NEUROTOXICOLOGY

DRUG AND CHEMICAL TOXICOLOGY

Series Editors

Frederick J. DiCarlo
Science Director
Xenobiotics Inc.
Denville, New Jersey

Frederick W. Oehme
Professor of Toxicology, Medicine and Physiology
Director, Comparative Toxicology Laboratories
Kansas State University
Manhattan, Kansas

Volume 1 Toxicity Screening Procedures Using Bacterial Systems, *edited by Dickson Liu and Bernard J. Dutka*

Volume 2 Chemically Induced Birth Defects, *James L. Schardein*

Volume 3 Neurotoxicology, *edited by Kenneth Blum and Luigi Manzo*

Other Volumes in Preparation

NEUROTOXICOLOGY

edited by

Kenneth Blum
University of Texas Health Science Center
at San Antonio
San Antonio, Texas

Luigi Manzo
University of Pavia Medical School
Pavia, Italy

MARCEL DEKKER, INC. New York and Basel

Library of Congress Cataloging in Publication Data

Main entry under title:

Neurotoxicology.

(Drug and chemical toxicology ; v. 3)
Includes bibliographies and index.
1. Nervous system--Diseases. 2. Neurotoxic agents.
3. Drugs--Side effects. 4. Toxicology. I. Blum,
Kenneth. II. Manzo, L. III. Series: Drug and chemical
toxicology (New York, N.Y. : 1984) ; v. 3.
[DNLM: 1. Nervous System--drug effects. 2. Nervous
System Diseases--chemically induced.
W1 DR513F v.3 / WL 100 N4967]
RC347.N488 1985 616.8 85-11605
ISBN 0-8247-7283-0

MARCEL DEKKER, INC.
270 Madison Avenue, New York, New York 10016

Current printing (last digit):
10 9 8 7 6 5 4 3 2 1

PRINTED IN THE UNITED STATES OF AMERICA

About the Series

Toxicology has come a long way since the ancient use of botanical fluids to eliminate personal and political enemies. While such means are still employed (often with more potent and subtle materials), toxicology has left the boiling-pots-and-vapors atmosphere of the "old days" and evolved into a discipline that is at the forefront of science. In this process, present-day toxicologists adopted a variety of techniques from other scientific areas and developed new skills unique to the questions asked and the studies being pursued. More often than not, the questions asked have never been thought about before, and only through the advances made in other disciplines (for example, in analytical chemistry) were the needs for answers raised. The compounding concerns of society for public safety, the maintenance of environmental health, and the improvement of the welfare of research animals have expanded the boundaries in which toxicologists work. At the same time, society has spotlighted toxicology as the science that will offer the hope of safety guarantees, or at least minimal and acceptable risks, in our everyday chemical encounters.

This *Drug and Chemical Toxicology* series was established to provide a means by which leading scientists may document and communicate important information in the rapidly growing arena of toxicology. Providing relevant and forward-looking subjects with diverse and flexible themes in an expedited and prompt publication format will be our goal. We will strive in this vehicle to provide fellow toxicologists and other knowledgeable and interested parties with appropriate new information that can be promptly applied to help answer current questions.

Of the several new specialty areas within toxicology, neurotoxicology is one of the most exciting. Understanding which of the multitude of environmental chemicals will adversely interact with the varying components of the central or peripheral nervous systems of specific animal species, establishing how the biochemical and morphological effects occur, documenting the clinical and laboratory diagnostic features of the toxicity to ease clinical recognition, studying the benefits of specific therapeutic agents to treat or manage the affected patient, and determining the ultimate long-term effect of the chemicals' involvements with the patients' nervous systems are complex, yet necessary and vital, challenges to overcome. Editors Blum and Manzo have dealt with all these aspects in *Neurotoxicology* and have generated a comprehensive but detailed volume that offers toxicologists and other interested scientists the best of many worlds. They have provided a book that gives state-of-the-art insights into the toxic effects, mechanisms of actions, and clinical features of the neurotoxicity produced by a range of chemicals and drugs. We hope *Neurotoxicology* will serve as an important building block for the documentation of continuing advances in understanding the adverse effects of chemicals on the nervous systems of animals and human beings throughout the world.

Frederick W. Oehme
Frederick J. DiCarlo

Foreword

Interest in nervous system toxicology has been growing in recent years, not only because of increased public concern over the impact of chemicals and toxicants on human health and the quality of life, but also because the nervous system has been shown to be especially vulnerable to chemical insult. This has led to the appearance and definition of a novel science, the science of neurotoxicology. The term and its domain, which borders on the domains of neuroscience and toxicology, are very recent.

Neurotoxicology studies the actions of chemical, biological, and certain physical agents that produce adverse effects on the nervous system and/or behavior in development and maturity. Toxic disorders of the nervous system of human beings and other animals may occur following exposure to abused substances (e.g., ethanol, inhalants, or narcotics), therapeutic drugs, products or components of living organisms, chemicals designed to affect certain organisms (e.g., pest-control products), industrial chemicals, chemical warfare agents, food additives and natural components of food, fragrance raw materials, and certain other types of chemicals encountered in the environment.

This book, edited by Kenneth Blum and Luigi Manzo, is at the very forefront of the progress in this field; it is designed to sensitize the reader to the new way to understand behavior, the functioning of the nervous system, how chemicals affect the mind, and the mechanisms involved in neurological and psychiatric disorders. The book should be of interest to anyone concerned with advances in neurotoxicology and it should be highly recommended also in view of the extensive international representation of the contributors.

G. Vettorazzi, M.D., Ph.D.
Senior Toxicologist
International Programme on Chemical Safety
World Health Organization
Geneva, Switzerland

Preface

The study of the nervous system has become increasingly important as a specialty area within toxicology. This domain, originally restricted to a few researchers, has now become a vast (although somewhat uncharted) common ground for scientists from very diverse disciplines including biochemistry, physiology, pharmacology, neurology, psychiatry, psychology, occupational health, and internal medicine. The increased interest stems principally from advances in neuroscience and from enhanced public awareness of the major problems arising from deliberate or unintentional exposure to potentially neurotoxic chemicals. Such exposure might involve substance abuse, neurological and psychiatric disorders caused by drugs, and health hazards linked with chemicals present in the work place or in the environment.

The research approach, based on the application of modern morphological techniques, has helped greatly to characterize toxic diseases of the nervous system and to establish neurotoxicology as a separate specialty. It is now also understood that toxicity of the nervous system comprises a variety of entities other than those defined in terms of anatomical evidence. In the last few years, more and more chemicals have been recognized as potential neurotoxins because of their ability to cause neurophysiological or neurohumoral alterations, behavioral abnormalities, or disruption of neurochemical activities (even in the absence of any obvious lesion in the nervous tissue).

Our incomplete understanding of structure/activity relationships and the mechanisms underlying most neurotoxic diseases prevents any systematic classification of neurotoxicants based on the functional or structural changes they can produce. Moreover, in most instances the action of neurotoxic chemicals is not restricted to a single target but is reflected by diffuse insults to various portions of the nervous system, resulting in complex clinical manifestations. Thus, at the present time neurotoxicology may be defined only in terms of distinct nosological entities that are often, but not invariably, associated with selective damage to neural substrates.

This volume provides an outline of biological bases and medical implications of neurotoxic phenomena. Both are discussed in terms of morphology, biochemistry, and physiology. The book's general approach is "problem-oriented," that is, attention is given primarily to chemicals causing neurotoxicity in man and to neurotoxic diseases currently regarded as relevant problems, either for epidemiologic reasons or because of their health and social implications.

The volume consists of four units. The first covers current concepts in the pathophysiology of toxic disorders of the human nervous system. This includes peripheral neuropathies and brain failures, plus dysfunctions of the neuroendocrine, sleeping and autonomic nervous systems. Genetic and developmental aspects of neurotoxicity and biological correlates of addictions and drug-seeking behaviors, regarded here as specific expressions of "motivational toxicity," are also discussed.

The second section surveys neurological side effects of selected classes of drugs (antiepileptic, anticancer, and psychotropic agents) that have neurotoxic potential that may reduce their efficacy or prevent their use in the treatment of major diseases in man.

The third section reviews nervous system toxicology as related to environmental chemicals including metals, pesticides, and biological toxins. Individual chapters are related to halogenated hydrocarbons, cyanide, and cyanogenetic compounds. While the acute neurotoxicity of these agents is firmly established, the effects of long-term, low-level exposure in humans are still undetermined, and the mechanisms underlying their neural toxicity need clarification.

The last section includes articles reviewing methodological issues. Many unsolved problems remain regarding the assessment of neurotoxicity. Most questions arise due to the enormous complexity of the nervous system, the uncertain definition of "normality" applied to highly integrated neural functions, and the potential influence of social or psychological factors on neurotoxic responses such as those manifested by behavioral, mood, and personality changes. Additionally, adaptive phenomena related to the functional plasticity of the nervous system may mask subtle effects of neurotoxicants, especially after prolonged low-dose exposure. It is apparent that no single screening method will suffice to explore nervous system responses or to establish suitable indexes of neurotoxicity. Moreover, most of the procedures currently used for detection of neurotoxic effects still need validation before they can be incorporated into standard protocols used routinely in any risk-assessment program (17). Recently developed methods (e.g., visualizing the mammalian brain without invasive procedures or the use of drugs to increase animal test sensitivity), promise to enhance our powers of investigation in the area of nervous system toxicology dramatically. For the present, sound methodology for studying neurological effects of toxicants and their mechanisms can rest only on the principle of multidisciplinarity, that is, on an integrated (physiological, biochemical, behavioral, and morphological) approach, and proper application of multiple test systems.

In a review volume of this size it is not possible to cover every aspect of the subject; however, the editors have attempted to compile a single-topic outline that is comprehensive but does not overlap extensively with other reference books (15,16,19,21-24), symposia proceedings (4,11,12,20), and volumes devoted to single topics (1-3,5-10,13, 14,18,25-27). Every effort has been made to provide an informative, basic text that presents as wide a view as possible of the current status of nervous system toxicology and shows the direction in which this emerging discipline is rapidly moving.

It is hoped that the contents of the volume will be useful in providing toxicologists with a greater appreciation of the general principles of their discipline, as applied to the nervous system. The volume also is addressed to basic scientists, clinicians, and other professionals who have a specialized interest in nervous system toxicology.

The editors are greatly indebted to all who have contributed to this manuscript, and to their colleagues, Professor F. Bertè and Professor A. Crema, for their encouragement and criticism during its preparation. We also wish to thank Dr. A. Baldry and Dr. Elly Xenakis-Gray for proofreading and index preparation. Finally, the editors acknowledge the wisdom and clear vision of Dr. Maurits Dekker for inviting us to edit such an important compendium.

Kenneth Blum
Luigi Manzo

REFERENCES

1. Breggin, P. R. 1983. *Psychiatric Drugs: Hazards to the Brain*, Springer, New York.
2. Ceccarelli, B., Clementi, F. (eds.). 1979. *Neurotoxins: Tools in Neurobiology*, Raven, New York.
3. Ecobichon, D. J., Joy, R. M. 1982. *Pesticides and Neurological Diseases*, CRC Press, Boca Raton.
4. Elmqvist, D. (ed.). 1984. Proceedings from the 1983 Swedish Neurotoxicology Symposium, *Acta Neurol. Scand. 70* (Suppl. 100): 1-225.
5. Englund, A., Ringen, K., and Mehlman, M. A. (eds.). 1982. *Occupational Health Hazards of Solvents*, Princeton Scientific Publ., Princeton.
6. Fuxe, K., Schwarcz, R. R. (eds.). 1983. *Excitotoxins*, Macmillan, London.

7. Gilioli, R., Cassitto, M. G., and Foà, V. (eds.). 1983. *Neurobehavioral Methods in Occupational Health*, Pergamon, Oxford.
8. Guilleminault, C., Lugaresi, E. (eds.). 1983. *Sleep/Wake Disorders: Natural History, Epidemiology, and Long-Term Evolution*, Raven, New York.
9. Hucho, F., Ovchinnikov, Y. A. (eds.). 1983. *Toxins as Tools in Neurochemistry*, Walter de Gruyter Publ., Hawthorne.
10. Iversen, L. L., Iversen, S. D., and Snyder, S. H. (eds.). 1984. *Drugs, Neurotransmitters, and Behavior*, Plenum, New York.
11. Kidman, A. D., Tomkins, J. K., and Cooper, M. N. A. (eds.). 1983. *Molecular Pathology of Nerve and Muscle. Noxious Agents and Genetic Lesions*, Humana Press, Clifton.
12. Manzo, L. (ed.). 1980. *Advances in Neurotoxicology*, Pergamon, Oxford.
13. Marotta, C. A. (ed.). 1983. *Neurofilaments*, University of Minnesota Press, Minneapolis.
14. Merigan, W. H., Weiss, B. (eds.). 1980. *Neurotoxicity of the Visual System*, Raven, New York.
15. Mitchell, C. L. (ed.). 1982. *Nervous System Toxicology*, Raven, New York.
16. Narahashi, T. (ed.). 1984. *Cellular and Molecular Neurotoxicology*, Raven, New York.
17. Office of Technology Assessment (1984). *Impacts of Neuroscience*, Washington, D.C.
18. Pfeifer, S. E. (1982-1983). *Neuroscience Approached Through Cell Culture*, CRC Press, Boca Raton.
19. Prasad, K. N., Vernadakis, A. (eds.). 1982. *Mechanisms of Action of Neurotoxic Substances*, Raven, New York.
20. Roizin, L., Shiraki, H., and Grcevic, N. (eds.). 1977. *Neurotoxicology*, Raven, New York.
21. Smith, W. T., Cavanagh, J. B. (eds.). 1979. *Recent Advances in Neuropathology*, Vol. I, Churchill Livingstone, Edinburgh.
22. Smith, W. T., Cavanagh, J. B. (eds.). 1982. *Recent Advances in Neuropathology*, Vol. II, Churchill Livingstone, Edinburgh.
23. Spencer, P. S., Schaumburg, H. H. (eds.). 1980. *Experimental and Clinical Neurotoxicology*, Williams & Wilkins, Baltimore.
24. Vinken, P. J., Bruyn, G. W. (eds.). 1979. *Intoxications of the Nervous System. Handbook of Clinical Neurology*, Vol. 36-37, North-Holland, Amsterdam.
25. Weiss, D. G. (ed.). 1982. *Axoplasmic Transport*, Springer, Berlin.
26. Yanai, J. (ed.). 1984. *Neurobehavioral Teratology*, Elsevier, New York.
27. Zbinden, G., Cuomo, V., Racagni, G., and Weiss, B. (eds.). 1982. *Application of Behavioral Pharmacology in Toxicology*, Raven, New York.

Contents

About the Series iii
Foreword (G. Vettorazzi) v
Preface vii
Contributors xv

Part I PATHOPHYSIOLOGY AND TARGETS OF NEUROTOXICITY

1 Peripheral Nervous System Toxicity: A Morphological Approach 1
J. B. Cavanagh

2 Toxic Brain Failure Syndromes 45
M. Bozza-Marrubini

3 Drug-Induced Dysfunctions of the Autonomic Nervous System 69
Virginia L. Zaratzian

4 Drug-Induced Neuroendocrine Disorders: Neuroleptics and Hormones 83
F. Brambilla

5 Drug-Induced Sleep Disorders 101
Christian Guilleminault

6 Actions of Abused Drugs on Reward Systems in the Brain 111
Roy A. Wise and Michael A. Bozarth

7 Neuropeptides and Addiction — 135
Jan M. van Ree and David de Wied

8 Developmental Neurotoxicology of Environmental and Industrial Agents — 163
B. K. Nelson

9 Genetic Aspects of Neurotoxicity — 203
Peter Propping

Part II SELECTED CLASSES OF DRUGS WITH NEUROTOXICITY POTENTIAL

10 Central Nervous System Toxicity of Psychopharmacological Agents — 219
Edmond H. Pi and George M. Simpson

11 Cerebellar Toxicity of Antiepileptic Drugs — 233
P. R. M. Bittencourt, E. Perucca, and A. Crema

12 Mechanisms of Neurotoxicity of Anticancer Drugs — 251
J. Alejandro Donoso and Fred Samson

13 Nitroimidazole Neurotoxicity — 271
Arthur J. Dewar and Geoffrey P. Rose

Part III NEUROTOXIC SUBSTANCES AND THE HUMAN ENVIRONMENT

14 Neurotoxicology of Lead — 299
Ellen K. Silbergeld

15 Neurotoxicity of Elemental Mercury: Occupational Aspects — 323
Vito Foà

16 Human Effects of Methyl Mercury as an Environmental Neurotoxicant — 345
Tadao Takeuchi

17 Methyl Mercury: Effects on Protein Synthesis in Nervous Tissue — 369
Saburo Omata and Hiroshi Sugano

18 Neurotoxicity of Selected Metals 385
Luigi Manzo, Kenneth Blum, and E. Sabbioni

19 Membrane Effects of Pesticides 405
John D. Doherty

20 Biochemical Toxicology of Organophosphorous Compounds 423
Mohamed B. Abou-Donia

21 Chlorinated Insecticides 445
S. C. Bondy

22 Halogenated Hydrocarbons 459
Anna Maria Seppäläinen

23 Alleged Neurotoxic Effects of Chronic Cyanide Exposure 473
Roger P. Smith

24 Toxicology of Selected Animal and Marine Neurotoxins 489
Lawrence Rodichok and Russell Mankes

Part IV ASSESSMENT OF NEUROTOXICITY

25 Neurotoxin-Induced Animal Models of Human Diseases 515
Edith G. McGeer and Patrick L. McGeer

26 Morphological Assessment of Neurotoxicity: Disulfiram Neuropathy as an Animal Model of Human Toxic Axonopathies 535
A. P. Anzil

27 Neural Culture: A Tool to Study Cellular Neurotoxicity 559
Antonia Vernadakis, David L. Davies, and Fulvia Gremo

28 Central Nervous System Toxicity Evaluation in Vitro: Neurophysiological Approach 585
Michael J. Rowan

29 Electrophysiological Methods for the in Vivo Assessment of Neurotoxicity 613
Yasuhiro Takeuchi and Yasuo Koike

30 Behavioral Toxicology: Animal Experimental Models 631
Zoltan Annau

Index 649

Contributors

Mohamed B. Abou-Donia, Ph.D., Department of Pharmacology, Duke University Medical Center, Durham, North Carolina

Zoltan Annau, Ph.D., Department of Environmental Health Sciences, The Johns Hopkins University, Baltimore, Maryland

A. P. Anzil, M.D.,* Department of Neuromorphology, Max Planck Institute for Psychiatry, Martinsried, Federal Republic of Germany

P. R. M. Bittencourt, M.D., Ph.D., Clinical Neurology Unit, Hospital Nossa Senhora das Graças, Curitiba, Brazil

Kenneth Blum, Ph.D., Department of Pharmacology, University of Texas Health Science Center at San Antonio, San Antonio, Texas

S. C. Bondy, Ph.D., Laboratory of Behavioral and Neurological Toxicology, National Institute for Environmental Health Sciences, Research Triangle Park, North Carolina

Michael A. Bozarth, Ph.D., Center for Studies in Behavioral Neurobiology, Department of Psychology, Concordia University, Montreal, Quebec, Canada

M. Bozza-Marrubini, M.D., Department of Anesthesiology, Resuscitation, and Intensive Care, Poison Control Center, Ente Ospedaliero Niguarda-Ca' Granda, Milan, Italy

---

**Present affiliation*:
Medical College of Pennsylvania, Philadelphia, Pennsylvania

F. Brambilla, M. D., Psychoendocrine Center, Ospedale Psichiatrico PINI, Milan, Italy

J. B. Cavanagh, M.D., Institute of Neurology, University of London, London, England

A. Crema, M.D., Department of Medical Pharmacology, University of Pavia, Pavia, Italy

David L. Davies, Ph.D., Department of Pharmacology, University of Colorado School of Medicine, Denver, Colorado

Arthur J. Dewar, M.Sc., D.T.C., Ph.D.,* Shell Toxicology Laboratory, Sittingbourne Research Centre, Sittingbourne, Kent, England

David de Wied, M.D., Rudolf Magnus Institute for Pharmacology, Medical Faculty, University of Utrecht, Utrecht, The Netherlands

John D. Doherty, Ph.D., Hazard Evaluation Division, Office of Pesticide Programs, Environmental Protection Agency, Washington, D.C.

J. Alejandro Donoso, Ph.D.,† Department of Neurology, and Ralph L. Smith Research Center, University of Kansas Medical Center, Kansas City, Kansas

Vito Foà, M.D., Industrial Hygiene and Occupational Toxicology Division, Institute of Occupational Health, Clinica L. Devoto, University of Milan, Milan, Italy

Fulvia Gremo, M.D., Department of Human Anatomy, University of Cagliari School of Medicine, Cagliari, Italy

Christian Guilleminault, M.D., Sleep Disorders Center, Stanford University School of Medicine, Stanford, California

Yasuo Koike, M.D.,‡ First Department of Internal Medicine, Nagoya University School of Medicine, Nagoya, Japan

Edith G. McGeer, Ph.D., Kinsmen Laboratory of Neurological Research, University of British Columbia, Vancouver, British Columbia, Canada

---

Present affiliations:
*Shell International Chemical Company, Shell Centre, London, England
†Veterans Administration Medical Center, Kansas City, Missouri
‡Nagoya University Hospital, Nagoya, Japan

Patrick L. McGeer, M.D., Ph.D., Kinsmen Laboratory of Neurological Research, University of British Columbia, Vancouver, British Columbia, Canada

Russell Mankes, Ph.D., Institute of Experimental Pathology and Toxicology, Albany Medical College, Albany, New York

Luigi Manzo, M.D., Department of Internal Medicine, Division of Pharmacology and Toxicology, University of Pavia Medical School, Pavia, Italy

B. K. Nelson, M.S., Division of Biomedical and Behavioral Science, National Institute for Occupational Safety and Health, U.S. Department of Health and Human Services, Cincinnati, Ohio

Saburo Omata, Ph.D., Department of Biochemistry, University of Niigata, Niigata, Japan

E. Perucca, M.D., Ph.D., Department of Medical Pharmacology, University of Pavia, Pavia, Italy

Edmond H. Pi, M.D.,* Department of Psychiatry and Behavioral Sciences, University of Southern California, School of Medicine, Los Angeles, California

Peter Propping, M.D.,† Institute of Human Genetics, University of Heidelberg, Heidelberg, Federal Republic of Germany

Lawrence Rodichok, M.D., Department of Neurology, Albany Medical College, Albany, New York

Geoffrey P. Rose, Ph.D., Shell Toxicology Laboratory, Sittingbourne Research Centre, Sittingbourne, Kent, England

Michael J. Rowan, Ph.D., Department of Pharmacology and Therapeutics, Trinity College, University of Dublin, Dublin, Ireland

E. Sabbioni, Radiochemistry and Nuclear Chemistry Division, EC Joint Research Center, Ispra, Varese, Italy

---

Present affiliations:
*The Medical College of Pennsylvania at Eastern Pennsylvania Psychiatric Institute, Philadelphia, Pennsylvania
†University of Bonn, Bonn, Federal Republic of Germany

Fred Samson, Ph.D., Ralph L. Smith Research Center, University of Kansas Medical Center, Kansas City, Kansas

Anna Maria Seppäläinen, M.D., Department of Occupational Medicine, Institute of Occupational Health, Helsinki, Finland

Ellen K. Silbergeld, Ph.D., Toxic Chemicals Program, Environmental Defense Fund, Washington, D.C.

George M. Simpson, M.D., Department of Psychiatry and Behavioral Sciences, University of Southern California, School of Medicine, Los Angeles, California

Roger P. Smith, Ph.D., Department of Pharmacology and Toxicology, Dartmouth Medical School, Hanover, New Hampshire

Hiroshi Sugano, Ph.D., Department of Biochemistry, University of Niigata, Niigata, Japan

Tadao Takeuchi, M.D.,* Department of Pathology, Kumamoto University, Kumamoto City, Japan

Yasuhiro Takeuchi, M.D., Department of Hygiene, Nagoya University School of Medicine, Nagoya, Japan

Jan M. van Ree, M.D., Rudolf Magnus Institute for Pharmacology, Medical Faculty, University of Utrecht, Utrecht, The Netherlands

Antonia Vernadakis, Ph.D., Departments of Psychiatry and Pharmacology, University of Colorado School of Medicine, Denver, Colorado

Roy A. Wise, Ph.D., Center for Studies in Behavioral Neurobiology, Department of Psychology, Concordia University, Montreal, Quebec, Canada

Virginia L. Zaratzian, Ph.D., Science Program, United States Department of Agriculture, Washington, D.C.

---

* *Present affiliation*:
Shokei-Gakuen University, Kumamoto City, Japan

Part I

PATHOPHYSIOLOGY AND TARGETS OF NEUROTOXICITY

1

Peripheral Nervous System Toxicity: A Morphological Approach

J. B. Cavanagh
Institute of Neurology, University of London, London, England

I. INTRODUCTION

It has long been realized by clinicians considering cases of peripheral nerve disease that several distinctly different forms exist that have different underlying pathological mechanisms. Each has its own natural history and individual pattern of symptoms and signs and its own rate of recovery. But predominant among peripheral nerve disorders is sensory and motor neuropathy showing symmetrical symptoms and signs affecting distal regions of the limbs, usually the legs more than the arms, and frequently leaving cranial nerves unaffected. Until recently this has been, perhaps, the most difficult disorder to understand; it certainly is among the most fundamental in its pathology because of the tantalizing similarities it has with the so-called systematized degenerations of nervous tissue which were one of the principal concerns of the classic authors of the nineteenth century. In order to understand this group of neuropathies we will inevitably be faced with understanding where lie the principal weaknesses in the neuron's capacity to maintain itself. It certainly has a huge task to supply and look after an axon of enormous length and volume over many years in the face of numerous environmental hazards and inherent weaknesses.

It is only about these that we are beginning to learn. To maintain a signalling system over relatively vast distances is the neuron's principal function, but only recently have we begun to learn how this might be done. It is becoming apparent that the supply system of the neuron, stretched as it has been in the course of evolution to extraordinary lengths, has developed points of weakness that render it susceptible to interference by a wide variety of environmental and genetic hazards. These points of weakness may not be apparent in the protected equilibrium state, but are exposed when a chemical intoxication or a deficiency state begins to place a stress on the system, such that it begins to break down. It is our role to try to determine at what point breakdown occurs.

This chapter tries to highlight some of the probable sites of weakness disclosed by chemical intoxication. Out of this arises an attempt at a rational classification of such disorders based on these putative critical sites of weakness. Virchow told us that the way toward understanding pathology was through the cell. There is no reason yet to doubt his advice.

II. MORPHOLOGICAL FACTORS IN DEVELOPMENT OF DAMAGE TO PERIPHERAL NERVES BY TOXIC CHEMICALS

A. Routes of Entry

1. Direct Entry

To produce a metabolic lesion in Peters' sense within the cellular complex of peripheral nerves a chemical substance must be able to enter fairly freely either the Schwann cell, the neuron, or its axon. Peripheral nerve, equally with the central nervous system, has built a privileged environment around itself for the protection of its own functional interests (the blood-nerve barrier) (7). The path through the vascular bed within nerves is closed by tight junctions and entry must therefore be through endothelial cells, unless these themselves are damaged by the toxic chemical. There is evidence that many drugs (2) can freely enter peripheral nerve, and it is probable that arsenic and thallium ions, isoniazid, 2,5-hexanedione, and so on, can directly reach the axon by this route.

The exact point of entry is not clear. The endoneurial space contains abundant glycoproteins which may impede movement of some types of material. The node of Ranvier has a layer of acidic glycoprotein, the "gap substance," on its external surface that could act as a cationic "sink." Nonetheless, it is probably that the node is a likely portal of entry if only because the alternative, through the Schwann cell and myelin, may well be less accessible.

2. *Entry from Axon Terminals*

Since the demonstration by Kristersson and Olsson (61) of the ascent of labeled proteins along axons from an intramuscular injection, it has been accepted that axon terminals are a ready site of access to both sensory and motor neurons. Indeed, once a chemical can enter the extracellular tissue space, it is able to enter a nerve cell by this route. It is an "open door" to the neuron's interior, but it has yet to be shown whether chemicals do in fact enter by this route and whether the nerve cell has any protective mechanisms against this.

3. *Entry Directly to the Sensory Cell Body*

Proteins in the bloodstream cannot normally enter the brain, but they can readily pass through the vascular bed of sensory ganglia (and autonomic ganglia) and can be shown (51), within a minute or two of intravenous injection, to be in the extracellular space of spinal or autonomic ganglia. Proteins can pass easily through fenestrations in this vascular bed and come into direct contact with the neuronal plasma membranes. How much satellite cells control the passage of such materials is unknown (77). It is particularly relevant that sensory rather than motor neurons are exposed in this way, and this could be one factor in the early and most frequent involvement of the primary sensory neuron in toxic neuropathies.

B. Permeability of the Vascular Bed

There are a number of chemicals and drugs which have the capacity to selectively damage the vascular bed of the brain, and thereby cause leakage of plasma constituents with accompanying neuronal necrosis (41,53,70). No peripheral neuropathy is known in which a toxic chemical appears to be having exactly the same kind of effect on the vascular bed of nerves. However, there is growing evidence that in lead intoxication there is a change in the ability of the vascular bed to maintain the equilibrium situation in the endoneurial space, although the evidence for a definite damage to the vascular bed is not beyond criticism (30).

C. Length and Frequency of Internodes Along Axons as a Factor in Toxic Neuropathy

It is now well established that internodal length is not uniform among peripheral nerve fibers, but is controlled by axon diameter and the growth of the part in which it finds itself. The time of onset of myelination during development determines its thickness and in any one fiber myelination begins earliest proximally and later distally.

Earlier myelinated fibers in limbs will be extended considerably more by the growth of the part than those in more proximal parts of the nerve. From this it can be concluded that the frequency of internodes and thus of nodes of Ranvier per unit length of nerve is often substantially less in limb nerves that have grown considerably since myelination commenced. In more proximal nerves, such as cranial nerves (e.g., oculomotor nerves), more nodes per unit of length is the rule.

These simple morphological relationships should not be neglected when considering functional disturbances from a process of apparently random segmental demyelination. In diphtheria intoxication, where functional changes, caused by slowing of conduction velocity, are a direct consequence of the number of affected nodes, nodal density will influence significantly the functional outcome.

D. Length and Diameter of Axons and Their Susceptibility to Toxic Neuropathic Processes

In toxic neuropathies symmetrical distal sensory loss is characteristic, and motor weakness is distal rather than proximal in the limb. Since such functional disturbances follow structural changes (i.e., wallerian degeneration), this can only mean that longer fibers are more at risk to degeneration of their distal regions than are shorter fibers, even though they have been equally exposed to the chemical. Larger-diameter fibers are also more at risk than small ones. Axon diameter does not, however, reflect the size of the projection field but is related to conduction velocity. Therefore it is not so much the specific signalling functions and the terminals that are relevant to susceptibility to degeneration in the toxic state, but the physical size of the axon, and the day-to-day domestic requirements for replacement of membrane materials or the provision of adequate amounts of energy to perform these maintenance operations. Such relatively large surface axons require constant provision, and from the effects of axon section we know (13,58) that loss of energy is one of the most important reasons for the ultimate breakdown of the axon.

III. GENERAL CLASSIFICATION OF TOXIC NEUROPATHIES ON MORPHOLOGICAL AND METABOLIC LINES

A. Vascular Bed Lesions with Secondary Myelin and Axon Degeneration

1. *Lead Neuropathy*

The classic signs and symptoms of neuropathy caused by chronic intake of relatively large amounts of lead salts have been well described by earlier authors (46,76). Typical is the predominantly motor nature of the condition, sensory disturbances being minor and not well defined. Frequent is the occurrence of motor weakness in an isolated limb or muscle group that may be determined by the work pattern of the individual. The term *mononeuritis multiplex* is often used here to demonstrate this apparent restriction and localization of the lesion. However, in other cases, and particularly in children, a symmetrical motor neuropathy may develop (72). Recovery on withdrawal from exposure is the rule unless evidence of encephalopathy is also present, when central nervous system symptoms may persist for long periods.

Postmortem studies on humans with chronic lead intoxication are few (65), but they have reported wallerian degeneration in peripheral nerves as well as segmental or primary demyelination. These changes, it should be noted, occur in many if not all nerves and are not restricted to those responsible for the paresis of limited muscle groups. The muscles may show denervation atrophy and the motor nerve cell body may show chromatolysis, but no involvement of the long tracts of spinal cord has ever been reported. The absence of systematized degeneration in this condition implies that the neuron is not the primary site of the metabolic lesion.

Experimental studies, particularly in the rat by Dyck and his group (74,95), have greatly helped our understanding of the condition. Earlier, Fullerton (34) had demonstrated in the guinea pig that both segmental myelin degeneration (Fig. 1) and axon degeneration were present which fully explained the marked early slowing of conduction

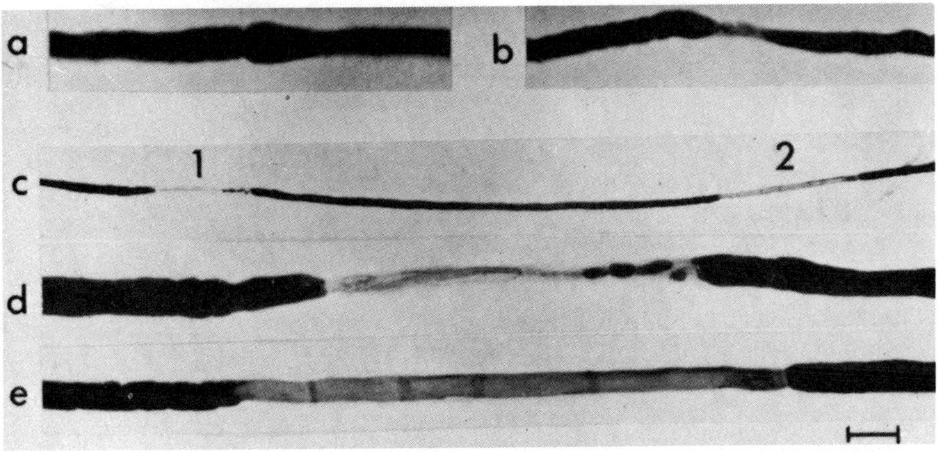

Figure 1 Guinea pig nerves teased and stained with osmium tetroxide to show the myelin sheaths: (a) normal node of Ranvier, (b, c) segmental paranodal myelin loss, and (d, e) thin, newly formed, regenerated myelin. (From Ref. 34; courtesy of Dr. P. M. Le Quesne.)

velocity in the affected nerves. Dyck and his colleagues (74) demonstrated the importance of edema of the endoneurium in the development of the lesion in the rat (Fig. 2), which raised an important question as to whether much of the apparent multifocal nature of the symptomatology might not be in part the result of increased pressure on the nerve fibers at critical sites along their course beneath ligaments and tendons. However, recent studies have shown that this edema of the endoneurium is not the result of a massive increase in the permeability of the vascular bed to horseradish peroxidase or to albumin (30). Yet despite this there is a steady increase in lead concentration within the endoneurium to reach a figure approaching 10 times that in the plasma (96). To account for this it must be concluded that protein-bound lead in some way becomes fixed to a component within the endoneurium. Any changes in vascular permeability that there might also be would intensify this steady incursion of lead into this tissue.

In the presence of this steady accumulation of lead within the Schwann cell's environment, it is perhaps not surprising that its metabolism becomes inadequate to maintain the myelin sheath intact and

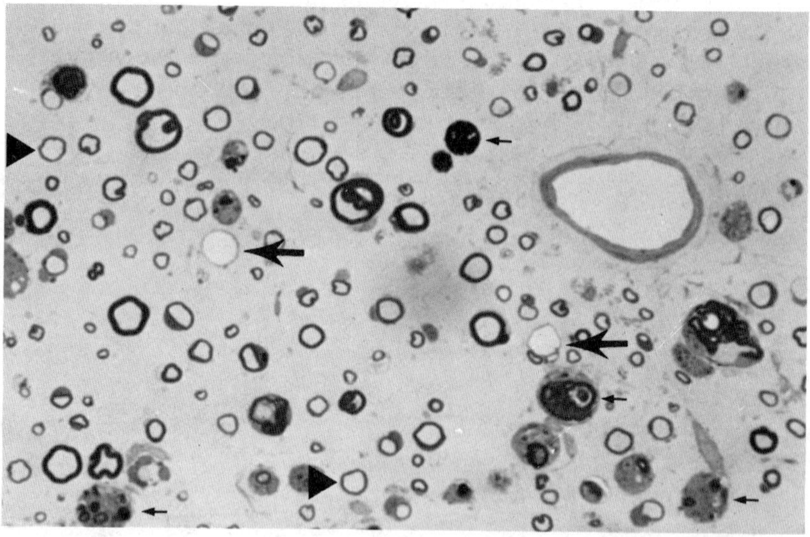

Figure 2 Section of nerve from a rat dosed with lead salts for 3 months. Note the marked endoneurial edema widely separating the nerve fibers from one another. Axons without myelin (large arrows) and axons with inappropriately thin myelin sheaths (arrowheads) can be seen. Occasional degenerating fibers may also be seen (small arrows) (toluidine blue, 420×). (Courtesy of Dr. P. J. Dyck.)

random breakdown of this occurs throughout the nerves. The metabolic mechanism responsible is as yet unknown, but other functions of the Schwann cell, such as its capacity to respond to the mitotic stimulus of axon degeneration, are also impaired (75).

B. Primary Schwann Cell Intoxications with Segmental Demyelination

1. *Diphtheria Intoxication*

Classic diphtheritic neuropathy is essentially motor in nature, with a relatively minor and irregular sensory component. Even the motor paresis is rarely complete in any region, but involvement of palatal, extraocular, and, not infrequently, respiratory muscles occurs early in the development of the disease and characterizes the clinical picture.

Morphologically the only detectable changes in peripheral nerves are segmental demyelination together with evidence of the cellular responses of myelin degradation by Schwann cells and macrophages leading to myelin removal and remyelination (Fig. 3) (14,68). The segmental demyelination is probably random, but may affect small-diameter fibers slightly earlier than large-diameter fibers. The whole internode only very rarely shows complete disintegration; indeed, most internodes show degeneration for only a few microns on either side of the node of Ranvier. The structure of the node of Ranvier itself, however, is destroyed by this process (Fig. 4) (4) so that the nerve is no longer capable of carrying impulses by saltatory conduction. Conduction velocity is thereby reduced down to rates of 1-2 m/s.

The metabolic mechanism responsible involves the early entry of the "nicked" toxin into the cell and its utilization of the available nicotinamide adenine dinucleotide (25). The result appears to be the temporary failure of protein synthesis, the myelin breakdown at the distal regions of the cell being secondary presumably to reduction in some (basic?) protein product required in the stabilization of the myelin membrane (81). The Schwann cell itself is not destroyed in the process and is able to recover fairly rapidly and to renew the damaged myelin and node of Ranvier. Conduction velocity returns as these are renewed, and no sequelae are known, other than the finding of many residual short intercalated internodes in previously affected nerves.

2. *Perhexilene Neuropathy*

Perhexilene maleate [2-(2,2-dicyclohexylethyl)piperidine maleate] is used as an agent for the reduction of attacks of angina pectoris in coronary disease. A number of toxic effects have been recorded from taking this drug, including peripheral neuropathy (Fig. 5) (66). This usually comes on after taking the drug (200 mg/day) over 12 months or more, and paresthesia and pain are frequent early symptoms. Later

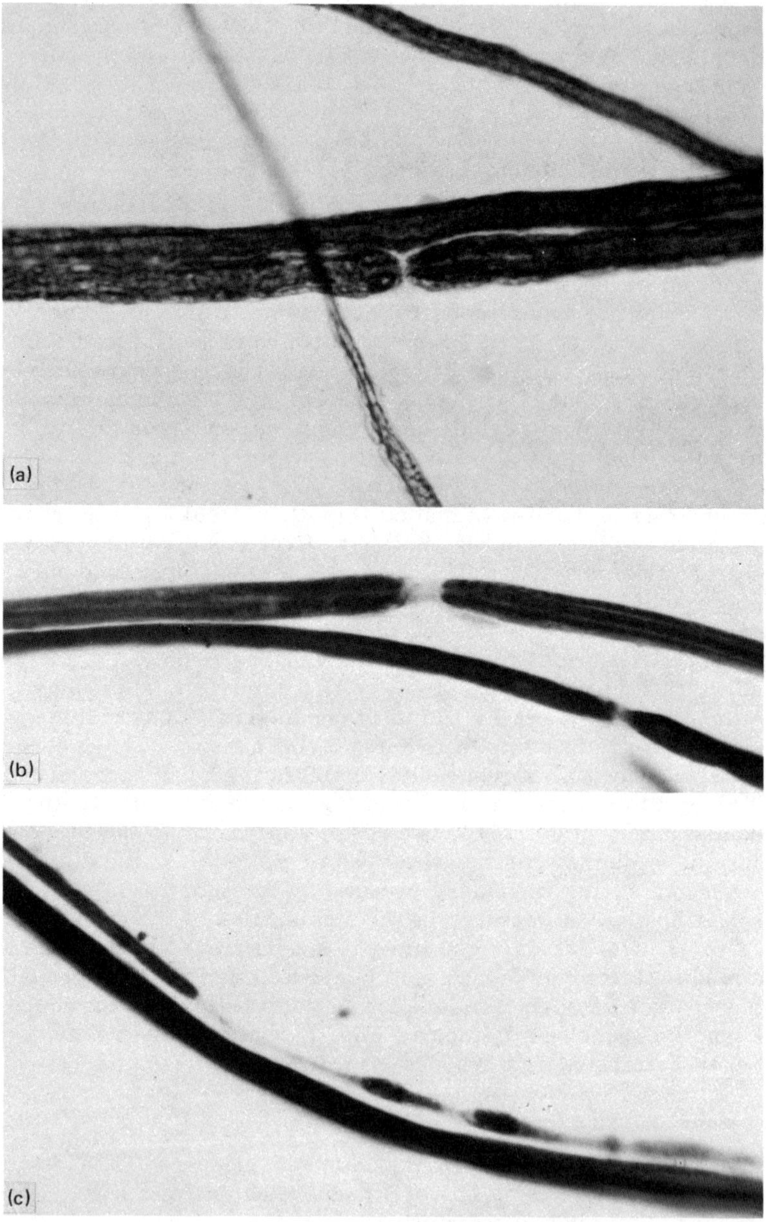

Figure 3 (a) Normal node of Ranvier, stained with Sudan black B, from the teased tibial nerve of a chicken (750×), (b) early widening of the nodes of Ranvier at the onset of flaccid paresis (Sudan black B, 530×), and (c) segmental myelin breakdown (Sudan black B, 530×).

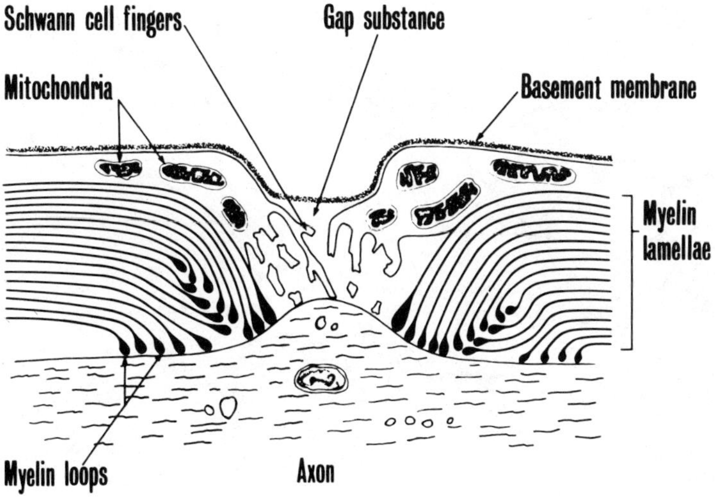

(a)

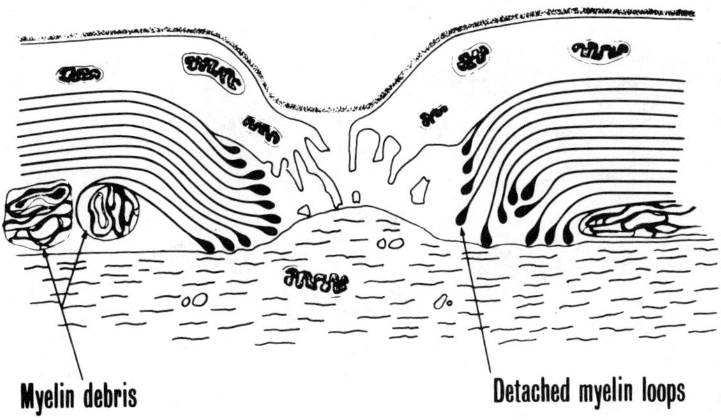

(b)

Figure 4 Diagrams of the disorganization of the paranodal region of myelin following injection of diphtheria toxin in the rat. (From Ref. 4.)

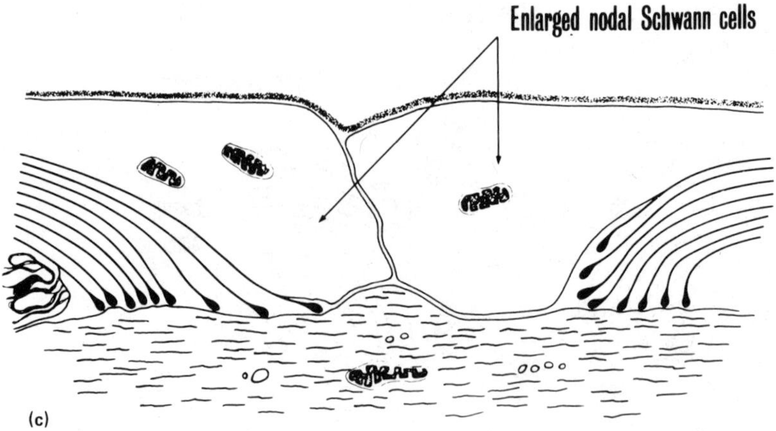

(c)

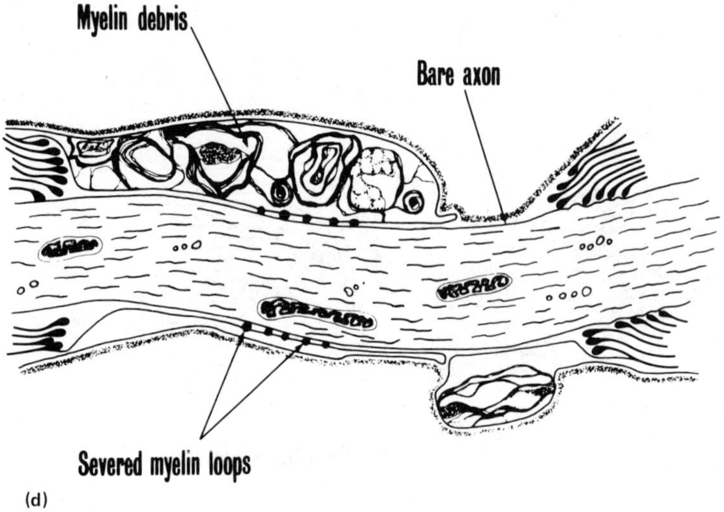

(d)

Figure 4 (Continued)

there may be some degree of weakness. Conduction velocity is found significantly slowed both in those with symptoms and also in many taking the drug without showing neurological signs (84).

Biopsy of peripheral nerves from such cases has shown striking selective myelin (internodal) degeneration and many shorter intercalated (remyelinated) internodes (83). A small number of axons also undergoing wallerian degeneration have been noted. The mechanism underlying this process is unknown, but it has been suggested that it may induce a disturbance of lipid metabolism in the Schwann cell. Metabolic breakdown and excretion of the drug may be markedly slower in those individuals showing this complication.

C. Primary Neuronal Intoxications

1. *The "Dying-Back" Process*

a. General Principles: Many toxic neuropathies have a clinical presentation in which sensory and motor dysfunction affects the limbs symmetrically to give a "glove-and-stocking" distal distribution of signs and symptoms. Short proximal lying nerves tend not to be affected in such conditions, although the long spinal tracts, sensory and motor, are not infrequently involved as well. The milder the intoxication, the more distal the distribution of changes; and the more severe the intoxication, the more proximal the functional changes may become. This pattern of involvement of the nervous system is not restricted to toxic conditions, but may be seen in deficiencies of several vitamins and in certain natural human conditions, such as Friedreich's disease. From these clinical and pathological observations the concept became generally acknowledged earlier this century that the degenerative changes affected the more distal regions of longer sensory and motor nerve cells earliest and, according to the severity of the condition, spread subsequently back toward the nerve cell body, and at the same time neurons with shorter axons also became involved in the process. The term *dying back* was applied to this neuronal degenerative process by Greenfield (38) as a purely descriptive term to separate it from other neuronal degenerations such as those caused by poliovirus, in which the perikaryon is destroyed first and the axon degenerated secondary to this, or Alzheimer's neurofibrillary change, where the perikaryon is obviously disordered. Spatz (87) drew attention to a number of natural human systematized diseases in which this type of slow retrograde degeneration of axons occurred ("centripetal degeneration"), and noted that the cell body usually remained intact, though became smaller, until the atrophy ascended beyond the last collateral. This is in accord with the general principle that the size of the cell body of a neuron is related to the size of its projection field. An experimental decrease or increase in this will be accompanied by parallel changes in perikaryon volume (78).

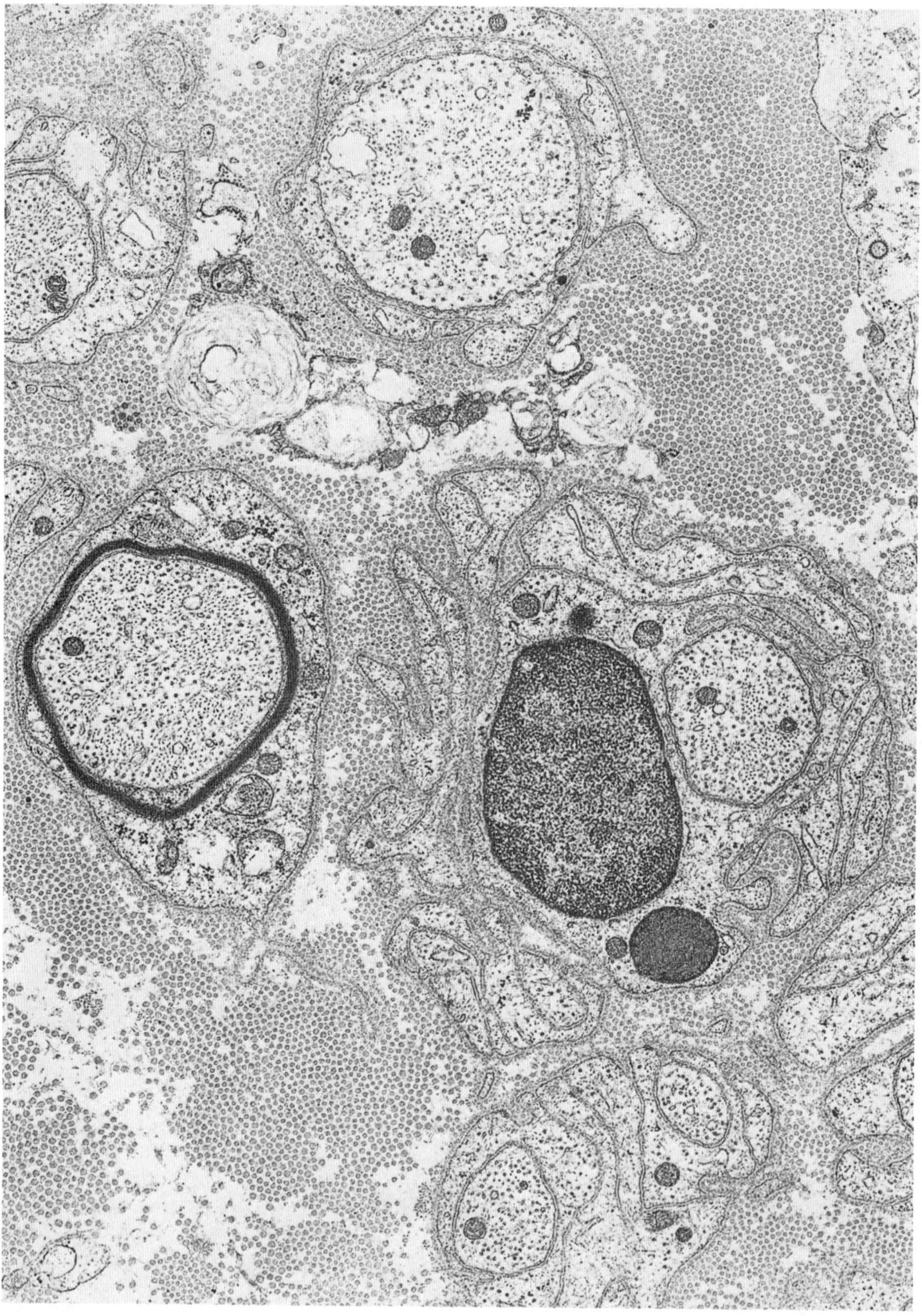

In toxic conditions, as in hereditary diseases, where a metabolic defect lies at the heart of a cellular disorder, a reduction in the cell's capacity to supply materials for the maintenance of the long axon and its branching terminal network will inevitably lead to degeneration and atrophy of the fiber until it "fits" the cell's capacity to continue to maintain it. The analogy with the branches of an old tree showing "heart rot" is close, where such a tree, being unable to maintain the full branching pattern of its prime, will show withering of its branches that can no longer be supplied. Nonetheless, the rest of the tree will continue to flourish and bear fruit for many years. In the case of the neurons the possible faults leading to incapacity to maintain its great volume and surface area of axon will be numerous because of the manifold steps in the synthesis, transport, and utilization chain of its maintenance functions. It is now becoming plain, as more toxic conditions are being adequately studied, that many different kinds of cellular tissues may be responsible for the breakdown of the system of supply. The present attempt to classify these conditions by metabolic or structural disturbances can be considered to be the beginning of a rational understanding of the many ways in which a nerve cell becomes incapable of keeping its domestic economy in a good and healthy state. Provided that the cell body is not totally destroyed by the metabolic lesion, the end result of all these processes is the same, namely, distal degeneration of longer and larger axons (Table 1).

In the following description of the morphological features of many of the known neurotoxic conditions, it will be seen that there are a growing number in which there is sufficient information available to discuss them in the light of their probable essential lesion. There are some, however, because less is known about them, where we cannot yet do that, and they are therefore placed in the category of "dying-back" conditions because this is a description that adequately fits their clinical picture. With time new information will allow them to be placed in a more precise etiological category. At present those in this holding category will probably be found to be all quite different in nature, one from the other. The chief one among this heterogeneous group is the delayed neuropathy due to certain organophosphates, and it is typical in all its clinical and morphological features of this class of neuronal disease process. While we know a great deal of its morphological and its biochemical basis, we still have not yet been able to pinpoint the type of intracellular lesion that leads to the distal axon degeneration in this instance.

Figure 5 Perihexilene neuropathy. Electron micrograph of a peripheral nerve showing the absence of myelin on larger-diameter axons and thin, probably recently formed myelin on one axon (9500×). (Courtesy of Dr. G. Said.)

Table 1 Classification of Primary Neuronal Intoxications: Toxic Conditions in Which a "Dying-Back" Type of Degeneration May Be Seen

A. Dying-back degeneration of unspecified mechanism
   1. Organophosphate intoxication
   2. Tetraethylthiuram disulfide (Antabuse)
   3. Clioquinol intoxication (subacute myelo-optico-neuropathy)
   4. Dapsone neuropathy

B. Due to probable defects induced in energy metabolism
   1. Antithiamine compounds, amprolium, pyrithiamine
   2. Trivalent arsenic intoxication
   3. Thallium poisoning
   4. Nitrofurantoin and related nitrofurans
   5. Nitroimidazole drugs, misonidazole, metronidazole

C. Due to probable defects in pyridoxal phosphate metabolism
   1. Isoniazid (isonicotinic acid hydrazide) neuropathy
   2. Hydralazine neuropathy
   3. Ethionamide neuropathy

D. Due to disturbance in protein synthesis
   1. Neuropathy of "pink" disease (calomel; mercurous chloride)
   2. Peripheral nerve degeneration in alkyl mercury intoxication

E. Disturbance to neurofilament behavior, caused by
   1. Certain hexacarbon compounds
   2. Carbon disulfide
   3. β,β'-Iminodipropionitrile and other propionitrile compounds

F. Disturbance to microtubular stability
   1. Vincristine and vinblastine neuropathy

G. Disturbance to smooth endoplasmic reticulum function (?)
   1. Acrylamide

b. Organophosphate-Delayed Neuropathy: Certain aryl and alkyl organophosphates have the capacity to produce symmetrical sensory and motor neuropathy of the dying-back type, which is also accompanied by analogous changes in long sensory and motor tracts in the spinal cord. The degeneration follows a delay period of 8-10 days or more, and they parallel the onset of the clinical features of the neuropathy. The chemical structure of the compounds capable of causing intoxication and the biochemical mechanisms that determine whether the neuropathy will occur or not have been well described (55; see also the chapter by Abou-Donia in this book), but the intimate intracellular lesion that prejudices the survival of the distal regions of the longer axons still eludes us.

Functional disturbances: In the early stages of intoxication various degrees of inhibition of acetylcholinesterase, according to the type of compound, lead to cholinergic symptoms that can be counteracted by pharmacological agents. These play no essential role in the neuropathy. With the onset of paresis there may be subjective sensory disturbances, but the prominent sign is flaccid paresis of the lower legs. This will spread to more proximal parts of the legs and to the hands in the ensuing 10 days or so (86). The extent of the paresis is dose dependent, but it never involves the proximal limb regions or the cranial nerves. Associated with these changes, there may be a variable amount of ataxia, but overt sensory disturbances are few in the early phase, though they may disclose themselves later as "glove-and-stocking" hypoesthesia in a proportion of cases.

Morphological features: Morphologically associated with these signs and symptoms, wallerian degeneration will be found in the relevant nerve bundles (Fig. 6) (9,15). This is more marked in more distal samples of nerve and no changes can be found in proximal nerve bundles, while the nerve cell bodies everywhere appear normal. Both motor and sensory nerve fibers are equally affected, but because of the predilection for larger-diameter axons, innervation to muscle spindles and Golgi tendon organs is particularly affected. Sensory end organs served by smaller axons, touch, temperature, pain, and so on, do not show degeneration. Where the intoxication is mild, only the distal parts of the nerves are damaged, and the nerve bundle emerging from a muscle may appear quite normal. This very distal distribution of the changes occasionally can lead to apparently poor correlation between axon degeneration and functional changes. For this reason frozen sections of muscle stained with Sudan black B showed more clearly than any other method the true extent of the degenerative changes (Fig. 7).

In the spinal cord degeneration of ascending dorsal columns and spinocerebellar tracts will be seen in their rostral regions, that is, in the upper cervical cord for the latter and in the anterior and posterior regions of the cerebellar vermis for the fibers of the former tract. Degeneration of descending corticospinal (Fig. 8) and probably rubro- and tectospinal tracts will be found in the lumbosacral regions. If the fibers of each tract are traced back toward their parent cell bodies, a point will come when no further degeneration can be found. Usually no degeneration of descending fibers will be found rostral to the lower cervical region in the descending tracts, and none caudal to the same level in the ascending tracts.

In addition to axon degeneration in both peripheral neurons and in spinal tracts, secondary myelin fragmentation will be seen, as well as invasion by macrophages from the bloodstream. Schwann cells in

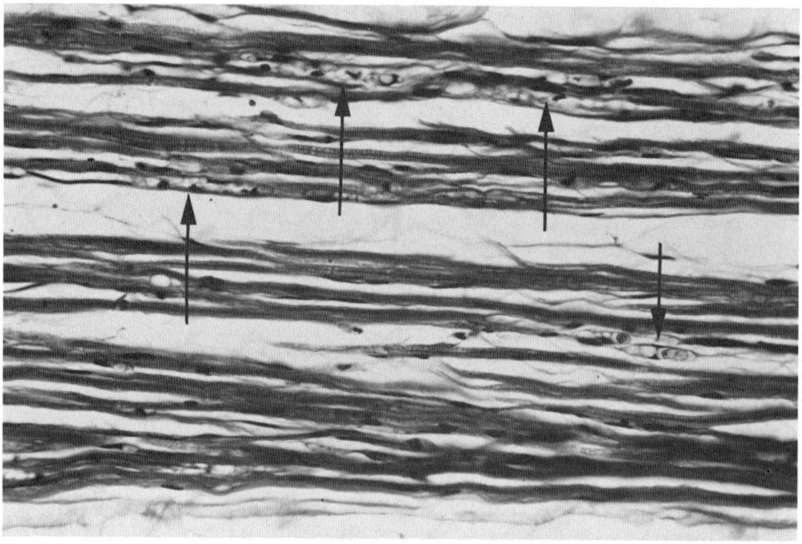

(a)

(b)

Figure 6 (a) Posterior tibial nerve from a cat poisoned with tri-*ortho*-cresyl phosphate. Note the numerous fragmenting fibers (arrows) with an increase in visible nuclei and vacuolation from degenerating myelin debris (hematoxylin and eosin, 230×). (b) Stain for axons showing a fragmenting (arrows) degenerating axon in TOCP intoxication in the posterior tibial nerve (Glees and Marsland stain, 338×).

(a)

(b)

Figure 7 (a) Normal spindle, from the foot muscle of a cat, stained with Sudan black B. Note the thick, well-stained myelin sheath running to the primary (annulospiral) sensory ending. The latter (arrows) is more faintly stained, being unmyelinated but having many mitochondria which take the stain (200×). (b) Spindle from a cat foot muscle after tri-*ortho*-cresyl phosphate intoxication. The primary sensory fiber has degenerated, and there are small lipid phagocytes in the lymph space of the spindle (arrows) containing fragments of the fiber (200×).

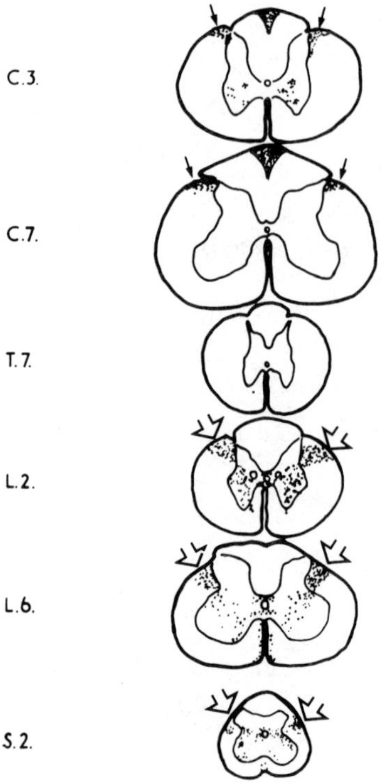

Figure 8 Pattern of degeneration in the long ascending and descending tracts of the cat spinal cord 55 days after tri-*ortho*-cresyl phosphate intoxication. The diagram shows the distribution of degeneration axons as shown by the Nauta-Gygax technique. The dorsal columns show degenerating fibers only in the medial (gracile) tracts in the cervical (C3 and C7) regions. The spinocerebellar tracts (small arrows) show degeneration in the same regions (the cell bodies of these fibers lie in the thoracic and upper lumbar levels: Clark's column). The corticospinal (descending) fibers are degenerating only distally in the lumbosacral levels (open arrows). A small number of uncrossed descending fibers are degenerating in the ventral tracts on either side of the ventral fissure. (From G. N. Patangia, Ph.D. thesis, University of London, 1964.)

peripheral nerves will undergo proliferation, while in the central nervous system astrocytes proliferate and surround the degenerating debris and take part in its degradation (Fig. 9). These cellular responses are typical of axon degeneration caused by whatever means and do not seem to be significantly changed in any intoxication so far studied, implying that the toxic substances at the doses that damage axons have no specifically deleterious effects on the functions of these supporting cells.

In the nerve cell bodies of the damaged axons no significant changes have been seen in organophosphate intoxication other than the subsequent cellular reaction to the axon degeneration known as chromatolysis, or the "axon reaction." This is a reparative response and indicates that the perikaryon is actively concerned with the synthesis of new axoplasm required for regeneration of the damaged axon.

Electron microscopy: Ultrastructural studies (6) of axons in the recurrent laryngeal nerve of the cat a few days before the onset of axon degeneration have revealed large numbers of smooth endoplasmic reticulum profiles occasionally associated with local vacuolation. Whether this is primarily associated with the local toxic process or perhaps a protective reaction to this putative lesion is unknown.

2. *Disturbances to Energy Supply to Axons*

a. Chronic Thiamine Deficiency (Beriberi): Chronic thiamine deficiency leads to a distal symmetrical sensory neuropathy with later motor involvement which affects legs more than arms. Morphologically, the changes of wallerian degeneration are found in the nerves, with denervation of muscles and chromatolysis (axon reaction) of sensory ganglion cells and of anterior horn cells (97). So far as it is known, there is no degeneration of long spinal tracts, except probably the dorsal columns, though postmortem reports are too few to be certain. Clinically and experimentally, there is no reason to believe that other spinal tracts are involved. It is generally recognized that depletion of the cofactor thiamine pyrophosphate in the tissues leads to impairment of pyruvate decarboxylation, which is a major source of energy in nervous tissue. Any chemical that interferes with thiamine metabolism, such as nitrofurans, could produce a neuropathy by this means (32).

b. Chronic Arsenic Intoxication: Chronic arsenic intoxication produces a peripheral neuropathy (54) which is indistinguishable from that seen in beriberi except by indirect evidence from the pigmented skin lesions, the hair and nail changes, and the presence of arsenic in the urine, tissues, and hair. Peripheral nerve biopsies show wallerian degeneration only (29), but because of the paucity of postmortem studies and a lack of another available sensitive species, the morphological basis of this intoxication is only inferential from the many clinical studies.

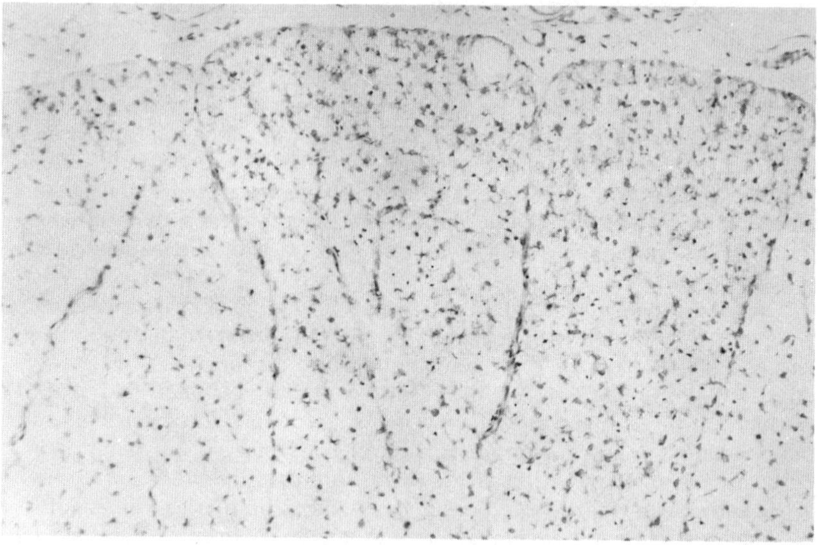

(a)

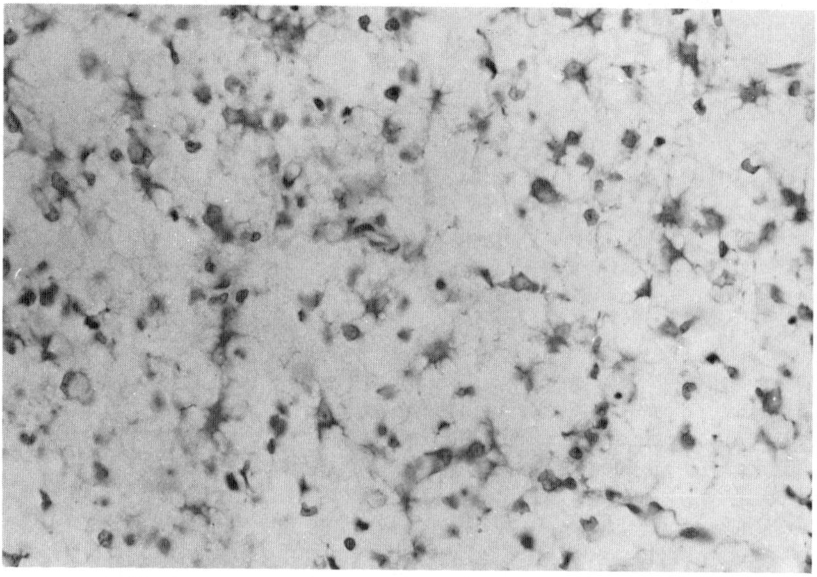

(b)

Figure 9 (a) Dorsal columns in the cervical level from a cat poisoned with tri-*ortho*-cresyl phosphate. Note the marked increase in cell population in the medial (gracile) tracts (cresyl fast violet, 96×). (b) Higher magnification to show the activated and hypertrophic astrocytes and increase in microglia (small nuclei) (cresyl fast violet, 309×).

c. Thallium Poisoning: Thallium poisoning has led to a number of chemical and morphological studies (18). The clinical picture in general is very closely similar to the above two conditions, it also being a distal symmetrical sensory neuropathy with motor involvement appearing slightly later.

Morphologically (60) there is severe wallerian degeneration in more distal parts of both motor and sensory nerve fibers, and sampling peripheral nerves at different levels shows that the numbers of degenerated axons increase in the more distal samples. Distal muscles are more severely denervated than proximal muscles. Anterior horn cells and spinal ganglion cells show chromatolysis (axon reaction) and this change may also be seen in brainstem motor nuclei (Fig. 10). The dorsal columns have shown severe loss of nerve fibers in postmortem cases of severe poisoning. No other spinal tract is involved.

While these postmortem cases show the severest degrees of damage, clinical examination of less severely poisoned cases indicate that the same but milder pattern of denervation has occurred, but which is capable of recovery over several months by regeneration of peripheral axons.

The reasons for believing that the neurotoxicity of thallium intoxication is due to impairment of energy supply are the following:

1. The clinical symptomatology is similar to that of beriberi in all essentials, differing only in severity and tempo.
2. Thallium intoxication in man and in monkey shows features similar to riboflavin deficiency, namely, peripheral neuropathy, characteristic perioral and facial skin lesions, and loss of hair (80).
3. Thallium forms an insoluble complex with riboflavin and would thus be expected to deplete the tissues of riboflavin-derived cofactors (62).
4. Flavoproteins and flavine adenine dinucleotide are required as cofactors at most of the steps in the use of several intermediates along the pathways of energy metabolism.

In view of the close morphological similarities between the neurological lesions of thallium intoxication and those of chronic thiamine deficiency, these metabolic conclusions gain substance.

d. Nitrofurantoin Neuropathy and the Neuropathy of 5-Nitroimidazole Drugs (Metronidazole and Misonidazole): Both these classes of substance are electron-affinic; they are both used as radiation-sensitizing agents for this reason, and they both lead to a persistent painful symmetrical peripheral neuropathy affecting the feet more than the hands. Electrophysiologically they show signs of motor nerve degeneration, with only mild slowing of conduction velocity. Sural nerve biopsy shows only the changes of wallerian degeneration, and

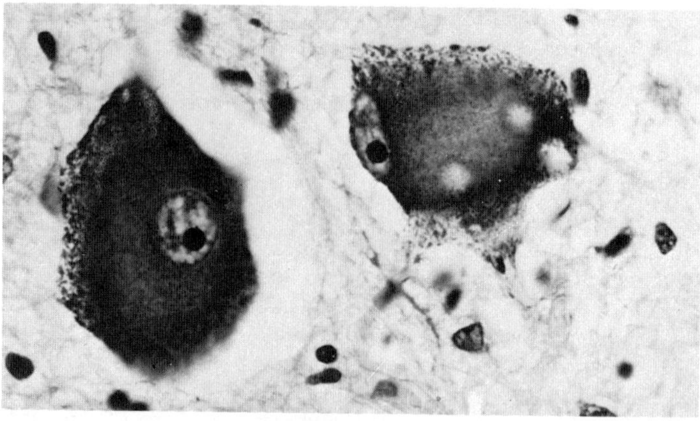

(a)

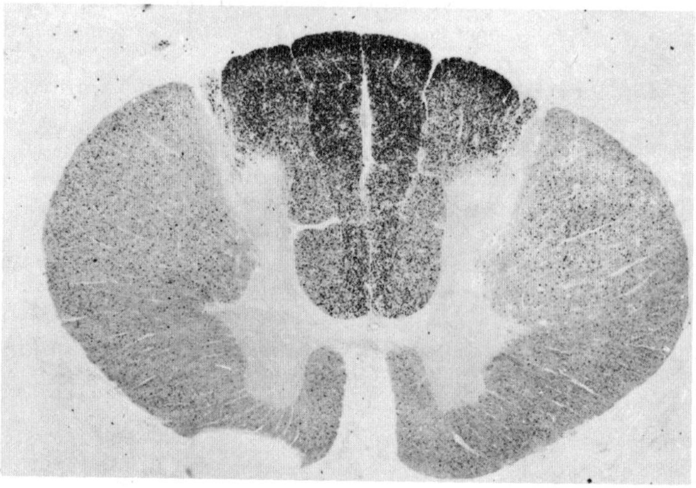

(b)

Figure 10 Thallium intoxication. (a) Chromatolysis of cervical anterior horn cells from a case published in Ref. 60 (cresyl fast violet, 450×). (b) Upper cervical spinal cord showing severe degeneration in the dorsal columns (more so in the gracile tracts than in the cuneate tracts). The small amount of degeneration in the lateral corticospinal tracts is caused by a recent brainstem lesion produced while in a respirator (Marchi stain, 4×).

at postmortem evidence of denervation atrophy can be found in muscles, as well as degeneration and regeneration in peripheral sensory nerves (unpublished observations). No definite changes in the central nervous system have been found in man.

Experimental studies on rats with misonidazole given 400 mg/kg per day for five doses have shown (41) that there is definite evidence of distal sensory denervation, but this is only present in the most distal part of the legs and feet. There are, however, also symmetrical and necrotic lesions in many brainstem nuclei of the rats, particularly in vestibular and dentate nuclei, which very closely resemble lesions occurring in the same regions in this species in acute experimental thiamine deficiency. It is a characteristic of the rat to be unusually resistant to the development of peripheral neuropathy, even when made severely deficient in vitamin B_1 (71).

It is thus on the basis of these morphological and clinical findings that it can be concluded that intoxication by these two classes of chemicals is probably producing the same metabolic lesion in the neurons as those better-defined conditions of thiamine deficiency and arsenic intoxication. However, neither in man nor in the rat can these changes be prevented by large doses of thiamine. In explanation, it might be suggested that these compounds could act as electron "sinks" (31) and thereby diminish the production of adenosine triphosphate in the cell normally formed by the actions of the electron transport chain.

3. *Disturbances to Pyridoxal Phosphate Metabolism*

A number of drugs produce peripheral neuropathies of the same general type which have in common the ability to interact with and deplete cells of the cofactor pyridoxal phosphate. Evidence exists that isonicotinic acid hydrazide (isoniazid), ethionamide, also used in the treatment of tuberculosis, and hydralazine, valuable in the treatment of hypertension, can chelate with pyridoxal phosphate (PPO_4) and block the enzyme pyridoxal phosphokinase. Only with isoniazid do we have morphological data as to their effects on the nervous system.

a. Isoniazid Intoxication: Isoniazid neuropathy in man is essentially a symmetrical distal sensory neuropathy affecting the feet rather than the hands. No other part of the nervous system is thus affected, but with higher doses it may be epileptogenic by reducing the synthesis of the inhibitory transmitter γ-aminobutyric acid, a PPO_4-dependent pathway. Biopsy studies have shown only wallerian degeneration in sensory nerves (73). While no postmortem reports are available, there is no clinical suggestion that axons other than peripheral, sensory and to a lesser degree motor, are involved.

Experimental studies (10) have confirmed this and have shown that in rats motor nerves may be very severely damaged when large doses of isoniazid are given. Many distal muscles may indeed be almost completely denervated. Sensory nerve fibers also show degeneration, but sometimes to a lesser extent in a way not fully understood (Fig. 11). Associated with this latter damage is some degree of degeneration of longer axons in the dorsal columns of the spinal cord, whose cells of origin lie in the dorsal root ganglia. No other tract or pathway has been found showing axon degeneration in this species.

The mechanism underlying this neuropathy is unknown. Predegenerative changes in axons following a large single dose of isoniazid in rats have been found (52) to consist of large amounts of smooth endoplasmic reticulum (SER). There is no way of determining yet whether these are a nonspecific protective or reparative response within the axon or whether they indicate a basic disturbance to the SER network within the axon.

Pyridoxal phosphate is not required for energy production mechanisms, and therefore we cannot invoke a metabolic mechanism like that in arsenical and analogous intoxications. This cofactor is required for amino decarboxylations, transaminations, and racemization reactions. It is also required for the synthesis of collagen, in a lysyl oxidase step, but so far we have no knowledge as to whether any of these mechanisms are essential to axon maintenance.

4. Disturbances to Protein Synthesis in the Perikaryon: Chronic Mercury Intoxication: There are three kinds of neurotoxic effects of chronic mercury poisoning known which superficially appear very different. However, closer examination of the morphological and chemical features of these conditions leads one to conclude that the differences lie probably in the degree of cellular damage produced by each kind of mercury and therefore its reversibility in each case, rather than to any qualitative differences in the toxicity of the different forms of mercury. These differences result almost certainly from the solubility characteristics of the three types of mercury responsible, and therefore their relative abilities to enter the nervous system and to be retained within it.

a. Inorganic (Mercurous Salt) Mercury Intoxication: Mercurous salts (calomel), when given to infants in teething powders, are generally considered to have been responsible for the now defunct condition "pink" disease (erythredema polyneuritis) (93). Mercury was found in quantity in the urine of such cases, and the condition disappeared when it ceased to be used for this purpose.

The infants showed a distal weakness and sensory irritation and marked mental disturbances, and at postmortem extensive nerve fiber degeneration was recorded in peripheral nerves (79). Sensory nerves were especially affected. No changes were noted elsewhere in the brain, but recent cases of chronic calomel intoxication in adults (26) have shown focal loss of granule cells in the cerebellum.

b. Inorganic (Mercuric Salt) Mercury Intoxication: Chronic intoxication due to either inhalation of elemental mercury or absorption of mercury salts, such as mercuric nitrate, leads to a characteristic

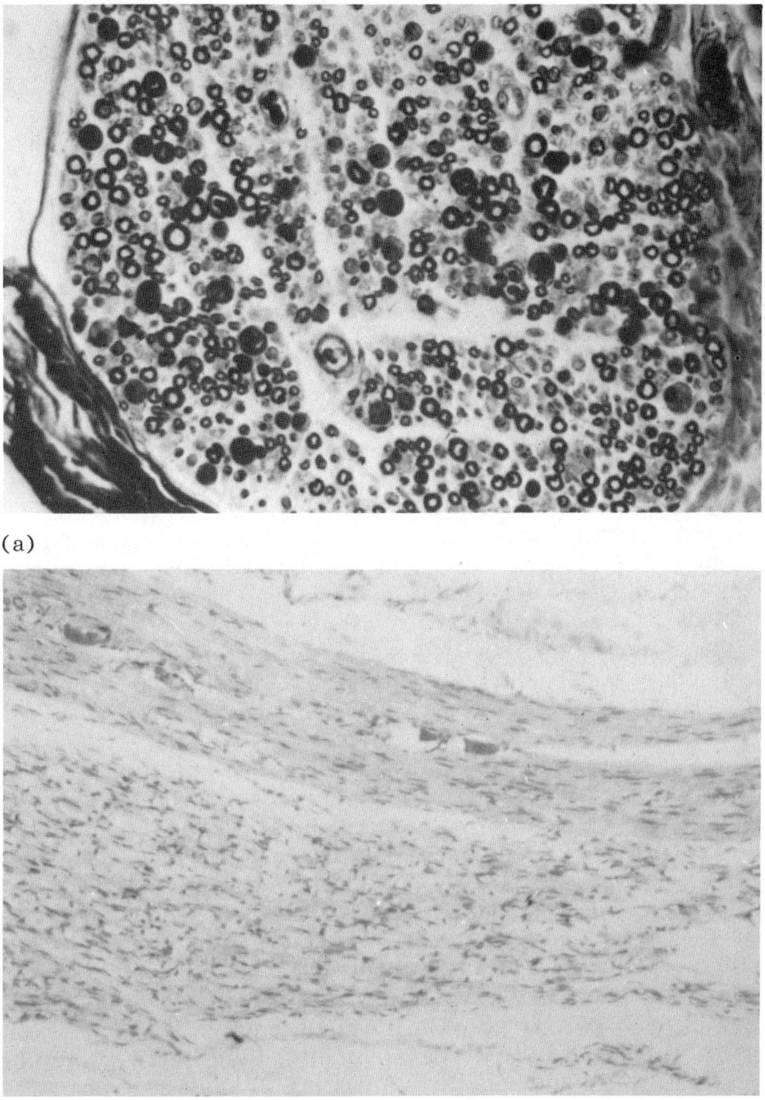

(a)

(b)

Figure 11 Isoniazid intoxication. (a) Rat posterior tibial nerve after 4 weeks of dosing with isoniazid. Note the severe reduction in the number of normal myelinated axons in this mixed motor and sensory nerve. There are many (solid black) degenerating fibers (iron hematoxylin, 330×). (b) Degeneration of the lumbar ventral root with greatly increased cell population by comparison with the normal dorsal root above (hematoxylin, 50×).

syndrome sometimes known as "hatters shakes" (46). Mental confusion, marked loss of motor control, and other disturbances to cerebral function are found, but these are all recoverable on withdrawal from the toxic source. No morphological studies are available for this reason, and it is probable that no irreversible cellular changes take place except in exceptional circumstances of severe intoxication. Recent electrophysiological studies suggest that workers exposed to mercury may develop a distal sensory and motor neuropathy (1). Experimental dosing of rats, however, with mercuric chloride has shown (33,49) that focal granule cell degeneration can be produced in the cerebellum after about 3 months of dosing. Secondly, it has been found that focal loss of ribosomes may be seen in dorsal root ganglion cells identical to that following dosing with methyl mercury, and this is also associated with a disturbance to protein synthesis.

c. *Organic Mercury Intoxication:* Human intoxication from organic (alkyl) mercury absorption has been well documented from industry (47), from Japan (32,90, see also T. Takeuchi in this book), and from Iraq (32). By comparison with inorganic mercury poisoning, the signs and symptoms are almost completely irreversible, and postmortem studies (48,91) have shown in the brain that areas of cortex and of cerebellum where small stellate nerve cells are numerous show extensive loss of these cells. The worst-damaged regions were therefore the visual cortex and cerebellum, the other regions of the granular cortex, for example, the auditory and sensory cortex, being next worst affected. It was noteworthy that large nerve cells in cerebral cortex, cerebellum, and elsewhere tended not to be killed. Sensory neuropathy was also a feature of these cases, though perhaps some of this symptom could be of cortical origin. Experimental studies (16, 50) have fully confirmed this pattern of central nervous system involvement, small nerve cells being conspicuously prone to necrosis while larger cells were frequently capable of recovery. Only with sensory ganglion cells was this rule broken, and experimental animals here consistently showed a severe sensory but not a motor neuropathy. Unlike other symmetrical peripheral neuropathies of the dying-back type, the axons degenerated rapidly back to a region close to the cell body (Fig. 12), for large numbers of degenerate axons were seen in the spinal roots, a feature rarely seen with dying-back types of neuropathy. It was found ultrastructurally (50) that the earliest changes in the cytoplasm in such cells were variable-sized foci of loss of polyribosomes and rough endoplasmic reticulum (Fig. 13), and these preceded the onset of axon degeneration by several days.

On the basis of these diverse morphological findings and the biochemical evidence of concomitant reduced incorporation of amino acids into proteins (17), it is probable that the primary metabolic lesion in nerve cells in mercury intoxication, whether organic or inorganic, is a disturbance to protein synthesis which may be sufficient to impair

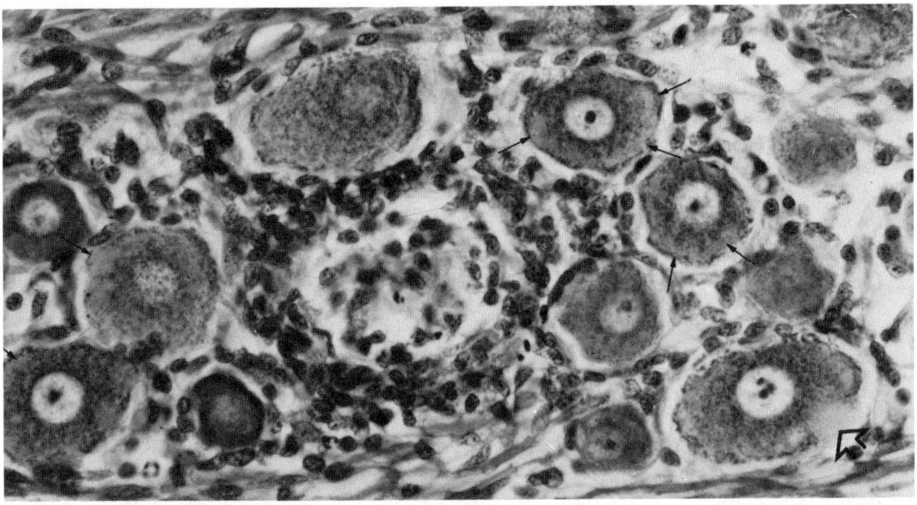

(a)

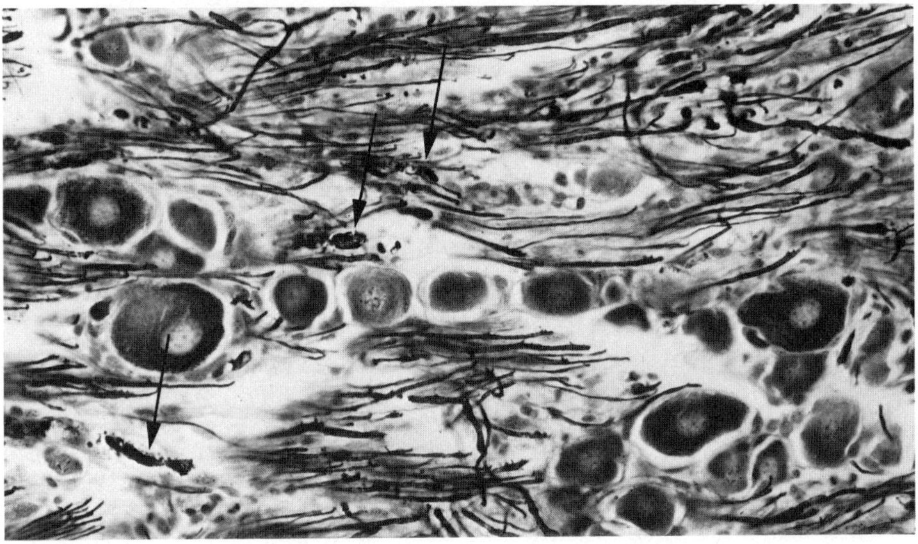

(b)

Figure 12 Organic mercury intoxication. (a) Degenerated rabbit lumbar dorsal root ganglion cell (center) with proliferation of satellite cells and macrophage infiltration. Other cells nearby show peripheral "clearing" of the Nissl material (arrows). This is not to be confused with the axon hillock (open arrow), which normally contains no stainable Nissl material (cresyl fast violet, 270×). (b) Rat dorsal ganglion showing several degenerating axons, but with the perikarya intact (Glees and Marsland stain, 340×).

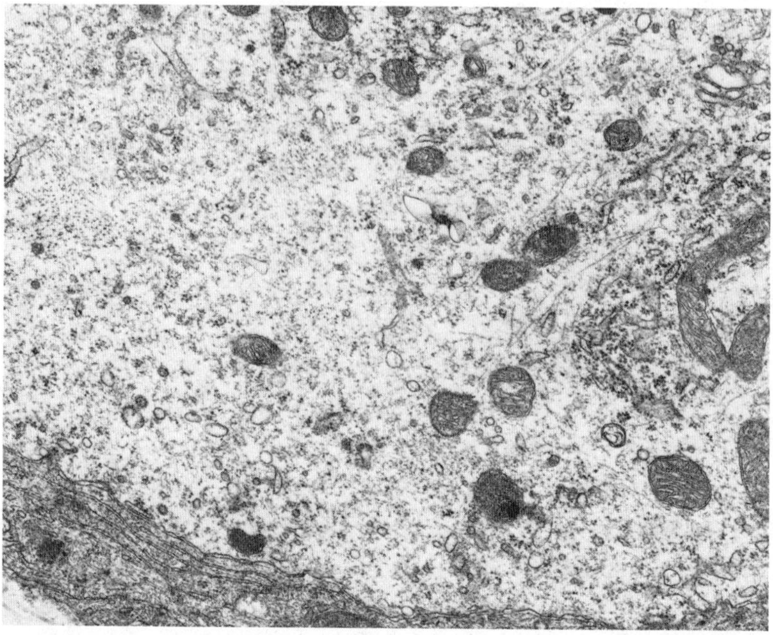

Figure 13 Dorsal root ganglion cell in methyl mercury poisoning showing focal loss of organized Nissl material (polyribosomes and rough endoplasmic reticulum) and many scattered single ribosomes. Mitochondria and adjacent satellite cells look normal (16,000×).

axon survival either distally or proximally or great enough to kill the cell. If mild enough, the cell is able to recover and probably regenerate its axon. In small cells of the central nervous system the amonts of ribosomal organelles are often extremely small, and for this reason such cells may have their whole ribosomal resources destroyed. With larger cells the amount of ribosomal material would be enough to allow cell survival and synthesis of new ribosomes. Only the large sensory cells were badly affected, in all likelihood because of the ready permeability of their vascular bed (51) and the inevitably easier access of mercury compounds from the bloodstream.

5. *Disturbances to Neurofilament Behavior*

An important component of axons are 10-nm filaments (intermediate filaments) which are synthesized in the cell body and move down the axon at 2-3 mm/day (64). Their precise functions are not known, but they may offer some skeletal support to the axon and they are probably concerned with axon growth in development and in regeneration. They must be able to slide easily one on another in order to negotiate the constrictions along axons at each node of Ranvier, for in larger-diameter axons the axon diameter may be reduced to as little as one-quarter of the diameter found in the internodal region (11). These paranodal constrictions are rather rigid and not easily distended owing to the abundance of myelin loops wound around the axon at these sites (63). At the axon terminals in the adult it is apparent that the filaments must be degraded, and the involvement of calcium-activated proteases is invoked for this (64).

Three intoxications are known in which the neurofilament appears to be the primary target and by cross-linking mechanisms their normal flow characteristics are impaired. In consequence of this, they now not only find it difficult to move along axons and pass through the paranodal constrictions, but by their bulk they may cause obstruction to the transport of other materials vital for the maintenance and survival of the more distal regions. A symmetrical sensory and motor distal neuropathy is thus indirectly produced by these substances.

a. Hexacarbon Intoxication: n-Hexane and the analog methyl n-butyl ketone on chronic exposure will cause a symmetrical sensory and/or motor distal neuropathy in man (3) and in experimental animals (88). This syndrome is associated with marked swelling of axons by accumulations of 10-nm filaments, and these can be readily seen on nerve biopsy. Teasing the nerve fibers shows (Figs. 14 and 15) that while the swellings may be seen along the myelin internodes, it is more usual to find them proximal to nodes of Ranvier. The overlying myelin may be greatly thinned or even appear nonexistent, but there is no true demyelination to be seen. The disorganization of the nodes of Ranvier caused by these filament masses (57) is responsible for the slowing of the conduction velocity that is commonly present.

Experimental studies (20) have shown, however, that this is no simple peripheral neuropathy, for axons in many regions of the central nervous system also show marked swelling (Fig. 16), but these seem to have little functional effect unless nerve fiber degeneration is also present.

Chronic intoxication will eventually lead to wallerian degeneration in the larger and longer axons of peripheral nerves and in the long fibers of spinal tracts. This degeneration does not appear to be a direct effect of the intoxication, but secondary to blockage of paranodal constrictions by filament masses (11). If the animal is allowed to recover, the filament masses in tracts with smaller axons slip down toward the terminals, where they are eventually removed (56); no degeneration occurs here. In larger axons of the lateral spinal tracts the filament masses may remain unmoved during the recovery period and a steady increase in axon degeneration can be found both here and in large-diameter peripheral axons. It is known that fast axoplasmic transport is increasingly slowed

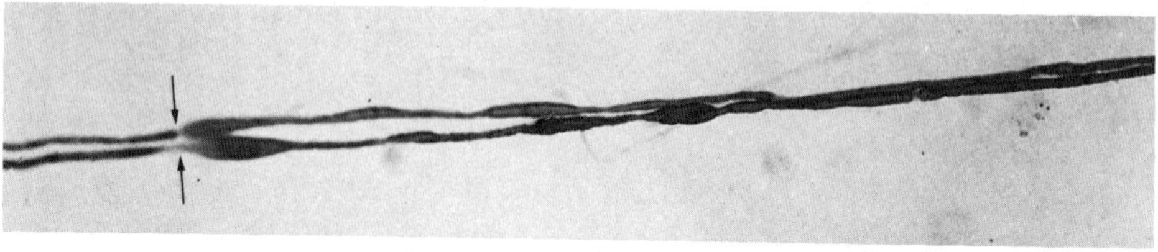

Figure 14 Teased nerve fibers from the posterior tibial nerve of a rat dosed with 2,5-hexanediol for 5 weeks. Note the markedly swollen regions both in the internodes and proximal to the nodes of Ranvier (arrows) (OsO_4 stained, teased in araldite, 200×).

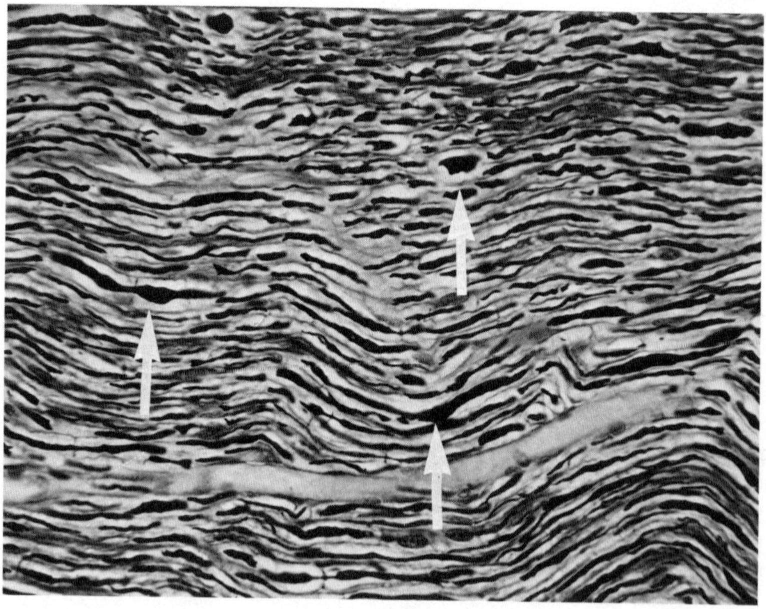

Figure 15 Sciatic nerve from a rat chronically exposed to 2,5-hexanediol for 5 weeks showing focal swelling of axons (arrows) (Glees and Marsland's stain, 360×).

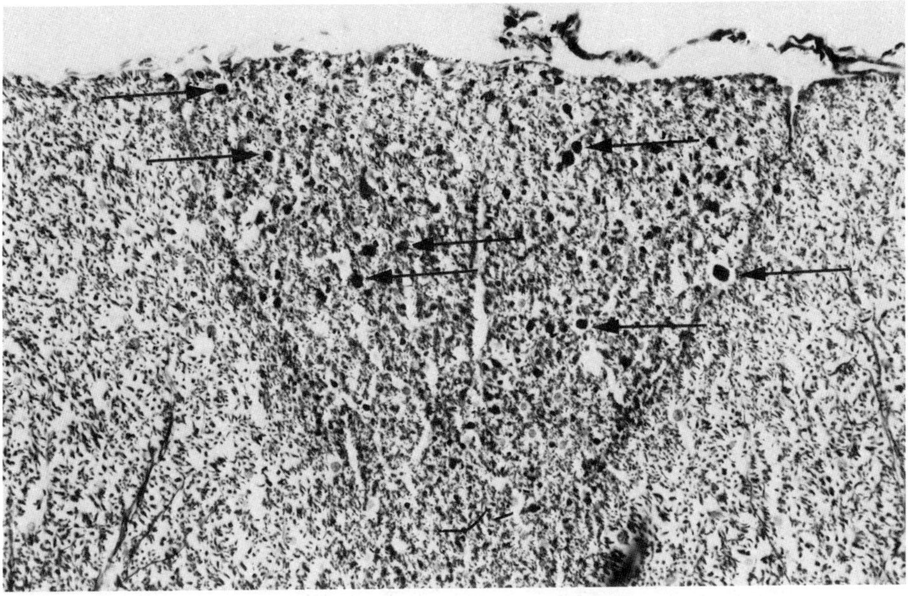

Figure 16 Dorsal spinal columns of a rat chronically intoxicated with 2,5-hexanediol. Note the numerous swollen argyrophilic axons (arrows) (Glees and Marsland stain, 250×).

(67) as the intoxication proceeds, and labeled materials can be found obstructed by the filament masses (40). The biochemical mechanism appears to result from metabolism of the hexacarbons to 2,5-hexanedione, which has the capacity to cross-link proteins, in this case axonal filaments (36,37).

b. Carbon Disulfide Intoxication: Exposure to carbon disulfide (CS_2) as an occupational hazard leads to a peripheral motor and sensory neuropathy with marked slowing of conduction velocity (92). As the result of animal exposure studies carried out in Finland (44), Poland (89), and Holland (23), it can now be stated that the morphological changes in the peripheral and central nervous systems are identical to those caused by chronic absorption of hexacarbon solvents (Figs. 17-19). So close is the morphological identity of these two toxic conditions that one can confidently expect that the precise primary lesion will be the same.

c. β,β'-Iminodipropionitrile Intoxication: β,β'-Iminodipropionitrile compounds derive from substances extracted from the sweet pea *Lathyrus odoratus*, and they have a marked effect on the stability of collagen. The above compound, when given to rats or cats, leads to a peripheral neuropathy in which the axons close to the nerve cell bodies in anterior

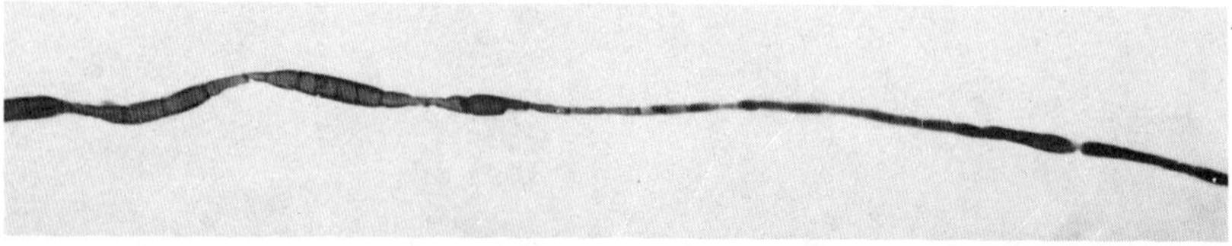

Figure 17 Teased ulnar nerve from a rat chronically exposed to carbon disulfide. Note the marked internodal swellings; the adjacent node of Ranvier is not swollen in this instance (OsO_4 stained, teased in araldite, 200×).

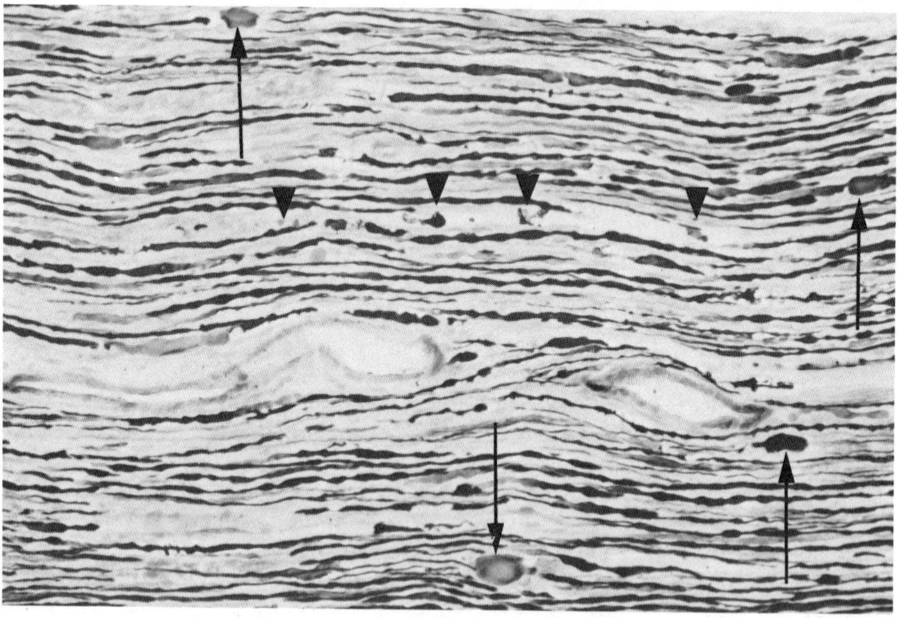

Figure 18 Sciatic nerve from rat chronically exposed to CS_2. Note the focal swellings of axons (arrows) and one fiber undergoing wallerian degeneration (arrowheads) (Glees and Marsland stain, 160×).

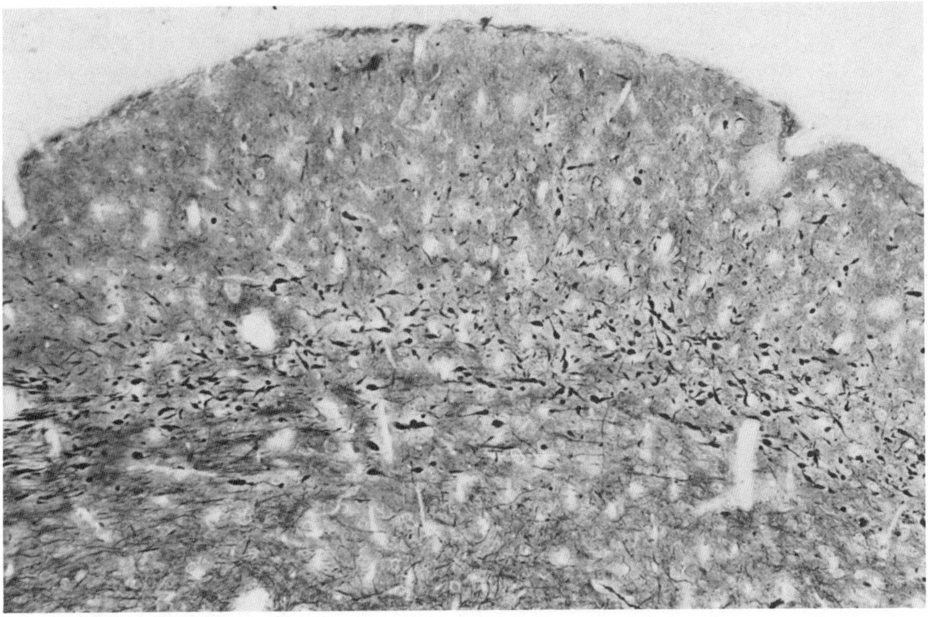

Figure 19 Superior colliculus of a rat chronically exposed to CS_2. Numerous grossly swollen and argyrophilic preterminal axons in the stratum opticum. The picture is the same as that produced in hexacarbon intoxication (Glees and Marsland stain, 100×).

horns, brainstem, Purkinje cells (24), and other neurons show great dilatation due to accumulation of whorled masses of 10-nm filaments. There basically appears to be no difference in this condition from hexacarbon or CS_2 intoxication, except that the filaments occur more markedly proximally rather than toward the distal regions of the axon (42,43). Filament masses may be found at all points, however, if carefully sought. In all other respects the same principles hold for the pathological changes associated with this phenomenon.

The reason for the different distribution of the filament masses may lie more in the nature of the chemical and its reactivities rather than in anything inherent in the nerve cell. It is of interest that dimethyl 2,5-hexanedione, an analog of the proximate toxic substance of hexacarbon poisoning, will also produce proximal swellings in axons (5).

6. *Disturbances to Microtubule Stability*

Microtubules form an essential quasi-stable structure within axons which probably play a vital role in the transport of organelles such as mitochondria and synaptic vesicles, and perhaps rapidly transported substances (28,64). Microtubules are synthesized in the cell body and slowly move distally along the axon, being presumably degraded distally in the same manner as filaments. They can be depolymerized by agents such as colchicine, vinblastine, and vincristine. The first does not pass through the blood-brain barrier, but the last two, when used in the treatment of lymphomas, particularly in children, has caused peripheral neuropathy (85).

a. Vincristine Neuropathy: Abdominal pain, sensory disturbances, distal weakness, and electrical evidence of denervation may follow treatment with vincristine in a dose-related manner. Biopsy of nerves shows wallerian degeneration with no special features (8).

Experimental administration to rabbits (85) has shown that wallerian degeneration will be found, but there are associated filament accumulations seen ultrastructurally in the affected nerve cells. These filaments probably represent products of excess tubulin dimer combined with vincristine in the cell. Absence of microtubules within the axon, whether produced by this way or by local injection into the axon, leads to block in axoplasmic transport and wallerian degeneration.

7. *Disturbance to Smooth Membrane Function*

There has been increasing evidence recently that a network of smooth endoplasmic reticulum (SER) exists throughout axons that plays an important role in the transport of proteins and other essentials and in the maintenance of axons (27). There is also growing evidence that accumulations of SER can be found within axons just before the onset of axonal degeneration in certain toxic neuropathies. This was first seen in organophosphorus toxicity in cats (6), then in isoniazid neuropathy in rats (52), and in a different context in acrylamide intoxication (21). In the first two cases it was suggested that it might be reactionary to some toxic metabolic lesion within the axon and could be in some way protective, counteracting the effects of the metabolic damage. In acrylamide the accumulations occurred in axons as an event preceding the curious and unexplained dose-dependent retrograde axon degeneration occurring proximal to a nerve ligature. In acrylamide intoxication there is also a remarkable disturbance (22) induced in the SER of Purkinje cells which may lead to cell death, and there are other factors of this intoxication which point to a rather special relationship that acrylamide seems to have with this important organelle. Until we know more about the functions of the axonal SER on the one hand and the metabolic lesions induced by acrylamide on the other, this relationship must be regarded as provisional but highly probable.

a. Acrylamide Intoxication: Acrylamide, a water-soluble substance, is widely used in industry, where the properties of its polymers are very valuable. It is readily absorbed and causes a mild sensory and motor distal neuropathy with two unusual features (35). There may be a marked truncal ataxia, occurring early in the condition, and there are striking signs of damage to the autonomic nervous system, namely, marked increase in sweating and red peeling changes in the palms and soles of the feet. Nerve biopsies show nerve fiber degeneration, and the electrophysiological changes confirm this as a generally distal event.

Studies in animals have clarified the pathological changes and have shown that this is a generalized condition of the nervous system not solely affecting peripheral nerves (12). In a sequential study in rats it has been found that the earliest change is loss of Purkinje cells from the cerebellum. Necrotic nerve cells begin within a few days of daily dosing with acrylamide, and continue so long as it is given. Death of these cells is preceded by some remarkable disturbances to the organization and arrangement of SER in this cell, and there is evidence that the cell attempts to discharge from itself the abnormal SER with the active aid of astrocytic intrusions entering close to where the SER masses have accumulated (22).

The next feature of this remarkable intoxication is that during the week or two before the occurrence of axon degeneration there are accumulations of filaments mixed with SER membranes and other organelles in the distal ends of axons (82). This leads to the swelling of the terminals and their increased staining with silver impregnation methods (Fig. 20), which, it seems, have an affinity for neurofilaments. This terminal axonal swelling is a generalized affair and can be seen peripherally in all sensory and motor nerves more or less equally wherever they may be found, and regardless of whether they will ultimately be found to degenerate. It also can be seen in the central nervous system, so that boutons terminaux become significantly enlarged and readily visible by light microscopy over anterior horn cells, neurons in cranial nerve nuclei, and any region where synapses are abundant. These are particularly well seen in the superior colliculus, where retinotectal fibers terminate. As the intoxication continues, the intra-axonal swelling extends away from the terminal region backward for some distance along the axon (Fig. 21). Swollen fibers may thus be seen in the sciatic nerve in the later stages of this predegenerative condition. According to the dose regimen employed, distal axon degeneration spreading proximally begins during the third or fourth week, and thereafter the changes in degenerating fibers are those of wallerian degeneration.

The third striking change in the predegenerative phase is the appearance of "chromatolysis" in sensory ganglion cells (Fig. 22), but here it occurs before axon degeneration can be found. This, in fact, is due to considerable accumulations of mitochondria, many of them grossly swollen, in the centers of the cells, degranulation of the rough endoplasmic reticulum and polyribosomes, and eccentricity of the nucleus. One can only conclude from these findings that acrylamide gravely disturbs the metabolism of the whole cell, but whether from a general metabolic lesion or from some local axonal defect with secondary repercussions throughout the cell cannot yet be stated.

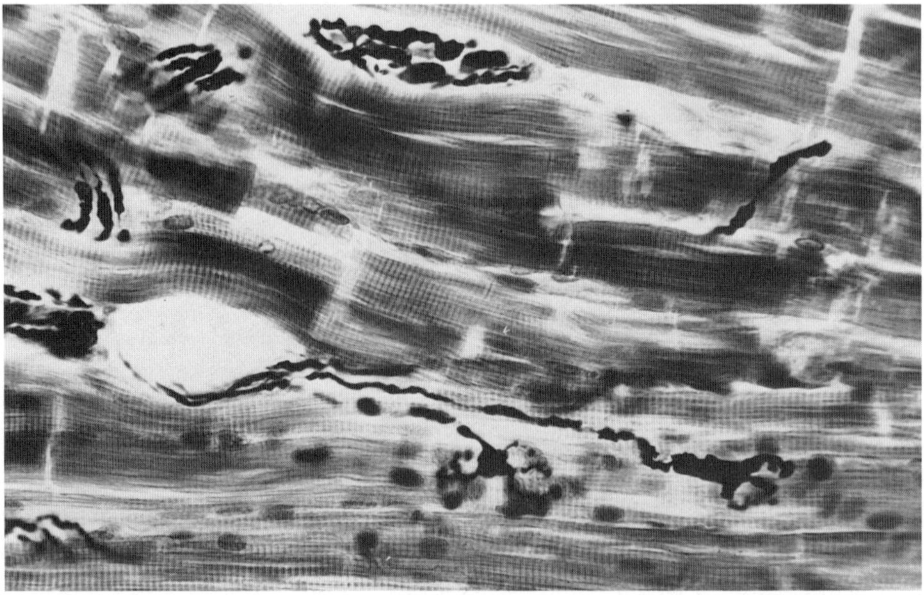

Figure 20 Tibialis anticus muscle showing swollen terminals and preterminal fibers of motor nerves. Swollen argyrophilic axons are also seen in the small nerve bundles above (acrylamide 30 mg/kg per day, 5 days/week, for 2 weeks; Glees and Marsland stain, 270×).

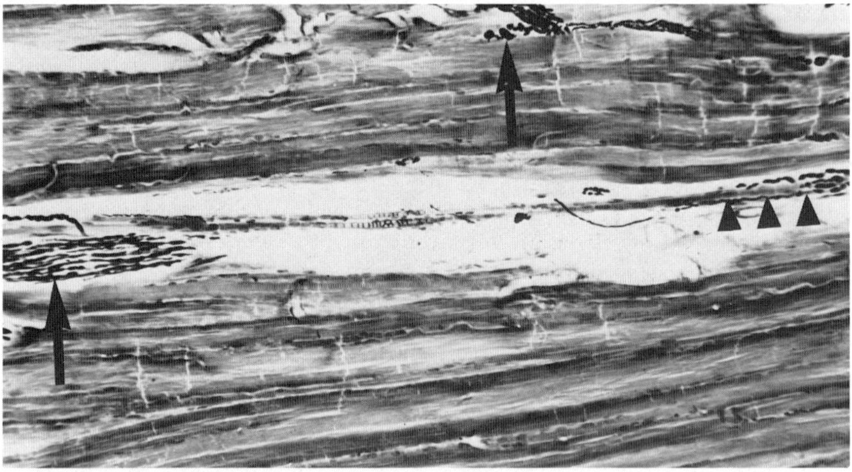

Figure 21 Muscle spindle of rat intrafusal muscle fibers showing nuclear bag and nuclear chain arrangement (center). No annulospiral formation is present because of its degeneration. Note the swelling and argyrophilia of secondary sensory fibers (arrowheads) and of adjacent intramuscular axons (arrows) (acrylamide 30 mg/kg per day, 5 days/week, for 21 days; Glees and Marsland stain, 108×).

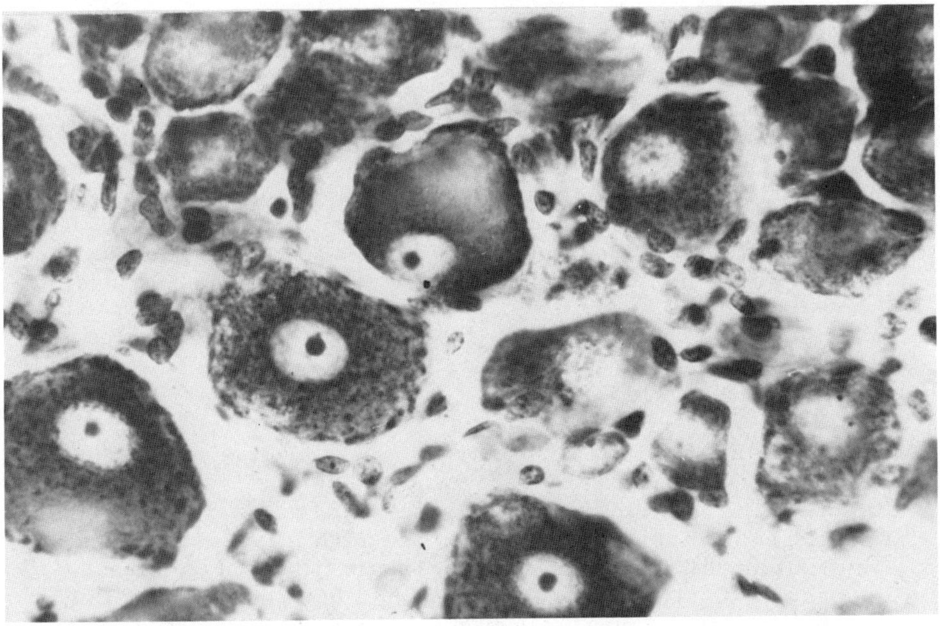

Figure 22 Lumbar spinal ganglion cells from a rat dosed with 30 mg/kg of acrylamide for 10 days. Note the chromatolytic neuron with an eccentric nucleus and central clearing of Nissl material from the cytoplasm (cresyl fast violet, 600×).

There are other morphological features of this intoxication which need to be emphasized if the whole picture is to be appreciated. Thus there is an interference with axonal regeneration after a nerve crush (69), inhibition of terminal sprouting after partial denervation and local botulinum intoxication (59), and retrograde axonal degeneration after nerve injury for several centimeters in a dose-dependent manner (19). There is also slowing of fast axonal transport (94) and, perhaps more relevant to the changes noted above, there is some failure of incorporation of radioactively labeled proteins into axon membranes (27,39). These labeled proteins are transported in relation to the SER, but appear to be unable to leave this on arrival distally.

The morphological features of acrylamide intoxication are thus complex and manifold, but generally leave an impression that the SER is markedly disturbed in a way that could interfere with axonal maintenance or repair. The reaction in the Purkinje cell is different from the changes in the sensory ganglion cell. But the general morphology (and the probable functions) of the SER in the two types of cell is strikingly different. In the former there is a loose subplasmalemmal network throughout the dendrites and insinuating into all the synaptic spines. In the latter the amount in the cell body is small, but the network in the axons would be anticipated to be concerned with quite different functions than in the Purkinje cell dendrites. It is not wholly surprising, therefore, that the intoxication would express itself differently in the two types of cell that are so greatly different in their forms.

The metabolic basis of the intoxication is unknown. It should be noted that acrylamide binds to –SH groups and persists in the tissues for a long time (45). However, no specific metabolic lesion has yet been identified that throws any light on these observed morphological changes.

IV. CONCLUSIONS

Because of the structural complexity of the nervous system and the specialization of functions that are found within it, it is impossible to begin to think constructively either about understanding how the clinical signs and symptoms of intoxication are produced or what the underlying pathological mechanisms are that lead to the condition without a thorough study of the morphological changes in each condition. Only on this basis are the pharmacological, electrophysiological, and the behavioral studies going to be interpretable. Without a full morphological analysis in a sensitive species, the rest is speculation, and this may be all too often misleading. Fortunately a sensitive species can usually (but not always) be found, and it is then that we can begin to make some kind of headway in understanding the mechanisms underlying neurotoxicity. Since diseases of the nervous system occur because of a breakdown of one or another complex and often stepwise intracellular processes, and since these complex processes have many points at which they can go wrong, it is becoming apparent that in order to analyze correctly the cellular basis of a neurotoxic event, we must have a fuller understanding of how the cell looks after itself and protects itself against defects in its daily economy. Neuropathology and likewise neurotoxicology can only advance as the processes of intracellular neurobiology become unraveled. Indeed, neurotoxic chemicals have a lot to offer to those that wish to understand how nerve cells work.

REFERENCES

1. Alber, J. W., Cavender, G. D., Levine, S. P., Langolf, G. D. 1982. Asymptomatic sensorimotor polyneuropathy in workers exposed to elemental mercury. *Neurology 32*:1168-1172.

2. Allen, B. W., Ellard, G. A., Gammon, P. T., King, R. C., McDougall, A. C., Rees, R. W., Weddell, A. G. M. 1975. The penetration of dapsone, rifampicin, isoniazid and pyrazinamide into peripheral nerves. *Br. J. Pharmacol.* *55*:151-155.
3. Allen, N., Mendell, J. R., Billmeier, D. J., Fontaine, R. E., O'Neill, J. 1975. Toxic polyneuropathy produced by the industrial solvent methyl n-butylketone. *Arch. Neurol. Chicago 32*: 209-218.
4. Allt, G., Cavanagh, J. B. 1969. Ultrastructural changes in the region of the node of Ranvier in the rat caused by diphtheria toxin. *Brain 92*:455-468.
5. Anthony, D. C., Boekelheide, K., Giangespero, F., Allen, J. C., Parks, H., Priest, J. W., Webster, D., Graham, D. G. 1982. The neurofilament neuropathies: A unifying hypothesis. *J. Neuropathol. Exp. Neurol. 41*:317.
6. Bouldin, T. W., Cavanagh, J. B. 1979. Organophosphorus neuropathy. II. A fine structural study of the early stages of axonal degeneration. *Am. J. Pathol. 94*:253-262.
7. Bradbury, M. 1979. *The Concept of the Blood-Brain Barrier*, Wiley, Chichester.
8. Bradley, W. G., Lassman, L. P., Pearce, G. W., Walton, J. N. 1970. The neuromyopathy of vincristine in man. Clinical, electrophysiological and pathological studies. *J. Neurol. Sci. 10*:107-131.
9. Cavanagh, J. B. 1964. Peripheral nerve changes in orthocresyl phosphate poisoning in the cat. *J. Pathol. Bacteriol. 87*:365-303.
10. Cavanagh, J. B. 1967. On the pattern of change in peripheral nerves produced by isoniazid intoxication in rats. *J. Neurol. Neurosurg. Psychiatry 30*:26-33.
11. Cavanagh, J. B. 1982. The pattern of recovery of axons in the nervous system of rats following 2,5-hexanediol intoxication: A question of rheology? *Neuropathol. Appl. Neurobiol. 8*:15-34.
12. Cavanagh, J. B. 1982. The pathokinetics of acrylamide intoxication: A reassessment of the problem. *Neuropathol. Appl. Neurobiol. 8*:315-336.
13. Cavanagh, J. B. 1984. The problems of neurons with long axons. *Lancet 1*:1284-1287.
14. Cavanagh, J. B., Jacobs, J. M. 1964. Some quantitative aspects of diphtheritic neuropathy. *Br. J. Exp. Pathol. 45*:309-322.
15. Cavanagh, J. B., Patangia, G. N. 1965. Changes in the nervous system of the cat as the result of tri-ortho-cresyl phosphate poisoning. *Brain 88*:165-180.
16. Cavanagh, J. B., Chen, F. C. K. 1971. The effects of methyl mercury dicyandiamide on the peripheral nerves and spinal cord of rats. *Acta Neuropathol. 19*:208-215.
17. Cavanagh, J. B., Chen, F. C. K. 1971. Amino acid incorporation into protein during the "silent phase" before organomercury and

p-bromophenylacetylurea neuropathy in the rat. *Acta Neuropathol.* *19*:216-224.
18. Cavanagh, J. B., Fuller, N. H., Johnson, H. M., Rudge, P. 1974. The effects of thallium salts with particular reference to the nervous system. *Q. J. Med.* *43*:293-319.
19. Cavanagh, J. B., Gysbers, M. F. 1980. "Dying back" above a nerve ligature produced by acrylamide. *Acta Neuropathol.* *51*: 169-177.
20. Cavanagh, J. B., Bennetts, R. J. 1981. On the pattern of changes in the rat nervous system produced by 2,5-hexanediol: A topographical study by light microscopy. *Brain* *104*:297-318.
21. Cavanagh, J. B., Gysbers, M. F. 1981. Ultrastructural changes in axons caused by acrylamide above a nerve ligature. *Neuropathol. Appl. Neurobiol.* *7*:315-326.
22. Cavanagh, J. B., Gysbers, M. F. 1983. Ultrastructural features of the Purkinje cell damage caused by acrylamide in the rat: A new phenomenon in cellular neurobiology. *J. Neurocytol.* *12*:413-437.
23. Cavanagh, J. B., de Groote, D. M. G. 1984. In preparation.
24. Chou, S.-M., Hartmann, H. A. 1964. Axonal lesions and waltzing syndrome after IDPN administration in rats with a concept of "axostasis." *Acta Neuropathol.* *3*:428-450.
25. Collier, R. J., Cole, H. A. 1969. Diphtheria toxin subunit active *in vitro*. *Science* *164*:1179-1181.
26. Davis, L. E., Nands, J. R., Weiss, S. A., Price, D. L., Girling, E. F. 1974. Central nervous system intoxication from mercurous chloride laxatives. *Arch. Neurol.* *30*:428-431.
27. Droz, B., Rambourg, A., Koenig, H. L. 1975. The smooth endoplasmic reticulum: Structure and role in renewal of axonal membrance and synaptic vesicles by fast axoplasmic transport. *Brain Res.* *93*:1-13.
28. Dustin, P. 1978. *Microtubules*, Springer-Verlag, Berlin.
29. Dyck, P. J., Gutrecht, J. A., Bastron, J. A., Karnes, W. E., Dale, A. J. D. 1968. Histologic and teased fibre measurements of sural nerve in disorders of lower motor and primary sensory neuron. *Mayo Clin. Proc.* *43*:82-123.
30. Dyck, P. J., Windebank, A. J., Low, P. A., Baumann, W. J. 1980. Blood-nerve barrier in rats and cellular mechanisms of lead induced segmental demyelination. *J. Neuropathol. Exp. Neurol.* *39*:700-709.
31. Edwards, D. I., Dye, M., Carne, H. 1973. The selective toxicity of anti-microbial nitroheterocyclic drugs. *J. Gen. Microbiol.* *76*: 135-145.
32. Elhassani, S. M. 1983. The many faces of methylmercury poisoning. *J. Toxicol. Clin. Toxicol.* *19*:875-906.

33. Enders, A., Noetzel, H. 1955. Spezifische Veränderungen im Kleinhirn bei chronischer oraler Vergiftung mit Sublimat. *Arch. Exp. Pathol. Pharmakol. 225*:346-351.
34. Fullerton, D. M. 1966. Chronic peripheral neuropathy produced by lead poisoning in guinea pigs. *J. Neuropathol. Exp. Neurol. 25*:214-236.
35. Garland, T. O., Patterson, W. H. 1967. Six cases of acrylamide poisoning. *Br. Med. J. 4*:134-138.
36. Graham, D. G. 1980. Hexane neuropathy: A proposal for the pathogenesis of a hazard of occupational exposure and inhalant abuse. *Chem. Biol. Interactions 32*:339-345.
37. Graham, D. G., Anthony, D. C., Boekelheide, K., Maschmann, N. A., Richards, R. G., Wolfram, J. W., and Shaw, B. R. 1982. Studies of the molecular pathogenesis of hexane neuropathy. II. Evidence that pyrrole derivatization of lysyl residues leads to protein crosslinking. *Toxicol. Appl. Pharmacol. 64*:415-422.
38. Greenfield, J. G. 1954. *The Spinocerebellar Degenerations*, Blackwell, Oxford.
39. Griffin, J. W., Price, D. L., Drachman, D. B. 1977. Impaired axonal regeneration in acrylamide intoxication. *J. Neurobiol. 8*: 355-370.
40. Griffin, J. W., Price, D. L., Spencer, P. S. 1979. Fast axoplasmic transport through giant axonal swellings in hexacarbon neuropathy. *J. Neuropathol. Exp. Neurol. 36*:603p.
41. Griffin, J. W., Price, D. L., Kuethe, D. O., Goldberg, A. M. 1979. Neurotoxicity of misonidazole in rats. I. Neuropathology, *Neurotoxicology 1*:299-312.
42. Griffin, J. F., Hoffman, P. N., Clark, A. W., Carroll, P. T., and Price, D. L. 1978. Slow axonal transport of neurofilament proteins. Impairment by β,β-iminodipropionitrile. *Science 202*: 633-635.
43. Griffin, J. W., Gold, B. G., Cork, L. C., Price, D. L., Lowndes, H. E. 1982. IDPN Neuropathy in the cat: Coexistence of proximal and distal axonal swellings. *Neuropathol. Appl. Neurobiol. 8*: 351-364.
44. Haltia, M., Linnoila, I. 1978. Distal axonopathy induced by carbon disulphide. Experimental studies. *J. Neuropathol. Exp. Neurol. 37*:621p.
45. Hashimoto, K., Aldridge, W. N. 1970. Biochemical studies on acrylamide, a neurotoxic agent. *Biochem. Pharmacol. 19*:2591-2604.
46. Hunter, D. 1955. *The Diseases of Occupations*, English Universities Press, London.
47. Hunter, D., Bomford, R. R., Russell, D. S. 1940. Poisoning by methyl mercury compounds. *Q. J. Med. 9N.S.*:193-213.

48. Hunter, D., Russell, D. S. 1954. Focal cerebral and cerebellar atrophy in a human subject due to organic mercury compounds. *J. Neurol. Neurosurg. Psychiatry 17*:235-241.
49. Jacobs, J. M., Cavanagh, J. B., Carmichael, N. G. 1975. The effect of chronic dosing with mercuric chloride on dorsal root and trigeminal ganglia of rats. *Neuropathol. Appl. Neurobiol. 1*:321-337.
50. Jacobs, J. M., Carmichael, N., Cavanagh, J. B. 1975. Ultrastructural changes in the dorsal root and trigeminal ganglia of rats poisoned with methyl mercury. *Neuropathol. Appl. Neurobiol. 1*:1-20.
51. Jacobs, J. A., McFarlane, R. M., Cavanagh, J. B. 1976. Vascular leakage in the dorsal root ganglia of the rat studied with horseradish peroxidase. *J. Neurol. Sci. 29*:95-107.
52. Jacobs, J. M., Miller, R. M., Whittle, A., Cavanagh, J. B. 1979. Studies on the early changes in acute isoniazid neuropathy in the rat. *Acta Neuropathol. 47*:85-92.
53. Jacobs, J. M., Ford, W. C. L. 1981. Neurotoxicity of 6-chloro-6-deoxyglucose, a chloro sugar with male antifertility properties. *Neurotoxicology 2*:405-418.
54. Jenkins, R. B. 1966. Inorganic arsenic and the nervous system. *Brain 89*:479-498.
55. Johnson, M. K. 1975. Organophosphorus esters causing delayed neurotoxic effects. *Arch. Toxicol. 34*:259-288.
56. Jones, H. B., Cavanagh, J. B. 1982. Recovery from 2,5-hexanediol intoxication of the retinotectal tract of the rat. *Acta Neuropathol. 58*:286-290.
57. Jones, H. B., Cavanagh, J. B. 1983. Distortion of the nodes of Ranvier from axon distension by filamentous masses in hexacarbon intoxication. *J. Neurocytol. 12*:459-473.
58. Joseph, B. S. 1973. Somatofugal events in wallerian degeneration: A conceptual overview. *Brain Res. 59*:1-18.
59. Kemplay, S., Cavanagh, J. B. 1983. Effects of acrylamide and some other sulphydryl reagents on "spontaneous" and pathologically-induced terminal sprouting from motor end-plates. *Muscle Nerve* 7:101-109.
60. Kennedy, P., Cavanagh, J. B. 1976. Spinal changes in the neuropathy of thallium poisoning. A case with neuropathological studies. *J. Neurol. Sci. 29*:295-301.
61. Kristersson, K., Olsson, T. 1971. Retrograde axonal transport of protein. *Brain Res. 29*:363-365.
62. Kuhn, R., Rudy, H., Wagner-Jauregg, T. 1933. Ueber Lactoflavin (vitamin B_2). *Ber. Dtsch. Chem. Ges. 66*:1950-1956.
63. Langley, O. K., Landon, D. N. 1968. A light and electron histochemical approach to the node of Ranvier and myelin of peripheral nerve fibres. *J. Histochem. Cytochem. 15*:722-731.
64. Lasek, R. J. 1981. The dynamic ordering of neuronal cytoskeleton. *Neurosci. Res. Program Bull. 19*:7-31.

65. Laslett, E. E., Warrington, W. B. 1898. The morbid anatomy of a case of lead paralysis. Condition of the nerves, muscles, muscle spindles and spinal cord. *Brain 21*:224-231.
66. Lhermitte, F., Fardeau, M., Chedru, F., Mallecourt, J. 1976. Polyneuropathy after perhexilene maleate therapy. *Br. Med. J. 1*:1256.
67. Mendell, J. R., Sahenk, A., Saida, K., Weiss, H. S., Savage, R., Courti, D. 1977. Alterations of fast axoplasmic transport in experimental methyl-n-butyl ketone neuropathy. *Brain Res. 33*:107-109.
68. Meyer, P. 1881. Anatomische Untersuchungen über diphtheritsche Lähmung. *Virchows Arch. Pathol. Anat. Physiol. 85*:181-226.
69. Morgan Hughes, J. A., Sinclair, S., Durston, J. H. 1974. The pattern of peripheral nerve regeneration induced by crush in rats with severe acrylamide neuropathy. *Brain 97*:235-250.
70. Morgan, K. T. 1974. Amprolium poisoning in preruminant lambs. An ultrastructural study of cerebral malacia and the nature of the inflammatory response. *J. Pathol. 112*:229-236.
71. North, J. D. K., Sinclair, H. M. 1956. Nutritional neuropathy. Chronic thiamine deficiency in the rat. *Arch. Pathol. 62*:341-353.
72. Nye, J. J. 1933. *Chronic Nephritis and Lead Poisoning*, Angus and Robertson, Sydney.
73. Ochoa, J. 1970. Isoniazid neuropathy in man. Quantitative electron microscopy study. *Brain 93*:831-850.
74. Ohnishi, A., Schilling, K., Brimijoin, W. S., Lambert, E. H., Fairbanks, V. G., Dyck, P. J. 1977. Lead neuropathy. I. Morphometry, nerve conduction and choline acetyl-transferase transport: New findings of endoneurial edema associated with segmental demyelination. *J. Neuropathol. Exp. Neurol. 36*:449-518.
75. Ohnishi, A., Dyck, P. J. 1981. Retardation of Schwann cell division and axonal regrowth following nerve crush in experimental lead neuropathy. *Ann. Neurol. 10*:469-477.
76. Oliver, T. 1902. *Dangerous Trades*, John Murray, London.
77. Pannese, E. 1963. Investigations on the ultrastructural changes of the spinal ganglion neurons in the course of axon regeneration and cell hypertrophy I. Changes during axon regeneration. *Z. Zellforsch. Mikrosk. Anat. 60*:74-740.
78. Pannese, E. 1981. The satellite cells of the sensory ganglia. *Adv. Anat. Embryol. Cell Biol. 65*:1-111.
79. Patterson, D., Greenfield, J. G. 1923-1924. Erythroedema polyneuritis (the so-called "pink disease"). *Q. J. Med. 17*:6-18.
80. Pentschew, A., Garro, F. 1969. Thallium encephalopathy in monkeys. *J. Neuropathol. Exp. Neurol. 28*:163.
81. Pleasure, D. B., Feldman, B., Prockop, D. J., 1973. Diphtheria toxin inhibits the synthesis of myelin proteolipid and basic proteins by peripheral nerve *in vitro*. *J. Neurochem. 20*:81-90.
82. Prinease, J. 1969. The pathogenesis of dying back polyneuropathies. Part III. An ultrastructural study of experimental

acrylamide intoxication in the cat. *J. Neuropathol. Exp. Neurobiol. 28*:598-621.
83. Said, G. 1978. Perhexilene neuropathy: A clinicopathological study. *Ann. Neurol. 3*:259-266.
84. Sebille, A. 1978. Prevalence of latent perhexilene neuropathy. *Br. Med. J.* 1321-1311.
85. Shelanski, M. L., Wisniewski, H. 1969. Neurofibrillary degeneration induced by vincristine neuropathy. *Arch. Neurol. Chicago 20*:199-206.
86. Smith, H., Spalding, J. M. K. 1959. Outbreak of paralysis in Morocco due to orthocresyl phosphate poisoning. *Lancet 2*:1015-1020.
87. Spatz, H. 1938. Die "Systemische Atrophien." *Arch. Psychiatr. 108*:1-55.
88. Spencer, P. S., Schaumburg, H. H. 1975. Experimental neuropathy produced by 2,5-hexanedione—A major metabolite of the neurotoxic industrial solvent methyl n-butyl betone. *J. Neurol. Neurosurg. Psychiatry. 33*:771-775.
89. Szendzikowski, S., Stetkiewicz, J., Wronska-Nofer, T., Zdrajkowska, I. 1973. Structural aspects of carbon disulphide neuropathy I. Development of neurohistological changes in chemically intoxicated rats. *Int. Arch. Arbeitsmed. 31*:135-149.
90. Takeuchi, T., Eto, N., Eto, K. 1979. Neuropathology of childhood cases of methylmercury poisoning (Minamata disease) with prolonged symptoms, with particular reference to the decortication syndrome. *Neurotoxicology 1*:1-20.
91. Takeuchi, T., Morikawa, N., Matsumoto, H., Shiraishi, Y. 1962. A pathological study of Minamata disease in Japan. *Acta Neuropathol. 2*:40-57.
92. Vasilescu, C. 1972. Motor nerve conduction velocity and electromyogram in carbon disulphide poisoning. *Rev. Roum. Neurol. 9*:63-71.
93. Warkany, J., Hubbard, D. M. 1948. Mercury in the urine of children with acrodynia. *Lancet 1*:829-830.
94. Weir, R. L., Glaubiger, G., Chase, T. N. 1978. Inhibition of fast axoplasmic transport by acrylamide. *Environ. Res. 17*:251-255.
95. Windebank, A. J., McCall, J. T., Hunder, H. G., Dyck, P. J. 1980. The endoneurial content of lead related to the onset and severity of segmental demyelination. *J. Neuropathol. Exp. Neurol. 39*:692-699.
96. Windebank, A. J., Dyck, P. J. 1981. Kinetics of ^{210}Pb entry into the endoneurium. *Brain Res. 225*:67-74.
97. Wright, H. 1901. Changes in the neuronal centres in beri beri neuritis. *Br. Med. J. 1*:1610-1616.

2

Toxic Brain Failure Syndromes

M. Bozza-Marrubini
Poison Control Center, Ente Ospedaliero Niguarda-Ca' Granda, Milan, Italy

I. DEFINITIONS: BRAIN FAILURE AND COMA

The term *coma* should be restricted to conditions of "unarousable unresponsiveness" according to Plum's definition (12); however, in resuscitation and intensive care the term *coma* is commonly used in a much broader sense to indicate a variety of neurological conditions whose main symptom is a disturbance of consciousness. When "coma" is caused by exogenous toxins, the range of possible variations is especially wide, including both depression of central nervous system activity and excitatory states.

A more correct term, analogous with terminology used in other conditions requiring resuscitation and intensive care, is *brain failure*. As stated by Plum and Posner (11), "impaired or decreased consciousness reflects severe brain dysfunction and coma means brain failure, just as uremia means renal failure."

II. CAUSES OF BRAIN FAILURE IN POISONED PATIENTS

In acute toxic pathology brain failure may be due to the following:

Direct toxic effects of one or more drugs on the central nervous system
A consequence of single or multiple organ or vital function failures due to toxins whose target organ is *not* the central nervous system
The coexistence of both the above mechanisms that may form a vicious circle leading to organic cerebral lesions

A frequent example is deep coma caused by hypnotic drugs that is the starting point of the sequence: respiratory depression, hypoxemia, further depression of the central nervous system, circulatory collapse, brain hypoxia/ischemia, and eventually neuronal damage. With few exceptions (for instance, methanol) central neurotoxins are functional poisons, that is, cause brain failure that is reversible without sequelae as soon as the poison is eliminated. But the same toxins may become "lesional" if brain oxygenation and perfusion are impaired—hence the importance of life support and intensive care in all the cases of toxic cerebral or extracerebral function failure.

III. DIAGNOSIS OF TOXIC BRAIN FAILURE

When a patient with brain failure is admitted to a casualty department, a full clinical and neurological examination is necessary to do the following:

1. Evaluate and classify the severity of brain failure and detect the presence or the imminent danger of vital function failures
2. Achieve a differential diagnosis between *organic* brain failure arising from cerebral hemorrhage, focal or global ischemia, encephalitis, brain injury, tumor, and so on, and *functional* brain failure due to exogenous or endogenous toxins or other metabolic causes (acidotic or hyperosmolar diabetes, hypoglycemia, uremia, liver failure, etc.)
3. Identify the poison or at least the class of poisons involved when the anamnesis and circumstances point to toxic brain failure, although the toxin is unknown or uncertain

The differential diagnosis between organic and functional brain failure is based on anamnestic and circumstantial data; the most important clinical signs pointing to organic brain failure are the *persistent asymmetries* in motor responses, muscular tonus, and reflexes (e.g., hemiplegia, unilateral extensor or flexor posturing, and/or unequal pupils). Bilateral spontaneous posturing, a frequent sign of organic coma, may also be observed in toxic brain failure, but only as a response to noxious stimuli and coupled with a dissociative behavior of brainstem reflexes (pupillary reactivity present, eye movements absent).

The clinical signs and laboratory data that indicate single or multiple organ and function failures fall within the scope of general medicine and resuscitation and will not be discussed here. It must be pointed out, however, that the diagnosis and correction of such failures has overall priority in all patients with brain failure, regardless of cause.

A. Specific Semiotics

In a patient with (ascertained) toxic brain failure, the relevant clinical signs may be grouped into four categories which, if carefully explored at short intervals, give all the data needed to assess the initial severity, to monitor the clinical course and to evaluate the indications of and responses to drugs or special procedures (dialysis, hemoperfusion, etc.). The continuous monitoring of brain electrical activity by electroencephalography or by devices especially designed for this purpose (see below) may be a very useful complement of the clinical examination in selected cases.

The four categories of clinical signs are the following:

Spontaneous motor activity
Verbal responses
Motor responses
Brainstem reflexes

Within each category the activity or responses that can be observed are ranked from best to worse. For recording or classification purposes a numerical score may be given to each sign of the sequence (see Sec. IV).

1. *Spontaneous motor activity* may be (Table 1)

Normal. The patient moves spontaneously in bed, seeking and changing comfortable positions and resting or sleep attitudes. He may also open his eyes spontaneously, blink, and orient his gaze toward surrounding objects and persons.

Absent, but arousable by adequate stimuli. The patient, if undisturbed, lies immobile in his bed, but a *normal* activity as above is observed after arousal by call or by pain or "visceral" stimuli (attempts to introduce a nasogastric or tracheal tube, laryngoscopy, vesical catheterism, etc.).

Abnormal. In toxic brain failure a variety of patterns of abnormal spontaneous motor activity can be observed.

Abnormal motor activity is considered *mild* when it does not interfere with the vital functions (breathing, heart rhythm, circulation,

Table 1 Spontaneous Motor Activity, Verbal Responses, and Motor Responses

Category	Numerical score	Clinical signs
Spontaneous motor activity and muscular tonus	5	Normal
	4	Absent (but present, normal, after stimuli)
	3	Abnormal, *mild*: tremors, myoclonus, choreoathetosis, hypertonus, single convulsive fits
	2	Abnormal, *severe*: generalized, repeated convulsions; status epilepticus; extensor spasms
	1	Absent even after stimulation; flaccidity
Verbal responses	5	Orientated speech
	4	Confused conversation
	3	Inappropriate speech
	2	Incomprehensible sounds
	1	No speech (intubation)
Motor responses to pain	5	Localizing
	4	Not localizing, grimacing
	3	Abnormal activity, mild
	2	Abnormal activity, severe
	1	No response

thermoregulation, or metabolic balance). Such is the case when the abnormal pattern consists of *tremors, myoclonus, choreoathetosis, or isolated epileptic seizures*.

On the other hand, *generalized and repeated convulsions or status epilepticus* are considered *severe*, because these types of abnormal motor activity are followed as a rule and within a short time by severe impairment of the vital functions.

When brain failure is severe, spontaneous motor activity is *absent* and remains so even after strong and repeated stimuli (pain, tracheal intubation or suction). In these cases there is generalized flaccidity or a marked reduction of muscular tonus.

2. The *verbal responses* are classified as in the Glasgow Coma Scale. This category of signs is very important in patients poisoned by drugs with central anticholinergic activity causing confusion, delirium, or hallucinations.

3. The *motor responses* are tested first by verbal orders. If the patient fails to *obey simple orders* such as "open/close your eyes," "push your tongue," "lift your hand," a painful stimulus is used to test the responses. Stimuli adequate for this purpose (and not producing unnecessary lesions even if often repeated) are pinching the trapezius muscle midway between the base of the neck and the shoulder, or firm pressure applied by a hard oblong object (e.g., a pencil) on the nail of a thumb or toe.

In response to such stimuli the patient may *localize pain* and try to remove its cause. Patients who *do not localize* pain may react by grimacing or withdrawing, with *mild* abnormal motor responses, or with *severe* abnormal motor responses. Abnormal motor responses include the same patterns as indicated for the spontaneous motor activity category, and are accordingly classified as mild or severe. Painful stimulation may either raise *ex novo* the abnormal responses or aggravate a spontaneous abnormal activity already present. Severe brain failure is characterized by *no response* either to pain or to laryngeal, tracheal, or carinal stimulation.

4. A highly important category for assessment of brain failure, however caused, involves *pupillary and eye movement reflexes*, because these have their pathways and synapses within the brainstem.

The *brainstem reflexes* giving the most valuable information in toxic brain failure are (a) the corneal reflex (CR), (b) the vestibulo-ocular reflex (VOR), and (c) the pupillary light reflex (LR) (Table 2). The CR is sensorimotor; the VOR is sensorimotor, but with multiple central synapses that probably are at least in part cholinergic. The LR has a sensory afferent limb and an efferent peripheral cholinergic synapse.

In organic brain diseases these reflexes indicate the depth of coma and have well-defined diagnostic and prognostic values (9; see other references in Ref. 1). In toxic brain failure the behavior of the brainstem reflexes is related not only to the severity of central nervous system depression, but also to the possible autonomic effects of the poison, in particular, its central or peripheral anticholinergic effects.

The serial study of the three reflexes (CR, VOR, and LR) may be a very useful tool for the differential diagnosis among different poisons and between toxic coma and coma due to cerebral organic lesions. They also give valuable diagnostic and prognostic information when toxic brain failure is complicated by cardiac arrest or by other vital function failures leading to hypoxic-ischemic damage of the brain. The pathways of the three reflexes and their range of responses are summarized in Table 2. With increasing severity of brain failure due to organic cerebral lesions or to central nervous system depressant drugs without autonomic effects, usually the VOR is the first reflex impaired and the CR the last to disappear.

The sequence usually observed is as follows. At the cold caloric test (VOR) *nystagmus* disappears practically in parallel with consciousness.

Table 2 Brainstem Reflexes

Stimulus to elicit reflex	Afferent neuron	Pathways connecting mechanisms	Efferent nerve or chain	Responses and scores	
Vestibulo-ocular reflex (VOR)					
Head flexed 30° on horizontal plane, irrigation of external ear with 20-60 ml of ice water in 10-60 sec.	Lateral semicircular canal VIII nerve	Vestibular nuclei, medial longitudinal fasciculus, Deiter's fasciculus, reticular system, III and VI nerve nuclei	III nerve motor fibers to rectus medialis muscle (one side), VI nerve motor fibers to rectus lateralis muscle (controlateral)	Nystagmus Conjugate deviation Disconjugate deviation No response	4 3 2 1
					1
Pupillary light reflex (LR)					
Light beam on open eye	Retina, optic nerve, chiasma, and optic tracts	Edinger-Westphal nuclei	III nerve (cholinergic fibers), ciliary ganglion, pupillary sphincter	Pupil constriction present Sluggish Absent	3 2 1
Corneal reflex (CR)					
Touching cornea with (sterile!) cotton or gauze wisp	Sensory terminals of V nerve, ophthalmic branch	VII nerve nucleus	VII nerve motor fibers to orbital and lid muscles	Active eye closing present Absent	2 1

When the *tonic conjugate* response of VOR is active, the LR and the CR are also still present. Disconjugate VOR (or atypical responses as vertical deviation) is inconstantly linked with impairment of the LR, which may be present but which is usually sluggish. Absence of the LR is coupled, as a rule, with absent or severely impaired VOR. The CR disappears only in very deep coma, and its absence is always linked with fixed pupils and immobile eyes. Exceptions to this general rule are due to the specific effects of some drugs and therefore have a diagnostic value. The VOR reflex is blocked by overdoses of *central anticholinergic drugs*, probably because, as stated before, some of its multiple central synapses are cholinergic. Therefore the VOR is impaired (no nystagmus or conjugate deviation) by doses of *atropine* that are only slightly higher than the therapeutic range.

In *barbiturate* overdoses VOR behavior is atypical. Hypnotic doses block nystagmus. At slightly higher doses (insufficient to abolish localized responses to pain) disconjugate deviation is observed, while with doses that abolish pain localization conjugate deviation is observed. In deep barbiturate coma (no response to pain) the VOR is always absent, while the LR may still be present (7).

It has been suggested that barbiturate facilitation of conduction in Deiter's fasciculus may be a possible explanation for the inversion of the two intermediate responses (disconjugate deviation with milder degrees of brain failure; conjugate deviation when the signs of central nervous system depression are more severe) (5). On the other hand, a fully active VOR is observed even after high doses of *benzodiazepines*. The reflex is impaired only when coma due to an overdose of these drugs is very deep.

The peripheral cholinergic synapses of the third cranial nerve (autonomic fibers) are blocked by all *anticholinergic drugs* (atropine and other belladonna alkaloids, quaternary ammonium compounds, tricyclic antidepressants, orphenadrine, phenothiazines, etc.). The doses of these drugs that block the LR are, however, higher than those that give other signs of a cholinergic block. For peripherally acting drugs, dry mouth, tachycardia, facial flushing, a block in sweating, and urinary retention appear before pupillary dilation with areflexia. For anticholinergic drugs that cross the blood-brain barrier (atropine, orphenadrine, antidepressants) the doses that block the LR are within the same range of the doses that produce a true coma with severe abnormal motor activity. In milder overdose cases, when only confusion and/or hallucination are present, the pupils may be dilated, but the LR is usually still present. The same occurs with mushrooms containing atropine-like toxins (Clitocybe, Psilocybe).

The block of LR is, however, the longer-lasting sign of anticholinergic poisoning, reaction to light and accommodation returning to normal long after (2-3 days for some drugs) the disappearance of all the other signs.

Severe cerebral hypoxia-ischemia and catecholamine excess also lead to pupillary areflexia. No effect on pupillary tonus and reflexes is seen with *neuromuscular blocking drugs* that instead block both the VOR and the CR even in conscious patients. The *botulinum toxin* constantly blocks VOR and CR, but areflexic pupillary dilation is an inconsistent sign, even in severe cases requiring artificial ventilation.

B. Cerebral Function Monitor Recording

The study of electroencephalogram patterns and of multisensory evoked potentials may perhaps give useful information in brain failure due to any cause. These studies, however, are seldom feasible in an emergency department or in a busy intensive care unit. Useful, even if less sophisticated, information can be obtained by continuous recording of cerebral activity with the Cerebral Function Monitor (CFM), an instrument especially devised for clinical use in surgical theaters and in intensive care units (13). The CFM is a small, portable apparatus with only two recording electrodes. It gives a compressed, 6-30 cm/hr, single-channel trace and is therefore suitable for continuous recording that may be prolonged for hours or days. Clear records can be obtained in the electrically hostile environment of intensive care units without interference with routine nursing or therapeutic procedures (e.g., mechanical/artificial ventilation) and with other types of monitoring. Artifacts are either automatically excluded or indicated on a separate trace. The most important advantage of the CFM is that the interpretation of the records is simple to learn even for doctors and nurses who are not electroencephalograph experts. The CFM trace is recorded on a special paper with a scale in microvolts and with time markings. The microvolt scale is linear from 0 to 6 μV, semilogarithmic between 8 and 20 μV, and logarithmic above 25 μV. This makes it possible to record large outputs when there is very little electrical activity arising in the brain, as in very deep coma or in brain death. Without range switching, however, it is also possible to record both the normal levels of cerebral activity and the exceptionally high levels which occur with epileptic seizures.

The key features of the trace to be considered are the following:

The distance of the lower level of the trace from the zero line (height of the base line)
The width of the band (difference in microvolts from peak to peak)
The variability of the trace, that is, its fluctuations with time, both spontaneous and in response to different stimuli

A normal "waking" trace is of middle width, with a base line fluctuating between 10 and 20 μV. In brain failure of progressive severity the trace may be high and fluctuating, of middle height and monotonous, with a low base line and resembling a comb with teeth pointing upward or downward, or very narrow with a base line very near the zero (13, p. 198). In toxic brain failure all these types of traces can be seen, as well as very high traces and/or high voltage peaks, indicative of clinical or subclinical convulsive activity.

Cerebral function monitoring during brain failure may give the clinician various useful types of information (3). For example, an early differential diagnosis between toxic and organic brain failure is facilitated. In two cases of *heroin* overdose, CFM traces were started 30 min after admission, that is, after resuscitation had been carried out and the vital signs had stabilized (Fig. 1). The first patient, a 31-year-old woman (Fig. 1a) was admitted in deep areflexic coma. At the time of recording, there was nonlocalizing response to pain. The patient had pin-point pupils, LR was absent, and CR was present. The CFM trace had a high base line, which was spontaneously fluctuating, especially in the right part of the trace, when the patient had fully regained consciousness 2 hr later. The second patient, a 25-year-old man (Fig. 1b), was admitted with cardiorespiratory arrest. Resuscitation restored vital signs, but no motor or reflex activity. The CFM trace, with a narrow band and a base line on 0 μV, remained unchanged 3 hr later, pointing to organic brain death due to cerebral ischemia-anoxia. There was no change until death 12 hr later.

During deep coma due to *hypnotic drugs* "awakening signs" can be detected on the CFM trace many hours before any clinical change indicative of recovery occurs, such as the return of motor or reflex responses. A barbiturate overdose patient, a 42-year-old woman, was found unconscious at home. She was admitted to the intensive care unit in a deep coma: There was no response to voice or pain, brainstem reflexes were absent, and she was suffering apnea and arterial hypotension (75/60 Torr). Artificial ventilation (IPPV) and circulatory support were provided. A CFM record was taken 24 hr after admission (Fig. 2, top). The patient was then still on IPPV, her arterial pressure was stable at 100/65, but there was no change in neurological status. The trace is monotonous here, the lower part almost on zero, but with a very broad band. At points 2 and 3 a painful stimulus was given; there was no motor response, but the trace shows a brisk reaction, the lower level suddenly rising for a prolonged period (10 and 5 min). Between points 2 and 3 (a $2\frac{1}{2}$-hour interval) there is also a slow progressive increase in the base line from 1 to 3-4 μV. The first clinical signs indicating a lightening of the brain failure level (nonlocalizing response to pain, LR and CR present) appear only 24 hr later. Figure 2 (bottom) shows the CFM record taken 3 days (left) and 4 days (right) later. At 3 days the patient was opening her eyes, and localizing motor response was present. At 4 days, at point 8, the patient was conscious, with normal spontaneous motor activity (indicated by the artifacts on the impedance line). In conclusion, the broad band with fluctuations in response to stimuli (a) predicted arousal long before the recovery of any

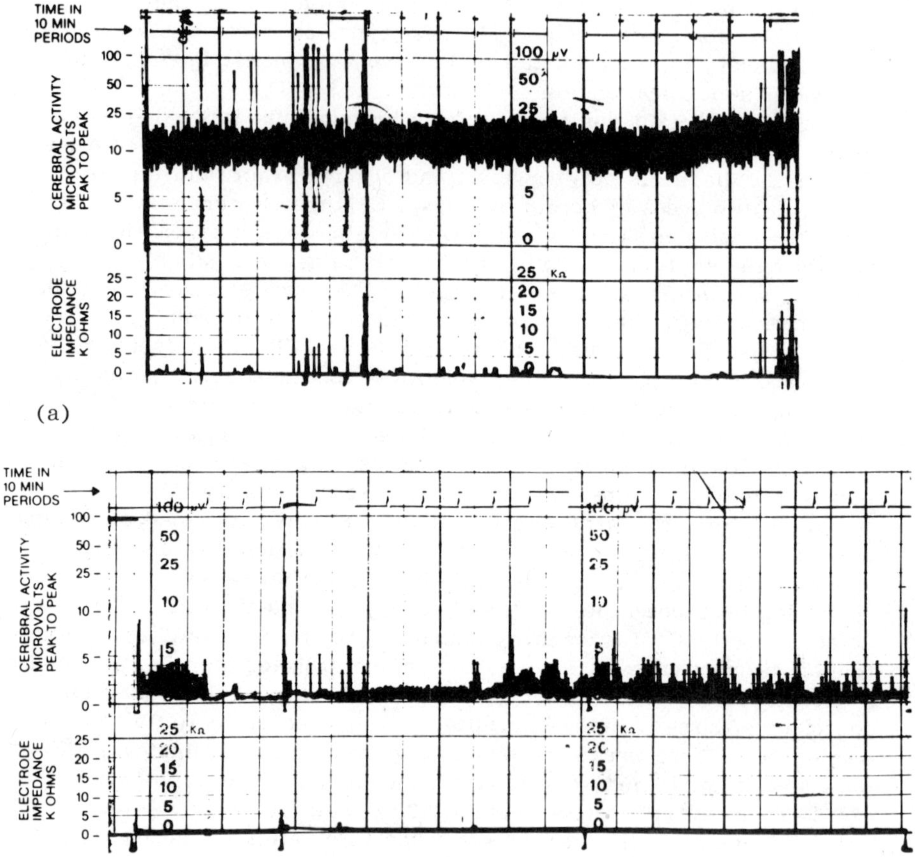

Figure 1 The CFM traces of two heroin overdose patients: (a) a 31 year-old woman, admitted in deep areflexic coma but regaining consciousness 2 hr later, and (b) a 25-year-old man, admitted with cardiorespiratory arrest and dying 12 hr later (see text for explanations).

Figure 2 The CFM traces of a barbiturate overdose patient admitted in deep coma. The top panel was recorded 24 hr after admission; the bottom panel was recorded at 3 days (left) and 4 days (right) after admission (see text for explanations).

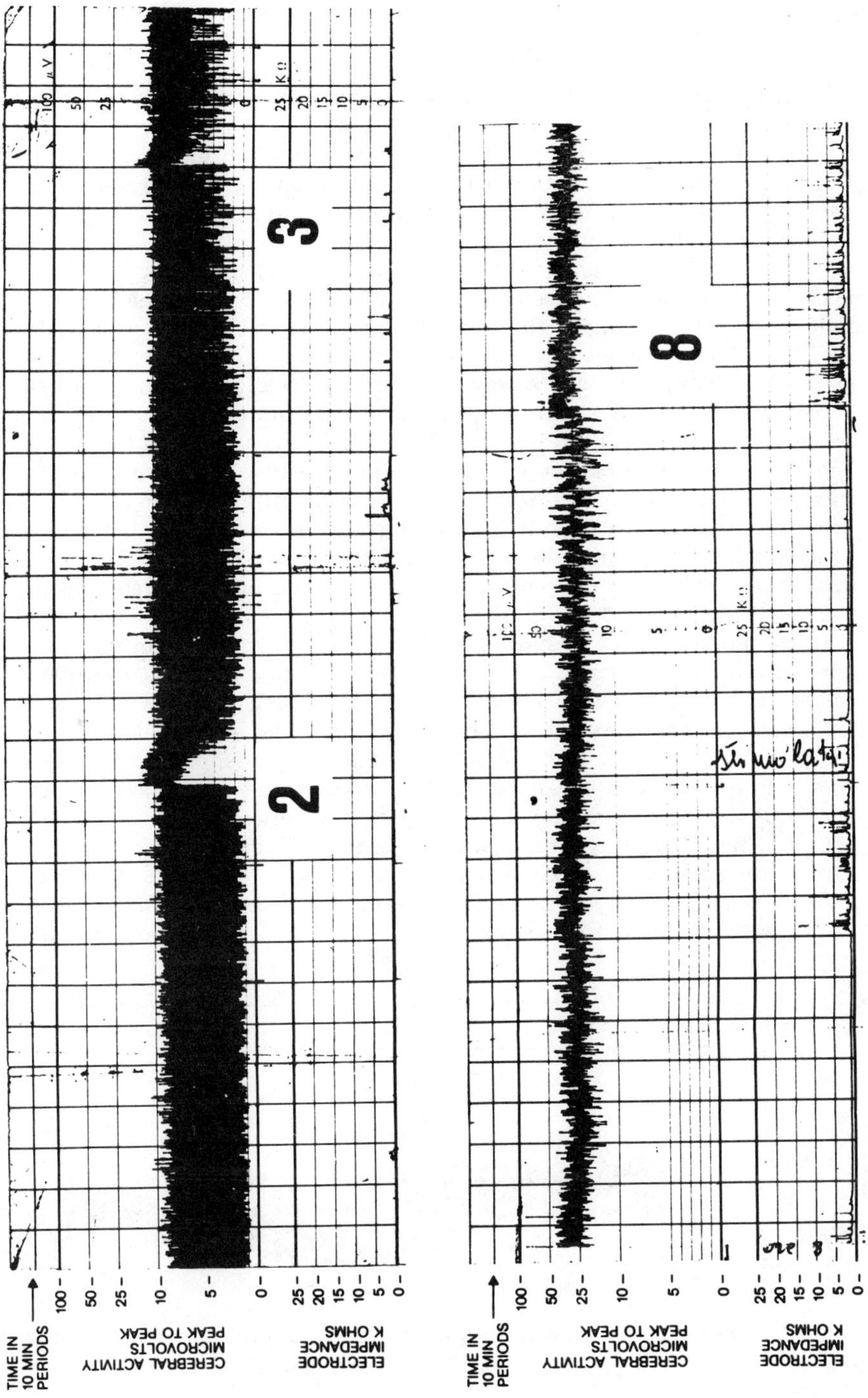
TIME IN 10 MIN PERIODS
CEREBRAL ACTIVITY MICROVOLTS PEAK TO PEAK
ELECTRODE IMPEDANCE K OHMS
2
3
8

detectable sign of reactivity and (b) excluded organic brain death as suggested by the clinical picture on admission. The narrow band with a high base line when the patient recovered to a lighter level of brain failure pointed to a barbiturate overdose. Upon recovering consciousness, the patient confirmed she had taken a pentobarbital overdose in an attempt to commit suicide.

A CFM trace may also give clear evidence of complicating factors, such as respiratory or circulatory insufficiency, that worsen brain failure. The following is the case of an 83-year-old woman who suffered a combined *tricyclic antidepressant-benzodiazepine* overdose. On admission the patient was reacting to but not localizing pain, and brainstem reflexes were present. An artificial ventilator was required to correct hypoxemia due to chronic respiratory disease. The CFM trace (not shown) had a relatively high base line (8-10 μV), but was unwavering and unreactive to stimuli. The trace in Figure 3, taken 24 hr later, has a lower base line, showing a deterioration in clinical conditions due to worsening of the respiratory failure (PO_2, 29 Torr) despite intensive life support. The first sudden but transient fall in the base line to the 0-μV level was due to a successfully resuscitated cardiac arrest. After restoration of spontaneous cardiac activity and adequate blood pressure, the patient remained in a very deep areflexic coma. The slow rise in the base line did not reach the previous day's level and was truncated by a second, irreversible cardiac arrest.

In some instances, when toxic brain failure is suspected but the poison is unknown, comparison of the clinical signs with the features of the CFM record can help in identifying the type of drug. Figure 4 shows typical traces of brain failure due to different types of drugs. The records were taken when the patients were all at about the same level of coma, that is, not obeying orders, not localizing pain, and reacting with mild abnormal motor activity. The top CFM recording in

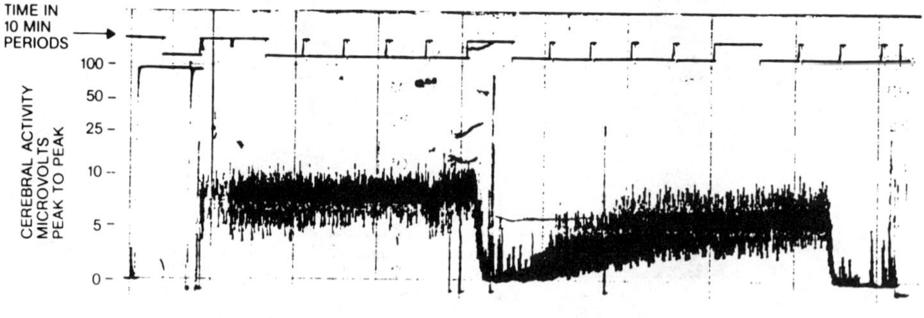

Figure 3 The CFM trace of a combined tricyclic antidepressant-benzodiazepine overdose patient taken 24 hr after admission (see text for explanations).

Figure 4 is typical of the *central anticholinergic syndrome* (in this case due to orphenadrine). The trace has a low base line (4-6 μV, lower than expected according to the level of coma) and a wide band with high voltage peaks; the peaks are higher and more frequent in the first part of the trace, when mild abnormal spontaneous motor activity was present (hypertonus, choreoathetosis), incompletely sedated by intravenous diazepam (arrow 1); and also at the end (extreme left) of the trace, when the patient recovered consciousness after 2 mg of intravenous *physostigmine* (arrow 2). The middle CFM recording in Figure 4 was taken at the same level of coma as above. This recording is typical of brain failure induced by *barbiturate and nonbarbiturate hypnotics and benzodiazepines*. The trace has a spontaneously fluctuating higher base line (between 8 and 15 μV) and is narrower, similar to the normal waking pattern. Very short periods of lower voltage (5 μV) occasionally give the trace the appearance of a comb with the teeth pointing downward ("suppression" phases). The bottom CFM recording in Figure 4 is typical of *lithium* toxicity during treatment or by voluntary overdose: Note the narrower band and the high baseline trace (25 μV) with peaks corresponding to clinical or subclinical seizures. In this case, the lithium plasma level was only slightly in excess of the therapeutic range (2.6 mEq/liter), but the patient showed signs of brain failure with continuous tremors and intermittent Jacksonian seizures (arrows 1 and 2), as well as a more severe and prolonged generalized seizure (arrows 3 and 4). At arrow 4 the interruption in the trace is due to the displacement of an electrode during the epileptic fit. Then 10 mg of diazepam were administered intravenously and were followed by prolonged sedation of the abnormal motor activity, with parallel lowering of the trace base line.

Following the CFM record may also be helpful for treatment. In the central anticholinergic syndrome, the reversible anticholinesterase agent physostigmine salicilate may have dramatic awakening effects, but at high doses, as required, for instance, in orphenadrine poisoning, this treatment may have convulsive effects. The CFM tract gives clear evidence of the electric stimulating effects of physostigmine and may guide the clinician in keeping the doses below the convulsive threshold. Take the case of a 16-year-old who poisoned herself with an overdose of tricyclic antidepressants. Figure 5 shows the CFM trace recorded 2 hr after admission: The patient was restless and confused and was not obeying orders or localizing pain. The trace shows the typical features of the central anticholinergic syndrome (compare with the upper trace in Fig. 4): Note the wide band with a low base line (6-7 μV), which remained unchanged even 12 hr after consciousness was regained. Two intravenous 2-mg doses of physostigmine injected rapidly (arrows 1 and 2) were followed within 2-5 min by a short-lasting recovery of consciousness, but also by a phase of high voltage peaks, especially noticeable after the second physostigmine injection.

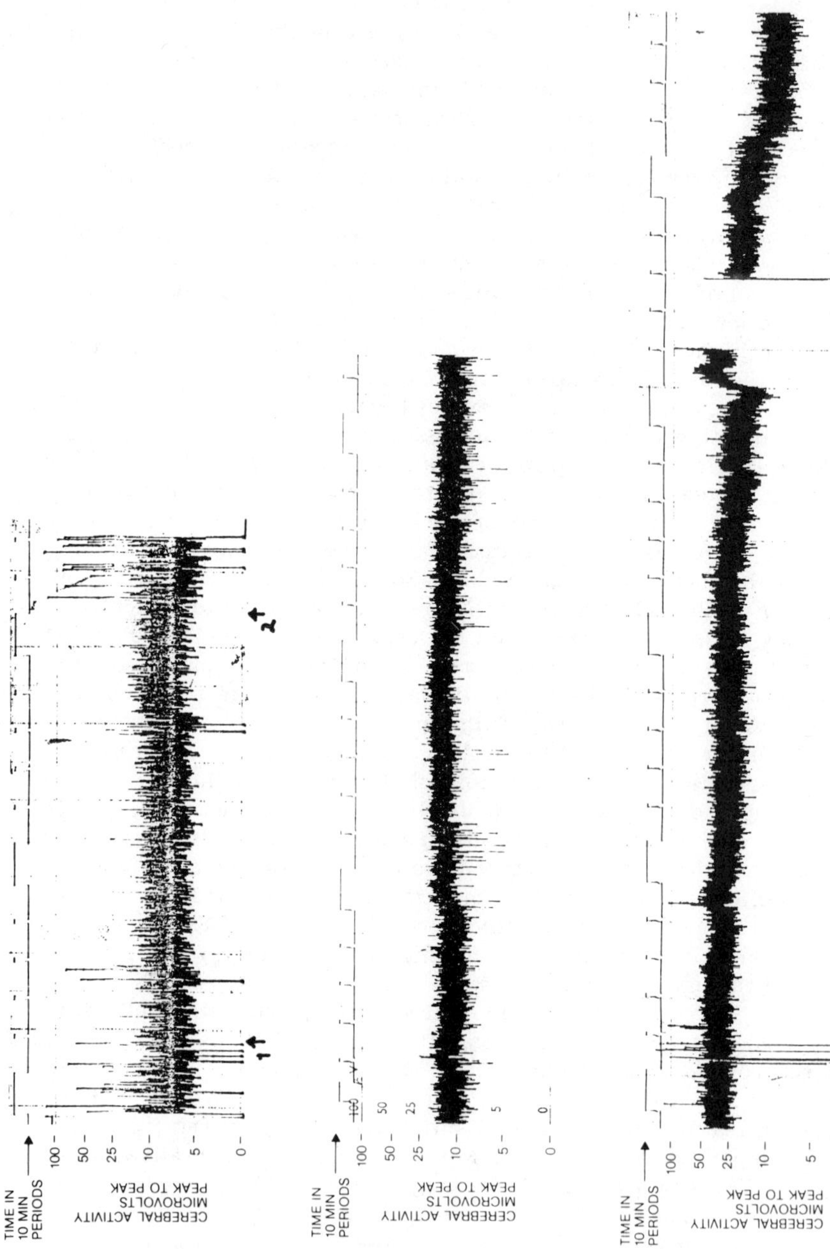
TIME IN 10 MIN PERIODS
CEREBRAL ACTIVITY MICROVOLTS PEAK TO PEAK
100
50
25
10
5
0
TIME IN 10 MIN PERIODS
CEREBRAL ACTIVITY MICROVOLTS PEAK TO PEAK
100
50
25
10
5
0
TIME IN 10 MIN PERIODS
CEREBRAL ACTIVITY MICROVOLTS PEAK TO PEAK
100
50
25
10
5
0

A third dose of physostigmine (3 mg, arrow 3) was then injected *very slowly*: The patient was fully alert within 5 min and did not relapse into coma; the trace remained stable and showed only two short, high voltage peak phases 20 and 80 min later. Similar information is given by the CFM trace when anticonvulsants must be used for the prevention or treatment of generalized seizures.

The CFM continuous recording should give useful information also for monitoring the effects of invasive procedures indicated to enhance the elimination of poisons (dialysis, hemoperfusion, etc.).

IV. CLASSIFICATION

The variety of clinical conditions corresponding to the definition of coma and, moreover, of those included within the broader term *brain failure* can be classified according to severity criteria.

In the last 35 years brain failure due to organic cerebral disease, especially brain injury, has been classified using different methods which fall into two main groups: coma scales and scoring systems.

Coma scales are based on *synthesis*. The clinical signs are assumed to be related and are therefore combined into syndromes called coma "levels," "degrees," or "stages." The levels are arranged in order of increasing severity to form a *scale* with steps indicating progressive brain dysfunction.

Scoring systems are based on *analysis*. The clinical signs are assumed to be all independent and are considered as separate categories. Within each category the responses observed are ordered following a hierarchy of progressive functional impairment and are accordingly graded by a numerical score. The best-known example of this method is the Glasgow Coma Scale for brain injuries (15), which, despite its name, is indeed a scoring system.

Classification of toxic brain failure meets a number of difficulties (4). Anatomical clinical correlations that can be recognized in organic brain failure do not exist here, the disturbance in consciousness being due to changes in neuronal activity and/or in synaptic transmission at various sites within the central nervous system. The neurological condition may range from excitatory/hallucinatory states to deep, areflexic coma. The clinical signs occur in apparently capricious combinations and change rapidly with time.

Figure 4 Typical CFM traces of brain failures caused by (top) central anticholinergic drugs, (middle) barbiturate and nonbarbiturate hypnotics and benzodiazepines, and (bottom) lithium poisoning (see test for explanations).

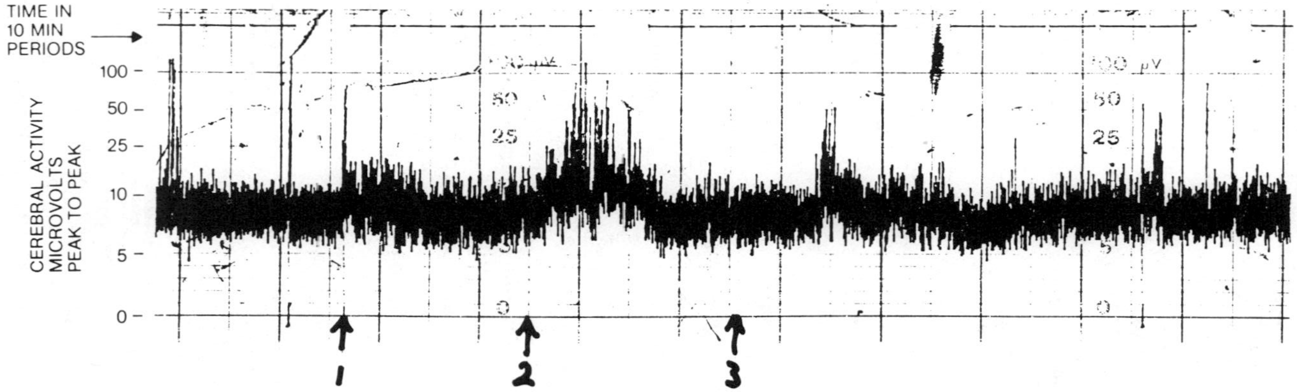

Figure 5 The CFM trace of a tricyclic antidepressant overdose patient (see text for explanations).

The most widely used classification, proposed in 1975 by Matthew and Lawson (10), is a coma scale with only four grades, each indicating the general reactivity of the patients from arousable drowsiness to generalized areflexia. This scale gives a very limited amount of information, owing to the attempt to include a very wide range of clinical conditions in a limited number of common patterns. Spontaneous or provoked abnormal motor activity, a common and relevant sign in toxic brain failure, is not taken into account.

The classification by Reed et al. (14) is more detailed, but was devised for barbiturate intoxication only. Classifications that are specific for a single class of poisons are of very limited value, since brain failure in most cases is due to mixed poisoning by two or more substances having different sites of action within the central nervous system and often opposite depressing, inhibitory, or excitatory actions. These drawbacks can, however, be overcome either by a scoring system or by a scale method, provided that a standardized semiology as previously described is adopted as a basis for the clinical examination.

A. Classification by a Scoring System

Tables 1 and 2 represent already a scoring system on four categories of signs. Within each category the signs are scored from best to worst by a numerical index. The four categories and their scores can be ranged vertically to form the axis of ordinates in a graph having as abscissa the course of time. Figure 6 is an example for a case of brain failure due to combined self-poisoning by phenobarbital and carbamazepine (dose unknown). The scoring system is used to record the clinical course of the case. Both the deepening of coma shortly after admission and the partial recovery of spontaneous motor activity and reactivity to pain after life support measures (intubation, artificial ventilation) were started are clearly visible on the flow sheet. The clinical signs showed no change in the following 36 hr, but during this period the progressive lightening of the coma level is demonstrated first by the return of a brisk LR (with transient anisocoria), then by the behavior of the VOR: Absent on the first day after admission, the reflex gave a conjugate response to stimulation of the right ear and a disconjugate response to stimulation of the left side on the second day. On the morning of the third day the reflex was conjugate on both sides, and this coincided with the return of a localizing response to the painful stimulus. Within the next 24 hr the patient was weaned from the respirator, extubated, and shortly thereafter recovered consciousness. The behavior of the clinical signs can thus be recorded on the graph to form a sequence of curves giving a visual profile of the clinical course of the case.

Scoring systems are primarily and best indicated for this purpose. Special events and treatments can also be registered on the graph that becomes therefore a most informative and comprehensive clinical document.

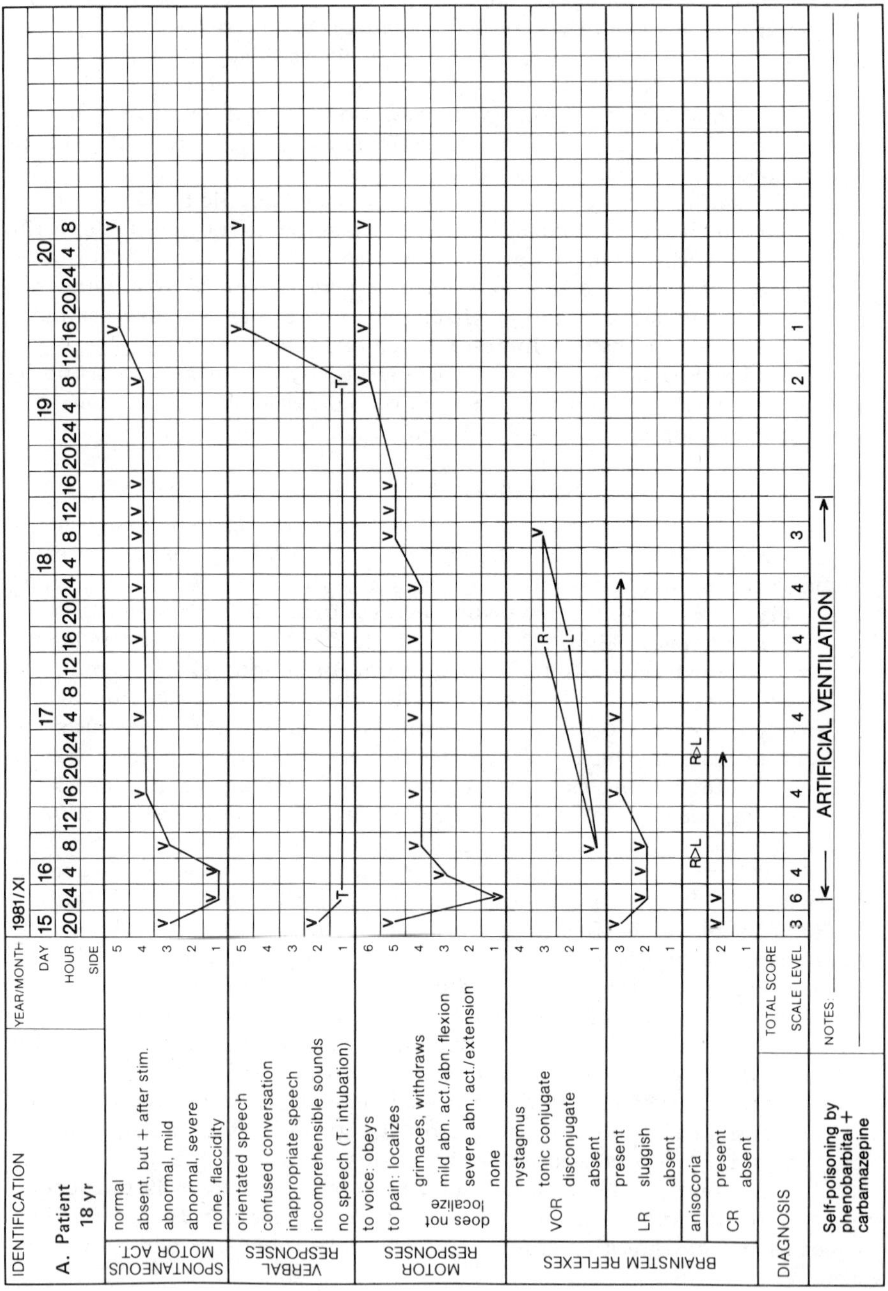

BRAIN FAILURE FLOW SHEET
IDENTIFICATION
A. Patient 18 yr
YEAR/MONTH 1981/XI
DAY 15 16 17 18 19 20
HOUR 20 24 4 8 12 16 20 24 4 8 12 16 20 24 4 8 12 16 20 24 4 8 12 16 20 24 4 8
SIDE
SPONTANEOUS MOTOR ACT
normal 5
absent, but + after stim. 4
abnormal, mild 3
abnormal, severe 2
none, flaccidity 1
VERBAL RESPONSES
orientated speech 5
confused conversation 4
inappropriate speech 3
incomprehensible sounds 2
no speech (T. intubation) 1
MOTOR RESPONSES
to voice: obeys 6
to pain: localizes 5
does not localize
grimaces, withdraws 4
mild abn. act./abn. flexion 3
severe abn. act./extension 2
none 1
BRAINSTEM REFLEXES
VOR
nystagmus 4
tonic conjugate 3
disconjugate 2
absent 1
LR
present 3
sluggish 2
absent 1
anisocoria
R>L
CR
present 2
absent 1
DIAGNOSIS
Self-poisoning by phenobarbital + carbamazepine
TOTAL SCORE
SCALE LEVEL 3 6 4 4 4 4 4 3 2 1
NOTES
ARTIFICIAL VENTILATION

When a global assessment and classification of a patient's condition is required, however, this analytical scoring system is of limited value. The total score obtained by adding up the numbers of the four categories is not a reliable clinical index. Each sign has a different "weight" or "value" as far as the severity of brain failure is concerned. The simple sum of their relative scores is therefore meaningless, almost as if one were adding up different foreign currencies by mere numerical value, without considering exchange rates. In some cases, besides, the total score cannot be obtained because one or more signs are missing or unreliable, as happens, for example, when a patient is intubated (no verbal responses) or has one or more brainstem reflexes blocked by autonomic side effects.

B. Classification by a Coma Scale

1. *Method*

Grouping of patients into classes of equivalent severity can be achieved more correctly through the synthesis given by a scale method. It must be emphasized that, to be of any use, a toxic brain failure scale should not be specific for every type of poison, since the majority of toxic comas are due to mixed poisonings.

The scale proposed here (Fig. 7) follows the same principles adopted for the classification of *organic* brain failure (1,2). Figure 7 evaluates the patient's responsiveness and the severity of his brain failure quite independently from the possible specific effects of a toxin. Nevertheless, atypical combinations of signs can be recognized and may have a diagnostic value (see below).

To define the level of brain failure only two categories of signs are used: the patient's *reactivity to voice or pain* and the *brainstem reflexes*.

Reactivity is ranged into seven possibilities, each defining a level or step of the scale:

Level 1. *Answers to questions* ("answer" meaning at least one elementary but oriented utterance, for instance, the correct patient's name).
Level 2. *Obeys orders*.
Level 3. *Localizes* pain.
Levels 4 and 5. *Does not localize* pain. At level 4 the patient responds to the painful stimulus by grimacing and/or with mild abnormal motor activity. At level 5 the nonlocalizing pain response consists mainly in severe abnormal motor activity, including extensor hypertonic posturing.

Figure 6 Scoring system for brain failure due to self-poisoning by phenobarbital and carbamazepine (see text for explanations).

Levels 6 and 7. *Has no response to pain.* The *brainstem reflexes* are present at levels 1-6, and globally absent at level 7. So level 6 is characterized by no response to pain and at least one brainstem reflex present; level 7 is characterized by total unresponsiveness and the absence of all brainstem reflexes.

In cases with variable responses or side differences, the principle to follow when allocating the patient to the correct level is to choose "the best response on the best side." Tests of reactivity and of reflexes should be repeated at least twice within a short time.

In Section III.A it was pointed out that drugs having peripheral and/or central anticholinergic effects may block the light reflex and especially the VOR, even at relatively superficial levels of brain failure. In evaluating the level of brain failure by this scale method, therefore, all three reflexes—VOR, light, and corneal—should always be tested.

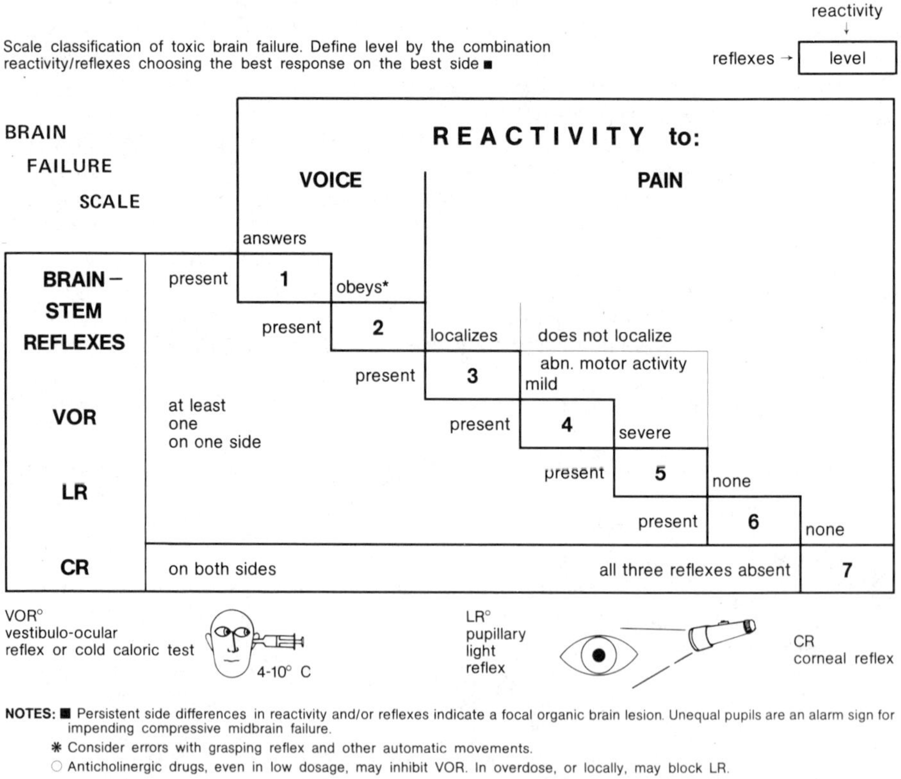

Figure 7 Toxic brain failure scale (see text for explanations).

2. Indications and Uses of the Scale Method

The main indication of the scale proposed are biostatistical and epidemiological studies. Patients with brain failure due to single drugs having different actions or to mixed poisoning may be grouped into homogeneous classes of severity. This is the necessary basis for studies on major aspects of clinical neurotoxicology such as epidemiology, mortality rates, and, within intensive care units, the cost/benefit ratio of invasive and complex treatments. For instance, in recent years hemoperfusion, an invasive, costly method to enhance drug elimination, has been the object of sharp contrasts in opinion (6,8) that might be more objectively settled if general agreement on uniform evaluation of severity of cases were reached. Similar considerations apply to the assessment of correlations between plasma levels and central nervous system effects of toxic substances and to the definition of drugs of choice in the management of poisoning.

The coma scale described may be also a help for diagnosis when atypical combinations of signs are observed. At brain failure levels 1-3 the absence of spontaneous motor activity, with deep tendon hypoareflexia and muscular flaccidity, points to benzodiazepines overdose. At level 1 the absence of the light reflex points to anticholinergic drugs that do not cross the blood-brain barrier. In this case other signs of peripheral cholinergic block will be very evident. At levels 1-5 the absence of VOR (and at levels 4 and 5 the absence also of LR) points to a central anticholinergic syndrome, which may be confirmed by CFM analysis.

Some hints about the identification of the specific drug are given here. Atropine and other belladonna alkaloids in adults mainly give a brain failure of level 2-3, with severe confusion and hallucinations. In children, however, a sudden drop to level 5 with generalized seizures may occur. Brain failure due to tricyclic antidepressants oscillates between levels 3 (with hallucinations) and 4. Abnormal motor activity at this level (both spontaneous and provoked) is represented mainly by tremor and by choreoathetosis. Severe orphenadrine poisoning gives a deeper coma level—usually level 5—with severe convulsions or status epilepticus. Cardiac arrhythmias, a life-threatening complication of both tricyclics and orphenadrine, are usually observed only at the deeper levels of coma (level 4 or below). Hypoxemia, hypercarbia, severe acidosis, circulatory collapse, or cardiac arrest mask all the specific signs and deepen the level of brain failure even to global areactivity/areflexia.

Both the levels and the specific syndromes can be identified, therefore, after restoration and stabilization of the vital functions. In these cases the persistence after resuscitation of an aspecific deep coma points to postanoxic-ischemic brain damage. The behavior of the brainstem reflexes may assist in defining the differential diagnosis and in predicting outcome. In very deep toxic coma motor responses may be totally absent while the brainstem reflexes are still active, at least the LR (with the sole exception of anticholinergic drugs). In irreversible postanoxic-ischemic encephalopathy the VOR is either severely impaired (disconjugate deviation) or absent, while some type of abnormal motor responses (grimacing, flexor or extensor posturing) may still be present. This diagnosis of organic brain failure and a prognosis of death or persistent vegetative state may be confirmed by a CFM monotonous trace of low voltage and with a base line near zero.

V. CONCLUSIONS AND SUMMARY

Brain failure and *coma* are not synonyms, the former indicating any condition whose main symptom is a disturbance of consciousness, the latter being restricted to unarousable patients, that is, patients who do not respond to call or orders even after a painful stimulus.

The clinical signs of toxic brain failure and coma can be grouped into four categories: spontaneous motor activity, verbal responses, motor responses, and the behavior of brainstem reflexes. Information obtained by monitoring the cerebral electrical activity is a valuable complement to the clinical examination, especially for differential diagnosis between toxic and organic brain failure, the detection of early signs of recovery or of worsening factors, and, in selected cases, for identifying the toxic agent.

Using the same clinical signs, brain failure may be classified by two methods with different basic assumptions and separate indications.

A scoring system is described by which the different signs of brain failure may be graphically represented on a flow sheet where the y axis is time. The main use for this method is the analytical monitoring and recording of the clinical course of brain failure.

When a synthetic evaluation of the patient's conditions is required for biostatistical studies, a scale method is preferable. With the brain failure scale proposed, patients can be grouped into classes of homogeneous severity, irrespective of the type or number of toxic agents involved. Besides the objective assessment of severity, the scale has also a diagnostic value, especially in cases with atypical combinations of signs.

REFERENCES

1. Bozza-Marrubini, M. 1982. Coma, in *Care of the Critically Ill Patient* (Tinker, J., Rapin, M., eds.), Springer, Berlin, pp. 719-740.
2. Bozza-Marrubini, M. 1984. Classifications of coma. *Intensive Care Med. 10*:217-226.

3. Bozza-Marrubini, M., Ghezzi, R., Moroni, C., Selenati, A. 1983. The cerebral function monitor (CFM) in toxic coma. *Hum. Toxicol. 2*:412.
4. Bozza-Marrubini, M., Ghezzi, R., Ruggeroni, M. Scaiola, A. 1983. Classification and monitoring of toxic coma. *Hum. Toxicol. 2*:410.
5. Cerchiari, E. L., Carugo, D., Della Puppa, T. 1984. Reliability of the vestibulo-ocular reflex as an index of the effects of hypnotic drugs on the central nervous system. *Br. J. Anaesth. 56*: 325-331.
6. Garella, S., Lorch, J. A. 1980. Haemoperfusion for acute intoxications. *Clin. Toxicol. 17*:515-527.
7. Greenberg, D. A., Simon, R. P. 1982. Flexor and extensor postures in sedative drug-induced coma. *Neurology 32*:448-451.
8. *Lancet*. 1979. Haemoperfusion for acute intoxication with hypnotic drugs. *Lancet 2*:1116.
9. Levati, A., Farina, M. L., Vecchi, G., Rossanda, M., Bozza-Marrubini, M. 1982. Prognosis of severe head injuries. *J. Neurosurg 57*:779-783.
10. Matthew, H., Lawson, A. A. H. 1975. *Treatment of Common Acute Poisonings*, Churchill Livingstone, Edinburgh, pp. 18-19.
11. Plum, F., Posner, J. B. 1966. *Diagnosis of Stupor and Coma*, Blackwell, Oxford, p. 2.
12. Plum, F. 1972. Organic disturbances of consciousness, in *Scientific Foundations of Neurology* (Critchely, M., O'Leary, J. L., Jennett, B., eds.), Heinemann, London, pp. 193-201.
13. Prior, P. 1979. *Monitoring Cerebral Fuction*, Elsevier/North-Holland, Amsterdam.
14. Reed, C. E., Driggs, M. F., Foote, C. C. 1952. Acute barbiturate intoxication: A study on 300 cases based on a physiologic system of classification of the severity of the intoxication. *Ann. Intern. Med. 37*:290-303.
15. Teasdale, G., Jennett, B. 1974. Assessment of coma and impaired consciousness. A practical scale. *Lancet 2*:81-84.

3

Drug-Induced Dysfunctions of the Autonomic Nervous System

Virginia L. Zaratzian
United States Department of Agriculture, Washington, D.C.

I. INTRODUCTION

This discussion focuses primarily on therapeutic drugs that induce adverse autonomic effects, especially on the peripheral autonomic nervous system, although drugs administered for specific autonomic effects will also be considered. Since drugs usually do not show specificity, numerous complex actions can occur at autonomic neuroeffector sites, where different classes of chemical mediators are physiologically involved. The agents can simulate the action of the mediators; alter the synthesis, storage, release, uptake, reuptake, or metabolism of the endogenous neurotransmitters (46).

II. GENERAL CONSIDERATIONS

The autonomic nervous system regulates the activity of smooth muscle structures of various organs, including the eye, gastrointestinal tract, respiratory tract, cardiac muscle, and exocrine glands (7,22,26,28,46).

Specific autonomic effects include cardiac abnormalities, achalasia, dryness of the mouth and skin, urinary retention, changes in gastric motility, changes in secretions, and ophthalamological and other effects. Although there are differences between autonomic and somatic nerves, there is an overlap of integration in the central nervous system between the autonomic or somatic centers, and the autonomic nervous system has both central and peripheral nervous system components (17). There are autonomic reflexes involved at the spinal cord and levels above the spinal cord at the hypothalamus and cortex.

III. AUTONOMIC NERVOUS SYSTEM TARGET ORGAN TOXICITY

A. Cardiovascular Effects

Adverse effects on the cardiovascular system are not readily detected when they are induced by drugs that are not used for cardiovascular purposes (21). There are reports of orthostatic hypotension, palpitations, circulatory collapse, and death (2,10,16,17,25,68,70). Orthostatic hypotension is a serious effect that can occur even at therapeutic doses. In cases of severe hypotension, patients with diseases of the vascular system are at risk of myocardial infarct, cardiovascular occusion, and fainting (64).

1. *Psychotropic Drugs*

a. Tranquilizers: The phenothiazines induce complex actions in the autonomic nervous system (25). These drugs can cause peripheral cholinergic, α-adrenergic, histaminergic, and tryptaminergic blockade. However, adrenergic activity can also be elicited. The major cardiovascular side effect of neuroleptics is postural hypotension and there are electrocardiographic disturbances (45,59). In some patients therapeutic doses of 50 mg of butaperazine taken three times daily induced hypotension (42). Reports in the literature indicate that other phenothiazines such as methotrimeprazine, acetophenazine, perphenazine, and trifluperazine also induce hypotension (32,43,49). The thioxanthenes and butyrophenones have also been implicated in incidences of hypotension. The paradoxical hypertension and tachycardia that have been reported with these agents may be attributed to compensatory mechanisms.

b. Antidepressant Drugs: Mixed cardiovascular responses are associated with the tricyclic antidepressants such as imipramine, desipramine, and amitriptyline (16). Tachyarrhythmias, alterations in cardiac conduction, and either hypertension or hypotension have been reported (27,29,45,55,56,67). Although there is a direct quinidine-like depressant activity on conduction of cardiac tissue, the tachyarrhythmias may be due in part to other mechanisms such as vagal blockade and/or blockade of norepinephrine reuptake (26,28,50,65). The toxicity to the autonomic nervous system is the result of the combined adrenergic and anticholinergic effects at various neuroeffector sites (53).

Both the α- and β-adrenoceptors are stimulated, as the tricyclic antidepressants block the reuptake of norepinephrine (37). In norepinephrine blockade, there is initially an increase in heart rate and mean arerial pressure; subsequently, there is rapid destruction of norepinephrine by monoamine oxidase and catechol-o-methyl transferase. When the norepinephrine is metabolized, the hypertensive activity is lost, and hypoxic damage to the myocardium may occur during the hypotension. The orthostatic hypotension caused by this drug class lasts a few weeks in patients overdosed with tricyclic antidepressants. The most frequent electrocardiographic abnormalities are sinus tachycardia with a prolonged PR interval and a widened QRS complex. Fatalities have been reported (40,55,61). Thioridazine most resembles the tricyclics in many pharmacological actions, including cardiac toxicity. The combination should be avoided (30).

The hydrazines are monoamine oxidase inhibitors and include, among others, phenelzine, an antidepressant, and procarbazine, an antineoplastic agent; tranylcypromine is a nonhydrazine antidepressant. The therapeutic agents prevent the destruction of catecholamines and indoleamines. Although the antidepressant drugs are administered primarily for their central effects, widespread adrenergic activity can be induced; the most serious complications are related to blood pressure (10,64). There is usually a vasopressor response with tachycardia, but when reflexes come into play, a persistent hypotension may occur. There is also a risk of severe cardiac disorders when antidepressant drugs, especially the monoamine oxidase inhibitors, are administered with adrenergic drugs and/or foods containing tyramine (21). In addition to the monoamine inhibition activity, tranylcypromine also has adrenergic activity similar to that of amphetamine.

Various nontricyclic nonmonoamine oxidase inhibitor antidepressants have been introduced in recent years, including agents with bicyclic (i.e., viloxazine) or tetracyclic (i.e., mianserin) ring nuclei and atypical compounds such as nomifensine and trazodone. In no case has the efficacy of these drugs been shown to exceed that of the tricyclics, but most have fewer cardiovascular side effects than amitriptyline (8).

Lithium ion is currently used as a therapeutic agent for the control of bipolar and unipolar depressions (53). In addition to replacing sodium in the nerve cell, lithium ion also alters the metabolism of biogenic amines; it inhibits the release of norepinephrine and serotonin and also increases the reuptake of norepinephrine at the synapses. Lithium is involved in the inhibition of the synthesis and release of acetylcholine and may contribute to the metabolism of the neurotransmitters glutamic and γ-aminobutyric acids. Cardiac function is altered; therapeutic levels of lithium ion induce changes in the T wave in patients with normal function; and there are reports of sinus and atrioventricular nodal arrhythmias, edema, congestive heart failure, diffuse myocarditis, and death.

2. *Antihypertensive Drugs*

Certain antihypertensive drugs have induced serious cardiovascular side effects (68). The rauwolfia alkaloids, reserpine in particular, have been implicated in a variety of cardiovascular disturbances (54). Reserpine administration may induce supine and orthostatic hypotension, bradycardia, and reduction in pressor responses, cardiac output, and systemic resistance. These reactions are attributed to the depletion of catecholamines and indoleamines. In addition, the antihypertensive drugs such as methyldopa, debrisoquin, and bethanidine can occasionally induce a rise in blood pressure. Minoxidil and hydralazine, direct peripheral arteriolar vasodilators, induce a reflex stimulation of the sympathetic nervous system (12,18). This effect is a consequence of the fall in blood pressure, which activates the carotid and aortic baroreceptors; as a result, there is an elevation in the cardiac rate and output.

3. *Other Drugs*

Such adrenergic agonists as adrenaline, isoproterenol, and metaproterenol, which are used therapeutically in aerosols for asthma, affect the cardiac β_1-adrenoceptors by stimulation and may cause adverse cardiac effects under certain conditions (e.g., coronary artery disease, hypertension, digitalis administration, hypoxia, and/or hypercarbia). Recently, a considerable debate was triggered on the possible involvement of inhaled sympathomimetic agents in cases of death in asthmatics. The adverse effects of these drugs, such as tachycardia, palpitations, hypertension, and ventricular arrhythmias, can be aggravated by the concurrent use of theophylline and by fluoroalkane propellants, which sensitize the myocardium to catecholamines. Metaproterenol is primarily a β_2-adrenergic agonist with little effect on the β_1 receptors of the heart. Thus, cardiac stimulation is sufficiently limited to give the drug a considerable therapeutic advantage over isoproterenol.

B. Esophageal Effects

Major tranquilizers, minor tranquilizers, and antidepressants cause impairment of speech and achalasia, but the deglutition abnormalities may be related to the mental disease (35). Drugs such as the phenothiazines and the tricyclic antidepressants decrease the tone of the esophageal and gastric sphincter. Amitriptyline, protriptyline, and imipramine were implicated in the development of a hiatal hernia in five patients taking these drugs (62). The effect may be attributed to the anticholinergic activity at the esophageal sphincter.

C. Gastrointestinal Effects

1. Psychotropic Drugs

A number of psychotropic drug classes, including the phenothiazines, tricyclic antidepressants, thioxanthenes, benzodiazepines, and the monoamine oxidase inhibitors, have been implicated in adverse gastrointestinal effects (36,63). The anticholinergic effects on the smooth muscle and secretions, in combination with the emergence of α- and β_2-adrenoceptor domination results in constipation in some individuals. The constipation caused by the tricyclic antidepressants can be attributed to a direct atropine-like effect on the smooth muscle of the colon. If the gastrointestinal motility and secretions are severely depressed, impaction and paralytic ileus may occur; this is more likely when combinations of drugs are administered with additive or synergistic anticholinergic effects (44,57). Amitriptyline overdosage also causes relaxation of the stomach and has been responsible for a high incidence of deaths from adynamic ileus (5). Chloropromazine and trifluperidol decrease the tone of the ileocecal valve; this is due to the α-adrenergic activity of these drugs. The anticholinergic effects of chlorpromazine are manifested as constipation accompanied by a decrease in intestinal secretions (17). On the other hand, most psychotropic drugs have on occasion caused diarrhea, and in some cases chlorpromazine and trifluperidol have been effective in the treatment of paralytic ileus, which did not respond to cholinergic drugs and vasopressin (52).

2. Antihypertensive Drugs

Ganglionic blocking agents, such as mecamylamine, can cause constipation. The tone and motility of the gastrointestinal tract are reduced, and propulsive movements of the small intestine may be completely blocked (60). The antihypertensive agents guanethidine, bethanidine, and debrisoquin, which depress the function of postganglionic sympathetic nerves, cause diarrhea (24). Methyldopa and the rauwolfia alkaloids also cause this effect, as they inhibit adrenergic function (23).

D. Ophthalmological Effects

The autonomic effects on the eye involve responses of the radial muscle, the sphincter muscle of the iris, and the ciliary muscle (46). Mydriasis and/or cycloplegia are induced by such drugs as the major tranquilizers, the tricyclic antidepressants, anticholinergic drugs, antihistamines, ganglionic blocking agents, and monoamine oxidase inhibitors (34,39). Narrow-angle glaucoma can be induced by drugs with anticholinergic effects (63).

1. Psychotropic Drugs

a. Antidepressants: Tricyclic antidepressants, amitriptyline in particular, may precipitate narrow-angle glaucoma (10). Acute glaucoma was observed in patients on maintenance doses of amitriptyline for period up to 6 months. Amitriptyline causes a narrowing of the visual field and a decrease in visual acuity. The severity of the narrow-angle glaucoma depends on the degree of anticholinergic activity of the drug. It appears that some of the new nontricyclic compounds ("second-generation" antidepressants) are less likely to produce ophthalmological anticholinergic effects than older drugs (31). The monoamine oxidase inhibitors cause blurred vision (17).

b. Tranquilizers: Chlorpromazine has also been reported to induce an increase in intraocular pressure (13). The butyrophenones and the thioxanthenes cause cholinergic blockade, but are less potent than the phenothiazines; the benzodiazepines are the least active with respect to the blurring of vision (7,22,43). The phenothiazines induce mydriasis by α-adrenergic stimulation as well as by cholinergic blockade. Phenothiazines such as chlorpromazine and prochlorperazine can also block α-adrenoceptors, and miosis has been induced in the human (38). Phenothiazines have also been implicated in bilateral Horner's syndrome.

2. Other Drugs

Drugs with anticholinergic properties, such as propantheline, and antispasmodic, and diphenhydramine, an antihistamine, have induced mydriasis (39).

The sympathetically innervated smooth muscle of the eyelid affects the position of the eyelid (20). Adrenergic neuron blocking agents or ganglionic blocking agents, such as bretylium tosylate and guanethidine, cause ptosis, as do anticholinergic agents.

E. Urological Effects

Urinary retention can be induced by adrenergic stimulation of the detrusor muscle, which relaxes while the trigone and sphincter contract. Cholinergic blocking agents achieve the same effects. Antidepressant drugs (imipramine, desipramine, etc.), antihistaminics (tripelennamine, diphenhydramine, etc.), antispasmodics (propantheline, hyoscine, etc.), and antiparkinsonian drugs (benztropine mesylate, trihexyphenidyl, etc.) have been reported to induce urinary retention (39,70).

The tricyclic antidepressants decrease the tone of the detrusor muscle and contract the trigone and sphincter in 62.5% of the patients (47). The severity of the urinary retention is dependent on the dose and duration of treatment, and the effect is induced transplacentally (58). With high doses the urinary retention may be observed in some patients within 2-4 weeks, but with low doses the effect may occur after years of therapy. In elderly men obstruction may develop into asymptomatic prostatic hypertrophy. Urinary retention has been induced by imipramine at daily doses of 25 mg in some patients. The anticholinergic effect of the tricyclic antidepressants has been applied to control enuresis in children (31). It has been suggested that imipramine activates regulatory autonomic mechanisms, which originate in the reticular formation. Imipramine also has a direct effect on the muscles of the bladder in children and in laboratory animals (3,67). There is an increase in the sphincter tone of the urinary bladder. In children on imipramine for the antienuretic effect, there is an increase in bladder capacity, which affects the desire to void.

F. Secretory Glandular Effects

The secretions of the salivary, bronchial, lacrimal, and sudoriferous glands, and of the stomach, intestine, and nose are under the influence of the autonomic nervous system (46). Phenothiazines and antiparkinsonian, antihistaminic, and other drugs with anticholinergic properties, such as benztropine and biperiden, decrease the secretions of the respiratory tract, stomach, and intestines. The effects can be reversed by the use of cholinergic drugs.

1. *Psychotropic Drugs*

The phenothiazines and the tricyclic antidepressants cause xerostomia because of cholinergic blockage (39,63,66). The tricyclic antidepressants also reduce the salivary flow (4) and increase the viscosity of sputum (6). The incidence of imipramine xerostomia varies from 10 to 50% in adults. When the salivary secretions are decreased, there is a possibility of dental caries, moniliasis, and parotitis (1,9). In healthy volunteers, amitriptyline, but not nomifensine, was effective in reducing the spontaneous palmar skin reflex, a physiologic response which is mediated by the sympathetic cervical ganglion and which reflects the action potential of the sweating glands (36). Excessive sweating may be observed in subjects treated with monoamine oxidase inhibitors, amphetamines, or methyl phenidate.

2. *Other Drugs*

The rauwolfia alkaloids, guanethidine, methyldopa, and α-adrenoceptor blocking agents, such as phentolamine, also cause a decrease in salivary secretion. Reserpine and methyldopa induce nasal congestion and obstruction as well (23). The antihistaminic and antiparkinsonian drugs possess anticholinergic properties and, as such, inhibit sudoriferous glandular secretions. Since the lacrimal glands are primarily under the influence of cholinergic systems, the antihistaminic and antiparkinsonian drugs also inhibit lacrimal secretions. There are reports that practolol, a β-adrenoceptor antagonist, caused an irreversible decrease in tear secretion (15). Propranolol and oxyprenolol have also induced a reduction in lacrimal glandular secretions, but this effect is probably not due to autonomic mechanisms but, rather, to a direct toxic effect on the lacrimal glands and conjunctiva.

G. Sexual Organ Effects

The autonomic nervous system has some regulatory functions in the physiology of the sex response. There are ejaculation disorders associated with the adrenergic blocking action of psychotropic and antihypertensive drugs (12,41,48,56). Since ejaculation is under sympathetic control, drugs having α-adrenergic blocking action, such as phenoxybenzamine, can inhibit this response. Delayed or retrograde ejaculation has been reported in patients receiving bethanidine or guanethidine (68; see also Ref. 33 for references). Moreover, there have been several reports of impotence related to β-adrenoreceptor antagonists (33). Impotence has also been described in patients receiving clonidine, likely as a result of the predominantly presynaptic α-adrenoreceptor-stimulating action of the drug and its tendency to produce sedation and depression (33).

Since erection is controlled by the anticholinergic segments of the autonomic nervous system, cholinergic blocking agents induce sexual dysfunction by inhibiting the achievement and the maintenance of erection; thus there are diminished libido and sexual function. These effects have been reported with atropine-like agents, tricyclic antidepressants, antiparkinson drugs (i.e., benztropine and benzhexol), as well as with the antiarrhythmic agent disopyramide (33). Moreover, impotence may develop in patients treated with antipsychotic drugs (phenothiazines and butyrophenones) owing to several factors, including stimulation of prolactin release (see Brambilla, in this volume), the sedative and anticholinergic actions of the drug, and α-adrenoreceptor antagonism (11,33,41). There have been mixed results with the monoamine oxidase inhibitor antidepressants. In some studies, there was an increase in spermatozoa and volume of ejaculate, and another report claims that the patient was impotent. The autonomic effects on ejaculation and erection may be mediated by stimulation of appropriate peripheral α-receptors in the pelvic plexus.

H. Pulmonary Effects

Cholinergic drugs that act at autonomic receptor sites in the lung cause obstruction of the airways (51). Direct cholinergic activation with drugs such as pilocarpine or carbachol or activation indirectly with anticholinesterases, such as neostigmine, may induce bronchoconstriction (14,19). Thiopental sodium and cyclopropane may also induce bronchoconstriction through cholinergic activation. The β-adrenoceptor blocking agents, such as propranolol, cause bronchoconstriction and aggravate preexisting asthma (69). Thus propranolol should not be administered to such individuals, as a death was reported from the intravenous administration of 5 mg of propranolol.

IV. SUMMARY

Drugs that suppress or activate autonomic neuroeffector sites can induce complex imbalances of the autonomic nervous system, which can lead to serious disturbances. The psychotropic drugs are the most prominent in this respect, but direct peripheral vasodilators, certain antihistamines, lithium carbonate, antiparkinsonian drugs, and gastrointestinal tract antispasmodics are also implicated in causing adverse autonomic effects. The drugs discussed induce serious cardiovascular, esophageal, gastrointestinal, ophthalmological, urological, secretory glandular, sexual, and pulmonary effects. Because of compensatory mechanisms and the variations in sensitivity of neuroeffector sites, as well as the nonspecificity of the autonomic effects of a particular agent, the actions cannot always be explained pharmacologically. Many biochemical reactions may be involved, which complicates the understanding of the mechanisms of such toxicities.

REFERENCES

1. Aldous, J. A. 1964. Induced xerostomia and its relation to dental caries. *J. Dent. Child. 31*:160-162.
2. Alexander, C. S., Nino, A. 1969. Cardiovascular complications in young patients taking psychotropic drugs. *Am. Heart J. 78*: 757-769.
3. Appel, P., Eckel, K., Harrer, G. 1971. Veranderungen des Blasen und Blasensphinktertonus durch Thymoleptic. *Int. Pharmacopsychiatry 6*:15-22.
4. Arnold, S. E., Kahn, R. J., Faldetta, L. L., Laing, R. A., McNair, D. M. 1981. Tricyclic antidepressants and peripheral anticholinergic activity. *Psychopharmacology 74*:325-328.
5. Ayd, F. J. 1965. Amitriptyline: Reappraisal after six years experience. *Dis. Nerv. Syst. 26*:719-727.
6. Baillie, R. M. 1967. Amitriptyline and sputum viscosity. *Lancet 2*:369-370 [Cited in Brewis, R. A. L. 1977. Respiratory disorders, in *Textbook of Adverse Drug Reactions* (Davies, D. M., ed.), Oxford University, New York, pp. 103-123.]

7. Baldessarini, R. J. 1980. Drugs and the treatment of psychiatric disorders, in *The Pharmacological Basis of Therapeutics* (Gilman, A. G., Goodman, L. S., Gilman, A., eds.), MacMillan, New York, pp. 391-447.
8. Ballinger, B. R., Feely, J. 1983. Antidepressants. *Br. Med. J. 286*:1885-1887.
9. Belfer, M. L., Shader, R. I. 1970. Autonomic effects, in *Psychotropic Drugs Side Effects* (Shader, R L., DiMascio, A., eds.), Williams and Wilkins, Baltimore, pp. 116-131.
10. Blackwell, B. 1981. Adverse effects of antidepressant drugs. *Drugs 21*:201-219.
11. Blair, J. H., Simpson, G. M. 1966. Effect of antipsychotic drugs on reproductive function. *Dis. Nerv. Syst. 27*:645-647.
12. Blaschke, T. F., Melmon, K. L. 1980. Antihypertensive agents and drug therapy of hypertension, in *The Pharmacological Basis of Therapeutics* (Gilman, A. G., Goodman, L. S., Gilman, A., eds.), MacMillan, New York, pp. 793-818.
13. Bock, R., Swain, J. 1963. Ophthalmologic findings in patients on long term chlorpromazine therapy. *Am. J. Ophthalmol. 56*: 808-810.
14. Brewis, R. A. L. 1977. Respiratory disorders, in *Textbook of Adverse Drug Reactions* (Davies, D. M., ed.), Oxford University Press, New York, pp. 103-123.
15. Bron, A. J. 1979. Mechanisms of ocular toxicity, in *Drug Toxicity* (Gorrod, J. W., ed.), Taylor and Francis, London, pp. 229-253.
16. Burgess, C. D. 1981. Effects of antidepressants on cardiac function. *Acta Psychiatr. Scand. 63* Suppl. 290: 370-379.
17. Byck, R. 1975. Drugs in the treatment of psychiatric disorders, in *The Pharmacological Basis of Therapeutics*, 5th ed. (Goodman, L. S., Gilman, A., eds.), MacMillan, New York, pp. 152-200.
18. Campese, V. M. 1981. Minoxidil: A review of its pharmacological properties and therapeutic use. *Drugs 22*:257-278.
19. Cancro, R., Davis, J. M., Klawans, H., Tancredi, L. 1981. Medical and legal implications of side effects from neuroleptic drugs. A round-table discussion, *J. Clin. Psychiatry 42*:78-82.
20. Crombie, A. L. 1977. Eye disorders, in *Textbook of Adverse Drug Reactions* (Davies, D. M., ed.), Oxford University, New York, pp. 313-323.
21. Davies, D. M., Gold, R. G. 1977. Cardiac disorders, in *Textbook of Adverse Drug Reactions* (Davies, D. M., ed.), Oxford University, New York, pp. 81-102.
22. Day, M. D. 1979. *Autonomic Pharmacology. Experimental and Clinical Aspects*, Churchill Livingstone, Edinburgh.
23. Diamond, C. 1977. Ear, nose, and throat disorders, in *Textbook of Adverse Drug Reactions* (Davies, D. M., ed.), Oxford University, New York, pp. 324-334.

24. Douglas, A. P. 1977. Gastrointestinal disorders, in *Textbook of Adverse Drug Reactions* (Davies, D. M., ed.), Oxford University, New York, pp. 134-145.
25. Ebert, M. D., Shader, R. I. 1970. Cardiovascular effects, in *Psychotropic Drug Side Effects Clinical and Theoretical Perspectives* (Shader, R. I., DiMascio, A., eds.), Williams and Wilkins, Baltimore, pp. 149-163.
26. Ellis, S. 1965. The autonomic nervous system—General considerations, in *Drill's Pharmacology in Medicine* (DiPalma, J. R., ed.), McGraw-Hill, New York, pp. 411-425.
27. Glassman, A. H., Bigger, J. T. 1981. Cardiovascular effects of therapeutic doses of tricyclic antidepressants. *Arch. Gen. Psychiatry 38*:815-820.
28. Green, A. F., Armstrong, J. M., Farmer, J. B., Fielden, R., Langer, S. Z., Maxwell, R. A., Natoff, I. L. 1979. Autonomic pharmacology: Report of the main working party, in *Pharmacological Therapy*, Vol. 5, *International Encyclopedia of Pharmacology and Therapeutics, Section 102, Pharmacological Methods in Toxicology* (Zbinden, G., Gross, F., eds.), Pergamon, Oxford, pp. 9-48.
29. Hayes, T. A., Panitch, M. L., Barker, E. 1975. Imipramine dosage in children: A comment on "imipramine" and EKG abnormalities in hyperactive children. *Am. J. Psychiatry 132*:546-547.
30. Heiman, E. M. 1977. Cardiac toxicity of thioridazine-tricyclic antidepressant combination. *J. Nerv. Ment. Dis. 165*:139-143.
31. Hollister, L. E. 1981. Current antidepressant drugs. *Drugs 22*: 129-152.
32. Honigfeld, G., Newhall, P. N. 1965. Hemodynamic effects of imipramine, acetophenazine and trifluoperazone in geriatric psychiatry. *Dis. Nerv. Syst. 26*:427-429.
33. Horowitz, J. D., Goble, A. J. 1979. Drugs and impaired male sexual function. *Drugs 18*:206-217.
34. Hsu, J. J., Yap, A. T. 1967. Autonomic reactions in relation to psychotropic drugs. *Dis. Nerv. Syst. 28*:304-310.
35. Hussar, A. E., Bragg, D. G. 1969. The effect of chlorpromazine on the swallowing function in chronic schizophrenic patients, *Am. J. Psychiatry 126*:570-573.
36. Ikeda, Y., Nomura, S., Sawa, Y., Nakazawa, T. 1982. The effects of antidepressants on the autonomic nervous system—A current investigation. *J. Neural Transm. 54*:65-73.
37. Jefferson, J. W. 1975. A review of cardiovascular effects and toxicity of tricyclic antidepressants. *Psychosom. Med. 37*:160-179.
38. Jonas, S. 1959. Miosis following administration of chlorpromazine and related agents. *Am. J. Psychiatry 115*:817-818.
39. Johnson, A. L., Hollister, L. E., and Berger, P. A. 1981. The anticholinergic intoxication syndrome: Diagnosis and treatment. *J. Clin. Psychiatry 42*:313-317.

40. Knudsen, K., Heath, A. 1984. Effects of self poisoning with maprotiline. *Br. Med. J. 288*:601-603.
41. Kotin, J., Wilbert, D. E., Verburg, D., Soldinger, S. M. 1976. Thioridazine and sexual dysfunction. *Am. J. Psychiatry 133*:82-85.
42. Kris, E. B. 1963. Five year community follow-up of patients discharged from a mental hospital. *Curr. Ther. Res. 5*:451-462.
43. Leuschner, F., Neuman, W., Hempel, R. 1980. Toxicology of antipsychotic agents, in *Psychotropic Agents. Handbook of Experimental Pharmacology*, Vol. 55/1 (Hoffmeister, F., Stille, G., eds.), Springer, Berlin, pp. 225-253.
44. Lyle, W. H. 1966. Adynamic ileus and nortriptyline. *Br. Med. J. 1*:980.
45. Martin, G. I., Zaug, P. J. 1975. Electrocardiographic monitoring of enuretic children receiving therapeutic doses of imipramine. *Am. J. Psychiatry 132*:540-542.
46. Mayer, S. E. 1980. Neurohumoral transmission and the autonomic nervous system, in *The Pharmacological Basis of Therapeutics*, 6th ed. (Gilman, A. G., Goodman, L. S., Gilman, A., eds.), MacMillan, New York, pp. 56-90.
47. Merrill, D. C., Markland, C. 1972. Vesical dysfunction induced by the major tranquilizers. *J. Urol. 107*:769-771.
48. Miller, R. A. 1976. Propranolol and impotence. *Ann. Intern. Med. 85*:682-683.
49. National Institute of Mental Health, Psychopharmacology Service Center Collaborative Study Group, 1964. Phenothiazine treatment in acute schizophrenia. *Arch. Gen. Psychiatry 10*:246-261.
50. Noble, J., Matthew, H. 1969. Acute poisoning by tricyclic antidepressants: Clinical features and management of 100 patients. *Clin. Toxicol. 2*:403-421.
51. Orme, M. L'E. 1979. Toxic effects of compounds on the pulmonary system, in *Drug Toxicity* (Garrod, J. W., ed.), Taylor and Francis, London, pp. 255-268.
52. Petri, G., Szenohradszky, J., Porszasz-Gibiszer, K. 1971. Sympatholytic treatment of paralytic ileus. *Surgery 70*:359-367.
53. Risch, S. C., Groom, G. P., Janowsky, D. S. 1981. Interfaces of psychopharmacology and cardiology—Part one. *J. Clin. Psychiatry 42*:23-34.
54. Risch, S. C., Groom, G. P., Janowsky, D. S. 1981. Interfaces of psychopharmacology and cardiology—Part two. *J. Clin. Psychiatry 42*:47-59.
55. Saraf, K. R., Klein, D. F., Gittelman-Klein, R., Groff, S. 1974. Imipramine side effects on children. *Psychopharmacologia 37*:265-274.
56. Shader, R. I. 1970. Ejaculation disorders, in *Psychotropic Drug Side Effects* (Shader, R. I., DiMascio, A., eds.), Williams and Wilkins, Baltimore, pp. 72-96.

57. Shader, R. I., Harmatz, J. C. 1970. Gastrointestinal effects, in *Psychotropic Drug Side Effects* (Shader, R. I., DiMascio, A., eds.), Williams and Wilkins, Baltimore, pp. 198-205.
58. Shearer, W. T., Schreiner, R. L., Marshall, R. E. 1972. Urinary retention in a neonate secondary to maternal ingestion of nortriptyline. *J. Pediatr. 81*:570-572.
59. Simpson, G. M., Pi, E. A., Smarek, J. J., Jr. 1981. Adverse effects of antipsychotic agents. *Drugs 21*:138-151.
60. Taylor, P. 1980. Ganglionic stimulating and blocking agents, in *The Pharmacological Basis of Therapeutics*, 6th ed. (Gilman, A. G., Goodman, L., Gilman, A., eds.), MacMillan, New York, pp. 211-219.
61. Thorstrand, C. 1976. Clinical features in poisoning by tricyclic antidepressants with special reference to the ECG. *Acta Med. Scand. 199*:337-344.
62. Tyber, M. A. 1975. The relationship between hiatus hernia and antidepressants: A report of five cases. *Am. J. Psychiatry 132*: 652-653.
63. Van der Kolk, B. A., Shader, R. I., Greenblatt, D. J. 1978. Autonomic effects of psychotropic drugs, in *Psychopharmacology: A Generation of Progress* (Lipton, M. A., DiMascio, A., Killam, K. F., eds.), Raven Press, New York, pp. 1009-1020.
64. Van Praag, H. M. 1978. *Psychotropic Drugs A Guide for the Practitioner*, Brunner/Mazel, New York.
65. Vohra, J. K. 1974. Cardiovascular abnormalities following tricyclic antidepressant drug overdosage. *Drugs 7*:323-325.
66. Von Knorring, L. 1981. Changes in saliva secretion and accommodation width during short-term administration of imipramine and zimelidine in healthy volunteers. *Int. Pharmacopsychiatry 16*:69-78.
67. Winsberg, B. G., Goldstein, S., Yepes, L. E., Perel, J. M. 1975. Imipramine and electrocardiographic abnormalities in hyperactive children. *Am. J. Psychiatry 132*:542-545.
68. Wollam, G. L., Gifford, R. W., Tarazi, R. C. 1977. Antihypertensive drugs: Clinical pharmacology and therapeutic use. *Drugs 14*:420-460.
69. Zaid, G., Beall, G. N. 1966. Bronchial response to beta-adrenergic blockade. *N. Engl. J. Med. 275*:580-584.
70. Zaratzian, V. L. 1980. Psychotropic drugs—Neurotoxicity. *Clin. Toxicol. 17*:231-270.

4

Drug-Induced Neuroendocrine Disorders: Neuroleptics and Hormones

F. Brambilla
Psychoendocrine Center, Ospedale Psichiatrico PINI, Milan, Italy

I. INTRODUCTION

In laboratory animals, the neuroendocrine system was shown to be a major target in the action of certain classes of neurotoxic chemicals, including pesticides, heavy metals, and excitotoxic amino acids (see Bondy, Manzo et al., and McGeer and McGeer in this book). In humans, the most obvious expressions of neuroendocrine toxicity apparently involve the exposure to psychotropic drugs.

Psychotropic drugs act by modifying the turnover of brain neurotransmitters and their receptor sensitivity in the central nervous system. Neurotransmitters in the tuberoinfundibular area physiologically influence the secretion of the releasing and inhibiting factors, regulators of the peripheral glands. Thus the prolonged and often massive administration of psychotropic drugs can lead to the appearance of clinical or subclinical endocrine disorders. In turn, peripheral hormones may interfere through feedback phenomena with neurotransmitter turnover and receptor sensitivity, with consequent changes in the individual response to psychopharmacological treatments. For these reasons, the study of the neuroendocrine function reaches particular significance during the course of psychotropic therapy.

Long-lasting administration of antianxiety (benzodiazepines and meprobamate) and antidepressant (monoamine oxidase inhibitors and tricyclics) drugs does not induce clinically evident neuroendocrine alterations, even though, occasionally, hormonal impairments have been observed in individual cases. Lithium therapy may cause disorders in thyroid and parathyroid function, electrolyte and water regulation, and in sugar metabolism. The most clinically relevant endocrine impairments are related to the administration of neuroleptic drugs.

Neuroleptics include phenothiazines, butyrophenones, diarylbutamines, thioxantenes, benzamides, and dibenzodiazepines. Since most of the endocrine effects of these drugs are similar, possibly because of their common antidopaminergic activity, they will be considered together.

In this chapter, we will first report data from animal studies. In this regard, we must mention that the animal investigations dealing with neuroendocrinology were done mostly to investigate physiological and biochemical aspects of the normal central nervous system, using the pharmacological stimulus only to clarify the biological dynamics of specific neuronal areas. As a result, the doses of the drugs have frequently been much higher than those administered therapeutically to humans, and therefore the hormonal effects observed in laboratory animals may differ from those elicited by the same drugs during the usual treatment of psychiatric patients. In addition, there is a large body of evidence indicating that the neuroendocrine effects in animals cannot always be extrapolated to humans, because of well-known species-related differences in the connections between neurotransmitter and neurohormone activities. Moreover, psychotic patients often exhibit hormonal disorders related to the inherent disease that may interfere with the typical neuroendocrine responses to therapy. All these elements must be taken into account in interpreting the data from the literature.

As mentioned before, neuroleptics act specifically on the dopamine system. Table 1 summarizes the main effects of dopamine on the neuroendocrine system.

II. EFFECT OF DRUGS ON THE INDIVIDUAL NEUROENDOCRINE FUNCTIONS

A. Prolactin

Acute or chronic administration of phenothiazines, haloperidol, pimozide, and sulpiride to experimental animals induced marked and sustained prolactin (PRL) hypersecretion (82).

In healthy volunteers, acute administration of neuroleptics stimulates PRL secretion. Peaks differ in amplitude from subject to subject,

Table 1 Effects of Dopamine on Neurohormones[a]

	Animals	Men
CRF/ACTH	↓	→
TRH/TSH	↓[b]	↓[c]
LHRH/FSH-LH	↑	↑
GHR-IF/GH	↓	↑
PIF-PRF/PRL	↓	↓

[a]CRF, corticotropin-releasing factor; ACTH, adrenocorticotropic hormone; TRH, thyrotropin-releasing hormone; TSH, thyroid-stimulating hormone; LHRH, luteinizing-hormone-releasing hormone; FSH, follicle-stimulating hormone; LH, luteinizing hormone; GHR-IF, growth-hormone-releasing and -inhibiting factors; GH, growth hormone; PIF-PRF, prolactin-inhibiting and -releasing factor; PRL, prolactin.
[b]Cold-induced TSH secretion.
[c]Hypothyroid subjects.

but are reproducible in the same subjects. Peaks also vary from one drug to another in parallel with the antipsychotic activity of the drug; they are higher in women than in men and exhibit dose dependency up to certain doses, above which PRL levels tend to plateau (13,36,49,54, 63,77,78,80,96,100).

Chronic administration of neuroleptics to psychotic patients increases PRL levels in both plasma and cerebrospinal fluid, with the values in the two compartments closely correlated (54,64). The values are generally lower than those obtained after acute administration of the drugs, but still higher in women than in men. According to Ghadirian et al. (30), hyperprolactinemia is significantly correlated with sexual dysfunction in both women and men. The hyperprolactinemia reaches its maximum 2-3 days after the start of therapy and returns to normal 48-72 hr after therapy is discontinued. During treatment, it maintains a plateau with minor and erratic fluctuations in the first 6-12 months, suggesting that tolerance to the drugs does not develop in the tuberoinfundibular area (3,4,19,28,32,36,42,49,57,60-63,67,68,90,96,98). However, when the therapy is prolonged beyond 2-3 years, PRL levels decrease in some patients, sometimes down to normal levels, suggesting that some tolerance develops (23,41,51). This could be due to several factors other than reduced target sensitivity to the specific effect of the drugs. For example, pharmacokinetic changes may intervene, or pituitary lactotrops may become exhausted after prolonged stimulation. However, in contrast with the last hypothesis, it has been observed that the acute thyrotropin-releasing hormone stimulation elicits PRL responses in patients with or without elevated basal PRL levels (67). Moreover, patients with normal basal PRL levels during chronic therapy show a brisk and marked lowering of the levels down to subnormal values after the washout of the therapy, together with a marked worsening of the schizophrenic symptomatology, and a substantial improvement when the drug treatment is reinstituted. These data suggest that during therapy a supersensitivity of the dopamine receptors had developed in these patients (51).

Since PRL levels tend to be correlated in individual patients with circulating concentrations of neuroleptics, it has been suggested that the PRL response may be a more convenient guide to neuroleptic treatment than measurement of plasma levels of the drugs (41). Moreover, PRL levels could serve as early biological markers for the patient's capacity to respond to a given dose of a given drug. However, the data in this regard have been contradictory (32,34,40,48,62-64,98). Ohman and Axelsson (71) used the Michaelis-Menten equation, based on the concept that dopamine receptors and neuroleptics interreact to produce the PRL increment, and proposed that it might be possible to predict the neuroleptic doses needed to obtain a significant clinical improvement for each patient when utilizing PRL levels to establish an individual pharmacokinetic constant. In other words, receptor and neuroleptic both act as substrate and the increment of PRL secretion represents the result. Of particular interest is the fact that clozapine, even though being an antischizophrenic drug, induces only a very modest increment in PRL levels, suggesting that, at least in the tuberoinfundibular area, the drug has a very mild antidopaminergic effect (45,69).

Prolonged phenothiazine therapy does not change the PRL circadian rhythm (93). The response to thyrotropin-releasing hormone stimulation is not modified by sulpiride or thioxantene therapy, but L-dopa inhibition is blocked (19,54).

Administration of long-acting drugs (fluphenazine decanoate and enanthate, perphenazine enanthate) increases PRL secretion, with a peak at 24-48 hr after injection and a successive plateau which persists in parallel with the clinical effect of the therapy (3,4,60). Prolactin levels remain elevated in some patients for up to a year following withdrawal of long-acting neuroleptics. In these subjects the gradual decline in PRL values and consequent gradual resensitization of dopamine receptors may act as a protective mechanism against development of supersensitivity-related reacutization of the disease.

Prolactin levels during chronic neuroleptic treatment have been suggested to be of value in establishing whether or not complete dopamine receptor blockade has been obtained. If PRL levels cannot be further increased in patients by an acute challenge or by increasing the dose of neuroleptics, this would mean that all the dopamine receptors available had been blocked by the therapeutic dose.

B. Pituitary-Adrenal Axis

Acute or chronic administration of phenothiazines and butyrophenones to experimental animals increases the secretion of adrenocorticotropic hormone (ACTH) and corticosterone, but the ACTH response to stimuli is blunted (24).

Acute administration of dihydrobenzperidol to healthy volunteers first stimulates and then immediately inhibits cortisol secretion. Since the stimulatory effect is blunted by previous administration of dexamethasone, it has been suggested that the drug acts on the pituitary and not on the adrenal glands (27). Sulpiride increases aldosterone blood levels without changing renin secretion (18).

The effects resulting from long-term treatments are contradictory. Psychiatric patients on prolonged therapy with chlorpromazine, levopromazine, and thioridazine show both reduced (20,27,97) and moderately increased (31) urinary levels of 17-ketosteroids and 17-hydroxycorticosteroids. The responses to the stimulation with ACTH, insulin shock, or pyrogens are sometimes blunted, sometimes normal, while that to metopyron is always blunted (3,4,15,27,31,79,81). Basal plasma cortisol levels, the circadian rhythm, and the responses to insulin and ACTH stimulation are unchanged during therapy with chlorpromazine, thioridazine, trifluoperazine, and fluphenazine decanoate (3,81). Aldosterone secretion and plasma renin activity are stimulated by chlorpromazine administration, possibly through the β-adrenergic activity of the drug (75).

C. Growth Hormone

Acute or chronic administration of phenothiazines to experimental animals does not change basal secretion or the circadian rhythm of growth hormone (GH), but blocks its response to insulin stimulation (82). Sulpiride administration blocks both the synthesis and the release of the hormone (55).

Acute administration of phenothiazines or haloperidol to healthy volunteers does not change basal GH levels (77), but the nocturnal peak is lowered or does not occur, and the responses to insulin-induced hypoglycemia or dopamine or L-dopa stimulation are blunted (39,48,65, 88,94). Instead, acute or prolonged administration of sulpiride to healthy volunteers or patients with psychosomatic disorders has no effect on either basal GH levels or the responses to arginine or L-dopa stimulation (1,54,95,100). Administration of pimozide in low doses and for a short time does not change basal GH levels, the nocturnal peak, or the response to insulin, glucagon, or L-dopa stimulation, only blunting the responses to apomorphine and γ-aminobutyric acid (13,16,58).

Chronic administration of chlorpromazine, perphenazine, thioridazine, trifluoperazine, fluphenazine decanoate, haloperidol, flupentixol, and pimozide to psychotic patients does not change basal GH levels. The response to insulin stimulation is reported as normal, blunted, and even enhanced (3,7,17,28,45,81,83). The nocturnal peak is also reported as both normal and blunted (33,45). The responses to stimulation with L-dopa, dopamine, methylphenidate, bromocriptine, and apomorphine are blunted by the administration of phenothiazines, pimozide, and clozapine (45,65,69). Instead, chronic sulpiride administration does not block the GH response to L-dopa or to insulin-induced hypoglycemia, except occasionally in psychotic children (8,66).

D. Gonadotropins

Delayed sexual maturation, decreases in ovarian and secondary sexual organ weight, ovulation blockade, depression of sexual activity, infertility, and abortion have been reported following acute or chronic administration of phenothiazines, haloperidol, pimozide, and sulpiride to experimental animals. Ovulatory peaks of follicle-stimulating hormone (FSH), luteinizing hormone (LH), and luteinizing-hormone-releasing hormone (LHRH) are also blunted (14,21,82).

The acute administration of chlorpromazine to healthy volunteers lowers LH, but not FSH secretion (57), while haloperidol in low doses does not modify the basal levels of either gonadotropin (77). Pimozide in low doses and given for a short period of time increases FSH and reduces LH secretion while blocking the response to LHRH stimulation (16,52). Sulpiride, when administered to either healthy volunteers or to patients with psychosomatic disorders, does not modify tonic FSH and LH secretion (except in postmenopausal women, in whom both gonadotropins are lowered to levels typical of those of fertile age), but increases either response to LHRH stimulation (1,22,54,56,101). The ovulatory peak and postovulatory secretion during the luteal phase are blunted (22).

Chronic administration of phenothiazines, haloperidol, flupentixol, sulpiride, or penfluridol delays or blocks ovulation in psychotic women, resulting in various menstrual disorders and often in amenorrhea (30, 85,93). In psychotic men they induce impotence, ejaculatio praecox, and several types of sexual impairments, including reduced sperm motility (53). Tonic secretion of FSH and LH and their responses to LHRH stimulation are generally unchanged, and only occasionally blunted (3,9-12,44,59). Only thioridazine induces a clear-cut reduction of LH secretion (11).

E. Testosterone

The prolonged administration of chlorpromazine to experimental animals induces atrophy of the testes and seminal vesicles (24). Sulpiride stimulates testosterone secretion and spermiogenesis (76). The acute or prolonged administration of thioridazine to healthy volunteers blunts the nocturnal secretion of testosterone (5). Sulpiride administration does not modify basal levels of the hormone, but increases its response to acute administration of chorionic gonadotropin (1).

In chronic schizophrenics, prolonged administration of chlorpromazine, fluphenazine decanoate, haloperidol, penfluridol, pimozide, and sulpiride causes no change in basal testosterone levels, which generally range around the lowest limits of normal, and in a few cases only lowers them. However, testosterone levels rise on withdrawal of medication and fall again on its resumption. This suggests that a decrease in hormone secretion caused by these drugs may be involved in the development of disorders such as impotence, ejaculatio praecox, and various sexual problems often seen in these patients (3,4,7,10,11,12,19,26,70, 85,91). Thioridazine induces a clear-cut fall in testosterone levels (11, 35,50).

F. Ovarian Steroids

In experimental animals, phenothiazine administration blocks ovulation, both spontaneous and induced, when administered at the critical time (24). In nonpsychotic patients prolonged sulpiride administration at low doses reduces progesterone secretion during the luteal phase of the cycle (22). Phenothiazines and haloperidol induce amenorrhea in a large percentage of schizophrenic women (44). However, the estrogen levels are not reduced, except in few cases (3,4,9). Progesterone is lowered in the luteal phase of the cycle (3,4).

G. Pituitary-Thyroid Axis

The effects induced by phenothiazine administration in experimental animals have been investigated with conflicting results, probably because of the different parameters used to assess thyroid function. Both the acute and chronic treatments seem, on the whole, to block the pituitary-thyroid axis (24).

Acute administration of chlorpromazine to healthy volunteers does not modify basal thyroid-stimulating hormone (TSH) levels, but blunts its response to thyrotropin-releasing hormone (TRH) stimulation (46, 98). Instead, short-term low-dose administration of pimozide reduces basal TSH levels (16). Sulpiride has no effect on TSH secretion in male individuals, but sometimes induces a net increment of basal TSH levels in women, which could be blunted by previous L-dopa administration.

Sulpiride also increases in both sexes the response of TSH to TRH stimulation (54,59,101).

In schizophrenics, chronic treatment with phenothiazines and haloperidol does not modify protein-bound iodine, cholesterol, basal metabolic rate or ^{131}I uptake, basal TSH, or response to TRH stimulation (6-98). However, increased ^{131}I uptake has been reported for some cases, possibly due to increased clearance of iodine (6). Chronically administered sulpiride increases basal TSH levels, expecially in women, while the response to TRH stimulation could be either normal or increased (59). Clozapine reduces protein-bound iodine and thyroxine levels, generally to the lowest limits of normal; however, values tend to fluctuate widely from day to day (89).

H. Melatonin

In experimental animals, chlorpromazine inhibits melatonin catabolism and increases its accumulation in tissues (99). In schizophrenics chronically treated with chlorpromazine, blood and cerebrospinal fluid levels of melatonin are higher than in nontreated patients, especially during the nocturnal hours (92).

I. Melanocyte-Stimulating Hormone

Treatment with phenothiazines increases melanocyte-stimulating hormone secretion in experimental animals (24). Acute administration of chlorpromazine to healthy volunteers results in little or no effect on β-melanocyte-stimulating hormone secretion (73).

J. Vasopressin

In experimental animals, the effects of acute phenothiazine administration on antidiuretic hormone secretion seem to depend partly on the state of hydration of the animal and partly on the dose of neuroleptic given (24). However, a transient stimulating action seems to prevail. Chronic treatment with high doses tends to decrease antidiuretic hormone levels. In schizophrenic patients on prolonged treatment, increased antidiuretic hormon secretion can occur occasionally (87).

K. Glucose-Insulin Metabolism

In experimental animals, phenothiazine administration causes hyperglycemia (86). In healthy volunteers, acute administration of chlorpromazine does not modify basal glucose levels or their response to a glucose load, while trifluoperazine blunts this response (47). Schizophrenic patients on chronic chlorpromazine treatment seem to develop diabetes mellitus more frequently (38,43), but this observation has not been confirmed (40). Haloperidol therapy does not modify basal glucose levels.

III. CONCLUSION

From the data we have reported, it appears that the long-term administration of neuroleptics to psychiatric patients may have extremely variable effects on the neuroendocrine system, sometimes inducing massive and prolonged alterations and sometimes none at all. This heterogeneity may be due to a variety of factors, including the peculiarities of the neuroendocrine substratum in each patient, the dose and duration of therapy, and abnormalities in receptor-neurotransmitter system functions linked to the inherent mental disorder. Though neuroendocrine alterations other than hyperprolactinemia are not constant corollaries of neuroleptic therapy, they must be considered for individual patients, in order to understand the development of specific symptoms and prevent serious drug-induced hormonal imbalances.

The pathogenesis of some of the endocrine impairments is clearly related to the specific pharmacological profile of neuroleptics. For instance, the hyperprolactinemia, which is the prominent effect elicited at therapeutic dosage, is obviously related to the antidopaminergic activity, the major effect of all neuroleptics. Other neuroendocrine impairments are related to the neuroleptic interactions with multiple neurotransmitter-neuromodulator systems, with effects reflecting these combined influences.

Most of the drug-induced abnormalities are subclinical and, at present, no therapeutic approaches are available to correct them. The only two syndromes that are clinically striking, amenorrhea and galactorrhea, have been the object of various studies. While amenorrhea can be and generally is easily corrected by substitution therapy, galactorrhea, or rather the hyperprolactinemia responsible for it, is still an unresolved problem, since dopamine agonists (L-dopa, bromocriptine, lisuride, and lergotrile) which block PRL hypersecretion cannot be given to psychiatric patients because they worsen the schizophrenic symptomatology (2,29) and might antagonize the antidopaminergic neuroleptic activity. Hyperprolactinemia can be an extremely important epiphenomenon, since PRL stimulates dopamine turnover in the central nervous system (25). Increased dopamine turnover has been considered a possible cause of schizophrenia, even though there is no clear-cut evidence for this (74). The increase in PRL levels might influence both the schizophrenic symptomatology and the response to therapy. This possibility warrants further investigation.

It has also been suggested that massive and prolonged drug-induced hyperprolactinemia might result in the development or stimulation of breast cancer. However, psychotic patients on chronic treatment do not seem to have an increased incidence of breast cancer as compared to the general population (37,72,84).

These data, even though still controversial, suggest that the neuroendocrine investigation in the fields of pharmacology and toxicology provides a promising approach to better understand both the mechanism of action and toxicity of psychotropic drugs in the context of a specific mental disorder. Moreover, these studies should provide some explanations for the heterogeneous responses to treatment in individual patients, possibly linked to the influence exerted by the hormonal impairments on the neurotransmitter-neuromodulator systems in the central nervous system.

REFERENCES

1. Ambrosi, B., Travaglini, P., Beck-Peccoz, P., Bara, R., Elli, E., Paracchi, A., Faglia, G. 1976. Effect of sulpiride-induced hyperprolactinemia on serum testosterone response to HCG in normal men. *J. Clin. Endocrinol. Metab.* *43*:700-703.
2. Angrist, B., Gershon, S. 1974. Dopamine and psychotic state: Preliminary remarks, in *Neuropsychopharmacology of Monoamines and Their Regulatory Enzymes* (Usdin, E., ed.), Raven Press, New York, pp. 211-219.
3. Beumont, P. J. V., Corker, C. S., Friesen, H. G., Kolakowska, T., Mandelbrote, B. M., Marshall, J., Murray, M. A. F., Wiles, D. H. 1974. The effects of phenothiazines on endocrine function: II. Effects in men and postmenopausal women. *Br. J. Psychiatry* *124*:420-430.
4. Beumont, P. J. V., Gelder, G. M., Friesen, G. H., Harris, G. V., MacKinnon, P. C. B., Mandelbrote, B. M., Wiles, D. H. 1974. The effects of phenothiazines on endocrine function. I. Patients with inappropriate lactation and amenorrhea. *Br. J. Psychiatry 124*: 413-419.
5. Bixler, E. O., Santen, R. J., Kales, A., Soldatos, C. R. 1977. Inverse effects of thioridazine (Melleril) on serum prolactin and testosterone concentrations in normal men, in *The Testis in Normal and Infertile Men* (Toen, P., Nankin, H. R., eds.), Raven Press, New York, pp. 403-407.
6. Blumberg, A. G., Klein, D. F. 1969. Chlorpromazine-procyclidine and imipramine: Effects on thyroid function in psychiatric patients. *Clin. Pharmacol. Ther.* *10*:350-354.
7. Brambilla, F., Guerrini, A., Guastalla, A., Rovere, C., Riggi, F. 1975. Neuroendocrine effects of haloperidol therapy in chronic schizophrenia. *Psychopharmacologia* *44*:17-22.

8. Brambilla, F., Guareschi-Cazzullo, A., Musetti, C., Musetti, L., Nobile, P. 1979. Growth hormone secretion in mentally disturbed children treated with sulpiride, in *Sulpiride and Other Benzamides* (Spano, P. F., Trabucchi, M., Corsini, U. G., Gessa, G. L., eds.), Italian Branch of the Research Foundation Press, Milan, pp. 221-229.
9. Brambilla, F., Guerrini, A., Guastalla, A., Riggi, F., Malagoli, G. 1979. Effect of chronic sulpiride treatment on basal and GnRH-stimulated gonadotropin secretion in male psychotic patients, in *Sulpiride and Other Benzamides* (Spano, F., Trabucchi, M., Corsini, U. G., Gessa, G. L., eds.), Italian Branch of the Research Foundation Press, Milan, pp. 231-241.
10. Brambilla, F., Riggi, F., Guerrini, A., Malagoli, G., Guastalla, A. 1979. The effect of prolonged sulpiride therapy on pituitary-gonadal function in male psychotic patients. *Clin. Ther.* *2*:91-97.
11. Brown, W. A., Laughren, T. P., Williams, B. 1981. Differential effects of neuroleptic agents on the pituitary-gonadal axis in men. *Arch. Gen. Psychiatry* *38*:1270-1272.
12. Casacchia, M., Meco, G., Caschedi, F., Di Ceglie, M., Falaschi, P., Rocco, A., Pompei, P., Frajese, G. 1979. Neuroendocrine side effects of antipsychotic therapy, in *Neuroendocrine Correlates in Neurology and Psychiatry* (Müller, E. E., Agnoli, A., eds.), Elsevier/North Holland, Amsterdam, pp. 211-224.
13. Cavagnini, F., Benetti, G. P., Invitti, C., Pinto, M., Dubini, A., Marelli, A. 1979. Effect of pimozide on growth hormone and prolactin secretion in response to gamma-amino-butyric acid in man. I. International Congress on Neuroactive Drugs in Endocrinology. Milan, abstracts.
14. Choudhury, S. A. R., Sharpe, R. M., Brown, P. S. 1974. The effect of pimozide, a dopamine antagonist, on pituitary gonadotropic function in the rat. *J. Reprod. Fertil.* *39*:275-283.
15. Christy, N. P., Longson, D., Worwitz, W. A., Knight, M. M. 1957. Inhibitory effect of chlorpromazine upon the adrenal cortical response to insulin hypoglycemia in man. *J. Clin. Invest.* *36*:543-549.
16. Collu, R., Jéquier, J. C., Leboeuf, G., Letartre, J., Ducharme, J. R. 1975. Endocrine effect of pimozide: A specific dopaminergic blocker. *J. Clin. Endocrinol. Metab.* *41*:981-984.
17. Cookson, J. C., Silverstone, T., Rees, L. 1982. Plasma prolactin and growth hormone levels in manic patients treated with pimozide. *Br. J. Psychiat.* *140*:274-279.
18. Costa, E., Frisina, M., De Pasquale, R. 1980. Increased aldosterone secretion induced by sulpiride. *J. Endocrinol.* *72*:273-277.
19. Cotes, M., Crow, J. T., Johnstone, E. C., Bartlet, W., Bourne, R. C. 1978. Neuroendocrine changes in acute schizophrenia as a function of clinical state and neuroleptic medication. *Psychol. Med.* *8*:657-665.

20. Co Tui, F., Brinitzer, W., Orr, A., Orr, E. 1960. Effect of chlorpromazine and of reserpine on adrenocortical function. *Psychiatr. Q. 34*:47-58.
21. Debeljuk, L., Khar, A., Jutisz, M. 1978. Effect of pimozide and sulpiride on the release of LH and FSH by pituitary cells in culture. *Mol. Cell. Endocrinol. 10*:159-162.
22. Delvoye, P., Taubert, H. D., Jurgensen, O., L'Hermite, M., Delogne, J., Robyn, C. 1975. Evolution des gonadotropines et de la progestérone sérique au cours d'une hyperprolactinémie induite par la sulpiride (Dogmatil) pendant la phase lutéale du cycle menstruel. *J. Pharmacol. Clin. 2*:121-124.
23. De Rivera, J., Lal, S., Ettigi, P., Hontela, S., Muller, H. F., Friesen, H. G. 1976. Effect of acute and chronic neuroleptic therapy on serum prolactin levels in men and women of different age groups. *Clin. Endocrinol. 5*:273-282.
24. De Wied, D. 1967. Chlorpromazine and endocrine function, in *Pharmacological Reviews* (De Wied, D., ed.), Williams and Wilkins, New York, pp. 251-288.
25. Drago, F. 1982. *Prolactin and Behavior* (Drago, F., ed.), Fidia Research Laboratory, Abano Terme, pp. 12-132.
26. Falaschi, P., Casacchia, M., Rocco, A., Meco, G., Caschedi, F., D'Urso, R., Toscano, V., Frajese, G. 1978. Sexual dysfunction during neuroleptic treatment, in the XIth CINP Congress, Abstract, p. 228.
27. Fotherby, K., Forrest, A. D., Laverty, S. G. 1959. The effect of chlorpromazine on adrenocortical function. *Acta Endocrinol. 32*:425-436.
28. Frantz, A. G., Kleinberg, D. L., Noel, G. L., Suh, H. K. 1972. Effects of neuroleptics on the secretion of prolactin and growth hormone, in *Endocrinology* (Scow, R. O., ed.), Excerpta Medica, Amsterdam, pp. 144-149.
29. Gershon, S., Angrist, B., Shopsin, B. 1975. Drugs, diagnosis and disease, in *Biology of Major Psychoses*, Vol. 54 (Freedman, D. X., ed.), Raven Press, New York, pp. 85-100.
30. Ghadirian, A. M., Chouinard, G., Annable, L. 1982. Sexual dysfunction and plasma prolactin levels in neuroleptic treated schizophrenic outpatients. *J. Nerv. Ment. Dis. 170*:463-467.
31. Gold, E. M., Di Raimondo, V. C., Forsham, P. H. 1960. Comparative effects of certain non-narcotic central nervous system analgesics and muscle relaxants on the pituitary-adrenocortical system. *Ann. N.Y. Acad. Sci. 86*:178-179.
32. Gruen, P. H., Sachar, E. J., Langer, G., Altman, N., Leifer, M., Frantz, A., Halpern, F. S. 1978. Prolactin response to neuroleptics in normal and schizophrenic subjects. *Arch. Gen. Psychiatry 35*:108-116.

33. Hall, J., Koulu, M., Pekkarinen, A., Aarima, M. 1978. Inhibiting effects of neuroleptic drug therapy on GH secretory peak in serum during sleep, in XIth CINP Congress, Abstract, p. 275.
34. Jorgensen, A., Andersen, J., Biorndal, N., Denker, S. J., Ludin, L., Malm, U. 1982. Serum concentrations of cis(Z)-flupentixol and prolactin in chronic schizophrenic patients treated with flupentixol and cis (Z)-flupentixol decanoate. *Psychopharmacology 77*:58-65.
35. Kales, A., Kales, J. D., Soldatos, C. R. 1975. Effects of thioridazine (Melleril) on anterior pituitary secretion: Changes in testosterone and prolactin. *Sleep Res. 4*:278.
36. Kato, Y., Ohgo, S., Chihara, K., Imura, H. 1975. Stimulation of human prolactin secretion by sulpiride. *Endocrinol. Jpn. 22*: 457-460.
37. Katz, J., Kunofsky, S., Patton, R. E., Allaway, N. C. 1967. Cancer mortality among patients in New York mental hospitals. *Cancer 20*:2194-2199.
38. Keskiner, A., El-Toumi, A., Bousquet, T. 1973. Psychotropic drugs, diabetes and chronic mental patients. *Psychosomatics 16*: 176-181.
39. Kim, S., Sherman, L., Kolodny, H., Benjamin, E., Singh, A. 1971. Attenuation by haloperidol of human serum growth hormone (HGH) response to insulin. *Clin. Res. 19*:718.
40. King, D. J. 1978. Lithium and neuroleptic drug combination: Lack of an effect on blood glucose. *Br. J. Clin. Pharmacol. 6*: 436-437.
41. Kolakowska, T., Wiles, D. H., Gelder, M. G., McNeilly, A. S. 1976. Clinical significance of plasma chlorpromazine levels. II. Plasma levels of the drug, some of its metabolites and prolactin in patients receiving long-term phenothiazine treatment. *Psychopharmacology 49*:101-107.
42. Kolakowska, T., Braddock, L., Wiles, D., Franklin, M., Gelder, M. 1981. Neuroendocrine tests during treatment with neuroleptic drugs. I. Plasma prolactin response to haloperidol challenge. *Br. J. Psychiat. 139*:400-412.
43. Korenyi, C., Lowenstein, B. 1968. Chlorpromazine-induced diabetes. *Dis. Nerv. Syst. 29*:827-828.
44. Kulcsar, S., Polishuck, W., Rubin, L. 1957. Aspect endocrinien du traitement à la chlorpromazine. *Presse Med. 65*:1288-1290.
45. Lal, S., Nair, N. P. V. 1979. Growth hormone and prolactin responses in neuropsychiatric research, in *Neuroendocrine Correlates in Neurology and Psychiatry* (Müller, E. E., Agnoli, A., eds.), Elsevier/North Holland, Amsterdam, pp. 179-194.
46. Lamberg, B. B., Linnoila, M., Fogelholm, R., Olkinuora, M., Kotilainen, P., Saarinen, P. 1977. The effect of psychotropic

drugs on the TSH response to thyroliberine (TRH). *J. Am. Med. Assoc. 181*:184-185.
47. Landolina, F., Chisari, C. 1962. Comportamento della curva glicemica da carico di glucosio dopo somministrazione di trifluoperazina. *Boll. Soc. Med. Chir. Suppl. 4*:227-236.
48. Langer, G., Sachar, E. J., Halpern, F. S. 1978. Effect of dopamine and neuroleptics on plasma growth hormone and prolactin in normal men. *Psychoneuroendocrinology 3*:165-169.
49. Langer, G. 1979. The prolactin model for the study of the pharmacodynamics of neuroleptic drugs in man, in *Neuroendocrine Correlates of Neurology and Psychiatry* (Müller, E. E., Agnoli, A., eds.), Elsevier/North Holland, Amsterdam, pp. 71-88.
50. Laughren, T. P., Brown, W. A., Petrucci, J. A. 1978. Effects of thioridazines on serum testosterone. *Am. J. Psychiatry 135*: 982-984.
51. Laughren, T. P., Brown, W. A., Williams, B. W. 1979. Serum prolactin and clinical state during neuroleptic treatment and withdrawal. *Am. J. Psychiatry 136*:108-110.
52. Leppaluoto, J., Mannisto, J., Rauta, T., Linnoila, M. 1976. Inhibition of midcycle gonadotropin release in healthy women by pimozide and fusaric acid. *Acta Endocrinol. 81*:455-459.
53. Levin, R. M., Amsterdam, J. D., Winokur, A., Wein, A. J. 1981. Effects of psychotropic drugs on human sperm motility. *Fertil. Steril. 36*:503-506.
54. L'Hermite, M., Denayer, P., Golstein, J., Virosoro, E., Vahaelst, L. Copinski, C., Robyn, C. 1978. Acute endocrine profile of sulpiride in the human. *Clin. Endocrinol. 9*:195-204.
55. MacLeod, R. M. Robyn, C. 1977. Mechanism of increased prolactin secretion by sulpiride. *J. Endocrinol. 72*:273-277.
56. Mancini, A. M., Guitelman, A., Debeljuk, L., Vargas, C. A. 1975. Effect of administration of sulpiride on serum follicle-stimulating hormone and luteinizing hormone levels in a group of postmenopausal women. *J. Endocrinol. 67*:127-128.
57. Mannisto, T. P., Korttila, K., Seppala, T. 1978. Serum prolactin levels after a single and subchronical oral administration of chlorpromazine and sulpiride. *Arzneim. Forsch. 28*:76-78.
58. Masala, A., Delitala, G., Alagna, S., Devilla, L., Ravasio, P. P., Lotti, G. 1978. Effect of dopaminergic blockade on the secretion of growth hormone and prolactin in men. *Metabolism 27*:921-926.
59. Massara, F., Camanni, F., Belforte, L., Vergano, V., Molinatti, G. M. 1978. Increased thyrotropin secretion induced by sulpiride in man. *Clin. Endocrinol. 9*:419-428.
60. Meltzer, H. Y., Sachar, E. J., Frantz, A. G. 1974. Serum prolactin levels in newly admitted psychiatric patients, in *Neuropsychopharmacology of Monoamines and Their Regulatory Enzymes* (Usdin, E., ed.), Raven Press, New York, pp. 299-315.

61. Meltzer, H. Y., Sachar, E. J., Frantz, A. G. 1975. Dopamine antagonism by thioridazine in schizophrenia. *Biol. Psychiatry* *10*:53-57.
62. Meltzer, H. Y., Fang, V. S. 1976. The effect of neuroleptics on serum prolactin in schizophrenic patients. *Arch. Gen. Psychiatry* *33*:279-286.
63. Meltzer, H. Y., Goode, D. G., Fang, V. S. 1977. Effects of chlorpromazine on plasma prolactin and chlorpromazine levels. *Psychopharmacology* *54*:59-60.
64. Meltzer, H. Y., Goode, D. J., Fang, V. S. 1978. The effects of psychotropic drugs on endocrine function. I. Neuroleptic precursors and agonists, in *Psychopharmacology: A Generation of Progress* (Lipton, M. A., Di Mascio, A., Killam, K. F., eds.), Raven Press, New York, pp. 508-530.
65. Mims, R. B. 1975. Inhibition of L-dopa-induced GH stimulation by pyridoxine and chlorpromazine. *J. Clin. Endocrinol. Metab.* *40*:256-259.
66. Mori, M., Kobayashi, Y., Shimoyama, S., Uehara, T., Nemoto, T., Fukuda, H., Kamio, N. 1977. Effect of sulpiride on serum growth hormone and prolactin concentrations following L-dopa administration in man. *Endocrinol. Jpn.* *24*:149-153.
67. Naber, D., Ackeneil, M., Laakman, G., Fischer, H. 1980. Basal and stimulated levels of prolactin, TSH and LH in serum of chronic schizophrenic patients, long-term treated with neuroleptics. *Pharmacopsychiatry* *13*:325-330.
68. Naber, D., Finkbeiner, C., Fischer, B., Zander, K. J., Ackeneil, M. 1980. Effect of long-term neuroleptic treatment on prolactin and norepinephrine levels in serum of chronic schizophrenics; relations to psychopathology and extrapyramidal symptoms. *Neuropsychobiology* *6*:181-189.
69. Nair, N. P. V., Lal, S., Cervantes, P. 1979. Clozapine on apomorphine-induced growth hormone secretion and serum prolactin concentration in schizophrenia. *Neuropsychobiology* *5*:136-142.
70. Nathan, R. S., Asnis, G. M., Dryenfurth, I. 1980. Plasma prolactin and testosterone during penfluridol treatment. *Lancet* *2*:94.
71. Ohman, R., Axelsson, R. 1978. Relationship between prolactin response and antipsychotic effect of thioridazine in psychiatric patients. *Eur. J. Pharmacol.* *14*:111-116.
72. Overall, J. E. 1978. Prior psychiatric treatment and the development of breast cancer. *Arch. Gen. Psychiatry* *35*:898-899.
73. Plummer, N. A., Thody, A. J., Burton, J. L., Godamali, S. K., Shuster, S., Cole, N. E., Boyns, A. R. 1975. The effect of chlorpromazine on the secretion of immunoreactive β-MSH and prolactin in man. *J. Clin. Endocrinol. Metab.* *41*:380-382.
74. Praag van, H. M. 1977. The significance of dopamine for the

mode of action of neuroleptics and the pathogenesis of schizophrenia. *Br. J. Psychiatry 130*:463-474.

75. Robertson, B. 1957. Breast changes in the male and female with chlorpromazine or reserpine therapy. *Med. J. Aust. 44*:239-241.
76. Rollet, J., Czyba, J. C., Cottinet, D. 1974. Retentissement endocrinien du sulpiride chez le hamster doré. *J. Pharmacol. Suppl. 2*:85.
77. Rubin, R., Poland, R. E., O'Connor, D., Gouin, P. R., Tower, B. B. 1976. Selective neuroendocrine effects of low-dose haloperidol in normal adult men. *Psychopharmacology 47*:135-140.
78. Rubin, R., Hays, S. 1978. Profiles of prolactin response to antipsychotic drugs: Some methodological observations. *Psychopharmacol. Bull. 14*:9-11.
79. Saarima, H. A., Sourander, C., Rinne, U. K. 1963. The effect of chlorpromazine on the secretion of corticotropin in psychiatric patients examined by the metopyron test. *Clin. Endocrinol. 1*: 173-180.
80. Sachar, E. J., Gruen, P. H., Karasu, T. B., Altman, N., Frantz, A. G. 1975. Thioridazine stimulates prolactin secretion in man. *Arch. Gen. Psychiatry 32*:885-886.
81. Saldanha, V. F., Havard, C. V. H., Bird, R., Gardner, R. 1972. The effect of chlorpromazine on pituitary function. *Clin. Endocrinol. 1*:173-180.
82. Scapagnini, U., Nisticò, G. 1978. Psychotropic drugs and neuroendocrine function, in *Perspective in Endocrine Psychobiology* (Brambilla, F., Bridges, P. K., Endroczi, E., Heuser, G., eds.), Akademiai Kiado, Budapest, pp. 229-286.
83. Schimmelbush, W. H., Mueller, P. S., Scheps, J. 1971. The positive correlation between insulin resistance and duration of hospitalization in untreated schizophrenics. *Br. J. Psychiatry 118*:429-436.
84. Schyve, P. M., Smithline, F., Meltzer, H. Y. 1978. Neuroleptic-induced prolactin level elevation and breast cancer. *Arch. Gen. Psychiatry 35*:1291-1301.
85. Shader, R. I., Bethesda, M. 1963. Sexual dysfunction associated with thioridazine hydrochloride. *J. Am. Med. Assoc. 188*:175-177.
86. Shader, R. I., Di Mascio, A. 1970. *Psychotropic Drug Side Effects*, Williams and Wilkins, Baltimore.
87. Shah, D. K., Wig, N. N., Chaudury, R. R. 1973. Antidiuretic hormone levels in patients with weight gain after chlorpromazine therapy. *Indian J. Med. 61*:771-776.
88. Shearman, R. P., Turtle, I. R. 1970. Secondary amenorrhea with inappropriate lactation: Observations on etiology, estrogen-gonadotropin interrelationship, the behaviour of human growth hormone and response to treatment. *Am. J. Obst. Gynecol. 106*: 818-827.

89. Simpson, G. M., Varga, E. 1974. Clozapine: A new antipsychotic agent. *Curr. Ther. Res. 16*:679-686.
90. Siris, S. G., Van Kammen, D. P., Fraitis, D. C. 1978. Serum prolactin and antipsychotic responses to pimozide in schizophrenia. *Psychopharmacol. Bull. 14*:12-14.
91. Siris, S. G., Siris, E. S., Van Kammen, D. P. 1980. Effects of dopamine blockade on gonadotropins and testosterone in men. *Am. J. Psychiatry 137*:211-214.
92. Smith, J. A., Mee, T. J. X., Barnes, J. L. C. 1978. Metabolism of the pineal gland in schizophrenia and the effect of neuroleptic drugs. *Neurosci. Lett. Suppl. 1*:213.
93. Story, N. L. 1974. Sexual dysfunction resulting from drug side effects. *J. Sex. Res. 10*:132-149.
94. Takahashi, Y., Knippis, D. M., Daughaday, W. H. 1968. Growth hormone secretion during sleep. *J. Clin. Invest. 47*: 2079-2090.
95. Thorner, M. O., McNeilly, A. S., Hagan, C., Besser, G. M. 1974. Long-term treatment of galactorrhea and hypogonadism with bromocryptine. *Br. Med. J. 2*:419-422.
96. Turkington, R. W. 1972. Prolactin secretion in patients treated with various drugs. *Arch. Intern. Med. 130*:349-354.
97. Vallade, L., Corsetti, R. 1963. Steroides urinaires et chemiotherapies psychotropes. *Ann. Med. Psychol. 2*:709-720.
98. Wode-Helgodt, B., Eneroth, P., Fyro, B., Gullberg, B., Sedval, G. 1977. Effect of chlorpromazine treatment on prolactin levels in cerebrospinal fluid and plasma of psychotic patients. *Acta Psychiatr. Scand. 56*:280-293.
99. Wurtman, R. J., Axelrod, J. 1966. Effect of chlorpromazine and other drugs on the disposition of circulating melatonin. *Nature 212*:312.
100. Zanoboni, A., Zanoboni-Muciaccia, W., Zanussi, C. 1979. Effect of sulpiride and L-dopa on anterior pituitary hormones in plasma of normal men and women. *Farmaco 2*:71-84.
101. Zanoboni, A., Zanoboni-Muciaccia, W., Zanussi, C., Mereghetti, E., Rognoni, L. 1979. Sulpiride as a research tool in studying dopaminergic control of anterior pituitary hormones, in *Sulpiride and Other Benzamides* (Spano, F., Trabucchi, M., Corsini, U.G., Gessa, G. L., eds.), Italian Branch of the Research Foundation Press, Milan, pp. 185-191.

5

Drug-Induced Sleep Disorders

Christian Guilleminault
Sleep Disorders Center, Stanford University School of Medicine, Stanford, California

I. INTRODUCTION

In this short chapter, we shall consider the paradoxical effects of drugs taken to relieve sleep complaints and then analyze the performances of a variety of drugs under objective testing. First, however, a brief overview of sleep physiology is useful.

Two main sleep states have been identified using polygraphic monitoring: rapid eye movement (REM) and nonrapid eye movement (NREM) sleep. Major physiological changes controlled by the central nervous system's neuronal network occur in these two sleep states. The autonomic nervous system, composed of a sympathetic component that accelerates heart rate and a parasympathetic component that inhibits heart rate, provides a good example. During quiet wakefulness there is usually a balance between the two systems that shifts when a subject falls asleep. The overall effect can be summarized as a progressive decrease in sympathetic activity and an increase in parasympathetic activity, which peaks during REM sleep, particularly during abrupt bursts of "phasic activity" (characterized by rapid, intense eye movement, an abrupt increase in heart rate, and an upsurge in blood pressure).

Drugs may interact with these neuronal networks and central controls, affecting a specific bodily function that results in a drug-induced derangement of sleep. The atropinic component of tricyclic medications, for example, moderately affects the autonomic nervous system during wakefulness, while its effects are greatly magnified during REM sleep. The sleep-related disorder leads to complaints of insomnia (disturbed nocturnal sleep) or hypersomnia (excessive daytime somnolence).

Very little sleep physiology is considered when drug trials are performed, so that drugs, directly or indirectly, may be responsible for significant sleep disturbances. Ironically, those administered for sleep disorders are the major culprits in disrupting sleep, but drugs prescribed for a number of disorders may also be disruptive.

II. DRUGS ADMINISTERED FOR SLEEP COMPLAINTS

Hypnotic sedatives, which account for the majority of drug sales in the Western world, are prescribed to provide symptomatic relief for "insomnia of unknown etiology." They fall into three main categories; listed in decreasing order on the basis of sales data, they are benzodiazepines, barbiturates, and others. In 1977, 25.5 million prescriptions for hypnotics (excluding hospital medications) were written in the United States alone (National Prescription Audit, Institute of Medical Statistics, Ambler, Pennsylvania). In 1979 the American National Academy of Sciences and the Institute of Medicine jointly sponsored a large study on "Sleeping Pills, Insomnia and Medical Practice" (13). Their report reviewed all published studies up to 1979, concluding that there was little known on the objective effects of "sleeping pills" on sleep. A single objective study with polygraphic recordings demonstrating that a hypnotic medication was helpful 6 weeks after initial drug ingestion did not exist. There was no information on objectively determined daytime somnolence following either one night or several nights of hypnotic ingestion. The committee concluded that it was "unable to endorse the efficacy of specific therapies, drugs, or classes of drugs in the management of insomnia complaints" (13). Data are still scarce.

Some data exist on short- to intermediate-acting barbiturate agents (pentobarbital, amobarbital, secobarbital). A study performed at Stanford showed that objective measurements of total sleep time were temporarily increased with pentobarbitol, but after 8 days of continuous ingestion total sleep time decreased to values lower than those noted during the predrug period. There were more sleep disruptions and wake time after sleep onset on drug night 8 than there had been on drug night 4 (7). Withdrawal from barbiturates induces a rebound insomnia with nightmares several nights following drug intake (15).

Rebound insomnia occurs with all hypnotic drugs. There was some controversy concerning this sleep-related problem and benzodiazepines. Short-lasting benzodiazepines were considered to be more prone to this side effect than long-acting ones such as flurazepam or nitrazepam (20). A review of data obtained at Stanford, both published and unpublished, indicates that the issue of rebound insomnia with benzodiazepines is not clear-cut. Long-acting benzodiazepines do not induce rebound insomnia 48 hr after their discontinuation because of the long half-life of their psychoactive metabolites (such as N-desalkylflurazepam). However, objective studies on total sleep time, wake after sleep onset, and other objective parameters performed 3-6 days after discontinuation of flurazepam, 15 or 30 mg, revealed rebound insomnia in some patients. Rebound insomnia can also be noted with short-lasting benzodiazepines. Several objective studies indicate that some patients present rebound insomnia while others do not. Very short-acting benzodiazepines have an obvious effect on the first third of the night, but, as with placebo, leave sleep structure unchanged for the second and third parts of the night in many patients (1,10,20). Early morning insomnia, a significant increase in the wakefulness during the final hours of drug nights, has been reported to occur after 1 or 2 weeks of nightly administration of benzodiazepine hypnotics with short elimination half-lives, such as midazolam and triazolam (16).

Nonbarbiturate/nonbenzodiazepine prescription hypnotics, which include chloral hydrate, glutethimide, methaqualone, ethchlorvynol, and methyprylon, can also lead to significant rebound insomnia after discontinuation. Objective efficacy studies with polygraphic monitoring have shown that several of these compounds, such as glutethimide or ethchlorvynol, initially only weakly affect total sleep time, sleep latency, and wake after sleep onset, but like barbiturates, may quickly worsen objective polygraphically monitored sleep parameters (14,17,18).

Benzodiazepines currently are the most popular hypnotic medication because of their relatively large efficacy/LD_{50} ratio. The most recent investigations of these compounds have tried to answer some of the questions raised by the joint study in 1979 (13). Comparative studies of several benzodiazepines and long-term placebo administration indicate that long-term administration (6 weeks) of these drugs leads to progressive loss of efficiency. If some of the benzodiazepines studied had initial positive effects compared to base-line placebo, none of them significantly changed total sleep time, sleep latency (i.e., time between lights out and sleep onset), or wake after sleep onset when the last week of placebo administration was compared to the last week of drug ingestion. There was no evidence that benzodiazepines worsened nocturnal sleep when compared to placebo, but the studies were short.

A joint 32-day study of 23 subjects from several American Sleep Disorders Centers found that when triazolam, 15 mg, and flurazepam, 30 mg, were administered double blind, using a counterbalance order and washout periods, these medications were virtually indistinguishable in their hypnotic effect on nocturnal sleep, but had dramatic effects on daytime sleep. We used the Multiple Sleep Latency Test (MSLT) (5) to evaluate daytime somnolence. This test involves polygraphically monitoring the latency between lights out and sleep onset five times during the day, usually at 10:00, 12:00, 14:00, 16:00, and 18:00. Each "scheduled" nap lasts 20 min, whether or not the subject sleeps. Flurazepam, 30 mg, significantly reduced the sleep latency on the MSLT, indicating objective daytime sleepiness, whereas triazolam did not (4). Carskadon et al. (6) evaluated the effects of 0.25 mg of triazolam and 15 mg of flurazepam in elderly insomniacs using nocturnal polygraphic recordings, daytime performance tests, and MSLT, and found similar results.

Systematic studies using the MSLT to investigate daytime sleepiness after benzodiazepine intake are still limited. The Stanford experience is presented in Table 1, which indicates the effects of several benzodiazepine hypnotics taken at bedtime on the following day's alertness. Worsening MSLT scores are usually correlated with the active hypnotic compounds' half-life and are thought to be the results of nightly drug intake (i.e., the longer the compound's half-life and the greater the dosage, the more likely daytime sleepiness will occur).

In conclusion, we know little about the long-term effects of sleeping pills on sleep parameters because of the dearth of studies. On a short-term basis, however, the effects are dubious. In the nonbenzodiazepine group, objective sleep parameters tended to deteriorate as the drugs were ingested, leading to frequent dosage increases, and ultimately drug dependency and addiction. In the benzodiazepine group, nocturnal sleep parameters did improve, but when sleep patterns during the 24-hr period and daytime somnolence were investigated, several benzodiazepines and other central nervous system depressant drugs were found to disturb sleep through secondary mechanisms in some patients.

III. EFFECTS OF MEDICATION TAKEN DURING THE DAY ON SLEEP

A. Cardiovascular Drugs

Diuretics and β-adrenoreceptor blockers, the most common antihypertensive medications, have an effect on the central nervous system. Bauer et al. (2) and Bengtsson et al. (3) have studied the side effects of antihypertensive treatments. Bengtsson's group studied sleep disturbances and nightmares in 1302 women, aged 44-66 years, who represented a cross-section of women in the Goteburg, Sweden, area. Sleep disturbances and nightmares were not significantly affected by antihypertensive drugs,

Table 1 Drug Effects on Daytime Sleep Tendencies (Daily MSLT Median)[a]

Drug	Dose (mg)	Short-term administration[b]			Long-term administration[c]		
		Better	Worse	Same	Better	Worse	Same
Triazolam	0.5	6	7	0	3	4	0
Oxazepam	30.0	2	4	1	—	—	—
Lormetazepam	1.0	1	5	1	3	3	1
Estazolam	2.0	0	5	0	—	—	—
Estazolam	1.0	0	5	0	—	—	—
Quazepam	30.0	0	3	2	0	4	1
Quazepam	15.0	1	5	4	2	3	1
Flurazepam	15.0	1	3	2	—	—	—
Flunitrazepam	1.0	—	—	—	0	7	3
Placebo		2	10	1	0	6	1

[a]The table compares patients' MSLT scores. To register as better or worse, postdrug MSLT scores had to deviate +6.0 min from base-line median scores.
[b]Short-term study, results after 1 week of medication.
[c]Long-term study, results after 21-22 consecutive nights of treatment.
Source: Reproduced from Seidel and Dement (1982) *Sleep* 5:188.

although there was a trend: 38% of the women using antihypertensive medication reported sleep disturbances, as compared to 33% in the total population; 13% of the drug group reported nightmares, versus 10% in the total population sample. Nightmares were more frequent with β-blockers than with diuretic medications (13 versus 7%). It appears that certain patients are more sensitive to them. In our own studies, the psychiatric interview and the Minnesota Multiphasic Personality Inventory taken after drug withdrawal and before the readministration of propranolol did not show significant psychiatric or psychological problems. As far as we know, most patients complaining of nightmares have been taking propranolol. With the recent introduction of longer-lasting β-blockers and the increasing popularity of this drug family, nightmares may become a more common complaint.

B. Antiarrhythmic Quinidine-Like Medications and Sleep Disturbances

Quinidine and drugs with quinidine-like effects may also lead to nightmares (REM sleep) and/or night terrors (NREM sleep). Some of the newer drugs (many not yet released in the United States but available elsewhere) such as lorcainide have this tendency. A Stanford study of lorcainide (unpublished data) indicated that this compound has a significant effect on sleep, with nightmares developing 5-7 days after initial ingestion and lasting, in most cases, 10-15 days. Nocturnal complaints decreased after 10-15 days, although significant insomnia was polygraphically documented after 30 days of drug intake. Patients using these medications may awaken frightened, having experienced a "weird" situation; they may be disoriented for several minutes and may appear confused. In some, the sleep disturbance does not disappear with time, so that drug withdrawal is necessary. For these patients, there is often a crossover effect with drugs closely related to quinidine.

C. Corticosteroids and Breathing Problems During Sleep

Corticosteroids, because of their anti-inflammatory property, are prescribed for a variety of conditions. Their adverse effects are well known: fulminating infections, osteoporosis, myopathy, diabetes mellitus, hypertension, psychotic reaction to weight gain, and hirsutism. These drugs can be life threatening and are difficult to control. Rarely discussed, however, are their sleep-related side effects, which we often note in our clinic at Stanford. The disturbed nocturnal sleep associated with the rare corticosteroid-related psychotic reaction is not as significant as the breathing abnormalities observed during sleep with their secondary cortege of insomnia or hypersomnia. The problem is more common in men than in women, but the sex difference decreases with age, particularly in postmenopausal women. Weight gain and fat accumulation, especially in the oropharynx, cause progressively heavier snoring at night, leading eventually to obstructive sleep apnea syndrome. We have seen 27 patients with significant breathing problems during sleep and daytime somnolence secondary to chronic corticosteroid medication intake (2). Obstructive sleep apnea syndrome most commonly disrupted sleep, but obesity hypoventilation syndrome during sleep was also responsible for daytime somnolence.

D. Tricyclic Medications and Sleep-Related Periodic Leg Movements (Nocturnal Myoclonus)

Tricyclic antidepressant medications are administered not only to depressed subjects, but also to narcoleptics to control their cataplectic attacks, one of the major auxiliary symptoms of narcolepsy. We first noted the development of periodic leg movements secondary to tricyclic intake while investigating the effects of certain tricyclic medications on cataplexy. In this initial study it looked as though the appearance of periodic leg movements was dose related, the threshold varying with the subject. Once the leg movements were induced, they disrupted nocturnal sleep much like nocturnal myoclonus syndrome (9).

The effect of tricyclic medications on sleep and sleep structure was also studied in normal volunteers and depressed patients. Passouant et al. carried out one of the most complete studies in 1972 (19). Most tricyclic medications will reduce total REM sleep time (18). However, as Passouant first observed and we have confirmed (9,19), the reduction is a disorganization of the three basic components used to score REM sleep (electroencephalogram, electro-oculogram, and chin electromyogram) rather than a disappearance of REM sleep per se. All the tricyclic medications we investigated—imipramine, desipramine, clomipramine, and protriptyline—led to a near or complete disappearance of the tonic muscle atonia seen during REM sleep. However, the typical electroencephalographic figures of REM sleep, such as sawtooth waves, are noted despite the high electromyographic activity (9,19). Tricyclic medications can also cause continuous eye movement, both slow and rapid, throughout the entire sleep period. Sometimes the electroencephalogram and electro-oculogram will be typical of REM sleep, but frequently the high electromyogram will cause "dissociated REM sleep" and an artifactual reduction of total REM sleep time and percent (9,19) in normal controls, as well as in depressed and narcoleptic patients.

The increased muscle tone activity induced by the medication may be responsible for the appearance of periodic leg movements and secondary nocturnal sleep disruption. Actually, the periodic leg movements problem was related to both the kind of drug and its dosage. In our study of the effect of tricyclic medication on periodic leg movements and sleep disturbance, we found that (a) dosages higher than 100 mg/24 hr of imipramine or a derivative, or 20 mg/24 hr of protryptyline, were necessary to induce the periodic leg movements; (b) periodic leg movements reappeared in less than four nights when the threshold initiating the disturbance was reached for the second or third time; and (c) there was no correlation between the appearance of continuous eye movements during sleep, increase in muscle tone during REM sleep, and the appearance of periodic leg movements.

E. Antiparkinsonian Medications and Sleep Disturbances

Parkinsonian patients' neurological problems can create sleep disturbances; their already precarious lives can be further jeopardized by drug-induced sleep problems. Popular and potent medications such as bromocriptine, amantadine, and the classic combination of levodopa and dopa decarboxylase inhibitors can induce not only nightmares and night terrors, but also mild toxic reactions manifested as confusion and wandering during the night. Hypnagogic hallucinations, sometimes terrifying, can also be seen in association with a panic reaction. Parkinsonian patients may also complain of insomnia caused by the transient appearance of a symptom, frequently during the late afternoon or at night, that leads to recurring awakenings.

F. Benzodiazepines and Breathing During Sleep

In sleep apneic patients, benzodiazepines (and all other hypnotic medications) can interact with the sleep-related control of breathing. As central nervous system depressants, benzodiazepines affect breathing during sleep through two different mechanisms: (a) They can work directly on the respiratory network, depressing respiration (19), or (b) they can modify the arousal threshold so that arousal responses are inhibited when abnormal respiratory events occur.

Twenty subjects, all over 50 years of age, participated in a study on the effects of a single bedtime dosage of flurazepam, 30 mg, a long-lasting benzodiazepine (21). These subjects were divided into three groups: 10 healthy elderly (mean age 73.3 years), 5 with chronic obstructive pulmonary disease (mean age 61 years), and 5 with obstructive sleep apnea syndrome (mean age 59 years). Flurazepam, 30 mg, worsened the oxygen saturation during sleep and increased the amount of sleep apnea in all subjects presenting an apnea index of 5 or above. In this same study, we found that the subjects whose apnea increased after benzodiazepine ingestion were also, based on their MSLT scores, more somnolent during the day.

Rudolph et al. (21) studied the effects of nitrazepam, taken as a hypnotic at bedtime, in chronic obstructive pulmonary disease patients. They found that, over 5-day periods, it induced moderate changes that were not clinically important, except in patients already hypercapnic before benzodiazepine intake.

In a study of five chronic obstructive pulmonary disease patients, we found that flurazepam, 15 mg, after 7 days of administration had no significant effect on the patients' nocturnal sleep disturbances when compared to placebo administration.

IV. CONCLUSIONS

Drugs can drastically impair sleep, causing nocturnal disruptions and daytime somnolence. Medications prescribed to counteract these complaints often aggravate them, so that a vicious cycle of sleep problems-hypersomnia-drug ingestion begins. Too often sleep-related side effects, ignored by industry, the pharmacist, and the prescribing physician, are responsible for a patient's discontinuing medication that is strongly needed. Medications are regularly prescribed without a real knowledge of their effects on sleep, and even less knowledge of their effects on vital functions, which alter with sleep-state shifts.

Recent pharmacological protocols have tried to investigate the sleep disturbances caused by cardiovascular, respiratory, and gastrointestinal medications, but there are few systematic objective evaluations of these phenomena. Also lacking are studies on the interaction of medications and sleep states. Systematic objective evaluation of medications are needed that include a complete sleep-wake protocol appreciating both the best time to administer the drug during the 24-hr cycle and the eventual changes on sleep and sleep-related central nervous system controls.

REFERENCES

1. Adams, K., Adarnson, L., Brezinova, V., Hunter, W. M., Oswald, I. 1976. Nitrazepam: Lastingly effective but trouble on withdrawal. *Br. Med. J. 1*:1558-1560.
2. Bauer, G. E., Baker, J., Hunyor, S. N., Marshall, P. 1978. Side effects of antihypertensive treatment: A placebo-controlled study. *Clin. Sci. Mol. Med.* 55:341-344.
3. Bengtsson, C., Lennartsson, J., Lindquist, O., Noppa, H., Sigurdsson, J. 1980. Sleep disturbances, nightmares and other possible central nervous system disturbances in a population sample of women, with special reference to those on antihypertensive drugs. *Eur. J. Clin. Pharmacol. 17*:173-177.
4. Bliwise, D. L., Seidel, W. F., Karacan, I., Mitler, M. M., Roth, T., Zorick, E., Dement, W. C. 1983. Daytime sleepiness as a criterion in hypnotic medication trials: Comparison of triazelam and flurazepam. *Sleep. 6*:156-163.
5. Carskadon, M. A., Dement, W. C. 1982. The multiple sleep latency test: What does it measure? *Sleep* 5:567-572.
6. Carskadon, M. A., Seidel, W. F., Greenblatt, D. J., Dement, W. C. 1982. Daytime carry over of triazolam and flurazepam in elderly insomniacs. *Sleep* 5:351-371.
7. Dement, W. C., Guilleminault, C. 1973. Sleep changes in drug dependency, hypersomnia, and insomnia: Causes manifestations and treatment. *Excerpta Med. 296*:42-43.
8. Dunleavy, D. L. F., Brezinova, V., Oswald, I., MacLean, A. W., Tinker, M. 1972. Changes during weeks in effects of tricyclic drugs on the human sleeping brain. *Br. J. Psychiatry 120*:663-672.
9. Guilleminault, C., Raynal, D., Takahashi, S., Carskadon, M., Dement, W. C. 1976. Evaluation of short term and long term treatment of the narcolepsy syndrome with clomipramine hydrochloride. *Acta Neurol. Scand.* 54:71-87.

10. Guilleminault, C., Seidel, W. F., Chernick, D. D., Dement, W. C. 1983. Objective long term nocturnal and daytime evaluation of flunitrazepam, in *Sleep 1982* (Koella, W. P., ed.), S. Karger, Basel, pp. 157-159.
11. Guilleminault, C., Silvestri, R. Aging, drugs and sleep. *Neurobiol. Aging 3*:379-386.
12. Guilleminault, C., Silvestri, R., Coburn, S., Dement, W.C., Cummiskey, J. Aging, breathing during sleep and long lasting benzodiazepines. *Br. J. Clin. Pharmacol.* (in press).
13. Institute of Medicine. 1979. *Report of a Study: Sleeping Pills, Insomnia and Medical Practice*, National Academy of Sciences, Washington, D.C.
14. Kales, A., Bixler, E. O., Kales, J. D., Scharf, M. B. 1977. Comparative effectiveness of nine hypnotic drugs: Sleep laboratory study. *J. Clin. Pharmacol. 17*:207-213.
15. Kales, A., Preston, T. A., Tan, T.C., Allen, C. 1970. Hypnotics and altered sleep-dream patterns. I: All night EEG studies of glutethimide, methyprylon and pentobarbital. *Arch. Gen. Psychiatry 23*:211-218.
16. Kales, A., Soldatos, C. R., Bixler, E. O., Kales, J. D. 1983. Early morning insomnia with rapidly eliminated benzodiazepines. *Science 220*:95-97.
17. Kripke, D. F., Lavie, P., Hernandez, J. 1978. Polygraphic evaluation of ethchlorvynol (14 days). *Psychopharmacology 56*: 221-223.
18. Mendelson, W. B. 1980. *The Use and Misuse of Sleeping Pills*, Plenum, New York.
19. Passouant, P., Cadilhac, J., Ribstein, M. 1972. Les privations de sommeil avec mouvements oculaires par les antidepresseurs. *Rev. Neurol. Paris 127*:173-192.
20. Roth, T., Hartse, K., Zorick, F. 1980. The differential effects of short and long acting benzodiazepines upon nocturnal sleep and daytime performance. *Drug Res. 30*:891-894.
21. Rudolf, M., Geddis, D. M., Turner, J. A. M., Saunders, K. B. 1978. Depression of central respiratory drive by nitrazepam. *Thorax 33*:97-100.

6

Actions of Abused Drugs on Reward Systems in the Brain

Roy A. Wise and Michael A. Bozarth
Center for Studies in Behavioral Neurobiology, Concordia University, Montreal, Quebec, Canada

I. INTRODUCTION

Drugs of abuse may be considered neuropoisons for a number of reasons other than the obvious one, that overdoses can be lethal. First, there are some direct neurotoxic effects of prolonged and excessive use that are attributable, at least, to opiates (20,39), ethanol (26,107), and psychomotor stimulants (40,79,99,106); that is to say, these agents can, under certain circumstances, alter behavior by causing structural abnormalities in the brain—abnormalities that are detectable in neurochemical or physiological tests. Second, there are physical dependence syndromes associated with some drugs of abuse. Barbiturates and benzodiazepine dependence can be severe; convulsions and sometimes death can result from discontinuation of drug use (60). Opiate dependence, on the other hand, is not so severe; withdrawal distress is unpleasant, but not life-threatening (60). In this case it is not easy to decide whether the dependence syndrome is sufficiently severe as to justify considering the drug a toxin in the traditional sense. Cell death or permanent changes in neural tissue are not known to be consequences of simple opiate dependence itself. If subjective reports are adequate grounds for defining toxicity; however, opiate withdrawal stress must be considered a mild toxic condition.

There is another sense in which drugs of abuse can be considered "neuropoisons." Experience with these agents changes behavior, almost certainly by changing underlying mechanisms in the nervous system. These changes result in neglect of personal hygiene and safety and they result in neglect of social relationships, which, in the long term, are equally critical for individual and species survival. There is an important measure of truth to the popular belief that drugs of abuse can poison the "mind." Drugs of abuse can cause what we might term "motivational toxicity"; under some circumstances, at least, the motivational priorities which normally ensure immediate health and long-term survival are reduced relative to increasingly dominant motivation for drug intake. Drugs with this capacity cause "motivational" toxicity in the sense that they lead to decreased incidence of essential activities; the physiological pathology that ultimately results is secondary to the behavioral changes, which in turn result from the alteration of the motivational hierarchy. The consequences to the body of motivational toxicity are varied; malnutrition, hepatitis, and gunshot wounds are some examples.

How can the motivation for a drug become so strong that the user forsakes good health and even life itself? A fully satisfying answer to this question awaits a complete understanding of the mechanisms of motivated behavior, but a partial answer is emerging from recent studies. While much remains to be learned about the neural mechanisms of motivated behavior, evidence has been accumulating to suggest that the mechanisms of "abnormal" behavior motivated by drugs and the mechanisms of "normal" behavior (motivated, for example, by food, shelter, and sex partners) involve common mechanisms. It now seems likely that drugs can come to dominate motivated behavior because they directly activate the endogenous motivational circuits of the brain, serving as hedonically effective substitutes for biologically important rewards and coming to be preferred to such biologically important rewards by virtue of their more powerful and direct activation of these brain circuits. The neural mechanisms underlying drug motivation and their relation to the neural mechanisms of normal motivated behavior are the major subjects of the present chapter.

II. BRAIN REWARD CIRCUITRY

The notion that there is specialized neural circuitry subserving the rewarding effects of such natural rewards as food, water, and sexual contact has been assumed by many since the demonstration that animals would work to electrically stimulate some but not all portions of their own brains (81). Brain stimulation is rewarding at various sites at all but the spinal levels of the central nervous system (53,83); however no single nucleus and its fibers can be supposed to account for all instances of rewarding brain stimulation. It is now generally assumed that stimulation of each of several anatomical pathways is rewarding; the neural elements at different sites may be connected to one another (connected in series) in some cases and be unrelated to one another (connected in parallel) in other cases (119). One of the major current problems for the reward specialist is to determine the anatomical relations of the various circuits and circuit elements.

Success in solving this problem depends on success with an even more fundamental problem: the problem of determining which of the dozens of intertwined pathways near a given stimulation site is critical for the rewarding property of stimulation. Despite almost three decades of active research, it is still not clear in a single instance (although major progress is being made; see Ref. 51) which of the multiple neural pathways near the site of rewarding stimulation is the directly activated substrate of reward.

The dominant theories of brain stimulation reward circuitry over the last decade have been based largely on circumstantial evidence suggesting direct activation of catecholaminergic fibers to be the common denominator of rewarding brain stimulation. Various workers proposed several noradrenergic and dopaminergic pathways in such roles (27,32, 38,53,69,89,105). It now seems unlikely that noradrenergic pathways are directly involved in reward function at all (25,27-29,36,46,63,91, 128,129,132), and, while dopaminergic pathways do seem to be involved (46-50,131), it seems that their involvement must be trans-synaptic rather than direct (51,100,116). Several lines of evidence have led to this current view of catecholamine involvement in reward.

The most damaging evidence for the various catecholamine theories is evidence that the medial forebrain bundle system, where the catecholamine fibers follow an overlapping and closely related trajectory, has refractory periods too short (127), conduction velocities too fast (8,100), and the wrong direction of conduction (101) to be those of any of the relevant catecholamine systems. These facts rule out either noradrenergic or dopaminergic elements as the reward-relevant fibers that are directly activated at the electrode tip.

Other evidence suggests, however, that dopaminergic systems do play a critical role, though it must be efferent to whatever cells are activated directly. Both anatomical and pharmacological evidence suggest the notion that dopaminergic systems are trans-synaptically activated by rewarding brain stimulation. Pharmacological blockade of dopamine receptors attenuates performance for brain stimulation reward, and it does so in a manner very different from the way noradrenergic blockers alter performance (46,132); the nature of the behavioral effects suggests the rewarding impact of the stimulation, and not merely the performance capacity of the animal, to be dependent on dopamine (46-50,131). The descending fiber system that represents the major directly activated component of the mechanism of rewarding medial forebrain bundle stimulation (101) appears to terminate in the region of the dopaminergic cells of the ventral tegmental area and substantia nigra (30), where the fibers and the dopamine cells share common mediolateral (117) and dorsoventral (30) boundaries. Because of this proximity of terminals to cell bodies, dopaminergic cells are one of the most likely synaptic targets of the descending system (116). Brain stimulation reward involving the dozens of other positive stimulation sites is, in cases thus far tested, also sensitive to dopaminergic receptor blockade and would thus also seem afferent to the same or some other dopaminergic link in brain reward circuitry. Only in the case of the medial forebrain bundle system, however, is there yet any strong reason to believe that the directly activated fibers project immediately to dopaminergic neurons.

In summary, the current picture regarding brain reward circuitry as identified from brain stimulation studies is that multiple directly activated circuit elements are implicated in reward involving different stimulation sites. One link in reward circuitry seems to involve dopamine as its transmitter, but this link appears not to be activated directly by rewarding stimulation, at least at the stimulation parameters that are usually tested (the dopaminergic fibers have an unusually high current threshold, seemingly well above the currents usually tested, see Ref. 8). The dopaminergic link seems to be efferent to most, if not all, brain stimulation reward sites, however. Activation of noradrenergic pathways, either directly or trans-synaptically, appears to be neither necessary nor sufficient for brain stimulation to be rewarding. Both anatomical mapping studies and lesion studies have attempted to determine which of the various dopaminergic projections might be involved, but no clear picture has yet emerged.

III. INTERACTION OF DRUGS OF ABUSE WITH BRAIN STIMULATION REWARD

One reason for believing that drugs of abuse act at some link in the system activated by rewarding brain stimulation is that such drugs are generally reported to facilitate brain stimulation reward. Amphetamine (105), cocaine (31), morphine (19,70,82), heroin (64), barbiturates (76,90), benzodiazepines (80,84), and ethanol (21,72,97,113) have all been reported to facilitate responding for lateral hypothalamic brain stimulation, although each of these agents also decreases responding under some circumstances (116,120). In each case, the doses that decrease responding are higher than the doses that increase responding; the high doses of stimulants seem to overstimulate the animals, while the doses of the other agents sedate them (116,120).

It is difficult to know that the doses which increase responding for brain stimulation reward do so because of rewarding actions of their own, let alone because they act in the same system as is activated by the rewarding brain stimulation (116). Anxiolytic, anticonvulsant, or analgesic actions of these drugs could, in theory, account for the increased responding for brain stimulation reward at doses or through mechanisms not involved in the rewarding effects of these agents (116). In the cases of amphetamine and morphine, however, there is direct evidence suggesting the reward-facilitating actions to involve the same brain sites and thus, in all likelihood, the same mechanisms as are involved in the direct rewarding effects. In the case of amphetamine, a site of reward-facilitating action (16) and a site of rewarding (77) action are both in the nucleus accumbens; in the case of morphine the site of the reward facilitating (17) and the site of the rewarding (11,13, 15,87,112) action are both in the ventral tegmental area. Inasmuch as the nucleus accumbens and the ventral tegmental area are bridged by a major dopaminergic fiber system (109), and inasmuch as dopaminergic circuit elements appear to play a role in the rewarding effects of both stimulants and opiates, it is tempting to conclude on this evidence alone that the rewarding effects of these agents and of brain stimulation reward are mediated by a common mechanism. However, while the interactions of drugs of abuse and brain stimulation reward are interesting, our knowledge regarding the mechanisms of action of these various agents comes more directly and importantly from studies of drugs as rewards in their own right. Until we know the individual mechanisms of brain stimulation reward and of drug reward we are unlikely to make much progress in determining the mechanisms of their interaction.

IV. MECHANISMS OF DRUG REWARD

A. Psychomotor Stimulants

The rewarding actions of the psychomotor stimulants amphetamine and cocaine, like the rewarding action of medial forebrain bundle electrical stimulation, involve a critical dopaminergic system; the rewarding neural consequences of these agents are initiated in a dopaminergic synapse. Amphetamine and cocaine have long been known to have actions in both the noradrenergic and the dopaminergic synapses. Both drugs block noradrenaline and dopamine reuptake from the synapse (3, 22,58) and amphetamine also causes presynaptic noradrenaline and dopamine release (3,22,58). It is only the dopaminergic actions that are critical for reward, however; selective blockade of dopaminergic but not noradrenergic receptors attenuates the rewarding impact of amphetamine and cocaine. This is clear from lever-pressing (36,91,92, 128,129) and place preference (103) experiments in animals and from subjective ratings of amphetamine euphoria in man (57).

Of the various dopaminergic projection systems, it appears to be limbic projections that are most importantly involved; injection of the neurotoxin 6-hydroxydopamine (71,93,94) or the dopamine receptor blocker spiroperidol (86) into the nucleus accumbens greatly reduces intravenous amphetamine and cocaine self-administration. Moreover, amphetamine appears to be rewarding when injected directly into the nucleus accumbens (77). Finally, selective dopaminergic but not noradrenergic agonists have amphetamine-like rewarding properties (6,34, 91,123,130). If dopamine is depleted from the dopaminergic neurons, the rewarding effects of the postsynaptic dopamine agonist apomorphine are not altered (5); thus the rewarding action of these agents reflects an action on postsynaptic receptors and not on autoreceptors (an autoreceptor action would only be expressed postsynaptically and thus would be blocked by depletion of synaptic dopamine). For these reasons it has been generally accepted for some years that the reward mechanism through which the psychomotor stimulants exert their behavioral control involves a dopamine synapse as its directly activated link (43,115).

B. Opiates

More recent evidence points to a common mechanism of rewarding action for opiates. Opiate receptors are found in a number of distinct brain regions, but direct injections into these various receptor fields has suggested that only one of them seems involved in the potent rewarding actions of opiates. This critical receptor field is localized in the region of the dopamine cell bodies of the ventral tegmental area (11,13, 87,88, 112). Rewarding unilateral injections of morphine into the region of the dopaminergic cells cause circling behavior (13) which is dopamine dependent (59). The fact that the animals turn away from the injected side suggests that the dopaminergic cells are activated (110) by morphine. The anatomical substrate for this effect has the same boundaries as the dopamine cell group (59). Opiate receptor antagonism restricted to this area of the brain by local microinjection reduces the rewarding impact of intravenous heroin (15), whereas local inhibition of enkephalinase causes conditioned place preference (56). Finally, dopamine receptor blockade seems to block the rewarding impact of systemic opiates, at least in the case of the higher doses thus far tested (42). Dopamine antagonists block place preferences established by rewarding opiate treatments (12,104), and in this paradigm it is clear that it is the rewarding impact of the opiates, and not the response capacity of the animal, that is disrupted by the tested doses of the dopamine antagonists (10). Current evidence thus suggests that morphine exerts its rewarding effects at the level of the dopamine cell bodies, whereas cocaine and amphetamine exert their rewarding effects at the level of the dopaminergic synaptic terminals. It would seem likely that the rewarding effects of opiates and the rewarding effects of psychomotor stimulants share a common dopaminergic mechanism; it is difficult to imagine how stimulants would fail to activate a subset of dopaminergic fibers that are selectively activated by opiates.

C. Other Drugs of Abuse

Studies of the mechanisms underlying rewarding effects of other drugs of abuse have not reached the same stage of development as have studies of psychomotor stimulants and opiates. It is possible that some will be found to activate the same system as seems to be involved in opiate, stimulant, and brain stimulation reward. It is also possible that different drugs of abuse will activate independent reward circuits (1). These possibilities are being actively explored in a number of laboratories at the present time. Only brief comments on various drugs can be offered on present evidence, however.

Ethanol has been the most intensively studied drug of abuse (23,67, 74), but most research on alcoholism deals with matters of tolerance and dependence (24,108), and ethanol's mechanisms of rewarding action are still only partially understood. Selective noradrenergic lesions and synthesis inhibition attenuate oral ethanol intake in rats (2, 18,62), and serotonergic uptake inhibition has similar effects (95). It is not yet known which anatomical projections of these transmitter systems are involved. There are known to be interactions between the three major central monoamines, but there is as yet little speculation as to how ethanol might interact with brain mechanisms involved in brain stimulation (1,119) or other drug reward phenomena.

Barbiturates are self-administered by lower animals (33,35,44), but little has been done to identify their mechanism of rewarding action. Benzodiazepine self-administration has also been demonstrated in higher animals (44,125,126), but, again, site and mechanism of action are not yet identified. Nicotine is self-administered both by smoking in man and monkeys (61) and by intravenous injection in monkeys and rats (37). While the site of rewarding action has not been determined, there is evidence suggesting that nicotine activates dopaminergic systems (54, 68) and may thus share a mechanism of action with opiates and psychomotor stimulants. Phencyclidine is self-administered by lower animals (4) and also appears to share its site of rewarding action with cocaine and amphetamine; local injections of phencyclidine into the nucleus accumbens are rewarding as shown by the conditioned place preference test (55).

Thus studies of cocaine, amphetamine, heroin, morphine, and phencyclidine suggest these agents to share the capacity for activating a common reward circuit. This circuit is activated synaptically by the directly stimulated descending fibers of medial forebrain bundle brain stimulation reward, and it appears to be activated synaptically by the sensory inputs which carry the signals of food and water rewards (52, 102,118,121,122). Morphine and heroin activate the system in the region of the dopaminergic cell bodies, while cocaine, amphetamine, and phencyclidine activate the system at the junction between the dopaminergic terminals and their efferent connections. The possibility seems good that at least some other habit-forming drugs, particularly nicotine (which has its own problems of associated, but indirect, toxicity), activate the same system in different ways. Firm knowledge as to which drugs activate common mechanisms of reward awaits more detailed understanding of the mechanisms of rewarding action of the individual agents.

V. ROLE OF DEPENDENCE IN DRUG REINFORCEMENT

The view of drug abuse which has, in one way or another, dominated recent Western thinking is that compulsive drug intake begins for frivolous or neurotic motivations or because of peer pressure, but that once drug dependence is established, drug intake becomes motivated by the need to avoid the physical distress of drug withdrawal. This view runs into several difficulties, but it nonetheless continues to influence the thinking and activities of scientists, clinicians, parents, and addicts. A question of importance in relation to the present topic is whether the rewarding impact of drugs of abuse derives primarily from their ability to relieve their own withdrawal symptoms, or rather whether their rewarding impact is a primary drug action. Do drugs come to dominate behavior because motivation to relieve withdrawal distress becomes stronger than motivation for food, shelter, and normal social activities, or do drugs come to dominate behavior because of more primary and positive effects of the drugs which are largely independent of dependence states?

There can be little doubt that addicts will "self-medicate" when withdrawal distress is acute and drug is available, but it is clearly not the case that all drug abuse is motivated by the need to relieve withdrawal stress. First, the view that drug is taken to relieve distress is a negative reinforcement model (7); the addict takes the drug to relieve the negative state of withdrawal stress, and not to produce the positive state of drug euphoria. This view is largely modeled after common-sense views of hunger and food intake. This model has not fared weli in attempts to understand normal hunger, however; experimental psychologists have largely abandoned such notions. The notion that we eat when we sense a need for nutrients (66) or that we drink when we sense a need for fluid (41) may be valid, but the view that such need states are the causal stimuli for all food or water intake is clearly wrong (45,65). Indeed, animals trained to lever-press for food when hungry will begin lever-pressing normally if put in the test box when not hungry; such lever-pressing decreases and ceases progressively, much as does lever-pressing in the motivated (hungry) animal that is tested under conditions of nonreward (78). Much eating and most drinking come long before there is significant tissue need (45,65), and it is likely that much drug intake, even in cases where it is compulsive and strongly motivated, comes under conditions where there is no major withdrawal stress.

Second, psychomotor stimulants are as habit forming as opiates or other sedative hypnotics, at least in the laboratory animal. There is, however, little evidence of a dependence syndrome with these compounds (111). There is a rebound depression following termination of chronic psychomotor stimulant intake, but the severity of this depression and its marked difference from the hyperexcitability of opiate or ethanol withdrawal make it clearly a different phenomenon from the prototypical withdrawal syndrome (60). An even weaker case can be made for physical dependence on cannabinoids (60). Even opiate self-administration patterns can be maintained at dose levels too low to produce significant dependence syndromes in laboratory animals (35,124). Third, attempts to break opiate self-administration habits in man have found treatment of withdrawal distress alone to be inadequate. Patients who have been detoxified for long periods of time experience strong desire for drug when returned to their old drug-associated environments (114). Alcoholics in a controlled environment can go through major periods of withdrawal stress without attempting to earn tokens for alcohol, only to begin working for the tokens after the distress symptoms have passed (75). Monkeys often show similar behavior, not self-administering alcohol during the period of withdrawal distress but beginning again when this distress has abated (125).

For these and other reasons, many workers have rejected the negative reinforcement model of drug abuse and some have offered positive reinforcement models in its place (9,67,73,74,98). In a positive reinforcement view it is the positive state produced by the drug (drug euphoria) and not the negative state relieved by the drug (withdrawal dysphoria) that is seen as the critical or primary factor in establishing and maintaining drug self-administration habits. From such a view, the ability to cause physical dependence and relieve withdrawal distress involves an independent mechanism from that associated with the ability to cause reinforcement and euphoria; only the latter ability is essential for the "motivational toxicity" of a drug. Two recent findings underscore the importance of this view, even for the case of opiate self-administration.

The first new finding is that heroin can establish a conditioned place preference on the very first pairing of an environmental location with a heroin injection (14). Other demonstrations of rewarding opiate effects, involving drug self-administration, require multiple opiate injections. With the self-administration paradigm, the first lever-press does not demonstrate that the drug injection is rewarding; rather, the first response is made before the animal has any opportunity to experience the rewarding effects of the drug. The first injection results from an accidental contact with the lever or a contact related to a prior habit, often one established with some other (training) reward. The next several injections may similarly reflect exploration or a previous habit. Only after several earned injections is it possible to infer that the drug is rewarding. In such a case it is not possible to rule out the possibilities that some degree of drug dependence has developed and that some degree of distress relief contributes to the reinforcement for lever-pressing. In any paradigm where the animal has had previous injections it is a logical possibility that some degree of withdrawal stress is relieved by all but the first injection.

In the one-trial conditioned place preference paradigm, however, no such possibility exists. The animal goes to the place where it has received heroin without ever having had the opportunity to learn that heroin relieves withdrawal distress. Thus the place preference that can be established by the very first heroin injection ever received by the animal (14) must reflect a positive, not a negative, reinforcement phenomenon. In this case the drug establishes a place preference because of the state it produces, not the state it relieves. In place preference paradigms where drug is given more than once, it remains possible that the second injection relieves some distress in the wake of the first injection; thus, at least with systemic drug injections, where both reward mechanisms and dependence mechanisms receive drug, only the one-trial place preference paradigm makes it possible to determine the uncontaminated contribution of positive and negative reinforcement mechanisms.

The second new finding is even more important; it suggests that the brain mechanisms of positive reinforcement and opiate dependence are anatomically distinct from one another. When animals are trained to lever-press for direct microinjection of morphine into the brain, they will work for very low doses injected into the ventral tegmental area (11). Such injections are strongly habit forming, even when restricted to sessions lasting only 4 hr/day. On such a regimen, however, no signs of physical dependence are evident (13). Moreover, even when much higher doses of drug are administered to the same region by chronic infusion, no morphine abstinence symptoms are seen following naloxone challenge (14). When the same infusions are given into the periaqueductal gray, a region some 2 mm removed from the ventral tegmental area, abstinence signs such as escape jumping and teeth chattering are observed (14). This suggests that opiates cause physical dependence because of neural events triggered at opiate receptors located in a different region than those involved in intracranial opiate self-administration (14). The effects of intracranial opiate antagonists in animals working for intravenous heroin suggest that the same ventral tegmental opiate receptors are involved in the rewarding effects of intravenous opiates; opiate antagonists in the same ventral tegmental area appear to reduce the rewarding effect of intravenous heroin (15). These recent studies suggest that at least one rewarding effect of opiates can be dissociated from the dependence-producing action of opiates.

This does not mean, of course, that the entire rewarding impact of opiates in drug-experienced subjects is attributable to positive reinforcement. It seems likely that experienced subjects take drugs for more than one reason. In humans the reasons are likely to include peer pressure and expectations conveyed through the use of language; in dependent animals these are likely to include both positive reinforcement (drug euphoria) and negative reinforcement (relief of withdrawal dysphoria). These studies do, however, point out that positive reinforcement is one of the consequences of opiate injections, and such positive reinforcement would seem to be the common denominator of the drug classes that are self-administered in lower animals. Psychomotor stimulants seem primarily to be positive reinforcers; it is difficult to make the case that these agents are self-administered for relief from their own after effects. The compulsive self-administration of drugs of abuse, and the motivational toxicity with which it is associated, seems best linked to the positive reinforcement effects shared by opiates and psychomotor stimulants, rather than to the relief of withdrawal distress which seems strongly associated with opiates and less reliably self-administered agents (in lower animals, at least) such as ethanol and barbiturates.

Indeed, if the negative reinforcing effects of drugs were the critical factor in motivational toxicity, then the potency of drugs as reinforcers, and their concomitant ability to disrupt essential behaviors, should be in proportion to their ability to produce a physical dependence syndrome. This would predict stronger reinforcing effects of opiates, barbiturates, benzodiazepines, and ethanol and weaker reinforcing effects of amphetamine and cocaine. Our experience with these agents in the laboratory rat would suggest psychomotor stimulants to be as strongly reinforcing as opiates and considerably more reinforcing than these other drugs of human abuse. This is reflected in the ease of training animals to lever-press, in regularity of responding for drug, and in the duration of sustained periods of active responding (Bozarth, M. A., Gerber, G. J., and Wise, R. A., unpublished observations, 1980). It is also reflected in the general health of animals allowed free access to drug; unlimited access to intravenous cocaine is much more likely to lead to self-starvation (85) and death than is unlimited access to intravenous heroin (Bozarth, M. A. and Wise, R. A., unpublished observations, 1984) or ethanol (Gerber, G. J. and Wise, R. A., unpublished observations, 1980). A similar self-starvation is seen when rats are given limited access to food and to rewarding brain stimulation (96); brain stimulation again shares the rewarding but not the dependence-producing capability of opiates. Thus the secondary toxicity resulting from drug-induced neglect of bodily needs—the motivational toxicity which qualifies drugs of abuse as social neuropoisons—seems clearly tied to primary reinforcing properties of drugs (properties reflecting the ability of drugs to activate endogenous reward mechanisms of the brain) and not to drug-induced physical dependence syndromes.

VI. SUMMARY AND CONCLUSIONS

Drugs of abuse can have direct toxic effects, but such effects are likely to discourage continued self-administration of the drug in question. The more insidious effects of drugs of abuse almost require the absence of major, direct, or immediate drug toxicity; these insidious effects lead to habits of repeated drug use and the neglect of bodily needs. The physical debilitation resulting from such habitual use is more frequently associated with lack of sleep and nutrients than with direct toxic effects of the drugs in question. The insidious toxicity of drugs of abuse is a "motivational" toxicity; it involves alterations of the motivational hierarchy of the animal such that drug self-administration replaces behaviors motivated by nutrients, sex objects, and safety. When there is also significant primary drug toxicity, it usually occurs only after long periods of repeated drug exposure or involves drug doses which are not usually or intentionally self-administered.

Several factors may contribute to the compulsive drug self-administration habits which can be established by opiates, psychomotor stimulants, ethanol, barbiturates, and benzodiazepines. Of these drug classes, opiates and psychomotor stimulants are the most reliably self-administered by lower animals; these classes share positive reinforcing properties which are likely correlates of subjectively experienced drug euphoria. The positive reinforcing properties of these two drug classes seem to involve a shared neural mechanism of action; psychomotor stimulants are reinforcing because they activate dopaminergic systems at the level of the synapses, probably in nucleus accumbens, whereas opiates are reinforcing because they activate dopaminergic systems at the level of the dopaminergic cell bodies, primarily in the ventral tegmental area. Opiates and psychomotor stimulants do not share the ability to produce physical dependence; the relief of dependence symptoms thus reflects a contribution to the reinforcing properties of opiates ("negative" reinforcement) which is not a critical determinant of abuse liability. The negatively reinforcing action of opiates would seem to involve the opiate receptor population of the periaqueductal gray, where opiate dependence phenomena are initiated and withdrawal symptoms are presumably relieved, and is independent of the positively reinforcing action, which involves the opiate receptor population of the ventral tegmental area.

The ability of drug self-administration and drug-seeking habits to dominate behavior, to the point that nutrition, rest, medical attention, personal safety, and normal social interactions are neglected, constitutes the most health-threatening consequence of drugs of abuse. It accounts for the fact that intelligent and often knowledgeable humans will subject themselves to severe health risks in the maintenance of drug self-administration habits. The health risks themselves—cirrhosis, anorexia, and brain damage in the extreme—are largely either secondary to forms of neglect which often accompany compulsive drug self-administration or are delayed until many years after the drug habits are engrained. The ability of pharmacological reinforcers to develop stronger impact and behavior control than the more natural reinforcers—food, rest, sex partners, and so on—which usually dominate behavior seems, in some instances at least, to derive from the fact that drugs can more powerfully and more directly influence the same reward systems of the brain as are less powerfully and directly activated by more natural stimuli.

The fact that reward (and the pleasure that usually accompanies it) can be produced centrally by drugs and electrical brain stimulation may prove to be a fact of considerable significance in the present phase of human evolution. The seeking of stimuli that activate reward systems in the brain has generally, over the course of evolution, led man to activities which have promoted survival of the individual and the species. The fact that the seeking of drugs reduces the seeking of such seemingly survival-promoting stimuli accounts for our temptation to call drugs "social neuropoisons" or "motivational toxins." Food, water, and sexual gratification appear to depend for their rewarding impact on their ability to activate the same dopaminergic system as is activated by rewarding brain stimulation, opiates, and psychomotor stimulants. Our technology makes it increasingly possible, however, to activate the reward systems of the brain in new ways which do not have the old consequences. The activation of brain reward systems by copulation had the consequence of reproduction; while the use of birth control devices may have a new form of survival value of its own, many people now opt for the pleasure of sexual gratification without its traditional survival-promoting correlate. Similarly, new technology has made it possible to obtain the pleasure of sweet tastes without the traditional survival-promoting correlate of caloric intake; again, there may be a new form of survival value associated with this practice, and we may no longer derive a net benefit from our appetite for sweet foods. In the current stage of civilization it seems likely that survival is better served by reduction of birth rate and caloric intake.

In the case of pharmacological activation of brain reward mechanisms it remains to be seen whether the likelihood of survival is enhanced or diminished. There are those who consider drug "abuse" to reflect a form of self-medication appealing largely to those individuals who are otherwise unable to cope with the stress and social estrangement of modern life. If this view is correct, then drug "abuse" may represent an adaptive response, for some individuals at least. The more sobering possibility is that drugs of abuse simply reflect pleasures so intense and so habit forming, because they can activate the brain circuits of reward so directly and so powerfully, that they are difficult to resist despite their health-damaging consequences. Such a possibility must be taken very seriously in light of the fact that lower animals will self-administer psychomotor stimulants and rewarding brain stimulation even when these activities lead to self-starvation. Both the intravenous stimulants and the intracranial stimulation that are capable of seducing animals away from food to the point of self-starvation should be considered neurotoxins of a very special type. We have chosen to term the toxicity which results from exposure to these agents "motivational toxicity," because it is the motivational hierarchy of the animal which is most directly altered by drug experience, and it is the change in the motivational hierarchy which is the seemingly critical determinant of tissue damage.

REFERENCES

1. Amit, Z., Brown, Z. W. 1982. Actions of drugs of abuse on brain reward systems: A reconsideration with specific attention to alcohol. *Pharmacol. Biochem. Behav.* *17*:233-238.
2. Amit, Z., Levitan, D. E., Lindros, K. O. 1976. Suppression of ethanol intake following administration of dopamine-beta-hydroxylase inhibitors in rats. *Arch. Int. Pharmacodyn.* *223*:114-119.
3. Axelrod, J. 1970. Amphetamine: Metabolism, physiological disposition and its effects on catecholamine storage, in *Amphetamines and Related Compounds* (Costa, E., Garattini, S., eds.), Raven Press, New York, pp. 207-216.
4. Balster, R. L., Woolverton, W. L. 1980. Continuous-access phencyclidine self-administered by rhesus monkeys leading to physical dependence. *Psychopharmacology* *70*:5-10.
5. Baxter, B. L., Gluckman, M. I., Scerni, R. A. 1976. Apomorphine self-injection is not affected by alpha-methylparatyrosine treatment: Support for dopaminergic reward. *Physiol. Behav.* *4*:611-612.
6. Baxter, B. L., Gluckman, M. I., Stein, L., Scerni, R. A. 1976. Self-injection of apomorphine in the rat: Positive reinforcement by a dopamine receptor stimulant. *Pharmacol. Biochem. Behav.* *2*:387-391.
7. Beach, H. D. 1957. Some effects of morphine on habit function. *Can. J. Psychol.* *11*:193-198.

8. Bielajew, C., Shizgal, P. 1982. Behaviorally derived measures of conduction velocity in the substrate for rewarding medial forebrain bundle stimulation. *Brain Res. 237*:107-119.
9. Bijerot, N. 1980. Addiction to pleasure: A biological and social-psychological theory of addiction, in *Theories on Drug Abuse: Selected Contemporary Perspectives* (Lettieri, D. J., Sayers, M., Pearson, H. W., eds.), National Institute on Drug Abuse, Rockville, Md., pp. 246-255.
10. Bozarth, M. A. 1983. Opiate reward mechanisms mapped by intracranial self-administration, in *Neurobiology of Opiate Reward Processes* (Smith, J. E., Lane, J. D., eds.), Elsevier, New York, pp. 331-359.
11. Bozarth, M. A., Wise, R. A. 1981. Intracranial self-administration of morphine into the ventral tegmental area in rats. *Life Sci. 28*: 551-555.
12. Bozarth, M. A., Wise, R. A. 1981. Heroin reward is dependent on a dopaminergic substrate. *Life Sci. 29*:1881-1886.
13. Bozarth, M. A., Wise, R. A. 1981. Localization of the reward-relevant opiate receptors, in *Problems of Drug Dependence 1980* (Harris, L. S., ed.), NIDA, Rockville, Md., pp. 158-164.
14. Bozarth, M. A., Wise, R. A. 1983. Dissociation of the rewarding and physical dependence-producing properties of morphine, in *Problems in Drug Dependence 1982* (Harris, L. S., ed.), NIDA, Rockville, Md., pp. 171-177.
15. Britt, M. D., Wise, R. A. 1983. Ventral tegmental site of opiate reward: Antagonism by a hydrophilic opiate receptor blocker. *Brain Res. 258*:105-108.
16. Broekkamp, C. L. E., Pijnenburg, A. J. J., Cools, A. R., Van Rossum, J. M. 1975. The effect of microinjections of amphetamine into the neostriatum and the nucleus accumbens on self-stimulation behavior. *Psychopharmacologia 42*:179-183.
17. Broekkamp, C. L. E., Van den Boggard, J. H., Heijnen, H. J., Rops, R. H., Cools, A. R., Van Rossum, J. M. 1976. Separation of inhibiting and stimulating effects of morphine on self-stimulation behavior by intracerebral microinjections. *Eur. J. Pharmacol. 36*: 443-446.
18. Brown, Z. W., Amit, Z., Levitan, D. E., Ogren, S.-O., Sutherland, A. 1977. Noradrenergic mediation of the positive reinforcing properties of ethanol: II. Extinction of ethanol drinking behavior in laboratory rats by inhibition of dopamine-beta-hydroxylase. Implications for treatment procedures for human alcoholics. *Arch. Int. Pharmacodyn. 230*:76-82.
19. Bush, H. D., Bush, M. F., Miller, A., Reid, L. 1976. Addictive agents and intracranial stimulation: Daily morphine and lateral hypothalamic self-stimulation. *Physiol. Psychol. 4*:79-85.

20. Carlson, K. R. 1977. Supersensitivity to apomorphine and stress two years after chronic methadone treatment. *Neuropharmacology 16*:795-798.
21. Carlson, R. H., Lydic, R. 1976. The effects of ethanol upon threshold and response rate for self-stimulation. *Psychopharmacology 50*:61-64.
22. Carlsson, A. 1970. Amphetamine and brain catecholamines, in *Amphetamines and Related Compounds* (Costa, E., Garattini, S., eds.), Raven Press, New York, pp. 289-300.
23. Cicero, T. J. 1980. Animal models of alcoholism, in *Animal Models in Alcohol Research* (Eriksson, K., Sinclair, D. J., Kiianmaa, K., eds.), Academic, New York, pp. 99-118.
24. Cicero, T. J. 1983. Endocrine mechanisms in tolerance to and dependence on alcohol, in *The Pathogenesis of Alcoholism: Biological Factors* (Kissin, B., Begleiter, H., eds.), Plenum, New York, pp. 285-357.
25. Clavier, R. M., Fibiger, H. C., Phillips, A. G. 1976. Evidence that self-stimulation of the region of the locus coeruleus in rats does not depend on noradrenergic projections to telencephalon. *Brain Res. 113*:71-81.
26. Cobb, C. F., Van Theil, D. J., Ennis, M. F., Gavaler, J. S., Lester, R. 1978. Acetaldehyde and ethanol are testicular toxins. *Gastroenterology 75*:958.
27. Cooper, B. R., Cott, J. M., Breese, G. R. 1974. Effects of catecholamine-depleting drugs and amphetamine on self-stimulation of brain following various 6-hydroxydopamine treatments. *Psychopharmacologia 37*:235-248.
28. Corbett, D., Skelton, R. W., Wise, R. A. 1977. Dorsal bundle lesions fail to disrupt self-stimulation from the region of locus coeruleus. *Brain Res. 133*:37-44.
29. Corbett, D., Wise, R. A. 1979. Intracranial self-stimulation in relation to the ascending noradrenergic fiber systems of the pontine tegmentum and caudal midbrain: A moveable electrode mapping study. *Brain Res. 177*:423-436.
30. Corbett, D., Wise, R. A. 1980. Intracranial self-stimulation in relation to the ascending dopaminergic systems of the midbrain: A moveable electrode mapping study. *Brain Res. 185*:1-15.
31. Crow, T. J. 1970. Enhancement by cocaine of intra-cranial self-stimulation in the rat. *Life Sci. 9*:375-381.
32. Crow, T. J. 1972. A map of the rat mesencephalon for electrical self-stimulation. *Brain Res. 36*:265-273.
33. Davis, J. D., Miller, N. E. 1963. Fear and pain: Their effect on self-injection of amobarbital sodium by rats. *Science 141*:1286-1287.
34. Davis, W. M., Smith, S. G. 1977. Catecholaminergic mechanisms of reinforcement: Direct assessment by drug self-administration. *Life. Sci. 20*:483-492.

35. Deneau, G., Yanagita, T., Seevers, M. H. 1969. Self-administration of psychoactive substances by the monkey. *Psychopharmacologia 16*:30-48.
36. DeWit, H., Wise, R. A. 1977. Blockade of cocaine reinforcement in rats with the dopamine receptor blocker pimozide but not with the noradrenergic blockers phentolamine or phenoxybenzamine. *Can. J. Psychol. 31*:195-203.
37. Dougherty, J., Miller, D., Todd, G., Kostenbauder, H. B. 1981. Reinforcing and other behavioral effects of nicotine. *Neurosci. Biobehav. Rev.* 5:487-495.
38. Dresse, A. 1966. Importance du système mésencéphalo-télencéphalique noradrénergique comme substratum anatomique du comportement d'autostimulation. *Life Sci.* 5:1003-1014.
39. Eibergen, R. D., Carlson, K. R. 1975. Dyskinesias elicited by methamphetamine: Susceptability of former methadone-consuming monkeys. *Science 190*:588-590.
40. Ellison, G., Eison, M. S., Huberman, H. S. 1978. Stages of constant amphetamine intoxication: Delayed appearance of abnormal social behaviors in rat colonies. *Psychopharmacology 56*:193-199.
41. Epstein, A. N. 1973. Epilogue: Retrospect and prognosis, in *The Neuropsychology of Thirst: New Findings and Advances in Concepts* (Epstein, A. N., Kissileff, H. R., Stellar, E., eds.), Winston, Washington, D.C., pp. 315-332.
42. Ettenberg, A., Pettit, H. O., Bloom, F. E., Koob, G. F. 1982. Heroin and cocaine intravenous self-administration in rats: Mediation by separate neural systems. *Psychopharmacology 78*:204-209.
43. Fibiger, H. C. 1978. Drugs and reinforcement mechanisms: A critical review of the catecholamine theory. *Annu. Rev. Pharmacol. Toxicol. 18*:37-56.
44. Findlay, J. D., Robinson, W. W., Peregrino, L. 1972. Addiction to secobarbital and chlordiazepoxide in the rhesus monkey by means of self-infusion preference procedure. *Psychopharmacologia 26*: 93-114.
45. Fitzsimons, J. T. 1971. The physiology of thirst: A review of the extraneuronal aspects of the mechanisms of drinking, in *Progress in Physiological Psychology*, Vol. 4 (Stellar, E., Sprague, J. M., eds.), Academic, New York, pp. 119-202.
46. Fouriezos, G., Hansson, P., Wise, R. A. 1978. Neuroleptic-induced attenuation of brain stimulation reward. *J. Comp. Physiol. Psychol. 92*:659-669.
47. Fouriezos, G., Wise, R. A. 1976. Pimozide-induced extinction of intracranial self-stimulation: Response patterns rule out motor or performance deficits. *Brain Res. 103*:377-380.
48. Franklin, K. B. J. 1978. Catecholamines and self-stimulation: Reward and performance deficits dissociated. *Pharmacol. Biochem. Behav.* 9:813-820.

49. Franklin, K. B. J., McCoy, S. N. 1979. Pimozide-induced extinction in rats: Stimulus control of responding rules out motor deficit. *Pharmacol. Biochem. Behav. 11*:71-75.
50. Gallistel, C. R., Boytim, M., Gomita, Y., Klebanoff, L. 1982. Does pimozide block the reinforcing effect of brain stimulation? *Pharmacol. Biochem. Behav. 17*:769-781.
51. Gallistel, C. R., Shizgal, P., Yeomans, J. 1981. A portrait of the substrate for self-stimulation. *Psychol. Rev. 88*:228-273.
52. Gerber, G. J., Sing, J., Wise, R. A. 1981. Pimozide attenuates lever pressing for water reinforcement in rats. *Pharmacol. Biochem. Behav. 14*:201-205.
53. German, D. C., Bowden, D. M. 1974. Catecholamine systems as the neural substrate for intracranial self-stimulation: A hypothesis. *Brain Res. 73*:381-419.
54. Giorguieff-Chesselet, M. F., Kemel, M. L., Wandscheer, D., Glowinski, J. 1979. Regulation of dopamine release by presynaptic nicotinic receptors in rat striatal slices: Effect of nicotine in a low concentration. *Life Sci. 25*:1257-1261.
55. Giovino, A. A., Glimcher, P. W., Mattei, C. A., Hoebel, B. G. 1983. Phencyclidine (PCP) generates conditioned reinforcement in the nucleus accumbens (ACC) but not in the ventral tegmental area (VTA). *Soc. Neurosci. Abstr. 9*:120.
56. Glimcher, P. D., Giovino, A. A., Margolin, D. H., Hoebel, B. G. 1984. Endogenous opiate reward induced by an enkephalinase inhibitor, thiorphan, injected into the ventral midbrain. *Behav. Neurosci. 98*:262-268.
57. Gunne, L. M., Anggard, E., Jonsson, L. E. 1972. Clinical trials with amphetamine-blocking drugs. *Psychiatr. Neurol. Neurochir. 75*:225-226.
58. Heikkila, R. E., Orlansky, H., Cohen, G. 1975. Studies on the distinction between uptake inhibition and release of (^{3}H)dopamine in rat brain tissue slices. *Biochem. Pharmacol. 24*:847-852.
59. Holmes, L. J., Bozarth, M. A., Wise, R. A. 1983. Circling from intracranial morphine applied to the ventral tegmental area in rats. *Brain Res. Bull. 11*:295-298.
60. Jaffe, J. H. 1980. Drug addiction and drug abuse, in *The Pharmacological Basis of Therapeutics* (Gilman, A. G., Goodman, L. S., Gilman, A., eds.), Macmillan, New York, pp. 535-584.
61. Jarvik, M. E. 1967. Tobacco smoking in monkeys. *Ann. N.Y. Acad. Sci. 142*:280-294.
62. Kiianmaa, K. 1980. Alcohol intake and ethanol intoxication in the rat: effect of a 6-OHDA-induced lesion of the ascending nordrenaline pathways. *Eur. J. Pharmacol. 64*:9-19.
63. Koob, G. F., Balcom, G. J., Myerhoff, J. L. 1976. Increases in intracranial self-stimulation in the posterior hypothalamus following unilateral lesions in the locus coeruleus. *Brain Res. 101*:554-560.

64. Koob, G. F., Spector, N. H., Meyerhoff, J. L. 1975. Effects of heroin on lever-pressing for intracranial self-stimulation, food and water in the rat. *Psychopharmacologia 42*:231-234.
65. Le Magnen, J. 1971. Advances in studies on the physiological control and regulation of food intake, in *Progress in Physiological Psychology*, Vol. 4 (Stellar, E., Sprague, J. M., eds.), Academic, New York, pp. 204-261.
66. Le Magnen, J. 1981. The metabolic basis of dual periodicity of feeding in rats. *Behav. Brain Sci. 4*:561-607.
67. Lester, D. 1966. Self-selection of alcohol by animals, human variation and the etiology of alcoholism: A critical review. *Q. J. Stud. Alcohol 27*:395-438.
68. Lichtensteiger, W., Hefti, F., Felix, D., Huwyler, T., Melamed, E., Schlumpf, M. 1982. Stimulation of nigrostriatal dopamine neurones by nicotine. *Neuropharmacology 21*:963-968.
69. Lippa, A. S., Antelman, S. M., Fisher, A. E., Canfield, D. R. 1973. Neurochemical mediation of reward: A significant role for dopamine? *Pharmacol. Biochem. Behav. 1*:23-28.
70. Lorens, S. A., Mitchell, C. L. 1973. Influence of morphine on lateral hypothalamic self-stimulation in the rat. *Psychopharmacologia 32*:271-277.
71. Lyness, W. H., Friedle, N. M., Moore, K. E. 1979. Destruction of dopaminergic nerve terminals in nucleus accumbens: Effect on *d*-amphetamine self-administration. *Pharmacol. Biochem. Behav. 11*:553-556.
72. Magnuson, D. J., Reid, L. C. 1977. Addictive agents and intracranial stimulation (ICS): Pressing for ICS under the influence of ethanol before and after physical dependence. *Bull. Psychonom. Soc. 10*:364-366.
73. McAuliffe, W. E., Gordon, R. A. 1974. A test of Lindesmith's theory of addiction: The frequency of euphoria among long-term addicts. *Am. J. Sociol. 79*:795-840.
74. Mello, N. K. 1973. A review of methods to induce alcohol addiction in animals. *Pharmacol. Biochem. Behav. 1*:89-101.
75. Mello, N. K., Mendelson, J. H. 1972. Drinking patterns during work-contingent and non-contingent alcohol acquisition. *Psychosom. Med. 34*:139-164.
76. Mogenson, G. J. 1964. Effects of sodium pentobarbital on brain self-stimulation. *J. Comp. Physiol. Psychol. 58*:461-462.
77. Monaco, A. P., Hernandez, L., Hoebel, B. G. 1980. Nucleus accumbens: Site of amphetamine self-injection: Comparison with the lateral ventricle, in *The Neurobiology of the Nucleus Accumbens* (Chronister, R. B., DeFrance, J. F., eds.), Haer Institute, New Brunswick, Maine, pp. 338-342.
78. Morgan, M. J. 1974. Resistance to satiation. *Anim. Behav. 22*:449-466.

79. Nwanze, E., Jonsson, G. 1981. Amphetamine neurotoxicity on dopamine nerve terminals in the caudate nucleus of mice. *Neurosci. Lett.* *26*:163-168.
80. Olds, J. 1966. Facilitatory action of diazepam and chlordiazepoxide on hypothalamic reward behavior. *J. Comp. Physiol. Psychol.* *62*:136-140.
81. Olds, J., Milner, P. M. 1954. Positive reinforcement produced by electrical stimulation of septal area and other regions of rat brain. *J. Comp. Physiol. Psychol.* *47*:419-427.
82. Olds, J., Tavis, R. P. 1960. Effects of chlorpromazine, meprobamate, pentobarbital and morphine on self-stimulation. *J. Pharmacol. Exp. Ther.* *128*:397-404.
83. Olds, M. E., Olds, J. 1963. Approach-avoidance analysis of rat diencephalon. *J. Comp. Neurol.* *120*:259-295.
84. Panksepp, J., Gandelman, R., Trowill, J. 1970. Modulation of hypothalamic self-stimulation and escape behavior by chlordiazepoxide. *Physiol. Behav.* 5:965-969.
85. Pickens, R., Harris, W. C. 1968. Self-administration of *d*-amphetamine by rats. *Psychoparmacologia* *12*:158-163.
86. Phillips, A. G., Broekkamp, C. L. E. 1980. Inhibition of intravenous cocaine self-administration by rats after microinjection of spiroperidol into the nucleus accumbens. *Soc. Neurosci. Abstr.* *6*:105.
87. Phillips, A. G., LePiane, F. G. 1980. Reinforcing effects of morphine microinjection into the ventral tegmental area. *Pharmacol. Biochem. Behav.* *12*:965-968.
88. Phillips, A. G., LePiane, F. G. 1982. Reward produced by microinjection of (d-ala^{2})-met^{5} enkephalinamide into the ventral tegmental area. *Behav. Brain Res.* 5:225-229.
89. Poschel, B. P. H., Ninteman, F. W. 1963. Norepinephrine: A possible excitatory neurohormone of the reward system. *Life Sci.* *10*:782-288.
90. Reid, L. D., Gibson, W. E., Gledhill, S. M., Porter, P. B. 1964. Anticonvulsant drugs and self-stimulation behavior. *J. Comp. Physiol. Psychol.* *7*:353-356.
91. Risner, M. E., Jones, B. E. 1976. Role of noradrenergic and dopaminergic processes in amphetamine self-administration. *Pharmacol. Biochem. Behav.* 5:477-482.
92. Risner, M. E., Jones, B. E. 1980. Intravenous self-administration of cocaine and norcocaine by dogs. *Psychopharmacology* *71*:83-89.
93. Roberts, D. C. S., Corcoran, M. E., Fibiger, H. C. 1977. On the role of ascending catecholaminergic systems in intravenous self-administration of cocaine. *Pharmacol. Biochem. Behav.* *6*:615-620.
94. Roberts, D. C. S., Koob, G. F., Klonoff, P., Fibiger, H. C. 1980. Extinction and recovery of cocaine self-administration following

6-OHDA lesions of the nucleus accumbens. *Pharmacol. Biochem. Behav. 12*:781-787.

95. Rockman, G. E., Amit, Z., Carr, G., Brown, Z. W., Ogren, S. O. 1979. Attenuation of ethanol intake by 5-hydroxytryptamine uptake blockade in laboratory rats. I. Involvement of brain 5-hydroxytryptamine in the mediation of the positive reinforcing properties of ethanol. *Arch. Int. Pharmacodyn. 241*:245-259.
96. Routtenberg, A. 1964. Self-starvation caused by "feeding-center" stimulation. *Am. Psychol. 19*:502-507.
97. St. Laurent, J., Olds, J. 1967. Alcohol and brain centers of positive reinforcement, in *Alcoholism Behavioral Research, Therapeutic Approaches* (Fox, R., ed.), Springer, New York, pp. 85-106.
98. Schuster, C. R., Thompson, T. 1969. Self-administration of and behavioral dependence on drugs. *Annu. Rev. Pharmacol. 9*:483-502.
99. Seiden, L. S., Fischman, M. W., Schuster, C. R. 1975/1976. Long-term methamphetamine induced changes in brain catecholamines in tolerant rhesus monkeys. *Drug Alcohol Dependence 1*:215-219.
100. Shizgal, P., Bielajew, C., Corbett, D., Skelton, R., Yeomans, J. 1980. Behavioral methods for inferring conduction velocity and anatomical linkage: I. Pathways connecting rewarding brain stimulation sites. *J. Comp. Physiol. Psychol. 94*:227-237.
101. Shizgal, P., Bielajew, C., Kiss, I. 1980. Anodal hyperpolarization block technique provides evidence for rostro-caudal conduction of reward related signals in the medial forebrain bundle. *Soc. Neurosci. Abstr. 6*:422.
102. Spiraki, C., Fibiger, H. C., Phillips, A. G. 1982. Attenuation by haloperidol of place preference conditioning using food reinforcement. *Psychopharmacology 77*:379-382.
103. Spiraki, C., Fibiger, H. C., Phillips, A. G. 1982. Dopaminergic substrates of amphetamine-induced place preference conditioning. *Brain Res. 253*:185-193.
104. Spiraki, C., Fibiger, H. C., Phillips, A. G. 1983. Attenuation of heroin reward in rats by disruption of the mesolimbic dopamine system. *Psychopharmacology 79*:278-283.
105. Stein, L. 1962. Effects and interactions of imipramine, chlorpromazine, reserpine and amphetamine on self-stimulation: Possible neurophysiological basis of depression, in *Recent Advances in Biological Psychiatry* (Wortis, J., ed.), Plenum, New York, pp. 288-308.
106. Steranka, L. R. 1981. Stereospecific long-term effects of amphetamine on striatal dopamine neurons in rats. *Eur. J. Pharmacol. 76*:443-446.
107. Streissguth, A. P., Martin, J. C. 1983. Prenatal effects of

alcohol abuse in humans and laboratory animals, in *The Pathogenesis of Alcoholism* (Kissin, B., Begleiter, H., eds.), Plenum, New York, pp. 539-589.

108. Tabakoff, B., Hoffman, P. L. 1983. Neurochemical aspects of tolerance to and physical dependence on alcohol, in *The Pathogenesis of Alcoholism* (Kissin, B., Begleiter, H., eds.), Plenum, New York, pp. 199-252.

109. Ungerstedt, U. 1971. Stereotaxic mapping of the monoamine pathways in the rat brain. *Acta Physiol. Scand. Suppl. 367*:1-48.

110. Ungerstedt, U. 1971. Striatal dopamine release after amphetamine or nerve degeneration revealed by rotational behavior. *Acta Physiol. Scand. Suppl. 367*:49-68.

111. Van Dyke, C., Byck, R. 1977. Cocaine: 1884-1974, in *Cocaine and Other Stimulants* (Ellinwood, E., Kilbey, M., eds.), Plenum, New York, pp. 1-30.

112. van Ree, J. M., de Wied, D. 1980. Involvement of neurohypophyseal peptides in drug-mediated adaptive responses. *Pharmacol. Biochem. Behav. Suppl. 1*:257-263.

113. Vrtunski, P., Murray, R., Wolin, L. R. 1973. The effect of alcohol on intracranially reinforced response. *Q. J. Stud. Alcohol 34*:718-725.

114. Wikler, A. 1973. Dynamics of drug dependence: Implications of a conditioning theory for research and treatment. *Arch. Gen. Psychiatry 28*:611-616.

115. Wise, R. A. 1978. Catecholamine theories of reward: A critical review. *Brain Res. 152*:215-247.

116. Wise, R. A. 1980. Action of drugs of abuse on brain reward systems. *Pharmacol. Biochem. Behav. Suppl. 1*:213-223.

117. Wise, R. A. 1981. Intracranial self-stimulation: Mapping against the lateral boundaries of the dopaminergic cells of the substantia nigra. *Brain Res. 213*:190-194.

118. Wise, R. A. 1982. Neuroleptics and operant behavior: The anhedonia hypothesis. *Behav. Brain Sci. 5*:39-53.

119. Wise, R. A., Bozarth, M. A. 1982. Action of drugs of abuse on brain reward systems: An update with specific attention to opiates. *Pharmacol. Biochem. Behav. 17*:239-243.

120. Wise, R. A., Routtenberg, A. 1983. Ethanol and brain mechanisms of reward, in *Pathogenesis of Alcoholism* (Kissin, B., Begleiter, H., eds.), Plenum, New York, pp. 77-106.

121. Wise, R. A., Spindler, J., deWit, H., Gerber, G. J. 1978. Neuroleptic-induced "anhedonia" in rats: Pimozide blocks the reward quality of food. *Science 201*:262-264.

122. Wise, R. A., Spindler, J., Legault, L. 1978. Major attenuation of food reward with performance-sparing doses of pimozide in the rat. *Can. J. Pharmacol. 32*:77-85.

123. Wise, R. A., Yokel, R. A., deWit, H. 1976. Both positive reinforcement and conditioned taste aversion from amphetamine and apomorphine in rats. *Science 191*:1273-1275.
124. Woods, J. H., Schuster, C. R. 1968. Reinforcing properties of morphine, cocaine and SPA as a function of unit dose. *Int. J. Addict. 3*:231-246.
125. Yanagita, T., Deneau, G. A., Seevers, M. H. 1965. Evaluation of pharmacologic agents in the monkey by long term intravenous self or programmed administration. *Twenty-Third International Congress on Physiological Science*, Tokyo.
126. Yanagita, T., Takahashi, S. 1973. Dependence liability of several sedative-hypnotic agents evaluated in monkeys. *J. Pharmacol. Exp. Ther. 185*:307-316.
127. Yeomans, J. S. 1979. Absolute refractory periods of self-stimulation neurons. *Physiol. Behav. 22*:911-919.
128. Yokel, R. A., Wise, R. A. 1975. Increased lever pressing for amphetamine after pimozide in rats: Implications for a dopamine theory of reward. *Science 187*:547-549.
129. Yokel, R. A., Wise, R. A. 1976. Attenuation of intravenous amphetamine reinforcement by central dopamine blockade in rats. *Psychopharmacology 48*:311-318.
130. Yokel, R. A., Wise, R. A. 1978. Amphetamine-type reinforcement by dopamine agonists in the rat. *Psychopharmacology 58*:289-296.
131. Zarevics, P., Setler, P. 1979. Simultaneous rate-independent and rate-dependent assessment of intracranial self-stimulation: Evidence for the direct involvement of dopamine in brain reinforcement mechanisms. *Brain Res. 169*:499-512.
132. Zarevics, P., Weidley, E., Setler, P. 1977. Blockade of intracranial self-stimulation by antipsychotic drugs: Failure to correlate with central alpha-noradrenergic blockade. *Psychopharmacology 53*:283-288.

7

Neuropeptides and Addiction

Jan M. van Ree and David de Wied
Rudolf Magnus Institute for Pharmacology, Medical Faculty, University of Utrecht, Utrecht, The Netherlands

I. INTRODUCTION

Psychoactive drugs have been used for many centuries to influence the brain function of healthy individuals as well as of mentally disturbed patients. Outside the medical setting, some of these drugs are self-administered because of their euphoric effects, which may ultimately lead to a "toxic" state, labeled addiction. In former days, naturally occurring drugs, mainly of plant origin, were used. During the last decades, however, a variety of psychoactive drugs has been synthesized and employed in and outside the medical practice. Many structurally related compounds were manufactured and it was recognized that the intrinsic activity of the drugs is very closely related to a particular structure. It was subsequently postulated that the body contains sites that recognize the structure of the drug in a rather specific way. Indeed, it has been shown that, for example, the brain contains specific binding sites for a number of psychoactive drugs. This observation led to the hypothesis that in the brain endogenous entities are present that serve as ligands for the specific binding sites. A number of these ligands have been isolated and some of them appeared to be peptide molecules. The present survey deals with the interaction of peptide molecules and addictive behavior.

II. ADDICTION

Repeated administration of psychoactive drugs may lead to a state of drug dependence characterized by the drug user performing substantial amounts of behavior leading specifically to further administration of the drug (57). Severe degrees of dependence are commonly labeled addiction, particularly in clinical practice. This "toxic" state is in several cases, especially with morphine and ethanol, accompanied by the occurrence of tolerance and physical dependence. In particular, physical dependence, characterized by a specific pattern of biological events which occur in response to withdrawal of the drug, has been considered as one of the most important mechanisms underlying drug addiction. In fact, drugs may be taken in order to alleviate or prevent withdrawal symptoms and by that stimulate drug-taking behavior. This idea was sustained because research has mainly been focused on morphinomimetics and ethanol, drugs where dependence coincides with physical dependence. However, other addictive drugs like cocaine induce little, if any, tolerance and physical dependence, and even with morphinomimetics it has become clear that neither tolerance nor physical dependence is a condition sine qua non for these drugs to be self-administered (91,100).

From the concepts of operational analysis of behavior as formulated by Skinner (84) a conceptual framework has emerged which allows to study the relation between drug administration by organisms and the behavioral consequences of drug administration. Application of these concepts led to the drug self-administration technique, in which drug administration is contingent on the occurrence of a prior response. In particular, intravenous self-administration has extensively been studied. In a typical experiment an animal is given access to a device when manipulated appropriately, delivers an intravenous injection of the drug. According to the principles of operational analysis of behavior, manipulation of the device is the response, while the subsequent injection is the reinforcement or reward. The animal can thus decide whether or not it proceeds to take the drug, which is more or less comparable to human addictive behavior. Using this technique, it has been shown that a variety of drugs from different pharmacological classes can serve as reinforcers in animal experiments (81,94). Interestingly, most of these drugs are abused by humans, whereas those drugs that do not initiate and maintain responding in animals are in general not abused by humans. Thus, empirically and theoretically, the self-administration technique is a useful method to predict the abuse potential of drugs. It establishes the reinforcing efficacy of drugs and is reliably applicable to evaluate the variables that interfere with drug-taking behavior (91).

External factors like the dose of the drug, the schedule of drug availability, and stimulus control are important in initiation, maintenance, and cessation of self-administration behavior (57,81). Drug-induced changes in the organism, including tolerance and physical dependence, may contribute more or less to the behavior associated by drug use. Although these alterations may change the pattern of drug intake, they seem not to play a primary role in the acquisition of drug self-administration. Internal factors, however, may be critical for initiation of self-administration and may contribute to the individual susceptibility to addictive drugs. Candidates for these factors are the neuropeptides, endogenous molecules in which specific information is encoded.

III. NEUROPEPTIDES

A. General Concepts

The pituitary gland produces a number of peptide molecules, that is, hormones, which play an essential role in the homeostasis of the organism. The hormones are secreted into the bloodstream and control a variety of endocrine processes in the body. During the last decades evidence was accumulated that these hormones are also implicated in brain processes. That pituitary hormones modulate brain functions was first suggested by behavioral disturbances that occurred following endocrine manipulations, that is, removal of the pituitary gland. These disturbances could be reversed by treatment with the hormones produced by the extirpated gland (18,20,23). Subsequent studies revealed that the behavior of intact animals is also sensitive to pituitary hormones. Interestingly, the peripheral endocrine action and the central action of these hormones are dissociated. Thus the classic endocrine effects need the whole or at least a major part for full intrinsic activity, whereas the central effects can be mimicked by small parts of the molecule which are devoid of the endocrine activity of the parent hormone (20,23,25,28,30,32,99). These data indicate that the brain is a target organ for pituitary hormones. Peptide molecules affecting the nervous system were designated as neuropeptides (26). At present we know that pituitary hormones and a large number of other peptide molecules are formed throughout the central nervous system. They are likely present in peptidergic pathways, which may function as an endocrine communication network of the brain.

The structure of hormones and neuropeptides makes these entities very suitable for coding specific information. They are small proteins made up of two or more amino acids linked by peptide bonds. The amino acids can be compared with letters of our alphabet, which can be sequenced to create words with a special meaning. The same holds for peptides, in that a certain sequence of amino acids contains specific information for its target organ. There are, however, more similarities between our language and information encoded in peptide molecules. As in spoken language, there are synonymous peptide words which may explain the similarity in the effects of structurally different peptides and could account for the redundance of information usually observed in peptide molecules. From words you can create new words with a different or even an opposite meaning by taking away one or more letters. Accordingly, enzyme systems are available which can generate bioactive neuropeptides from other neuropeptides or from inactive precursor molecules by cleaving a certain peptide bond.

Neuropeptides are thus derived from precursor molecules. These molecules may be pituitary hormones with classic endocrine effects or inactive storage molecules. Enzymatic processing of these precursor molecules results in the generation of neuropeptides of the first order. Further enzymatic processing of these neuropeptides may yield neuropeptides of the second order and possibly of the third order. All these subsequently generated neuropeptides have specific information of brain processes. For example, pro-opiomelanocortin (Fig. 1) is the precursor molecule for the adrenocorticotropic hormone (ACTH) and the fat-mobilizing hormone β-lipotropin (β-LPH). In the intermediate lobe of the pituitary and the brain, ACTH can be further processed to α-melanocyte-stimulating hormone (α-MSH) and in the brain to a number of other fragments. The non-opiate-like β-lipotropin is the precursor molecule for β-endorphin, a neuropeptide with an opiate-like action. In the brain particularly, β-endorphin is the precursor mole-

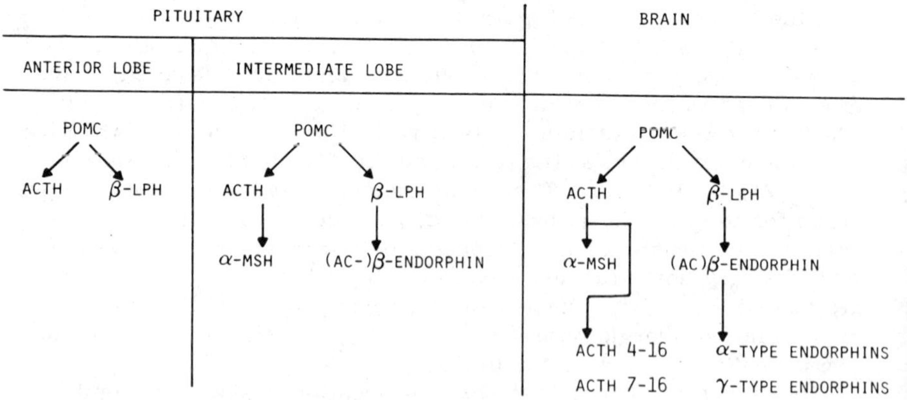

Figure 1 Processing of pro-opiomelanocortin (POMC) in pituitary and brain (ACTH, adrenocorticotrophic hormone; β-LPH, β-lipotropin; α-MSH, α-melanotropic hormone).

cule for the neuroleptic-like γ-type endorphins and the psychostimulant-like α-type endorphins. Thus enzymatic processing may control the bioavailability of the various neuropeptides which may function as neurotransmitters or neuromodulators and induce or inhibit activity in other neuronal systems, or modulate ongoing activity (2). In this way the peptidergic systems play a unique role in the regulation of brain homeostatic mechanisms.

B. ACTH and Related Peptides

Adrenocorticotropic hormone is derived from a precursor molecule of molecular weight 31,000 named pro-opiomelanocortin (64,70,76). This molecule is present in the pituitary and, among others, in a peptidergic pathway with cell bodies in the nucleus arcuatus and terminals in structures of the midbrain limbic system and of the lower brainstem (114). Proteolytic enzymatic processing of pro-opiomelanocortin results in the formation of ACTH and β-LPH. The ACTH can be further cleaved to form α-MSH and other ACTH-related peptides. The N-terminal portion of pro-opiomelanocortin contains an amino acid sequence which resembles α-MSH and has been designated as γ_2-MSH (see Table 1).

It has been shown that ACTH-like neuropeptides and especially $ACTH_{4-10}$ are effective in a variety of behavioral animal test procedures, including active and passive avoidance behavior, appetitively and sexually motivated behavior, social behavior, and experimentally induced amnesia (23,32,41). In addition, intracerebroventricular administration of these peptides induces excessive grooming, yawning, and stretching (49). Structure-activity relationship studies using extinction of pole-jumping avoidance behavior revealed that (a) $ACTH_{4-7}$ is the shortest active sequence in delaying extinction; (b) the phenylalanine moiety—position 7—is important for this activity, since peptides with $D\text{-}Phe^7$ produce an opposite effect, that is, facilitation of extinction; and (c) various substitutions in $ACTH_{4-9}$ and $ACTH_{4-16}$ yield peptides with, respectively, 1000- (Org 2766) or 1,000,000-fold (Org 5042) potentiation (50). Later studies have shown that Org 2766, in addition to its $ACTH_{4-10}$-mimicking action, induces at higher dose levels effects on passive avoidance and social behavior which are opposite to that of $ACTH_{4-10}$ (38,41,71).

The influence of ACTH neuropeptides on conditioned behavior has been interpreted to result from a temporary increase in motivation and selective attention, although some involvement of these peptides in learning and memory processes cannot be excluded (8,22,32,59). Accordingly, single doses of $ACTH_{4-10}$ and Org 2766 administered to humans increase attention and improve task-oriented motivation, but do not consistently affect learning and memory (45,78). Subchronic treatment with Org 2766 in aged people improves mood, reduces depression and anxiety, and increases social contacts and feelings of

Table 1 Amino Acid Sequences of Peptides Related to ACTH, Vasopressin, and β-Endorphin

	1 2 3 4 5 6 7 8 9 10 11 12 13 14 15 16 17 18 19 20 21 22 23 24 25 26 27 28 29 30 31
$ACTH_{1-24}$	H–Ser–Tyr–Ser–Met–Glu–His–Phe–Arg–Trp–Gly–Lys–Pro–Val–Gly–Lys–Lys–Arg–Arg–Pro–Val–Lys–Val–Tyr–Pro–OH
$ACTH_{4-10}$	H–Met–Glu–His–Phe–Arg–Trp–Gly–OH
Org 2766	H–Met(→O)–Glu–His–Phe–^{D}Lys–Phe–OH
γ_2-MSH	Tyr–Val–Met–Gly–His–Phe–Arg–Trp–Asp–Arg–Phe–Gly–OH
β-Endorphin (βE)	H–Tyr–Gly–Gly–Phe–Met–Thr–Ser–Glu–Lys–Ser–Gln–Thr–Pro–Leu–Val–Thr–Leu–Phe–Lys–Asn–Ala–Ile–Val–Lys–Asn–Ala–His–Lys–Lys–Gly–Glu–OH
γ-Endorphin (βE_{1-17})	H–Tyr–Gly–Gly–Phe–Met–Thr–Ser–Glu–Lys–Ser–Gln–Thr–Pro–Leu–Val–Thr–Leu–OH
DTγE (βE_{2-17})	H–Gly–Gly–Phe–Met–Thr–Ser–Glu–Lys–Ser–Gln–Thr–Pro–Leu–Val–Thr–Leu–OH
DEγE (βE_{6-17})	Thr–Ser–Glu–Lys–Ser–Gln–Thr–Pro–Leu–Val–Thr–Leu–OH
α-Endorphin (βE_{1-16})	H–Tyr–Gly–Gly–Phe–Met–Thr–Ser–Glu–Lys–Ser–Gln–Thr–Pro–Leu–Val–Thr–OH
Met-enkephalin (βE_{1-5})	H–Tyr–Gly–Gly–Phe–Met–OH
Vasopressin	Cys–Tyr–Phe–Gln–Asn–Cys–Pro–Arg–Gly–NH_2 (Cys1–Cys6 bridged)
DG-Vasopressin	Cys–Tyr–Phe–Gln–Asn–Cys–Pro–Arg–OH (Cys1–Cys6 bridged)
Oxytocin	Cys–Tyr–Ile–Gln–Asn–Cys–Pro–Leu–Gly–NH_2 (Cys1–Cys6 bridged)

competence. These latter effects may be related to the above-mentioned effects of this peptide on animal behavior, and which are not seen after treatment with $ACTH_{4-10}$.

γ_2-Melanocyte-stimulating hormone, whose core structure is only one amino acid different from the $ACTH_{4-9}$ sequence, induces a number of behavioral effects which are opposite to those of ACTH/MSH-like neuropeptides (74,106). The influence of γ_2-MSH is also opposite to that of β-endorphin, and interestingly resembles that of opiate antagonists. Accordingly, γ_2-MSH attenuated several effects of intracerebroventricularly injected β-endorphin, including antinociception, hypothermia, hypoactivity, and α-MSH release. In addition, injection of γ_2-MSH into the periaqueductal gray of opiate-naive rats elicited opiate withdrawal-like symptoms. These and other in vivo experiments, but also in vitro studies, suggest that the effects of γ_2-MSH resemble those of opiate antagonists and that this peptide acts as a functional antagonist of brain opioids.

C. Endorphins and Related Peptides

The discovery of specific opiate binding sites (77,83,86) was followed by the isolation and characterization of a number of endogenous peptide molecules with morphinelike activity (12,51,53,63). Three families of these so-called endorphins can be distinguished. These families stem from different precursor molecules: firstly, proenkephalin, whose processing yields the pentapeptides (Met)-enkephalin and (Leu)-enkephalin, and which is present in short neuronal pathways widespread throughout the brain but also in the periphery (e.g., the adrenal cortex) (72,114); secondly, pro-opiomelanocortin, whose processing yields α-, β-, and γ-endorphins. This system is located predominantly in the pituitary, but also in a neuronal pathway in the brain (see before) (70,114). And thirdly, there is prodynorphin, yielding dynorphins, which is located in the hypothalamic pituitary neuronal system, but also in other parts of the brain such as the hippocampus (56,115). Some evidence is available that the peptides belonging to a certain family can activate a specific subclass of opioid receptors (120) (enkephalin-δ; β-endorphin-μ; dynorphin-κ).

Intracerebroventricular injection of β-endorphin into the cerebrospinal fluid of animals induces a variety of behavioral effects (antinociception, temperature changes, excessive grooming, immobilization, muscular rigidity), which are also observed after morphine treatment (4,48,52,96). These effects are mediated by opiate receptor systems, since they are readily blocked by opiate antagonists. The other endorphins produce similar effects, but they are less potent in this respect, probably because of a more rapid in vivo degradation of the shorter endorphins. β-Endorphin mimics morphine not only after a single injection, but also after repeated injections. In fact, repeated treatment with β-endorphin leads to development of tolerance and physical dependence and induction of experimental addiction (96,102,117).

In addition to their morphinomimetic effects, the endorphins induce behavioral effects which are apparently not mediated by opiate receptor systems. Thus subcutaneous injection of α- and β-endorphin and (Met)-enkephalin delays extinction of pole-jumping avoidance behavior, an effect that persists in the presence of naltrexone (29). Subsequent studies revealed that γ-endorphin has unique properties as compared to the other endorphins, in that it facilitates rather than attenuates extinction of avoidance behavior (30). This effect of γ-endorphin could also be dissociated from the action of γ-endorphin on opiate receptor systems. Removal of the N-terminal amino acid tyrosine, which eliminates the opiate-like action of γ-endorphin, even potentiates the influence on avoidance behavior. The effects of the peptide des-Tyr^1-γ-endorphin (DTγE, βE_{2-17}) resembles that of neuroleptic drugs in a number of animal test procedures, including active and passive avoidance behavior, induction of the grasping response, and electrical brain self-stimulation behavior (30,34). The reduction in locomotion activity and the sedation seen with haloperidol was, however, less pronounced with DTγE, and the peptide did not displace neuroleptics from their specific binding sites (30,101). From these data it was suggested that DTγE or a closely related peptide is an endogenous substance with neuroleptic-like action (24). α-Endorphin, which is only one amino acid shorter than γ-endorphin, induces behavioral effects just opposite to those of γ-endorphin. The effects of α-endorphin and related peptides are in several aspects comparable with those of amphetamine (103). Thus α-type endorphins may function as endogenous substances with a psychostimulant-like action. Subsequent studies have shown that α- and γ-endorphin and their nonopiate fragments can be formed when β-endorphin is incubated with brain synaptic plasma membrane fractions in vitro (13,14). Structure-activity studies revealed that desenkephalin-γ-endorphin (DEγE, βE_{6-17}) is the shortest sequence with full neuroleptic-like activity (31). Concerning the mode of neuroleptic-like action of γ-type endorphins (DTγE, DEγE), it appeared that these peptides, like classic and atypical neuroleptics, function as antagonists of a certain dopaminergic receptor system, which is present in nucleus accumbens, a terminal area of the mesolimbic dopaminergic pathways (108, 109). It has been proposed that the physiological available γ-type endorphins exert a control over this dopaminergic system and that a chronic deficiency of these peptides may lead to a sustained increase in dopaminergic activity (109).

The opiate-like action led to several postulates in which these peptides were implicated in chronic pain, as well as in various psychopathological disorders like psychotic depression, mania, and addiction. The data collected so far in clinical studies, in which patients were treated with opioid antagonists, β-endorphin, or a highly potent (Met)-

enkephalin analog, do not permit to draw definite conclusions in this respect. The postulate that psychosis of the schizophrenic type may result from a relative deficiency of γ-type endorphins (24) stimulated research on the purported antipsychotic action of these peptides. Both DTγE and DEγE appeared to possess antipsychotic effects in a number of schizophrenic patients (110,113).

D. Vasopressin, Oxytocin, and Related Peptides

The neurohypophyseal hormones vasopressin and oxytocin are synthesized predominantly in cell bodies of the hypothalamus and from there transported within the axons to terminals. These terminals are present in the posterior pituitary, where the hormones are stored and can be released into the bloodstream, and in several structures of the limbic system and the lower brainstem. Vasopressin is involved in maintaining the osmolarity and volume of body fluids via an action on the kidney and is therefore designated as antidiuretic hormone. Oxytocin acts on the mammary gland, causing ejection of milk and may play an important role in the expulsion of the fetus at parturition. But these hormones also have profound effects on the central nervous system. This was first demonstrated by removal from rats of the posterior part of the pituitary, which led to behavioral disturbances, and the fact that these disturbances could be restored by treatment with vasopressin (19,20). Subsequent studies in intact rats showed that a single subcutaneous injection with vasopressin results in a long-term resistance to extinction of one-way active avoidance behavior (21). This long-term effect and a lot of other studies suggested that vasopressin is involved in memory processes. Vasopressin neuropeptides facilitate consolidation of acquired information and play a role in retrieval processes or the expression of stored information (20,99). Oxytocin, which differs by only two amino acids from vasopressin (see Table 1), exerts an opposite effect, in that both consolidation and retrieval processes are attenuated by oxytocin treatment (9,10,99). Hence oxytocin can be regarded as an amnesia-producing peptide. Evidence has been presented that both vasopressin and oxytocin are physiologically involved in memory processes (10,27,111). Also, with these hormones, a dissociation exists between the peripheral and central effects. Thus removal of the C-terminal amino acid (glycinamide) from vasopressin almost completely eliminates the blood pressure-increasing and antidiuretic action of the hormone, but not the behavioral activities (25). Structure-activity studies suggested that the effects of these peptide molecules on consolidation and retrieval reside in different parts of the molecule, while the amnestic effects of oxytocin are mainly a feature of the whole molecule (11,99). Recent findings show that both vasopressin and oxytocin can be converted by brain synaptosomal plasma membranes to highly potent specific molecules, with respect to their effects on consolidation and retrieval (15,16).

Vasopressin and related peptides seem to improve learning and memory processes of both healthy individuals and patients with disturbances in these processes. Thus, in cognitively unimpaired or impaired adults, in men aged from 50 to 65 years, and in patients suffering from diabetes insipidus or from amnesia, an improvement of several aspects of memory was observed following treatment with vasopressin and related peptides (54,61,62). It has been reported that oxytocin induced an amnestic effect in women, particularly with respect to retrieval functions (40).

Development of tolerance and physical dependence on opiates and ethanol has been regarded as adaptive processes which in some ways resemble learning and memory (104,107). Thus vasopressin and oxytocin and related peptides could also be involved in the development of opiate- and ethanol-induced adaptive changes. Indeed, it was found that des-glycinamide^9-vasopressin and the C-terminal fragment of oxytocin (prolyl-leucyl-glycinamide, PLG) affect the development of tolerance and physical dependence of morphine, β-endorphin, and ethanol (for references see Refs. 79 and 107). However, the direction and magnitude of the effects seem to depend on the test procedure. It has been argued that maybe the different neuronal processes involved in development of tolerance and physical dependence are differentially affected by neurohypophyseal hormones and their fragments (107).

IV. NEUROPEPTIDES AND EXPERIMENTAL ADDICTION

A. Animal Test Procedures

As outlined before, the drug self-administration models reliably predict abuse potential of psychoactive drugs. These models involve a variety of species (rats, cats, dogs, monkeys) and routes of drug administration (e.g., intravenous, intragastric, oral, inhalation, intracerebroventricular, intracerebral) (57,81). The most commonly applied technique is the intravenous one, with the exception in the case of ethanol, which has frequently been offered via the oral route.

For our studies concerning the influence of neuropeptides, we have selected intravenous heroin self-administration in rats. With this drug self-injecting behavior develops relatively quickly and is rather reproducible, at least under standard conditions (97,100). We focused on acquisition of the behavior, since neuropeptides are involved in adaptation and adaptive processes play an important role in the mechanism by which drug injection gains and maintains control over behavior. Thus rats were tested on five consecutive days for 6 hr per day for self-injecting behavior. Pressing one of the two available levers, marked by an illuminated light, resulted in an intravenous injection with heroin. Using this test procedure, it was found that self-administering behavior was readily acquired and increased during the first test days, reaching a ceiling level on the fourth and fifth test days (92,97). For intracerebroventricular and intracerebral self-administration a similar test procedure was used. Rats were then equipped with stainless steel cannula in the lateral brain ventricles or at specific brain sites using a stereotaxic instrument.

Oral consumption of morphine and ethanol solutions have been studied in relation to the effectiveness of neuropeptides. When acquisition of a morphine-drinking habit was investigated, the rats could continuously choose between a sucrose solution without morphine and a sucrose solution with a fixed concentration of morphine. Acquisition and retention of ethanol drinking was studied in rats which were found to drink increasing concentrations of ethanol in water (43).

Brain electrical self-stimulation is widely used to explore the significance of certain brain structures with respect to reward (116). In these experiments, rats can press a lever which is then followed by the delivery of a train of electrical stimuli via an electrode placed in a specific brain site. There are behavioral similarities between drug self-administration and intracranial electrical self-stimulation (ICSS). Because of these similarities it has been suggested that morphinelike drugs are reinforcing because they interact with central reward structures (36,66). Moreover, it has been argued that brain dopamine systems are critically involved in both ICSS and drug self-administration (118,119; see also Chapter 6 by Wise and Bozarth). It has indeed been shown that the excitability of the neural systems which mediate self-stimulation is affected by morphine treatment (65,66,112). Brain endorphins seem to be involved in ICSS, since the opioid antagonist naloxone suppresses responding and increases the threshold of ICSS (3,112). Thus the effect of neuropeptides on ICSS was analyzed to gain insight into the mechanism of interaction between neuropeptides and addictive drugs.

B. ACTH and Related Peptides

Motivational factors have been implicated in both drug self-administration and ICSS (46,87,119). As mentioned before (see Sec. III.B), the behavioral effects of ACTH neuropeptides have been interpreted as mediated by motivational and attention processes. Others have proposed that ACTH neuropeptides may affect behavior by assuming an intrinsic reinforcing effect of these peptides (55). It has indeed been reported that $ACTH_{1-24}$ and $ACTH_{4-10}$ are self-administered by some but not all rats tested. The self-administration of ACTH was markedly decreased by naloxone, which may implicate opioid receptor systems in ACTH self-administration (55).

We tested a number of ACTH neuropeptides for their possible modulating role in acquisition of heroin self-administration of rats. Neither $ACTH_{1-24}$ nor $ACTH_{4-10}$ significantly influenced the pattern of heroin self-injecting behavior, but (D-Phe^7)-$ACTH_{4-10}$ and a high dose of the potent peptide Org 2766 decreased heroin intake, especially after some days of testing (93). The closely related peptide γ_2-MSH markedly decreased acquisition of heroin self-injecting behavior, which may be related to the naloxone-like effects of this peptide, as suggested from other experiments (see Sec. III.B). That ACTH neuropeptides may influence processes involved in drug addiction was also concluded from experiments showing that $ACTH_{4-10}$ changed the performance of rats which were trained to dissociate the opiate fentanyl from saline (17). This test procedure requires the rat to evaluate the subjective effects produced by drug administration, and which are likely involved in the initiation and maintenance of drug-seeking behavior.

Intracranial electrical self-stimulation for certain brain regions has been reported to be sensitive to ACTH-like neuropeptides (37,60,73). Low doses of Org 2766 and $ACTH_{4-10}$ enhanced ICSS behavior elicited from the medial septum at low current intensities and attenuated this behavior at high current intensities. (D-Phe^7)-$ACTH_{4-10}$ and high doses of Org 2766 decreased ICSS behavior of threshold currents. This inhibitory effect of Org 2766 was prevented by the opioid antagonist naltrexone (39). ACTH neuropeptides also affect ICSS elicited from the medial forebrain bundle (60,73). Further experiments suggest that the influence of ACTH neuropeptides may be mediated by an effect on motivational processes involved in ICSS rather than that these peptides directly changed the reward of the stimulation (37). Nevertheless, the decreasing effect of (D-Phe^7)-$ACTH_{4-10}$ and high doses of Org 2766 on ICSS may be related to the attenuating influence of these peptides on acquisition of heroin self-administration.

C. Endorphins and Related Peptides

The presence of endorphins in the pituitary and brain led to the postulate that these peptides may be involved in the functioning of physiological systems that are implicated in reward and which are susceptible to narcotic drugs. Thus the endorphins may have intrinsic reinforcing effects of opiate and electrical self-stimulation. Enkephalins, and especially (Leu)-enkephalin, were found to support self-administration in rats when injected intracerebroventricularly (3). Although we could not reproduce self-administration of (Met)-enkephalin, we found that β-endorphin shares the reinforcing properties of heroin (102). Rats worked for a response-contingent intracerebroventricular injection when they received relatively low doses of β-endorphin. The dose used was much lower than that needed to produce antinociception (96). Self-administration of β-endorphin was accompanied with a low level of physical dependence (102). This is in accord with other data suggesting that physical dependence is not an essential condition for the acquisition of drug self-administration (110,121). Synthetic enkephalin analogs that are resistant to enzymatic breakdown, for example, (D-Ala^2-

Met5)-enkephalinamide and FK33-824 have also been demonstrated to serve as reinforcers in naive and opiate-dependent rats and monkeys (67,75,80,88). These peptides were administered either intravenously, intracerebroventricularly, or intrahypothalamically. Thus β-endorphin and enkephalin analogs have been shown to serve as positive reinforcers. Other data suggest that these peptides also possess discriminative internal stimulus properties similar to those of narcotic drugs (102,122). Both the positive reinforcing and the discriminative stimulus properties of β-endorphin indicate that this peptide may exert powerful control over behavior. This action of β-endorphin may be mimicked by narcotic drugs, and in this way organisms may become dependent on these drugs. It may be hypothetized that an altered bioavailability of endorphins may be a critical factor in the process of addiction. The involvement of endorphins in reward is supported by the influence of opiate antagonists on ICSS. Moreover, the endorphins may also be implicated in dependence on other drugs. Blockade of opiate receptor systems with naltrexone decreased the reinforcing effects of ethanol in monkeys self-administering ethanol via the intravenous route (1). In alcoholics, the concentration of β-endorphin in the cerebrospinal fluid was markedly decreased as compared to that of normal individuals, while that of ACTH was increased (47). Naloxone has been reported to reduce the amount of cigarettes smoked by chronic smokers (58). Thus β-endorphin may be a common factor in drug addiction. In addition, other brain endorphins, that is, enkephalins, have been implicated in drug addiction, especially ethanol consumption of experimental animals (5-7).

Using the test procedure for acquisition of heroin self-administration as outlined before, it was found that systemic injection with the opioid peptides α-, γ-, and β-endorphin did not significantly change heroin intake. However, the nonopiate, neuroleptic-like fragments of γ-endorphin, that is, DTγE and DEγE, decreased acquisition of heroin self-administration (93). A similar effect was observed with the neuroleptic drug haloperidol. These data further support the neuroleptic-like action of γ-type endorphins and agree with the involvement of dopaminergic systems in opiate self-administration, since neuroleptic drugs are potent antidopaminergic substances. In addition, it has been found that DTγE, like haloperidol, attentuated ICSS behavior elicited from the ventrotegmental-medial substantia nigra area and the nucleus accumbens, where the cell bodies and terminals of the mesolimbic dopaminergic system are respectively located (34,105). These data indicate that γ-type endorphins interfere with brain reward processes associated with stimulation of the mesolimbic dopaminergic system, which may be related to the effects of these peptides on acquisition of heroin self-administration.

α-Endorphin and related peptides slightly but not significantly increased heroin intake of rats when tested for acquisition of self-administration (93). However, α-endorphin has been shown to increase ICSS behavior elicited from the ventrotegmental-medial substantia nigra area (34). That this effect is not reflected in a more marked effect on heroin self-administration may be due to the nearly maximal rate of acquisition under the conditions used. This agrees with the suggestion that neuropeptides primarily modulate the initiation and not the maintenance of drug self-administration. Accordingly, both DTγE and α-endorphin change ICSS behavior when threshold current are delivered, but these peptides are less effective when high currents are used.

D. Vasopressin, Oxytocin, and Related Peptides

Learning and memory processes play an important role in the mechanism by which drug injection gains and maintains control over behavior. Since vasopressin and oxytocin are implicated in memory processes (see Sec. III.B), we have tested these peptides and their fragments on acquisition of heroin self-administration. We found that subcutaneous treatment with desglycinamide^{9}-arginine^{8}-vasopressin (DG-AVP) decreased heroin intake in a dose-dependent fashion (97). A similar effect was observed when DG-AVP was administered orally or intracerebroventricularly (92,98). The observation that intracerebroventricularly applied antivasopressin serum increased heroin self-administration suggests that vasopressin plays a physiological role in this behavior (98). Antiserum against oxytocin and growth hormone were not effective in this respect, but antiprolactin serum also increased heroin intake. This may indicate that more pituitary hormones are involved in experimental addiction. Structure-activity relationship studies with vasopressin and oxytocin peptides (Fig. 2) revealed that the ring structure of vasopressin (pressinamide) and, to a minor degree, the C-terminal part of vasopressin are important for attenuating heroin self-administration. Interestingly, oxytocin slightly and the C-terminal tripeptide PLG markedly increased heroin intake (97). A similar opposing action of DG-AVP and PLG has been observed in rats tested for ICSS behavior elicited from the ventrotegmental substantia nigra area (33) and for acquisition of fentanyl self-administration when this opiate was administered directly in the ventrotegmental substantia nigra area (104). Both these behaviors were decreased by DG-AVP and stimulated by PLG. These findings suggest that the interaction between neurohypophyseal hormones-related peptides and heroin reward may be due to an interference with dopaminergic systems, especially the mesolimbic pathways. Others have reported that ICSS behavior elicited from the lateral hypothalamic structures is also attenuated and increased by vasopressin and oxytocin, respectively (82).

Treatment with DG-AVP may decrease heroin intake by attenuating the reinforcing efficacy of the heroin injection. This is supported by

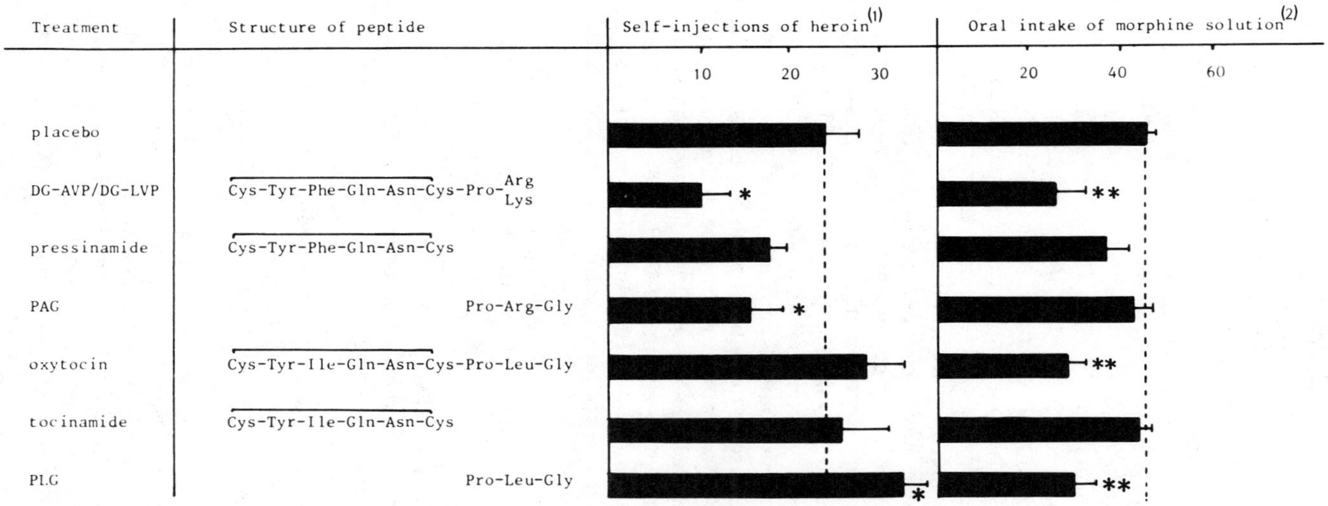

Figure 2 Influence of neuropeptides related to neurohypophyseal hormones on acquisition of heroin self-administration and on oral morphine intake in rats. (1) Rats were subjected to a 6-hr experimental session daily for five consecutive days. During these sessions the animals were allowed to self-administer heroin (0.15 mg/kg per injection) intravenously. Treatment was performed subcutaneously 1 hr prior to each session. Groups of animals (n = 8-11) were injected with 1 μg of the peptides (5 μg in the case of PAG). The mean ± SEM number of self-injections on the fifth test day is presented. (From Refs. 88 and 93.) (2) Rats received a free choice between two bottles of drinking fluid for 11 days. The bottles contained 5% sucrose solution and 5% sucrose solution with 0.02% morphine. Treatment was performed subcutaneously every day. Groups of rats (n = 10) were injected with 10 μg of the peptides. The mean ± SEM intake of the morphine solution expressed as the percentage of total fluid intake is presented. (*,Different from placebo-treated control rats; *, < 0.05; **, p < 0.01.)

the finding showing that when rats treated with DG-AVP have to perform more effort to obtain heroin, they refused to do so (92). The effectiveness of DG-AVP in decreasing reward may depend on the degree of reinforcement control over behavior. Indeed, it has been shown that DG-AVP did not decrease opiate intake during maintenance of heroin and morphine self-administration in rats and monkeys, respectively (68,92). Moreover, ICSS is attenuated by DG-AVP at threshold but not at high current intensities (33).

The data reviewed so far consistently show that vasopressin-related peptides attenuate brain and heroin reward. However, using other models of experimental addiction and testing phenomena related to addiction, like tolerance and physical dependence, the emerged picture is complicated. Firstly, the development of tolerance to and physical dependence on opiates is enhanced by DG-AVP and more potently by oxytocin and PLG (95). Others have reported just opposite effects with PLG (107). Similar contradictory effects have been reported on the influence of vasopressin- and oxytocin-like neuropeptides on development of ethanol tolerance and physical dependence (79,85,104). These effects of the peptides may interfere with the interaction of these peptides and self-administration of opiates and ethanol.

Secondly, when rats were forced to drink increasing concentrations of ethanol in water, they stabilized at a certain acceptance concentration of ethanol. Treatment with DG-AVP and PLG during acquisition of this ethanol-drinking habit led to acceptance of higher concentrations of ethanol (43,69). It was demonstrated that DG-vasopressin was effective when administered during the training period and not after a first level of acceptance had been reached.

Thirdly, we have tested the influence of vasopressin- and oxytocin-related peptides on oral morphine intake (unpublished data). Rats received a free choice between two bottles of drinking fluid for 11 days. One bottle contained a solution of 5% sucrose and the other a solution of 5% sucrose and 0.02% morphine. Groups of rats were subcutaneously treated with various peptides every day. The total fluid intake did not differ among the various treatment groups. The intake of morphine solution, expressed as the percentage of total fluid intake, increased during the test period from about 20% (day 1) to about 60% (day 11). The mean intake of morphine solution over the 11 days was about 44% (Fig. 2). Desglycinamide^{9}-lysine^{8}-vasopressin (DG-LVP), oxytocin, and PLG attenuated morphine intake, while the peptides pressinamide (ring structure of vasopressin), tocinamide (ring structure of oxytocin), and prolyl-arginyl$_2$-glycinamide (PAG, tail structure of vasopressin) were ineffective. Especially rats treated with DG-LVP and PLG did not increase their morphine intake during the test period, which may support that particularly acquisition of this drinking habit is decreased, as was reported for heroin self-administration and oral ethanol consumption.

These data indeed show that the interaction between drug-taking behavior and neurohypophyseal hormones-related peptides is rather complex. Maybe the effects of these peptides on oral morphine and ethanol consumption are somehow related to their effects on development of tolerance and physical dependence, since in all these test systems DG-vasopressin and PLG exert a similar action. Assuming this, it is, however, puzzling that oral morphine consumption is decreased rather than increased. Nevertheless, yet available data indicate that neurohypophyseal hormones and their fragments profoundly influence drug-taking behavior.

V. NEUROPEPTIDES AND HUMAN ADDICTION

In view of the attenuating effect of DG-AVP on intravenous heroin self-administration in rats, the effects of this neuropeptide have been studied in heroin addicts. Since DG-AVP may be especially effective in situations in which the reinforcement control over self-injecting behavior is changed and/or the behavior is less strictly controlled, the influence of DG-AVP was studied on drug-seeking behavior in heroin addicts during the initial phase of methadone detoxification (44,89,90). The ambulant methadone detoxification program offers short-term detoxification to mild or moderate addicts. Patients receive their calibrated methadone dose once daily for 2 weeks. After this period the methadone dose was gradually reduced, but only when heroin use was more or less stopped. In two double-blind, placebo-controlled studies, 15 patients received 1 mg of DG-AVP sublingually per day for at least 5 days and 15 patients received placebo. It was observed that DG-AVP facilitates the methadone detoxification, in that the percentage of patients succeeding in reducing the methadone dose and the percentage of patients with successful detoxification was higher in the DG-AVP group as compared to the placebo group. Also the time course of attending the clinic was longer when patients were treated with DG-AVP, suggesting a beneficial effect on the methadone detoxification. The percentage of urine samples with detectable morphine and cocaine appeared to be lower for the DG-AVP group than for the placebo group in the first and both studies, respectively. Moreover, the medical attendant judged the detoxification of the DG-AVP treated patients to have been more successful than that of the patients on placebo. The beneficial effect of DG-AVP was present not only during treatment with this neuropeptide, but also in the period following cessation of DG-AVP administration. These preliminary findings indicate once more the predictive value of experimental drug self-administration for human addiction. The decrease of heroin intake in patients treated with DG-AVP supports the postulate that this neuropeptide attenuates the reinforcing efficacy of heroin. The decrease of cocaine use may suggest that the peptide has changed drug-seeking behavior in general.

Smokers subjected to aversive conditioning sessions designed to assist them to stop smoking were intranasally treated with LVP for five consecutive days following a double-blind, placebo-controlled design (35). During the week of aversive conditioning those subjects receiving LVP smoked significantly fewer cigarettes than the saline-treated group. During the follow-up period of 6 weeks the LVP group had significantly more urges to smoke and smoked more cigarettes than the placebo group. The authors concluded that LVP was associated with a facilitation of the acquisition of the avoidance response followed by an apparent acceleration of the extinction of the learned response as compared to saline. However, in view of the effects of vasopressin-like peptides on heroin addiction, it may also be that LVP diminished the reinforcing action of nicotine. These data suggest that vasopressin-like peptides can beneficially change addictive behavior not only of experimental animals, but also of humans.

VI. CONCLUDING REMARKS

The reviewed data indicate that pituitary hormones and their fragments are present in the brain and are involved in a variety of behavioral performances. Besides these peptide molecules, a large number of other peptides have been found in brain tissue. The so-called neuropeptides are derived from precursor molecules by enzymatic activity and form the brain endocrine communication network that plays a unique role in the regulation of brain homeostatic mechanisms. The neuropeptide systems are also implicated in opiate and brain reward processes. Thus the endorphins, especially β-endorphin, exert reinforcing effects similar to those of opioids, and neuropeptides related to ACTH/MSH, β-endorphin, and neurohypophyseal hormones modulate both acquisition of heroin self-administration and intracranial electrical self-stimulation. Preliminary data suggest that neuropeptides may also change addictive behavior of humans. Maybe disturbances in brain and/or pituitary hormone-containing systems or in the generation of neuropeptides from these hormones lead to a state in which drug-seeking behavior is easily elicited. Demonstration and subsequent treatment of these disturbances in the human addicts may yield a more goal-directed treatment of addiction in the future.

REFERENCES

1. Altshuler, H. L., Phillips, P. E., Feinhandler, D. A. 1980. Alteration of ethanol self-administration by naltrexone. *Life Sci. 26*:679-688.

2. Barchas, J. D., Akil, H., Elliott, G. R., Holamn, R. B., Watson, S. J. 1978. Behavioral neurochemistry: Neuroregulators and behavioral states. *Science 200*:964-973.
3. Belluzzi, J. D., Stein, L. 1977. Enkephalin may mediate euphoria and drive-reduction reward. *Nature 266*:556-558.
4. Bloom, F., Segal, D., Ling, N., Guillemin, R. 1976. Endorphins: Profound behavioral effects in rats suggest new etiological factors in mental illness. *Science 194*:630-632.
5. Blum, K., Briggs, A. H., Elston, S. F. A., Delallo, L., Sheridan, P. J., Sar, M. 1982. Reduced leucine-enkaphaline-like immunoreactive substance in hamster basal ganglia after long-term ethanol exposure. *Science 216*:1425-1427.
6. Blum, K., Briggs, A. H., Delallo, L., Elston, S. F. A., Ochoa, R. 1982. Whole brain methionine-enkephalin of ethanol-avoiding and ethanol-preferring C57B1 mice. *Experientia 38*:1469-1470.
7. Blum, K., Elston, S. F. A., Delallo, L., Briggs, A. H., Wallace, J. E. 1983. Ethanol acceptance as a function of genotype amounts of brain (met)enkephalin. *Proc. Nat. Acad. Sci. 80*:6510-6512.
8. Bohus, B., De Wied, D. 1978. Pituitary-adrenal system hormones and behavior, in *Neuro-psychopharmacology*, Proceedings of the 10th Congress Collegium International Neuro-Psychopharmacologicum, Quebec, July 1976 (Deniker, P., Radouco-Thomas, C., Villeneuve, A., eds.), Pergamon, Oxford, pp. 987-993.
9. Bohus, B., Kovács, G. L., De Wied, D. 1978. Oxytocin, vasopressin and memory: Opposite effects on consolidation and retrieval processes. *Brain Res. 157*:414-417.
10. Bohus, B., Urban, I., Van Wimersma Greidanus, T. B., De Wied, D. 1978. Opposite effects of oxytocin and vasopressin on avoidance behavior and hippocampal theta rhythm in the rat. *Neuropharmacology 17*:239-247.
11. Bohus, B., Kovács, G. L., Greven, H. M., De Wied, D. 1978. Memory effects of arginine-vasopressin (AVP) and oxytocin (OXT): Structural requirements. *Neurosci. Lett. Suppl. 1*: S80.
12. Bradbury, A. F., Smyth, D. G., Snell, C. R., Birdsall, N. J. M., Hulme, E. C. 1976. C-Fragment of lipotropin has a high affinity for brain opiate receptors. *Nature 260*:793-795.
13. Burbach, J. P. H., Loeber, J. G., Verhoef, J., Wiegant, V. M., De Kloet, E. R., De Wied, D. 1980. Selective conversion of β-endorphin into peptides related to γ- and α-endorphin. *Nature 283*:96-97.
14. Burbach, J. P. H., De Kloet, E. R., Schotman, P., De Wied, D. 1981. Proteolytic conversion of β-endorphin by brain synaptic membranes: Characteristization of generated β-endorphin fragments and proposed metabolic pathway. *J. Biol. Chem. 256*:12,463-12,469.

15. Burbach, J. P. H., Kovács, G. L., De Wied, D., Van Nispen, J. W., Greven, H. M. 1983. A major metabolite of arginine vasopressin in the brain is a highly potent neuropeptide. *Science 221*:1310-1312.
16. Burbach, J. P. H., Bohus, B., Kovács, G. L., Van Nispen, J. W., Greven, H. M., De Wied, D. 1983. Oxytocin is a precursor of potent behaviourally active neuropeptides. *Eur. J. Pharmacol. 94*:125-131.
17. Colpaert, F. C., Niemegeers, C. J. E., Janssen, P. A. J., Van Ree, J. M., De Wied, D. 1978. Selective interference of $ACTH_{4-10}$ with discriminative responding based on the narcotic cue. *Psychoneuroendocrinology 3*:203-210.
18. De Wied, D. 1964. Influence of anterior pituitary on avoidance learning and escape behavior. *Am. J. Physiol. 207*:255-259.
19. De Wied, D. 1965. The influence of the posterior and intermediate lobe of the pituitary and pituitary peptides on the maintenance of a conditioned avoidance response in rats. *Int. J. Neuropharmacol. 4*:157-167.
20. De Wied, D. 1969. Effects of peptide hormones on behavior, in *Frontiers in Neuroendocrinology* (Ganong, W. F., Martini, L., eds.), Oxford University Press, New York, pp. 97-140.
21. De Wied, D. 1971. Long term effect of vasopressin on the maintenance of a conditioned avoidance response in rats. *Nature 232*:58-60.
22. De Wied, D. 1974. Pituitary-adrenal system hormones and behavior, in *The Neurosciences, Third Study Program* (Schmitt, F. O., Worden, F. G., eds.), MIT Press, Cambridge, pp. 653-666.
23. De Wied, D. 1977. Peptides and behavior. *Life Sci. 20*:195-204.
24. De Wied, D. 1978. Psychopathology as a neuropeptide dysfunction, in *Characteristics and Function of Opioids* (Van Ree, J. M., Terenius, L., eds.), Elsevier/North-Holland, Amsterdam, pp. 113-122.
25. De Wied, D., Greven, H. M., Lande, S., Witter, A. 1972. Dissociation of the behavioral and endocrine effects of lysine vasopressin by tryptic digestion. *Br. J. Pharmacol. 45*:118-122.
26. De Wied, D., Van Wimersma Greidanus, T. B., and Bohus, B. 1974. Pituitary peptides and behavior: Influence on motivational, learning and memory processes, in *Excerpta Medica International Congress Series No. 359, Proceedings of the IXth Congress of the Collegium International Neuropsychopharmacology*, Excerpta Medica, Amsterdam, pp. 653-658.
27. De Wied, D., Bohus, B., Van Wimersma Greidanus, T. B. 1975. Memory deficit in rats with hereditary diabetes insipidus. *Brain Res. 85*:152-156.
28. De Wied, D., Witter, A., Greven, H. M. 1975. Behaviorally active ACTH analogues. *Biochem. Pharmacol. 24*:1463-1468.

29. De Wied, D., Bohus, B., Van Ree, J. M., Urban, I. 1978. Behavioral and electrophysiological effects of peptides related to lipotropin (β-LPH). *J. Pharmacol. Exp. Ther. 204*:570-580.
30. De Wied, D., Kovács, G. L., Bohus, B., Van Ree, J. M., Greven, H. M. 1978. Neuroleptic activity of the neuropeptide β-LPH_{62-77} ([Des-Tyr^1]-γ-endorphin; DTγE). *Eur. J. Pharmacol. 49*:427-436.
31. De Wied, D., Van Ree, J. M., Greven, H. M. 1980. Neuroleptic-like activity of peptides related to (Des-Tyr^1)-γ-endorphin: structure activity studies. *Life Sci. 26*:1575-1579.
32. De Wied, D., Jolles, J. 1982. Neuropeptides derived from pro-opiocortin: Behavioral, physiological and neurochemical effects. *Physiol. Rev. 62*:976-1059.
33. Dorsa, D. M., Van Ree, J. M. 1979. Modulation of substantia nigra self-stimulation by neuropeptides related to neurohypophyseal hormones. *Brain Res. 172*:367-371.
34. Dorsa, D. M., Van Ree J. M., De Wied, D. 1979. Effects of [Des-Tyr^1]-γ-endorphin and α-endorphin on substantia nigra self-stimulation. *Pharmacol. Biochem. Behav. 10*:899-905.
35. Ehrensing, R. H., Michell, G. F., Baker, R. P. 1982. Vasopressin's effects on acquisition and extinction of conditioned avoidance response to smoking. *Peptides 3*:527-530.
36. Esposito, R., Kornetsky, C. 1977. Morphine lowering of self-stimulation thresholds: Lack of tolerance with long-term administration. *Science 195*:189-191.
37. Fekete, M., Bohus, B., Van Wolfswinkel, L., Van Ree, J. M., De Wied, D. 1982. Comparative effects of the ACTH-(4-9) analog (Org 2766), ACTH-(4-10) and [D-Phe^7]ACTH-(4-10) on medial septal self-stimulation behavior in rats. *Neuropharmacology 21*:909-916.
38. Fekete, M., De Wied, D. 1982. Naltrexone-insensitive facilitation and naltrexone-sensitive inhibition of passive avoidance behavior of the ACTH-(4-9) analog (ORG 2766) are located in two different parts of the molecule. *Eur. J. Pharmacol. 81*:441-448.
39. Fekete, M., Drago, F., Van Ree, J. M., Bohus, B., Wiegant, V. W., De Wied, D. 1983. Naltrexone-sensitive behavioral actions of the ACTH 4-9 analog (ORG 2766). *Life Sci. 32*:2193-2204.
40. Ferrier, B. M., Kennett, D. J., Devlin, M. C. 1980. Influence of oxytocin on human memory processes. *Life Sci. 27*:2311-2317.
41. File, S. E. 1979. Effects of $ACTH_{4-10}$ in the social interaction tests of anxiety. *Brain Res. 171*:157-160.
42. File, S. E. 1981. Contrasting effects of ORG 2766 and α-MSH on social and exploratory behavior in the rat. *Peptides 2*:255-260.
43. Finkelberg, F., Kalant, H., LeBlanc, A. E. 1978. Effect of vasopressin-like peptides on consumption of ethanol by the rat. *Pharmacol. Biochem. Behav. 9*:453-458.

44. Fraenkel, H. M., Van Beek-Verbeek, G., Fabriek, A. J., Van Ree, J. M. 1983. Desglycinamide^9-arginine^8-vasopressin and ambulant methadone-detoxification of heroin addicts. *Alcohol Alcoholism 10*:331-335.
45. Gaillard, A. W. K. 1981. ACTH analogs and human performance, in *Endogenous Peptides and Learning and Memory Processes* (Martinez, J. L., Jensen, R. A., Messing, R. B., Rigter, H., McGaugh, J., eds.), Academic, New York, pp. 181-196.
46. Gallistel, C. R. 1966. Motivating effects in self-stimulation. *J. Comp. Physiol. Psychol. 62*:95-101.
47. Genazzani, A. R., Nappi, G., Facchinetti, F., Mazzella, G. L., Parrini, D., Sinforiani, E., Petraglia, F., Savoldi, F. 1982. Central deficiency of β-endorphin in alcohol addicts. *Clin. Endocrinol. Metab. 55*:583-586.
48. Gispen, W. H., Wiegant, V. M., Bradbury, A. F., Hulme, E. C., Smuth, D. G., Snell, C. R., De Wied, D. 1976. Induction of excessive grooming in the rat by fragments of lipotropin. *Nature 264*:794-795.
49. Gispen, W. H., Isaacson, R. L. 1981. ACTH-induced grooming in the rat. *Pharmacol. Ther. 12*:209-246.
50. Greven, H. M., De Wied, D. 1977. Influence of peptides structurally related to ACTH and MSH on active avoidance behaviour in rats. A structure-activity relationship study, in *Frontiers of Hormone Research*, Vol. 4 (Tilders, F. J. H., Swaab, D. F., and Van Wimersma Greidanus, T. B., eds.), S. Karger, Basel, pp. 140-152.
51. Guillemin, R., Ling, N., Burgus, R. 1976. Endorphins. Hypothalamic and neurohypophyseal peptides with morphinomimetic activity. Isolation and primary structure of α-endorphin. *C.R. Acad. Sci. Paris Ser. D 282*:783-785.
52. Holaday, J. W., Loh, H. H., Li, C. H. 1978. Unique behavioral effects of β-endorphin and their relationship to thermoregulation and hypothalamic function. *Life Sci. 22*:1525-1536.
53. Hughes, J., Smith, T. W., Kosterlitz, H. W., Fothergill, L. A., Morgan, B. A., Morris, H. R. 1975. Identification of two related pentapeptides from the brain with potent opiate agonist activity. *Nature 258*:577-579.
54. Jolles, J. 1983. Vasopressin-like peptides and the treatment of memory disorders in man. *Prog. Brain Res. 60*:169-182.
55. Jouhaneau-Bowers, M., Le Magnen, J. 1979. ACTH self-administration in rats. *Pharmacol. Biochem. Behav. 10*:325-328.
56. Kakidani, H., Furutani, Y., Takahashi, H., Noda, M., Morimoto, Y., Hirose, T., Asai, M., Inayama, S., Nakanishi, S., Numa, S. 1982. Cloning and sequence analysis of cDNA for porcine β-neo-endorphin/dynorphin precursor. *Nature 298*:245-249.

57. Kalant, H., Engel, J. A., Goldberg, L., Griffiths, R. R., Jaffe, J. H., Krasnegor, N. A., Mello, N. K., Mendelsohn, J. H., Thompson, T., Van Ree, J. M. 1978. Behavioral aspects of addiction—Group report, in *The Bases of Addiction*, Report of the Dahlem Workshop on the Bases of Addiction, Berlin 1977, September 26-30 (Fishman, J., ed.), Dahlem Konferenze, Berlin, pp. 463-496.
58. Karras, A., Kane, J. M. 1980. Naloxone reduces cigarette smoking. *Life Sci.* *27*:1541-1545.
59. Kastin, A. J., Sandman, C. A., Stratton, L. O., Schally, A. V., Miller, L. H. 1975. Behavioral and electrographic changes in rat and man after MSH. *Prog. Brain Res.* *42*:143-150.
60. Katz, R. J. 1980. Effects of an $ACTH_{4-9}$ related peptide upon intracranial self stimulation and general activity in the rat. *Psychopharmacology* *71*:67-70.
61. Laczi, F., Van Ree, J. M., Wagner, A., Valkusz, Zs., Járdánházy, T., Kovács, G. L., Telegdy, G., Szilárd, J., László, F. A., De Wied, D. 1983. Effects of desglycinamide-arginine-vasopressin (DG-AVP) on memory processes in diabetes insipidus patients and non-diabetic subjects. *Acta Endocrinol.* *102*:205-212.
62. Legros, J. J., Gilot, P. 1979. Vasopressin and memory in the human, in *Brain Peptide: A New Endocrinology* (Gotto, Jr., A. M., Peck, Jr., E. J., Boyd III, A. E., eds.), Elsevier/North-Holland, Amsterdam, pp. 347-363.
63. Li, C. H., Chung, D. 1976. Isolation and structure of an untriakontapeptide with opiate activity from camel pituitary glands. *Proc. Nat. Acad. Sci. U.S.A.* *73*:1145-1148.
64. Mains, R., Eipper, B. A., Ling, N. 1977. Common precursor to corticotropins and endorphins. *Proc. Nat. Acad. Sci. U.S.A.* *74*: 3014-3018.
65. Marcus, R., Kornetsky, C. 1974. Negative and positive intracranial reinforcement thresholds: effects of morphine. *Psychopharmacologia* *38*:1-13.
66. Maroli, A. N., Tsang, W. K., Stutz, R. M. 1978. Morphine and self-stimulation: Evidence for action on a common neural substrate. *Pharmacol. Biochem. Behav.* *8*:119-123.
67. Mello, N. K., Mendelson, J. H. 1978. Self-administration of an enkephalin analog by rhesus monkey. *Pharmacol. Biochem. Behav.* *9*:579-586.
68. Mello, N. K., Mendelson, J. H. 1979. Effects of the neuropeptide DG-AVP on morphine and food self-administration by dependent rhesus monkey. *Pharmacol. Biochem. Behav.* *10*:415-419.
69. Mucha, R. F., Kalant, H. 1979. Effects of desglycinamide$^{(9)}$-lysine$^{(8)}$-vasopressin and prolyl-leucyl-glycinamide on oral ethanol intake in the rat. *Pharmacol. Biochem. Behav.* *10*:229-234.

70. Nakanishi, S., Inoue, A., Kita, T., Nakamura, M., Chang, A. C., Cohen, S. N., Numa, S. 1979. Nucleotide sequence of cloned cDNA for bovine corticotropin-β-lipotropin precursor. *Nature* *278*:423-427.
71. Niesink, R. J. M., Van Ree, J. M. 1984. Analysis of the facilitatory effect of the ACTH 4-9 analog ORG 2766 on active social contact in rats. *Life Sci.* *34*:961-970.
72. Noda, M., Furutani, Y., Takahashi, H., Toyosato, M., Hirose, T., Inayama, S., Nakanishi, S., Numa, S. 1982. Cloning and sequence analysis of cDNA for bovine adrenal proenkephalin. *Nature* *295*:202-206.
73. Nyakas, C., Bohus, B., De Wied, D. 1980. Effects of $ACTH_{4-10}$ on self-stimulation behavior in the rat. *Physiol. Behav.* *24*:759-764.
74. O'Donohue, T. L., Handelmann, G. E., Loh, Y. P., Olton, D. S., Leibowitz, J., Jacobowitz, D. M. 1981. Comparison of biological and behavioral activities of alpha- and gamma-melanocyte stimulating hormones. *Peptides* *2*:101-104.
75. Olds, M. E., Williams, K. N. 1980. Self-administration of D-Ala^2-met-enkephalinamide at hypothalamic self-stimulation sites. *Brain Res.* *194*:155-170.
76. Orth, D. N., Nickolson, W. E. 1977. Different molecular forms of ACTH. *Ann. N.Y. Acad. Sci.* *297*:27-46.
77. Pert, C. B., Snyder, S. H. 1973. Properties of opiate-receptor binding in rat brain. *Proc. Nat. Acad. Sci. U.S.A.* *70*:2243-2247.
78. Pigache, R. M., Rigter, H. 1981. Effects of peptides related to ACTH on mood and vigilance in man, in *Frontiers of Hormone Research*, Vol. 8 (Van Wimersma Greidanus, T. B., Rees, L. H., eds.), Karger, Basel, pp. 193-207.
79. Rigter, H., Crabbe, J. C. 1980. Neurohypophysial peptides and ethanol, in *Hormones and the Brain* (De Wied, D., Van Keep, P. A., eds.), MTP Press, Lancaster, pp. 263-275.
80. Roemer, D., Buescher, H. H., Hill, R. C., Pless, J., Bauer, W., Cardinaux, F., Closse, A., Hauser, D., Huguenin, R. 1977. A synthetic enkephalin analogue with prolonged parenteral and oral analgesic activity. *Nature* *268*:547-549.
81. Schuster, C. R., Thompson, T. 1969. Self-administration of and behavioral dependence on drugs. *Annu. Rev. Pharmacol.* *9*:483-502.
82. Schwarzberg, H., Hartmann, G., Kovács, G. L., Telegdy, G. 1976. Effect of intraventricular oxytocin and vasopressin on self-stimulation in rats. *Acta Physiol. Acad. Sci. Hung.* *47*:127-131.
83. Simon, E. J., Hiller, J. M., Edelman, I. 1973. Stereospecific binding of the potent narcotic analgesic (^{3}H)etorphine to rat brain homogenate. *Proc. Nat. Acad. Sci. U.S.A.* *70*:1947-1949.

84. Skinner, B. F. 1938. *The Behavior of Organisms*, Appleton-Century-Crofts, New York.
85. Szabó, G., Kovács, G. L., Telegdy, G. 1983. The effect of oxytocin and an oxytocin fragment (prolyl-leucyl-glycinamide) on the development of ethanol tolerance. *Acta Endocrinol. Suppl. 256*:242.
86. Terenius, L. 1973. Characteristics of the "receptor" for narcotic analgesics in synaptic plasma membrane fraction from rat brain. *Acta Pharmacol. Toxicol. 33*:377-384.
87. Thompson, T., Pickens, R. 1969. Drug self-administration and conditioning, in *Scientific Basis of Drug Dependence* (Steinberg, H., ed.), Churchill, London, pp. 177-198.
88. Tortella, F. C., Moreton, J. E. 1980. D-Ala^2-Methionine-enkephalinamide self-administration in the morphine-dependent rat. *Psychopharmacology 69*:143-147.
89. Van Beek-Verbeek, G., Fraenkel, M., Geerlings, P. J., Van Ree, J. M., De Wied, D. 1979. Des-glycinamide-arginine-vasopressin in methadone detoxification of heroin addicts. *Lancet 2*:738-739.
90. Van Beek-Verbeek, G., Fraenkel, H. M., Van Ree, J. M. 1983. Des-Gly^9-[Arg^8]-vasopressin may facilitate methadone detoxification of heroin addicts. *Substance Alcohol Actions/Misuse 4*:375-382.
91. Van Ree, J. M. 1979. Reinforcing stimulus properties of drugs. *Neuropharmacology 18*:963-969.
92. Van Ree, J. M. 1982. Neurohypophyseal hormones and addiction, in *Advances in Pharmacology and Therapeutics II*, Vol. 1 (Yoshida, H., Hagihara, Y., Ebashi, S., eds.), Pergamon, Oxford, pp. 199-209.
93. Van Ree, J. M. 1983. The influence of neuropeptides related to pro-opiomelanocortin on acquisition of heroin self-administration of rats. *Life Sci. 33*:2283-2289.
94. Van Ree, J. M., Slangen, J. L., De Wied, D. 1974. Self-administration of narcotic drugs in rats: Dose-response studies, in *Neuropsychopharmacology: Proceedings of the IXth Congress of the Collegium International Neuropsychopharmacology*, Paris, July 7-12, 1974, Excerpta Medica International Congress Series, No. 359, Excerpta Medica, Amsterdam, pp. 231-239.
95. Van Ree, J. M., De Wied, D. 1976. Prolyl-leucyl-glycinamide (PLG) facilitates morphine dependence. *Life Sci. 19*:1331-1340.
96. Van Ree, J. M., De Wied, D., Bradbury, A. F., Hulme, E. C., Smyth, D. G., Snell, C. R. 1976. Induction of tolerance to the analgesic action of lipotropin C-fragment. *Nature 264*:792-794.
97. Van Ree, J. M., De Wied, D. 1977. Modulation of heroin self-administration by neurohypophyseal principles. *Eur. J. Pharmacol. 43*:199-202.
98. Van Ree, J. M., De Wied, D. 1977. Heroin self-administration is under control of vasopressin. *Life Sci. 21*:315-320.

99. Van Ree, J. M., Bohus, B., Versteeg, D. H. G., De Wied, D. 1978. Neurohypophyseal principles and memory processes. *Biochem. Pharmacol.* *27*:1793-1800.
100. Van Ree, J. M., Slangen, J. L., De Wied, D. 1978. Intravenous self-administration of drugs in rats. *J. Pharmacol. Exp. Ther.* *204*:547-557.
101. Van Ree, J. M., Witter, A., Leysen, J. E. 1978. Interaction of des-tyrosine-γ-endorphin (DTγE, β-LPH_{62-77}) with neuroleptic binding sites in various areas of rat brain. *Eur. J. Pharmacol.* *52*:411-413.
102. Van Ree, J. M., Smyth, D. G., Colpaert, F. 1979. Dependence creating properties of lipotropin C-fragment (β-endorphin): Evidence for its internal control of behavior. *Life Sci.* *24*:495-502.
103. Van Ree, J. M., Bohus, B., De Wied, D. 1980. Similarity between behavioral effects of des-tyrosine-γ-endorphin and haloperidol and of α-endorphin and amphetamine, in *Endogenous and Exogenous Opiate Agonists and Antagonists* (Leong Way, E., ed.), Pergamon, New York, pp. 459-462.
104. Van Ree, J. M., De Wied, D. 1980. Involvement of neurohypophyseal peptides in drug-mediated adaptive responses. *Pharmacol. Biochem. Behav. Suppl.* *1*:257-263.
105. Van Ree, J. M., Otte, A. P. 1980. Effects of (Des-Tyr^1)-γ-endorphin and α-endorphin as compared to haloperidol and amphetamine on nucleus accumbens self-stimulation. *Neuropharmacology* *19*:429-434.
106. Van Ree, J. M., Bohus, B., Csontos, K. M., Gispen, W. H., Greven, H. M., Nijkamp, F. P., Opmeer, F. A., De Rotte, A. A., Van Wimersma Greidanus, T. B., Witter, A., De Wied, D. 1981. Behavioral profile of γ-MSH: Relationship with ACTH and β-endorphin action. *Life Sci.* *28*:2875-2888.
107. Van Ree, J. M., De Wied, D. 1981. Vasopressin, oxytocin and dependence on opiates, in *Endogenous Peptides and Learning and Memory Processes* (Martinez, J. L., Jensen, R. A., Messing, R. B., Rigter, H., McGaugh, J. L., eds.), Academic, New York, pp. 397-411.
108. Van Ree, J. M., Caffé, A. M., Wolterink, G. 1982. Non-opiate β-endorphin fragments and dopamine. III. γ-Type endorphins and various neuroleptics counteract the hypoactivity elicited by injection of apomorphine into the nucleus accumbens. *Neuropharmacology* *21*:1111-1117.
109. Van Ree, J. M., De Wied, D. 1982. Neuroleptic-like profile of γ-type endorphins as related to schizophrenia. *Trends Pharmacol. Sci.* *3*:358-361.
110. Van Ree, J. M., Verhoeven, W. M. A., De Wied, D., Van Praag, H. M. 1982. The use of the synthetic peptides γ-type endorphins

in mentally ill patients. *Ann. N.Y. Acad. Sci. 398*:478-495.

111. Van Wimersma Greidanus, T. B., Dogterom, J., De Wied, D. 1975. Intraventricular administration of anti-vasopressin serum inhibits memory consolidation in rats. *Life Sci. 16*:637-644.
112. Van Wolfswinkel, L., Van Ree, J. M. 1982. Effects of morphine and naloxone on ventral tegmental electrical self-stimulation, in *Drug Discrimination: Application in CNS Pharmacology* (Colpaert, F. C., Slangen, J. L., eds.), Elsevier, Amsterdam, pp. 391-397.
113. Verhoeven, W. M. A., Van Praag, H. M., Van Ree, J. M., De Wied, D. 1979. Improvement of schizophrenic patients treated with [Des-Tyr^1]-γ-endorphin (DTγE). *Arch. Gen. Psychiatry 36*:294-298.
114. Watson, S. J., Akil, H., Berger, P. A., Barchas, J. D. 1979. Some observations on the opiate peptides and schizophrenia. *Arch. Gen. Psychiatry 36*:35-41.
115. Watson, S. J., Akil, H., Ghazarossian, V. E., Goldstein, A. 1981. Dynorphin immunocytochemical localization in brain and peripheral nervous system: Preliminary studies. *Proc. Nat. Acad. Sci. U.S.A. 78*:1260-1263.
116. Wauquier, A., Rolls, E. T., eds. 1976. *Brain-Stimulation Reward*, Elsevier/North-Holland, Amsterdam.
117. Wei, E., Loh, H. 1976. Physical dependence on opiate-like peptides. *Science 193*:1262-1263.
118. Wise, R. A. 1978. Catecholamine theories of reward: A critical review. *Brain Res. 152*:215-247.
119. Wise, R. A. 1980. Action of drugs on abuse of brain reward systems. *Pharmacol. Biochem. Behav. Suppl. 1*:213-223.
120. Wood, P. L. 1982. Multiple opiate receptors: Support for unique mu, delta and kappa sites. *Neuropharmacology 21*:487-497.
121. Woods, J. H., Schuster, C. R. 1971. Opiates as reinforcing stimuli, in *Stimulus Properties of Drugs* (Thompson, T., Pickens, R., eds.), Appleton-Century-Crofts, New York, pp. 162-175.
122. Young, A. M., Woods, J. H., Herling, S., Hein, D. W. 1983. Comparison of the reinforcing and discriminative stimulus properties of opioids and opioid peptides, in *The Neurobiology of Opiate Reward Processes* (Smith, J. E., Lane, J. D., eds.), Elsevier, Amsterdam, pp. 147-174.

8

Developmental Neurotoxicology of Environmental and Industrial Agents

B. K. Nelson
Division of Biomedical and Behavioral Science, National Institute for Occupational Safety and Health, U.S. Department of Health and Human Services, Cincinnati, Ohio

I. INTRODUCTION

Life forms and processes are often more susceptible to toxic action when they are in a developmental stage than when they have reached maturity (66,67). Consequently, developmental neurotoxicology, encompassing neurotoxic effects resulting from pre- and perinatal exposures to toxic agents, is of increasing importance in the overall field of neurotoxicology. This chapter reviews what is known of the developmental neurotoxicity of industrial and environmental agents based on animal research. Included herein are agents to which animals have been exposed during prenatal or early postnatal development; subsequently, they were examined for behavioral or other postnatal effects.

A. Complexity of Prenatal Development

A more intricate meshing of specific biochemical potentialities and precision timing than occurs in prenatal development is difficult to imagine. That a single fertilized cell can differentiate into a complete organism possessing countless components and abilities is nothing short of amazing! Continuous activation and repression of specific genetic programming at critical developmental stages are vital. As development proceeds, individual cells and certain regions of embryonic mass are destined to become specific organs. Though not yet well understood, this process is modulated through genetic nuclear components, with precision timing of critical importance. This induction must occur in appropriate sequence relative to surrounding events or that particular organ will likely be abnormal. Once a particular mass of cells is induced to a particular fate (i.e., destined to become a particular organ), it loses competence to form a different organ. Consequently the specificity of fates beginning early in development reaches its apex during organogenesis. Following a zygote through the processes of cleavage, gastrulation, organogenesis, growth and histological differentiation, and maturation, one is awed, not with the occasional flaws in development, but with the *lack* of errors. Using even the most liberal estimates of defects in development, some 90% of human births are completely normal; the majority of the 10% that are defective exhibit postnatal functional deficits, the subject of this chapter (67).

B. Historical Perspective

In retrospect, it is somewhat surprising that interest in human birth defects, at least sufficient interest to develop the area into a science (teratology), has been quite recent.* The evolution of teratological testing (119) has focused primarily on the testing of drugs, but as recently as 25 years ago, only effects on fertility and on particular sex organs of the parental animals were observed. Following the thalidomide tragedy of the early 1960s (116), the need for better testing in teratology became apparent.

This resulted in a closer scrutiny of the offspring, including internal examination for soft tissue and skeletal malformations in teratological research. Contemporary with this, Werboff and Gottlieb initiated a series of studies in which the *functional* capacity of the offspring was tested (117,118). However, after the genesis of functional teratology testing in the early 1960s, there was little interest in this new discipline until the early 1970s.

While the concept of the teratological effects mediated by exposure of pregnant females is well established, research on the effects of *paternal* exposures to environmental agents on subsequent offspring functional development is more limited. In one of the few studies

---

*The history of teratology is fascinating, as reviewed by Warkany (116). [Indeed, the very life of this great man is a history of teratology (10).]

completed, lead was found to produce behavioral deviations in offspring after paternal exposure prior to mating that were similar to those found after maternal exposure during pregnancy (8). In another investigation, injection of male mice with morphine prior to mating was shown to alter offspring behavior (34). More recently, cyclophosphamide has been demonstrated to alter offspring behavior after treatment of males prior to mating (2). Also, inhalation exposure of male rats, which were then mated with unexposed females, to 25 ppm 2-methoxyethanol for 7 hr/day and 7 days/week for 6 weeks resulted in offspring exhibiting a number of neurochemical deviations, though no behavioral changes were detected (73). Whereas additional studies are urgently needed to confirm such effects with other agents, it appears established that such transfer can take place in experimental animals, and raises the possibility of such effects in humans as well.

A point of continuing concern is our ability to provide risk assessment in humans based on animal research. In order to have a solid basis for such assessment, one would hope to have some degree of consistency in effects: (a) after doses which can conceivably be encountered by man, (b) after and along environmentally relevant routes of exposure, (c) across species (not that the same type of effect should be observed across species, but that some adverse effects are noted in more than one species, with several species tested), (d) in the *pattern* of effects seen (within the context of the species being tested and the type and number of observations being made; a sporadic deviation from controls may suggest the need for further testing, but would not justify great concern), and (e) which persist for an extended period of time, as into adulthood (that is, transient effects which are not life threatening would not be of as great a concern as permanent effects).

Other items could be added to this list, but the above are sufficient to illustrate the point that in nearly all cases of agents reviewed in this chapter, the data base is sufficiently weak that risk assessment for humans is tenuous at best. In many cases, there has been only one study, with a single species and limited dose levels. Bearing in mind that there is no ideal model test system, one sees important methodological differences between studies even for the best-investigated agents (e.g., different species, with accompanying differences in central nervous system ontogeny and pharmacokinetic patterns; differences in dose, duration, and timing of exposure; differences in sex and ages of animals examined, as well as in methodologies and types of postnatal evaluations utilized). Hence it is not surprising that there are often inconsistent effects reported in animal studies. For example, methyl mercury has been extensively studied in experimental animals, and has shown developmental neurotoxicity in humans; but, as can be observed in Table 1, the animal results are not at all consistent. There are many tests which have not shown differences from controls. Consequently, extrapolation of animal results reviewed in this chapter to possible effects in humans can only be made with caution.

C. What Is Normal?

Primarily as a result of the events surrounding the thalidomide tragedy, there emerged more clearly the need to define normality, and investigators began to appreciate that there was no simple line separating normal from abnormal. As investigators from different backgrounds and disciplines became interested in teratology, the "gray area" separating normal from abnormal began expanding. As illustrated in Figure 1, scientists in the pre-thalidomide era might have considered offspring normal if their outward appearance was reasonably typical of others of their species and they could survive and reproduce. However, as teratologists looked more closely at the offspring, different levels of examination began to reveal internal or functional abnormalities which were missed by other means of examination.

One subtle point that relates to the concept of normality as it is used in this chapter is the direction of changes that may be observed. Most changes from normal can be in either direction (e.g., an animal can be either *hypo*active or *hyper*active). The direction of change is unimportant; that is, a deviation in either direction is equivalent and indicates that the treatment has produced some deviation from normal. Reverting to an example from classic teratology, an agent that induced the development of extra ribs (or digits) would be classified as a teratogen, just as would be an agent which induced the absence of ribs (or digits). Thus, when preparing this chapter, I viewed changes from control in either direction as deviations, indicating that the treatment has produced some effect.

D. Developmental Neurotoxicology

As with its parent disciplines of teratology and neurotoxicology, developmental neurotoxicology initially focused on the testing of drugs (21,22,45,114). Its development can be followed by examining several

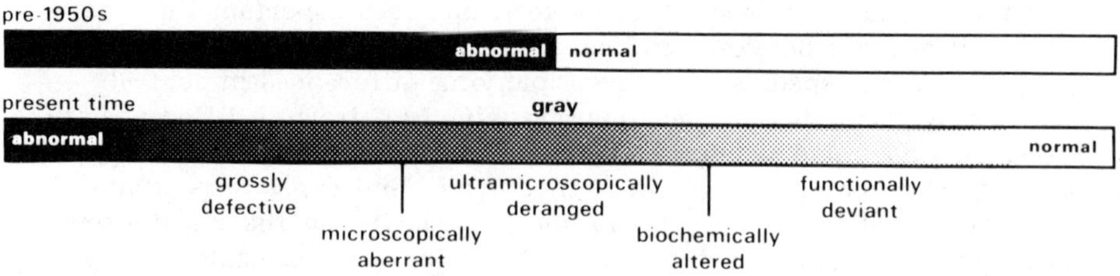

Figure 1 Though specifically applied to teratology and the increasing recognition of additional types of alterations which can be detected in offspring after a developmental insult, this somewhat simplified illustration emphasizes, more generally, the ever-increasing "gray area" separating normal from abnormal.

reviews (4,35,48,53,66,100,115,117,123), with more recent discussion centering on the choice of which behavioral tests are most appropriate (1,13,15,67,68,84,124; see also Chapter 30 by Annau). As the discipline has expanded, investigators from different backgrounds have become interested, and there has been a branching from perdominantly behavioral testing (e.g., sensory, activity, and learing measures) to other areas of postnatal evaluation. These include neurochemical assays, morphological (including neuropathological) examinations, and immune system evaluations, among the more important ones.

Second to behavioral testing, neurochemical teratology has historically received more attention than the other areas. Typically, either whole brain samples or general brain regions have been examined for steady-state levels of classic putative neurotransmitters (or their accompanying enzymes) at limited developmental periods; there is need for more temporal studies of a wider variety of transmitters, including turnover rates and receptor binding characteristics in discrete regions of the brain. Once again, most work has involved the testing of drugs (58,61,65,80,113,125), and several reviews have been published (66,94,100,112). There have been even fewer studies in which offspring were examined for microscopic or ultramicrospic pathology after developmental exposures. However, the majority of this research has involved the testing of environmental agents, particularly heavy metals (18). Such studies have been reported for methyl mercury (17,82,83), lead (54), radiation (9,42,92,121), hyperthermia (31,60), asphyxia (64), and halothane (77). Evaluations of the immune system and the few other tests of functional integrity have rarely been included in research on environmental contaminants. Consequently these will not be discussed here, but notation of their effect is included as appropriate in the tables in this chapter.

II. ENVIRONMENTAL AND INDUSTRIAL AGENTS EXAMINED FOR DEVELOPMENTAL NEUROTOXICITY

Owing to the recent birth of developmental neurotoxicology as a science, and with its infancy focusing on the testing of drugs, it should be no surprise that only a few industrial and environmental agents have been tested for such effects. Following are summaries of studies reported in the literature on the postnatal effects in animals after exposure during prenatal or early postnatal development to industrial chemical (Table 1) and physical (Table 2) agents. The tabular data are categorized by commonly accepted classes of chemical and physical agents, but with the recognition that the classes are not completely parallel (e.g., "heavy metals" is a chemical class, but "solvents" is more of a usage class). Included in the tabular presentation are the agent, species tested, dose and route of administration, time of exposure, effects under investigation, whether or not such effects were

Table 1 Chemical Agents for Which Developmental Exposure Has Been Evaluated for Behavioral Teratogenicity

	Species	Dose	Time[a] (days)	Time and effect examined for[b]	Effects reported[c]	References
A. Heavy metals 1. Mercury (organic)	Mice	1-10 mg/kg, gavage	g, 8	8 weeks: open field	No	43
				Two-way avoidance	-	
				Conditioned suppression	No	
				"Punished behavior"	No (+ on extinction)	
				Water escape	No	
				Body weight	- (not at 8 weeks)	
	Rats	0.05-5 mg/kg, gavage	g, 0, 7, or 14	26 weeks: operant (base line)	No	44
				Amphetamine challenge	-	
	Rats	0.05-5 mg/kg, gavage	g, 6-9	12 weeks: operant (differential reinforcement of high rates)	- (dose dependent	63
	Mice	6-12 mg/kg, subcutaneous	g, 10	3,5 weeks: activity (open field)	+ (both ages)	105
				3,6,9 weeks: activity (photocell)	- (all ages)	
				10 weeks: "convulsive behavior" threshold	-	
				Body weight	-	
	Mice	8 mg/kg, intraperitoneal	g, 7 or 9	Week 4: open field	-	103
				Week 5: swimming	-	
				Week 9: choline acetyl-transferase	No	
				Cholinesterase	No	
	Rats	2.5 mg/kg, drinking water	g, 1-21; L, 1-21	Weeks 4,5: water escape T maze	-	126

	Rats	0.1-2.5 mg/kg, gavage	g, 6-15	Weeks 1-4: surface righting	No	98
				Inclined screen	+	
				Body weight	- (2.5 mg/kg)	
				Week 4: cholinesterase	- (at or above 0.5 mg/kg)	
				5HT	- (2.5 mg/kg)	
				5HIAA	No	
				NE	- (0.5, but not 2.5 mg/kg)	
				DA	No (regional differences observed)	
	Mice	8 mg/kg, injection	g, 9	55-60 weeks: electron microscopy (cerebellum)	Yes	17
				Body weight	No	
	Mice	0.001-5 mg/kg,	g, 6-17	Weeks 0-4: light microscopy	Yes (transient at 1 mg/kg; apparently not present at 5 mg/kg)	51
				Body weight	No	
	Hamster	2-10 mg/kg, oral	g, 10 or 10-15	Weeks 0-4: light microscopy	No	83
				Electron microscopy	Yes	
				Week 40: light microscopy	Yes	
				Electron microscopy	Yes	
2. Lead	Mice	0.13%, drinking water	All g + L, + post	4 weeks: social behavior	-	28
				Activity (jiggle cage)	No (prenatal)	
				15 weeks: social behavior	+	
				Activity (jiggle cage)	No (postnatal)	
				3,13 weeks: body weight	- (postnatal)	
	Rats	0.5-250 ppm, diet	pre-g, all g + L, + 6-9 months post	Begun at birth:		37
				physical landmarks	-	
				Week 1: surface righting	-	
				Weeks 2-4: air righting	-	
				Weeks 1,2: "locomotor development"	No	

Table 1 (Continued)

Species	Dose	Time[a] (days)	Time and effect examined for[b]	Effects reported[c]	References
			Week 1-3: auditory startle	No	
			Weeks 2-4: visual placing	No	
			Weeks 3-4: activity (open field)	No	
			Weeks 3,4,5,9,14: Activity (photocell)	No	
			Weeks 8-12: Activity (running wheel)	No	
			Week 6: rotorod	No	
			Body weight	-	
Rats	25-200 mg/kg, gavage	p, 3-30	Spatial alternation	No	62
Rats	0.02-2%, drinking water	All g + L	Week 1: negative geotaxis	+ (?)	75
			Week 2: forepaw suspension	No	
			Auditory startle	-	
			Air righting	-	
			Weeks 1-3: body weight	-	
Rats	9-81 mg/kg, gavage	p, 2-21	4 weeks: activity (photocell)	No	97
			4-8 weeks: passive avoidance	No	
			One-way avoidance	No	
			20 weeks: two-way avoidance	-	
			8 weeks: lever press avoidance/escape	- (reversal phase)	
			Weeks 3-4:		
			AchE	- (?)	
			butyrocholinesterase	- (?)	
			NE	No	

				DA	No	
				5HT	No	
				5HIAA	No	
				Body weight	No	
	Rats	1.38 g/kg, diet	Pre, all g + L	Weeks 13,16,20,24: activity (open field)	-	120
				Lashley maze discrimination	-	
				Body weight	+	
	Rats	1.0 g/kg, drinking water	Pre, all g + L	5-7 weeks: T-maze learning	-	127
				Body weight	- (not at 50 days)	
	Rats	109 ppm, drinking water	All L	Weeks 3,4,13: activity (photocell)	No	78
				Week 14: DA	No	
				NE	No	
				Week 13: activity (photocell, in response to amphetamine challenge)	+	
				DA (after challenge)	No	
				NE (after challenge)	-	
	Rats	4%, drinking water	All L	Week 4: myelination	Yes	54
				Morphometric analyses	Yes	
				Light microscopic	Yes	
				Electron microscopic	Yes	
	Rats	25, 50 ppm, drinking water	Pre, all g + L, post until assay	Week 8: antibody levels	-	59
				Immunoglobin levels	-	
				Light microscopy	No	
				Body weight	No	
3. Cadmium	Rats	4.3-34 μg/ml, drinking water	Pre, all g	5-9 weeks: activity (running wheel)	+ (-17 μg/ml)	20
				30 weeks: visual discrimination	No	
				Body weight	No	

Table 1 (Continued)

	Species	Dose	Time[a] (days)	Time and effect examined for[b]	Effects reported[c]	References
	Rats	17.2 μg/ml, drinking water	Pre, all g	5-10 weeks: activity (running wheel)	-	40
				19 weeks: spatial discrimination	No	
				Body weight	- (birth only)	
	Rats	1-100 μg/100 g, gavage	p, 1-30	4 weeks: activity (photocell)	+	79
				Tyrosine hydroxylase activity	+ striatum (10 μg)	
				4 weeks: tryptophan hydroxylase	+ midbrain (10 μg)	
				MAO	No (midbrain)	
				COMT	- (midbrain)	
				TRP	No (100 μg)	
				NE	No (10 μg) + (100 μg)	
				DA	No (10 μg) + (100 μg)	
				5HT	No (10 μg) + (100 μg)	
				5HIAA	+ (100 μg) (differential effects by brain region)	
				Body weight	- (100 μg)	
4. Tin (organic)	Rats	1.5-3 mg/kg, subcutaneous	p, 5	Weeks 3,4,9,13:		39
				Forelimb grip strength	No	
				Hindlimb grip strength	No	
				Negative geotaxis	-	
				Auditory startle	-	
				Activity (photocell)	+ (- at week 13)	
				Body weight	No	
				Week 9: two-way avoidance conditioning	-	

	Rats	3-12 mg/kg, intraperitoneal	p, 5	Weeks 1-2: negative geotaxis	No	81
				Placing reaction	No	
				Ascending on wire mesh	No	
				Auditory startle	No	
				Air righting	No	
				Weeks 2-3: descent on rope	-	
				Weeks 1-4, 16: activity (open field)	No	
				Activity (figure 8 maze)	+	
				Weeks 1-50: body weight	- (until weaning)	
	Rats	3-9 mg/kg, intraperitoneal	p, 5	Week 9: visual evoked potential	-	30
				Hippocampal after discharges	No	
				Kindling	-	
				Seizure induction	No	
5. Manganese	Rats	1-20 μg-g, gavage	p, 1-24	4 weeks: 5HT	+ (hypothalmus)	26
					No (corpus striatum)	
				AChE	No (hypothalmus)	
					+ (corpus striatum); (all after 20 μg/g)	
	Rats	10-20 μg/g, gavage	p, 1-24	4 weeks: DA	+ (hypothalmus)	27
					No (corpus striatum)	
				NE	- (hypothalmus)	
					No (corpus striatum)	
				Tyrosine hydroxylase activity	- (hypothalmus)	
					No (corpus striatum)	
				MAO activity	+ (hypothalmus)	
					No (corpus striatum)	
	Rats	15 mg/kg, gavage	All L	2 weeks: succinic dehydrogenase	-	88
				Adenosine triphosphatase	-	
				Adenosine deaminase	-	

Table 1 (Continued)

	Species	Dose	Time[a] (days)	Time and effect examined for[b]	Effects reported[c]	References
				AChE	-	
				MAO	+	
				4 weeks: same chemicals	Similar effects	
				Microscopic pathology	Yes	
				Body weight	No	
	Guinea pigs	10 mg/kg, intraperitoneal	p, 2-30	5 weeks: TYR	No (- in serum)	90
				TRP	-	
				DA	+	
				NE	No	
				5HT	-	
				HVA	No	
				5HIAA	-	
B. Pesticides 1. Diazinon	Mice	0.18, 9 mg/kg, diet	All g	(Neonate) reflex testing	- (low level, not high)	102
				Week 5: visual cliff	?	
				Week 5: auditory startle	No	
				Week 7: swimming ability	No	
				Week 9: rotorod	No	
				Week 10: inclined plane	-	
				Week 11: open field	No	
				Week 12: Lashley III maze	-	
				Body weight	- (through 3 weeks only)	
	Mice	0.18-9 mg/kg, diet	All g	Weeks 14,57,114: serum immunoglobins	- (female)	5
	Mice	0.18-9 mg/kg, diet	All g	Week 14: corticosterone	+ (at lower dose but not at higher dose)	23
2. Carbofuran	Mice	0.01-0.5 mg/kg, diet	All g	Weeks 14,57,114: serum immunoglobins	+ (male) - (female at lower dose but not at higher dose)	5
				Body weight	-	

	Mice	0.01-5 mg/kg, diet	All g	Week 14: corticosterone	+ (at lower dose but not at higher dose)	23
3. Kepone	Rats	1 mg/pup, subcutaneous	Neonatal	Operant habituation	- (female)	110
				Discrimination	No (but, - in reversal phase)	
4. Chlordane	Mice	0.16-8 mg/kg, diet	All g	Week 14: corticosterone	+ (males at lower dose, but not at higher dose)	23
5. DDE	Mallard duck	3 ppm, diet	3-5 months before mating	Week 1: "approach behavior"	+	41
				"Avoidance behavior"	-	
	Black duck	10 ppm, diet	7 months before mating	Reproductive success	-	55
				Shell thickness	-	
6. 2,4,5-T	Rats	100 mg/kg, gavage	g, 7, 8, or 9	13 weeks: open field	+ (males only)	93
	Chickens	7-225 mg/kg, 75-150 mg/kg	(Egg) day 8 or 15, (neonatal day 2	Week 1: activity (photocell)	+	97
				Visual discrimination	-	
7. Baygon	Rats	1000 ppm, diet	g, 6-15	Week 1: auditory startle	-	86
				Week 2: righting reflex	No	
				Week 3: encephalography	-	
				Visual evoked response	-	
				Weeks 1,2: body weight	-	
8. Carbaryl	Rats	1-5 mg/kg	g, 11-21	Week 1: AchE	No	25
9. Maneb	Rats	0.5-10 ppm, diet	All L	Week 4: activity (open field)	- (10 ppm)	96
				Passive avoidance	-	
				Week 14: lever press avoidance/escape	+ ($\geq$1 ppm)	
				Week 26: cholinesterase	-	
				Body weight	No	

Table 1 (Continued)

	Species	Dose	Time[a] (days)	Time and effect examined for[b]	Effects reported[c]	References
10. Mirex	Rats	1 mg/kg	g, 15-21	Preparturition: "heart blocks"	+	36
				"Regulatory distress"	+	
				Cataracts	+	
				Week 0: electrocardiogram	+	
11. Nitrofen	Rats	75 mg/kg, oral	g, 11	Weeks 0-4: activity (figure 8 maze)	No	50
				Harderian gland	Yes	
				Hydropenia	+	
	Mice	0-200 mg/kg	g, 7-17	Harderian gland	- (⩾50 mg/kg)	38
	Rats	0-25 mg/kg	g, 7-16	Harderian gland	- (⩾12.5 mg/kg)	
	Hamster	0-40 mg/kg	g, 7-11	Harderian gland	- (⩾100 mg/kg)	
C. Solvents 1. Benzene	Rats	550 mg/kg, subcutaneous	p, 9,11,13	Weeks 6,9,14: forelimb grip strength	No	111
				Hindlimb grip strength	No	
				Auditory startle	No	
				Air puff startle	No	
				Negative geotaxis	No	
				Weeks 14-19: activity (photocell)	+	
				Week 23: activity (operant chamber, touch sensors)	No	
				Weeks 0-14: body weight	No	
2. t-Butanol	Mice	0.5-1%, diet	g, 6-20	Week 1: surface righting	+	24
				Negative geotaxis	-	
				Weeks 1-3: rotorod	-	
				Activity (open field)	+	
				Cliff avoidance	-	
				Body weight	-	

3. Carbon disulfide	Rats	17-67 ppm inhalation, 8 hr/day	All g	Weeks 3,4,13: open field	-	106
				Oxygen consumption	+	
				Liver lipid levels	-	
				Body weight	-	
	Rats	0.1-3 ppm, inhalation	All g	Week 1: reflex development (eyelid, pinna)	No	107
				Week 1: cliff avoidance	No	
				Week 2: olfaction (homing response)	No	
				Hearing (startle)	-	
				Week 3: visual placing	-	
				Weeks 2,3: open field	+ (by week 3, no)	
				Week 3: air righting	-	
				Week 3: rotorod	No	
				Week 4: T-maze learning	-	
4. Chloroform	Mice	31.1 mg/kg, gavage	Pre, all g + L	Week 1: surface righting	No	14
				Forelimb placing	-	
				Forepaw grasp	No	
				Rooting reflex	No	
				Visual cliff	No	
				Auditory startle	No	
				Bar holding	No	
				Week 2: inverted screen	No	
				Week 3: passive avoidance	No	
				Weeks 1-3: body weight	No	
5. Methyl chloroform	Rats	2100 ppm inhalation, 6 hr/day	All g	Week 3: activity (open field)	No	122
				Weeks 6-16: activity (running wheel): no challenge	No	
				Amphetamine challenge	No	
				Weeks 0-17: body weight	No	

Table 1 (Continued)

	Species	Dose	Time[a] (days)	Time and effect examined for[b]	Effects reported[c]	References
6. 2-Ethoxyethanol	Rats	100 ppm inhalation, 7 hr/day	g, 7-13; g, 14-20	Week 2: ascent on wire mesh screen	No	71
				Weeks 3,4: rotorod	-	
				Weeks 2,3,4,6,8: activity (openfield)	-	
				Week 5: activity (running wheel)	-	
				Weeks 5,9: two way avoidance conditioning	-	
				Week 6: operant (progressive fixed ratio)	No	
				Weeks 0-5: body weight	No	
				Weeks 0,3: ACh	+	
				Weeks 0,3: DA	+	
				Weeks 0,3: NE	+	
				Weeks 0,3: 5HT	+ (differential effects by brain region)	
Plus ethanol	Rats	100 ppm (2-ethoxy-ethanol) inhalation, 7 hr/day, plus 10% ethanol, drinking water	g, 7-13	Week 2: ascent on wire mesh screen	No	69,70
			g, 14-20	Weeks 3,4: rotorod	No	
				Weeks 2,4,6,8: activity (open field)	-	
				Week 5: activity (running wheel)	No	
				Weeks 5,9: two-way avoidance conditioning	-	
				Week 6: operant (progressive fixed ratio)	No	
				Weeks 0-5: body weight	No	
				Weeks 0,3: ACh	+	
				Weeks 0,3: DA	No	

				Weeks 0,3: NE	No	
				Weeks 0,3: 5HT	- (differential effects by brain region)	
7.	Formaldehyde	Rats	0.4-4 ppm inhalation, 4 hr/day	g, 1-19	Weeks 4,8: "neuromuscular excitability:		89
					Activity	+	
8.	2-methoxy-ethanol	Rats	25 ppm inhalation, 7 hr/day	g, 7-13; g, 14-20	Week 2: ascent on wire mesh screen	No	73
					Weeks 3,4: rotorod	No	
					Weeks 2,4,6,8: activity (open field)	No	
					Week 5: activity (running wheel)	No	
					Weeks 5,9: two-way avoidance conditioning	+	
					Week 6: operant (progressive fixed ratio)	No	
					Weeks 0-5: body weight	No	
					Weeks 0,3: ACh	+	
					Weeks 0,3: DA	-	
					Weeks 0,3: NE	+	
					Weeks 0,3: 5HT	+ (differential effects by brain region)	
9.	Methyl n-butyl ketone	Rats	500-2000 ppm, 6 hr/day	g	Weeks 4,8,18,80: righting reflex	?	76
					Inclined screen	+	
					Activity (open field)	-	
					Activity (running wheel)	+ (in young), - (in geriatrics)	
					"Food maze behavior"	- (in young), + (in geriatrics)	
					"Swimming stress test"	No	
					Two-way avoidance	-	

Table 1 (Continued)

	Species	Dose	Time[a] (days)	Time and effect examined for[b]	Effects reported[c]	References
				Pentobarbitol sleep time	+	
				Pathology	No	
				Blood chemistry	Sporadic	
				Body weight	-	
10. Methylene chloride	Rats	4500 ppm	Pre, g, 1-17	1,16 weeks: activity (jiggle cage)	No	6
				2 weeks: activity (photocell)	+	
				8-15 weeks: activity (running wheel)	No	
				Weeks 1,2: one-way avoidance	No	
				Body weight	No	
11. Paint thinner	Rats	50,000 ppm inhalation (2 × 10 min/day)	p, 1-30	1-5 weeks: activity (photocell)	+	56
	Rats	50,000 ppm inhalation (2 × 10 min/day)	p, 1-33	1-5 weeks: swimming ability	-	85
				Body weight	No (except on day 30)	
12. Perchloro-ethylene	Rats	100, 1000 ppm inhalation, 7 hr/day	g, 7-13; g, 14-20	Week 1: olfaction	No	74
				Week 2: ascent on wire mesh screen	-	
				Weeks 3,4: rotorod	-	
				Week 4: activity (open field)	+	
				Week 5: activity (running wheel)	No	

				Two-way avoidance conditioning	No	
				Week 6: progressive fixed ratio	No	
				Weeks 0-5: body weight	No	
				Weeks 0,3: ACh	No	
				Weeks 0,3: DA	-	
				Weeks 0,3: NE	No	
				Weeks 0,3: 5HT	No	
				Weeks 0,3: light microscopy	No	
D. Other 1. Halogenated hydrocarbons a. PCBs	Monkeys	2.5 ppm, diet	All g + L	44 months: activity (photocell)	-	7
	Mice	0.3-32 mg/kg, gavage	g, 10-16	Weeks 3,6,24: activity	+	57
				Light microscopy	Yes	
	Mice	11-82 ppm, diet	All g + L, post	Week 3: two-way avoidance	-	104
				Week 4: activity (open field)	+	
				Week 1: negative geotaxis	No	108
				Week 4: surface righting	No	
				Weeks 1-3: visual cliff	No	
				Weeks 1-2: swimming	-	
				Week 2: auditory startle	No	
				Weeks 2,3: hindlimb support	-	
				Week 3: visual placing	-	
				Weeks 3,10,34: activity (open field)	+	
				Weeks 6,20: water-filled T maze	-	
				Week 34: two-way avoidance conditioning	-	
				Weeks 0-10: body weight	-	

Table 1 (Continued)

	Species	Dose	Time[a] (days)	Time and effect examined for[b]	Effects reported[c]	References
	Mice	32 mg/kg, gavage	g, 10-16	Weeks 5,9: reflex tests	-	109
				Activity (photocell)	No	
				Visual placement	-	
				Grip strength	-	
				One-way avoidance	+	
				Wire rod traversing	-	
				Body weight	No	
b. PBBs	Mice	0.3-32 mg/kg, gavage, 8 hr/day	g, 10-16	Weeks 3,6,24: activity	No	57
				Light microscopy	No	
2. Carbon monoxide	Rats	150 ppm inhalation, 24 hr/day	All g	Weeks 0-3: activity (jiggle cage)	-	32
				DA	-	
				NE	No	
				Protein	-	
	Rats	150 ppm inhalation, 24 hr/day	All g	Week 1: surface righting	No	33
				Week 1: negative geotaxis	-	
				Week 1: homing response	-	
				Week 0: DA	No	
				Week 0: NE	No	
				Body weight	-	

3. Halothane	Rats	1.2-2.5% inhalation (2 hr)	g, 3, 10, or 17	Week 11: Y-maze discrimination	+	95
				Week 12: foot shock sensitivity	+	
				Activity (running wheel)	No	
				Body weight	No	
	Rats	10 ppm inhalation (8 hr/day, 5 days/week)	p, 1-60	Week 20: Y-maze	-	77
				Spatial discrimination	-	
				"Adults": electron microscopy	Yes	
4. Solvent-refined coal hydrocarbons	Rats	0.17-0.99 ml/kg, gavage	g, 12-16	Weeks 1-3: surface righting	No	3
				Air righting	No	
				Visual placing	No	
				Swimming activity	No	
				Forelimb strength	No	
				Auditory startle	No	
				Weeks 0-6: clinical chemistry	No	
				Body weight	-	

[a] g = gestation, L = lactation, p = postnatal (followed by days of exposure).
[b] In most cases the time of testing (in weeks) is given; the following abbreviations are used: ACh, acetylcholine; AChE, acetylcholinesterase; DA, dopamine; NE, norepinephrine; 5HT, 5-hydroxytryptamine; 5HIAA, 5-hydroxyindole acetic acid; MAO, monoamine oxidase; COMT, catechol-O-methyltransferase; TRP, tryptophane, TYR, tyrosine; HVA, homovanillic acid; AMP, adenosine monophosphate; ADP, adenosine diphosphate; ATP, adenosine triphosphate.
[c] No = no effect; - = significant effect lower than or retarded from controls; + = significant effect higher than in controls; yes = some effect observed (used for pathology examination). (In some cases, doses above which the effects were noted are given.)

observed, and the reference citation. Generally these studies involve administration of a test agent to pregnant test animals and, after allowing the animals to litter, offspring are either reared by their biological mother or are fostered to untreated controls. Litters are usually culled to a standardized number of offspring, often six or eight. Testing of the offspring usually begins within the first weeks of life, but can extend throughout the lifetime of the animals. In some cases the test agent is administered to neonatal animals rather than to pregnant animals, and these animals are tested later in life.

A. Chemical Agents

1. *Heavy Metals*

There has been extensive experimental research on the developmental neurotoxic effects of methyl mercury and lead. Space does not permit a thorough review of either compound. Consequently a few representative studies have been included in Table 1, Section A. Methyl mercury at 1-10 mg/kg during gestation has caused postnatal disturbances in rats and mice after administration by injection, gavage, and in the drinking water (17,43,44,51,63,93,98,103,105,126). Based upon these and other reports in the literature (see Chapter 30 by Annau), it appears that levels above 8-10 mg/kg typically induce behavioral alterations in the developing animals (generally hypoactivity, but changes in learning measures are also reported); however, neurochemical and ultramicroscopic deviations may be induced by lower concentrations. Thus there is sufficient research to be relatively certain that methyl mercury is developmentally neurotoxic to experimental animals; it is also a known behavioral teratogen in humans, producing effects typically referred to as cerebral palsy-like symptoms (101; see also Chapter 16 by T. Takeuchi). The questions that remain focus on the lower end of the dose-effect curve, addressing the question, What is a safe concentration for humans?

Lead at concentrations above 25 ppm in the drinking water or diet, often throughout gestation and lactation in rats and mice, consistently produces postnatal alterations (28,37,52,54,59,62,75,97,120,127). Based upon these and other studies, several areas of concern can be noted: (a) Most studies have involved administration of lead for relatively long periods of time, often throughout gestation and lactation; some have extended treatment for weeks or months after weaning. Such chronic treatment is atypical of most developmental neurotoxicity studies, complicating methodological comparisons. A closely related point is the (b) postnatal effects have often been reported only at concentrations which also produce weight reductions. Such reductions complicate analysis of test results. That is, are the results dependent upon or incident to weight reductions? (c) It appears that parenteral administration requires a smaller dose of lead to produce alterations than does oral administration. Consequently, reasons for the typical use of dietary administration are unclear (except for ease of administration). Whereas none of these concerns detract from the apparent ability of lead to produce developmental neurotoxic effects in experimental animals (often behavioral deviations such as hyper- or hypoactivity and pathological changes), they do add to aforementioned concerns about risk assessment. Since lead *can* induce behavioral disorders in children [typically including hyperactivity and minimal brain dysfunction (91); see also Chapter 14 by Silbergeld], the question is not, Is lead a hazard? but, rather, as stated before, *How much* lead is a hazard?

Pre- and/or perinatal exposure of experimental animals to cadmium ($\geq$ 1 μg/ml in drinking water) (20,40,79), tin ($\geq$ 3 mg/kg) (30,39,81), or manganese ($\geq$ 1 μg/g) (26,27,88,90) has resulted in some behavioral and/or neurochemical alterations in the developing animals. It appears that learning and activity measures are typically affected by heavy metals, and neurochemical and pathological changes are also frequently observed. However, the lack of consistency in either behavioral or neurochemical effects makes a judgment as to possible risks in humans impossible.

2. *Pesticides*

As exemplified in Table 1, Section B, a variety of pesticides, including organochlorines, organophosphates, and carbamates, have been shown to alter offspring behavior, neurochemistry, or immune function after exposure in prenatal or neonatal life. Dietary administration of Diazinon (0.18-9 mg/kg) produced few behavioral changes (102) but did alter serum immunoglobin (5) and corticosterone (23) levels. Carbofuran (0.01-0.5 mg/kg) altered serum immunoglobin (5) and corticosterone (23) levels, but no studies for behavioral effects have been reported. Chlordane produced a paradoxical increase in corticosterone in mice (23). Both DDE (3-10 ppm) and 2,4,5-trichlorophenoxyacetic acid (7-225 mg/kg) produced hyperactivity in rats (93), and functional alterations in avians as well (41,55,87). There are also isolated reports of a number of other pesticides which have produced effects; these are reported in Table 1. Given the wide variety of pesticides which have produced some functional alterations following perinatal exposure in experimental animals, albeit not a consistent pattern of effects in any which have been looked at in more than one study, it would appear that further testing of these chemicals is needed. Changes which have most consistently been observed include those in the immune system, in activity levels, and in pathological alterations. No such effects have been documented for humans.

3. *Solvents*

Table 1, Section C, shows that one dozen solvents have been examined for behavioral teratogenicity; the majority of those tested have shown significant behavioral or neurochemical differences from controls. Neonatal administration of benzene (550 mg/kg) produced hyperactivity in rats, though the other tests with these animals did not show differences from controls (111). Dietary administration of tertiary butanol (0.5-1%) to pregnant mice produced a number of alterations in young (preweanling) mice; the authors concluded that t-butanol was five times more potent as a behavioral teratogen than ethanol (24). Carbon disulfide (airborne concentrations from 3 to 67 ppm; U.S. Permissible Exposure Limit (PEL) = 20 ppm) produced alterations in rats (106,107). Neither chloroform (31 mg/kg) nor methyl chloroform (2100 ppm in air, or six times the U.S. PEL) showed effects on a battery of behavioral tests (14,122). Inhalation of 100 ppm 2-ethoxyethanol (one-half the PEL) by pregnant rats produced a number of behavioral and neurochemical deviations (71); furthermore, an interaction was observed when pregnant rats were exposed to 100 ppm 2-ethoxyethanol for 7 hr/day and 10% ethanol was given concomitantly in the drinking water overnight (69,70). A closely related glycol ether, 2-methoxyethanol, when administered at the PEL of 25 ppm, was also found to produce a number of neurochemical alterations, though few behavioral changes were observed (73). [It is also possible, by inference, that the acetates of these two glycol ethers would produce similar effects, though they have not been tested for functional alterations (72).] Exposure of pregnant rats to 0.4 or 4 ppm (PEL = 3 ppm) formaldehyde throughout pregnancy resulted in behavioral alterations in offspring at both exposure levels (89). Methyl n-butyl ketone at airborne concentrations of 1000 and 2000 ppm (PEL = 100 ppm) produced effects on most of the behavioral tests included in the battery (76). Methylene chloride (4500 ppm in air, or nine times the PEL) produced hyperactivity, but had no effect on the other tests (6). Inhalation of 50,000 ppm paint thinner for 10 min once or twice per day in neonatal rats resulted in behavioral alterations (56,85). Perchloroethylene at 1000 ppm produced sporadic behavioral and neurochemical differences from controls, but no differences were detected after exposure to 100 ppm, the PEL (74).

As illustrated above, one-third of the solvents tested have shown functional alterations (typically including alterations in activity levels and/or neurochemistry) after exposure to currently acceptable concentrations of carbon disulfide, 2-ethoxyethanol, 2-methoxyethanol, and formaldehyde. Two (chloroform and methyl chloroform) showed no effects at the concentrations tested, however, most of the data are from single studies. Consequently, there is a need to replicate these studies and to examine other solvents for developmental neurotoxicity.

Table 2 Physical Agents for Which Developmental Exposure Has Been Evaluated for Behavioral Teratogenicity

	Species	Dose	Time[a]	Time and effect examined for[b]	Effects reported[c]	References
A. Ionizing radiation						
1. X-rays have been extensively studied in several species with effects reviewed						9,42,92,121
2. Cobalt-60 irradiation	Squirrel monkeys	100-200 rads	g	Neonatal: reflex testing	-	11
				Visual acuity	-	
				Activity (open field)	-	
				Week 0: brain morphometry	-	
				Week 4: choline acetyltransferase	-	
				AchE	+	
				Tyrosine hydroxylase	No	
				MAO	+	
				DA	No	
				NE	No	
				5HT	No	
				Weeks 4, 25: serum immunoglobins	+	
				Body weight	-	
B. Nonionizing radiation						
1. 915 mHz	Rats	10 mW/cm^2, 8 hr/day	g	Week 1: surface righting	+	47
				Negative geotaxis	No	
				Week 2: air righting	-	
				Auditory startle	No	

Table 2 (Continued)

	Species	Dose	Time[a]	Time and effect examined for[b]	Effects reported[c]	References
				Week 3: visual placing	-	
				Week 9: activity (open field)	No	
				Activity (running wheel)	No	
				Water T maze	No	
				Swimming	No	
				Hanging from rope	No	
				Two-way avoidance conditioning	No	
				Weeks 0-13: body weight	No	
2. 918 mHz	Rats	5 mW/cm^2, 20 hr/day	g, 1-18	13 weeks: two-way avoidance conditioning	-	49
3. 2450 mHz	Mice	38 mW/g, 600 sec	g, 14	5 weeks: Lashley III maze learning	No	19
	Rats	8.26 mW/cm^2	g, 10	1,2,3,5 weeks:		46
				ATP	+ all ages	
				ADP	- weeks 1,2, + week 3	

				AMP	- weeks 1,2, + week 3 (no changes in any at 5 weeks)	
C. Other						
1. Noise	Rats	"high intensity," "composite frequency"	g, 5-18	11 weeks: maze learning	-	99
	Sheep	130 db broadband 4 hr/day	g (?)	5 weeks: hearing Cochlea pathology	No Yes (?)	29
2. Ultrasound	Mice	250 mW/cm^2 1 mHz	g, 9	0-3 weeks: 11 developmental tests	- (two tests)	12

[a]g = gestation (followed by days of exposure).
[b]In most cases the time of testing (in weeks) is given; the following abbreviations are used: AchE, acetylcholinesterase; DA, dopamine; NE, norepinephrine; 5HT, 5-hydroxytryptamine; MAO, monoamine oxidase; AMP, adenosine monophosphate; ADP, adenosine diphosphate; ATP, adenosine triphosphate.
[c]No = no effect; - = significant effect lower than in or retarded from controls; + = significant effect higher than in controls; yes = some effect observed (used for pathology examination).

4. *Other Chemicals*

A number of chemicals which cannot be readily classified as above have produced behavioral teratogenic effects in experimental animals (Table 1, Section D). Several investigators have observed consistent effects with polychlorinated biphenyls (PCBs) (7,57,104,108,109); in contrast, the only study to examine polybrominated biphenyls (PBBs) did not find effects (57). Carbon monoxide at 150 ppm (PEL = 50 ppm) resulted in behavioral and neurochemical deviations from control (32,33). Low levels of the anesthetic gas halothane have produced behavioral and electron-microscopic deviations in experimental animals (77,95). Finally, solvent-refined coal hydrocarbons were tested but not found to produce behavioral or clinical chemistry deviations (3).

B. Physical Agents

1. *Ionizing Radiation*

The deleterious effects of x-irradiation on nervous system development are well documented in several reviews (e.g., Refs. 9, 42, 92, and 121) and need not be elaborated here; doses above 25-30 rads during prenatal development often cause behavioral alterations (hyperactivity, "emotionality," and decreased learning ability) in the offspring of experimental animals (9,16,42,92,121). Since other kinds of ionizing radiation would be expected to produce similar effects (11), exposure to ionizing radiation above approximately 10 rads during perinatal development is contraindicated.

2. *Nonionizing Radiation*

As illustrated in Table 2, Section B, exposure to nonionizing radiation during development has not shown a clear pattern of effects in the offspring. Some studies have found behavioral and neurochemical effects, but additional research is needed to determine the conditions which lead to consistent effects even in experimental animals before risk to humans can be assessed.

3. *Other Physical Agents*

Exposures of pregnant experimental animals to noise (29,99) and to ultrasound (12) have caused behavioral effects in the offspring. However, neither these nor other physical agents have been well investigated.

III. SUMMARY AND CONCLUSIONS

This chapter has reviewed the studies which have examined industrial and environmental agents for developmental neurotoxicology. It has endeavored to assemble relevant studies on a wide variety of agents and thus, indirectly, to indicate where additional investigations are most urgently needed. As is illustrated herein, most agents examined for developmental neurotoxicity have had inadequate testing. Many have been tested in only one study, in one species, and at limited dose levels—frequently only one. Not uncommonly, relatively high concentrations (approaching the maternally toxic levels) are used, and a no-effect level is not included. This is a major concern, since available teratology literature indicates that there is an exposure level threshold for inducing teratogenic and developmental neurotoxic effects below which the offspring are not adversely affected. In addition, some studies have examined only very young animals; hence whether or not effects persist into adulthood, or what effects might be found as the offspring reach senescence, cannot be predicted. Taking all of these factors into consideration, and bearing in mind the usual lack of consistency in effects observed, much research must yet be completed before stating that the agents reviewed in this chapter—with the exception of methyl mercury, lead, and ionizing radiation—pose a substantive risk in humans. However, the animal data do raise some degree of concern for pregnant women (and in some cases men) who may be exposed to chemicals producing effects in animals and do suggest the need to consider maternal and paternal exposure to environmental agents when attempting to determine the etiology of childhood behavioral disorders. It is the author's hope that this review will serve as a stimulus for much needed research in this area in order to reduce or prevent developmental neurotoxicity in humans.

ACKNOWLEDGMENTS

The author acknowledges numerous colleagues whose insights and experience have formed the base on which developmental neurotoxicology is being built. He thanks those who have played a role in the preparation of this chapter, particularly Mrs. Nadine Dickerson for her uncomplaining preparation of the manuscript.

REFERENCES

1. Adams, J., Buelke-Sam, J. 1981. Behavioral assessment of the postnatal animal: Testing and methods development, in *Developmental Toxicology* (Kimmel, C. A., Buelke-Sam, J., eds.), Raven Press, New York, pp. 233-258.
2. Adams, P. M., Fabricant, J. D., Legator, M. S. 1980. Cyclophosphamide-induced spermatogenic effects detected in the F_1 generation by behavioral testing. *Science 211*:80-82.
3. Andrew, F. D., Lytz, P. S., Buschbom, R. L., Springer, D. L. 1982. Postnatal effects following prenatal exposure of rats to solvent

refined coal (SRC) hydrocarbons. *Teratology 25*:26A.

4. Barlow, S. M., Sullivan, F. M. 1975. Behavioral teratology, in *Teratology: Trends and Applications* (Berry, C. L., Poswillo, D. E., eds.), Springer-Verlag, New York, pp. 103-120.
5. Barnett, J. B., Spyker-Cranmer, J. M., Avery, D. L., Hoberman, A. M. 1980. Immunocompetence over the lifespan of mice exposed in utero to carbofuran or diazinon: 1. Changes in serum immunoglobin concentrations. *J. Environ. Pathol. Toxicol. 4*:53-63.
6. Bornschein, R. L., Hastings, L., Manson, J. M. 1980. Behavioral toxicity in the offspring of rats following maternal exposure to dichloromethane. *Toxicol. Appl. Pharmacol. 52*:29-37.
7. Bowman, R. E., Heironimus, M. P. 1981. Hypoactivity in adolescent monkeys perinatally exposed to PCBs and hyperactive as juveniles. *Neurobehav. Toxicol. Teratol. 3*:15-18.
8. Brady, K., Herrera, Y., Zenick, H. 1975. Influence of parental lead exposure on subsequent learning ability of offspring. *Pharmacol. Biochem. Behav. 3*:561-565.
9. Brent, R. L. 1977. Radiation and other physical agents, in *Handbook Teratology 1. General Principles and Etiology* (Wilson, J. G., Fraser, F. C., eds.), Plenum, New York, pp. 153-223.
10. Brent, R. L. 1982. Biography of Josef Warkany. *Teratology 25*: 137-151.
11. Brizzee, K. R., Ordy, J. M., Kaack, B., Martin, L., Walker, L. 1978. Prenatal cobalt-60 irradiation effects on early postnatal development of the squirrel monkey offspring, in *Developmental Toxicology of Energy-Related Pollutants* (Mahlum, D. D., Sikov, M. R., Hackett, P. L., Andrew, F. D., eds.), NTIS Conference 771017, Springfield, Virginia, pp. 204-227.
12. Brown, N., Galloway, W. D., Monohan, J. C., Fisher, B. 1980. Postnatal behavior and development following in utero exposure of mice to ultrasound and microwave radiation, in *Effects of Foods and Drugs on the Development and Function of the Nervous System: Methods for Predicting Toxicity* (Gryder, R. M., Frankos, V. H., eds.), HHS Publication No. (FDA) 80-1076, U.S. Dept. of Health and Human Services, Rockville, Md., pp. 161-163.
13. Buelke-Sam, J., Kimmel, C. A. 1979. Development and standardization of screening methods for behavioral teratology. *Teratology 20*:17-30.
14. Burkhalter, J. E., Balster, R. L. 1979. Behavioral teratology evaluation of trichloromethane in mice. *Neurobehav. Toxicol. Teratol. 1*:199-205.
15. Butcher, R. E., Wootten, V., Vorhees, C. V. 1980. Standards in behavioral teratology testing: test variability and sensitivity. *Teratogenesis Carcinogenesis Mutagenesis 1*:49-61.
16. Cabe, P. A., McRee, D. I. 1980. Behavioral teratological effects

of microwave radiation in Japanese quail (*Coturnix coturnix japonica*): An exploratory study. *Neurobehav. Toxicol. Teratol.* *2*:291-296.

17. Chang, L. W., Reuhl, K. R., Spyker, J. M. 1977. Ultrastructural study of the latent effects of methylmercury on the nervous system after prenatal exposure. *Environ. Res.* *13*:171-185.
18. Chang, L. W., Wade, P. R., Pounds, J. G., Reuhl, K. R. 1980. Prenatal and neonatal toxicology and pathology of heavy metals. *Adv. Pharmacol. Chemotherap.* *17*:195-231.
19. Chernovetz, M. E., Justesen, D. R., King, N. W., Wagner, J. E. 1975. Teratology, survival, and reversal learning after fetal irradiation of mice by 2450-MHz microwave energy. *J. Microwave Power* *10*:391-409.
20. Cooper, G. P., Choudhury, H., Hastings, L., Petering, H. G. 1978. Prenatal cadmium exposure: Effects on essential trace metals and behavior in rats, in *Development Toxicology of Energy-Related Pollutants* (Mahlum, D. D., Sikov, M. R., Hackett, P. L., Andrew, F. D., eds.), NTIS Conference 771017, Springfield, Virginia, pp. 627-637.
21. Coyle, I., Wayner, M. J., Singer, G. 1976. Behavioral teratogenesis: A critical evaluation. *Pharmacol. Biochem. Behav.* *4*: 191-200.
22. Coyle, I., Wayner, M. J., Singer, G. 1980. Behavioral teratogenesis: A critical evaluation, in *Advances in the Study of Birth Defects*, Vol. 4 (Persuad, T. V. N., ed.), University Park Press, Baltimore, pp. 111-133.
23. Cranmer, J. S., Avery, D. L., Grady, R. R., Kitay, J. I. 1978. Postnatal endocrine dysfunction resulting from prenatal exposure to carbofuran, diazinon, or chlordane. *J. Environ. Pathol. Toxicol.* *2*:357-369.
24. Daniel, M. A., Evans, M. A. 1982. Quantitative comparison of maternal ethanol and maternal tertiary butanol diet on postnatal development. *J. Pharmacol. Exp. Ter.* *222*:294-300.
25. Declume, C., Cambon, C., Derache, R. 1979. The effects on newborn rats of repeated Carbaryl administration during gestation. *Toxicol. Lett.* *3*:191-196.
26. Deskin, R., Bursian, S. J., Edens, F. W. 1981. Neurochemical alterations induced by manganese chloride in neonatal rats. *Neurotoxicology* *2*:65-73.
27. Deskin, R., Bursian, S. J., Edens, F. W. 1981. The effect of chronic manganese administration on some neurochemical and physiological variables in neonatal rats. *Gen. Pharmacol.* *12*:279-280.
28. Donald, J. M., Cutler, M. G., Moore, M. R., Bradley, M. 1981. Development and social behavior in mice after prenatal and postnatal administration of low levels of lead acetate. *Neuropharmacology* *20*:1097-1104.

29. Dunn, D. E., Lim, D. J., Ferraro, J. A., McKinley, R. L., Moore, T. J. 1981. Effects on the auditory system from *in utero* noise exposure in lambs, as cited in *Prenatal Effects of Exposure to High-Level Noise* (1982), National Academy Press, Washington, D.C.
30. Dyer, R. S., Howell, W. E., Reiter, L. W. 1981. Neonatal triethyltin exposure alters adult electrophysiology in rats. *Neurotoxicology 2*:609-623.
31. Edwards, M. J., Penny, R. H. C., Zevnik, I. 1971. A brain cell deficit in newborn guinea pigs following prenatal hyperthermia. *Brain Res. 28*:341-345.
32. Fechter, L. D., Annau, Z. 1977. Toxicity of mild prenatal carbon monoxide exposure. *Science 197*:680-682.
33. Fechter, L. D., Annau, Z. 1980. Prenatal carbon monoxide exposure alters behavioral development. *Neurobehav. Toxicol. Teratol. 2*:7-11.
34. Friedler, G., Wheeling, H. S. 1979. Behavioral effects in offspring of male mice injected with opioids prior to mating. *Pharmacol. Biochem. Behav. 11*:23-28.
35 Golub, M. S., Golub, A. M. 1981. Behavioral teratogenesis, in *Advances in Perinatal Medicine*, (Milunsky, A., Friedman, E., and Gluck, L., eds.), Plenum, New York, pp. 231-293.
36. Grabowski, C. T., Payne, D. B. 1983. The causes of perinatal death induced by prenatal exposure of rats to the pesticide, Mirex. Part 1: parturation observations of the cardiovascular system. *Teratology 27*:7-11.
37. Grant, L. D., Kimmel, C. A., West, G. L., Martine-Vargas, C. M., Howard, J. L. 1980. Chronic low-level lead toxicity in the rat. II. Effects on postnatal physical and behavioral development. *Toxicol. Appl. Pharmacol. 56*:42-58.
38. Gray, L. E., Kavlock, R. J., Chernoff, N., Ferrell, J. 1982. The effects of prenatal exposure to the herbicide TOK on the postnatal development of the Harderian gland of the mouse, rat, and hamster. *Teratology 25*:45A.
39. Harry, G. J., Tilson, H. A. 1981. The effects of postpartum exposure to triethyltin in the neurobehavioral functioning of rats. *Neurotoxicology 2*:283-296.
40. Hastings, L., Choudhury, H., Petering, H. G., Cooper, G. P. 1978. Behavioral and biochemical effects of low level prenatal cadmium exposure in rats. *Bull. Environ. Contam. Toxicol. 20*: 96-101.
41. Heinz, G. H. 1976. Behavior of mallard ducklings from parents fed 3 ppm DDE. *Bull. Environ. Contam. Toxicol. 16*:640-645.
42. Hicks, S. P., D'Amato, C. J. 1978. Effects of ionizing radiation on developing brain and behavior, in *Studies on the Development of Behavior and the Nervous System*, Vol. 4 (Gottlieb, G., ed.), Academic, New York, pp. 35-72.

43. Hughes, J. A., Annau, Z. 1976. Postnatal behavioral effects in mice after prenatal exposure to methylmercury. *Pharmacol. Biochem. Behav.* *4*:385-391.
44. Hughes, J. A., Sparber, S. B. 1978. d-Amphetamine unmasks postnatal consequences of exposure to methylmercury in utero: Methods for studying behavioral teratogenesis. *Pharmacol. Biochem. Behav.* *8*:365-375.
45. Hutchings, D. E. 1978. Behavioral teratology: Embryopathic and behavioral effects of drugs during pregnancy, in *Studies on the Development of Behavior and the Nervous System*, Vol. 4 (Gottlieb, G., ed.), Academic, New York, pp. 7-34.
46. Jamakosmanovic, A., Nakas, M., Drecan, M., Shore, M. L. 1981. The levels of ATP, ADP, and AMP in rat brain during postnatal development following *in utero* 2450 MHz microwave irradiation. *Period. Biol.* *83*:151-152.
47. Jensh, R. P. 1980. Behavioral teratology: Application in low dose chronic microwave irradiation studies, in *Advances in the Study of Birth Defects*, Vol. 4 (Persuad, T. V. N., ed.), University Park Press, Baltimore, pp. 135-162.
48. Joffe, J. M. 1969. *Prenatal Determinants of Behavior,* Pergamon, New York.
49. Johnson, R. B., Mizumori, S., Lovely, R. H. 1978. Adult behavioral deficit in rats exposed prenatally to 918 MHz microwaves, in *Developmental Toxicology of Energy-Related Pollutants* (Mahlum, D. D., Sikov, M. R., Hackett, P. L., Andrew, F. D., eds.), NTIS Conference 771017, Springfield, Virginia, pp. 281-299.
50. Kavlock, R. J., Gray, J. A. 1983. Morphometric, biochemical, and physiological assessment of perinatally induced renal dysfunction. *J. Toxicol. Environ. Health* *11*:1-13.
51. Khera, K. S., Tabacova, S. A. 1973. Effects of methylmercury chloride on the progeny of mice and rats treated before or during gestation. *Food Cosmet. Toxicol.* *11*:245-254.
52. Kimmel, C. A. 1983. Critical periods of exposure and developmental effects of lead, in *Toxicology and the Newborn* (Kachew, S., Reasor, M. J., eds.), Elsevier, New York, pp. 218-287.
53. Kornetsky, C. 1970. Psychoactive drugs in the immature organism. *Psychopharmacology* *17*:105-136.
54. Krigman, M. R., Hogan, E. L. 1974. Effect of lead intoxication on the postnatal growth of the rat nervous system. *Environ. Health Perspect.* 5:187-199.
55. Longcore, J. R., Stendell, R. C. 1977. Shell thinning and reproductive impairment in black ducks after cessation of DDE dosage. *Arch. Environ. Contam. Toxicol.* *6*:293-304.
56. Lorenzana-Jimenez, M., Salas, M. 1980. Effects of neonatal exposure to paint thinner on the development of swimming in rats. *Neurobehav. Toxicol. Teratol.* *2*:1-6.

57. Lucier, G. W., Davis, G. J., McLachlan, J. A. 1978. Transplacental toxicology of the polychlorinated and polybrominated biphenyls, in *Developmental Toxicology of Energy-Related Pollutants* (Mahlum, D. D., Sikov, M. R., Hackett, P. L., Andrew, F. D., eds.), NTIS Conference 771017, Springfield, Virginia, pp. 188-203.
58. Lundborg, P. Engel, J. 1978. Neurochemical brain changes associated with behavioral disturbances after early treatment with psychotropic drugs, in *Maturation of Neurotransmission* (Vernadakis, A., Giacobini, E., Filogama, G., eds.), S. Karger, Basel, pp. 226-235.
59. Luster, M. I., Faith, R. E., Kimmel, C. A. 1978. Depression of humoral immunity in rats following chronic developmental lead exposure. *J. Environ. Pathol. Toxicol. 1*:397-402.
60. Lyle, J. G., Edwards, M. J., Jonson, K. M. 1977. Critical periods and the effects of prenatal heat stress on the learning and brain growth of mature guinea pigs. *Biobehav. Rev. 1*:1-13.
61. Middaugh, L. D., Blackwell, L. A., Santos III, C. A., Zemp. J. W. 1974. Effects of d-amphetamine sulfate given to pregnant mice on activity and on catecholamines in the brains of offspring. *Dev. Psychobiol. 7*:429-438.
62. Milar, K. S., Krigman, M. R., Grant, L. D. 1981. Effects of neonatal lead exposure on memory in rats. *Neurobehav. Toxicol. Teratol. 3*:369-373.
63. Musch, H. R., Bornhausen, M., Kriegel, H., Greim, H. 1978. Methylmercury chloride induces learning deficits in prenatally treated rats. *Arch. Toxicol. 40*:103-108.
64. Myers, R. D. 1977. Experimental models of perinatal brain damage: Relevance to human pathology, in *Intrauterine Asphyxia and the Developing Fetal Brain* (Gluck, L., ed.), Year Book Medical, Chicago, pp. 37-97.
65. Nair, V. 1974. Prenatal exposure to drugs: Effect on the development of brain monoamine systems, in *Drugs and the Developing Brain* (Vernadakis, A., Weiner, N., eds.), Plenum, New York, pp. 171-197.
66. Nair, V., DuBois, K. P. 1968. Prenatal and early postnatal exposure to environmental toxicants. *Chicago Med. Sch. Q. 27*: 75-89.
67. Nelson, B. K. 1978. Behavioral assessment in the developmental toxicology of energy-related industrial pollutants, in *Developmental Toxicology of Energy-Related Pollutants* (Mahlum, D. D., Sikov, M. R., Hackett, P. L., Andrew, F. D., eds.), NTIS Conference 771017, Springfield, Virginia, pp. 410-424.
68. Nelson, B. K. 1981. Dose/effect relationships in developmental neurotoxicology. *Neurobehav. Toxicol. Teratol. 3*:255.
69. Nelson, B. K., Brightwell, W. S., Setzer, J. V. 1982. Prenatal interactions between ethanol and the industrial solvent 2-ethoxyethanol in rats: Maternal and behavioral teratogenic effects.

Neurobehav. Toxicol. Teratol. 4:387-394.

70. Nelson, B. K., Brightwell, W. S., Setzer, J. V., O'Donohue, T. L. 1982. Prenatal interactions between ethanol and the industrial solvent 2-ethoxyethanol in rats: Neurochemical effects in the offspring. *Neurobehav. Toxicol. Teratol. 4*:395-401.
71. Nelson, B. K., Brightwell, W. S., Setzer, J. V., Taylor, B. J., Hornung, R. W. 1981. Ethoxyethanol behavioral teratology in rats. *Neurotoxicology 2*:231-249.
72. Nelson, B. K., Setzer, J. V., Brightwell, W. S., Mathinos, P. R., Kuczuk, M. H., Weaver, T. E., Goad, P. T. 1984. Comparative inhalation teratogenicity of four glycol ether solvents and an amino derivative in rats. *Environ. Health Perspect. 57*:261-271.
73. Nelson, B. K., Brightwell, W. S., Burg, J. R., Massari, V. J. 1984. Behavioral and neurochemical alterations in the offspring of rats after maternal or paternal inhalation exposure to the industrial solvent 2-methoxyethanol. *Pharmacol. Biochem. Behav. 20*: 269-279.
74. Nelson, B. K., Taylor, B. J., Setzer, J. V., Hornung, R. W. 1979. Behavioral teratology of perchloroethylene in rats. *J. Environ. Pathol. Toxicol. 3*:233-350.
75. Overmann, S. R., Zimmer, L., Woolley, D. E. 1981. Motor development, tissue weights and seizure susceptibility in perinatally lead-exposed rats. *Neurotoxicology 2*:725-742.
76. Peters, M. A., Hudson, P. M., Dixon, R. L. 1981. The effect totigestational exposure to methyl n-butyl ketone has on postnatal development and behavior. *Ecotoxicol. Environ. Safety 5*:291-306.
77. Quimby, K. L., Aschkenase, L. J., Bowman, R. E., Katz, J., Chang, L. W. 1974. Enduring learning deficits and cerebral synaptic malformation from exposure to 10 parts of halothane per million. *Science 185*:625-627.
78. Rafales, L. S., Greenland, R. D., Zenick, H., Goldsmith, M., Michaelson, I. A. 1981. Responsiveness to d-amphetamine in lead-exposed rats as measured by steady state levels of catecholamines and locomotor activity. *Neurobehav. Toxicol. Teratol. 3*:363-367.
79. Rastogi, R. B., Merali, Z., Singhal, R. L. 1977. Cadmium alters behavior and the biosynthetic capacity for catecholamines and serotonin in neonatal rat brain. *J. Neurochem. 28*:789-794.
80. Rech, R. H., Lomuscio, G., Algeri, S. 1980. Methadone exposure in utero: Effects on brain biogenic amines and behavior. *Neurobehav. Toxicology 1*:75-78.
81. Reiter, L. W., Heavner, G. B., Dean, K. F., Ruppert, P. H. 1981. Developmental and behavioral effects of early postnatal exposure to triethyltin in rats. *Neurobehav. Toxicol. Teratol. 3*:285-293.
82. Reuhl, K. R., Chang, L. W. 1979. Effects of methylmercury on the development of the nervous system: a review. *Neurotoxicology 1*:21-55.

83. Reuhl, K. R., Chang, L. W., Townsend, J. W. 1981. Pathological effects of in utero methylmercury exposure on the cerebellum of the golden hamster. 1. Early effects upon the neonatal cerebellar cortex. *Environ. Res. 26*:281-306. 2. Residual effects on the adult cerebellum. *Environ. Res. 26*:307-327.
84. Rodier, P. M. 1978. Behavioral teratology, in *Handbook of Teratology*, Vol. 4 (Wilson, J. G., Fraser, F. C., eds.), Plenum, New York, pp. 397-428.
85. Rodriguez, R., Lorenzana-Jimenez, M., Manjarrez, A., Gomez-Ruiz, H. 1978. Behavioral effects from the acute and chronic inhalation of thinner in rats of various ages, in *Voluntary Inhalation of Industrial Solvents* (Sharp, C. W., Carroll, L. T., eds.), DHEW Publication No. (ADM) 79-779, Rockville, Md., pp. 213-225.
86. Rosenstein, L., Chernoff, N. 1978. Spontaneous and evoked EEG changes in perinatal rats following in utero exposure to Baygon: A preliminary investigation. *Bull. Environ. Contam. Toxicol. 20*:624-632.
87. Sanderson, C. A., Rogers, L. J. 1981. 2,4,5-Trichlorophenoxyacetic acid causes behavioral effects in chickens at environmentally relevent doses. *Science 211*:593-595.
88. Seth, P. K., Husain, R., Mushtaq, M., Chandra, S. V. 1977. Effects of manganese on neonatal rat: Manganese concentration and enzymatic alterations in brain. *Acta Pharmacol. Toxicol. 40*:553-560.
89. Sheveleva, G. A. 1971. Specific action of formaldehyde on the embryogeny and progeny of white rats. *Toksikol. Nov. Prom. Khim. Veshchestv 12*:78-96, as reviewed in Barlow, S. M., Sullivan, F. M. 1982. *Reproductive Hazards of Industrial Chemicals*, Academic, New York, pp. 339-340.
90. Shukla, G. S., Chandra, S. V. 1979. Species variation in manganese induced changes in brain biogenic amines. *Toxicol. Lett. 3*:249-253.
91. Silbergeld, E. K., Goldberg, A. M. 1974. Hyperactivity: A lead-induced behavior disorder. *Environ. Health Perspect. 7*:227-232.
92. Sikov, M. R., Mahlum, D. D. (eds.). 1969. *Radiation Biology of the Fetal and Juvenile Animal*, AEC Symposium Series, Richland, Wash., NTIS Conference 690501, Springfield, Virginia.
93. Sjoden, P. O., Soderberg, U. 1972. Sex dependent effect of prenatal 2,4,5-trichlorophenoxyacetic-acid on rats open-field behavior. *Physiol. Behav. 9*:357-360.
94. Slotkin, T. A., Thadani, P. V. 1980. Neurochemical teratology of drugs of abuse, in *Advances in the Study of Birth Defects*, Vol. 4 (Persuad, T. V. N., ed.), University Park Press, Baltimore, pp. 199-234.

95. Smith, R. F., Bowman, R. E., Katz, J. 1978. Behavioral effects of exposure to halothane during development in the rat: Sensitive period during pregnancy. *Anesthesiology 49*:319-323.
96. Sobotka, T. J., Brodie, R. E., Cook, M. P. 1972. Behavioral and neuroendocrine effects in rats of postnatal exposure to low dietary levels of Maneb. *Dev. Psychobiol. 5*:137-148.
97. Sobotka, T. J., Brodie, R. E., Cook, M. P. 1975. Psychophysiologic effects of early lead exposure. *Toxicology 5*:175-191.
98. Sobotka, T. J., Cook, M. P., Brodie, R. E. 1974. Effects of perinatal exposure to methylmercury on functional brain development and neurochemistry. *Biol. Psychiatry 8*:307-320.
99. Sontag, L. W. 1970. Effect of noise during pregnancy upon foetal and subsequent adult behavior, in *Physiological Effects of Noise* (Welch, B. L., Welch, A. S., eds.), Plenum, New York, pp. 131-141.
100. Sparber, S. B. 1972. Effects of drugs on the biochemical and behavioral responses of developing organisms. *Fed. Proc. 31*: 74-80.
101. Spyker, J. M. 1975. Behavioral teratology and toxicology, in *Behavioral Toxicology* (Weiss, B., Laties, V. G., eds.), Plenum, New York, pp. 311-344.
102. Spyker, J. M, Avery, D. L. 1977. Neurobehavioral effects of prenatal exposure to the organophosphate Diazinon in mice. *J. Toxicol. Environ. Health 3*:989-1002.
103. Spyker, J. M., Sparber, S. B., Goldberg, A. M. 1972. Subtle consequences of methylmercury exposure: Behavioral deviations in offspring of treated mothers. *Science 177*:621-623.
104. Storm, J. E., Hart, J. L., Smith, R. F. 1981. Behavior of mice after pre- and postnatal exposure to Arochlor 1254. *Neurobehav. Toxicol. Teratol. 3*:5-9.
105. Su, M. Q., Okita, G. T. 1976. Behavioral effects on the progeny of mice treated with methylmercury. *Toxicol. Appl. Pharmacol. 38*:195-205.
106. Tabacova, S., Hinkova, L. 1979. Neurotoxicological screening of early effects of prenatal carbon disulphide exposure. *Act. Nerv. Super. 21*:268-269.
107. Tabacova, S., Hinkova, L., Balavaeva, L. 1978. Carbon disulphide teratogenicity and postnatal effects in rat. *Toxicol. Lett. 2*:129-133.
108. Tanimura, T., Ema, M., Kihara, T. 1980. Effects of combined treatment with methylmercury and polychlorinated biphenyls (PCBs) on the development of mouse offspring, in *Advances in the Study of Birth Defects*, Vol. 4 (Persuad, T. V. N., ed.), University Park Press, Baltimore, pp. 163-198.
109. Tilson, H. A., Davis, G. J., McLachlan, J. A., Lucier, G. W. 1979. The effects of polychlorinated biphenyls given prenatally

on the neurobehavioral development of mice. *Environ. Res. 18*: 466-474.

110. Tilson, H. A., Squibb, R. E., Burne, T. A. 1981. Effects of postnatal exposure to Kepone on neurobehavioral development of rats. *Teratology 24*:58A.
111. Tilson, H. A., Squibb, R. E., Meyer, O. A., Sparber, S. B. 1980. Postnatal exposure to benzene alters the neurobehavioral functioning of rats when tested in adulthood. *Neurobehav. Toxicol. Teratol. 2*:101-106.
112. Timiras, P. S. 1974. Developmental changes in the responsivity of the brain to endogenous and exogenous factors, in *Drugs and the Developing Brain* (Vernadakis, A., Weiner, N., eds.), Plenum, New York, pp. 417-427.
113. Tonge, S. R. 1973. Permanent alterations in catecholamine concentrations in discrete areas of brain in the offspring of rats treated with methylamphetamine and chlorpromazine. *Br. J. Pharmacol. 47*:425-427.
114. Vorhees, C. V., Brunner, R. L., Butcher, R. E. 1979. Psychotropic drugs as behavioral teratogens. *Science 205*:1220-1225.
115. Vorhees, C. V., Butcher, R. E. 1982. Behavioral teratogenicity, in *Developmental Toxicology* (Snell, K., ed.), Croom Helm, London, pp. 247-298.
116. Warkany, J. 1971. *Congenital Malformations—Notes and Comments*, Year Book Medical, Chicago.
117. Werboff, J. 1970. Developmental psychopharmacology, in *Principles of Psychopharmacology* (Clark, W. G., del Giudice, J., eds.), Academic, New York, pp. 343-353.
118. Werboff, J., Gottlieb, J. S. 1963. Drugs in pregnancy: Behavioral teratology. *Obstet. Gynecol. Surv. 18*:420-423.
119. Wilson, J. G. 1979. The evolution of teratological testing. *Teratology 20*:205-212.
120. Winneke, G., Brockhaus, A., Baltissen, R. 1977. Neurobehavioral and systemic effects of longterm blood-lead elevation in rats. 1. Discrimination learning and open field behavior. *Arch. Toxicol. 37*:247-263.
121. Yamazaki, J. N. 1966. A review of the literature on the radiation dosage required to cause manifest central nervous system disturbances from in utero and postnatal exposure. *Pediatrics 37*:877-903.
122. York, R. G., Sowry, B. M., Hastings, L., Manson, J. M. 1982. Evaluation of teratogenicity and neurotoxicity with maternal inhalation exposure to methylchloroform. *J. Toxicol. Environ. Health 9*:251-266.
123. Young, R. D. 1967. Developmental psychopharmacology: A beginning. *Psychol. Bull. 67*:73-86.

124. Zbinden, G. 1981. Experimental methods in behavioral teratology. *Arch. Toxicol. 48*:69-88.

125. Zemp, J. W., Middaugh, L. D. 1975. Some effects of prenatal exposure to d-amphetamine sulfate and phenobarbital on developmental neurochemistry and on behavior. *Addict. Dis. Int. J. 2*:307-331.

126. Zenick, H. 1974. Behavioral and biochemical consequences in methylmercury chloride toxicity. *Pharmacol. Biochem. Behavior 2*:709-713.

127. Zenick, H., Padich, R., Tokarek, T., Aragon, P. 1978. Influence of prenatal and postnatal lead exposure on discrimination learning in rats. *Pharmacol. Biochem. Behav. 8*:347-350.

9

Genetic Aspects of Neurotoxicity

Peter Propping*
Institute of Human Genetics, University of Heidelberg, Heidelberg, Federal Republic of Germany

I. INTRODUCTION

The genetic diversity of individuals is an established fact in human genetics. Less well established is our knowledge of genetic effects on neurotoxicity in humans, which is something of an unknown quantity. Recent evidence points overwhelmingly to different responses to xenobiotics from individual to individual because of an interplay between chemicals and genetic factors.

Genetic factors causing increased susceptibility to neurotoxic disorders may play a role in both the metabolism of potentially harmful chemicals and their interaction with the target tissue, that is, the nervous system.

Studies on protein polymorphism based mainly on electrophoretic analysis of blood proteins have provided some insight into average genetic variation. These data suggest that man is likely to exhibit an average heterozygosity per genetic locus of about 20% (28). On the other hand, electrophoretic separation of brain proteins, both in humans and in different strains of mice, has only revealed a very limited degree of possibly genetically based variations (47). This applies even

*Present affiliation:
University of Bonn, Bonn, Federal Republic of Germany

to two-dimensional electrophoretic methods (12). Furthermore, studies aimed at finding genetic diversity in the brain mostly used nonspecific protein staining procedures so that precise knowledge on the function of the few polymorphic proteins known is not yet available.

Since genetic diversity has been found in various tissues, one may reasonably assume that it also exists in the nervous system. Experimental approaches for detecting such a diversity should include assessment of differential reactions to neurotoxicants based on genetic differences established by means of clinical data analysis, family studies, or animal models using inbred strains.

II. GENETIC PHENOMENA IN NEUROTOXICITY WITH A COMPLEX ETIOLOGY

A. Side Effects After Chronic Treatment with Neuroleptics

About 15% of patients receiving phenothiazine drugs in high therapeutic dosages develop extrapyramidal symptoms within a few weeks. It has been shown that the risk of developing these side effects is increased by a factor of about 3 in individuals who have a relative with parkinsonism (46).

An additional side effect of neuroleptics that may develop after prolonged drug administration is tardive dyskinesia (see Pi and Simpson in this volume). This syndrome consists of dystonic movements of the tongue, face, and neck, including buccofacial movements with salivation, oculogyric crisis, and opisthotonus. Involuntary choreiform movements involving the trunk and extremities may also be observed. The incidence of tardive dyskinesia in a population of psychiatric patients continuously exposed to neuroleptics for 4 years was found to be 12% (34).

The risk factors relating a predisposition to the development of tardive dyskinesia are far from clear, but the results of recent studies suggest that genetic determinants may be involved (69). Among 500 inpatients receiving neuroleptics, those with first-degree relatives who had received the same therapy were identified. The eight patient pairs identified matched as regards either the presence or absence of tardive dyskinesia.

The results of animal experiments are also consistent with the hypothesis that genetic factors are involved in the drug-induced tardive dyskinesia. Long-term administration of neuroleptic drugs to rats has been found to produce an increase in the number of caudate dopamine receptors. Since this increase apparently goes hand in hand with sensitivity to behavioral effects of dopamine receptor agonists (48), the rise in the number of dopamine receptors has been proposed as an animal model for tardive dyskinesia. Studies performed on inbred mice have demonstrated remarkable strain-dependent differences in the rise of striatal dopamine receptor numbers after treatment with haloperidol (5, 11). Genetic influences on the behavioral response to both dopaminergic agents and dopamine antagonists have also been reported in mice (58).

Recent data suggest that genetic factors seem to contribute to the interindividual variation in the number of platelet α_2-adrenoreceptors in humans (55). Similar genetic differences might exist with respect to other types of receptors (including those for dopamine in the striatum). This would provide an explanation for the differing individual likelihood of developing tardive dyskinesia after chronic treatment with neuroleptics.

B. Psychoses Induced by Chemical Agents

There is overwhelming empirical evidence for the involvement of genetic factors in the etiology of "endogenous" schizophrenia. A wide variety of factors, including exposure to toxic chemicals, may lead to psychopathological changes indistinguishable from "endogenous" schizophrenia (see Ref. 54). Since not all individuals exposed to chemicals capable of inducing a schizophrenia-like psychosis do in fact develop such a condition, the question arises as to whether interindividual differences exist in the occurrence of toxically induced psychosis. In this respect, family studies carried out on agents acting on the central nervous system, such as methamphetamine, ethanol, and LSD, have clearly demonstrated that the risk is increased in the relatives of the index cases (see Table 1). Thus a psychotic predisposition may be discovered as the result of the exposure to certain chemicals.

As regards organ-specific end points of alcoholism, a study of the medical histories of 15,924 male twin pairs was examined (29). The

Table 1 Findings Supporting the Hypothesis of a Relationship Between Pharmacologically Induced Psychosis and "Endogenous" Schizophrenia

	Findings in relatives
Methamphetamine psychosis	Incidence of schizophrenia in sibs, 4.6%
Alcohol hallucinosis	Lifetime risk of schizophrenia in sibs, 2.6-2.9%
Schizophrenia-like symptoms after LSD intake	Many first-degree relatives of schizophrenics exhibited psychotic symptoms after LSD intake

Source: From Ref. 54.

concordance rate of alcoholic psychosis (delirium tremens, hallucinosis, or Korsakoff's syndrome) was 21.1% for monozygotic and 6.0% for dizygotic twins. The Korsakoff syndrome is a confabulatory psychosis with a disproportionate loss of retentive memory in an otherwise clear sensorium. This condition develops only in a small percentage of alcoholics and is thought to be related to thiamine deficiency in most cases. Recent evidence suggests that a biochemical peculiarity related to the activity of the thiamine-dependent enzyme transketolase plays a role as a factor predisposing to the Korsakoff psychosis (7,40). Thus in some Korsakoff patients red blood cell transketolase shows a reduced activity or an abnormally low affinity for thiamine pyrophosphate (7,39).

C. Tolerance to and Dependence on Alcohol

Among addictive agents, ethanol is the only one for which there is clear-cut evidence that genetic factors are involved in pharmacological effects in humans. Genetic variation may exist in relation to metabolism of alcohol or in relation to alcohol's effect on the central nervous system. Genetic factors are thought to be responsible for the well-known interindividual differences in neurosensitivity (initial tolerance), functional (acquired) tolerance, and physical dependence (4). The analysis of genetic influences on the interaction of alcohol with biological systems is of particular interest because family, twin, and adoption studies indicate that it is very probable that genetic factors are involved in the etiology of alcoholism (25).

Alcohol is predominantly metabolized by two liver enzymes, alcohol dehydrogenase (ADH) and acetaldehyde dehydrogenase (ALDH). The accumulation of the major metabolite of alcohol degradation, acetaldehyde, is apparently responsible for a number of symptoms occurring in acute or chronic intoxication. The role of genetic factors is suggested by the fact that more than 80% of Orientals but only 5% of whites exhibit facial flushing combined with an increase in skin temperature and pulse rate and a feeling of discomfort after ingestion of small amounts of alcoholic beverages (66,67). These symptoms can also be elicited by the administration of acetaldehyde to volunteers (2), as well as by the ingestion of alcohol in subjects pretreated with the aldehyde dehydrogenase inhibitor disulfiram.

Both ADH and ALDH show genetic polymorphism. At the ADH_2 locus, which is responsible for most of the liver ADH activity in adults, two alleles are known (see Ref. 70). These alleles differ remarkably in their racial distribution: More than 90% of whites only have the usual enzyme ADH_2^1, whereas 85% of Orientals have the "atypical" enzyme ADH_2^2. At the physiological pH the atypical ADH is many times more active than the usual one, and it has been hypothesized that the racial difference in alcohol sensitivity could be due to different rates of acetaldehyde formation (60), even though there is no apparent difference in the rate of alcohol elimination between carriers of ADH_2^1 and ADH_2^2.

Liver ALDH also differs between whites and Orientals. Virtually all whites have two major isoenzymes, ALDH I and ALDH II, while about 50% of Orientals have only ALDH II. Since ALDH I has a high affinity for acetaldehyde, the absence of this isoenzyme might well be responsible for initially high concentrations of acetaldehyde formed after alcohol intake. This suggests that ALDH I deficiency may play a role in the "flushing reaction" mechanism and in the higher alcohol sensitivity among Orientals occurring as the result of the development of abnormally high acetaldehyde levels (24).

Acetaldehyde dehydrogenase polymorphism attracted particular interest when detected in fibroblasts and hair roots, opening up the way for genetic studies in populations (23). Since individuals with ALDH I deficiency experience discomfort and symptoms of intoxication at low alcohol intake levels, the possible relevance of this condition to susceptibility to alcoholism must be considered. In Orientals, alcoholism is much less frequent than in whites. Harada et al. (27) observed very few subjects with ALDH I deficiency among Japanese alcoholics as compared with 50% deficiency in healthy controls. This finding would suggest that Orientals, or at least approximately half of them, have metabolic protection from overusing alcoholic beverages and thus developing alcoholism. Since Europeans do not have a genetic polymorphism, this process is presumably of little significance in their vulnerability to alcohol effects.

Among the various procedures suggested for studying genetically dependent variations in acute effects of alcohol on the central nervous system, electroencephalography has several advantages. The resting electroencephalogram exhibits a remarkable interindividual variation within the normal range which is genetically determined (63). Moreover, the electroencephalogram is the most sensitive means of measuring the effects of centrally acting drugs objectively.

In a twin study, it was demonstrated that individuals differ remarkably in their electroencephalographic reactions to alcohol and that the differential response is under genetic control (52). There is also evidence that certain electroencephalographic patterns may reflect a degree of vulnerability for chronic alcohol abuse (19,51,56) and possibly other psychiatric disorders (32).

D. Malignant Hyperthermia

Malignant hyperthermia (or hyperpyrexia) is an uncommon but often fatal complication of general anesthesia, characterized by a dramatic rise in core temperature and generalized muscle rigidity in 70-80% of reported cases. This condition has been associated with exposure to a variety of anesthetic agents, including halothane, nitrous oxide, methoxyflurane, cyclopropane, and their combinations. It is more common when succinylcholine is used as premedication. The incidence in North America has been estimated at 1 in 14,000 general anesthesias (8). Susceptibility to malignant hyperthermia can apparently be triggered by factors other than anesthesia, since this condition has also been related to heat stroke or sudden infant death (13,14).

Though a positive family history can be established in only one-third of all patients, 80% of families investigated after the occurrence of an index case were found to contain susceptible relatives, suggesting the implication of genetic factors and the existence of an individual predisposition that may be revealed by general anesthetics (44).

The increase in plasma creatine phosphokinase in affected families may represent the mildest form of myopathy associated with susceptibility to malignant hyperthermia and may be helpful in identifying persons at risk. On the other hand, about one-third of all affected patients have a history of previously uncomplicated anesthesia, making genetic analysis difficult. Screening tests developed to determine susceptibility to disease yielded abnormal findings in many but not all cases (44). Muscle biopsies from patients who survived malignant hyperthermia provided evidence of variable and nonspecific changes. An in vitro pharmacological assay on isolated skeletal muscle has been developed. The test consists of measuring the contraction tension of muscle fibers when exposed to various concentrations of caffeine both in the presence and absence of the anesthetic agent. It has been suggested that in malignant hyperthermia an abrupt loss in control of calcium movements in the presence of triggering agents causes a rise in intracellular calcium, muscle contracture, metabolic stimulation, and permeability changes. This is reflected in the lower concentrations of caffeine or halothane needed to produce contracture in vitro in muscle biopsy specimens from susceptible patients (26). In fact, preparations from patients with rigid malignant hyperthermia show an increased sensitivity to caffeine (33), but this kind of response is less consistent in the nonrigid form as well as in relatives of index cases.

A total of 80% of subjects who survived malignant hyperthermia and some of their relatives have high plasma creatine phosphokinase levels. Electromyography is another test that reveals abnormalities in some of the cases (42). In the original description of malignant hyperthermia as a pharmacogenetic disorder, an autosomal dominant inheritance with incomplete penetrance was postulated (15). A Mendelian mode of inheritance with "incomplete penetrance" has frequently been proposed when the observed family data were not compatible with the results expected on theoretical grounds. In the meantime, malignant hyperthermia proved to be genetically heterogeneous (44): Half of the families studied demonstrated autosomal dominant transmission; about 20% of cases were sporadic. In a few families, recessive inheritance cannot be excluded, but most of the remaining pedigrees suggest multifactorial inheritance (44).

Malignant hyperthermia has also occurred in certain muscle disease such as Duchenne muscular dystrophy, myotonia congenita, and central core disease (9,44). In the majority of such cases general anesthesia is uneventful. Similarities in the clinical signs of malignant hyperthermia and the neuroleptic malignant syndrome suggest that these disorders may have a common pathophysiology (10).

III. NEUROTOXIC PHENOMENA WITH A SIMPLE GENETIC BASIS

Analytical understanding is usually better in cases of the Mendelian mode of inheritance of a trait which points to one single factor being responsible.

A. N-Acetyltransferase Polymorphism

Isoniazid and several other monosubstituted hydrazines such as phenelzine, hydralazine, dapsone, sulfamethazine, procainamide, and the amino metabolite of nitrazepam are acetylated in the liver by a soluble N-acetyltransferase which is known to exist in two distinct forms exhibiting different catalytic activity. The two enzyme forms are determined by two different alleles which are responsible for either "rapid" or "slow" elimination of the drug. N-Acetyltransferase polymorphism follows a Mendelian mode of inheritance, with "rapid" elimination being an autosomal dominant trait (see Ref. 53). Several methods have been developed giving reliable phenotyping of an individual as belonging to the group of either "rapid" or "slow" acetylators. In whites and blacks "rapid" and "slow" acetylators have the same frequency, about 50% each, whereas up to 80-90% of Orientals are known to be rapid acetylators (53). The physiological function of polymorphic N-acetyltransferase is unknown. The two types of acetylators are healthy and can only be differentiated on the basis of the rate of elimination of the drugs that are substrates for the enzyme.

The N-acetyltransferase polymorphism must be considered in relation to potential side effects of the tuberculostatic agent isoniazid and of other drugs which are potentially neurotoxic and metabolized by N-acetylation. Slow acetylators develop higher plasma levels of isoniazid and are more prone to developing a toxic peripheral neuropathy, as compared with rapid acetylators (41). Occasionally hydralazine (61) and dapsone (36) have also been described as inducing peripheral nervous system disorders in slow acetylators. Pyridoxine deficiency is apparently involved in these side effects, and clinical data suggest that isoniazid-induced polyneuritis can be prevented by the simultaneous administration of pyridoxine (30). It seems significant that a very high incidence of slow acetylators has been observed among diabetics with peripheral neuropathy both in juvenile or maturity onset diabetes (43). These patients were not exposed to drugs serving as substrates for the polymorphic N-acetyltransferase. It would also appear that slow acetylators are more likely to develop the idiopathic lupus erythematosus-like syndrome, even when not treated with hydralazine (53).

Isoniazid may occasionally induce schizophrenia-like psychoses (49) and convulsions (57). The monoamine oxidase inhibitor phenelzine has also been reported to induce psychosis (59). Slow acetylators may thus be at a higher risk for these central neurotoxic disorders.

B. Serum Paraoxonase Polymorphism

Paraoxon, the metabolite of the neurotoxic organophosphorous insecticide parathion, is a direct irreversible inhibitor of acetylcholinesterase. The arylesterase paraoxonase (EC 3.1.1.2) of human serum promotes the paraoxon hydrolysis leading to formation of p-nitrophenol (Fig. 1). Methyl paraoxon and chloromethylparaoxon are additional substrates for this enzyme (18).

In whites paraoxonase activity has been reported to be polymorphic (22) and to exhibit either a bimodal (17,50) or trimodal distribution (21,22). Paraoxonase phenotypes have been classified on the basis of serum enzyme activity measured by different procedures, and evidence has been obtained for the involvement of a dominant gene in the regulation of the enzyme (31). It seems that the low phenotype represents homozygosity for a low allele with a gene frequency of about 0.7 (17).

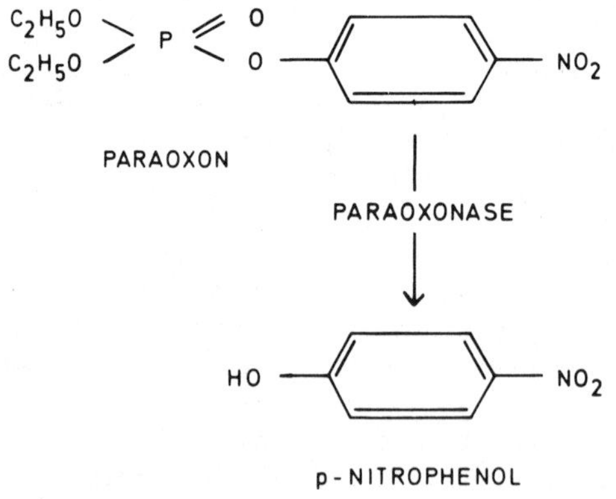

Figure 1 Reaction catalyzed by paraoxonase.

Recent data indicate that the quantitative differences in paraoxonase activity reflect a qualitative difference in a two-allele system (16): A high-activity allele can be stimulated by NaCl or KCl, while a low-activity allele is not stimulated by salts. Studies are needed to clarify whether the different paraoxonase phenotypes have different neurotoxic risk rates following exposure to organophosphorous insecticides (20).

C. Pseudocholinesterase Variants

Pseudocholinesterase (EC 3.1.1.8) is a nonspecific esterase located in plasma and tissues which catalyzes the hydrolysis of a variety of choline esters, including certain agents of pharmacological interest. A typical substrate for pseudocholinesterase is the short-acting neuromuscular blocking drug succinylcholine, which is commonly used in anesthesia. There is a good inverse correlation between duration of the neuromuscular blockade by succinylcholine and pseudocholinesterase activity in plasma. The pharmacological effect of this compound, which usually lasts about 2 min, may last for up to 2-3 hr in individuals whose capacity to metabolize it is reduced. At least two loci of pseudocholinesterase have been identified, one of which ($\underline{E}_{\underline{1}}$) is genetically well established.

Variation in plasma pseudocholinesterase activity has been observed in a variety of conditions (64). Reduced enzyme activity associated with abnormally high sensitivity to succinylcholine has been related to the occurrence of rare enzyme variants. At least four allelic genes exist at the $\underline{E}_{\underline{1}}$ locus and have been designed as follows: $\underline{E}_{\underline{1}}^{\underline{u}}$, unusual; $\underline{E}_{\underline{1}}^{\underline{a}}$, atypical, dibucaine-resistant; $\underline{E}_{\underline{1}}^{\underline{f}}$, fluoride-resistant; and $\underline{E}_{\underline{1}}^{\underline{s}}$, silent. The alleles have been classified according to the enzyme sensitivity to selected inhibitors such as dibucaine and fluoride. The four alleles give rise to 10 different genotypes, all of which have been recognized, with various sensitivities to succinylcholine (64) (see Table 2). Exceptionally, in some families, additional alleles ($\underline{E}_{\underline{i}}^{\underline{j}}$, $\underline{E}_{\underline{1}}^{\underline{k}}$) have been described (53).

Evidence exists that approximately 70% of patients with prolonged apnea following succinylcholine administration have detectable variants of pseudocholinesterase, which may be an explanation as to why this side effect occurs. In addition, 13% of the relatives of individuals sensitive to succinylcholine are apparently themselves sensitive to the drug (64). The sensitivity depends on the genotype of both the proband and his relative (see Table 2).

Studies carried out using highly purified pseudocholinesterase pointed to a structural abnormality in the active center of the variant forms (45). In the enzyme obtained from the $\underline{E}_{\underline{1}}^{\underline{u}}$ and $\underline{E}_{\underline{1}}^{\underline{a}}$ forms the amino acid

Table 2 Pseudocholinesterase types, percent inhibition of enzyme activity by dibucaine and fluoride, and succinylcholine sensitivity

Genotype of clinical significance	Dibucaine number	Fluoride number	Phenotype frequency in "European" populations	Succinylcholine sensitivity
$E_1^a\ E_1^a$	22	27	1:3,200	+++
$E_1^s\ E_1^s$	0	0	1:170,000	++++
$E_1^f\ E_1^f$	66	35	1:28,000	++
$E_1^a\ E_1^s$	22	27	1:11,000	+++
$E_1^a\ E_1^f$	49	33	1:2,500	+++
$E_1^f\ E_1^s$	67	43	1:33,000	++
$E_1^u\ E_1^u$	80	59	95%	None
$E_1^u\ E_1^a$	62	48	3%	(+)
$E_1^u\ E_1^f$	74	50	1%	(+)
$E_1^u\ E_1^s$	80	59	1:200	+

Source: From Refs. 53 and 65.

sequence of the active site was found to differ only in the exchange of a single amino acid unit (68).

D. Diphenylhydantoin Hydroxylation

The primary and rate-limiting step in the metabolism of diphenylhydantoin (DPH) is p-hydroxylation to 5-phenyl-5'-p-hydroxyphenylhydantoin, which then undergoes excretion in urine as the glucuronide acid conjugate. Individuals differ remarkably in their steady-state plasma DPH concentrations (53) and the rate of elimination of the drug is genetically determined (1). In subjects with a reduced rate of drug elimination, therapeutic doses of DPH may give rise to excessive plasma drug concentrations associated with a high incidence of neurological side effects such as nystagmus, ataxia, and lethargy (3; see also Chapter 11 by Bittencourt, Perucca, and Crema).

Three two-generation families with subjects of both generations showing deficient DPH p-hydroxylation have been described (37,38). In another four-generation family studied by Vasko et al. (62) four subjects in three distinct generations exhibited a defective metabolism of the drug. The reported pedigrees seem to be compatible with autosomal dominant inheritance of deficient DPH hydroxylation, but polygenic inheritance cannot be ruled out. The hydroxylation of phenytoin is inhibited by isoniazid. This was found to occur more frequently, with increased risk of phenytoin toxicity, in slow isoniazid acetylators (37,38).

E. Porphyrias

These metabolic disorders are characterized by inherited enzyme defects or acquired abnormalities in the regulation of heme synthesis with excretion of large amounts of heme precursors (6). A wide variety of drugs and chemicals have stimulatory effects on heme synthesis leading to the accumulation of porphyrin precursors. They include barbiturates, sulfonamides, antiepileptics, imipramine, oral hypoglycemic agents, estrogens, and oral contraceptives (35). In subjects with latent porphyria these drugs may precipitate an acute porphyric attack, that is, an endogenous intoxication characterized by abdominal pain, vomiting, hypertension, tachycardia, mental disturbances, loss of consciousness, and convulsions.

Porphyrias are generally regarded as rare diseases. Since the enzyme disorder usually has no consequences in the absence of exposure to drugs, the number of undiagnosed cases may be not so small. Indeed, many previously undiagnosed cases were discovered during screening studies on selected groups of patients treated with porphyrogenic drugs (54).

ACKNOWLEDGMENTS

This work was supported by the Deutsche Forschungsgemeinschaft.

REFERENCES

1. Andreasen, P. B., Froland, A., Skovsted, L., Andersen, S. A., Hauge, M. 1973. Diphenylhydantoin half-life in man and its inhibition by phenylbutazone: The role of genetic factors. *Acta Med. Scand. 193*:561-564.
2. Asmussen, E., Hald, J., Larsen, V. 1948. The pharmacological action of acetaldehyde on the human organism. *Acta Pharmacol. 4*:311-320.

3. Atkinson, A. J., Shaw, J. M. 1973. Pharmacokinetic study of a patient with diphenylhydantoin toxicity. *Clin. Pharmacol. Ther. 14*:521-528.
4. Belknap, J. K. 1980. Genetic factors in the effects of alcohol: Neurosensitivity, functional tolerance and physical dependence, in *Alcohol Tolerance and Dependence* (Rigter, H. and Crabbe, J. C., eds.), Elsevier/North-Holland, Amsterdam, pp. 157-180.
5. Belmaker, R. H., Bannet, J., Brecher-Fride, E., Yanai, J., Ebstein, R. P. 1981. The effect of haloperidol feeding on dopamine receptor number. *Clin. Genet. 19*:353-356.
6. Bickers, D. R. 1982. Environmental and drug factors in hepatic porphyria. *Acta Derm. Venereol. Suppl. 100*:29-41.
7. Blass, J. P., Gibson, G. E. 1979. Genetic factors in Wernicke-Korsakoff syndrome. *Alcohol. Clin. Exp. Res. 3*:126-134.
8. Britt, B. A, Kalow, W. 1970. Malignant hyperthermia: A statistical review. *Can. Anaesth. Soc. J. 17*:293-315.
9. Brownell, A. K. W., Paasuke, R. T., Elash, A., Fowlow, S. B., Seagram, C. G. F., Diewold, R. J., Friesen, C. 1983. Malignant hyperthermia in Duchenne muscular dystrophy. *J. Anesthesiol. 58*:180-182.
10. Carof, S., Rosenberg, H., Gerber, J. C. 1983. Neuroleptic malignant syndrome and malignant hyperthermia. *Lancet 1*:244.
11. Ciaranello, R. D., Boehme, R. E. 1981. Biochemical genetics of neurotransmitter enzymes and receptors: Relationships to schizophrenia and other major psychiatric disorders. *Clin. Genet. 19*:358-372.
12. Comings, D. E. 1981. Pc 1 Duarte—A polymorphism associated with depression, in *Genetic Research Strategies in Psychobiology and Psychiatry* (Gershon, E. S., Matthysse, S., Breakefield, X. O., Ciaranello, R. D., eds.), Boxwood, Pacific Grove, California, pp. 59-64.
13. Denborough, M. A. 1982. Heat stroke and malignant hyperpyrexia. *Med. J. Aust. 1*:204-205.
14. Denborough, M. A., Galloway, G. J., Hopkinson, K. C. 1982. Malignant hyperpyrexia and sudden infant death. *Lancet 2*:1068-1069.
15. Denborough, M. A., Lovell, R. R. H. 1960. Anaesthetic deaths in a family. *Lancet 2*:45.
16. Eckerson, H. W., Romson, J., Wyte, C., La Du, B. N. 1983. The human serum paraoxonase polymorphism: Identification of phenotypes by their response to salts. *Am. J. Hum. Genet. 35*:214-227.
17. Eiberg, H., Mohr, J. 1981. Genetics of paraoxonase. *Ann. Hum. Genet. 45*:323-330.
18. Flügel, M., Geldmacher-von Mallinckrodt, M. 1978. Zur Kinetik des Paraoxon-spaltenden Enzyms in menschlichen Serum (EC

3.1.1.2). *Klin. Wochenschr.* *56*:911-916.
19. Gabrielli, W. F., Mednick, S. A., Volavka, J., Pollock, V. E., Schulsinger, F., Itil, T. M. 1982. Electroencephalograms in children of alcoholic fathers. *Psychophysiology 19*:404-407.
20. Geldmacher-von Mallinckrodt, M. 1978. Polymorphism of human serum paraoxonase. *Hum. Genet. Suppl. 1*:65-68.
21. Geldmacher-von Mallinckrodt, M., Hommel, G., Dumbach, J. 1979. On the genetics of the human serum paraoxonase (EC 3.1.1.2). *Hum. Genet. 50*:313-327.
22. Geldmacher-von Mallinckrodt, M., Lindorf, H. H., Petenyi, M., Flügel, M., Fischer, T., Hiller, T. 1973. Genetisch determinierter Polymorphismus der menschlichen Serum-Paraoxonase (EC 3.1.1.2). *Humangenetik 17*:331-335.
23. Goedde, H. W., Agarwal, D. P., Harada, S., Meier-Tackmann, D., Ruofu, D., Bienzle, U., Kroeger, A., Hussein, L. 1983. Population genetic studies on aldehyde dehydrogenase isozyme deficiency and alcohol sensitivity. *Am. J. Hum. Genet. 35*:769-772.
24. Goedde, H. W., Harada, S., Agarwal, D. P. 1979. Racial differences in alcohol sensitivity: A new hypothesis. *Hum. Genet. 51*: 331-334.
25. Goodwin, D. W. 1979. Alcoholism and heredity. *Arch. Gen. Psychiatry 36*:57-61.
26. Gronert, G. A. 1980. Malignant hyperthermia. *Anesthesiology 53*:345-423.
27. Harada, S., Agarwal, D. P., Goedde, H. W., Ishikawa, B. 1983. Aldehyde dehydrogenase isoenzyme variation and alcoholism in Japan. *Pharmacol. Biochem. Behav. 18*:151-153, Suppl. 1.
28. Harris, H. 1980. *The Principles of Human Biochemical Genetics*, Elsevier/North-Holland, Amsterdam.
29. Hrubec, Z., Omenn, G. S. 1981. Evidence of genetic predisposition to alcoholic cirrhosis and psychosis: Twin concordances for alcoholism and its biological end points by zygosity among male veterans. *Alcohol. Clin. Exp. Res. 5*:207-215.
30. Irskens, K. J. 1964. Neurologische und psychotische Komplikationen bei tuberkulostatischer Behandlung. *Nervenarzt 35*:415-416.
31. Iselius, L., Evans, D. A. P., Playfer, J. R. 1982. Genetics of plasma paraoxonase activity. *J. Med. Genet. 19*:424-426.
32. Itil, T. M., Marasa, J., Saletu, B., Davis, S., Mucciardi, A. N. 1975. Computerized EEG: Predictor of outcome in schizophrenia. *J. Nerv. Ment. Dis. 160*:188-203.
33. Kalow, W., Britt, B. A., Richter, A. 1977. The caffeine test of isolated human muscle in relation to malignant hyperthermia. *Can. Anaesth. Soc. J. 24*:678-694.
34. Kane, J. M., Woerner, M., Weinhold, P., Wegner, J., Kinon, B. 1982. A prospective study of tardive dyskinesia development:

Preliminary results. *J. Clin. Psychopharmacol.* *2*:345-349.
35. Kappas, A., Sassa, S., Anderson, K. E. 1983. The porphyrias, in *The Metabolic Basis of Inherited Disease* (Stanbury, J. B., Wyngaarden, J. B., Fredrickson, D. S., Goldstein, J. L., Brown, M. S., eds.), McGraw-Hill, New York, pp. 1301-1384.
36. Koller, W. C., Gehlmann, L. K., Malkinson, F. D., Davis, F. A. 1977. Dapsone-induced peripheral neuropathy. *Arch. Neurol.* *34*:644-646.
37. Kutt, H. 1971. Biochemical and genetic factors regulating dilantin metabolism in man. *Ann. N.Y. Acad. Sci.* *179*:704-722.
38. Kutt, H., Wolk, M., Scherman, R., McDowell, F. 1964. Insufficient parahydroxylation as a cause of diphenylhydantoin toxicity. *Neurology* *14*:542-548.
39. Leigh, D., McBurney, A., McIlwain, H. 1981a. Erythrocyte transketolase activity in the Wernicke-Korsakoff syndrome. *Br. J. Psychiatry* *139*:153-156.
40. Leigh, D., McBurney, A., McIlwain, H. 1981b. Wernicke-Korsakoff syndrome in monozygotic twins: A biochemical peculiarity. *Br. J. Psychiatry* *139*:156-159.
41. Lunde, P. J. M., Frislid, K., Hansteen, V. 1977. Disease and acetylation polymorphism. *Clin. Pharmacokinet.* *2*:182-197.
42. Mamoli, B., Sporn, B., Steinbereithner, Dal-Bianco, P. 1980. Elektrophysiologische Untersuchungen bei einer Familie mit maligner Hyperthermie. *Fortschr. Neurol. Psychiatr.* *48*: 314-323.
43. McLaren, E. H., Burden, A. C., Moorhead, P. J. 1977. Acetylator phenotype in diabetic neuropathy. *Br. Med. J.* *2*:291-293.
44. McPherson, E., Taylor, C. A. 1982. The genetics of malignant hyperthermia: Evidence for heterogeneity. *Am. J. Med. Genet.* *11*:273-285.
45. Muensch, H., Yoshida, A., Altland, K., Jensen, W., Goedde, H.-W. 1978. Structural difference at the active site of dibucaine resistant variant of human plasma cholinesterase. *Am. J. Med. Genet.* *30*:302-307.
46. Myrianthopoulos, N. C., Kurland, A. A., Kurland, L. T. 1962. Hereditary predisposition in drug-induced parkinsonism. *Arch. Neurol.* *6*:5-9.
47. Omenn, G. S. 1981. Protein polymorphism, in *Genetic Research Strategies in Psychobiology and Psychiatry* (Gershon, E. S., Matthysse, S., Breakefield, X. O., Ciaranello, R. D., eds.), Boxwood, Pacific Grove, California, pp. 65-77.
48. Owen, F., Cross, A. J., Waddington, J. L., Poulter, M., Gamble, S. J., Crow, T. J. 1980. Dopamine-mediated behavior and ^{3}H-spiperone binding to striatal membranes in rats after nine months of haloperidol administration. *Life Sci.* *26*:55-59.
49. Pauleikhoff, B. 1957. Über psychopathologische Besonderheiten

einer unter Isonicotinsäurehydrazid (INH)–Behandlung aufgetretenen Psychose. *Arch. Psychiatr. Nervenkr. 195*:489-501.
50. Playfer, J. R., Eze, L. C., Bullen, M. F., Evans, D. A. P. 1976. Genetic polymorphism and interethnic variability of plasma paroxonase activity. *J. Med. Genet. 13*:337-342.
51. Pollock, V. E., Volavka, J., Goodwin, D. W., Mednick, S. A., Gabrielli, W. F., Knop, J., Schulsinger, F. 1983. The EEG after alcohol administration in men at risk for alcoholism. *Arch. Gen. Psychiatr. 40*:857-861.
52. Propping, P. 1977. Genetic control of ethanol action on the central nervous system. An EEG study in twins. *Hum. Genet. 35*:309-334.
53. Propping, P. 1978. Pharmacogenetics. *Rev. Physiol. Biochem. Pharmacol. 83*:123-173.
54. Propping, P. 1983. Genetic disorders presenting as "schizophrenia." Karl Bonhoeffer's early view of the psychoses in the light of medical genetics. *Hum. Genet. 65*:1-10.
55. Propping, P., Friedl, W. 1983. Genetic control of adrenergic receptors on human platelets. A twin study. *Hum. Genet. 64*:105-109.
56. Propping, P., Krüger, J., Mark, N. 1981. Genetic disposition to alcoholism. An EEG study in alcoholics and their relatives. *Hum. Genet. 59*:51-59.
57. Ramakrishna, T., Smith, B. H. 1970. Seizures secondary to isoniazid ingestion. *Neurology 20*:1142-1145.
58. Severson, J. A., Randall, P. K., Finch, C. E. 1981. Genotypic influences on striatal dopaminergic regulation in mice. *Brain Res. 210*:201-215.
59. Sheehy, L. M., Maxmen, J. S. 1978. Phenelzine-induced psychosis. *Am. J. Psychiatry 135*:1422-1423.
60. Stamatoyannopoulos, G., Chen, S.-H., Fukui, M. 1975. Liver alcohol dehydrogenase in Japanese: High population frequency of atypical form and its possible role in alcohol sensitivity. *Am. J. Hum. Genet. 27*:789-796.
61. Tsujimoto, G., Horai, Y., Ishizaki, T., Itoh, K. 1981. Hydralazine-induced peripheral neuropathy seen in a Japanese slow acetylator patient. *Br. J. Clin. Pharmacol. 11*:622-625.
62. Vasko, M. R., Bell, R. D., Daly, D. D., Pippenger, C. E. 1980. Inheritance of phenytoin hypometabolism: A kinetic study of one family. *Clin. Pharmacol. Ther. 27*:96-103.
63. Vogel, F. 1970. The genetic basis of the normal human electroencephalogram (EEG). *Humangenetik 10*:91-114.
64. Whittaker, M. 1980. Plasma cholinesterase variants and the anaesthetist. *Anaesthesia 35*:174-197.
65. *WHO Tech. Rep. Ser. No. 524.* 1973. Pharmacogenetics.
66. Wolff, P. H. 1972. Ethnic differences in alcohol sensitivity. *Science 175*:449-450.

67. Wolf, P. H. 1973. Vasomotor sensitivity to alcohol in diverse mongoloid populations. *Am. J. Hum. Genet.* *25*:193-199.

68. Yamamoto, K., Huang, I.-Y., Muensch, H., Yoshida, A., Goedde, H.-W., Agarwal, D. P. 1983. Amino acid sequence of the active site of human pseudocholinesterase. *Biochem. Genet.* *21*:135-145.

69. Yassa, R., Ananth, J. 1981. Familial tardive dyskinesia. *Am. J. Psychiatry* *138*:1618-1619.

70. Yoshida, A. 1982. Molecular basis of difference in alcohol metabolism between Orientals and Caucasians. *Jpn. J. Hum. Genet.* *27*:55-70.

Part II

SELECTED CLASSES OF DRUGS WITH NEUROTOXICITY POTENTIAL

10

Central Nervous System Toxicity of Psychopharmacological Agents

Edmond H. Pi and George M. Simpson
University of Southern California, School of Medicine, Los Angeles, California

I. INTRODUCTION

During the past three decades there has been steady progress in the field of psychopharmacology, serving to greatly alleviate human suffering from various types of psychiatric illnesses. This has led to deinstitutionalization of previously chronically hospitalized patient populations, more rapid improvement, and in some cases even total remission. However, these psychopharmacological agents may produce many adverse effects, particularly on the central nervous system. These range from mild sedation to sometimes irreversible toxic reactions. Adequate knowledge of the pharmacology of these medications is therefore clinically imperative. In this chapter the clinical aspects of adverse central nervous system effects produced by psychopharmacological agents are discussed; this includes side effects encountered with therapeutic use as well as those from misuse or abuse, including withdrawal reactions.

II. NEUROLEPTICS

Unwanted effects on the central nervous system are one of the important adverse effects produced by neuroleptics. These central nervous system side effects were reported soon after the introduction of chlorpromazine in the 1950s. Since then, many of these effects have been systematically studied, some have been successfully prevented and treated, while others, like tardive dyskinesia and the neuroleptic malignant syndrome, remain puzzles.

A. Sedation

Sedation is the most common central nervous system side effect of neuroleptics. It is present in many subjects at low doses, but is observed more often when an excessive amount is ingested or in combined treatment when other sedatives are given, particularly to elderly populations. The initial mild sedation generally disappears within 1-2 weeks; if not, reducing the dosage most often eliminates this effect. Since neuroleptics suppress central nervous system function in general, extra caution must be exercised when treating patients with respiratory disorders or any precomatose states such as severe liver disease or head injury.

In cases of overdose, patients may present acute onset drawsiness, agitation, extrapyramidal side effects, decreased deep tendon reflex, seizures, delirium, or even fatal hyperpyrexia due to disturbance of temperature regulation. The basic principles of treatment are supportive and symptomatic.

B. Behavioral Side Effects

1. *Acute Exacerbation of Psychosis*

Usually this occurs within the first few days of neuroleptic administration. The clinical picture consists of acute onset and dramatic worsening of preexisting psychotic symptoms, sometimes associated with extrapyramidal side effects and tremor. It can be managed by increasing the dose if no extrapyramidal side effects are present; if no improvement is observed, decreasing or discontinuing the neuroleptic or prescribing antiparkinsonian agents usually provides good relief (48,57, 58). Neuroleptics in some cases can produce delirium, particularly if combined with antiparkinsonian agents; in such cases, it is best to stop medications and observe for a few days.

2. *Catatonic-Like States*

These usually consist of acute onset of extrapyramidal side effects and catatonic features, that is, mutism, waxy flexibility, negativism, withdrawal, and posturing. Certain organic factors and major psychiatric disorders which may cause the condition must be carefully ruled out; otherwise the treatment involves reduction or elimination of neuroleptics. The addition of an anticholinergic may also be helpful (1,2,14,29,66).

3. *Klüver-Bucy-Like Syndrome*

There have been two cases of a neuroleptic-induced "Klüver-Bucy syndrome" reported (60) which respond well to antiparkinsonian medication.

4. *Electroencephalographic Changes and Epileptogenic Effects*

Neuroleptics are well known to cause generalized slowing of alpha rhythm, increased synchronization and amplitude, with superimposed sharp, fast activity (13). These electroencephalographic changes usually persist between 6 and 10 weeks after cessation of neuroleptics (54).

Most neuroleptics are epileptogenic (20,28,47), the more potent ones less so than the less potent ones, such as chlorpromazine or thioridazine; these effects are more likely when low-potency neuroleptics are prescribed at high doses with rapid increases in dose level. Although there is no evidence that prescribing anticonvulsants prevents seizure attacks, it would appear beneficial to carefully monitor dosage of anticonvulsants with known epileptic patients who need neuroleptics for concomitant psychosis.

5. *Neuroleptic Malignant Syndrome*

This is a rare form of potentially fatal central nervous system toxicity produced by neuroleptics, particularly certain high-potency neuroleptics. This syndrome, which is still being defined, is characterized by muscular rigidity, hyperthermia, altered consciousness, and autonomic dysfunction (4); it may progress to coma and death. It is best to discontinue all neuroleptics and provide patients with supportive treatments. Antiparkinsonian agents are reported as ineffective in reducing the duration or the mortality of the neuroleptic malignant syndrome. A recent report claims that one case was successfully treated with amantadine (30).

6. *Hyperpyrexia*

Whether this is a separate entity or part of the neuroleptic malignant syndrome remains unresolved. It is attributed to the effect of neuroleptics on the hypothalamic temperature regulation center, resulting in rapid elevation of body temperature in hot climates (3). Since the death rate from hyperpyrexia is high, it requires vigorous medical treatment, as in the treatment of heat stroke. In cold climates hypothermia may occur, since the effect in the temperature control system leaves ambient temperature in control of body temperature.

7. Extrapyramidal Side Effects

All the neuroleptics currently available cause various degrees of extrapyramidal side effects, with the possible exception of clozapine (50).

a. Acute Dystonic Reactions: These reactions consist of acute, bizarre muscular spasms mainly affecting the head, neck, and facial muscles—oculogyric crises, torticollis, and lockjaw. In more dramatic cases, the muscles of the back and extremities are affected. These reactions are more likely to affect males than females, the young rather than the old, and to occur with higher dosages more often than with lower dosages. Onset usually occurs in the first 24-48 hr after initiating neuroleptics or, less frequently, after increase of the dose. Parenteral administration of antiparkinsonian agents can dramatically interrupt reactions.

b. Akathisia: The onset is usually within a few days of initiation, but it can also occur later in neuroleptic treatment. Clinical manifestations include restlessness, agitation, inability to sit still or tolerate inactivity, rocking, and shifting of the legs while in a sitting position. Sometimes akathisia may be misdiagnosed as exacerbation of psychosis and result in an increase of neuroleptics which may worsen the akathisia. Antiparkinsonian agents or benzodiazepines may alleviate the symptoms. If there is no symptom relief, the dosage should be reduced or the neuroleptic withdrawn.

c. Pseudoparkinsonism: Indistinguishable from classic parkinsonism, it usually appears after a few days to a few weeks. Symptom triads are the following: (a) akinesia, defined as loss of voluntary movement of the extremities, flexion of the arms, a stooped and shuffling gait, pill-rolling tremor of the fingers, slow monotonous speech, and depressive-like manner with dysphoria recently described as akinetic depression (40,59); (b) resting tremor, which usually presents as a regular rhythmic oscillation of the fingers and hand, 4-8/sec, that improves when hands are in motion; and (c) cogwheel and other rigidity. All these symptoms can be alleviated by either reducing the dosage of neuroleptics or adding antiparkinsonian agents. They are totally reversible after stopping neuroleptics (50).

8. Tardive Dyskinesia

This late onset neurological condition usually occurs after several years of taking neuroleptics, although there have been a few early onset cases reported. The elderly tend to be affected more than the young, and women more than men. Other putative predisposing factors include individual susceptability, antiparkinsonian agent usage, duration and doses of treatment with neuroleptics, affective disorder, history of alcohol abuse, organic brain syndrome, and brain injury. Overall prevalence is quoted as 0.5-65% (51). Recently the syndrome has been expanded from the initial buccolingual-masticatory syndrome to include various choreoathetoid movements of the extremities and trunk (34,51, 52). There are, as a rule, no subjective complaints from the patient. Dyskinetic movements disappear during sleep. Some of the dyskinetic movements worsen on withdrawal of neuroleptics and then gradually disappear after discontinuation. Considerable efforts have been made to distinguish withdrawal tardive dyskinesia (reversible) from persistent tardive dyskinesia (irreversible). However, as of now it is impossible to predict reversibility. Differential diagnoses include Huntington's chorea, Wilson's disease, Tourette's syndrome, schizophrenic mannerism and stereotypical behavior, and senile dyskinesia (15). Of these conditions, spontaneous senile dyskinesia is gaining particular attention in relation to the role of neuroleptics in the etiology of tardive dyskinesia. One study found little difference in the incidence of dyskinetic movements between neuroleptic-treated versus nonneuroleptic-treated schizophrenic groups (22), and another two found a high incidence of dyskinetic movements, ranging from 9 to 37%, in elderly populations who never received neuroleptics (9,61).

Pharmacological treatment of tardive dyskinesia is difficult, unsatisfactory, and inconsistent. Most of the available pharmacological interventions are based on one hypothesis of dopamine receptor hypersensitivity and are all in an experimental stage. Therefore, prevention is the most important medical strategy in dealing with tardive dyskinesia (36).

9. *Other Central Nervous System Side Effects*

All neuroleptics suppress release of the prolactin-inhibiting factor from the hypothalamus which is mediated by dopaminergic pathways, and consequently, produce an elevation of circulating serum prolactin level (32). Therefore galactorrhea and gynecomastia may be seen in patients treated with neuroleptics, whether males or females (see also Chapter 4 by Brambilla).

III. ANTIDEPRESSANTS

A. Tricyclic Antidepressants

1. *Overdose*

Both accidental and purposeful overdose of tricyclic antidepressants have shown a rapid increase in recent years and are an important public health problem. A correlation between toxicity and total tricyclic dose and central anticholinergic side effects has been claimed; that is doses of 1-3 mg/kg are therapeutic, those of 10-20 mg/kg cause moderate to severe toxicity, and those of 30-40 mg/kg are usually fatal (21). Clinical manifestations of central nervous system toxicity include acute onset of agitation, restlessness, delirium with visual hallucinations, hyperreflexia, and seizures, possibly progressing to coma and death (45).

2. *Sedation*

The tertiary amines, for example, amitriptyline, imipramine, and doxepin, cause more drowsiness or sedation than secondary amines, for example, desipramine and nortriptyline. This side effect usually disappears within a few weeks. Dizziness may also occur, most likely secondary to orthostatic hypotension. Once-a-day dosage at bedtime may help the patient avoid daytime sedation. Some of the new antidepressants have fewer anticholinergic effects and less toxic effects on the cardiovascular system and may in the long run prove safer, particularly for the elderly (53).

3. *Cognitive Impairment*

Recently Cole and Schatzberg (6) claimed that tricyclic antidepressants which possess more anticholinergic effects can complicate patients' recent memory functions and storage. Even with the least anticholinergic tricyclic antidepressants, such as desipramine, similar effects can be observed.

4. *Precipitate Manic Episode*

Tricyclic antidepressants can trigger manic/hypomanic episodes in patients with bipolar affective disorders (62), and perhaps exacerbate psychotic symptoms in schizophrenics.

5. *Akathisia or Dyskinesia*

Three patients who developed an akathisia-like syndrome after the abrupt withdrawal of imipramine were reported. A link between this phenomenon and dopamine turnover in the central nervous system was postulated (42). Two cases of dyskinesia similar to neuroleptic-induced tardive dyskinesia were reported and attributed to the anticholinergic action from long exposure to tricyclic antidepressants (12). A case of akathisia and cogwheel rigidity induced by amoxapine was reported (41). This suggests further study to determine the risk of tardive dyskinesia associated with amoxapine.

B. Monoamine Oxidase Inhibitors

1. *Behavioral Side Effects*

The central stimulating properties can cause insomnia, irritability, restlessness and agitation; they can also precipitate manic episodes in bipolar patients.

2. *Hypertensive Crises*

Hypertensive crises with associated severe occipital headaches result from ingestion of tyramine, for example, in aged cheese. Associated central nervous system toxic effects include neuromuscular excitability, restlessness, and hyperpyrexia attributed to excessive adrenergic stimulation. Rarely, intracranial bleeding may occur during a hypertensive crisis and prove fatal (63).

IV. LITHIUM SALTS

A. Neurotoxicity

The increased usage of lithium salts for prophylaxis and treatment of affective disorders has resulted in increased reporting of early and late side effects, as well as of severe neurotoxicity.

Severe lithium-induced neurotoxicity is usually associated with serum lithium levels above 1.5 mEq/liter. Neurotoxicity in patients with therapeutic serum lithium levels may be related to the patient's age; that is, older subjects are more vulnerable than younger ones. Other possible factors include tissue retention of lithium due to negative sodium balance caused by diuretics, fever with excessive sweating or diarrhea, seizure diathesis, acute concurrent medical illness, and drug synergism, particularly the combination of lithium and neuroleptics (37,56). The report of irreversible brain damage caused by this latter combination (5) has been critically reviewed, and the findings fail to support significant increased risk from a neuroleptic-lithium combination (23,25).

Central nervous system neurotoxic symptoms include drowsiness, muscular fasciculation, ataxia, hyperactive deep tendon reflexes, choreoathetoid movements, urinary incontinence, slurred speech, delirium, and seizures. If untreated, some cases will progress to death (19). The treatment of lithium neurotoxicity is supportive and symptomatic, with forced diuresis, alkalization of urine, and dialysis in severe cases (43).

B. Hand Tremor

Hand tremor, a common nuisance rather than a serious side effect of lithium treatment, occurs with therapeutic lithium levels but may be related to dosage and serum lithium levels nevertheless, and one may benefit from lowered dose. Propranolol or other β-blockers may help this side effect (27,37).

C. Cognitive Impairment

Mild, reversible impairment of memory and cognition may be produced by lithium in both normal controls and affective disorders, but the clinical significance of this remains uncertain (26,64).

V. ANTIANXIETY AGENTS

A. Sedation or Drowsiness

This is the most common side effect caused by antianxiety agents, and this tends to disappear or decrease in severity within days to weeks as the patient develops tolerance or with further careful dosage titration.

Symptoms of acute, mild to moderate intoxication with antianxiety agents resemble alcohol intoxication, including lateral nystagmus, drowsiness, slurred speech (dysarthria), ataxia, hyporeflexia, and stupor.

Severe intoxications, mostly caused by either intentional or accidental overdoses, require emergency medical treatment because of serious complications such as coma and hypotension. Although high oral doses of benzodiazepines have not caused fatalities if taken alone, extreme caution must be exercised when treating these patients because of the high possibility of ingestion of multiple drugs, particularly alcohol (16,39).

The basic principle of treatment is symptomatic and supportive.

B. Withdrawal Reaction

Onset and severity of withdrawal reactions are dependent upon the type of medication involved, mainly the elimination half-life, the dose, and the length of time for which it was used. Wikler (65) reported that when patients took benzodiazepines in amounts equivalent to 800 mg/day of short-acting barbiturates for longer than 6 weeks, 20% of the patients experienced major symptoms of withdrawal such as convulsions within 72 hr after abrupt discontinuation of medications. Grand mal seizures usually began 3-7 days after stopping the drug, and psychomotor symptoms and delirium appeared any time between days 4 and 6.

Dosage and duration of various nonbarbiturate antianxiety agents, including ethchlorvynol, glutethimide, methaqualone, meprobamate, chlordiazepoxide, and diazepam, which cause withdrawal symptoms have been reported (44).

Compared with other types of antianxiety agents, benzodiazepines are safer; that is, it is more difficult to kill oneself; they cause a lesser degree of hepatic microsomal enzymatic induction, and they are more consistently effective with a wider range of therapeutic effects. However, withdrawal symptoms are more prevalent than previously postulated (17,39), particularly with abrupt withdrawal after high dosage, prolonged administration, or if other sedative-hypnotics or alcohol are involved.

Minor benzodiazepine withdrawal symptoms are anxiety, insomnia, tremor, diaphoresis, lethargy, weakness, and gastrointestinal symptoms.

Sometimes these symptoms are difficult to differentiate from the recurrence of the original anxiety symptoms. Major benzodiazepine withdrawal symptoms are psychosis, delirium, hyperthermia, and seizures (7,10,38,55,67).

Pharmacokinetics of antianxiety agents have significant clinical implications in dealing with withdrawal symptoms; for example, after abrupt discontinuation of medications with long elimination half-lives, such as diazepam and chlordiazepoxide, symptoms will not occur for days or even weeks, but with short elimination half-life medications, such as barbiturates or lorazepam, withdrawal symptoms may occur within 24 hr. Therefore for patients taking sedatives for long periods of time, particularly the short-acting ones, dosage must be gradually reduced (18,35).

Such management will prevent major withdrawal symptoms and minimize sleep disturbances, for example, rapid eye movement rebound phenomenon.

Sedative-hypnotics tend to unmask suppressed emotion and disinhibit behavior (8). Indications for prescribing these agents should be carefully considered.

In summary, antianxiety agents are used to treat anxiety disorders, muscular spasm, insomnia, seizure disorders, and alcohol withdrawal reaction. Among all the available antianxiety agents, the benzodiazepines have proved to be the safest and most effective. Despite the reported tendency for abuse or misuse of these agents, the benefits far outweigh the risks, particularly if careful monitoring is exercised.

VI. ANTICHOLINERGIC AGENTS

A. Sedation

This central nervous system side effect can be alleviated by adjusting the dosage or time of administration.

B. Overdose

This manifests as the central anticholinergic syndrome: excitement, agitation, confusion, worsening of mental symptoms, or toxic psychosis, hallucinations, especially visual, dizziness, muscle weakness, ataxia, mydriases, numbness of extremities, and headache. In severe cases patients will experience delirium, coma, convulsions, hyperthermia, and respiratory arrest (11).

Treatment is symptomatic and supportive; administration of physostigmine salicylate, 1-2 mg subcutaneously or intravenously, may reverse symptoms of central anticholinergic intoxication (see Chapter 2 by Bozza-Marrubini). It is also a diagnostic procedure.

C. Withdrawal of Anticholinergic Agents

Anticholinergic agents are effective treatments for extrapyramidal side effects when they occur (46). However, controversy arises over the practice of giving these medications to prevent the appearance of extrapyramidal side effects. A few studies showed that when patients receiving anticholinergic agents were abruptly withdrawn, a considerable number had a recurrence of extrapyramidal side effects requiring the restarting of anticholinergic agents (24,31,33). It is recommended that anticholinergic agents be decreased gradually in order to avoid withdrawal symptoms such as agitation or delirium.

VII. CONCLUSION

Despite the central nervous system side effects of psychopharmacological agents, the benefits of pharmacotherapy still commonly outweigh the risks. An adequate knowledge of the potential central nervous system side effects of these agents will serve to significantly improve their risk/benefit ratio and lead to safer and more effective clinical use.

REFERENCES

1. Angus, J. W. S., Simpson, G. M. 1970. Hysteria and drug induced dystonia. *Acta Psychiatr. Scand. Suppl. 212*:25-58.
2. Behrman, S. 1972. Mutism induced by phenothiazines. *Br. J. Psychiatry 121*:599-604.
3. Borbely, A. A., Leopfe-Hinkkanen, M. 1979. Phenothiazines, in *Modern Pharmacology-Toxicology*, Vol. 16 (Schoenbaum, L. P., ed.), Marcel Dekker, New York, pp 403-426.
4. Caroff, S. N. 1980. The neuroleptic malignant syndrome. *J. Clin. Psychiatry 41*:79-83.
5. Cohen, W. J., Cohen, N. H. 1974. Lithium carbonate, haloperidol and irreversible brain damage. *J. Am. Med. Assoc. 230*: 1283-1287.
6. Cole, J. O., Schatzberg, A. F. 1980. Memory difficulty and tricyclic antidepressants, in *Psychopharmacology Update* (Cole, J. O., ed.), Collamone, Lexington, Mass., pp. 189-195.
7. Cole, J. O., Haskell, D. S., Orzack, M. H. 1981. Problems with the benzodiazepines: An assessment of the available evidence. *McLean Hosp. J. 6*:46-74.
8. Cook, L., Sepinwall, J. 1975. Behavioral analysis of the effects and mechanisms of action of benzodiazepines, in *Mechanisms of Action of Benzodiazepines* (Greengard, P., ed.), Raven Press, New York, pp. 1-28.
9. Delwaide, P. J., Desseilles, M. 1977. Spontaneous buccolinguofacial dyskinesia in the elderly. *Acta Neurol. Scand. 56*:256-262.

10. Dysken, M. W., Chan, C. H. 1977. Diazepam withdrawal psychosis: A case report. *Am. J. Psychiatry 134*:573-574.
11. Dysken, M. W., Merry, W., Davis, J. M. 1978. Anticholinergic psychosis. *Psychiatr. Ann. 8*:452-456.
12. Fann, W. E., Sullivan, J. L., Richman, B. W. 1976. Dyskinesias associated with tricyclic antidepressants. *Br. J. Psychiatry 128*: 49-493.
13. Fink, M. 1976. Electroencephalograms, mental state and psychoactive drugs. *Pharmacol. Physicians 3*:1-5.
14. Gelenberg, A. J., Mandel, M. R. 1977. Catatonic reactions to high-potency neuroleptic drugs. *Arch. Gen. Psychiatry 34*:947-950.
15. Granacher, R. P. 1981. Differential diagnosis of tardive dyskinesia: An overview. *Am. J. Psychiatry 138*:1288-1297.
16. Greenblatt, D. J., Allen, M. D., Noel, B. J., Shader, R. I. 1977. Acute overdosage with benzodiazepine derivatives. *Clin. Pharmacol. Ther. 21*:497-514.
17. Greenblatt, D. J., Shader, R. I. 1978. Dependence, tolerance, and addiction to benzodiazepines: clinical and pharmacokinetic considerations. *Drug Metab. Res. 8*:13.
18. Greenblatt, D. J., Shader, R. I. 1980. Pharmacokinetic aspects of anxiolytic drug therapy. *Can. J. Neurol. Sci. 7*:269-270.
19. Horowitz, L. C., Fisher, C. U. 1969. Acute lithium toxicity. *N. Engl. J. Med. 281*:1369.
20. Itil, T. M., Soldatos, C. 1980. Epileptogenic side effects of psychotropic drugs. *J. Am. Med. Assoc. 244*:1460-1463.
21. Jackson, J. E., Bressler, R. 1982. Prescribing tricyclic antidepressants: Part III: Management of overdose. *Drug Ther. 12*: 49-63.
22. Owen, D. G. C., Johnstone, E. C., Frith, C. D. 1982. Spontaneous involuntary disorders of movement, the prevalence, severity, and distribution in chronic schizophrenics with and without treatment with neuroleptics. *Arch. Gen. Psychiatry 39*:452-461.
23. Juhl, R. P., Tsuang, M. T., Perry, P. J. 1977. Concomitant administration of haloperidol and lithium carbonate in acute mania. *Dis. Nerv. Syst. 38*:675-676.
24. Klett, C. J., Point, P., Caffey, E. 1972. Evaluating the long-term need for antiparkinson drugs by chronic schizophrenics. *Arch. Gen. Psychiatry 26*:374-379.
25. Krishina, N. R., Taylor, M. A., Abrams, R. 1978. Combined haloperidol and lithium carbonate in treating manic patients. *Comp. Psychiatry 19*:119-120.
26. Kysumo, K. S., Vaughn, M. 1977. Effects of lithium salts on memory. *Br. J. Psychiatry 131*:453-457.
27. Lapierre, Y. D. 1976. Control of lithium tremor with propranolol. *Can. Med. Assoc. J. 51*:622.

28. Logothetis, J. 1967. Epileptic seizures in the course of phenothiazine therapy. *Neurology 17*:869-875.
29. May, R. H. 1959. Catatonic-like states following phenothiazine therapy. *Am. J. Psychiatry 115*:1119-1120.
30. McCarron, M. M., Boettger, M. L., Peck, J. J. 1982. A case of neuroleptic malignant syndrome successfully treated with amantadine. *J. Clin. Psychiatry 43*:381-382.
31. McClelland, H., Blessed, A., Bhate, S., Ali, N., Clarke, P. 1974. Abrupt withdrawal of antiparkinsonian drugs in schizophrenic patients. *Am. J. Psychiatry 124*:151-159.
32. Meltzer, H. Y., Sachar, E. J., Frantz, A. G. 1974. Serum prolactin levels in unmedicated schizophrenic patients. *Arch. Gen. Psychiatry 31*:564-569.
33. Orlov, P., Kasparian, G., DiMascio, A., Cole, J. 1971. Withdrawal of antiparkinson drugs. *Arch. Gen. Psychiatry 25*:410-412.
34. Pi, E. H., Simpson, G. M. 1981. Tardive dyskinesia and abnormal tongue movements. *Br. J. Psychiatry 139*:526-528.
35. Pi, E. H., Simpson, G. M. 1982. The use and misuse of benzodiazepines: An update. *J. Cont. Educ. Family Physician 16*: 102-106.
36. Pi, E. H., Simpson, G. M. 1984. Prevention of tardive dyskinesia, in *Neurobehavioral Dysfunction Induced by Psychotherapeutic Agents: Neurophysiological, Neuropharmacological Bases and Clinical Implications* (Shah, N. S., Donalds, A. G., eds.), Plenum, New York (in press).
37. Pi, H. T., Surawicz, F. G. 1978. Severe neurotoxicity and lithium therapy. *Clin. Toxicol. 13*:479-486.
38. Preskorn, S. H., Denner, L. J. 1977. Benzodiazepines and withdrawal psychosis. Report of 3 cases. *J. Am. Med. Assoc. 237*: 36-38.
39. Rickels, K. 1981. Benzodiazepines: Use and Misuse, in *Anxiety: New Research and Changing Concepts* (Klern, D. F., Rabkin, J., eds.), Raven Press, New York, pp. 1-26.
40. Rifkin, A., Quitkin, F., Klein, D. F. 1975. Akinesia, a poorly recognized drug-induced extrapyramidal behavioral disorder. *Arch. Gen. Psychiatry 32*:672-674.
41. Ross, D. R., Walker, J. I., Peterson, J. 1983. Akathisia induced by amoxapine. *Am. J. Psychiatry 140*:115-116.
42. Sathananthan, G. L., Gershon, S. 1973. Imipramine withdrawal: An akathisia-like syndrome. *Am. J. Psychiatry 130*:1286-1287.
43. Schou, M., Amidsen, A., Trap-Jensen, J. 1968. Lithium poisoning. *Am. J. Psychiatry 125*:520-527.
44. Shader, R. I., Caine, E. D., Meyer, R. D. 1975. Treatment of dependence on barbiturates and sedative-hypnotics, in *Manual of Psychiatric Therapeutics* (Shader, R. I., ed.), Little, Brown, Boston, p. 196.

45. Sholomskas, A. J. 1980. An old side effect revisited: Visual hallucination. *Psychiatr. Ann. 10*:47-59.
46. Simpson, G. M. 1970. Controlled studies of antiparkinsonian agents in the treatment of drug-induced extrapyramidal system disorders. *Acta Psychiatr. Scand. Suppl. 212*:44-51.
47. Simpson, G. M. 1975. CNS effects of neuroleptic agents. *Psychiatr. Ann. 5*:11-14.
48. Simpson, G. M., Varga, E., Haher, E. J. 1976. Psychotic exacerbations produced by neuroleptics. *Dis. Nerv. Syst. 37*: 367-369.
49. Simpson, G. M., Pi, E. H. 1981. Pharmacokinetics of antipsychotic agents, in *Recent Advances in Neuropsychopharmacology*, Vol. 31 (Angrist, B., ed.), Pergamon, Oxford, pp. 365-372.
50. Simpson, G. M., Pi, E. H., Sramek, J. J. 1981. Adverse effects of antipsychotic agents. *Drugs 21*:138-151.
51. Simpson, G. M., Pi, E. H., Sramek, J. J. 1982. Management of tardive dyskinesia: Current update. *Drugs 23*:381-383.
52. Simpson, G. M., Pi, E. H., Sramek, J. J. 1983. The current status of tardive dyskinesia. *J. Psychiatr. Treat. Eval. 5*:127-133.
53. Simpson, G. M., Pi, E. H., White, K. 1983. Plasma drug levels and clinical response to antidepressants. *J. Clin. Psychiatry 44(Sec. 2)*:27-34.
54. Steiner, W. G., Pollack, S. L. 1965. Limited usefulness of EEG as a diagnostic and in psychiatric cases receiving tranquilizing drug therapy. *Prog. Brain Res. 16*:97-105.
55. Stewart, R. B., Salem, R. B., Springer, P. K. 1980. A case report of lorazepam withdrawal. *Am. J. Psychiatry 137*:1113-114.
56. Strayhorn, J. M., Jr., Nash, J. L. 1977. Severe neurotoxicity despite serum therapeutic lithium levels. *Dis. Nerv. Syst. 38*: 107-112.
57. Tornatore, F. L., Lee, D., Sramek, J. J. 1981. Psychotic exacerbation with haloperidol. *Drug Intelligence Clin. Pharm. 15*:209-213.
58. Van Putten, T., Mutalipassi, L. R. 1975. Fluphenazine enanthate induced decompensations. *Psychosomatics 16*:37-40.
59. Van Putten, R., May, P. 1978. Akinetic depression in schizophrenia. *Arch. Gen. Psychiatry 35*:1101-1107.
60. Varga, E., Haler, E. J., Simpson, G. M. 1974. Neuroleptic-induced Klüver-Bucy syndrome. *Biol. Psychiatry 10*:65-68.
61. Varga, E., Sugerman, A. A., Varga, V., Zomorodi, A., Zomorodi, W., Menken, M. 1982. Prevalence of spontaneous oral dyskinesia in elderly. *Amer. J. Psychiatry 139*:329-331.
62. Wehr, T. A., Goodwin, F. K. 1979. Rapid cycling in manic-depressives induced by tricyclic antidepressants. *Arch. Gen. Psychiatry 36*:555-559.

63. White, K., Simpson, G. M. 1981. Combined MAOI-tricyclic antidepressant treatment: A re-evaluation. *J. Clin. Psychopharmacol. 1*:264-282.
64. White, K., Bohart, R., Whipple, K., Boyd, J. 1979. Lithium effect on normal subjects relationships to plasma and RBC lithium levels, *Int. Pharmacopsychiatry 14*:176-183.
65. Wikler, A. 1968. Diagnosis and treatment of drug dependence of the barbiturate type. *Am. J. Psychiatry 125*:758-765.
66. Williams, P. 1972. An unusual response to chlorpromazine therapy. *Br. J. Psychiatry 121*:439-440.
67. Winokur, A., Rickels, K., Greenblatt, D. J., Snyder, P. J., Schatz, N. J. 1980. Withdrawal reaction from long-term low-dosage administration of diazepam. A double-blind, placebo-controlled case study. *Arch. Gen. Psychiatry 37*:101-105.

11

Cerebellar Toxicity of Antiepileptic Drugs

P. R. M. Bittencourt
Hospital Nossa Senhora das Graças, Curitiba, Brazil

E. Perucca and A. Crema
University of Pavia, Pavia, Italy

I. INTRODUCTION

The most common manifestations of toxicity of antiepileptic drugs are those affecting the central nervous system (68,88). Among the various symptoms and signs described (see Table 1), many suggest involvement of structures within the posterior fossa (23). Some appear in relation to high serum drug levels and resolve when the amounts are decreased. Occasionally there is no resolution even after drugs are withdrawn altogether, which has led to the long-standing argument on the pathogenesis of "cerebellar syndromes" in these patients.

This chapter will attempt to review the acute and chronic effects of antiepileptic drugs on posterior fossa structures. A brief review of the anatomy and physiology of the cerebellum and its connections will also be presented, since it is relevant to the discussion of the clinical findings.

Table 1 Manifestations of Acute Central Nervous System Toxicity of Antiepileptic Drugs[a]

Ataxia, dysarthria	Psychiatric and behavioral disorders
Vertigo, nystagmus	Paradoxical excitation
Sedation, drowsiness	Paradoxical increase in fit frequency
Impairment of cognitive function	Choreoathetoid movements
Diplopia, blurring of vision	Other diskynetic movements, dystonia
Tremors	Coma
	Headache

[a]The list includes the most common manifestations seen with clinically prescribed doses (not overdose) of phenytoin, phenobarbitone, primidone, carbamazepine, and benzodiazepines. Adverse effects may differ somewhat from one drug to another (88).

II. ANATOMY AND PHYSIOLOGY

The cerebellum is composed of the vermis and two lateral lobes, or hemispheres. A morphological subdivision based on phylogeny and physiology shows it to be divided into a flocculonodular lobe (archicerebellum), an anterior lobe (paleocerebellum), and the remaining part, made of most of the hemispheres (neocerebellum). The archicerebellum deals mostly with vestibular signals, while the paleocerebellum is concerned with proprioceptive information from the spinal cord. The neocerebellum is related strongly to the cerebral cortex.

One afferent system to the cerebellum is via the mossy fibers, which synapse with granule and parallel fibers and reach the Purkinje cells in the cerebellar cortex. A second afferent pathway is represented by the climbing fibers, which are excitatory and synapse directly with the Purkinje cells. The mossy fibers bring vestibular and proprioceptive information from higher cerebral areas as well as from the spinal cord, via the vestibular and pontine nuclei. The climbing fibers bring information from the cerebral cortex, mesencephalon, and inferior olivary complex (16).

The only inhibitory efferent pathway from the cerebellum originates from the Purkinje cells. The great majority of the axons from these cells relay in the intrinsic nuclei. The fastigial nuclei receive fibers from the anterior and posterior parts of the vermis, the globose and emboliform nuclei from paravermal areas, and the dentate nuclei from lateral hemispherical areas. Fibers from the dentate, globose, and emboliform nuclei go to the red nucleus and parts of the thalamus.

Fibers relaying in the fastigial nuclei and the few fibers from the Purkinje cells that do not relay in the intrinsic nuclei go mostly to the vestibular nuclei (16).

The cerebellum is an important integrative center for the coordination of movements. It receives continuous information about the length of the muscle spindles and utilizes it to correct the position of the body in space, acting as a position servomechanism. The archicerebellum is mainly concerned with equilibrium, the paleocerebellum with the stretch reflex and antigravity posture, and the neocerebellum with the regulation of voluntary movements. Lesions within the archicerebellum usually affect the vermis and result in truncal ataxia. Lesions within the neocerebellum produce dysarthria, limb ataxia, and oculomotor disturbances. Many of the disturbances commonly attributed to cerebellar dysfunction, however, may also result from alterations in cerebellar connections in the brainstem (25).

Studies of oculomotor function have considerably improved our knowledge of cerebellar and brainstem pathophysiology (4,37). It is known that the flocculus receives direct information from the retina and monitors the gain of the vestibulo-ocular reflex. Thus floccular dysfunction leads to decreased visual suppression and altered gain control of the vestibulo-ocular reflex. This could explain the broken smooth-pursuit eye movements in patients with cerebellar lesions (8). After cerebellectomy there is absence of smooth pursuit and incapacity to maintain eccentric gaze. If the subject is asked to maintain gaze on an eccentric target, the result will be paretic nystagmus; this is distinctive in that the slow phase is directed away from the target and toward the central position, while the fast phase is a saccade which takes the eyes back to the target. In this type of nystagmus, which is strongly suggestive of brainstem or cerebellar dysfunction, the amplitude of the movement increases with the eccentricity of the gaze. Unfortunately, most reports of "cerebellar syndromes" in epileptic patients are not detailed enough to be accepted as indicative of cerebellar dysfunction, as they do not provide details on broken smooth pursuit, gaze-paretic nystagmus, altered gain of the vestibulo-ocular reflex, or its visual suppression (23,29,36,53).

III. ACUTE TOXICITY OF ANTIEPILEPTIC DRUGS

A. Phenytoin

Already in the very first report on the clinical use of phenytoin, Merritt and Putnam (53) described toxic symptoms such as "dizziness, ataxia, tremors, blurring of vision, diplopia and slight nausea." Finkelman and Arieff (29) found that ataxia, tremor, and nystagmus were the most frequent signs of central nervous system toxicity, severe enough to warrant withdrawal of the drug in 6 out of 44 patients. These symptoms, which are usually reversible on reduction of drug dosage, do not originate from the labyrinth (60) and are generally attributed to cerebellar dysfunction (23). In line with this hypothesis, drug-treated epileptic patients have been shown to have prominent abnormalities in smooth pursuit, which appear to be linearly related to decreased ability to suppress the vestibulo-ocular reflex (8). Saccadic velocity and duration do not seem to be impaired in these patients, suggesting that the flocculonodular lobe of the cerebellum is the site responsible for much of the observed vestibular/oculomotor disorders (9).

Although there is general agreement that clinically relevant signs of cerebellar toxicity usually occur at serum phenytoin concentrations above 80-100 μmol/liter (10-20 μg/ml), individual patients may show them at levels higher or lower than these (62,63,69). Several studies have examined the relationship between nystagmus and serum phenytoin concentration. Kutt et al. (44), from purely clinical observations, suggested the existence of a strong inverse relationship. Thus at 60 μmol/liter (15 μg/ml) nystagmus could be observed at extreme lateral gaze, while at 120 μmol/liter (30 μg/ml) it could be observed at 45° lateral gaze, and at concentrations of 200 μmol/liter (50 μg/ml) or higher nystagmus could be observed in the mid-position. These findings have not been confirmed in subsequent studies. Lehtovaara (45), for example, failed to find any correlation between clinically observed postrotational gaze nystagmus and serum phenytoin levels. Riker et al. (70) examined the situation more carefully, trying to correlate both free and total serum phenytoin levels with nystagmus recorded by electro-oculography. Nystagmus was observed in only 7 out of 29 patients, and its severity was not correlated with the concentration of either free or total drug.

A partial explanation for the frequently reported lack of correlation between phenytoin concentration and some indices of brainstem-cerebellar function in different patients is provided by a series of experiments by Bittencourt et al. (7-9,14). These authors reported decreased smooth-pursuit eye movement velocity in patients with severe epilepsy on polytherapy; in patients with epilepsy of moderate severity on monotherapy with phenytoin, carbamazepine, or sodium valproate; and in healthy subjects acutely exposed to therapeutic concentrations of carbamazepine and benzodiazepines. No abnormality of smooth pursuit was observed in the healthy subjects at therapeutic concentrations of sodium valproate and at phenytoin concentrations below 42 μmol/liter. It was concluded that the smooth-pursuit abnormality in phenytoin-treated epileptic patients is due to acute phenytoin toxicity as well as to underlying permanent brainstem and/or cerebellar damage, which may or may not be related to long-term exposure to the drug (14).

B. Carbamazepine

Dizziness, diplopia, slurred speech, and ataxia, with drowsiness, headache, and feelings of drunkeness, have been observed in patients exposed to carbamazepine in therapeutic doses or in overdose (40,82,83). Bittencourt and Richens (11,14) described a linear relationship between serum carbamazepine concentration and the velocity of smooth-pursuit eye movements, an objective index of brainstem-cerebellar function. Umida et al. (85) found that carbamazepine-induced gaze nystagmus was dyskinetic and pendular in severely intoxicated patients but that these cerebellar signs disappeared on lower doses, whereas convergence weakness and gaze-paretic nystagmus persisted. These findings, together with the observed depression of optokinetic nystagmus, are suggestive of upper brainstem involvement. It has been commented that the nystagmus caused by carbamazepine is different from that caused by phenytoin (48). Evidence has also been provided of concurrent labyrinthine abnormalities, especially on overdosage (see Ref. 48 for a review).

It has been repeatedly reported that adverse effects are usually more prominent at the onset of therapy and tend to subside during the next few days. This phenomenon partly reflects a decrease in serum carbamazepine levels, due to an autoinduction effect (28). Clinically relevant signs of central nervous system toxicity usually appear at levels above 34 μmol/liter (8 μg/ml), even though the degree of interindividual variability may be considerable. Höppener et al. (36) and Albani et al. (1) reported that the severity of toxic effects may wax and wane during the day, probably in relation with the fluctuations in serum concentration.

C. Primidone

Gallagher et al. (30) described the occurrence of nystagmus and ataxia in four patients first exposed to primidone. In some studies nystagmus, ataxia, and drowsiness were seen in up to 40-80% of patients given primidone for the first time, requiring discontinuation of treatment in up to a third of cases (47,50,74,75). These signs usually appear soon after drug administration, at a time when phenobarbitone levels are very low (30,46) and are thus considered to be due to the parent compound. The serum concentration at which they occur is around 23-46 μmol/liter (5-10 μg/ml) (46). It is clear, however, that their incidence can be reduced by starting the treatment with very low doses and increasing the dosage gradually. There is also some evidence that previous exposure to barbiturates decreases the severity of early primidone toxicity.

There are few details about the type of nystagmus caused by primidone, but this is claimed to be similar to that seen with other antiepileptic drugs (46). There are reports of downbeat nystagmus and external ophthalmoplegia during primidone treatment, even though these patients were also on phenytoin (3,61). The latter signs are suggestive of brainstem rather than cerebellar dysfunction (4).

D. Phenobarbitone and Other Barbiturates

The most important adverse effects of barbiturates include sedation, drowsiness, impairment in cognitive function, and (especially in children) behavioral problems. Signs of brainstem-cerebellar dysfunction may also be observed, especially at relatively high doses (49). Bergman et al. (6) described the occurrence of nystagmus after administration of 100 mg of amobarbital in healthy subjects. The more accurate observations of Rashbass (66,67) confirmed the gaze-paretic nature of the nystagmus and provided evidence for the occurrence of disturbances in smooth-pursuit eye movements. The latter were abolished 8 min after intravenous thiopentone (100 mg) administration and 100 min after oral amilobarbitone (100 mg). Together with earlier reports of disturbances of optokinetic nystagmus (6), these findings suggest that the observed effects may be mediated by the cerebellar flocculus (14). Quinalbarbitone, pentobarbitone, and phenobarbitone have all been shown to affect eye movements in a similar way (35,59). There are no reliable studies correlating brainstem-cerebellar effects with serum phenobarbitone levels.

E. Benzodiazepines

Although sedation is the most prominent adverse effect of benzodiazepine therapy, ataxia, dysarthria, dizziness, and nystagmus are also reported (5,27,72). In healthy volunteers a negative relationship has been described between the velocity of saccadic and smooth pursuit eye movements (parameters of brainstem-cerebellar function) and serum concentrations of diazepam, nitrazepam, and temazepam (12,13). During chronic treatment tolerance develops to some of the adverse effects of benzodiazepines and this may provide an explanation for the observed lack of relationship between central nervous system effects and serum concentration in this situation (10).

F. Ethosuximide and Sodium Valproate

Unlike other antiepileptic drugs, ethosuximide and sodium valproate do not usually induce major signs of cerebellar dysfunction during acute intoxication.

IV. CHRONIC TOXICITY OF ANTIEPILEPTIC DRUGS

The possibility of organic damage in the cerebellum of epileptic patients is documented by three separate lines of evidence: (a) the occasional clinical finding of a permanent cerebellar syndrome, irreversible irrespectively of alterations in drug treatment; (b) presence of cerebellar atrophy at radiological (air encephalography or computerized tomography) examination; and (c) marked histological changes with loss of Purkinje cells in cerebellar bioptic or autopsy clinical specimens. While there is general agreement on the morphological and clinical picture associated with these findings, interpretation of the data has led to continuing controversy. Opinions differ in respect to the possible role of chronic phenytoin therapy in the pathogenesis of the cerebellar degeneration (there are at present no reports implicating antiepileptic drugs other than phenytoin)(23). Because the cerebellum is particularly sensitive to hypoxia and because convulsions alone can cause cerebellar degeneration by this mechanism, it has been difficult to establish whether phenytoin itself can have an independent toxic effect.

A. Human Studies

Descriptions of abnormal anatomy in the brains of epileptic patients were made as early as 1825 (58), and neuronal death in the cerebellar cortex was reported in detail by Spyelmeyer's studies in the 1920s (78-81). Necrosis of recent onset was noted in the cerebral and cerebellar cortices of patients dying soon after a bout of seizures. In the cerebellum the Purkinje cells and granular layer were affected, the frequency and severity of the lesions there being second only to those in the Ammon horns. Spyelmeyer considered these changes to be a consequence of functional vascular disturbances related to generalized tonic-clonic seizures. Anatomical changes in the same areas were also observed by Norman (56,57) in children dying in status epilepticus and by Zimmerman (89) in nonepileptic children who died during or after severe convulsions of various etiologies. These authors also considered the cerebellar changes to be due to ischemia. Scholz (73) showed that neuronal death leading to gliosis and scarring in the hippocampus and cerebellum were only part of widespread damage affecting also the thalamus and the diencephalon. Since there were more acute changes in patients who had died 1 or 2 days after a series of seizures and more chronic changes in those who had died longer after the seizures, he concluded that these changes were related to the seizures themselves.

It was 20 years after the introduction of phenytoin that Utterback and collaborators (86,87) linked phenytoin to widespread loss of Purkinje cells in a patient who died in status epilepticus after receiving an unspecified amount of the drug. Hoffmann (34) also blamed phenytoin for the histological changes observed in the brain of a 28-year-old epileptic patient who developed frequent tonic-clonic generalized and complex partial seizures after a febrile illness and had been given six hourly doses of phenytoin (250 mg) and phenobarbitone (65-130 mg) daily for 17 days prior to her death. The postmortem showed severe cerebellar damage, with disappearance of the Purkinje cells, and only mild microscopic abnormalities in the basal ganglia. The causative role of phenytoin was suggested by the fact that in the authors' view other possible causes of Purkinje cell damage (hypoxia, hypoglycemia, etc.) could be excluded either by lack of involvement of other parts of the brain or by lack of exposure. Haberland (33) reported three cases of similar cerebellar changes, but blamed phenytoin in only one patient. A further case of permanent cerebellar damage presumably caused by phenytoin was reported by Kokenge et al. (42). Their patient was an 18-year-old girl with intractable partial seizures who had received for many years up to 1000 mg of phenytoin daily in combination with barbiturates and had on admission a putative serum phenytoin level (calculated in retrospect) in excess of 300 μmol/liter (75 μg/ml). This patient never recovered from the severe brainstem-cerebellar signs she had on admission.

Against this background, a large amount of data by Dam's group appeared in the literature (18-22,24,55). Most of the work was carried out in animals and will be discussed in the next section. The data from epileptic patients (21,22) corroborated the animal studies and suggested that frequent generalized tonic-clonic seizures rather than phenytoin was the cause of cerebellar degeneration. To a large extent, these conclusions were based on the study of 32 brains of patients who had "grand mal" epilepsy without evidence of other disease that could induce cerebellar degeneration. Based on dose/weight considerations, it was estimated that 12 patients must have had serum phenytoin levels around 60 μmol/liter (15 μg/ml) and 18 levels lower than these (2 patients never received phenytoin). Ten control brains from subjects dying in accidents or cardiac arrest were included in the study. After classifying the patients into groups according to the severity of the epilepsy or the dosage of phenytoin, the authors observed that the number of Purkinje cells was decreased in (a) the group with severe seizures and (b) the group treated with large doses of phenytoin. In the latter group, however, only half of the patients had low Purkinje cells counts and also in half (which half not specified) of the cases the high dose of phenytoin was associated with severe epilepsy. As mentioned above, these data were considered as evidence that the cerebellar damage was caused by the seizures and not by the drug. This conclusion is of course supported by the data from studies before the advent of phenytoin (81,89).

Although the role of epileptic seizures is established beyond doubt, a number of the more recent studies have led to partial criticism of Dam's conclusion, reinforcing the view that phenytoin has a toxic effect of its own. Horne (38) reported two cases of cerebellar degeneration (demonstrated by air encephalography) following chronic therapy with phenytoin and barbiturates. One of these cases, a 35-year-old female with an "irreversible" cerebellar syndrome, is particularly relevant in that there was no history of status epilepticus, and the severity of the epilepsy was comparatively mild. Ghatak et al. (32) reported the case of a 78-year-old phenytoin-treated epileptic female who developed a progressive cerebellar syndrome during the last 10 years of her life, at a time when she had only about one seizure per year. On autopsy the major change was a marked cerebellar atrophy with loss of the Purkinje cells and increased number of Bergmann's astrocytes. Because of the lack of lesions elsewhere in the central nervous system, the authors concluded that their findings could not be explained on the basis of generalized hypoxia and considered phenytoin as the most likely cause. This conclusion is supported further by the study of Rapport and Shaw (65), who reported the occurrence of phenytoin-induced degeneration without seizures.

An interesting observation, even though not particularly relevant to the problem of the etiology of cerebellar degeneration, is the chance finding by Cooper et al. (17) that all five of their biopsied cerebellar stimulation patients had severe decrease or absence of Purkinje cells. In a similar study on bioptic material obtained during implantation of cerebellar electrodes, Salcman et al. (71) found changes similar to those reported by Dam (23). These changes were blamed on a variety of factors associated with seizures, such as hypoglycemia, hypotension, and hypoxia. These same factors, however, would be unlikely to explain the cerebellar degeneration (documented by computerized tomography) observed by Koller et al. (43) in another group of eight patients without cerebellar symptoms, all of whom had "good seizure control."

Further evidence for a toxic action of phenytoin was provided by Iivanainen et al. (39), who examined in a retrospective study the air-encephalographic findings of 131 mentally retarded epileptic patients and found significant correlations between high serum phenytoin levels and degree of cerebellar atrophy. Interestingly, patients with cerebellar atrophy had higher serum phenytoin levels than patients with large ventricles or temporal lobe atrophy. Although the patients with high serum phenytoin levels were more likely to have severe epilepsy, cerebellar atrophy was better correlated with the concentration of the drug than with the seizures. Tsukamoto et al. (84) correlated the computer-tomographic findings in 47 epileptic patients with seizure frequency and duration of phenytoin treatment. More marked cerebellar atrophy was found in patients with tonic-clonic and partial complex seizures. Cerebellar atrophy seemed to be better correlated with the duration of phenytoin treatment, whereas cerebral atrophy was better correlated with seizure frequency.

The existence of an independent toxic effect of phenytoin is also supported by a study by McLain et al. (54) in five epileptic patients who developed a cerebellar syndrome while on phenytoin. All five patients had high plasma phenytoin levels and none were considered to have seizures of a type that could have caused systemic hypoxia at the time the cerebellar signs appeared. Cerebellar degeneration was confirmed by atrophy on computer-tomographic scans and permanence of the syndrome when plasma phenytoin levels were decreased. In three of the patients signs of cerebellar degeneration appeared after several years of treatment at serum drug levels that had previously been well tolerated (in the other two patients, the onset of the permanent cerebellar syndrome was unmasked by acute phenytoin intoxication). One explanation for these findings can be based on the hypothesis that cerebellar degeneration is a slowly progressive disease and that a prolonged period may elapse before its clinical manifestations appear (indeed, clinical manifestations may never become manifest). This hypothesis is of course supported by the several reports of asymptomatic cerebellar atrophy in these patients (17,43,71).

In a more recent computer-tomographic scan study, Bittencourt et al. (15) confirmed the occurrence of marked brainstem and cerebellar atrophy in 40 patients with chronic severe epilepsy whose medical and drug histories had been carefully recorded for up to 10 years. Analysis of these data is in progress and may contribute to the elucidation of some aspects of the etiology of cerebellar degeneration.

At present it would seem wise to conclude with McLain et al. (54) that "the question of whether phenytoin or the cumulative effect of hypoxia from repeated convulsions causes cerebellar degeneration should not be posed as one of exclusive alternatives, as hypoxia is a well known cause of cerebellar atrophy. Instead, the question should be posed of whether or not phenytoin can be responsible." Although much of the evidence reviewed above suggests that it can, no conclusive statements can be made as yet.

B. Animal Studies

The effect of severe seizure activity on the cerebellum was clearly documented by Gastaut et al. (31), who injected 14 cats with alumina cream in the left temporal lobe. The five cats who developed severe "psychomotor seizures" with frequent secondary generalization showed loss of Purkinje cells and Bergmann's glial proliferation in the cerebellum (presumably secondary to ischemia), while the nine cats who did not develop generalized convulsions did not show cerebellar changes.

An independent toxic effect of phenytoin on the cerebellum was suggested by Utterback and co-workers (86,87). These authors reported that in cats phenytoin doses of 8 mg/kg produced a severe cerebellar syndrome, while doses of 20 or 30 mg/kg caused "widespread" destruction of the Purkinje cells, without changes in the rest of the brain. Doses of 20 mg/kg also produced a severe bleeding disorder that could be averted by vitamin supplements, suggesting that the animals had previously been undernourished (22). In the study by Kokenge et al. (42) rats were injected with 100-200 mg/kg of phenytoin daily. Only doses of 200 mg/kg produced chronic cerebellar signs, even though these doses caused death in 50% of the animals if introduced suddenly, and a comatose state if introduced gradually. Loss of Purkinje cells and other cytological changes were seen only after at least 14 days of treatment. In the same study, cats tolerated only doses up to 30 mg/kg, with a 50% death rate if this dose was maintained for longer than 5 days. This tolerability problem was circumvented by lowering the dose when the animals became so ataxic that they could not stand. Similarly to rats, only animals treated for longer than 14 days developed cerebellar changes. It was concluded that prolonged periods of high phenytoin concentrations in the bloodstream are required to produce cerebellar damage. In a more detailed study based on optical as well as electron-microscopic analysis, Del Cerro and Snider (26,76,77) described in rats given 200 mg/kg of phenytoin daily not only changes similar to those already reported (i.e., reduced numbers of Purkinje cells, proliferation of Bergmann's astrocytes), but also membranous cytoplasmic bodies in or near the dendrites of the damaged Purkinje cells, which were also claimed to show sprouting.

Dam (22) critically reviewed the data of Utterback et al. (86,87), Kokenge et al. (42) and Snider and Del Cerro (26,76,77) and argued that cerebellar changes were found in animals who became comatose because of phenytoin intoxication and could thus be hypoxic. Furthermore, in Dam's view the histological changes described could have been caused by fixation artifacts or by incorrect evaluation of Purkinje cells numbers. This criticism applied to his own previous report of Purkinje cell loss in pigs chronically exposed to phenytoin (18). The basis for Dam's arguments was provided by a number of studies in pigs and monkeys fed for 55-182 days with doses of phenytoin (51-300 mg/kg) which produced serum levels of 44-292 μmol/liter (11-73 μg/ml) (19-22, 24,55). Compared with controls, the animals did not show reduction of Purkinje cell counts or proliferation of Bergmann's astrocytes. Dam (22) explained these negative findings on the basis that his animals did not develop coma and that his fixation techniques and Purkinje cell quantitation were more accurate than those of previous investigators. This view was partly supported by a later study by Puro and Woodward (64), who found no cerebellar damage in phenytoin-treated rats fixated by perfusion.

The more recent studies on the cerebellar toxicity of phenytoin have added relatively little to preexisting evidence. Karkos (40) found "degenerative" changes in the cerebellum of rats chronically exposed to phenytoin; changes, albeit less prominent, were also seen with other antiepileptic drugs. The noxious role of severe epileptic seizures was reemphasized by Meldrum and Brierley (51), who described, in the brains of baboons with bicuculline-induced status epilepticus, Purkinje and basket cell changes restricted to the boundary zone between the territories of the superior and posterior inferior cerebellar arteries. These changes were considered to be due to a combination of hypoglycemia, hypoxia, hypotension, and hyperpyrexia. In another study (52), 2 out of 13 baboons with allylglycine-induced status epilepticus developed proliferation of Bergmann's astrocytes. Although these results clearly indicate cerebellar damage caused by seizures, the pathological changes were obvious only at arterial boundary zones (51), while in the reported epileptic patients (see Sec. IV) the cerebellar changes were generalized.

The ultrastructural changes caused by chronic phenytoin treatment have been more recently investigated by Alcala et al. (2). In this study the cerebellum of three rats exposed chronically to phenytoin (serum levels ranging from 1 to 30 μmol/liter, or 0.25-7.5 μg/ml) was found to show minimal or no changes on light microscopy, but definite changes on electron microscopy.

V. CONCLUSIONS

Phenytoin, carbamazepine, barbiturates, primidone, and benzodiazepines induce reversible signs of brainstem-cerebellar dysfunction during episodes of acute toxicity. Only phenytoin has been implicated in the development of permanent, irreversible cerebellar damage, even though opinions differ on this point. There is clear evidence, in fact, that factors associated with seizures (hypoxia, hypoglycemia, and hyperthermia) can cause by themselves pathoanatomical changes similar to those described in the cerebellum of epileptic patients. Nevertheless, permanent cerebellar damage appears to be reported mainly in patients exposed to high blood levels of phenytoin, and some of them did not have a great number of generalized tonic-clonic seizures. This suggests that the drug may have an independent noxious effect of its own.

It is likely that many of the issues presently in doubt will be settled in the future, owing to the progressively greater number of patients on single-drug therapy and to the wider availability of noninvasive neurophysiological and neuroradiological techniques.

REFERENCES

1. Albani, G., Riva, R., Ambrosetto, G., Contin, M., Cortelli, P., Perucca, E., Baruzzi, A. 1984. Diurnal fluctuations in free and total steady-state plasma concentrations of carbamazepine and correlation with intermittent side effects. *Epilepsia* 25:476-481.

2. Alcala, H., Lertratanangkoon, K., Steinbach, W., Kellaway, P., Horning, M. G. 1978. The Purkinje cell in phenytoin intoxication: Ultrastructural and Golgi studies. *Pharmacologist 20*:240.
3. Alpert, J. 1978. Downbeat nystagmus due to anti-convulsant toxicity. *Ann. Neurol. 4*:471-473.
4. Baloh, R. W., Honrubia, V. 1979. *Clinical Neurophysiology of the Vestibular System*, F. A. Davis, California.
5. Baruzzi, A., Michelucci, R., Tassinari, C. A. 1982. Benzodiazepines: nitrazepam, in *Antiepileptic Drugs*, 2nd ed. (Woodbury, D. M., Penry, J. K., Pippenger, C. E., eds.), Raven Press, New York, pp. 753-770
6. Bergman, P. S., Nathanson, M., Bender, M. B. 1952. Electrical recordings of normal and abnormal eye movements modified by drugs. *Arch. Neurol. Psychiatry 67*:357-374.
7. Bittencourt, P. R. M. 1981. The effects of some centrally-acting drugs on saccadic and smooth pursuit eye movements in man, Ph.D. thesis, University of London, London.
8. Bittencourt, P. R. M., Gresty, M. A., Richens, A. 1980. Quantitative assessment of smooth pursuit eye movements in healthy and epileptic subjects. *J. Neurol. Neurosurg. Psychiatry 43*:1119-1124.
9. Bittencourt, P. R. M., Smith, A. T., Richens, A. 1980. Quantitative eye movement measurements in epileptic patients on monotherapeutic regimens. *Acta Neurol. Scand. Suppl. 79*:92.
10. Bittencourt, P. R. M., Dhillon, S. 1981. Benzodiazepines: Clinical aspects, in *Therapeutic Drug Monitoring* (Richens, A., Marks, V., eds.), Churchill Livingstone, Edinburgh, pp. 255-271.
11. Bittencourt, P. R. M., Richens, A. 1981. Serum drug concentrations and effects on smooth pursuit and saccadic eye movements, in *Abstracts of the 12th World Congress of Neurology*, Excerpta Medica, Amsterdam, p. 523.
12. Bittencourt, P. R. M., Wade, P., Smith, A. T., Richens, A. 1981. The relationship between peak velocity of saccadic eye movements and serum benzodiazepine concentration. *Br. J. Clin. Pharmacol. 12*:523-533.
13. Bittencourt, P. R. M., Wade, P., Smith, A. T., Richens, A. 1983. Benzodiazepines impair smooth pursuit eye movements. *Br. J. Clin. Pharmacol. 15*:259-262.
14. Bittencourt, P. R. M., Richens, A. 1982. Assessment of antiepileptic drug toxicity by eye movements, in *Kyoto Symposia, 10th International Congress of EEG and Clinical Neurophysiology (EEG Supplement 36)* (Buser, P. A., Cobb, W. A., Okuma, T., eds.), Elsevier Biomedical, Amsterdam, pp. 467-481.
15. Bittencourt, P. R. M. 1983. Cerebral and cerebellar atrophy in patients with severe epilepsy: A preliminary report, in *Chronic*

Toxicity of Antiepileptic Drugs (Oxley, J., Janz, D., Meinardi, H., eds.), Raven Press, New York, pp. 237-246.
16. Carpenter, R. H. S. 1977. *Movements of the Eyes*, Pion, London.
17. Cooper, I. S., Amin, I., Riklan, M., Waltz, J. M., Poon, T. P. 1976. Chronic cerebellar stimulation in epilepsy. *Arch. Neurol.* *33*:559-570.
18. Dam, M. 1966. Organic changes in phenytoin-intoxicated pigs. *Acta Neurol. Scand.* *42*:491-494.
19. Dam, M. 1970. Number of Purkinje cells after diphenylhydantoin intoxication in pigs. *Arch. Neurol.* *22*:64-67.
20. Dam, M. 1970. The number of Purkinje cells after diphenylhydantoin intoxication in monkeys. *Epilepsia* *11*:199-206.
21. Dam, M. 1970. Number of Purkinje cells in patients with grand mal epilepsy treated with diphenylhydantoin. *Epilepsia* *11*:313-320.
22. Dam, M. 1972. The density and ultrastructure of the Purkinje cells following diphenylhydantoin treatment in animals and man. *Acta Neurol. Scand. Suppl.* *49*:48.
23. Dam, M. 1982. Phenytoin toxicity, in *Antiepileptic Drugs*, 2nd ed. (Woodbury, D. M., Penry, J. K., Pippenger, C. E., eds.), Raven Press, New York, pp. 247-256.
24. Dam, M., Nielsen, M. 1971. Purkinje's cell density after diphenylhydantoin intoxication in rats. *Arch. Neurol.* *23*:555-557.
25. Daroff, R. B., Troost, B. T. 1978. Supranuclear disorders of eye movements, in *Neuro-ophthalmology* (Glaser, J., ed.), Harper and Row, Hagerstown, Md., pp. 201-218.
26. Del Cerro, M. P., Snider, R. S. 1967. Studies on dilantin intoxication: Ultrastructural analogies with the lipoidoses. *Neurology* *17*:452-466.
27. Dreifuss, F. E., Sato, S. 1982. Benzodiazepines: Clonazepam, in *Antiepileptic Drugs* (Woodbury, D. M., Penry, J. K., Pippenger, C. E., eds.), Raven Press, New York, pp. 737-752.
28. Eichelbaum, M., Kothe, K. W., Hoffmann, F., Von Unruh, G. E. 1979. Kinetics and metabolism of carbamazepine during combined antiepileptic drug therapy. *Clin. Pharmacol. Ther.* *26*:366-371.
29. Finkelman, I., Arieff, A. J. 1942. Untoward effects of phenytoin sodium in epilepsy. *J. Am. Med. Assoc.* *118*:1209-1212.
30. Gallagher, B. B., Baumel, I. P., Mattson, R. H., Woodbury, S. G. 1973. Primidone, diphenylhydantoin and phenobarbital: Aspects of acute and chronic toxicity. *Neurology* *23*:145-149.
31. Gastaut, H., Naquet, R., Meyer, A., Cavanagh, J. B., Beck, E. 1959. Experimental psychomotor epilepsy in the cat. Electro-clinical and anatomo-pathological correlations. *J. Neuropathol. Exp. Neurol.* *18*:270-293.
32. Ghatak, N. R., Santoso, R. A., McKinney, W. M. 1976. Cerebellar degeneration following long-term phenytoin therapy. *Neurology* *26*:818-820.

33. Haberland, C. 1962. Cerebellar degeneration with clinical manifestation in chronic epileptic patients. *Psychiatr. Neurol. 142*: 29-44.
34. Hoffmann, W. W. 1958. Cerebellar lesions after parenteral dilantin administration. *Neurology 8*:210-214.
35. Holzman, P. S., Levy, D. L., Uhlenhuth, E. H., Proctor, L. R., Friedman, D. X. 1975. Smooth pursuit eye movements and diazepam, CPZ and secobarbital. *Psychopharmacologia 44*:111-115.
36. Höppener, R. J., Kuyer, A., Meijer, J. W. A., Hulsman, J. 1980. Correlation between daily fluctuations of carbamazepine serum levels and intermittent side-effects. *Epilepsia 21*:341-350.
37. Hotson, J. R. 1982. Cerebellar control of fixation eye movements. *Neurology 32*:31-36.
38. Horne, P. D. 1973. Long term anticonvulsant therapy and cerebellar atrophy. *J. Ir. Med. Assoc. 66*:147-152.
39. Iivanainen, M., Viukari, M., Helle, E. P. 1977. Cerebellar atrophy in phenytoin-treated mentally retarded epileptics. *Epilepsia 18*:375-386.
40. Karkos, J. 1973. Effect of antiepileptic drugs on the morphological picture of rat cerebellum. *Neuropatol. (Pol.) 11*:427-439.
41. Killian, J. M., Fromm, G. H. 1968. Carbamazepine in treatment of neuralgia. Use and side-effects. *Arch. Neurol. 19*:129-136.
42. Kokenge, R., Kutt, H., McDowell, F. 1965. Neurological sequelae following dilantin overdose in a patient and in experimental animals. *Neurology 15*:823-829.
43. Koller, W. C., Glatt, S. L., Fox, J. H. 1980. Phenytoin-induced cerebellar degeneration. *Ann. Neurol. 8*:203-204.
44. Kutt, H., Winters, W., Kokenge, R., McDowell, F. 1964. Diphenylhydantoin metabolism, blood levels and toxicity. *Arch. Neurol. 11*:642-648.
45. Lehtovaara, R. 1973. Post-rotational nystagmus: an unreliable sign of diphenylhydantoin toxicity. *Epilepsia 14*:447-449.
46. Leppik, I. E., Cloyd, J. C. 1982. Primidone toxicity, in *Antiepileptic Drugs* (Woodbury, D. M., Penry, J. K., Pippenger, C. E., eds.), Raven Press, New York, pp. 441-447.
47. Livingston, S., Petersen, D. 1956. Primidone (Mysoline) in the treatment of epilepsy *N. Engl. J. Med. 254*:327-329.
48. Masland, R. L. 1982. Carbamazepine neurotoxicity, in *Antiepileptic Drugs* (Woodbury, D. M., Penry, J. K., Pippenger, C. E., eds.), Raven Press, New York, pp. 521-532.
49. Mattson, R. J., Cramer, J. A. 1982. Phenobarbital toxicity, in *Antiepileptic Drugs* (Woodbury, D. M., Penry, J. K., Pippenger, C. E., eds.), Raven Press, New York, pp. 351-364.
50. Matzke, G. R., Cloyd, J. C., Sawchuk, R. J. 1981. Acute phenytoin and primidone intoxication: A pharmacokinetic analysis. *J. Clin. Pharmacol. 21*:92-99.

51. Meldrum, B. S., Brierley, J. B. 1973. Prolonged epileptic seizures in primates. *Arch. Neurol. 28*:10-17.
52. Meldrum, B. S., Horton, R. W., Brierley, J. B. 1974. Epileptic brain damage in adolescent baboons following seizures induced by allylglycine. *Brain 97*:407-418.
53. Merritt, H. H., Putnam, T. J. 1938. Sodium diphenylhydantoin in the treatment of convulsive disorders. *J. Am. Med. Assoc. 111*:1068-1073.
54. Mc Lain, L. W., Jr., Martin, J. T., Allen, J. H. 1980. Cerebellar degeneration due to chronic phenytoin therapy. *Ann. Neurol. 7*:18-23.
55. Nielsen, M. H., Dam, M., Klinken, L. 1971. The ultrastructure of Purkinje cells in diphenylhydantoin intoxicated rats. *Exp. Brain Res. 12*:447-456.
56. Norman, R. M. 1962. Neuropathological findings in acute hemiplegia in childhood, in *Clinics in Developmental Medicine N. 6* (Bax, M., Mitchell, R., eds.), Heinemann, London.
57. Norman, R. M. 1964. The neuropathology of status epilepticus. *Med. Sci. 4*:46-51.
58. Norman, R. M., Sandry, S., Corsellis, J. A. N. 1974. The nature and origin of patho-anatomical changes in the epileptic brain, in *Handbook of Clinical Neurology*, Vol. 15 (Vinken, P. J., Bruyn, G. W., eds.), Elsevier-North Holland, Amsterdam, pp. 611-620.
59. Norris, H. 1968. The time course of barbiturate action in man investigated by measurement of smooth tracking eye movement. *Br. J. Pharmacol. Chemother. 33*:117-128.
60. Nozue, M., Mizuno, M., Kaga, K. 1973. Neurotological findings in diphenylhydantoin intoxication. *Ann. Otol. Rhinol. Laryngol. 82*:389-394.
61. Orth, D. N., Almedia, H., Walsh, F. B., Honda, M. 1967. Ophthalmoplegia resulting from diphenylhydantoin and primidone intoxication. *J. Am. Med. Assoc. 201*:485-487.
62. Perucca, E., Richens, A. 1981. Antiepileptic drugs. Clinical aspects, in *Therapeutic Drug Monitoring* (Richens, A., Marks, V., eds.), Churchill Livingstone, Edinburgh, pp. 321-348.
63. Perucca, E., Crema, A. 1983. Therapeutic monitoring of serum antiepileptic drug levels, in *Epilepsy: an Update in Research and Therapy* (Nisticó, G., Di Perri, R., Meinardi, H., eds.), Alan R. Liss, Inc., New York, pp. 267-283.
64. Puro, D. G., Woodward, D. G. 1973. Effects of diphenylhydantoin on activity of rat cerebellar Purkinje cells. *Neuropharmacology 12*:433-440.
65. Rapport, R. L., Shaw, C. M. 1977. Phenytoin related cerebellar degeneration without seizures. *Ann. Neurol. 2*:437-439.
66. Rashbass, C. 1959. Barbiturate nystagmus and the mechanism of visual fixation. *Nature 183*:897-898.

67. Rashbass, C. 1961. The relationship between saccadic and smooth tracking eye movements. *J. Physiol. 159*:326-338.
68. Reynolds, E. H. 1975. Chronic antiepileptic toxicity: A review. *Epilepsia 16*:319-352.
69. Richens, A. 1982. Clinical pharmacology and medical treatment, in *Textbook of Epilepsy* (Laidlaw, J., Richens, A., eds.), Churchill Livingston, London, pp. 292-347.
70. Riker, W. W., Downes, H., Olsen, G. D., Smith, B. 1978. Conjugate lateral gaze nystagmus and free phenytoin concentrations in plasma: lack of correlation. *Epilepsia 19*:93-98.
71. Salcman, M., Defendini, R., Correll, J., Gilman, S. 1978. Neuropathological changes in cerebellar biopsies of epileptic patients. *Ann. Neurol. 3*:10-19.
72. Schmidt, D. 1982. Benzodiazepines: Diazepam, in *Antiepileptic Drugs*, 2nd ed. (Woodbury, D. M., Penry, J. K., Pippenger, C. E., eds.), Raven Press, New York, pp. 711-736.
73. Scholz, W. 1959. The contribution of patho-anatomical research to the problem of epilepsy. *Epilepsia 1*:36-55.
74. Sciarra, D., Carter, S., Vicale, C. T., Merritt, H. H. 1954. Clinical evaluation of primidone (Mysoline), a new anticonvulsant drug. *J. Am. Med. Assoc. 154*:827-829.
75. Smith, B. H., McNaughton, F. L. 1952. Mysoline, a new anticonvulsant drug. Its value in refractory cases of epilepsy. *Can. Med. Assoc. J. 68*:464-467.
76. Snider, R. S., Del Cerro, M. P. 1966. Membranous cytoplasmic spirals in dilantin intoxication. *Nature 212*:536-537.
77. Snider, R. S., Del Cerro, M. P. 1967. Drug induced dendritic sprouts on Purkinje cells in the adult cerebellum. *Expl. Neurol. 17*:466-480.
78. Spyelmeyer, W. 1920. Uber einige Beziechungen zwischen Ganglienzellveranderungen und gliosen Erscheinungen, besonders in Kleinhirm. *Z. Gesamte Neurol. Psychiatr. 54*:1-38.
79. Spyelmeyer, W. 1927. Pathogeneses des epileptischen Krampfes. *Z. Gesamte Neurol. Psychiatr. 109*:501-520.
80. Spyelmeyer, W. 1930. The anatomic substratum of the convulsive state. *Arch. Neurol. Psychiatry 23*:869-875.
81. Spyelmeyer, W. 1933. Funktionelle Kreislaufstorungen und Epilepsie. *Z. Gesamte Neurol. Psychiatr. 148*:285-298.
82. Strandjord, R. E., Johannessen, S. I. 1975. A preliminary study of serum carbamazepine levels in healthy subjects and in patients with epilepsy, in *Clinical Pharmacology of Antiepileptic Drugs* (Schneider, H., Janz, D., Gardner-Thorpe, C., Meinardi, H., Sherwin, A. L., eds.), Springer-Verlag, Berlin, pp. 181-188.
83. Sullivan, J. B., Rumack, B. H., Peterson, R. G. 1981. Acute carbamazepine toxicity from overdose. *Neurology 31*:621-624.
84. Tsukamoto, Y., Kondo, K., Yoshioka, M. 1980. Antiepileptic drug and cerebellar atrophy. *Acta Neurol. Scand. Suppl. 79*:89.

85. Umida, Y., Sakata, E. 1977. Equilibrium disorders in carbamazepine toxicity. *Ann. Otol. Rhinol. Laryngol. 86*:318-322.
86. Utterback, R. A. 1958. Parenchymatous cerebellar degeneration complicating diphenylhydantoin (dilantin) therapy. *Arch. Neurol. Psychiatry 80*:180-181.
87. Utterback, R. A., Ojeman, R., Malek, J. 1958. Parenchymatous cerebellar degeneration with dilantin intoxication. *J. Neuropathol. Exp. Neurol. 17*:516-519.
88. Woodbury, D. M., Penry, J. K., Pippenger, C. E. (eds.) 1982. *Antiepileptic Drugs*, Raven Press, New York.
89. Zimmerman, H. M. 1938. The histopathology of convulsive disorders in children. *J. Pediatr. 13*:859-890.

12
Mechanisms of Neurotoxicity of Anticancer Drugs

J. Alejandro Donoso* and Fred Samson
University of Kansas Medical Center, Kansas City, Kansas

I. INTRODUCTION

The growing list of chemotherapeutic agents includes some whose use is unfortunately limited by their neurotoxic actions. Although neurotoxicity from anticancer drugs does not appear to be a frequent consequence of chemotherapy, central and peripheral nervous system disturbances can be serious side effects and dose limiting. The most common nervous system-related problems are nausea, vomting, sensory disturbances, motor incoordination, depression, dizziness, headache, hallucinations, convulsions, and hyperexcitability. Nausea and vomiting are prevalent debilitating reactions to many anticancer drugs, but these reactions may be normal adaptive physiological responses to toxic materials rather than toxicity, strictly defined as an inability of tissues to carry on normal metabolic responses. Many of these disturbances may be the result of imbalances of neuroactive substances such as neurotransmitters, neuropeptides, and extracellular components that are

**Present affiliation*: Veterans Administration Medical Center, Kansas City, Missouri readily reversible when the chemotherapy is discontinued. Since cancer patients are now living longer as a result of improved therapies, there is a greater possibility of neurotoxicity appearing in a larger proportion of chemotherapy patients either as a delayed effect or following repeated dosing (107,109). In many clinical schedules multiple agents are used, increasing the difficulty in identifying the agents or combination of agents causing the underlying toxic events. The attempt here is to review the probable mechanisms of neurotoxicity of those anticancer drugs whose neurotoxicity is at least moderately common.

Acute neurotoxicity occurs with a few anticancer drugs. More frequently neurotoxicity is seen only after repetitive exposure, either as the cumulative effect on several processes or an effect on processes with slow temporal courses. Whether the mechanism of the neurotoxicity is related to the antineoplastic action of the drug is sometimes disputable, but certainly likely. Also, it is difficult to extrapolate from cytotoxic reactions in animal cells or cultured human cells to human cells in situ. Anticancer drugs are used with the aim of impairing the capability of rapid division of the cancer cells, but, by hitting the same molecular target in slowly dividing or nondividing cells, they may induce a cytotoxic effect, usually on a slower temporal course.

One approach in the development of anticancer drugs has been to take advantage of metabolic and special requirement differences between cancer and normal cells. Thus the antimetabolites to the various vitamins, amino acids, purines, pyrimidines, and nucleosides have been developed with the intent of interfering with rapid metabolism characteristics of tumor cells. As might be predicted from such designs, toxicity commonly occurs in organs with high rates of cell division, such as bone marrow, the gastrointestinal tract, and the hair follicle. Since the mature neuron does not divide and the blood-brain barrier limits access of drugs to the central nervous system, it might be anticipated that neurotoxicity from anticancer drugs would be rare. However, neurons are continuously renewing their molecular constituents, have a large surface-to-volume ratio to maintain, and are highly dependent upon an ongoing energy flow with relatively little reserve. Those neurons which support extensive axonal and dendritic processes are particularly vulnerable to insults, such as disruption of axoplasmic transport. Some anticancer agents depolymerize the microtubules of the mitotic spindle apparatus and this is believed to be the basis of their cancer therapeutic value. Since microtubules are abundant organelles in nerves, this type of anticancer drug frequently alters the normal functioning of nerves.

Anticancer drugs have various degrees of neurotoxicity, depending on the following variables:

1. Access to the cell. The permeability of a drug through the blood-brain barrier and the integrity of the barrier have important roles in neurotoxic actions in the central nervous system.

As a general rule, lipid-soluble drugs readily pass the blood-brain barrier; water-soluble drugs or drugs that are highly charged do not. Some analogs may enter by the normal substrate transport systems.

2. Access to the target molecule. If the target molecule is within the cell, the permeability or transport of the drug into the cell can be a major factor.
3. The importance of the target molecule to the cell's maintenance and functions.
4. The cell's capability to develop tolerance, renew, or repair damage.
5. Modifications of the drug which may either increase or decrease its toxicity.
6. The biological stage or physiological state of the cell at the time it is exposed to the drug.

II. VULNERABLE TARGETS OF NEUROTOXICITY

The targets for neurotoxicity may be considered in terms of cytological structure, such as mitochondria, Nissl substance, fibrillar components (microtubules, microfilaments, neurofilaments), myelin, synapse, or physiological processes (protein synthesis, axoplasmic transport, action potentials, synaptic transmission, etc.), or in terms of larger brain systems. These identifications are undoubtedly valuable, but a more powerful identification of targets would be at the molecular level, that is to say, the target molecules of the anticancer drug, the character of the molecular interaction, and how it leads to the cytotoxic perturbation of function or structure. Although this is an ultimate aim, currently the target molecules for many of the drugs are conjectured. The current status of cytological targets has been recently reviewed by Hirano and Llena (53), and that of the physiological targets by Price and Griffin (82).

III. DRUGS

A. Vinca Alkaloids

The vinca alkaloids, vincristine (VCR), vinglastine (VBL), and vindesine (VDS), are well-established cancer chemotherapeutic agents which interfere with the functions of microtubules, especially those that are constituents of the spindle apparatus in dividing cells. Vinca alkaloids disrupt the mitotic spindle apparatus, arrest cell division in metaphase, and eventually cause cell death (5,24,25,70). Although these vinca alkaloids have in common their interaction with microtubules (30,52,75,76), they have dissimilar chemotherapeutic actions and toxic side effects (22,24,57,101,106-108). With some overlap, there is a specificity in the oncolytic activity of these agents against particular cell lines. For example, VDS is much more effective than VCR and VBL in prolonging the life of mice inoculated intraperitoneally with B16 melanoma cells (3,72,101). Furthermore, their actions on nonmalignant cells differ; VCR toxicity is most frequently manifested by sensory and motor peripheral neuropathy, whereas VBL toxicity produces marrow suppression, with neuropathy being a relatively infrequent occurrence (22,24,57,101,106-108). VDS (still under going clinical trial) may cause both a neuropathy and marrow suppression, but of a milder form than with VCR or VBL (37,106). The neurological disorders induced by the vinca alkaloids, especially VCR, have been associated with their interaction with microtubules and the role that microtubules may play in the transport of macromolecules and vesicular elements along the axons (axoplasmic transport), materials necessary for the maintenance of the structural and functional integrity of axons. The lack of macromolecular synthesizing capacity in axons makes these neuronal extensions dependent upon arrival via axoplasmic transport of material synthesized in the cell body.

Studies on the comparative neurotoxicity of vinca alkaloids have shed some light on these differential effects. It has been shown that axoplasmic transport was blocked to a greater extent by VCR than VBL or VDS (18) when desheathed nerves were incubated in vitro with each of these drugs. Furthermore, it has been reported that VCR is taken up in a greater amount than VDS or VBL (55). On the other hand, in similar experiments on intact cat vagus nerves incubated in vitro with concentrations no greater than 100 μm, clear differences were not found in the blockage of fast axoplasmic transport caused by these three vinca alkaloids. Ultrastructural examination of nerves incubated in vitro for 2 hr with the vinca alkaloids showed at low concentrations (1-50 μM) a decrease in the microtubule number per axon, and at high concentrations (50-100 μM) paracrystalline structures concomitant with a decrease of microtubule number (30).

In vivo experiments have investigated the correlation between VCR neurotoxicity and axoplasmic transport. In one report on cats treated with doses of VCR comparable to those used in human chemotherapy, fast axoplasmic transport was studied in vivo after the animals had developed signs of peripheral neuropathy (42). A faster peak of transported proteins, absent in controls, was always present in the VCR-treated animals. Furthermore, the transport of proteins was partially blocked in a few preparations of cats with VCR-induced neuropathy. Other investigators have reported no changes in axonal transport in animals treated with VCR and presenting a VCR-induced peripheral neuropathy (11).

Microtubule stability in the axons or other cellular systems is regulated by, among other factors, the presence and/or type of microtubule-associated proteins. The ionic environment may cause depolymerization of microtubules by an increase or decrease in calcium concentration. Changes in these factors will affect the integrity of the microtubules and, as a consequence, their functions, in this case axoplasmic transport, would be affected. In this connection it has been shown that the interaction of VCR with the microtubule-tubulin system is influenced by the presence of microtubule-associated proteins (MAPs), or polycation components (47). In the absence of microtubule-associated proteins, the microtubules are completely depolymerized by VCR, whereas in their presence microtubules are only partially depolymerized, forming spirals and eventually forming paracrystal structures (31,47).

The microtubule stability varies even within regions of individual cells; for example, under conditions in which microtubules are structurally stable in axons, they are not seen in terminal regions. The stability of axonal microtubules in regions distant from or close to the cell bodies has been studied. The number of microtubules per axon was decreased by low temperatures in both distal and proximal regions of the nerve, being more pronounced in the distal region (32). Concomitant with the change in microtubule number per axon, changes in the axonal cross-sectional area were observed even when the tissue was immediately fixed in cold fixative. The fact that the microtubules and the axonal cross-sectional area appear to be more sensitive to cold in the distal regions of the nerve suggests that axonal regions far from their cell bodies are more vulnerable to neurotoxins. The mechanism of this differential reaction of axonal microtubules could be explained by differential stability of microtubules along the axons or differential distribution of agents such as microtubule-associated proteins or local ionic composition.

Until recently it was believed that microtubules do not enter the presynaptic terminal region. However, experiments with modified fixation techniques have revealed that microtubules are not only abundant in these regions, but also associated with a large number of synaptic vesicles (39,41). Further evidence for the presence of the microtubules is the fact that a large amount of tubulin has been isolated from nerve terminals (14,60,114).

The difficulty in detecting microtubules in the terminal regions suggests that they have a greater susceptibility than those in the main length of the axon to antitubulin drugs. Thus the low concentration of chronically administered VCR may produce clinical effects owing to an effect on intracellular domains with actively depolymerizing-polymerizing tubulin subunits, such as the mitotic apparatus of rapidly dividing cells and possibly synaptic and afferent nerve terminals.

B. Etoposide

Etoposide (VP-16-213), a derivative of podophyllotoxin (a toxic component in the roots of the mandrake herb), is an antitumor compound with significant activity in a broad spectrum of tumors, including lymphoma, leukemia, small-cell lung carcinoma, and ovarian and testicular cancers (56,83,105). Etoposide stops cell division in a way similar to the "spindle poison" colchicine and the parent compound podophyllotoxin. However, unlike podophyllotoxin, etoposide does not interfere with microtubule assembly (68), but instead may induce DNA breaks directly and arrest cells in the late S or G_2 phase of the cell cycle (69). Hematological toxicity is dose limiting with 43% of patients experiencing leukopenia when treated with etoposide alone (105). Peripheral neuropathy may result from a cumulative effect when other neurotoxins such as cisplatin and vinca alkaloids are given concomitantly (83,103). This type of neuropathy is characterized by an insidious onset, reversibility by reduction of the dose or cessation of the drug administration, and by "dying-back" type of neuronal degeneration. The fragmentation of DNA in neurons in the central nervous system would alter the synthesis of proteins or of macromolecules necessary to maintain neuronal structure and functions. The peripheral axonal regions would be affected earlier than axonal regions close to the cell bodies, since these regions would be expected to obtain enough material from the cell bodies when an inadequate synthesis of macromolecules occurs. That is, peripheral neuropathy may result when material is consumed as it moves along the axons, leaving a progressively smaller amount to reach the nerve endings and the distal regions of the nerves would be the first and most affected. Furthermore, a lack of materials essential for renewal would make the distal regions of the nerve more vulnerable to toxic substances.

C. Methotrexate

Methotrexate (MTX) is a folic acid analog and antagonist that inhibits cell division. It is a highly ionized, lipid-insoluble drug and poorly penetrates the blood-brain barrier. However, when it is given intrathecally or in conjunction with irradiation (which leads to a change in the blood-brain barrier permeability), there is commonly a spectrum of complications, with neuronal involvement. MTX causes primarily long-term injury to the central nervous system; that is, massive doses of MTX alone may lead to leukoencephalopathy (leukodystrophy), or sclerosis of white matter, degeneration of white matter with demyelinization and glial reaction. Again, it should be emphasized that these neuropathologies are more likely to occur if the MTX is given with irradiation. Deficits in short-term memory which develop several years later have been reported from long-term studies on children given intrathecal MTX and irradiation.

When given intrathecally or with irradiation, common complications are meningeal irritation, transverse myelitis, dementia, seizures, and coma (1,74,95). A high concentration of methotrexate in the cerebrospinal fluid leads to severe, acute neurotoxicity (7,9). Chronic encephalopathy occurs in some children with intrathecal administration (usually in combination with irradiation), with personality changes leading to dementia. Ataxia and spasticity may occur and multifocal infarcts of white matter secondary to a fibrinoid degeneration and thrombosis of small vessels have been reported (94,95). In a study with monkeys, intrathecal MTX along with irradiation led to a profound neurotoxicity some $3\frac{1}{2}$ months later. The necrotizing encephalopathy found in this study was not seen with either irradiation or MTX alone.

The mechanism of neurotoxicity is related to folic acid metabolism. The deficiency of folic acid leads to an inability to synthesize purines and pyrimidine thymine. The conversion of dihydrofolate into tetrahydrofolate is required for the synthesis of thymidylic acid and then nucleic acids. Thus fast-growing cells are inhibited by limiting the folic acid. Folic acid deficiency affects virtually every organ and system (45). Folic acid and derivatives have a more generalized function concerned with the transfer of one-carbon compounds. Folic acid has no direct coenzyme activity, but is converted by reduction into an active form, tetrahydrofolic acid by dihydrofolic acid reductase. This enzyme is strongly and specifically inhibited by certain analogs of folic acid such as aminopterin.

It has been suggested that the enzyme dehydrofolate reductase is important for the production of biopterins, which are cofactors for tyrosine hydroxylase (96). The brain contains this enzyme, although it is present only at about 15% of the activity in liver. Since the reduced folates easily enter the brain by a transport system from the plasma to the cerebrospinal fluid in the choroid plexus (97), the importance and role of this enzyme in brain are not entirely clear.

According to Kaplan and Wiernik (58), there is no clear biochemical explanation for chronic MTX neurotoxicity. It has been speculated that MTX causes neurotoxicity by demyelinization, which is often a feature of the myelopathic syndrome. The chemical "arachnoiditis," paraparesis, acute arachnoiditis, and myelopathic reactions appear to be caused by the MTX itself and not solely by the diluent, volume, and preservatives involved in delivery (9,33,107,108). Subarachnoid block can lead to elevated drug levels and might explain encephalopathic toxicity in some brain tumor patients. Another mechanism for MTX neurotoxicity could be abnormalities in the cerebrospinal fluid by bulk flow which may be impaired by MTX (8,9). Thus, neoplastic infiltration of meninges and arachnoid granulations may decrease the cerebrospinal fluid reabsorption and the transport of the MTX to the circulation. When intrathecal MTX is given with intravenous vincristine, the latter may lead to an increase in the concentration of MTX in the cerebrospinal fluid (4,102).

In summary, the consistent character of lesions is disseminated necrotizing leukoencephalopathy, demyelinization, multifocal white matter necrosis, and astrocytic reaction, but without inflammation or axonal damage. Strangely, the gray matter is spared (58). The acute or subacute reactions to MTX are meningeal irritation and arachnoiditis, which begin about 2 hr after the injection and last about 72 hr. These reactions lead to symptoms of a stiff neck, headache, nausea, vomiting, and lethargy and occur in about 60% of the patients. The paraplegia may be transient or permanent and is related to intrathecal MTX. It may be a severe direct toxic reaction or related to the demyelinization (90). The chronic reactions are an encephalopathy related most closely to the intrathecal or intraventricular route. It may be subacute, chronic, transient, or permanent.

Since the biochemical "target" of MTX is a deficit of folic acid with the consequent blockade of nucleic acid synthesis, mature neurons as nondividing cells would not be expected to be especially vulnerable to MTX, as indeed it appears that they are not. However, neurons are very dependent on other "support" systems such as glial cells, cerebral capillary integrity, and regulation of the brain cell microenvironment by the blood-brain barrier. Thus it is possible that neurotoxicity, especially the chronic clinical signs of nervous system damage, is secondary to malfunctions of certain nonneuronal cells.

D. Cytosine Arabinoside

Cytosine arabinoside (ARA-C) is an analog of deoxycytidine and is believed to inhibit DNA synthesis. It is used in therapy for acute myelocytic leukemia. It is also used by the intrathecal route for meningeal neoplasm when methotrexate has not been effective. Neurotoxic effects from intrathecal ARA-C have been reported. In one case reported, the patient (2,12) developed a myelopathy spreading a number of segments up into the thoracic cord and chiefly affecting white matter (12). Two patients who had received systemic chemotherapy, cranial irradiation, and prophylactic monthly intrathecal ARA-C developed blindness long after total remission in acute lymphocytic leukemia (35).

Peripheral neuropathy can also occur with ARA-C (89,93). The cases fail to form a coherent pattern, but are important nevertheless. Intrathecal antimetabolites may damage the nervous system by relatively nonspecific means. In the case of paraplegia seen after intrathecal therapy the intrathecal injection of a number of compounds has been followed by arachnoidal scarring and damage to nerve roots or spinal cord. Perhaps this is not related to the compound given, but the procedure itself. Withdrawal and repeated reinjection of cerebrospinal fluid alone can cause a demyelinating lesion of cat spinal cord (12).

E. 5-Fluorouracil

5-Fluorouracil is the 5-fluoro analog of uracil, which is effective against several types of solid tumors and is used topically for solar keratoses and multiple superficial basal cell carcinomas (110). When converted in vivo to 5-fluoro-2-deoxyuridine-5-monophosphate (FdUMP), it inhibits thymidylate synthetase and thus blocks thymidine monophosphate formation (see the review in Ref. 80). It is usually given intravenously, since gastrointestinal absorption is incomplete and unpredictable. It distributes to all body compartments, including the cerebrospinal fluid, probably by simple diffusion (10,21,98) and is more toxic to proliferating than to nonproliferating cells (80).

Since 5-fluorouracil must be converted in vivo to be active, a major determinant of tumor sensitivity would be dependent upon the biochemical pathways responsible for its conversion to FdUMP. However, this relationship between FdUMP production and the clinical response has not been established in human cancers (16). The side effects are nausea, vomiting, anorexia, and diarrhea. Also, there is a bone marrow depression, alopecia, and thinning of the skin from its toxicity toward proliferating cells. Its neurotoxic actions seem to be relatively uncommon; however, acute cerebellar syndrome does appear in about 1% of patients (107,108). This syndrome is characterized by ataxia, nystagmus, slurred speech, and dizziness, all of which are claimed to be reversible within 1-6 weeks after the therapy is discontinued. Owing to its toxicity for proliferating cells, 5-fluorouracil is teratogenic (20, 45); however, it does not seem to lead to any irreversible neurocytotoxic effect. The acute cerebellar syndrome has been proposed to be a consequence of an inhibition of the tricarboxylic acid cycle; that is, the fluorouracil is converted by the affected cells to fluorocitrate, which in turn inhibits the tricarboxylic acid cycle by blocking the normal conversion of citrate to isocitrate (61). In view of the ongoing energy requirements of nervous tissue, a perturbation of the normal events of the tricarboxylic acid cycle would be expected to lead to central nervous system disturbances. However, since it is reversible (107,108), it seems that such an inhibition is not pronounced enough to produce long-term structural damage such as occurs with severe hypoxia. It is possible that the production of fluorocitrate in many neurons is not rapid enough or extensive enough in most neurons to cause any substantial interference with the supply of energy for normal central nervous system function.

In addition to fluorouracil, floxuridine is also available for clinical use in the United States. When incubated with cells in culture, floxuridine is generally more potent than fluorouracil, probably because it is more an immediate precursor of the active form of FdUMP. In humans, however, fluorouracil and floxuridine are essentially equally toxic on a molar basis because floxuridine is rapidly degraded to fluorouracil in the body.

F. Cisplatin

Cisplatin (cis-dichlorodiamine platinum) is a potent anticancer platinum compound which has the property of inhibiting cell division but not cell growth (81,85,87,88,112). It has been proposed that the mechanism for the antitumor activity of cisplatin involves "binding" of the platinum to cellular DNA (46,48). Tissue culture and simultaneous in vivo and in vitro studies using concentrations of the platinum drugs having a therapeutic effect on animals with cancer have determined that the primary action at the cellular level is a severe and prolonged inhibition of new DNA synthesis with some effect upon RNA and protein synthesis (54).

Cisplatin has been found particularly beneficial in patients with testicular, ovarian, head, and neck tumors (50,51,111,113) and is now recomended as first-line chemotherapy in stabilized combination regimens for testicular tumors and ovarian adenocarcinomas. Trials of the drug are also in progress for head and neck squamous carcinoma and other tumors (19,85,112). The major types of clinical toxic effects of cisplatin are gastrointestinal, hematopoietic, immunosuppressive, otological, and renal. The first three of these side effects are claimed to be relatively mild (50,67), but the gastrointestinal distress has been found to be significant by some clinicians (R. Stephens, personal communication, 1983). Ototoxicity, tinnitus, and loss of hearing in the high-frequency range (4000-8000 Hz) has been reported as the most frequent neurological disorder caused by cisplatin (19,26,34,48). Ototoxicity appeared to be dose related and cumulative and occurred both unilaterally and bilaterally (81). This hearing loss was detected within 4 days after treatment and persisted for up to 6 months in surviving patients (50,63,78). The degree of reversibility of cisplatin-induced ototoxicity has not been established (40,84). In some cases the hearing losses have proved permanent over a 12- to 18-month period, confirming an earlier report on the irreversible nature of the hearing loss (19,64). No vestibular symptoms appear to be associated with the audiological toxicities.

In the treatment of cancer patients with cisplatin the auditory function should be monitored. If the toxicity is diagnosed and the drug is continued, clinical hearing loss at the spoken-voice range (100-4000 Hz) will occur. Partially reversible, high-frequency nerve deafness is usually the earliest sign of toxicity. The postulated mechanism of ototoxicity is a direct effect on the hair cells of the organ of Corti (67). Animal studies have shown that the damage occurs mainly to the outer hair cells in the basal turns of the cochlea, while labyrinth damage was not seen (36,73). The preferential damage of outer hair cells versus inner hair cells may be explained by the higher metabolic activity of the outer hair cells, as suggested by Koide et al. (62). The outer hair cells were shown to have a higher protein turnover than inner hair cells in an autoradiographic study. Another possibility is that the blood circulation is not as abundant in the lower turn as in the apical turns in the cochlea (36). Finally, the action of cisplatin could be increased if the drug were accumulated in the perilymph of the inner ear in a manner similar to the aminoglycoside antibiotics. The presence of the drug in the cell will interfere with DNA, RNA, and protein synthesis, damaging the cells with higher metabolic rate. The onset of ototoxicity by cisplatin occurs approximately 5 days after initiation of treatment. No direct damage to the auditory nerve has been reported.

Because high-frequency sounds are processed in the basal end of cochlea, these frequencies are the first to be affected. As the ototoxic lesion advances, it progressively involves the hair cells toward the apex, resulting in losses at progressively lower frequencies.

Other neurotoxic reactions caused by cisplatin are peripheral neuropathies, including paresthesia in both upper and lower extremities, tremor, leg weakness, and loss of taste (23,44,49,59,81). Seizures were also reported in some patients (71), as well as cortical blindness (6). When cisplatin therapy was discontinued, these neurological signs improved.

G. Procarbazine

Procarbazine is a monoamine oxidase inhibitor currently used as a chemotherapeutic agent in combination with other active agents such as vincristine, methotrexate, and bleomycin in the therapeutic treatment of lymphomas (99), small-cell lung carcinoma (79), melanoma (100), and brain tumors (66,77). The greatest therapeutic effectiveness of procarbazine is in Hodgkin's disease, particularly in combination with mechlorethamine, vincristine, and prednisone (28,29). The mechanism(s) by which procarbazine inhibits tumor growth has not yet been determined. Procarbazine induces chromatid breaks and translocations (92,104) in Ehrlich ascites and leukemia cells. In addition, procarbazine can inhibit RNA, DNA, and protein synthesis in vivo and in vitro (43,65), and these may be responsible for the disturbances in cell division. Cytological studies indicate suppression of mitosis as a result of prolongation of interphase. The most common toxic side effects include leukopenia, thrombocytopenia, nausea, and vomiting. About 10-15% of the patients treated with procarbazine showed some signs of peripheral neuropathy several weeks after the treatment had started (13,91), but these authors believe the disturbances usually are not serious enough to discontinue the drug. A reduction of the dose can reverse the neurotoxic side effects. The peripheral neuropathy caused by procarbazine may be a consequence of depleting the plasma of pyridoxal phosphate (15,17,27). Other neurotoxic side effects induced by procarbazine are ataxia, confusion, and predisposition to a variety of drug interactions common to monoamine oxidase-inhibiting compounds (107,108).

IV. SUMMARY

In the process of stopping cell division and/or killing malignant cells, anticancer drugs also cause some form of cytotoxicity in normal cells. Although the most frequent toxic side effects occur in cell renewal systems such as skin, bone marrow, and small intestine, central and peripheral nervous system disturbances do occur with some chemotherapeutic agents. Despite the fact that mature neurons are nondividing and protected by barriers such as the blood-brain barrier, these agents can reach the neuron and their processes and affect their functions. The neuronal cell body is a site of high protein synthesis, and chemotherapeutic agents that inhibit protein synthesis will impair the process of maintenance in the nervous tissue. A lack of proteins essential to the structural and functional integrity of the neuron will eventually lead to pathological reactions in those areas with a high rate of turnover. Furthermore, neurons are characterized by a high degree of asymmetry and functional segregation with an almost complete restriction of protein synthesis and organelle formation to the neuronal cell body. The major fraction of axonal proteins, then, has to be translocated from the cell body to the neuronal processes and nerve endings. The neuronal translocation (axoplasmic transport) is dependent upon the integrity of the neuronal cytoskeleton, including microtubules. Since some anticancer agents bind tubulin and depolymerize microtubules (antimitotic agents), they can initiate a cascade of pathological events, especially in the peripheral nervous system. The nervous system is highly dependent on a continuous energy flow from the oxidation of glucose and anticancer agents may cause changes in the blood and/or energy supply or alter functions in nonneuronal cells that will indirectly affect the nervous system. There is a need for better understanding of the neurotoxic mechanisms of anticancer agents, not only because of their potentially serious debilitating effects, but also because of the relatively limited capability of the nervous system for repair.

REFERENCES

1. Allen, J. C. 1978. The effects of cancer therapy on the nervous system. *J. Pediatr.* *93*:903-909.
2. Bagshawe, K. D., Magrath, I. T., Golding, P. R. 1959. Intrathecal methotrexate. *Lancet* *2*:1258.
3. Barnett, C. J., Cullinan, G. J., Gerzon, K., Hoying, R. C. Jones, W. E., Newlon, W. M., Poore, G. A., Robinson, R. L., Sweeney, M. J., Todd, G. C. 1978. Structure-activity relationships of dimeric *Catharanthus* alkaloids. 1. Deacetylvinblastine amide (vindesine) sulfate. *J. Med. Chem.* *21*:88-96.
4. Bender, R. A., Nichols, A. P., Norton, L., Simon, R. M. 1978. Lack of therapeutic synergism between vincristine and methotrexate in L1210 murine leukemia *in vivo*. *Cancer Treat. Rep.* *62*:997-1003.
5. Bensch, K. G., Malawista, S. E. 1969. Microtubular crystals in mammalian cells. *J. Cell Biol.* *40*:95-107.
6. Berman, I. J., Mann, M. P. 1980. Seizures and transient cortical blindness associated with cisplatinum (II) diamminedichloride (PDD) therapy in a thirty-year old man. *Cancer* *45*:764-766.
7. Bertino, J. R. 1981. Clinical use of methotrexate—With emphasis on use of high doses. *Cancer Treat. Rep. Suppl.* *1*:131-135.
8. Bleyer, W. A., Dedrick, R. L. 1977. Clinical pharmacology of intrathecal methotrexate I. Pharmacokinetics in nontoxic patients after lumbar injection. *Cancer Treat. Rep.* *61*:703-708.
9. Bleyer, W. A., Drake, J. C., Chabner, B. A. 1973. Neurotoxicity and elevated cerebrospinal-fluid methotrexate concentration in meningeal leukemia. *N. Engl. J. Med.* *289*:770-773.
10. Bourke, R. A., West, C. R., Chheda, G., Tower, D. B. 1973. Kinetics of entry and distribution of 5-fluorouracil in cerebrospinal fluid and brain following intravenous injection in a primate. *Cancer Res.* *33*:1735-1746.
11. Bradley, W. G., Williams, M. H. 1973. Axoplasmic flow in axonal neuropathies: Axoplasmic flow in cats with toxic neuropathies. *Brain* *96*:235-246.
12. Breuer, A. C., Pitman, S. W., Dawson, D. M., Schoene, W. C. 1977. Paraparesis following intrathecal cytosine arabinoside: A case report with neuropathological findings. *Cancer* *40*:2817-2822.
13. Brunner, K. W., Young, C. W. 1965. A methylhydrazine derivative in Hodgkin's disease and other malignant neoplasma: Therapeutic and toxic effects studied in 51 patients. *Ann. Intern. Med.* *63*:69-86.
14. Burke, B. E., De Lorenzo, R. J. 1982. CA^{2+} and calmodulin-regulated endogenous tubulin kinase activity in presynaptic nerve terminal preparations. *Brain Res.* *236*:393-415.

15. Chabner, B. A., DeVita, V. T., Considine, N., Oliverio, V. T. 1969. Plasma pyridoxal phosphate depletion by the carcinostatic procarbazine. *Proc. Soc. Exp. Biol. Med.* *132*:1119-1122.
16. Chabner, B. A., Myers, C. E., Coleman, C. N., Johns, D. G. 1975. The clinical pharmacology of antineoplastic agents (first of two parts). *N. Engl. J. Med.* *292*:1107-1113.
17. Chabner, B. A., Sponzos, R., Hubbard, S., Canellos, G. P., Young, R. C., Schein, P. S., DeVita, V. T. 1973. High-dose intermittent intravenous infusion of procarbazine (NSC-77213). *Cancer Chemother. Rep.* *57*:361-363.
18. Chan, S. W., Worth, R., Ochs, S. 1980. Block of axoplasmic transport *in vitro* by vinca alkaloids. *J. Neurobiol.* *11*:251-264.
19. Chapman, P. 1982. Rapid onset hearing loss after cisplatinum therapy: Case reports and literature review. *J. Laryngol. Otol.* *96*:159-162.
20. Chaube, S., Murphy, M. L. 1968. The teratogenic effects of the recent drugs active in cancer chemotherapy. *Adv. Teratol.* *3*:181-237.
21. Clarkson, B., O'Connor, A., Winston, L., Hutchison, D. 1965. The physiologic disposition of 5-fluorouracil and 5-fluoro-2'-deoxyuridine in man. *Clin. Pharmacol. Ther.* *5*:581-610.
22. Conrad, R. A., Cullinan, G. J., Gerzon, K., Poore, G. A. 1979. Structure-activity relationships of dimeric catharanthus alkaloids. 2. Experimental antitumor activities of N-substituted deacetylvinblastine amide (vindesine) sulfates. *J. Med. Chem.* *22*:391-400.
23. Cowan, J. D., Kies, M. S., Roth, J. L., Joyce, R. P. 1980. Nerve conduction in patients treated with cis diaminedichloroplatinum (II): A preliminary report. *Cancer Treat. Rep.* *64*:1119-1122.
24. Creasey, W. A. 1981. The vinca alkaloids and similar compounds. *Cancer Chemother.* *3*:79-96.
25. Cutts, J. H. 1961. The effect of vincaleukoblastine on dividing cells *in vivo*. *Cancer Res.* *21*:168-172.
26. DeConti, R. C., Toftness, B. R., Lange, R. C., Creasey, W. A. 1973. Clinical and pharmacological studies with cis-diamminedichloroplatinum (II). *Cancer Res.* *33*:1310-1315.
27. DeVita, V. T., Hahn, M. A., Oliverio, V. T. 1965. Monoamine oxidase inhibition by a new carcinostatic agent, N-isopropyl-α-(2 methylhydrazino)-p-toluamide (MIH). *Proc. Soc. Exp. Biol. Med.* *120*:561-565.
28. DeVita, V. T., Jr., Serpick, A. A., Carbone, P. P. 1970. Combination chemotherapy in the treatment of advanced Hodgkin's disease. *Ann. Intern. Med.* *73*:881-895.
29. DeVita, V. T., Jr., Simon, R. M., Hubbard, S. M., Young, R. C., Berard, C. W., Moxley, S. H., Frei III, E., Carbone, P. P.,

Canellos, G. P. 1980. Curability of advanced Hodgkin's disease with chemotherapy. *Ann. Intern. Med.* *92*:587-595.

30. Donoso, J. A., Green, L. S., Heller-Bettinger, I. E., Samson, F. E. 1977. Action of the vinca alkaloids vincristine, vinblastine, and desacetyl vinblastine amide on axonal fibrillar organelles *in vitro*. *Cancer Res.* *37*:1401-1407.
31. Donoso, J. A., Haskins, K. M., Himes, R. H. 1979. Effect of microtubule associated proteins on the interaction of vincristine with microtubules. *Cancer Res.* *39*:1604-1610.
32. Donoso, J. A., Samson, F. E. 1981. Differential stability of microtubules in cervical vagus nerve unmyelinated axons. *J. Cell Biol.* *91*:88A.
33. Duttera, M. J., Bleyer, W. A., Pomeroy, T. C., Leventhal, C. M., Leventhal, B. G. 1973. Irradiation, methotrexate toxicity and the treatment of meningeal leukemia. *Lancet* *2*:703-707.
34. Einhorn, L. H., Williams, S. D. 1979. Current concepts in cancer: The role of cisplatinum in solid tumor therapy. *N. Engl. J. Med.* *300*:289-291.
35. Fishman, M. L., Bean, S. C., Cogan, D. G. 1976. Optic atrophy following prophylactic chemotherapy and cranial radiation for acute lymphocytic leukemia. *Am. J. Ophthalmol.* *82*:571-576.
36. Fleischman, R. W., Stadnicki, S. W., Ethier, M. F., Schaeppi, U. 1975. Ototoxicity of cis dichlorodiammine platinum (II) in the guinea pig. *Toxicol. Appl. Pharmacol.* *33*:320-332.
37. Focan, C., Olivier, R., Le Hung, S., Bays, R., Claessens, J. J., Debruyne, H. 1981. Neurological toxicity of vindesine used in combination chemotherapy of 51 human solid tumors. *Cancer Chemother. Pharmacol.* *6*:175-181.
38. Geiser, C. F., Bishop, Y., Jaffe, N., Furman, L., Traggis, D., Frei, E. 1975. Adverse effects of intrathecal methotrexate in children with acute leukemia in remission. *Blood* *45*:189-195.
39. Gordon-Weeks, P. R., Burgoyne, R. D., Gray, E. G. 1982. Presynaptic microtubules: Organisation and assembly/disassembly. *Neuroscience* *7*:739-749.
40. Gottlieb, J. A., Drewinko, B. 1975. Review of the current clinical status of platinum coordination complexes in cancer chemotherapy. *Cancer Chemother. Rep.* *59*:621-628.
41. Gray, E. G. 1975. Presynaptic microtubules and their association with synaptic vesicles. *Proc. R. Soc. London Ser. B* *190*:369-372.
42. Green, L. S., Donoso, J. A., Heller-Bettinger, I. E., Samson, F. E. 1977. Axonal transport disturbances in vincristine-induced peripheral neuropathy. *Ann. Neurol.* *1*:255-262.
43. Gutterman, J., Huang, A. T., Hochstein, P. 1969. Studies on the mode of action of N-isopropyl-α-(2-methylhydrazino)-p-toluamide (MIH) (33658). *Proc. Soc. Exp. Biol. Med.* *130*:797-802.

44. Hadley, D., Herr, H. W. 1979. Peripheral neuropathy associated with cis-dichlordiammine platinum (II) treatment. *Cancer 44*: 2026-2028.
45. Harbison, R. D. 1980. Teratogens, in *Casarett and Doull's Toxicology*, 2nd ed. (Doull, J., Klaassen, C. D., Amdur, M. O., eds.), Macmillan, New York, pp. 158-175.
46. Harder, H. C., Rosenberg, B. 1970. Inhibitory effects of anti-tumor platinum compounds on DNA, RNA and protein synthesis in mammalian cells *in vitro*. *Int. J. Cancer 6*:207-216.
47. Haskins, K. M., Donoso, J. A., Himes, R. H. 1981. Spirals and paracrystals induced by vinca alkaloids: Evidence that microtubule-associated proteins act as polycations. *J. Cell Sci. 47*:237-247.
48. Helson, L., Okonkwo, E., Anton, L., Cvitkovic, E. 1978. Cis-platinum ototoxicity. *Clin. Toxicol. 13*:469-478.
49. Hemphill, M., Pestronk, A., Walsh, T., Parhad, I., Clark, A., Rosenshein, N. 1980. Sensory neuropathy in cis-platinum chemotherapy. *Neurology 30*:429.
50. Higby, D. J., Wallace, H. J., Albert, D., Holland, J. F. 1974. Diaminedichloroplatinum in the chemotherapy of testicular tumors. *J. Urol. 192*:100-104.
51. Hill, J. M., Loeb, E., MacLellan, A., Hill, N. O., Khan, A., King, J. J. 1975. Clinical studies of platinum coordination compounds in the treatment of various malignant diseases. *Cancer Chemother. Rep. 59*:647-659.
52. Himes, R. H., Kersey, R. N., Heller-Bettinger, I., Samson, F. E. 1976. Action of vinca alkaloids, vincristine, vinblastine and desacetyl vinblastine amide on microtubules *in vitro*. *Cancer Res. 36*:3798-3802.
53. Hirano, A., Llena, J. F. 1980. The central nervous system as a target in toxic-metabolic states, in *Experimental and Clinical Neurotoxicology* (Spencer, P. S., Schaumburg, H. H., eds.), Williams and Wilkins, Baltimore, pp. 24-34.
54. Howle, J. A., Gale, G. R. 1970. Cis-dichlorodiamine platinum (II): Persistent and selective inhibition of deoxyribonucleic acid synthesis *in vivo*. *Biochem. Pharmacol. 19*:2757-2762.
55. Iqbal, L., Ochs, S. 1980. Uptake of vinca alkaloids into mammalian nerve and its subcellular components. *J. Neurochem. 34*:59-68.
56. Issell, B. F., Crooke, S. T. 1979. Etoposide (VP-16-213). *Cancer Treat. Rev. 6*:107-124.
57. Johnson, L. S. 1982. Plant alkaloids, in *Cancer Medicine* (Holland, J. F., Frei, E., eds.), Lea and Febiger, Philadelphia, pp. 910-920.
58. Kaplan, R. S., Wiernik, P. H. 1984. Neurotoxicity of antitumor agents, in *Toxicity of Chemotherapy* (Perry, M. C., Yarbro, J. W., eds.), Grune and Stratton, New York, pp. 365-431.

59. Kedar, A., Cohen, M. E., Freeman, A. I. 1978. Peripheral neuropathy as a complication of cis-dichlorodiammineplatinum (II) treatment: A case report. *Cancer Treat. Rep. 62*:819-821.
60. Kelly, P. T., Cotman, C. W. 1978. Synaptic proteins: Characterization of tubulin and actin and identification of a distinct postsynaptic density polypeptide. *J. Cell Biol. 79*:173-183.
61. Koenig, H., Patel, A. 1970. The acute cerebellar syndrome in 5-fluorouracil chemotherapy: A manifestation of fluoroacetate intoxication. *Neurology 20*:416.
62. Koide, Y., Hata, A., Hando, R. 1966. Vulnerability of the organ of Corti in poisoning. *Acta Otolaryngol. 61*:332-344.
63. Kovach, J. S., Moertel, C. G., Shutt, A. J., Reitemeier, R. G., Hahn, R. G. 1978. Phase II study of cis-diamminedichloroplatinum (NSC-119875) in advanced carcinoma of the large bowel. *Cancer Chemother. Rep. 57*:357-359.
64. Krakoff, I. H., Lippman, A. J., 1974. Clinical trials of cis-platinum (II) diamine-dichloride (PDD) in patients with advanced cancer, in *Platinum Coordination Emphasis in Cancer Chemotherapy* (Connors, T. A., Roberto, J. J., eds.), Springer-Verlag, Heidelberg, pp. 183-190.
65. Kreis, W. 1966. Metabolism and reaction mechanism of a methylhydrazine derivative in BDF P-815 leukemic mice. *Proc. Am. Cancer Res. 7*:39.
66. Kumar, A. R. V., Renandin, J., Wilson, C. B., Boldrey, E. B., Enot, K. J., Levin, V. A. 1974. Procarbazine hydrochloride in the treatment of brain tumors. Phase 2 Study. *J. Neurosurg. 40*:365-371.
67. Lippman, A. J., Helson, C., Helson, L., Krakoff, I. H. 1973. Clinical trials of cis-diamminedichloroplatinum (NSC-119875). *Cancer Chemother. Rep. 57*:191-200.
68. Loike, J. D., Brewer, C. F., Sternlicht, H., Gensler, W. J., Horwitz, S. B. 1978. Structure-activity study of the inhibition of microtubule assembly *in vitro* by podophyllotoxin and its congeners. *Cancer Res. 38*:2688-2693.
69. Loike, J. D., Horowitz, S. B. 1976. Effects of VP-16-213 on the intracellular degradation of DNA in HeLa cells. *Biochemistry 15*:5443-5448.
70. Malawista, S. E., Sato, H., Bensch, K. G. 1968. Vinblastine and griseofulvin reversibility disrupt the living mitotic spindle. *Science 160*:770-772.
71. Mead, G. M., Arnold, A. M., Green, J. A., Macbeth, F. R., Williams, C. J., Whitehouse, J. M. 1982. Epileptic seizures associated with cisplatin administration. *Cancer Treat. Rep. 66*:1719-1722.
72. Miller, J. C., Gutowski, G. E., Poore, G. A., Boder, G. B. 1977. Alkaloids of vinca rosea L. (*Catharanthus roseus* G. Don). 38. 4'-dehydrated derivatives. *J. Med. Chem. 20*:409-413.

73. Nakai, Y., Konishi, K., Chang, K. C., Ohashi, K., Morisaki, N., Minowa, Y., Morimoto, A. 1982. Ototoxicity of the anticancer drug cisplatin. *Acta Otolaryngol. 93*:227-232.
74. Nelson, R. W., Frank, J. R. 1981. Intrathecal methotrexate-induced neurotoxicities. *Am. J. Hosp. Pharm. 38*:65-68.
75. Owellen, R. J., Hartke, C. A., Dickerson, R. M., Hains, F. O. 1976. Inhibition of tubulin-microtubule polymerization by drugs of the vinca alkaloid class. *Cancer Res. 36*:1499-1502.
76. Owellen, R. J., Owens, A. H., Jr., Donigian, D. W. 1972. The binding of vincristine, vinblastine and colchicine to tubulin. *Biochem. Biophys. Res. Commun. 47*:685-691.
77. Pezzotta, S., Knerich, R., Butti, G. 1982. Chemotherapy for medulloblastoma in children. Current status and future prospects. *Child's Brain 9*:294-298.
78. Piel, I. J., Meyer, D., Perlia, C. P., Wolf, V. I. 1974. Effects of cis-diamminechloroplatinum (NSC-119875) on hearing function in man. *Cancer Chemother. Rep. 58*:871-875.
79. Poplin, E. A., Aisner, J., Van Echo, D. A., Whitacre, M., Wiernik, P. H. 1982. CCNV, vincristine, methotrexate, and procarbazine treatment of relapsed small cell lung carcinoma. *Cancer Treat. Rep. 66*:1557-1559.
80. Pratt, W. B., Ruddon, R. W. 1979. *The Anticancer Drugs*, Oxford University Press, New York, pp. 122-131.
81. Prestayko, A. W., D'Aoust, J. C., Issel, B. F., Crooke, S. T. 1979. Cisplatin (cis-diammine dichloro platinum II). *Cancer Treat. Rev. 6*:17-39.
82. Price, D. L., Griffin, J. W. 1980. Neurons and ensheathing cells as targets of disease processes, in *Experimental and Clinical Neurotoxicology* (Spencer, P. S., Schaumburg, H. H., eds), Williams and Wilkins, Baltimore, pp. 2-23.
83. Radice, P. A., Bunn, P. A., Jr., Ihde, D. C. 1979. Therapeutic trials with VP-16-213 and VM-26: Active agents in small cell lung cancer, non-Hodgkin's lymphomas, and other malignancies. *Cancer Treat. Rep. 63*:1231-1239.
84. Reddel, R. R., Kefford, R. F., Grant, J. M., Coates, A. S., Fox, R. M., Tattersall, M. H. N. 1982. Ototoxicity in patients receiving cisplatin: Importance of dose and method of drug administration. *Cancer Treat. Rep. 66*19-23.
85. Rose, W. C., Schurig, J. E., Huftalen, J. B., Bradner, W. T. 1982. Antitumor activity and toxicity of cisplatin analogs. *Cancer Treat. Rep. 66*:135-146.
86. Rosenberg, B. 1975. Possible mechanisms for the antitumor activity of platinum coordination complexes. *Cancer Chemother. Rep. 59*:589-598.
87. Rosenberg, B., VanCamp, L., Grimley, E. B., Thomson, A. J. 1967. The inhibition of growth or cell division in *Escherichia coli*

by different ionic species of platinum (IV) complexes. *J. Biol. Chem.* *242*:1347-1352.

88. Rosenberg, B., VanCamp, L., Trosko, J. E., Mansour, V. H. 1969. Platinum compounds: A new class of potent antitumor agents. *Nature* *222*:385-386.
89. Russell, J. A., Powless, R. L. 1974. Neuropathy due to cytosine arabinoside. *Br. Med. J.* *2*:652-663.
90. Saiki, J. H., Thompson, S., Smith, F., Atkinson, R. 1972. Paraplegia following intrathecal chemotherapy. *Cancer* *29*:370-374.
91. Samuels, M. L., Leary, W. B., Alexanian, R., Howe, C. D., Frei III, E. 1967. Clinical trials with N-isopropyl-α-(2-methylhydrazino)-p-toluamide hydrochloride in malignant lymphoma and other disseminated neoplasia. *Cancer* *20*:1187-1194.
92. Sartorelli, A. C., Tsunamura, S. 1965. Metabolic alteration in L5178Y lymphoma cells induced by N-isopropyl-α-(2-methylhydrazino)-p-toluamide (procarbazine). *Proc. Am. Assoc. Cancer Res.* *6*:55.
93. Sawicka, J., Dawson, D. M., Blum, R. 1978. Neurologic aspects of the treatment of cancer. *Curr. Neurol.* *1*:301-335.
94. Shapiro, W. R., Allen, J. C., Horten, B. C. 1980. Chronic methotrexate toxicity to the central nervous system. *Clin. Bull.* *10*:49-52.
95. Shapiro, W. R., Chernick, N. L., Posner, J. B. 1973. Necrotizing encephalitis following intraventricular instillation of methotrexate. *Arch. Neurol.* *28*:96-102.
96. Spector, R., Levy, P., Abelson, H. T. 1977. Identification of dihydrofolate reductase in rabbit brain. *Biochem. Pharmacol.* *26*:1507-1511.
97. Spector, R. 1977. The effect of pronase on choroid plexus transport. *Brain Res.* *134*:573-576.
98. Sterman, A. B., Shaumburg, H. H. 1980. Neurotoxicity of selected drugs, in *Experimental and Clinical Neurotoxicology* (Spencer, P. S., Schaumburg, H. H., eds.), Williams and Wilkins, Baltimore, pp. 593-612.
99. Stolinsky, D. C., Solomon, J., Pugh, R. P., Stevens, A. R., Jacobs, E. M., Irwin, L. E., Wood, D. A., Steinfold, J. L., Bakeman, J. R. 1970. Clinical experience with procarbazine in Hodgkin's disease, reticulum cell sarcoma, and lymphosarcoma. *Cancer* *26*:984-900.
100. Stragand, J. J., Drewinko, B., Barlogie, B., Papadopulos, N., White, R. A. 1982. Serial analysis of melanoma growth kinetics in a patient receiving bleomycin and procarbazine therapy: Comparison of tumor volume, flow cytometry and thymidine labeling techniques. *Cancer Treat. Rep.* *66*:529-534.

101. Sweeney, M. J., Boder, G. B., Cullinan, G. J., Culp, H. W., Daniles, W. D., Dyke, R. W., Gerzon, K., McMahon, R. E., Nelson, R. L., Poore, G. A., Todd, G. C. 1978. Antitumor activity of deacetyl vinblastine amide sulfate (vindesine) in rodents and mitotic accumulation studies in culture. *Cancer Res.* *38*:2886-2891.

102. Tejada, F., Zubrod, C. G. 1979. Vincristine effect on methotrexate cerebrospinal fluid concentration. *Cancer Treat. Rep.* *63*:143-145.

103. Thant, M., Hawley, R. J., Smith, M. T., Cohen, M. H., Minna, J. D., Bunn, P. A., Ihde, D. C., West, W., Matthews, M. J. 1982. Possible enhancement of vincristine neuropathy by VP-16. *Cancer 49*:859-864.

104. Therman, E. 1972. Chromosome breakage by 1-methyl-2-benzylhydrazine in mouse cancer cells. *Cancer Res. 32*:1133-1136.

105. Vogelzang, N. J., Raghavan, D., Kennedy, B. J. 1982. VP-16-213 (etoposide): The mandrake root from Issyk-Kul. *Am. J. Med. 72*:136-144.

106. Weber, W., Tackmann, H. J., Freund, H. J., Kaeser, H., Obrist, R., Obrecht, J. P. 1981. The evaluation of neurotoxicity in cancer patients treated with vinca alkaloids with special reference to vindesine. *Anticancer Res. 1*:31-34.

107. Weiss, H. D., Walker, M. D., Wiernik, P. H. 1974. Neurotoxicity of commonly used antineoplastic agents (first of two parts). *N. Engl. J. Med. 291*:75-81.

108. Weiss, H. D., Walker, M. D., Wiernik, P. H. 1974. Neurotoxicity of commonly used antineoplastic agents (second of two parts). *N. Engl. J. Med. 291*:127-133.

109. Wierzba, K., Wankowicz, B., Piekarczyk, A., Danysz, A. 1983. Cytostatic and immunosuppressive drugs, in *Side Effects of Drugs* (Dukes, M. N. G., ed.), Excerpta Medica, Amsterdam, pp. 425-449.

110. Williams, A. C., Klein, E. 1970. Experiences with local chemotherapy and immunotherapy in premalignant and malignant skin lesions. *Cancer 25*:450-462.

111. Wiltshaw, E., Kroner, T. 1976. Phase II study of cis-dichlorodiamineplatinum (II) (NSC 119875) in advanced adenocarcinoma of the ovary. *Cancer Treat. Rep. 60*:55-60.

112. Wiltshaw, E. 1979. Cis-platinum in the treatment of cancer. *Platinum Met. Rev. 23*:90-98.

113. Wiltshaw, E., Subramarian, S., Alexopoulos, C., Barker, G. H. 1979. Cancer of the ovary: A summary of the experience with cis-dichlorodiamineplatinum (II) at the Royal Marsden Hospital. *Cancer Treat. Rep. 63*:1545-1548.

114. Zisapel, N., Levi, M., Gozes, D. 1980. Tubulin: An integral protein of mammalian synaptic vesicle membranes. *J. Neurochem. 34*:26-32.

13

Nitroimidazole Neurotoxicity

Arthur J. Dewar* and Geoffrey P. Rose
Shell Toxicology Laboratory, Sittingbourne Research Centre, Sittingbourne, Kent, England

I. INTRODUCTION

2'-Nitroimidazole (azomycin) was discovered in the early 1950s and was found to possess trichomonacidal properties (48). This stimulated the synthesis and screening of many nitroimidazoles as potential therapeutic agents for treating protozoal infections. One of the most potent compounds for this purpose was found to be the 5'-nitroimidazole 1-(β-hydroxyethyl)-2-methyl-5-nitroimidazole (29) (Fig. 1). Under the name metronidazole (Flagyl) this drug was introduced into clinical practice for the treatment of trichomoniasis and intestinal and extra-intestinal amebiasis. Metronidazole is generally regarded as one of the most significant advances in the treatment of protozoal infections and, in addition to two other 5'-nitroimidazoles, tinidazole (Fasigyne) and nimorazole (Naxogin), remains widely used.

Metronidazole at the dosage required as a trichomonacide has proved to be a safe drug. Side effects have been rarely sufficiently severe to

* *Present affiliation*: Shell International Chemical Company, Shell Centre, London, England

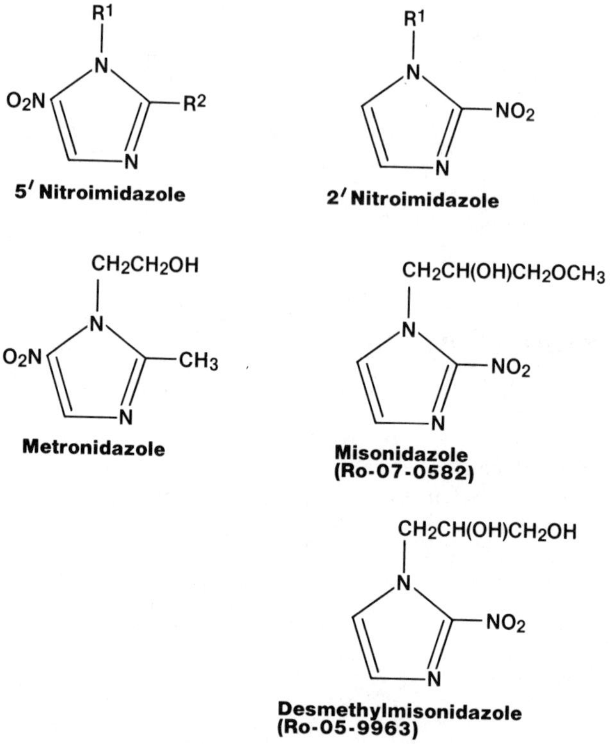

Figure 1 Structures of 5'- and 2'-nitroimidazoles.

necessitate the discontinuation of therapy. Nevertheless, there were some indications that metronidazole could, in larger doses, have adverse effects on the nervous system. Dizziness, vertigo, and, albeit rarely, incoordination, ataxia, and paresthesia have been reported (72), and treatment with metronidazole is contraindicated in patients with active diseases of the central nervous system or with renal impairment.

In the 1970s nitroimidazoles began to be used in larger doses for controlling infections by anaerobic bacteria and for treating Crohn's disease. When nitroimidazoles were used for this purpose, sometimes for prolonged periods, a number of cases of neurotoxicity were reported (30,49,79). The clinical and electrophysiological evidence suggested a sensory distal axonopathy (10,30,67), and this was confirmed in reports of nerve biopsy (10,79). Morphological examination of a sural nerve biopsy from a patient being treated with metronidazole for Crohn's disease showed axonal degeneration of myelinated fibers. There was no evidence of primary demyelination (79). Recent developments in radiation oncology have increased the significance and importance of the neurotoxic properties of nitroimidazoles.

II. NITROIMIDAZOLE RADIOSENSITIZERS

It is believed that hypoxic cells occur in a significant proportion of human tumors and, because they are resistant to radiotherapy (43), constitute a major obstacle to improvement in local tumor control by radiotherapy (4). The radiosensitivity of these resistant tumor cells may be partially restored by use of hyperbaric oxygen (101), but this is laborious and requires excessively long treatment times. There has therefore been a search for chemicals which could diffuse into poorly vascularized tumors and mimic the effects of oxygen, thus serving to sensitize hypoxic cells more cheaply. There are now a number of compounds which have been identified as effective hypoxic cell sensitizers (for a review, see Ref. 2). The largest class of these compounds is the electron-affinic group whose efficiency of sensitization is directly related to their electron affinity (4). However, although demonstrably effective in vitro, many of these compounds lack appreciable sensitizing capability in vivo because of metabolic instability or toxicity.

The first electron-affinic nitro compound to show appreciable radiosensitization of hypoxic mammalian cells in vitro was para-nitroacetophenone (3). This led to a search for similar properties in other nitro-containing aromatics and nitro heterocyclic compounds. Nitrofurans were found to be potent hypoxic sensitizers in vitro (18) and, because of their existing clinical use as antibacterials, were initially looked upon as potentially promising candidates for in vivo use. However, in the high doses required for sensitization, the toxic effects of nitrofurans were seen to be a problem. High doses of nitrofurans had been shown to produce a severe and disabling polyneuropathy—in some cases so severe that narcotic analgesia was required to control the pain (24) and in some cases leading to death (41).

In 1973 metronidazole was tested for its sensitizing ability in hypoxic Chinese hamster cells and found to be effective (18). Because of its past successful use in clinical medicine, its relatively well-known pharmacology and toxicology, and because it was considered sufficiently metabolically stable to diffuse intercellularly to hypoxic cells situated in poorly vascularized areas of tumors distant from functional capillaries and close to areas of necrosis, metronidazole was introduced on an experimental basis into the clinical treatment of cancer with radiotherapy.

A. Clinical Studies

Phase 1 clinical trials* commenced in 1974 (32,90) and were followed by a phase 2 randomized trial involving 36 patients and using ^{60}Co γ

*Radiosensitizer clinical trials are generally divided into three phases: Phase 1, establishment of dose-limiting toxicity; phase 2, radiation and sensitizer used in combination, with study of specific tumor sites; and phase 3, large-scale investigations which are controlled and randomized.

radiation and multiple doses of metronidazole to treat supratentorial glioblastoma. Glioblastoma was chosen because of the very high failure rate (91). There was a significant improvement, but the treatment involved using a nonstandard fractionation regime. In the in vitro studies enhancement ratios of up to 1.9 had been demonstrated (18), but data from in vivo studies indicated that in doses tolerable by man enhancements of less than 1.3 were to be expected (8,68). Consequently the search continued for more effective sensitizers, although clinical research on metronidazole has continued, albeit on a limited scale to the present day (53). Although neurotoxicity was not observed in the early phase 1 trials (90), the use of high doses of metronidazole for an extended period in cancer therapy was found to precipitate peripheral neuropathy, primarily of a sensory nature (81).

On theoretical grounds 2'-nitroimidazoles would be expected to have electron affinities greater than that of 5'-nitroimidazole compounds (2), and consequently the 2'-nitroimidazoles would be expected to be more efficient hypoxic cell sensitizers than the 5-substituted analogs. Studies in a number of different animal tumor systems demonstrated that this was indeed the case (5). The 2'-nitroimidazole of particular interest was 1-(2-nitro-1-imidazolyl)-3-methoxy-2-propanol (Ro-07-0582, misonidazole), a drug originally developed by Roche Products Ltd. as a trichomonacide (Fig. 1) It was found to give in vitro enhancement ratios of up to 2.5 (7), and in vivo studies in mouse tumors gave ratios of 2.1 for high doses and ratios of up to 1.8 for doses likely to be tolerated by man (31,83). After the sensitizing properties of misonidazole had been demonstrated experimentally in man, misonidazole was introduced into clinical trials (35).

Dose tolerance and drug distribution studies established that misonidazole was a readily diffusible molecule, not appreciably protein bound, and penetrated readily into a wide range of histologically different tumors in a variety of tissues. Patients given single oral doses of misonidazole attained serum concentrations expected to be of value in radiosensitization. Tumor concentration relative to serum was also judged to be satisfactory (between 64 and 92% in the biopsy samples that could be assayed reliably) (44). Observation of tumor response in patients provided evidence of enhancement of effect by misonidazole, and it was concluded that the drug might be of promise as a radiosensitizer in clinical radiotherapy (89).

The immediate symptoms which were considered to limit the dose that could be administered were anorexia, nausea, and vomiting. These become troublesome above 140 mg/kg and limited the practical dose to a maximum of 120 mg/kg. No signs or symptoms of neurotoxicity were detected in spite of the fact that they were particularly looked for in view of the findings of some earlier toxicological studies of drugs in the 2-nitroimidazole series.

Daily doses of 50-200 mg/kg of the 2-nitroimidazole Ro-07-1051 given to dogs had been found to produce ataxia, convulsions, and death after 3-20 days (depending on the dose) (80). If, however, the drug was stopped at the onset of symptoms and the animal sedated, recovery occurred. At autopsy histological examination revealed degeneration of Purkinje cells in the cerebellum. Administration of the 5-nitroimidazoles Ro-07-0207 and metronidazole gave identical effects; however, these clinical and histological findings could not be reproduced in mice, rats, guinea pigs, or rabbits at doses of 400 mg/kg per day or in rats at doses of 1000 mg/kg per day. It was concluded that the cerebellar changes found in the dog were probably an acute sensitivity reaction, and consequently of little practical significance for humans (80). In toxicity studies of misonidazole in baboons (62) it was found that doses of up to 200 mg/day were tolerated for up to 28 days, but doses in excess of 400 mg/kg produced severe muscular incoordination and incipient convulsions by the 11th day. Cessation of dosing resulted in recovery of normal motor function within 24 hr. There was no evidence of cerebellar lesions, but small bilateral lesions were found ventral to the fourth ventricle in the region of the vestibular nucleus and in the corpus pontobulbare. These lesions were also found in the animals given lower doses and which had exhibited no neurological signs. These changes were not produced by doses of the order required to produce sensitization, provided that they were given only three times per week.

In view of the promise of the drug and the lack of conclusive adverse toxicological findings, more extensive clinical trials involving multiple doses were commenced in several centers. In one study (36) 16 patients received a total of 15-51 g in 3-20 doses. Although immediate tolerance was good, 11 of the patients developed peripheral neuropathy. Symptoms variously described as pins and needles, numbness, dead feeling, or loss of feeling were reported in the hands and feet. The ability to perform fine movements such as the fastening of buttons and walking was also adversely affected. In some cases the paresthesia was sufficiently painful to warrant strong analgesia to control it. The patient given the largest dose suffered from mental confusion and episodes of sustained muscular spasm of the limbs and eventually suffered grand mal convulsions. Neurological signs of a gross bilateral disturbance sited in the upper brainstem appeared.

Other independent phase 1 studies being conducted at the same time yielded similar findings, that is, evidence of a distal symmetrical sensory neuropathy and, at higher doses, convulsions and encephalopathy (51,92,93). Electron-microscopic examination of a sural nerve biopsy from an affected patient confirmed distal axonal degeneration, but some indications of segmental demyelination were also observed (93) (Fig. 2). Although the neuropathy was primarily sensory, it was

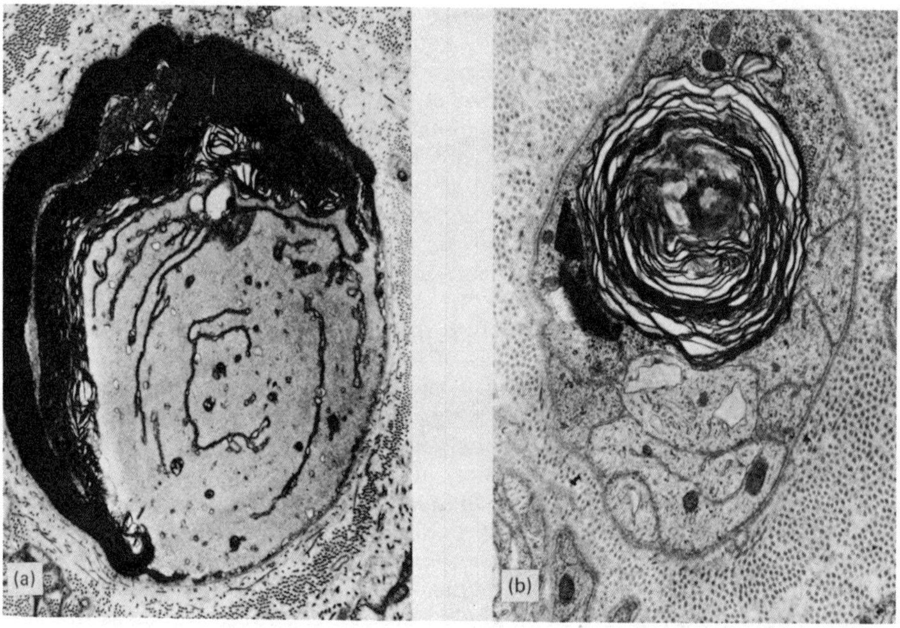

Figure 2 Electron micrographs of a sural nerve biopsy from a misonidazole-treated patient with symptoms of peripheral neuropathy. (Reproduced from Ref. 93.) (a) Damaged large myelinated nerve fiber showing axonal swelling, an increase in neurofilaments, and partial denudation of the myelin sheath. (b) Myelin debris and small myelinated nerve fiber. (c) Remyelinated small myelinated nerve fiber showing axonal distortion, thinning and irregularity of the myelin sheath, and a periaxonal deposit of collagen. Adjacent nonmyelinated nerve fibers are of normal appearance.

noticed in some centers that patients given high doses experienced motor weakness and cramps in addition to sensory changes (64). In addition, ototoxicity was detected. The loss of hearing was significant and could be readily measured audiographically (28). Although most of the neurotoxic effects (particularly those involving the central nervous system) were transient and generally reversed within 2 weeks, some cases of peripheral neuropathy were found to persist for periods of over 2 years. Recently a fatal encephalopathy involving focal cortical neuronal necrosis has been reported (54).

In 1978 all the clinical groups having experience in the administration of misonidazole to patients pooled their data (from a total of 177 patients) and submitted them to statistical analysis. The data revealed a relationship between the incidence and severity of peripheral neuro-

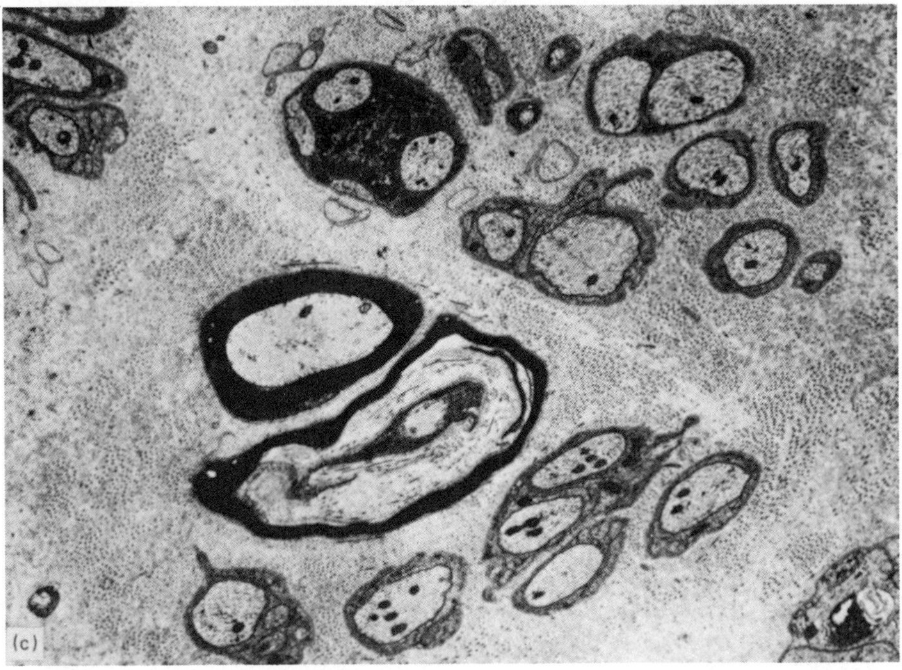
(c)

pathy and dose (37). No patients were found to have suffered from neuropathy at cumulative doses of less than 13.5 g (Fig. 3). In an evaluation of phase 1 studies, the Radiation Therapy Oncology Group concluded that the maximum tolerable dose of misonidazole was 6 g/m^2 in 1 week, 10.5 g/m^2 in 3 weeks, and 12 g/m^2 in 6 weeks (65).

These findings imposed constraints on the design and execution of phase 2 and phase 3 trials. Nevertheless, in the 7 years since the introduction of misonidazole into phase 1 trials, over 2000 patients have been treated with the drug. In Radiation Therapy Oncology Group phase 2 studies, 28% of the 466 treated patients showed peripheral neuropathy, 7% showed central nervous system toxicity, and 5% ototoxicity. Of the 428 patients involved in phase 3 trials, 16% exhibited peripheral neuropathy, 3% central nervous system toxicity, and 1% ototoxicity. A detailed discussion of the clinical findings in these trials and the conclusions regarding the value of misonidazole as a sensitizer in radiotherapy is beyond the scope of this chapter, and the reader is referred to a number of recent reviews on the subject (1,38, 65). However, the data available to date indicate that the benefits of misonidazole in clinical radiotherapy have not yet been unequivocally demonstrated, principally because its neurotoxicity prevents the drug's use at a dosage necessary to give the full degree of hypoxic cell sensitization. The recommended dose limitation of 12 g/m^2 (36) means that

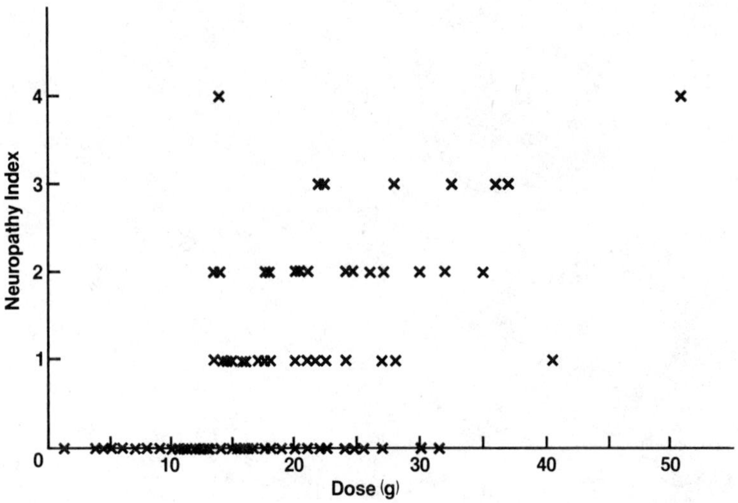

Figure 3 A scatter diagram showing the relationship between neurotoxicity and total dose of misonidazole given. Symptoms due to neurotoxicity are graded: 0 = none, 1 = slight, 2 = moderate, 3 = severe, and 4 = convulsion. (From Ref. 37.)

where the drug is given at low doses (e.g., 0.6 g/m^2), with each fraction of a conventional multifraction radiotherapy regime the resulting enhancement ratios are unlikely to exceed 1.2-1.3 (100).

The finding that the clinical effectiveness of misonidazole, like that of the nitrofurans and metronidazole, is seriously limited by its neurotoxicity has led to both a search for second-generation sensitizers clinically more effective and less neurotoxic than misonidazole and for means whereby the neurotoxicity of misonidazole itself can be minimized. These objectives have stimulated studies aimed at an understanding of the nature and mechanism of nitroimidazole neurotoxicity and the development of animal models capable of predicting the likely neurotoxic effects of new candidate sensitizers.

B. Neurotoxicity in Animals

The activities of the lysosomal enzymes β-glucuronidase and β-galactosidase increase dramatically in nerves undergoing wallerian degeneration (47,56). Degeneration of the wallerian type is by far the most common type encountered in chemically induced peripheral neuropathy (14), and consequently the peripheral nerve degeneration produced by many neurotoxic chemicals such as acrylamide and methyl mercury is accompanied by substantial increases in the activity of these two enzymes (34,52,75). In fact, the measurement of β-glucuronidase and β-galactosidase has proved a useful (and quantitative) method for detecting chemically induced peripheral neuropathy (6,33).

When applied to rats dosed orally for 7 days with 400 mg/kg per day misonidazole, the assay of lysosomal enzymes provided evidence consistent with distal axonal degeneration in peripheral nerves (73). Transient increases in β-glucuronidase and β-galactosidase activity were found in the sciatic/posterior tibial nerve. These were maximal 4 weeks after dosing. Enzyme activities returned to control values within 8 weeks (73,75) (Fig. 4a). The largest increases in enzyme activity were observed in the more distal section of the nerve. At maximum the increases were approximately 170% of control value—increases considerably less than those found with neurotoxic doses of such toxicants as acrylamide and methyl mercury, where changes of the order of 300-600% are found (34,52). This suggests that, in comparison with these agents, the peripheral nerve damage produced by misonidazole is of a lesser order. Subsequent studies revealed that the increases in lysosomal enzyme activities in peripheral nerve were related to the dose of misonidazole administered (73) (Fig. 4b) and that very similar dose-related changes could be produced by both metronidazole and nitrofurantoin (76).

This method has been extended to the mouse, but instead of using a fluorimetric assay in nerve homogenates for measuring β-glucuronidase, a quantitative cytochemical method using microdensitometry was employed. Intraperitoneal misonidazole administration resulted in 100-160% increase in β-glucuronidase and acid phosphatase activities in the sciatic nerve (19,20). As in the rat, the increases were greater in the distal than in the proximal part of the nerve, reached a maximum three to four weeks after dosing, and were dose related. Enzyme increases were detectable at doses which were comparable to clinically relevant doses in humans (0.2 mg/kg total).

Biochemical changes have also been observed in the central nervous system of misonidazole-treated rats and mice (20,73). There is evidence that certain types of degeneration in the central nervous system are accompanied by increases in β-glucuronidase activity. This is particularly true of degeneration in which cellular proliferation is a feature, for example, the encephalopathy produced by cuprizone (9). There is also evidence that β-galactosidase is a neuronal marker in the central nervous system (84) and that neuronal damage or loss is accompanied by a reduction in the activity of this enzyme (9). Misonidazole administration in the rat produced a dose-related decrease in cerebellar β-galactosidase activity and a dose-related increase in β-glucuronidase (73). Cytochemical measurement of β-glucuronidase in misonidazole-treated mice revealed increased enzyme activity in the granular regions of the cerebellum (20).

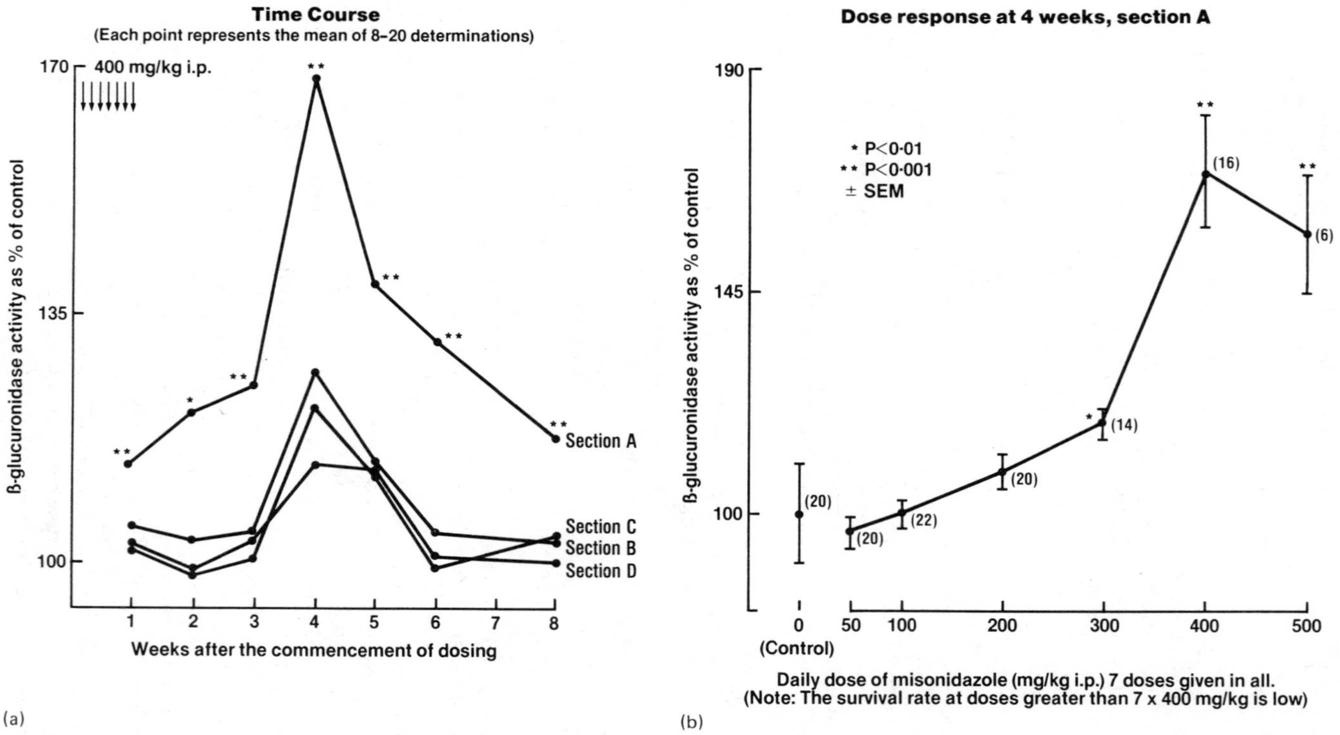

Figure 4 The effect of subacute administration of misonidazole on β-glucuronidase activity in sections of the rat sciatic/posterior tibial nerve (section A = most distal, section D = most proximal). (Data from Ref. 75.)

In a subchronic study in rats, the animals dosed intraperitoneally with misonidazole in doses of 260 or 340 mg/kg for 3 weeks to 5 months showed neurological signs within 11-19 days (at the higher dose) or 26-31 days (at the lower dose) (45). The onset of signs was rapid and preceded by poor feeding and weight loss. There was a loss of righting reflexes, reduced activity, and abnormal stance and gait. The tail was held in a rigid dystonic posture, and severely affected animals tilted their head to one side or fell to one side. These signs indicative of central nervous system involvement were not accompanied by any distinctive clinical evidence of peripheral neuropathy.

Pathological examination of perfusion-fixed tissues revealed lesions in both the central and peripheral nervous systems (45). The lesions in the peripheral nervous system were milder and less frequent than those in the central nervous system and consisted of distal axonal degeneration. The most severely affected structures were the sensory terminals on intrafusal fibers, although there was some loss of motor nerve terminals in extrafusal fibers. Axons undergoing degeneration were present within the intramuscular nerve branches of the hand and foot at a time when the tibial and sciatic nerves and the dorsal and ventral roots appeared normal. No evidence of segmental demyelination was detected, but edema was found in the dorsal root ganglia and in the endoneurium of affected intramuscular nerve branches. Morphological examination of the peripheral nervous system of mice dosed intraperitoneally with misonidazole revealed essentially similar findings (27).

In the central nervous system the major findings were localized degeneration of specific nuclear groups in the medulla. Particularly affected were the lateral vestibular nuclei, the superior vestibular nuclei, the spinal vestibular nuclei, and the superior olives. Mild lesions were characterized by spongy changes (due primarily to extracellular edema) and axonal swellings. Advanced lesions showed necrosis associated with a marked infiltration of phagocytes. No abnormalities were seen in the Purkinje cells of the cerebellum, although occasional lesions were seen in root nuclei of the cerebellum. It is not clear whether these changes could account for the observed dose-related decrease in cerebellar β-galactosidase activity and increase in cerebellar β-glucuronidase (73).

The findings of the pathological examination of misonidazole-treated rats resemble those reported for metronidazole in an earlier study. Here histological sections of the perfused central nervous system of rats treated with 800 mg/kg metronidazole showed sharply bounded symmetrical lesions in the nuclei vestibularis and cochlearis, in the nuclei of the cerebellar root, and in the superior olive (70).

These findings are interesting because the clinical and pathological alterations are qualitatively and topographically comparable to lesions seen in Wernicke's encephalopathy in adults and children (70) and in rats rendered thiamine deficient either by dietary restriction (69) or by the administration of thiamine antogonists such as pyrithiamine (25). The similarity of neuropathy produced by nitrofuran drugs and that produced by thiamine deficiency has also been noted (82,85).

C. Mechanisms of Neurotoxicity

The biochemical basis for the neurotoxicity of nitroimidazoles is, as yet, not understood, although a number of possible mechanisms have been advanced. Evidence is currently accumulating that suggests that an interference in energy metabolism could be involved.

On the basis of findings on the effects of metronidazole on the acetylcholine responses of the frog rectus muscle and rat diaphragm, it was suggested that nitroimidazoles exhibit anticholinesterase activity (50), but the drug concentration used in this study was unrealistically high and would prove lethal in vivo in both experimental animal and humans. Subsequent measurement of acetylcholinesterase activity in the presence of concentrations of misonidazole and metronidazole within the clinically tolerated dose range failed to demonstrate any significant inhibition (98). Furthermore, measurement of acetylcholinesterase activity in the cerebellum of mice treated with misonidazole in the clinically tolerated dose range failed to show any inhibitory effect of the drug (22). It is clear from these findings that the neurotoxicity of the nitroimidazoles cannot be attributed to an effect on cholinesterase activity.

In addition to their radiation-sensitizing effects on hypoxic cells, nitroimidazoles are also cytotoxic in their own right, particularly under hypoxic conditions (46). They also have the ability to sensitize cells to other cytotoxic agents (71). The mechanisms of nitroimidazole cytotoxicity and chemosensitization are currently areas of active research (for a recent review see Ref. 11), and since there is a correlation between the cytotoxicity and neurotoxicity of the nitroimidazoles, some of this work could help to throw light on the possible mechanism of nitroimidazole neurotoxicity.

Recent studies have shown that misonidazole depletes intracellular nonprotein thiols (such as glutathione) in hypoxic cells (97) and that the cytotoxic action of the drug to mammalian cells in vitro can be reduced by the addition of exogenous thiols (88). Since an interaction with sulfhydryl groups is associated with certain types of neurotoxicity (e.g., those produced by some heavy metals), it has been suggested that an interaction with sulfhydryl groups could be involved in nitroimidazole neurotoxicity (55). There is, however, no experimental evidence to support this suggestion.

Nitroimidazoles have been shown to inhibit DNA synthesis in mammalian cells in vitro (61), inhibit the uptake of nucleotides (61), and, after nitro reduction, bind to DNA, RNA, and protein in vivo (95). Interference with RNA and protein metabolism is a potential mechanism of neurotoxicity (33), but it is doubtful whether this has direct relevance to nitroimidazole neurotoxicity, since a study on RNA synthesis in the nervous system of rats dosed with metronidazole failed to demonstrate any significant inhibition (10).

It will be recalled that the neuropathology associated with nitroimidazole neurotoxicity is strongly reminiscent of that found in thiamine deficiency. Thiamine, in the form of its pyrophosphate, is an impor- cofactor required for the oxidative decarboxylation of pyruvate to acetyl coenzyme A and therefore plays a crucial role in energy metabolism. For this reason recent reports of effects of nitroimidazoles on energy metabolism are probably very relevant to their neurotoxic properties.

Thiamine supplementation has proved ineffective in preventing the development of nitroimidazole neuropathy both in man (36) and experimental animals (45,74). However, an early in vitro study (albeit with nonnervous tissue) demonstrated that nitrofuran drugs act competitively against thiamine pyrophosphate in the oxidation of pyruvate (63).

Hypoxic cells depend on the glycolysis pathway for energy, and consequently the effects of nitroimidazoles on this pathway have been studied in some depth, lactate production being taken as an index of activity. Under conditions of prolonged anaerobic incubation with misonidazole and other nitroimidazoles, Ehrlich ascites tumor cells were shown to have lowered lactate production and decreased glucose utilization. This inhibition of glycolysis could also be demonstrated in other cell lines and was found to be reversed by reincubation in a buffer containing cysteine (96). A suggested explanation for this phenomenon is that under hypoxic conditions nitroreduction products of nitroimidazoles react with −SH groups in several of the glycolytic enzymes.

The effects of nitroimidazoles on glycolysis under aerobic conditions are more controversial. One research group has claimed that under aerobic conditions misonidazole actually stimulates glucose utilization, possibly because of a stimulation of the hexose monophosphate shunt (96). In contrast, another group was able to show that prolonged incubation of CHO cells with misonidazole resulted in a time- and dose-dependent inhibition of lactate production (17). The observation that the ability of nitroimidazoles to inhibit glycolysis was not only restricted to hypoxic cells led to the suggestion that an inhibition of glycolysis could be a potential mechanism for the neurotoxicity of these compounds.

To investigate this possibility, the effects of misonidazole and another nitroimidazole sensitizer, desmethylmisonidazole, on lactate production and glucose utilization in brain tissue of C3H mice were investigated in vivo and in vitro. Mice given neurotoxic doses of the nitroimidazoles were found to produce significantly less lactate in the brain. In vitro brain slices showed a 40% reduction in lactate production and a 15% reduction in glucose utilization in the presence of the nitroimidazoles (16). In a later in vivo study lactate and pyruvate were measured in a number of brain areas from mice treated for 1-3 weeks with misonidazole. Misonidazole treatment resulted in a 30% decrease in lactate (coupled with a 30-35% rise in pyruvate) in the brainstem after 1 week of dosing, with no further changes over the succeeding 2 weeks. No comparable changes were detected in the cerebral cortex, cerebellum, or serum (26).

There are therefore some grounds for believing that a contributing factor to the neurotoxicity of nitroimidazoles in the central nervous system is their interference with energy metabolism. However, the precise mechanism is as yet unclear, although glycolysis and/or pyruvate decarboxylation appear to be involved. A further reason for believing this is that there is a large body of indirect evidence associating abnormal energy metabolism with peripheral neuropathies of the distal axonopathy type (14,78,82,85; see also Cavanagh and Manzo, Blum, and Sabbioni, this volume).

III. METHODS FOR REDUCING THE NEUROTOXICITY OF NITROIMIDAZOLE RADIOSENSITIZERS

The neurotoxicity of misonidazole has been related to both the total dose of drug administered and total tissue exposure (31). The degree of radiosensitization depends only on the concentration of sensitizer in the tumor at the time of irradiation (57). A method for improving the drug concentration is to use intravenous infusion, thereby eliminating a slower absorption phase.

Another potential approach to reduce neurotoxicity is to develop less lipophilic analogs of misonidazole. Lipophilicity, as measured by the octanol:water partition coefficient P, can greatly influence the absorption, distribution, excretion, and toxicity of a drug and analogs, with a lower value of P expected to have a diminished ability to cross the lipid blood-brain barrier and blood-neural barrier. There is evidence from studies in rodents that the neurotoxic potential of nitroimidazoles is a function of their lipophilic properties (1,21), but lipophilicity does not affect radiosensitizing efficiency in vitro (4).

The possibility of decreasing toxicity by attempting to exclude nitroimidazoles from the dose-limiting tissue has been explored intensively. A series of 2'-nitroimidazoles of similar electron affinity but with P values covering a range of 0.014-1.5 were compared to misonidazole with respect to the acute LD_{50} in mice, plasma half-life, tumor penetration, brain penetration, and radiosensitizing ability. The P value did not influence the radiosensitizing ability or the peak tumor levels, but did influence the degree of brain penetration and the LD_{50}. The acute LD_{50} in the mouse varied from 1 to 23 mmol/kg. The compounds with lower P values (i.e., the less lipophilic compounds) exhibited the lowest brain penetration and the lowest toxicity (12,103).

On this evidence, therefore, it should be possible to develop a misonidazole analog with as good, if not better, a radiosensitizing ability as misonidazole, but with lower toxicity. However, the assumption behind these experiments on animals is that brain is the dose-limiting tissue. As we shall see, this assumption, while possibly valid for some experimental animals, does not necessarily apply to man. Furthermore, it is always dangerous to draw conclusions regarding chronic neurotoxicity (i.e., the type that is of particular concern in the case of nitroimidazoles) from data from acute dosing only. Experience with many classes of neurotoxic chemicals has shown that large, potentially lethal single doses yield little information about chronic neurotoxicity. Repeated dosing is nearly always required to provide this information (33).

IV. PROTECTION AGAINST MISONIDAZOLE NEUROTOXICITY

The earliest strategy aimed at the prevention or reduction of misonidazole neurotoxicity was stimulated by the similarity of nitroimidazole neurotoxicity to that produced by thiamine deficiency. However, as was mentioned earlier, attempts to use thiamine supplementation as a means of ameliorating symptoms or as a prophylactic to prevent the development of nitromidazole neurotoxicity have been unsuccessful both in man (36) and experimental animals (41,75). Attempts to use other B vitamins such as riboflavin, nicotinamide, and pyridoxine for this purpose have also met with little success (36). There is, however, one report that thiamine and niacin delay (but do not prevent) the onset of desmethylmisonidazole-induced neuropathy (15), and there is some evidence from biochemical studies that intramuscular administration of thiamine pyrophosphate has a protective effect against misonidazole-induced neuropathy in rats (77),

An alternative approach has been to attempt to manipulate the pharmacokinetics of misonidazole and in particular to reduce the AUC (total area under the drug level-time curve) by shortening the half-life. Pretreatment of mice with microsomal enzyme-inducing agents like phenytoin and phenobarbitone has been found to affect the pharmacokinetics of misonidazole profoundly (102); the half-life of misonidazole being reduced by 20-67% and the AUC being decreased by 23-49%. The decrease in half-life was associated initially with a 1.5- to 2.0-fold increase in the circulating concentration of the O-demethylated metabolite desmethylmisonidazole, nevertheless, the AUC for total 2'-nitroimidazole was reduced overall by 20-50%. Although they markedly change the half-life of misonidazole, neither phenytoin nor phenobarbitone significantly alter the peak misonidazole concentration in tumors, brain, or plasma.

Phenytoin was subsequently found to reduce the half-life of misonidazole in man (99) and in patients receiving daily radiotherapy plus daily or weekly misonidazole (60). Phenobarbitone and phenytoin are frequently given to patients with brain tumors for sedative and anticonvulsant therapy. It has been found that such patients have a low incidence of peripheral neuropathy (99). It would appear on preliminary evidence, therefore, that the use of these drugs could be a practical and effective method of reducing the neurotoxicity of misonidazole while not interfering with its capabilities as a radiosensitizer. However, the possible effects on the antitumor cytotoxicity of the drug remain uncertain.

Another promising method currently under investigation is the use of the glucocorticoid dexamethasone. Brain tumor patients are frequently maintained on this drug to control cerebral edema, and in a limited clinical study this drug was found to have a protective effect against misonidazole neurotoxicity (99). The results of a later randomized clinical study involving 14 patients undergoing radiation therapy corroborated this finding (94). One suggestion to explain the reduction of misonidazole neurotoxicity by dexamethasone is that it acts at the nerve cell membrane level, restoring and stabilizing cell surface properties (94). There is some evidence that suggests that vitamin E, another membrane stabilizer, exerts a protective effect against misonidazole-induced neuropathy in rats (77).

There are, however, problems associated with the use of dexamethasone. One problem is side effects such as hypertension and diabetes insipidus; another is that there are indications from in vitro experiments with Chinese hamsters that this agent, although not affecting hypoxic cell radiosensitizing properties of misonidazole, can reduce the radiation sensitivity of cells (58). It is possible, therefore, that the inclusion of dexamethasone in radiotherapy/chemotherapy regimes could reduce the therapeutic gain. Recently other agents have been examined in an attempt to reproduce the beneficial effects of dexamethasone without the problem of the reduction of radiation sensitivity. There are indications from in vitro studies that the nonsteroidal anti-inflammatory agent flurbiprofen is promising in this respect (58,59), but, as yet, its efficacy in vivo has not been demonstrated.

To conclude this section it may be said that the strategy of reducing the neurotoxicity of misonidazole has one further potential problem. If the reduction of neurotoxicity permits the administration of misonidazole in doses in excess of the present maximum of 12 g/m^2, there is the possibility that other side effects of misonidazole, for instance, damage to the bone marrow (6), could become limiting.

V. METHODS FOR ASSESSING NITROIMIDAZOLE NEUROTOXICITY

A prerequisite of any program designed to develop new more effective and less neurotoxic nitroimidazole radiosensitizers is a reliable method for detecting and quantifying neurotoxic effects of new compounds. The most desirable method would be a rational mechanism-based test that could be employed in vitro and allowing direct prediction of neurotoxicity to man. This could be used conveniently in conjunction with the various methods (detailed in Refs. 86 and 87) for screening candidate compounds for their radiosensitizing ability.

Although the example of the organosphosphates and the neurotoxic esterase assay demonstrate that it is possible to devise such methods for ranking the neurotoxic potential of a series of related compounds (33,35), the knowledge of the mechanism underlying nitroimidazole neurotoxicity is unfortunately not yet at a level that permits the design and implementation of equivalent tests for nitroimidazoles. Consequently, at present, reliance has to be placed on in vivo animal models and the development of methods for quantifying the various manifestations of nitroimidazole neurotoxicity.

One of the first attempts to quantify the magnitude of nitroimidazole neurotoxicity was by using electrophysiological techniques. A sensitive electrophysiological technique for measuring impaired axonal conduction (nerve train analysis) has been found to be capable of detecting neuropathy in rats dosed subchronically with misonidazole before overt clinical signs appear (40). Since this technique is noninvasive, it can be used clinically in patients undergoing misonidazole therapy, as well as in experimental animals. It therefore constitutes a potentially promising method for evaluating nitroimidazole-induced peripheral neuropathy and is currently being used in clinical studies. Electrophysiological methods have also been applied with success to the detection of the effects of misonidazole on the central nervous system. Of the techniques investigated to date brainstem auditory evoked potentials appear to be the most sensitive indicator of misonidazole-induced toxicity (66).

In testing for peripheral neuropathy it is often useful to use tests for assessing neuromuscular dysfunction. These functional tests involve giving the animal a task requiring good neuromuscular coordination and scoring its performance. Examples include the rotarod test (in which an animal is placed on a rotating rod and the time it takes to fall off is measured) and the narrowing bridge test (in which the animal has to traverse a progressively narrowing bridge and the number of times it slips is counted). Experience with such neurotoxins as acrylamide has shown that such tests offer very sensitive methods for detecting subclinical neuropathy, quantifying disability in a clinically expressed neuropathy and monitoring recovery (33,34,52).

These tests have the disadvantage that they are relatively nonspecific, and this has proved a problem when using them to assess misonidazole-induced peripheral neuropathy. However, in the mouse both the rotarod and the narrowing bridge tests have been used with success to quantify the functional impairment produced by misonidazole neuropathy and that produced by metronidazole. With both tests it has proved possible to construct dose-response curves to compare the neurotoxicities of these compounds (21,26,27). The measurement of the modulation of the startle reflex in mice has also proved an effective method for ranking the ototoxicity of nitroimidazoles (26).

Fundamental to an understanding of the effects of a chemical on the nervous system is neuropathological examination. Thus the development of new nitroimidazole sensitizers will ultimately involve rigorous neuropathological examination of dosed animals using modern fixation, embedding, and staining techniques. Although such techniques (including electron microscopy) have been used in an integrated testing strategy for testing nitroimidazoles (26), they are demanding and not ideally suited to the preliminary screening of large numbers of compounds. Simpler histological methods such as simple paraffin histology are insufficiently sensitive. Quantification of pathology is also problematic, since quantitative histological techniques are notoriously laborious and time-consuming.

An alternative approach (discussed earlier) for rapidly detecting and quantifying nitroimidazole-induced nerve damage is to measure nerve β-glucuronidase activity. With either biochemical or cytochemical methods, the neurotoxicities of various nitroimidazoles can be compared by evaluating dose-response curves of lysosomal enzyme activities in distal sections of peripheral nerve. This has been done successfully in both the rat and the mouse (21,76). In a mouse study, β-glucuronidase measurements were used to evaluate the neurotoxicities of members of a homologous series of 1-substituted 2'-nitroimidazole compounds. It was found that the increase in peripheral nerve β-glucuronidase activity produced by a compound was correlated positively with the number of methylene groups in its side chain and hence its lipophilicity (Fig. 5).

VI. THE NEUROTOXICITY OF SECOND-GENERATION SENSITIZERS

The search for second-generation sensitizers more potent than misonidazole as radiosensitizers, but less neurotoxic, has been proceeding apace (87), but from a neurotoxicological viewpoint the results have so far been disappointing.

To date most attention has been focused on the drug desmethylmisonidazole (Ro-05-9963), a major metabolite of misonidazole (23,42)

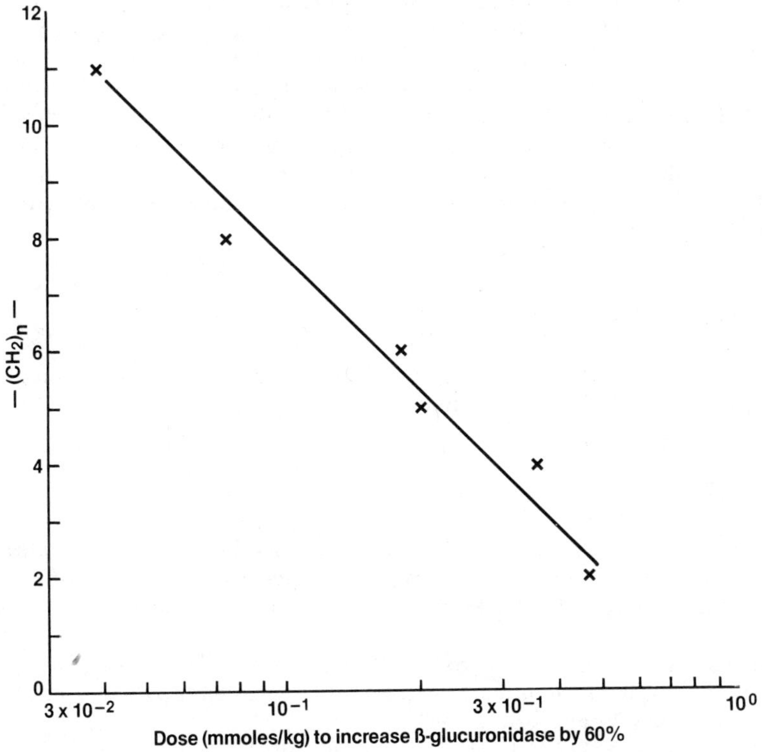

Figure 5 A correlation between neurotoxicity, expressed in terms of elevated β-glucuronidase (arbitrarily set at 60%), and the number of methylene groups in the side chain of a homologous series of 1-substituted 2-nitroimidazole compounds. (From Ref. 23.)

(Fig. 1). This is comparable in radiosensitizing efficiency to the parent compound, but less lipophilic (P = 0.13). On the evidence of experiments in dogs, its penetration into the cerebrospinal fluid is lower than that of misonidazole (39,100). The drug also has a reduced half-life in blood, both in animals and humans (39,100), and there is some evidence that it is less neurotoxic than misonidazole in mice (26).

When clinical toxicity trials were undertaken using multiple doses up to 12 g/m^2, the incidence of peripheral neuropathy was comparable to that produced by misonidazole, although central nervous system toxicity was absent (39). It appears, therefore, that this drug offers no major advantages over misonidazole. It is also apparent that although lipophilicity has a bearing on central nervous system toxicity in humans, it has considerably less influence on peripheral nervous system toxicity.

At present there are two other compounds at the stage of clinical evaluation: SR-2508 and Ro-03-8799. The compound SR 2508 is considerably less lipophilic than misonidazole, with a P value of only 0.046 (i.e., nearly 10 times less than misonidazole) (13,21). The compound Ro-03-8799 is as lipophilic as misonidazole (at physiological pH), but is less neurotoxic to primates (1). The performance of these two compounds in clinical trials will be of importance in assessing the validity of the hypothesis that the neurotoxicity of nitroimidazoles is related to their lipophilicity.

ACKNOWLEDGMENT

The authors wish to thank Dr. I. J. Stratford for helpful discussions during the preparation of this chapter.

REFERENCES

1. Adams, G. E. 1982. Accomplishments, problems and prospects: a conference summary, in *Chemical Modification, Radiation and Cytotoxic Drugs* (Sutherland, R. M., ed.), Pergamon, Oxford, pp. 805-808.
2. Adams, G. E. 1978. Hypoxic cell sensitizers for radiotherapy, in *Cancer Comprehensive Treatise*, Vol. 6 (Becker, F. F., ed.), Plenum, New York, pp. 181-223.
3. Adams, G. E., Asquith, J. C., Dewey, D. L., Foster, J. L., Michael, B. D., Willson, R. L. 1971. Electron affinic sensitization 11. Paranitroacetophenone, a radiosensitizer for anoxic bacterial and mammalian cells. *Int. J. Radiat. Biol. 19*:575-585.
4. Adams, G. E., Dawson, K., Stratford, I. J. 1980. Electron-affinic radiation sensitizers for hypoxic cells: Prospects and limitations with present and future drugs, in *Progress in Radio-Oncology International Symposium, Baden (Austria), May 1978* (Karcher, K. H., Kogelnik, H. D., Meyer, H. J., eds.), Academic, New York, pp. 84-95.
5. Adams, G. E., Fowler, J. F. 1976. Nitroimidazoles as hypoxic cell sensitizers *in vitro* and *in vivo*, in *Modification of Radiosensitivity of Biological Systems*, Vienna, Publishers I.A.E.A., pp. 103-117.
6. Allalunis, J. M., Turner, R. A., Partington, I. P., Urtasun, R. C. 1980. Effects of misonidazole therapy on human granulopoietic stem cells. *Cancer Treat. Rep. 64*:1097-1102.
7. Asquith, J. C., Watts, M. E., Patel, K., Smithen, C. E., Adams, G. E. 1974. Electron affinic sensitisation. V. Radiosensitisation of hypoxic bacteria and mammalian cells *in vitro* by some nitroimidazoles and nitropyrazoles. *Radiat. Res. 60*:108-118.

8. Begg, A. C., Sheldon, P. W., Foster, J. L. 1974. Demonstration of hypoxic cell radiosensitization in solid tumour by metronidazole. *Br. J. Radiol.* *47*:399-404.
9. Bowen, D. M., Flack, R. H. A., Martin, R. O., Smith, C. B., White, P., Davison, A. N. 1974. Biochemical studies on degenerative neurological disorders 1: Acute experimental encephalitis. *J. Neurochem.* *22*:1099-1107.
10. Bradley, W. G., Karlsson, I. J., Rassol, C. G. 1977. Metronidazole neuropathy. *Br. Med. J.* *2*:610-611.
11. Brown, J. M. 1982. The mechanisms of cytotoxicity and chemosensitisation by misonidazole and other nitroimidazoles. *Int. J. Radiat. Oncol. Biol. Phys.* *8*:675-682.
12. Brown, J. M., Workman, P. 1980. Partition coefficient as a guide to the development of radiosensitizers which are less toxic than misonidazole. *Radiat. Res.* *82*:171-190.
13. Brown, J. M., Yu, N. Y., Brown, D. M., Lee, W. 1981. SR 2508: A 2 nitroimidazole amide which should be superior to misonidazole as a radiosensitizer for clinical use. *Int. J. Radiat. Oncol. Biol. Phys.* 7:695-703.
14. Cavanagh, J. B. 1984. The problems of neurons with long axons. *Lancet* *1*:1284-1287.
15. Chao, C. F., Subjeck, J. R., Johnson, R. J. R. 1980. Interference with desmethylmisonidazole induced neurotoxicity, thiamine and niacin. *J. Cell Biol.* *87*:321a.
16. Chao, C. F., Subjeck, J. R., Johnson, R. J. R. 1981. Neurotoxicity of nitroimidazoles and glycolysis in brain. *J. Cell Biol.* *91*:95a.
17. Chao, C. F., Subjeck, J. R., Johnson, R. J. R. 1982. Nitroimidazole inhibition of lactate production in CHO cells. *Int. J. Radiat. Oncol. Biol. Phys.* *8*:729-732.
18. Chapman, J. D., Reuvers, A. P., Borsa, J., Henderson, J. S., Migliore, R. D. 1974. Nitroheterocyclic drugs as selective radiosensitizers of hypoxic mammalian cells. *Cancer Chemother. Rep. Part 1*, *58*:559-570.
19. Clarke, C., Dawson, K. B., Sheldon, P. W., Chaplin, D. J., Stratford, I. J., Adams, G. E. 1980. Quantitative cytochemical method for assessing the neurotoxicity of misonidazole, in *Radiation Sensitizers: Their Use in the Clinical Management of Cancer* (Brady, L. W., ed.), Masson, New York, pp. 245-249.
20. Clarke, C., Dawson, K. B., Sheldon, P. W. 1982. Quantitative cytochemical assessment of the neurotoxicity of misonidazole in the mouse. *Br. J. Cancer* *45*:582-587.
21. Clarke, C., Dawson, K. B., Sheldon, P. W., Ahmed, I. 1982. Neurotoxicity of radiation sensitizers in the mouse. *Int. J. Radiat. Oncol. Biol. Phys.* *8*:787-789.

22. Clarke, C., Hobbiger, F., Sheldon, P. W. 1981. Acetylcholinesterase and cholinesterase activities in the mouse cerebellum following misonidazole treatment. *Eur. J. Pharmacol. 69*:209-211.
23. Coleman, C. N., Wasserman, T. H., Phillips, T. L., Strong, J. M., Urtasun, R. C., Schwade, J. G., Johnson, R. J., Zagars, G. 1982. Initial pharmacology and toxicology of intravenous desmethylmisonidazole. *Int. J. Radiat. Oncol. Biol. Phys. 8*:371-375.
24. Collings, H. 1960. Polyneuropathy associated with nitrofuran therapy. *Arch. Neurol. 3*:65-68.
25. Collins, G. 1976. The morphology of myelin degeneration in thiamine deficiency, in *Thiamine* (Gubler, C., Fujiwara, M., Dreyfus, P., eds.), Wiley, New York, pp. 261-269.
26. Conroy, P. J., McNeill, T. H., Passalacqua, W., Merritt, J., Reich, K. R., Walker, S. 1982. Nitroimidazole neurotoxicity: Are mouse studies predictive? *Int. J. Radiat. Oncol. Biol. Phys. 8*:799-803.
27. Conroy, P. J., Van Burg, R., Passalacqua, W., Penney, D. P., Sutherland, R. M. 1979. Misonidazole neurotoxicity in the mouse. Evaluation of functional, pharmacokinetic, electrophysiologic and morphologic parameters. *Int. J. Radiat. Oncol. Biol. Phys. 5*:983-991.
28. Cooper, J. S., Fife, K. D., Borok, T. L., Waltzman, J. B. 1980. Detection of toxicity from the radiosensitizing drug misonidazole by serial audiograms, in *Radiation Sensitizers* (Brady, L. W., ed.), Masson, New York, pp. 490-494.
29. Cosar, C., Julou, L. 1959. Activité de l'(hydroxy-2'-ethyl)-1-methyl-2-nitro-5-imidazole (8, 823 R.P.) vis-à-vis des infections expérimentales à *Trichomonas vaginalis*. *Ann. Inst. Pasteur 96*: 238-241.
30. Coxon, A., Pallis, C. A. 1976. Metronidazole neuropathy. *J. Neurol. Neurosurg. Psychiatry 39*:403-405.
31. Denekamp, J., Harris, S. R. 1975. Tests of two electron-affinic radiosensitisers *in vivo* using regrowth of an experimental carcinoma. *Radiat. Res. 61*:191-203.
32. Deutsch, G., Foster, J. L., McFadzean, J. A., Parnell, M. 1975. Human studies with 'high dose' metronidazole: A nontoxic radiosensitiser of hypoxic cells. *Br. J. Cancer 31*:75-80.
33. Dewar, A. J. 1980. Neurotoxicity testing—With particular reference to biochemical methods, in *Testing for Toxicity* (Garrod, J., ed.), Taylor Francis, London, pp. 199-217.
34. Dewar, A. J., Moffett, B. J. 1979. Biochemical methods for detecting neurotoxicity—A short review. *Pharmacol. Ther. 5*:545-562.
35. Dische, S., Gray, A. J., Zanelli, G. D. 1976. Clinical testing of the radiosensitizer Ro-07-0582 11. Radiosensitization of normal and hypoxic skin. *Clin. Radiol. 27*:159-166.

36. Dische, S., Saunders, M., Lee, M., Adams, G. E., Flockhart, L. 1977. Clinical testing of the radiosensitizer Ro 07-0582: Experience with multiple doses. *Br. J. Cancer* 35:567-579.
37. Dische, S., Saunders, M. I., Anderson, P., Urtasun, R. C., Karcher, K. H., Kogelnik, N. D., Bleehen, N., Phillips, T. L., Wasserman, T. H. 1978. The neurotoxicity of misonidazole: Pooling of data from five centres. *Br. J. Radiol.* 51:1023-1024.
38. Dische, S., Saunders, M. I., Anderson, P., Stratford, M. R. L., Minchinton, A. 1982. Clinical experience with nitroimidazoles as radiosensitizers. *Int. J. Radiat. Oncol. Biol. Phys.* 8:335-338.
39. Dische, S., Saunders, M. I., Stratford, M. R. L. 1981. Neurotoxicity with desmethylmisonidazole. *Br. J. Radiol.* 54:156-157.
40. Edwards, M. S., Bolger, C. A., Levin, V. A., Phillips, T. L., Jewett, D. L. 1982. Evaluation of misonidazole peripheral neurotoxicity in rats by analysis of nerve trains evoked response. *Int. J. Radiat. Oncol. Biol. Phys.* 8:69-74.
41. Ellis, F. G. 1962. Acute polyneuritis after nitrofurantoin therapy. *Lancet* 2:1136-1138.
42. Flockhart, I. R., Sheldon, P. W., Stratford, I. J., Watts, M. E. 1978. A metabolite of the 2-nitroimidazole misonidazole with radiosensitizing properties. *Int. J. Radiat.* 34:91-94.
43. Gray, L. H., Conger, A. D., Ebert, M., Hornsey, S., Scott, O. C. A. 1953. The concentration of oxygen dissolved in tissues at the time of irradiation as a factor in radiotherapy. *Br. J. Radiol.* 26:638-648.
44. Gray, A. J., Dische, G. E., Adams, G. E., Flockhart, I. R., Foster, J. L. 1976. Clinical testing of the radiosensitizer Ro-07-0582 1. Dose tolerance, serum and tumour concentrations. *Clin. Radiol.* 27:151-157.
45. Griffin, J. W., Price, D. L., Kuetha, D. O., Goldberg, A. M. 1980. Neurotoxicity of misonidazole in rats 1. Neuropathology. *Neurotoxicology* 1:653-666.
46. Hall, E. J., Miller, R., Astor, M., Rini, F. 1978. The nitroimidazoles as radiosensitizers and cytoxic agents. *Br. J. Cancer Suppl.* 3:120-123.
47. Hollinger, D. M., Rossiter, R. J. 1952. Chemical studies of peripheral nerve during wallerian degeneration 5: β-Glucuronidase. *Biochem. J.* 52:659-663.
48. Horie, H. 1956. Anti-*Trichomonas* effect of azomycin. *J. Antibiot.* 9:168.
49. Ingham, H. R., Selkon, J. B., Hale, J. H. 1975. Treatment with metronidazole of 3 patients with serious infections due to *Bacteroides fragilis*. *J. Antimicrob. Chemother.* 1:235-242.
50. Jadhov, H., Balsara, J., Joshi, V., Salunkhe, D. 1974. The effect of metronidazole on striated muscle. *Eur. J. Pharmacol.* 25:263.

51. Jentsch, K., Karcher, K. H., Kogelnik, H. D., Maida, E., Mammoli, B., Wessley, P., Zaunbauer, F., Nitsche, V. 1977. Initial clinical experience with the radiosensitizing nitroimidazole Ro-07-0582. *Strahlentherapie 153*:825-831.
52. Kaplan, M. L., Murphy, S. D. 1972. Effect of acrylamide on rotarod performance and sciatic nerve β-glucuronidase activity of rats. *Toxicol. Appl. Pharmacol. 22*:259-268.
53. Kapp, D. S., Wagner, F. C., Lawrence, R. 1982. Glioblastoma multiforme: Treatment by large dose fraction irradiation and metronidazole. *Int. J. Radiat. Oncol. Biol. Phys. 8*:351-355.
54. Kun, L. E., Ho, K. C., Moulder, J. E. 1982. Fatal misonidazole-induced encephalopathy. An RTOG case report. *Cancer 49*:423-426.
55. Mamoli, B., Wessely, P., Kogelnik, H. D., Muller, M., Rathkolb, O. 1979. Electroneurographic investigations of misonidazole polyneuropathy. *Eur. Neurol. 18*:405-414.
56. McCamen, R. E., Robbins, E. 1959. Quantitative biochemical studies of wallerian degeneration in the peripheral and central nervous systems II: Twelve enzymes. *J. Neurochem.* 5:32-42.
57. McNally, N. J., Denekamp, J., Sheldon, P., Flockhart, I. R., Steward, F. A. 1979. Radiosensitation by misonidazole (Ro 07-0582). The importance of timing and tumour concentration of sensitizer. *Radiat. Res. 73*:568-580.
58. Millar, B. C., Jinks, S., Powles, R. J. 1981. Flurbiprofen, a non-steroid anti-inflammatory agent, protects cells against hypoxic cell radiosensitizers in vitro. *Br. J. Cancer 44*:733-739.
59. Millar, B. C., Jinks, S., Powles, R. J. 1982. Protection against misonidazole-induced toxicity *in vitro* by Flurbiprofen, a non steroidal anti-inflammatory agent, in *Prostaglandins and Cancer First International Conference*, (Powles, R. J., Bockman, R. S., Mann, K. V., Ramwell, T., eds.), A. R. Liss, Inc., New York, pp. 793-797.
60. Moore, J. L., Paterson, I. C. M., Dawes, P. J. D., Henk, J. M. 1982. Misonidazole in patients receiving radical radiotherapy: Pharmacokinetic effects of phenytoin, tumour response and neurotoxicity. *Int. J. Radiat. Oncol. Biol. Phys. 8*:361-364.
61. Olive, P. L. 1979. Inhibition of DNA synthesis by nitroheterocycles II. Mechanisms of cytotoxicity. *Br. J. Cancer 40*:94-104.
62. Parkes, M. W. 1974. Personal communication cited in Adams (Ref. 2) and Gray et al. (Ref. 47).
63. Paul, M. F., Paul, H. E., Kopko, F., Bryson, M. J., Harrington, C. 1954. Inhibition by furacin of citrate formation in testes preparations. *J. Biol. Chem. 206*:491-497.
64. Phillips, T. L., Wasserman, T. H., Johnson, R. J., Levin, V. A., Van Raalte, G. 1981. Final report on the United States phase 6 clinical trial of the hypoxic cell radiosensitizers, misonidazole (Ro 07-0582; NSC #261 037). *Cancer 48*:1697-1704.

65. Phillips, T. L., Wasserman, T. H., Stetz, J., Brady, L. W. 1982. Clinical trials of hypoxic cell sensitizers. *Int. J. Radiat. Oncol. Biol. Phys.* *8*:327-334.
66. Powers, S. K., Edwards, M. S. B., Baringer, R. J., Phillips, T. 1982. Evoked potentials in rats with misonidazole neurotoxicity. *Int. J. Radiat. Oncol. Biol. Phys.* *8*:814.
67. Ramsey, I. D. 1968. Endocrine ophthalmopathy. *Br. Med. J.* *4*:706.
68. Rauth, A. M., Kaufman, K. 1975. In vivo testing of hypoxic radiosensitizers using the KHT murine tumour assayed by the lung-colony technique. *Br. J. Radiol.* *48*:209-220.
69. Robertson, E., Wasan, S., Skinner, D. 1968. Ultrastructural features of early brain stem lesions of thiamine deficient rats. *Am. J. Pathol.* *52*:1081-1097.
70. Rogulja, P., Kovac, W., Schmid, H. 1973. Metronidazole-Encephalopathie der Ratte. *Acta Neuropathol.* *25*:36-45.
71. Roizin-Towle, L. A., Hall, E. J., Glynn, M., Biaglow, J. E., Varnes, M. E. 1982. Enhanced cytotoxicity of melphalan by prolonged exposure to nitroimidazoles. The role of endogenous thiols. *Int. J. Radiat. Oncol. Biol. Phys.* *8*:757-760.
72. Rollo, I. M. 1975. Miscellaneous drugs used in the treatment of protozoal infections, in *The Pharmacological Basis of Therapeutics*, 5th ed. (Goodman, L. S., Gilman, A., eds.), Macmillan, New York, pp. 1081-1089.
73. Rose, G. P., Dewar, A. J., Stratford, I. J. 1980. A biochemical method for assessing the neurotoxic effects of misonidazole in the rat. *Br. J. Cancer* *42*:890-899.
74. Rose, G. P., Dewar, A. J., Stratford, I. J. 1980. A neurotoxicity study on the radiosensitizing drug misonidazole: The effect of thiamine on misonidazole neurotoxicity in the rat. *Toxicol. Lett. Suppl.* *1*:127A.
75. Rose, G. P., Dewar, A. J., Moffett, B. J., Stratford, I. J. 1981. A biochemical assessment of the neurotoxicity of the radiosensitizing drug misonidazole in the rat. *Clin. Toxicol.* *18*:1411-1426.
76. Rose, G. P., Dewar, A. J., Stratford, I. J. 1982. A biochemical neurotoxicity study relating the neurotoxic potential of metronidazole, nitrofurantoin and misonidazole. *Int. J. Radiat. Oncol. Biol. Phys.* *8*:781-785.
77. Rose, G. P., Dewar, A. J., Stratford, I. J. 1983. Protection against misonidazole-induced neuropathy in rats: A biochemical assessment. *Toxicol. Lett.* *17*:181-185.
78. Sabri, M. I., Spencer, P. S. 1980. Toxic distal axonopathy: Biochemical studies and hypothetical mechanisms, in *Experimental and Clinical Neurotoxicology* (Spencer, P. S., Schaumburg, H. H., eds.), Williams and Wilkins, Baltimore, pp. 206-219.

79. Said, G., Goasguen, J., Laverdant, C. 1978. Polynévrites au cours des traitements prolongés par le metronidazole. *Rev. Neurol. 134*:515-521.
80. Schärer, K. 1972. Selective alterations of Purkinje cells in the dog after oral administration of high doses of nitroimidazole derivatives. *Verh. Dtsch. Ges. Pathol. 56*:407-410.
81. Schipper, H., Beale, F. A., Bush, R. S., Fryer, C. 1976. Metronidazole as a radiosensitizer (letter to the editor). *N. Engl. J. Med. 295*:901.
82. Schoental, R., Cavanagh, J. B. 1977. Mechanisms involved in the "dying back" process—An hypothesis implicating coenzymes. *Neuropathol. Appl. Neurobiol. 3*:145-157.
83. Sheldon, P. W., Foster, J. L., Fowler, J. F. 1974. Radiosensitization of C3H mouse mammary tumours by a 2-nitroimidazole drug. *Br. J. Cancer 30*:560-565.
84. Sinha, A. K., Rose, S. P. R. 1972. Compartmentation of lysosomes in neurons and neuropil and a new neuronal marker. *Brain Res. 39*:181-196.
85. Spencer, P. S., Sabri, M. I., Politis, M. 1980. Methyl N-butyl ketone, carbon disulfide and acrylamide. Putative mechanism of neurotoxic damage, in *Advances in Neurotoxicology* (Manzo, L., ed.), Pergamon, Oxford, pp. 173-180.
86. Stratford, I. J. 1980. The development of hypoxic cell sensitizers for clinical use, in *Scientific Foundation of Oncology* (Syrington, T., Carter, R., eds.), Heinemann, London, pp. 116-130.
87. Stratford, I. J. 1982. Mechanisms of hypoxic cell radiosensitization and the development of new sensitizers. *Int. J. Radiat. Oncol. Biol. Phys. 8*:391-398.
88. Taylor, Y. C., Rauth, A. M. 1980. Sulphydryls, ascorbate and oxygen as modifiers of the toxicity and metabolism of misonidazole *in vitro*. *Br. J. Cancer 41*:892-900.
89. Tomlinson, R. H., Dische, S., Gray, A. J., Errington, L. M. 1976. Clinical testing of the radiosensitizer Ro 07-0582 III. Response of tumours. *Clin. Radiol. 27*:167-174.
90. Urtasun, R. C., Chapman, J. D., Band, P., Rabin, H., Fryer, C., Sturmwind, J. 1975. Phase 1 study with high dose metronidazole: a specific *in vivo* and *in vitro* radiosensitizer of hypoxic cells. *Radiology 117*:129-133.
91. Urtasun, R. C., Band, P., Chapman, J. D., Feldstein, M. L., Mielke, B., Gryer, C. 1976. Radiation and high dose metronidazole (Flagyl) in supratentorial glioblastomas. *N. Engl. J. Med. 293*:1364-1367.
92. Urtasun, R. C., Band, P., Chapman, J. D., Rabin, H. R., Wilson, A. F., Fryer, C. G. 1977. Clinical phase 1 study of the hypoxic cell radiosensitizer Ro-07-0582, a 2-nitroimidazole derivative. *Radiology 122*:801-804.

93. Urtasun, R. C., Chapman, J. D., Feldstein, M. L., Band, R. P., Rabin, H. R., Wilson, A. F., Marynowski, B., Starreveld, E., Shnitka, T. 1978. Peripheral neuropathy related to misonidazole: Incidence and pathology. *Br. J. Cancer Suppl.* *111*:271-275.
94. Urtasun, R. C., Tanasichuk, H., Fulton, D., Agboola, O., Turner, A. R., Koziol, D., Raleigh, J. 1982. High dose misonidazole with dexamethasone rescue: A possible approach to circumvent neurotoxicity. *Int. J. Radiat. Oncol. Biol. Phys.* *8*:365-369.
95. Varghese, A. J., Whitmore, G. F. 1980. Binding to cellular macromolecules as a possible mechanism for the cytotoxicity of misonidazole. *Cancer Res.* *40*:2165-2169.
96. Varnes, M. E., Biaglow, J. E. 1982. Misonidazole-induced biochemical alterations of mammalian cells: Effects on glycolysis. *Int. J. Radiat. Oncol. Biol. Phys.* *8*:683-686.
97. Varnes, M. E., Biaglow, J. E., Koch, C. J., Hall, E. J. 1980. Depletion of non protein thiols of hypoxic cells by misonidazole and metronidazole, in *Radiation Sensitizers* (Brady, L., ed.), Masson, New York, pp. 121-126.
98. Von Burg, R., Conroy, P. J. 1979. Evaluation of the anti-cholinesterase activity of metronidazole and misonidazole. *Eur. J. Pharmacol.* *55*:417-420.
99. White, R. A. S., Workman, P. 1980. Pharmacokinetic and tumour-penetration properties of the hypoxic cell radiosensitizer desmethylmisonidazole (Ro 05-9963) in dogs. *Br. J. Cancer* *41*:268-276.
100. Windeyer, B. 1978. Hyperbaric oxygen radiotherapy. The Medical Research Council's Working Party. *Br. J. Radiol.* *51*:875.
101. Workman, P. 1979. Effects of pretreatment with phenobarbitone and phenytoin on the pharmacokinetics and toxicity of misonidazole in mice. *Br. J. Cancer* *40*:835-345.
102. Workman, P., Brown, J. M. 1981. Structure-pharmacokinetic relationships for misonidazole analogues in mice. *Cancer Chemother. Pharmacol.* *6*:39-49.
103. Wasserman, T. H., Phillips, T. L., Van Raalte, G., Urtasun, R., Partington, J., Koziol, D., Schwade, J. G., Gangji, D., Strong, J. M. 1980. The neurotoxicity of misonidazole: Potential modifying role of phenytoin sodium and dexamethasone. *Br. J. Radiol.* *53*:172-173.

Part III

NEUROTOXIC SUBSTANCES AND THE HUMAN ENVIRONMENT

14

Neurotoxicology of Lead

Ellen K. Silbergeld
Environmental Defense Fund, Washington, D.C.

I. INTRODUCTION

Research on the toxicology of lead has increased greatly over the past decade. This interest has been stimulated by several factors: First, lead exposure is widespread in both industrial societies and some less-developed countries where lead-containing ceramic glazes, cosmetics, and medicines are still common (22,49). Second, lead acts on the central and peripheral nervous systems at very low dose; it is an agent which can explicate both normal and abnormal functionings of the nervous system. Third, a consistent set of findings is emerging from clinical and experimental research on lead, such that the various animal models developed over the past 15 years provide an exceptionally reliable basis for understanding this illness and for evaluating current methods of treatment (52). Fourth, the relatively simple nature of lead, as a stable element, means that extensive metabolic changes are not involved in its toxicity [although a simplistic hypothesis that lead acts solely as ionic lead in causing toxicity, particularly in the nervous system, is probably inadequate (73)].

This chapter will review information on the nature and extent of lead exposure in selected human populations, with emphasis on those at increased risk for toxicity. The evidence for lead as a neurotoxin will be reviewed, and parallels between clinical and experimental data will be drawn, with discussion of areas of agreement and disagreement. For complete reviews of all aspects of lead toxicity, the reader is referred to several recent monographs (49,50,59,85).

II. CLINICAL TOXICOLOGY OF LEAD EXPOSURE

The toxicity of lead has been known for centuries, although the molecular identity of the toxic agent has obscured until very recently consistent diagnosis and regulation. Thus, in antiquity, while the use of leaded solder (metallic lead or lead oxides) was known to be hazardous, "sugar of lead" (lead subacetate) was recommended as a sweetening agent for wines. Only centuries later was this practice of adulteration outlawed. Leaded ceramic glazes were recognized as hazardous early in this century (39). However, understanding that this hazard also affected consumers using lead-glazed ware was delayed by some 50 years. Lead compounds continued to be used in paints until the mid-1900s. In the United States, lead in paint was not effectively restricted until 1972. In the 1920s, alkyllead compounds were introduced as fuel additives for internal combustion engines. No government has adequately regulated this large-scale use of lead, although by now (1984) most countries have placed some limits on the maximum amount of alkyllead in gasoline.

This brief history indicates the length of time it has taken to understand the elemental identity of lead as a toxic substance. With the exception of alkyllead compounds, which behave differently from inorganic lead in the environment and within organisms, how lead enters the environment only slightly alters its toxicity to the receptor organism. The major part of this review concerns inorganic lead; however, where the organic compounds significantly differ in toxicity will be indicated. For a more complete review of the toxicity of alkyllead compounds, the reader is referred to Ref. 29.

The early studies of lead toxicity, although fragmented, as noted above, nevertheless identified some important aspects: the insidious and cumulative nature of its effects, the nervous system and reproductive system as specific targets of toxic action, the progression of neurological disease from subtle psychological effects to death, and the importance of exposure—dose and rate—in determining toxicity. Not until the twentieth century was an additional important aspect of lead toxicity noted: the special sensitivity of young children. This was first appreciated in those cases where workers bringing home on their clothing large amounts of lead-containing dusts thus exposed their families (pregnant wives and young children). Since that time medical attention has focused on the young child as the most critical target for lead poisoning (40). This concern is increased by the fact that children usually cannot prevent their exposure to ubiquitous environmental toxins like lead.

A. Childhood Lead Toxicity

Like all forms of lead intoxication, the nature and course of the neurotoxic disease in children depend upon duration and amount of exposure. Low-dose chronic exposures cannot be equated with high-dose acute exposures. High-dose acute exposures are associated with the so-called signs of lead encephalopathy, encompassing such central nervous system signs as lethargy, seizures, brain edema, and eventual coma and death. Progression can be rapid, and these symptoms are not well correlated with indices of exposure (15). Under such conditions, lead neurotoxicity is accompanied by damage to other organ systems, notably the kidney, hematopoietic system, and liver. Both central and peripheral neuropathy can be present. In a study of brains at postmortem, extensive demyelination, gliosis, and loss of cortical cells have been noted (36). Peripherally, the neuropathology of high-dose lead exposure is characterized by progressive segmental demyelination, swelling of the endoneurium, and axonal destruction proceeding in a retrograde fashion to the motor nerve cell bodies in the spinal cord (21, see also Cavanagh in this book).

At lower doses, the signs and indices of lead neurotoxicity and exposure are less readily observed. Toxicity may be subtle, particularly in young children, where detection of adverse effect may require sophisticated measurement of complex behaviors (see below). Aminoaciduria and other indicators of nonneural toxicity may not be present, thus indicating the relatively greater sensitivity of the nervous system. However, biochemical inhibition of heme synthesis is observed in children with blood lead concentrations as low as 10 μg/100 ml (56). With low level exposure, it has been assumed that central nervous system toxicity occurs in children in the absence of peripheral neuropathy. However, few studies have been directed to resolving this issue. In a study of children exposed to lead from a smelter, both types of neurotoxicity, with decreases in IQ scores and slowing of nerve conduction velocity, have been reported (38). Similarly, Feldman et al. (21) found slowed peripheral nerve conduction velocity in children with long-term exposure to lead paint; however, assessment of central nervous system dysfunction was not done. Needleman et al. (51), in a study of children several years after undetected intoxication, reported significant decrements in the performance of tests which involve the peripheral as well as the central nervous systems. This may indicate multisite neurotoxicity, but the nature of these effects (on reaction time) suggest that attentiveness, mediated by central nervous system centers, was probably the site of lead-induced dysfunction, rather than peripheral parameters related to the performance of the reaction task.

Neurotoxic aspects of childhood lead toxicity do not appear to be readily reversible when external exposure ceases. This could reflect both the induction of persistent damage and the continued presence of lead in the nervous system. Even with therapeutic intervention lead neurotoxicity is relatively irreversible (21,52). Experimental data indicate that chelation therapy with conventional agents does not remove lead from the brain (26,28). As shown in Fig. 1, in a group of children restudied several months after extensive chelation, neurochemical abnormality was still present, even though blood lead levels showed significant decreases toward the values of controls (16).

It is now agreed that low level of lead exposure of young children has long-term neurotoxic sequelae, which are dependent upon the nature of the original exposure. Although debate continues over the level, in terms of blood lead, at which adverse effects on the nervous system occur, this debate is centered on the range of 10-25 μg/100 ml (52). High-dose exposures have been well documented to be followed by severe mental retardation and significant behavior disorders, including dementia and psychosis, and, less severely, many of the symptoms which were formerly clustered under the diagnostic description of minimal brain dysfunction (13,17,18). Lower lead exposures, not associated with classic lead encephalopathy, are now recognized to entail significant and persistent decrements in neurological function (64).

Low-level lead exposure early in childhood appears to affect later measures of central nervous system function associated with decreases in learning (both acquisition and retention) (see Ref. 52). Sensory processes, such as auditory and visual acuity, may also be adversely affected. Spontaneously expressed behavior is also detectably different as measured in children in classrooms long after exposure, as shown in Table 1.

Nervous system function has also been studied biochemically and electrophysiologically in lead-exposed children. Biochemically, the urinary excretion of the major catecholamine metabolites has been investigated with respect to lead dose. As shown in Table 2, good statistical correlation was found between blood lead and 24-hr urinary excretion of the dopamine metabolite homovanillic acid (HVA). Urinary HVA is not well correlated with other indices of lead burden or physiological effect, such as urinary lead (spontaneous or chelatable), urinary d-aminolevulinic acid, or red cell protoporphyrin (16). In an earlier report we found similar increases in urinary vanillyl mandelic acid, a metabolite of norepinephrine (78); however, this was not found

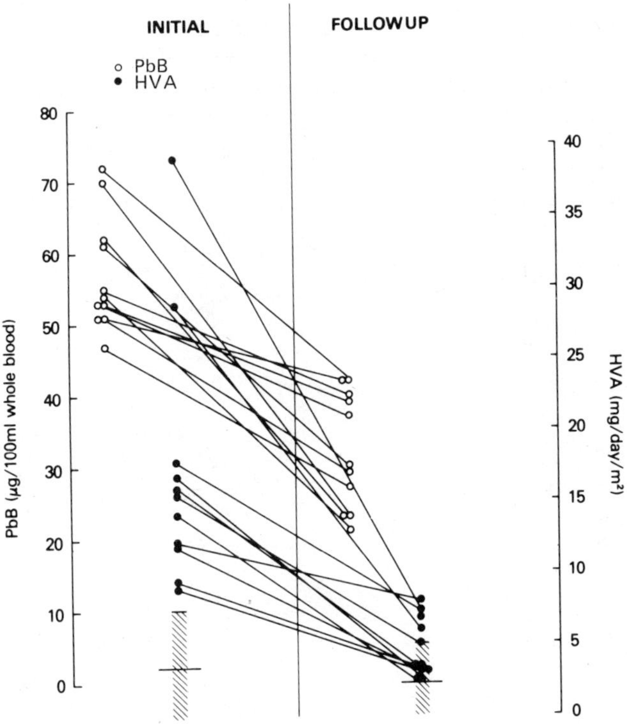

Figure 1 Blood lead levels (PbB) (in micrograms per 100 ml, shown on the left axis) and urinary homovanillic acid (HVA) levels (in mg/day per square meter of body surface area, shown on the right axis) in 11 children studied before and after extensive chelation therapy and removal from lead-contaminated homes. Lines and shaded areas indicate the means and standard deviations of urinary HVA excretion in controls of comparable ages. Although all values decreased from initial to follow-up studies, all children continued to have blood lead levels above 20 μg/100 ml; four children showed persistent elevations in HVA as well (16).

Table 1 Classroom Behavior in Children Exposed to Lead

	Percentage of intervals in which behavior was noted	
	High[a]	Low[b]
Out of seat	0.8 ± 1.4	0.3 ± 0.7
Glances at classmate	9.3 ± 7.4	6.5 ± 2.4
Glances at observer	10.1 ± 4.0	5.6 ± 3.3
Glances away	4.2 ± 3.1	2.2 ± 1.9
Interacts with classmate	12.0 ± 8.1	7.9 ± 8.4

[a]Dentine lead, 20 ppm.
[b]Dentine lead, 10 ppm.
Source: Ref. 3.

Table 2 Simple Linear Correlation Coefficients Between Indicators of Body Lead Burden and Several Biochemical Effects

	PbB	PbU/C	PbU	HVA	ALA	EP
Indicators						
Blood lead (PbB)	1.00	0.46[a]	0.13	0.73[b]	0.47[a]	-0.31
Chelatable lead (PbU/C)		1.00	0.32	0.13	0.65[b]	-0.11
Spontaneously excreted lead (PbU)			1.00	0.13	0.37	0.01
Effects						
Homovanillic acid (HVA)				1.00	0.19	-0.20
d-Aminolevulinic acid (ALA)					1.00	0.03
Erythrocyte protoporphyrin (EP)						1.00

[a]$p < 0.05$.
[b]$p < 0.001$.
Source: Ref. 16.

in subsequent studies, nor were consistent changes found in other major catecholamine metabolites measured in urine. The relative stability of HVA production as compared to the other products may be responsible for this apparent selectivity of neurochemical effect. Alternatively, lead may have highly specific actions on catecholaminergic neurotransmission and/or metabolism.

Several electrophysiological studies have been done on children with various exposures to lead. Studies in children who had suffered encephalopathic episodes have reported many types of irregularities in electroencephalograms, including paroxysmal spike and wave complexes (20,63). Electrophysiological changes have more recently been described in children with much lower levels of exposure. Indeed, such measurements may provide sensitive indicators of neurotoxicity. These studies are consistent in finding persistent alterations at relatively low dose. Spectral analysis of electroencephalograms on some of the children studied by Needleman et al. (51) demonstrated significant changes in the patterns of cortical slow and fast waves (10). A selected type of electroencephalogram (the visual-evoked response) has also been used to study children with relatively low blood leads at the time of exposure and in a 1-year follow-up (54). Statistically significant alterations were reported in the visual-evoked response pattern, with deviations from controls at blood lead as low as 15 μg/100 ml. Similar changes in evoked responses and other electrophysiological measures have been reported in another group of lead-exposed children whose exposure was estimated using hair lead analysis (89).

B. Adult Lead Neurotoxicity

Lead is also neurotoxic to adults. Like the childhood disease, lead encephalopathy in adults is now relatively rare, but cases still occur. Among the cases are abusers of lead-contaminated illicit whiskey, who present with multiple seizures, mania, delirium, mental obtundation, blindness, and aphasia (93). In such exposures, other organ systems are also affected, so that anemia, basophilic stippling, porphyria, and aminoaciduria accompany the signs of encephalopathy.

Recent *prospective* studies undertaken in occupational settings indicate that well before other indicators of toxic damage, workers experience distinct neuropsychiatric changes, which can be most sensitively seen using mental status examination (46). Performance of tasks requiring quick reactions and good eye-hand coordination are also impaired at relatively low exposures (60).

Peripheral neurotoxicity has been proposed as a sensitive measure of occupational neurotoxicity (67). Lead is a well-established example of predominantly motor neuropathy (36). Moreover, undetected lead exposure has been proposed to produce conditions similar or identical to motor neuron disease or amyotrophic lateral sclerosis (37).

At lower doses, associated with blood levels in the range of 40-80 μg/100 ml, motor nerve conduction velocity is slowed (21,67,90). It is not yet clear whether such electrophysiological changes follow or precede cellular damage to elements of the peripheral system. Conduction abnormalities may be due to biochemical changes preceding morphological damage (81). Slight slowing may occur in cases where very little axonal pathology is detected by biopsy (8), although endoneurial swelling may have already been induced (96).

Some of the neurotoxic effects of lead may be related to nonneural effects of lead. One occupational study strongly reported that decrements observed in performance of complex neurobehavioral tasks are better related to the degree of altered hematopoiesis, rather than to increase in body lead burden (90). The similarities between lead neurotoxicity and hereditary porphyria (a disease which involves changes in the synthesis of heme similar to those induced by lead) have been remarked upon elsewhere (73,83).

In adults, lead toxicity also affects the reproductive system (74). This is noted here, in a chapter on neurotoxicity, because it has been suggested that there is an increased incidence of mental retardation in children of lead-intoxicated parents (47). At present there are few data on this subject, except in the experimental literature (23,98), but the special sensitivity to lead of both germ cells and the fertilized zygote should be indicated. There may be reason to suspect that developing neurons are highly sensitive to lead, as evidenced in studies of hyperinnervation of noradrenergic systems induced by lead in the transplanted embryonic cerebellum (4).

C. Extent of Lead Poisoning

Recent data from the United States demonstrate that exposure to lead is endemic, a situation which probably is not different from countries which use alkyllead in gasoline (44). If Table 3 is taken as representative of such national incidences, and it is assumed that neurotoxic sequelae can occur in young children with blood lead values above 20 μg/100 ml; then it is apparent that millions of children may be suffering neurotoxic damage every year as a consequence of lead exposure.

D. Special Sensitivity

Sensitivity to lead as a neurotoxin is clearly modulated by age. In addition, the state of trace metal nutrition is very important (45). Lead competes with zinc, calcium, and phosphorus for uptake and retention after ingestion. Moreover, lead and calcium may interact at sites of neurotoxicity, and possibly lead and iron may compete at several steps in the regulation of heme synthesis, as well as in availability

Table 3 Blood Lead Levels of Persons 6 Months to 74 Years Old, with Mean, Standard Error of the Mean, and Percent Distribution, by Race and Age, United States, 1976-1980[a]

	Estimated population in thousands (1978)	Blood lead level (μg/100 ml)		Percent distribution				
		Mean	Standard error of the mean	Less than 10	10-19	20-29	30-39	39
All races								
All ages	203,554	13.9	0.24	22.1	62.9	13.0	1.6	0.3
6 months to 5 years	16,852	16.0	0.42	12.2	63.3	20.5	3.6	0.4
6-17 years	44,964	12.5	0.30	27.6	64.8	7.1	0.5	—
18-74 years	141,728	14.2	0.25	21.2	62.3	14.3	1.8	0.4
White								
All ages	174,528	13.7	0.24	23.3	62.8	12.2	1.5	0.3
6 months to 5 years	13,641	14.9	0.43	14.5	67.5	16.1	1.8	0.2
6-17 years	37,530	12.1	0.30	30.4	63.4	5.8	0.4	—
18-74 years	123,357	14.1	0.25	21.9	62.3	13.7	1.8	0.4
Black								
All ages	23,853	15.7	0.48	13.3	63.7	20.0	2.3	0.6
6 months to 5 years	2,584	20.9	0.61	2.5	45.4	39.9	10.2	2.0
6-17 years	6,529	14.8	0.53	12.8	70.9	15.6	0.7	—
18-74 years	14,740	15.5	0.54	14.7	62.9	19.6	2.0	0.9

[a]Numbers may not add to totals because of rounding.
Source: Ref. 44.

of tetrahydrobiopterin for catecholamine synthesis (58). In addition, lead affects the transport of certain amino acids, such as tyrosine and tryptophan, from blood to brain (41). Dietary undersupply of these substances may also increase sensitivity to lead, although this has not been investigated.

It has been proposed that females are more sensitive than males to lead, on the basis that occupationally exposed women show slightly greater inhibition of red cell ALA dehydrase activity as compared to men (62). But this could result from differences in male and female porphyrin metabolism (a biochemical pathway sensitive to sex steroids) or real differences in sensitivity to toxic effects. The literature on *neurotoxicity* has not been examined for differential sensitivity based on sex in children or adults.

There are no clinical data which suggest genetic differences in neurotoxic response to lead. Conditions of coexisting lead exposure and sickle cell anemia have been described, but insufficiently to construe any differences in neurotoxicity compared to external dose or body burden (1). The experimental literature suggests some intraspecies differences dependent upon strain, such as those described in rodent strains bred for differences in brain weight (11). However, these strains may be more sensitive to many types of neuroactive substances, and not specifically vulnerable to lead.

Repeated exposure to lead appears to increase toxic consequences of subsequent exposures. Rodents with prior semichronic exposure concentrate more lead in their brains after an acute exposure, as compared to naive animals (25). Preliminary studies indicate that children with repeated episodes of lead toxicity have higher excretion rates of urinary HVA, at the same blood lead levels as children with no previous history of exposure (J. J. Chisolm and E. K. Silbergeld, unpublished data). As indicated earlier, this may be due to the long half-life of lead in the nervous system, so that sequential exposures are essentially additive in terms of dose at critical sites of action.

III. NATURE AND MECHANISMS OF LEAD NEUROTOXICITY

The evidence that lead is a neurotoxin is very strong. As reviewed above, the clinical history of lead poisoning amply indicates that exposure over a range of doses is associated with neuropsychiatric sequelae ranging in intensity and duration. However, neurological signs may be associated with many types of intoxication, and general changes in mood frequently accompany damage to other organs (92). There is extensive evidence that lead truly is a specific neurotoxin.

First, neurotoxic effects are induced prior to the onset of more general or other organ system toxicity. With tests of sufficient sensitivity (such as electrophysiology and neurochemistry), significant alterations in nervous system function occur well before demonstrated damage to other organs, decreases in body weight, or other indicators of toxicity. However, inhibition of heme synthesis may occur at equally low dose. Also alkyllead compounds possess mutagenic properties at very low dose which may precede or occur at the same time as their toxic actions on the nervous system (29).

Second, lead is neurotoxic at levels of exposure below those associated with cell loss or morphological alterations to intracellular organelles. Lead can alter biochemical and electrophysiological properties of nerve cells, as well as the behavior of animals, in the absence of morphological damage. Even with sophisticated neuroanatomical techniques—indices of synaptic complexity and quantitative measurements of myelination—it is clear that cellular damage is not the precedent factor in lead neurotoxicity (7). Nevertheless, at higher exposure, lead can cause large decreases in synaptic complexity and even reduce laminar thickness, particularly of the hippocampus (42,55). In the case of alkyllead compounds, glia may be more sensitive than neurons (61).

The mechanisms of low-level lead neurotoxicity are therefore likely to be neurochemical. At such levels of exposure lead produces extensive neurochemical changes (81). Some of the more consistent findings are summarized in Table 4.

Lead exposure interferes with neurotransmission using acetylcholine, dopamine, norepinephrine, and γ-aminobutyric acid (GABA) as transmitters. These effects are regionally specific and depend upon the nature of exposure, in that some effects which are seen at low dose in young animals are not produced by higher doses (19,68,81).

In the central nervous system both presynaptic and postsynaptic aspects of neurotransmission are affected. Presynaptically, in vivo lead exposure of rodents alters patterns of transmitter release, reducing the amount of some transmitters (acetylcholine and GABA) and increasing release of others (norepinephrine and dopamine). In vitro, lead has more restricted actions, augmenting release of dopamine, but with no effect on other transmitters studied (81). Another presynaptic process sensitive to lead is the sodium-dependent high-affinity uptake of several neurotransmitters and neurotransmitter precursors. In vivo lead exposure produces decreases in the uptake of dopamine, choline, glycine, and GABA (81). In vitro, the uptake of both dapamine and choline is sensitive to inorganic and alkyllead compounds (5,70). Transport of amino acids not specifically involved in the synthesis of transmitters (such as leucine) is not affected (79).

Postsynaptically, inorganic and organic lead in vivo and in vitro directly inhibits transmitter-sensitive adenylate cyclase in striatal tissue (48,94,95). Inorganic lead blocks some classes of dopamine receptors which are labeled with neuroleptics, particularly those in the limbic system (43). In vivo lead exposure, semichronically but not acutely,

Table 4 Some Neurochemical Effects of Lead Exposure[a]

Species	In vivo exposure	Effects
Mouse	2-10 mg/ml lead acetate starting at birth	↓ Dopamine (DA) uptake ↓ Choline uptake ↑ Tyrosine uptake ↓ Acetylcholine (ACh) release ↑ Norepinephine (NE) ↑ Brain HVA ↑ Monoamine oxidase activity; No change in Ach
Rat	40-50 mg/ml lead acetate starting at birth	No change in ACh, 5-hydroxytryptamine or GABA levels No change in tyrosine hydroxylase activity ↑ NE ↓ DA
	10 mg/ml lead acetate starting at birth	↑ ACh ↑ Choline acetyltransferase activity ↓ ACh esterase activity ↓ GABA uptake and release ↑ GABA transaminase activity

[a]See Refs. 68 and 81 for details and references.

also increases the density of cerebellar GABA receptors (27,82) and affects responsiveness of noradrenergic receptors (88). As shown in Table 5, some of these postsynaptic effects are only observed after in vivo exposure, while for others—notably the inhibition of dopamine-sensitive cyclase—effects have been found in both in vivo and in vitro experiments (see Ref. 72 for a review).

In the peripheral nervous system, lead blocks the release of acetylcholine at neuromuscular junction and superior cervical ganglia (33). This action is reversible by calcium and by washout.

In developing hypotheses as to how lead produces these neurochemical effects, it is assumed that lead has no physiological function. There is no process yet identified whose function is enhanced by or dependent upon lead at any level. Thus its accumulation, or binding within cells, may be assumed to represent localization at sites of toxicity. As a pure toxin, lead might be expected to act nonspecifically, producing dysfunction at any site to which it gains access. However, lead is not

Table 5 Postsynaptic Effects of Lead in Brain

	Effect of lead[a]		
		In vitro	
Parameter	In vivo (inorganic)	Inorganic	Alkyl
Enzymes			
Adenylcyclase			
Basal activity	—	—	—
Dopamine stimulated	↓	↓	↓
Receptors			
[^{3}H]Spiroperidol	—	—	↓[b]
[^{3}H]Sulpiride	↓	Not done	Not done
[^{3}H]GABA	↑	—[c]	—
[^{3}H]QNB (quinuclidinyl benzylate)	—	—	↓
[^{3}H]Naloxone	—	—	Not done

[a] —, no effect; ↓, decrease; ↑, increase (in density of binding sites).
[b] Inhibition observed only when lead tri(n-butyl)acetate concentration is above 10^{-6} M.
[c] Inhibition reported by Bondy et al. (5); not found by Silbergeld et al (82).

uniformly distributed within nervous tissue, so some selective processes are involved in its distribution and these processes are, presumably, related to its toxic effects. Several types of studies have been undertaken on the subcellular localization of lead within nervous tissue, but none is without methodological problems (see Ref. 75 for a discussion). Nevertheless, the results of these studies consistently suggest that the sites of lead's action are relatively restricted to membranes and mitochondria.

A. Membrane-Mediated Neurotoxicity of Lead

Actions of lead on neural membranes include interference with ion fluxes and inhibition of membrane-associated enzymes. Lead alters movement of calcium across neuronal and other cell membranes (57,76). In peripheral nervous tissues, lead in vitro blocks the neuronal uptake of ^{45}Ca and may alter some properties of membrane conductance (2). In the central nervous system both indirect and direct evidence suggest that lead increases transmembrane calcium flux (32,76). Two important enzymes associated with membranes, Na-K adenosinetriphosphatase and adenyl cyclase, are inhibited by lead in brain and other tissues (48, 69,87). Inhibition of Na-K adenosinetriphosphatase can affect cellular ion flux because of the important role of this pump in maintaining ion homeostasis.

Alkyllead compounds may act on glial cells through membrane-mediated actions associated with inhibition of synthesizing enzymes (34) and possibly also with inhibition of glial porphyrin biochemistry (65).

B. Mitochondrial Toxicity of Lead

In circumstances under which lead gains access inside cells, which are not yet understood, lead is preferentially and rather tightly bound to mitochondria, possibly to the Ca-Mg adenosine triphosphate complex within this organelle (77; E. K. Silbergeld, D. Joy, D. Maher, and J. Costa, unpublished observations). The consequences of this are several. First, mitochondrial energy metabolism is affected (9,30). Since neurons are highly active cells, their dependence upon cellular energy is high, and decrements in available adenosine triphosphate may rapidly compromise function and cause cell death. Second, the ability of mitochondria to regulate intracellular calcium concentrations is inhibited. The neuronal mitochondrion is an important part of calcium regulation, which controls neuronal secretory processes and the activity of some synthesizing enzymes of importance to neurotransmission, such as tyrosine hydroxylase (91). Lead reduces the lability of the mitochondrial pool (Fig. 2) and therefore increases its function as a "sink" for both extra- and intracellular calcium (57,76). Removal of mitochondrial pools of calcium from participation in neuronal calcium cycling has been proposed to be a factor in increasing exocytotic events, which may be associated with increased release of catecholaminergic neurotransmitters (76).

C. Molecular Mechanisms

The molecular nature of lead neurotoxicity is not fully understood. Lead does not appear to directly compete with calcium, although it may occlude some important calcium sites, particularly in the periphery. Lead does not compete with other ions, although some of its effects can be diminished by sodium (Fig. 3). Lead is hypothesized to act through its affinity with sulfhydryl groups, but the relevance of this physicochemical interaction to the neurotoxicity of lead is unknown. Lead is not particularly toxic for enzymes or receptors known to be rich in sulfhydryl components, such as acetylcholinesterase or muscarinic cholinergic receptors, nor is lead concentrated in areas of the brain richest in such groups. Alkyllead compounds may show a clearer

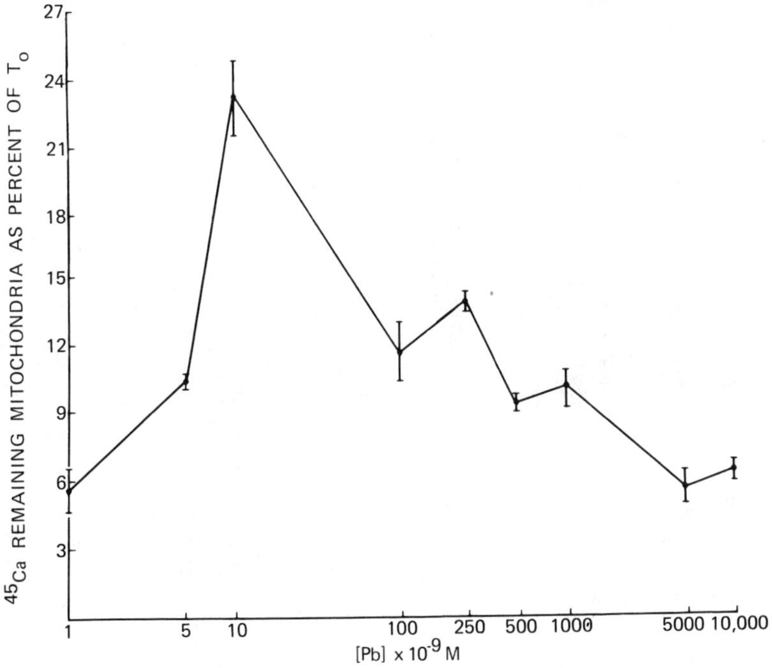

Figure 2 Effects of lead in vitro on ^{45}Ca release from preloaded mitochondria derived from rat caudate synaptosomes. Concentrations of lead greater than 1 nM were associated with an inhibition in this efflux, measured as the percent of that loaded into mitochondria. Apparent return to base line at concentrations higher than 10 nM probably reflects inhibition of adenosinetriphosphatases by lead (76).

affinity for cysteine-containing lipoproteins, particularly the sphingomyelin components of the central nervous system. Inorganic lead can bind with adenosine triphosphate through complexing with phosphate. This may explain its affinity for adenosinetriphosphatase and for mitochondria, although the exact sites for its binding to these components are not known.

The neurotoxicity of lead is affected by a combination of factors which result in the distribution of lead to sites within the nervous system, as well as the sensitivity of these sites to lead. Understanding these factors is important to determine the threshold, if any, for lead as a neurotoxin. Once lead gains access to the nervous system, and to specific sites within neurons, the "threshold" for its effects is very low, probably in the submicromolar range. In vitro experiments on preparations of neuronal tissue demonstrate significant effects on

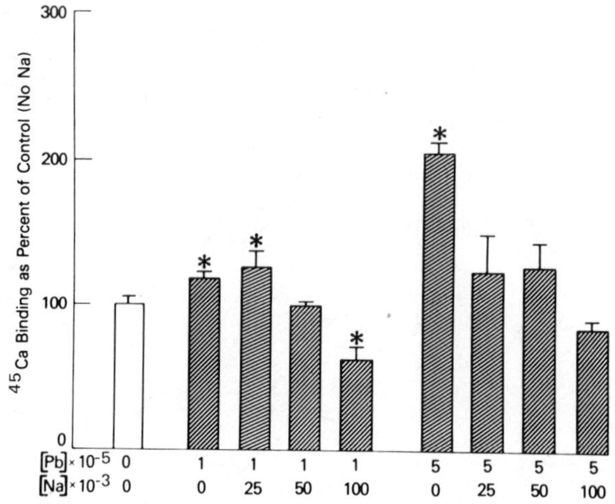

Figure 3 Effects of added sodium (Na) on the increased uptake of ^{45}Ca by rat caudate synaptosomes exposed to lead in vitro at either 10 or 50 μM. As can be seen, Na was able to antagonize these effects of lead at concentrations of 25-50 mM (76).

enzymes, receptor binding, and ion transport when lead concentrations are as low as 5 nM (Fig. 2). Clearly, at the level of exogenous, environmental exposure, thresholds exist. However, these may be related almost entirely to factors governing tissue distribution. The ability of red cells to bind lead and the existence of nonneuronal binding sites within liver, bone, kidney, and spleen, for example, may prevent lead from equilibrating rapidly with sensitive neuronal sites (53).

IV. A HYPOTHESIS OF LEAD NEUROTOXICITY

Given the complex neurobehavioral effects described in lead-exposed children and animals, and our incomplete understanding of lead neurotoxicity at the cellular level, it may be premature to advance a hypothesis for lead neurotoxicity which integrates cellular mechanisms and organism-level dysfunctions. It has been proposed by clinical and experimental investigators that lead may cause, or constitute an animal model of, the childhood behavior disorder of minimal brain dysfunction (17,79). In fact, testing for increased lead burden is now recommended for minimal brain dysfunction diagnosis and treatment (66).

However, this hypothesis does not greatly advance understanding of lead toxicity. Minimal brain dysfunction has been proposed to involve overactivation of biogenic amine systems, imbalance among such systems, or inhibition of dopamine function (71). Nevertheless, the analogy with minimal brain dysfunction caused many researchers (including this author) to direct attention to the effects of lead on catecholamine pathways and to consider lead neurotoxicity primarily as a derangement of neural processing in these systems. As discussed above, there is evidence that such changes do occur. Moreover, the frequent finding of altered responses in lead-exposed animals to drugs which act as agonists on catecholamine pathways (amphetamine and apomorphine) has remained a continuing stimulus of research in lead (14,35, 43,79,86).

Recently we have revised this hypothesis and directed our attention to the effects of lead on other pathways of neurotransmission. GABAergic neurotransmission is very sensitive to lead exposure, under both acute and chronic conditions of exposure (72,82,84). Unlike other neurotoxic effects of lead, acute exposure, even in adult animals, can rapidly decrease uptake and release of GABA (73).

It is therefore our current hypothesis that this system is the most sensitive to lead neurotoxicity, and the site involved in the earliest expression of lead neurotoxicity. Such a hypothesis is consistent with the other neurochemical effects of lead exposure, since reduced GABAergic function can by itself be associated with disinhibition of adrenergic and dopaminergic function, although this can also be produced by lead added in vitro to neural tissue (5).

This hypothesis that lead is a GABAergic neurotoxin provides an interesting perspective on its behavioral effects which have been described experimentally and clinically. The importance of GABA as a regulator of overall brain activity and mood is well known, particularly in light of its major role in regulating cortical electrical activity and its intricate involvement with benzodiazepine-mediated pathways.

Many of the behavioral effects of lead can be characterized as failures of appropriate inhibition of response. This may range from the general level of motor activity, an increase in which is frequently (although not always) observed in animals (80), to the constellation of off-task and inattentive behaviors described in classroom observation of lead exposed children (3). The paradigms of passive response and measurements of time to extinction/reversal of operant tasks are frequently noted as being particularly sensitive to lead (6,12,31,97). The clinical effects of low-level lead exposure on reaction time, which depends upon the time lapse between conditioned and unconditioned stimuli, are also characteristic of a disinhibited state where the subject is distractable and inattentive. Whether the electrophysiological effects of lead are consistent with disinhibition of the nervous system has not been investigated. However, an ontogenic study of low-level lead exposure has reported changes in minimal electroshock seizures, which is consistent with altered GABAergic function (24).

In summary, this chapter has provided a description of both the clinical and experimental neurotoxicity of lead, a discussion of current theories of mechanisms of lead neurotoxicity, and finally an integration of mechanistic studies with the behavioral toxicology. This review has empahsized lead neurotoxicity in terms of its age dependence, the sensitivity of certain pathways in the central nervous system, and the possibility that lead at low dose may act specifically on GABAergic systems. Clearly much remains to be elucidated in terms of molecular mechanisms of lead toxicity, and the full picture of low-level neurotoxicity (particularly in adults) requires more clinical description. Nevertheless, among the neurotoxins of significance in clinical disease, a great deal is known about lead, sufficient to guide both the important political decisions necessary to reduce exposure to this toxin, and the future studies on its actions in the nervous system.

ACKNOWLEDGMENTS

The author wishes to thank Joan Dolby for editorial and secretarial assistance, and colleagues at the Environmental Protection Agency review meeting, particularly Dr. David Weil, for many insights and references.

REFERENCES

1. Anku, V. D., Harris, J. W. 1975. Peripheral neuropathy and lead poisoning in a child with sickle-cell anemia. *J. Pediatr.* *85*:337-340.
2. Atchison, W. D., Narahashi, T. 1982. Mechanism of action of lead on neuromuscular junctions. *Neurotoxicology* *3*:122-124.
3. Bellinger, D., Needleman, H. 1983. Low level lead exposure and psychological deficit in children, in *Advances in Developmental and Behavioral Pediatrics* (Vol. 3) (Wolraich, M., Routh, D. K., eds.), JAI Press, Greenwich, Ct., pp. 523-527.
4. Bjorklund, H., Olson, L., Sieger, A., Hoffer, B. 1980. Chronic lead and brain development. *Environ. Res.* *22*:224-236.
5. Bondy, S. C., Anderson, C. L., Harrington, M. E., Prasad, K. N. 1979. The effects of organic and inorganic lead and mercury on neurotransmitter high-affinity transport and release mechanisms. *Environ. Res.* *19*:102-111.
6. Bornschein, R., Pearson, D., Reiter, L. 1980. Behavioral effects of moderate lead exposure in children and animal models. *CRC Crit. Rev. Toxicol.* *8*:43-152.
7. Bouldin, T. W., Krigman, M. R. 1975. Acute lead encephalopathy in the guinea pig. *Acta Neuropathol.* *33*:185-190.

8. Buchthal, F., Behse, F. 1979. Electrophysiology and nerve biopsy in men exposed to lead. *Br. J. Ind. Med. 36*:135-147.
9. Bull, R. J., Lutkenhoff, S. D., McCarty, G. E., Miller, R. G. 1979. Delay in the postnatal increase of cerebral cytochrome concentrations in lead-exposed rats. *Neuropharmacology 18*:93-102.
10. Burchfiel, J., Duffy, F., Bartels, P. H., Needleman, H. L. 1980. The combined discriminating power of quantitative electroencephalography and neuropsychologic measures in evaluating central nervous system effects of lead at low levels, in *Low Level Lead Exposure* (Needleman, H. L., ed.), Raven, New York pp. 75-90.
11. Burright, R. G., Donovick, P. J., Michels, K., Fanelli, R. J., Dolinsky, Z. 1982. Effect of amphetamine and cocaine on seizure in lead treated mice. *Pharmacol. Biochem. Behav. 16*:631-655.
12. Bushnell, P. J., Bowman, R. E. 1979. Reversal learning deficits in young monkeys exposed to lead. *Pharmacol. Biochem. Behav. 10*:733-742.
13. Byers, R. K., Lord, E. E. 1943. Late effects of lead poisoning on mental development. *Am. J. Dis. Child. 66*:471-494.
14. Carter, R. B., Leander, J. D. 1980. Chronic low-level lead exposure: Effects on schedule-controlled responding revealed by drug challenge. *Neurobehav. Toxicol. 2*:345-353.
15. Chisolm, J. J. 1971. Lead poisoning. *Sci. Am. 224*:15-23.
16. Chisolm, J. J., Silbergeld, E. K. 1981. The effects of lead exposure in young children on urinary excretion of homovanillic acid, in *Proc. Int. Symp. on Heavy Met.*, WHO-EEC-EPA, Amsterdam, pp. 546-549.
17. David, O., Clarke, J., Voeller, K. 1972. Lead and hyperactivity. *Lancet 2*:900-903.
18. de la Burde, B., Choate, M. S. 1972. Does asymptomatic lead exposure in children have latent sequelae? *J. Pediatr. 81*:1088-1091.
19. Dubas, T. C., Hrdina, P. D. 1978. Behavioral and neurochemical consequences of neonatal exposure to lead in rats. *J. Environ. Pathol. Toxicol. 2*:473-484.
20. Fejerman, N., Gimenez, A., Vallejo, N. E., Medina, C. S. 1973. Lennox's syndrome and lead intoxication. *Pediatrics 52*:227-234.
21. Feldman, R. G., Hayes, M. K., Younes, R., Aldrich, F. D. 1977. Lead neuropathy in adults and children. *Arch. Neurol. 34*:481-488.
22. Fernando, N. P., Healy, M. A., Aslam, M., Davis, S. S., Hussein, A. 1981. Lead poisoning and traditional practices: The consequences for world health. A study in Kuwait. *Public Health (London) 95*:250-260.
23. Flynn, J. C., Flynn, E. R., Patton, J. H. 1979. Effects of pre- and postnatal lead on affective behavior and learning in the rat. *Neurobehav. Toxicol. 1*:93-103.

24. Fox, D. A., Overmann, S. R., Woolley, D. E. 1978. Early lead exposure and ontogeny of seizure responses in the rat. *Toxicol. Appl. Pharmacol.* *45*:270-282.
25. Goldberg, A. M., Cohen, S. 1977. Does lead enhance its own toxicity? *Abstracts of the International Congress on Toxicology (Toronto)*, 11-14.
26. Goldstein, G. W., Asbury, A. K., Diamond, I. 1974. Pathogenesis of lead encephalopathy. *Arch. Neurol.* *31*:382-389.
27. Govoni, S., Memo, M., Lucchi, L., Spano, P. F., Trabucchi, M. 1980. Brain neurotransmitter systems and chronic lead intoxication. *Pharmacol. Res. Commun.* *12*:447-460.
28. Goyer, R. A., Cherian, M. G. 1979. Ascorbic acid and EDTA treatment of lead toxicity in rats. *Life Sci.* *24*:433-438.
29. Grandjean, P., Nielsen, T. 1979. Organolead compounds: Environmental health aspects. *Residue Rev.* *72*:98-148.
30. Holtzman, D., Shen Hse, J. 1976. Early effects of inorganic lead on immature rat brain mitochondrial respiration. *Pediatr. Res.* *10*:70-75.
31. Jason, K. M., Kellogg, C. K. 1980. Behavioral neurotoxicity of lead, in *Lead Toxicity* (Singhal, R., Thomas, J. A., eds.), Urban and Schwartzenberg, Baltimore, pp. 241-272.
32. Kim, C. S., O'Tauma, L. A., Cookson, S. L., Mann, J. D. 1980. The effects of lead poisoning on calcium transport by brain in 30-day old albino rabbits. *Toxicol. Appl. Pharmacol.* *52*:491-496.
33. Kober, T. E., Cooper, G. P. 1976. Lead competitively inhibits calcium-dependent synaptic transmission in the bullfrog sympathetic ganglion. *Nature* *262*:704-705.
34. Konat, G., Clausen, J. 1976. Triethyllead-induced hypomyelination in the developing rat forebrain. *Exp. Neurol.* *50*:124-133.
35. Kostas, J., McFarland, D. J., Drew, W. G. 1978. Lead-induced behavioral disorders in the rat: Effects of amphetamine. *Pharmacology* *16*:226-236.
36. Krigman, M. R., Bouldin, T. W., Mushak, P. 1980. Lead, in *Clinical and Experimental Neurotoxicology* (Spencer, P. S., Schaumburg, H. H., eds.), Williams and Wilkins, Baltimore, pp. 490-507
37. Kurlander, H. M., Patten, B. M. 1978. Metals in spinal cord tissue of patients dying of motor neuron disease. *Ann. Neurol.* *6*:21-24.
38. Landrigan, P., Baker, E., Whitworth, R., Feldman, R. G. 1980. Neuroepidemiological evaluation of children with chronic increased lead absorption, in *Health Effects of Lead at Low Dose* (Needleman, H. L., ed.), Raven, New York, pp. 17-34.
39. Legge, T. M. 1901. Industrial lead poisoning. *J. Hyg.* *1*:96-108.

40. Lin-fu, J. S. 1973. Vulnerability of children to lead exposure and toxicity. *N. Engl. J. Med. 289*:1229-1233.
41. Lorenzo, A. V., Gewirtz, M. 1977. Inhibition of ^{14}C-tryptophan transport into brain of lead-exposed neonatal rabbits. *Brain Res. 132*:386-392.
42. Louis-Ferdinand, R. T., Brown, D. R., Fiddler, S. F., Daughtrey, W. C., Klein, A. W. 1978. Morphometric and enzymatic effects of neonatal lead exposure in the rat brain. *Toxicol. Appl. Pharmacol. 43*:351-360.
43. Lucchi, L., Memo, M., Airaghi, M. L., Spano, P. F., Trabucchi, M. 1981. Chronic lead treatment induces in rat a specific and differential effect on dopamine receptors in different brain areas. *Brain Res. 213*:397-404.
44. Mahaffey, K. R., Annest, J. L., Roberts, J., Murphy, R. S. 1982. National estimates of blood lead levels: U.S. 1976-1980; association with selected demographic and socioeconomic factors. *N. Engl. J. Med. 307*:575-579.
45. Mahaffey, K., Michaelson, I. A. 1980. Interaction between lead and nutrition, in *Low Level Lead Exposure* (Needleman, H. L., ed.), Raven Press, New York, pp. 159-200.
46. Mantere, F., Hanninen, H., Hernberg, S. 1983. Subclinical neurotoxic effects of lead. *Neurobehav. Toxicol. Teratol. 4*:725-728.
47. Moore, M., Meredith, P. A., Goldberg, A. 1977. A retrospective analysis of blood lead in mentally retarded children. *Lancet 1*:717-719.
48. Nathanson, J. A., Bloom, F. E. 1976. Heavy metals and adenosine cyclic 3', 5'-monophosphate metabolism: Possible relevance to heavy metal toxicity. *Mol. Pharmacol. 12*:390-398.
49. National Academy of Sciences. 1980. *Lead in the Human Environment*, National Academy of Sciences, Washington, D.C.
50. Needleman, H. L (ed.). 1980. *Low Level Lead Exposure: The Clinical Implications of Current Research*, Raven Press, New York.
51. Needleman, H. L., Gunnoe, C., Leviton, A., Reed, R., Peresis, H., Maker, C., Barrett P. 1979. Deficits in psychological and classroom performance of children with elevated lead levels. *N. Engl. J. Med. 300*:689-695.
52. Needleman, H. L., Landrigan, P. L. 1981. The health effects of low level exposure to lead. *Annu. Rev. Public Health 2*:277-298.
53. Oskarsson, A., Squibb, K. S., Fowler, B. A. 1982. Intracellular binding of lead in the kidney. *Biochem. Biophys. Res. Commun. 104*:290-298.
54. Otto, D. A., Benignus, V. A., Muller, K. E., Barton, C. N. 1981. Effects of age and body lead burden on CNS function in young children. I: Slow cortical potentials. *Electroencephal. Clin. Neurophysiol. 52*:229-239.

55. Petit, T. L., LeBoutillier, J. C. 1979. Effects of lead exposure during development on neocortical dendritic and synaptic structure. *Exp. Neurol. 64*:482-492.
56. Piomelli, S., Seaman, C., Zullow, D., Curran, A., Davidow, B. 1982. Threshold for lead damage to heme synthesis in urban children. *Proc. Nat. Acad. Sci. U.S.A. 79*:3335-3339.
57. Pounds, J. G., Wright, R., Morrison, D., Casciano, D. A. 1982. Effect of lead on calcium homeostasis in the isolated rat hepatocyte. *Toxicol. Appl. Pharmacol. 63*:389-401.
58. Purdy, S. E., Blair, J. A., Leeming, R. J., Hilburn, M. E. 1981. Effect of lead on tetrahydrobiopterin synthesis and salvage: A cause of neurological dysfunction. *Int. J. Environ. Stud. 17*: 141-145.
59. Ratcliffe, J. 1981. *Lead in Man and the Environment*, Ellis Horwood, London.
60. Repko, J. D., Morgan, B. B., Nicholson, J. A. 1974. Final report on the behavioral effects of occupational exposure to lead. NIOSH, Cincinnati, Technical Report no. ITR-74-27.
61. Reyners, H., Granfelici de Reyners, E., Vander Parren, J., Maisin, J. R. 1978. Evolution de l'équilibre des populations gliales dans le cortex cerebral du rat intoxiqué au plomb. *C. R. Soc. Biol. 172*:998-1002.
62. Roels, H. A., Balis-Jacques, M. N., Bucket, J. P., Lauwerys, R. 1979. The influence of sex and of chelation therapy on erythrocyte protophyrin and urinary ALA in lead-exposed workers. *J. Occup. Med. 21*:527-539.
63. Romano, C., Grossi-Bianchi, M. L., Tortorolo, G., Verde, J. 1960. L'encefalopatia cronica da piombo. *Minerva Pediatr. 18*: 416-425.
64. Rutter, M. 1980. Raised lead levels and impaired cognitive/ behavioral function. *Dev. Med. Child Neurol. Suppl. 42*:1-36.
65. Sassa, S., Whetsell, W. J., Kappas, A. 1979. Studies on porphyrin-heme biosynthesis in organotypic cultures of chick dorsal root ganglia. II. The effect of lead. *Environ. Res. 20*: 513-525.
66. Schain, R. J. 1980. Medical and neurological differential diagnosis, in *Handbook of Minimal Brain Dysfunctions* (Rie, H. E., Rie, S. D., eds.), Wiley, New York, pp. 338-406.
67. Seppalainen, A., Hernberg, S. 1972. A sensitive technique for detecting subclinical lead neuropathy. *Br. J. Ind. Med. 29*: 443-449.
68. Shih, T. M., Hanin, I. 1978. Chronic lead exposure in immature animals: Neurochemical correlates. *Life Sci. 23*:877-888.
69. Siegel, G. J., Fogt, S. K. 1977. Inhibition by lead ions of electrophorous electroplax Na+K-adenosine triphosphatase and K^+-p-nitrophenylphosphatase. *J. Biol. Chem. 252*:5201-5205.

70. Silbergeld, E. K. 1977. Interactions of lead and calcium on the synaptosomal uptake of dopamine and choline. *Life Sci. 20*:309-318.
71. Silbergeld, E. K. 1977. The neuropharmacology of hyperkinesis, in *Current Developments in Psychopharmacology* (Essman, W., Valzelli, L., eds.), Spectrum, New York, pp. 181-214.
72. Silbergeld, E. K. 1982. Neurochemical and ionic mechanism of lead neurotoxicity, in *Mechanisms of Actions of Neurotoxic Substances* (Prasad, K. N., Vernadakis, A., eds.), Raven Press, New York, pp. 1-25.
73. Silbergeld, E. K., Hruska, R. E., Bradley, D., Lamon, J., Frykholm, B. C. 1982. Neurotoxic aspects of porphyrinopathies: lead and succinylacetone. *Environ. Res. 29*:459-471.
74. Silbergeld, E. K. 1983. Effects of lead on reproduction: Review of experimental studies, in *Lead versus Health* (Rutter, M., Russell-Jones, R., eds.), Wiley, London, pp. 217-227.
75. Silbergeld, E. K. 1983. Localization of lead: implications for mechanism of neurotoxicology. *Neurotoxicology 4*:193-200.
76. Silbergeld, E. K., Adler, H. S. 1978. Subcellular mechanisms of lead neurotoxicity. *Brain Res. 148*:451-467.
77. Silbergeld, E. K., Adler, H. S., Costa, J. 1977. Subcellular localization of lead in synaptosomes. *Res. Commun. Chem. Pathol. Pharmacol. 17*:715-725.
78. Silbergeld, E. K., Chisolm, J. J. 1976. Lead poisoning: Altered urinary catecholamine metabolites as indicators of intoxication in mice and children. *Science 192*:153-155.
79. Silbergeld, E. K., Goldberg, A. M. 1975. Pharmacological and neurochemical investigations of lead-induced hyperactivity. *Neuropharmacology 14*:431-444.
80. Silbergeld, E. K., Goldberg, A. M. 1979. Problems in experimental studies of lead poisoning, in *Lead Toxicity* (Singhal, R., Thomas, J. A., eds.), Urban and Schwarzenberg, Baltimore, pp. 19-42.
81. Silbergeld, E. K., Hruska, R. E. 1980. Neurochemical investigations of low level lead exposure, in *Low Level Lead Exposure* (Needleman, H. L., ed.), Raven Press, New York, pp. 135-157.
82. Silbergeld, E. K., Hruska, R. E., Miller, L. P., Eng, N. 1980. Effects of lead in vivo and in vitro in GABAergic neurochemistry. *J. Neurochem. 34*:1712-1718.
83. Silbergeld, E. K., Lamon, J. M. 1980. The role of altered heme synthesis in the neurotoxicity of lead. *J. Occup. Med. 22*:680-684.
84. Silbergeld, E. K., Miller, L. P., Kennedy, S., Eng. N. 1979. Lead, GABA and seizures: Effects of subencephalopathic lead exposure on seizure sensitivity and GABAergic function. *Environ. Res. 18*:371-382.

85. Singhal, R., Thomas, J. A. (eds.). 1980. *Lead Toxicity*, Urban and Schwartzenberg, Baltimore.
86. Sobotka, T. J., Cook, M. P. 1974. Postnatal lead acetate exposure in rats: Possible relationship to minimal brain dysfunction. *Am. J. Ment. Defic.* *79*:5-9.
87. Suketa, Y., Ujiie, M., Okada, S. 1982. Alteration of Na and K mobilization and of adrenal function by long-term ingestion of lead. *Biochem. Pharmacol.* *31*:2913-2919.
88. Taylor, D., Nathanson, J., Hoffer, B., Olson, L., Sieger, A. 1978. Lead blockade of norepinephrine-induced inhibition of cerebellar Purkinje neurons. *J. Pharmacol. Exp. Ther.* *206*:371-381.
89. Thatcher, R. W., Lester, M. L., McAlaster, R., Horst, R. 1982. Effects of low levels of cadmium and lead on cognitive functioning in children. *Arch. Environ. Health* *37*:159-166.
90. Valciukas, J. A., Lilis, R., Eisinger, J., Blumberg, W. E., Fischbein, A., Selikoff, J. 1978. Behavioral indices of lead neurotoxicity: Results of a clinical field survey. *Int. Arch. Occup. Environ. Health* *41*:217-236.
91. Vickers, G. R., Dowdall, M. J. 1976. Calcium uptake in preterminal central synapses: Importance of mitochondria. *Exp. Brain Res.* *25*:429-445.
92. Weiss, B. 1980. Conceptual issues in the assessment of lead toxicity, in *Low Level Lead Exposure: The Clinical Implications of Current Research* (Needleman, H. L., ed.), Raven Press, New York, pp. 127-134.
93. Whitfield, C. L., Ch'ien, L. T., Whitehead, J. D. 1972. Lead encephalopathy in adults. *Am. J. Med.* *52*:289-298.
94. Wilson, W. E. 1982. Dopamine sensitive adenylate cyclase inactivation by organolead compounds. *Neurotoxicology* *3*:100-107.
95. Wince, L. C., Donovan, C. A., Azzaro, A. J. 1980. Alterations in the biochemical properties of central dopamine synapses following chronic postnatal $PbCO_3$ exposure. *J. Pharmacol. Exp. Ther.* *214*:642-650.
96. Windebank, A., McCall, J. T., Hunder, H. E., Dyck, P. J. 1980. The endoneurial content of lead related to the onset and severity of segmental demyelinization. *J. Neuropathol. Exp. Neurol.* *39*:692-699.
97. Winneke, G., Brockhaus, A., Baltissen, R. 1977. Neurobehavioral and systemic effects of long term blood lead elevation in rats. *Arch. Toxicol.* *37*:247-263.
98. Zenick, H., Rodriguez, W., Ward, J., Elkington, B. 1978. Deficits in fixed-interval performance following prenatal and postnatal lead exposure. *Dev. Psychobiol.* *12*:509-514.

15

Neurotoxicity of Elemental Mercury: Occupational Aspects

Vito Foà
Institute of Occupational Health, Clinica L. Devoto, University of Milan, Milan, Italy

I. INTRODUCTION

The industrial use of mercury right from ancient times led to descriptions of the clinical aspects of intoxication as long as 20 centuries ago: Plinius described clinical aspects in goldsmiths using mercury, and Galen identified the intoxication that caused slow death in the miners of the mercury mines in Almaden in Spain. But is was not until the seventeenth century that Ellenborg and Paracelsus recognized mercury intoxication as a typical occupational disease, and it was Ramazzini who wrote the first systematic treatise on the classic clinical picture due to excessive mercury absorption in the already large number of categories of artisans exposed.

II. PRODUCTION, USE, AND CONSUMPTION OF MERCURY

Mercury and its inorganic compounds are involved in a vast number of industrial processes (Table 1). In the medical field, only the amalgam used for dental fillings and certain skin disinfectants contain mercury.

Table 1 Occupations Involving Mercury Exposure

Mining, distillation, and processing of mercury
Production of mercury compounds and amalgams
Mining and processing of gold and silver ores
Production of filament lamps, lighting tubes, fluorescent tubes, transmitting and receiving lamps, tubes and bulbs for radio and television equipment, x-ray emitters
Production, maintenance, and repair of current rectifiers, mercury vapor lamps, and mercury electrical conductors
Production, maintenance, and repair of thermometers, barometers, gauges, and other measuring instruments and laboratory instruments
Production and use of mercury pumps and mercury blower engines
Use of mercury and its compounds as catalyzers in the chemical industry
Electrolysis with mercury cathodes
Production and packaging of pharmaceuticals and phytopharmaceuticals containing mercury or its compounds
Production and use of mercury-based dyes, varnishes, paints, and glues (ceramica, ship keels)
Gold-plating (especially for lighting conductors), silver-plating, tin-plating, sweeping, and damasking using mercury or its compounds
Preparation of raw materials for the hat industry
Work on felt; treatment and preservation of animal furs
Use of amalgams in dental surgeries
Work in photographic laboratories and in clinicochemical and histological analysis laboratories
Production and use of detonators containing mercury fulminate (explosives)
Preservation of wood by "chromization" (telegraph poles and railway cross-ties)
Agriculture: disinfection and conservation of seeds, soil treatment; use of bactericides, fungicides, and pesticides containing mercury

Until the 1950s, severe cases of occupational mercury intoxication occurred mostly in the hat industry, where mercury was used in solution with nitric acid in the felt hat process (39,60). At present, however, the main industrial use of mercury is in chloralkali production by mercury cathode electrolytic processes.

Mercury is also widely used in the production of electrical equipment including under this heading a wide range of applications, such as in mercury vapor lamps for industrial and street illumination. Other uses include silent electric switches for the home and mercury batteries which are long lasting and can be used at high temperature and humidity.

The use of mercury as a fungicide and general pesticide should no longer pose any problem, at least in countries where all organic mercury compounds have been banned for use in agriculture.

In 1972 the United States banned organic mercury compounds in the paper industry (75), where they had been used to prevent mold developing on the wet paste during processing or storage.

III. METABOLISM

Absorption, distribution, and excretion of mercury vary considerably according to its chemical form (11). In addition to the elemental state (Hg^0), mercury also exists in the +1 and +2 oxidation states. Mercury (II) forms salts such as nitrates, sulfates, and chlorides, but is also the basis of an important class of organometallic compounds, the toxicity of which will be dealt with in other chapters of this volume.

It is usual to classify mercury compounds as "organic" or "inorganic." The inorganic compounds include metallic mercury, monovalent and bivalent mercury salts, and complexes in which bivalent mercury binds reversibly to the thiol groups and proteins in tissues. The compounds in which mercury is combined directly to a carbon atom with a covalent bond are classified as organic mercury compounds. The distinction between "organic" and "inorganic" has mainly chemical value, rather than any toxicological basis, because the toxicological properties of elemental mercury vapors are different from those of inorganic mercury salts, and, moreover, the alkyl mercurials are dramatically different from the other organic mercury compounds regarding their adverse effects. In fact, elemental mercury vapors and the short-chain alkyl mercurials are the forms which have posed the greatest risk for human health.

A. Absorption and Transport

Metallic mercury (Hg^0) is the form most widely involved in occupational exposure. Mercury vapors and dusts or aerosols of inorganic mercury are absorbed via the respiratory tract in amounts or at sites dependent upon their particle size and solubility in biological fluids.

Absorption by inhalation is a highly efficient process. Equilibrium between mercury in air and in plasma is reached within a very short time from beginning of exposure, and 20 hr after cessation of exposure the lung contains practically no more metal.

The percentage of lung retention (subjects exposed to concentrations of 100 $\mu g/m^3$ of labeled mercury) varies from 74-76%, when inspiration is through the nose, to 50% when inspiration and expiration is through the mouth; the retention percentage remains constant over time even if exposure continues (40).

In occupational exposure, absorption by ingestion may be considered only as an accidental event. Nevertheless, if the temperature of the inhaled vapors is higher than the body temperature, they may condense on the mucosa of the mouth and upper respiratory tract and then be swallowed. When elemental mercury is ingested, the oxidation process in the intestinal tract is usually too slow to be completed before the mercury is eliminated with the feces (21,22). Absorption of inorganic mercury and its salts occurs to some extent in the intestinal tract (20). In recent studies using ligated gastrointestinal segments, the duodenum was shown to be the major absorption site of Hg^0 and $HgCl_2$. The alkalinity of bile promotes the absorption of mercury from the alimentary tract (27).

The importance of skin as a route of entry of metallic mercury is unclear. Studies in animals indicate that inorganic salts of mercury can cross the skin barrier (22).

After absorption, a fraction of 50% or more of inorganic mercury is transported in the plasma preferentially bound to the albumin, while about 90% of the organic compounds is transported inside the red blood cells. The distribution pattern is determined by several factors, including the relative density of sulfhydryl groups, the affinity of mercury for cellular components, and the rates of association and dissociation of mercury-protein complexes.

B. Distribution and Biotransformation

The metabolic patterns of mercurials have been reviewed in a recent paper by Mushak (60). When mercury vapors are inhaled, a catalase-dependent oxidation of elemental mercury to mercuric ion occurs in erythrocytes and in various organ tissues. A fraction of the absorbed mercury is also present in blood in a nonionized state (52,53). The diffusion of the latter portion into the tissues is facilitated by factors such as liposolubility and lack of electrical charge which also allow the passage of the elemental mercury across the blood-brain barrier. Within brain tissue elemental mercury is oxidized to mercury ion in a coupled oxidation reaction involving the catalase-hydrogen peroxide complex I. The brain concentrations of mercury are 10 times higher following exposure to elemental mercury vapors than after absorption of identical quantities of ionized mercury (62).

Chang et al. have demonstrated that even minute amounts of mercury ions (<1.0 ppm) are capable of impairing within hours the blood-brain barrier function, leading to extravasation of normally barred plasma solutes (18,82). It has been proposed (63) that mercury ions cause damage to the membrane structure by forming cross-linkages with the protein moiety of the cell membrane, which in turn impair the selective permeability membrane functions ("leaky membrane" phenomenon). Electron-microscopic and histochemical studies have demonstrated mercury deposits upon biological membranes after mercury intoxication (15), and the early damage by mercury ion seems to occur at the level of endothelial and glial components (82).

Since the blood-brain barrier is an active site for regulation of the metabolite, uptake processes from blood to the nervous system (10), the disruption of the blood-brain barrier and the inhibition of certain associated enzymes may account for the reduced cerebral uptake of amino acids and other metabolites after mercury intoxication (14).

Studies on intracerebral distribution of the elemental mercury in rats and mice revealed higher mercury concentrations in the gray matter than in the white matter, the maximum levels being reached in areas of cerebellum, spinal cord, medulla, pons, and mid-brain. In the cerebellum, selective localization was noted in the Purkinje cells and in neurons of the dentate nucleus. Moreover, there is histochemical evidence (16) that Hg^{2+} appears first in the neurons of the dorsal root ganglia, then in the Purkinje cells, and finally in the elements of the ventral horns, calcarine cortex, and granular layer of the cerebellum. These latter cells, however, are those exhibiting the highest susceptibility to mercury toxicity (13,14,17).

There is also adequate evidence of mercury tropism for the peripheral nervous system. Mercury has been found in the peripheral motor and sensory fibers, as well as in the central neurons (16,17). Within the nerve fibers mercury was predominantly found in myelin sheaths, mitochondria, as well as in axonal components, including neurofilaments.

Mercury ion is very rapidly distributed and accumulated in other organs and tissues, including kidney, liver, myocardium, mucosa of the digestive and respiratory tracts, testis, skin, bone marrow, and placenta. Conspicuous accumulation occurs in the renal cortex. The biological half-life of inorganic mercury in the kidney (64 days) is longer than that in the whole body (58 days), but considerably shorter than the half-life in the brain (as long as 1 year) (5,6,77).

It has not yet been possible to establish any definite relationship between the concentration of inorganic mercury in the whole body or in individual organs and the concentration in biological fluids such as blood and urine. On the other hand, significant correlation was demonstrated between extent of exposure (product of duration by concentration in the air) and the quantity of mercury in the occipital cortex of monkeys exposed to mercury vapors (7).

C. Excretion

The elimination of mercury occurs by different routes, including kidney, bile, intestinal mucosa, sudorific and salivary glands, and skin (65). The kidney acts as a multicompartmental model in which at least one compartment is characterized by a high retention time (57). The physiological mechanisms involved in the renal excretion of mercury are still unclear. Generally, the rate of excretion becomes maximal after the kidney has accumulated a certain quantity of mercury. In intermittent exposure (as in the case in many occupational exposures), this latency mechanism can give rise to the appearance of an excretion peak during periods of nonexposure.

The persistence of urinary elimination of mercury even for considerably long periods of time after cessation of exposure is consistent with the hypothesis that the metal is incorporated irreversibly into cell components and that delayed excretion occurs at rates depending on the metabolic turnover of proteins.

Excretion of mercury with the feces represents a significant fraction of total body elimination, at least after exposure to moderate doses (67). The other nonrenal routes of elimination have not been sufficiently studied. In three workers at a chloroalkali plant (50) the concentration of mercury in the sweat was found to be 50-200% higher than that in the urine. This finding assumes significance if one considers the high temperatures (with consequent excessive sweating) usually present in the environment of chloroalkali plants, which are not built in the open.

IV. CLINICAL ASPECTS OF ABNORMAL MERCURY ABSORPTION

A. Acute Intoxication

Acute occupational intoxication by elemental mercury or its inorganic compounds is rather rare, but can occur following inhalation of large quantities of mercury vapor which may develop in accidental situations (fires, breakdowns in extraction furnaces, explosion of mercury fulminate, breakage of mercury vapor lamps, explosion of pipes or receptacles containing heated mercury), or by accidental ingestion of mercury salts.

Cases of severe lung lesions have been described following exposure to elevated concentrations of mercury vapor. The clinical picture of acute intoxication shows acute anuric tubular nephritis similar to other toxic tubular diseases. The nervous system is not involved except for the symptomatology accompanying the terminal picture of anuric nephropathy. However, in a case of acute exposure following an accident, in addition to the mainly pulmonary symptoms, signs of irritability and a permanent decrease in libido were observed (55).

B. Chronic Intoxication

The clinical picture of chronic intoxication has without doubt received the most attention in view of the number of cases reported. Multiple and variable individual responses have been observed at the same mercury exposure conditions, raising the question of the existence of an individual "susceptibility" to the metal (24,76,81). This individual susceptibility has been related to life-style factors such as personal hygiene, smoking (39), diet (9), and alcohol consumption (39), which may alter mercury metabolism by affecting absorption, cellular conversion of Hg^0 to Hg^{2+}, distribution in tissues, and the rate of excretion (58,64). Moreover, complex interactions may occur between mercury and certain trace elements such as cobalt and selenium, resulting in marked changes in physiological effects and toxicity (44).

The earliest and most typical symptoms in chronic mercury poisoning are personality changes denominated "psychic erethism" (39). Irritability, anxiety, insomnia, hyperreactivity, shyness, and emotional instability, sometimes with excitation in response to unmotivated stimuli, have been described. In the most severe cases there is considerable decrease in attention and subsequently in memory, eventually leading to a general depersonalization.

The insidious change in mood and character, which the subject himself does not perceive at first, may cause increasing difficulties in social relationships, revealed by intolerance toward any criticism, quarrelsomeness, impressionability, tendency to melancholy, and signs of hypochondria. Psychodiagnostic tests showed a positive correlation between certain psychological parameters, such as neuroticism and shyness in the sense of introversion, and length of exposure in apparently healthy workers exposed to mercury (7,30).

In the absence of significant clinical signs, the finding of a dysthymic personality involution may provide useful information as far as the degree of adverse effect of mercury in exposed workers is concerned (29,30,31,46,59,81).

The value of electroencephalographic examination is controversial. Electroencephalographic abnormalities were described in workers with manifest signs of mercury intoxication. However, in working environments with mercury concentrations below the threshold limit value (TLV) (0.05 mg/m^3), no abnormalities were observed that were substantially different from those found in the general population (36). In a population of Polish workers nonspecific encephalographic alterations were observed in 44.5% of the subjects exposed to concentrations of mercury between 0.02 and 0.05 mg/m^3, compared to 15.3% observed in the controls (71).

Tremor is the most conspicuous clinical sign, and has been observed in all reported cases of overt mercury intoxication. There was a 68% frequency of tremor in the mercury-exposed workers of five Italian hat factories (2). The tremor is static and intentional, and is seriously aggravated by emotion; it is initially slight and unnoticeable, but becomes evident when intricate movements involving the hands, such as threading a needle, writing, and doing up buttons on clothing, are hindered. Tremor generally starts at the corners of the lips. At rest, fine tremors can be observed involving the eyelids, tongue, and extremities which are clearly accentuated by fatigue, emotion, and alcohol. Later tremor becomes gross and is alternated by violent tonal and clonal spasms, which may wake the patient from sleep during the night and even cause him to fall out of bed. When the upper and lower limbs are involved, tremor can hinder normal habitual movements like eating, dressing, and walking (80). Tremor is the main symptom of a series including asynergy, adiadochokinesia, and nystagmus. The voice becomes monotonous and faltering, with changes in speech, stammering, and difficulty in pronounciation, especially of sibilants (mercurial psellism).

In the more severe cases there are also alterations of basal nuclei with rigidity and chorea, myoclonus, and "facies parkinsoniana." Painful contractions are now only rarely seen, and the flaccid paralyses observed in the past are even rarer. The following classification of tremor has been proposed (2):

1. Slight static tremor with arms outstretched that does not interfere with muscular activity
2. Static tremor associated with early intentional tremor influencing more intricate muscular movements
3. Tremor that seriously disturbs actions requiring particular manual dexterity, such as writing, shaving, and drinking
4. Tremor that makes wide movements difficult
5. Intense generalized tremor.

The neurological picture of chronic inorganic mercury intoxication is consistent with the previously mentioned involvement of areas and structures of the brain and cerebellum which physiologically control muscular contractions and movements. Similarly, the action of mercury on the temporal lobe of the brain could explain not only loss of memory, but also the psychological and character changes observed in intoxicated workers.

Cases of polyneuritis have been described in mercury-intoxicated workers. In addition to significant electroneurographic evidence of motor conduction velocity impairment in workers (37,81), clinical findings reflecting alterations in both sensory and motor nerve fibers have been reported (47,81). Distal paresthesias of the limbs with alterations in reflexes, cramps, and fasciculation have been described in workers making thermometers. These findings have been explained as the result of the toxic effect of mercury to the anterior horn motor neurons with axonal degeneration. In workers with past exposure to mercury, nerve fibers were degenerating even in the absence of the metal, indicating that adverse effects may also appear after cessation of exposure (16,17).

In one study, there was a prevalence and early appearance of sensory alterations in less-exposed workers, while motor alterations were prominent in workers with a longer period of exposure (37). These data might be explained by an early selective damage to the sensory nerve fibers that could be transitory and eventually reversible; however, it is also conceivable that these clinical manifestations are expressions of different stages of the same toxic process involving both the sensory and motor systems. In fact, sensory disorders in the toxic peripheral neuropathies generally precede the motor deficit (see Cavanagh in this volume).

Cochlear and vestibular lesions (84) have also been reported.

V. DOSE-RESPONSE AND DOSE-EFFECT RELATIONSHIPS

Determination of the dose-response and dose-effect relationships is of utmost importance for the establishment of metal concentrations that can be tolerated by the human body (34,61,68). In practice, it is not possible to establish the "dose" defined as the quantity or the concentration of a particular compound in the critical organ,* or, anyway, at the site where its presence leads to a definite effect (7,61). Therefore dose is evaluated indirectly by means of "external" parameters, such as the concentration in the air and duration of exposure and by biological parameters obtained by measuring the amount of metal in the accessible biological compartments, such as blood, urine, feces, sweat, and hair. However, it is uncertain to what extent levels of

---

*The "critical organ" is defined as that particular organ in which the critical concentration of a metal is first attained under specific circumstances of exposure and for a given population; the "critical organ concentration" is defined as the mean concentration in the organ at the time any of its cells reaches the "critical concentration," defined as the concentration at which undesirable functional changes, reversible or irreversible, occur in the cell (78).

mercury in air reflect the actual exposure and how much recent exposure and the body burden of mercury from earlier exposure influence mercury concentrations in blood and urine.

Attempts have been made to evaluate the early central nervous system changes in persons chronically exposed to inorganic mercury in relation to "dose" indices (72). Significant correlation was found (59, 66) between levels of mercury in blood and urine, and occurrence of abnormalities in neuromuscular or psychomotor parameters, such as increased frequency of forearm tremor, increased bandwidth of the surface electromyogram, decreased speed of hand- and foot-tapping tests with a parallel tendency to increased response times, and moderate impairment in ability to consistently move the hand quickly to a discrete position (eye-hand coordination tests). All these findings were noted at mercury levels between 10 and 20 μg/liter of blood and were always reversible at mercury values below 60 μg/liter of blood.

The above results were substantiated in two other investigations. In the first study (31), a correlation was found between the results of mechanical and visual memory tests, psychomotor ability tests, and the number of times per year, for the previous 10 years, that the urinary concentrations of mercury had exceeded 100 μg/liter. The second study (46) found a correlation between the results of psychometric tests and the number of urinary mercury peaks above 500 μg/liter observed in the previous 3, 6, and 12 months.

An important aspect of the latter investigations is that both the populations studied worked in environmental conditions that were on the whole acceptable, and no single worker showed clinical signs of intoxication.

Mental dynamism and reaction speed also seem to be influenced by exposure to low levels of mercury. Table 2 reports the significant correlations obtained between the results of mental dynamism tests (Raven, block design), visual memory tests (Benton), psychomotor ability test (digit symbol), and reactions times, and the urinary mercury values in a population of workers employed on mining and the transformation of mercury (12).

There are very few data on mercury concentrations in the human brain following exposure to mercury vapors. Concentrations of 6-9 mg/kg were found in the brains of two persons who, several years before death, had shown a complete clinical picture of mercury intoxication. Mercury concentrations in the brain exceeding 0.5 μg/kg are unlikely to occur in man in the absence of known mercury exposure (8).

Watanabe (83) reported mercury concentrations in various tissues and brain areas from the autopsy of two miners who had died 10 years after cessation of exposure to high concentrations of mercury vapor. Relatively high concentrations were found in the occipital and parietal

Table 2 Significant Correlations Obtained Between Test Variables and Urinary Mercury Levels (n = 33)

	r	t	p
Raven PM 38	-0.49	3.14	.01
Block design	-0.48	3.03	.01
Benton VRT (correct reproduction)	-0.47	2.98	.01
Benton VRT (errors)	-0.57	3.87	.001
Digit symbol	-0.36	2.16	.05
Simple RT × (visual signals)	0.48	3.03	.01
Simple RT δ (visual signals)	0.62	5.56	.001
Simple RT δ (auditory signals)	0.39	2.36	.05
Choice RT δ	0.47	2.98	.01

PM, Progressive matrices; VRT, visual retention test; RT, retention times.

cortex and in substantia nigra. High concentrations were also found in the cerebellum. The thalamus and pallidum showed mercury concentrations similar to those found in the kidneys.

VI. ENVIRONMENTAL SURVEILLANCE

Sampling and analysis of mercury in the air are particularly complex owing to the numerous chemical species of mercury that may be present in the working environment. The possibility of interconversion of the various chemical forms further complicates sampling and analysis.

Several attempts have been made to develop models, instruments, and sampling techniques suitable for personal monitoring of the individual TLV (3). A personal dosimeter with a piezoelectric crystal has been developed which permits measurement of total integrated exposure to mercury vapors (68). A number of studies have been made to establish correlations between environmental concentrations of mercury and the incidence and severity of the signs and symptoms of intoxication (2,38,73).

Following the studies made by Smith et al. (73) and a review of the previous literature, in 1971 the American Conference of Governmental Industrial Hygienists (25) proposed lowering the TLV for elemental and inorganic mercury and nonalkyl compounds from 0.10 to 0.05 mg/m^3 for exposures of 8 working hours per day for 5 days per week during the whole of the working life. The American National Standards Institute (1) also adopted the same value in 1972, followed by other countries, like Italy in 1975 (74). In the USSR a value of 0.01 mg/m^3 has been in force for over 30 years (51).

Regarding the possibility of contamination of the food chain, the joint Food and Agriculture Organization/World Health Organization Expert Committee on food additives has temporarily fixed the permissible weekly dose of total mercury at 0.3 mg per person, of which not more than 0.2 mg is to be in the form of methyl mercury. This latter portion is equivalent to a value of 0.03 MeHg per 70 kg per day, which is obtained by multiplying the minimum concentration required to cause appearance of symptoms in more highly susceptible subjects (0.3 mg MeHg per 70 kg per day) (4,35) by 10^{-1}, which is the correction factor due to individual variability (23). As a percentage of probability of intoxication, this limit corresponds to 0.02% risk.

For organic compounds, the International Committee (41) proposed a level of 0.01 mg Hg/m^3 of methyl and ethyl mercury salts for an exposure of 8 consecutive hours per day for the whole of the working life, which corresponds to a concentration of 10 μg Hg/100 ml blood.

VII. BIOLOGICAL MONITORING OF EXPOSURE TO ELEMENTAL AND INORGANIC MERCURY

A. Mercury in Urine

Many authors have stressed the importance and usefulness of this biological parameter to evaluate mercury exposure and absorption (26,33,42). Nevertheless, in spite of all the attempts in this direction, it is now generally accepted that urinary mercury is not a reliable indicator of internal dose. In fact, there is marked individual variability in occupationally exposed groups. Moreover, urinary excretion fluctuates very widely over periods of 24 hr (65) and a week (53), with three- and even five-fold variations.

The correlation between daily exposure (evaluated on the environmental mercury concentration) and mercury in urine (and also in blood) on the same day is generally rather poor and in some cases even nonexistent (48), which proves that the concentration of mercury in urine and blood at the end of the work shift does not reflect the exposure that has occurred on that day.

The measurement of urinary mercury can, however, be used as an index of "exposure" for a group of exposed workers. The ratio between mean values of urinary mercury (in micrograms per liter) in exposed workers and mercury in the air (in milligrams per cubic meter) was previously evaluated at about 2 (73), but has now been recalculated to 1 (3,56). Therefore at 50 μg/m^3 Hg (TLV for working environments), the corresponding mean urinary excretion, evaluated on all exposed workers, is 50 μg/liter.

The mean levels of mercury in the urine in nonoccupationally exposed populations have been evaluated at less than 10 μg/liter (19,70, 79). No permissible levels for urinary mercury in exposed workers have, however, been established. A permissible limit of between 200 and 250 μg/liter has been proposed (26), based on the reported frequency of appearance of clinical signs of intoxication in relation to the urinary mercury values. However, using the value of 1 as a mean ratio between urinary and atmospheric mercury, the excretion in a group of workers exposed to an atmospheric concentration equal to the TLV (50 μg/m^3) should be closer to 50 μg/liter than to 100 μg/liter. Therefore it has been proposed (49) that in a group of exposed workers, a level of 50 μg/liter, that is, the concentration which, on the basis of the above, corresponds to an exposure of about the TLV, should be used as a comparative index for evaluation of the efficacy of the environmental surveillance measures.

B. Mercury in Whole Blood

A correlation between the blood concentrations of mercury and the incidence of subjective symptoms was found in exposed groups (65). However, the individual correlation was very low.

Again with the exposed group considered as a whole, blood mercury was used as an index of "exposure" (38,73). In this way a correlation was determined between blood and urinary levels of mercury, with a ratio calculated at about 0.3 (34,73). The mean levels of mercury in the blood in nonexposed subjects were calculated to be lower than 1.0 μg/100 ml (19,79). No permissible blood levels for mercury have been established. A value of 10 μg/100 ml has been proposed as a "health evaluation action level" (59): If the level of 10 μg/100 ml were exceeded, the person would be routinely examined to determine if any alteration in his health status had taken place. The outcome of these functionally based health evaluations would assist in determining if the individual should be removed from exposure or retested at frequent intervals. The author stresses, however, that the basis for choosing between these alternatives should be the subject of further investigations.

With a 0.3 ratio between blood and urinary levels of mercury, a value of 10 μg Hg/100 ml blood corresponds to about 35 μg Hg/liter urine. Therefore, in a group of exposed workers, the value of 10 μg Hg/100 ml blood, together with the measurement of urinary mercury, can be used as the basis for establishing that exposure is lower than 50 μg Hg/m^3 air.

Studies have been performed with labeled mercury vapor (48). When extrapolated to stable exposure conditions, the results indicated that about half the mercury in the blood (and also in the urine) reflects recent exposure (last 1 or 2 weeks), and that the other half reflects exposure that occurred in the previous 5-6 months.

Therefore, it is advisable to wait until the concentrations of mercury measurable in blood and urine of exposed workers can be significantly correlated with the atmospheric levels of mercury, but not until the worker has been exposed for at least 6 months or more.

C. Biochemical Indicators of Effect

On the basis of experimental data, a number of biochemical indicators has been extensively studied in occupationally exposed populations. However, at the present time, no biochemical tests can be identified which permit biological monitoring of the exposed individual, especially in relation to possible lesions occurring at early stages when toxic changes may be still reversible.

Epidemiological research on groups of exposed workers has shown that significant correlations exist between mercury exposure and inhibition of erythrocyte cholinesterase. However, the role played by the erythrocyte cholinesterase in the toxicity of mercury is controversial (28).

There has also been evidence of inhibition of membrane Na^+-K^+ adenosinetriphosphatase of erythrocytes, with increase in blood and urinary lactic dehydrogenase following exposure to elemental mercury (69), but the wide dispersion of individual results precludes using the measurement of these enzymes as individual indices of early effects.

In this respect, interest is aroused by clinical studies which, in agreement with the results of animal investigations (32), have indicated preferential accumulation of mercury at the lysosomal level. In fact, significant increase in the activity of some plasma acid hydrolases was found in subjects exposed to mercury vapors, even to concentrations below 50 $\mu g/m^3$ (28).

These data, although preliminary, enable the search for individual indices of "early reversible effect" of mercury to be continued.

VIII. TREATMENT

The most common form of treatment of mercury intoxication is to move the individual away from the site and source of exposure; however, although such action is correct and necessary, it is not always efficacious (5,63). The need for active measures led to the use of various chelating agents, including penicillamine and its derivative N-acetylpenicillamine.

Comparative studies on the effects of the various chelating agents have shown that N-acetylpenicillamine is the most effective (43). Patients with neurological symptoms due to inorganic mercury intoxication who were treated with this compound (375-1000 mg/day given in three or four oral doses) showed improvement of the clinical picture, together with increased urinary excretion of mercury. Side effects may consist of cutaneous manifestations, blood dyscrasias, and nephrotic syndrome. Although N-acetylpenicillamine is at present the drug of choice for inorganic mercury intoxication, it is ineffective in intoxications by methyl mercury and other organic mercury compounds (4).

Effective extracorporeal removal of mercury by hemoperfusion through agarose-polymercaptal microsphere beads has been described in a recent study (54). Moreover, animal experiments have provided promising results with a new method called "extracorporeal complexing dialysis" (45), which consists of mixing the blood with a mercury-chelating agent (cysteine) and then removing the dialysable complex.

ACKNOWLEDGMENTS

The author wishes to thank Ms. Kathleen White for reviewing the English text and Ms. Carla Antonini and Ms. Giovanna Radaelli for Assistance in its preparation.

REFERENCES

1. *American National Standards Allowable Concentrations of Mercury*, Vol. 237, ANSI, New York, 1972, p. 8.
2. Baldi, G., Vigliani, E. C., Zurlo, N. 1953. Mercurialismo cronico nei cappellifici. *Med. Lav. 44*:161.
3. Bell, Z. G., Lovejoy, H. B., Vizena, T. R. 1973. Mercury exposure evaluations and their correlations with urine mercury excretion. 3. Time-weighted average (TWA) mercury exposure and mercury urine levels. *J. Occup. Med. 15*:501.
4. Berglund, F., Berlin, M., Birke, G., Cederlöf, R., Von Euler, U., Friberg, L., Holmstedt, B., Johnsson, E., Luning, K. G., Ramel, C., Skerfving, S., Swensson, A., Tejning, S. 1971. Methylmercury in fish: A toxicologic-epidemiologic evaluation of risks. Report from an expert group. *Nord. Hyg. Tidskr. Suppl. 4.*
5. Berlin, M., Ullberg, S. 1963. Accumulation and retention of mercury in the mouse. I. An autoradiographic study after a single intravenous injection of mercuric chloride. *Arch. Environ. Health 6*:589.

6. Berlin, M., Ullberg, S. 1963. Accumulation of mercury in the mouse. II. An autoradiographic comparison of phenylmercuric acetate with inorganic mercury. *Arch. Environ. Health 6*:602.
7. Berlin, M. 1976. Dose-response relations and diagnostic indices of mercury concentrations in critical organs upon exposure to mercury and mercurials, in *Effects and Dose-Response Relationships of Toxic Metals* (Nordberg, G. F., ed.), Elsevier, Amsterdam, pp. 235-245.
8. Berlin, M. 1979. Mercury, in *Handbook on the Toxicology of Metals* (Friberg, L., Nordberg, G. F., Vouk, V. B., eds.), Elsevier/North Holland, Amsterdam, pp. 503-526.
9. Blackstone, S., Hurley, R. J., Hughes, R. E. 1974. Some interrelation between vitamin-C (L-ascorbic acid) and mercury in the guinea pig. *Food Cosmet. Toxicol. 12*:511-516.
10. Broman, T., Steinwall, O. 1967. Blood-brain barrier, in *Pathology of the Nervous System* (Mincker, J., ed.), Blakiston, New York.
11. Brown, J. R., Kulkarni, M. V. 1967. A review of the toxicity and metabolism of mercury and its compounds. *Med. Serv. J. Can. 23*:786-808.
12. Camerino, D., Cassitto, M. G., Desideri, E., Angotzi, G. 1981. Behaviour of some psychological parameters in a population of a Hg extraction plant. *Clin. Toxicol. 18*:1299-1309.
13. Cavanagh, J. B., Chen, F. C. K. 1971. The effects of methylmercury-dicyandiamide on the peripheral nerves and spinal cord of rats. *Acta Neuropathol. 19*:208-215.
14. Cavanagh, J. B., Chen, F. C. K. 1971. Aminoacid incorporation in protein during the "silent phase" before organo-mercury and p-bromophenylacetylurea neuropathy in the rat. *Acta Neuropathol. 19*:216-224.
15. Chang, L. W., Hartmann, H. A. 1972. Electron microscopic histochemical study on the localization and distribution of mercury in the nervous system after mercury intoxication. *Exp. Neurol. 35*:122.
16. Chang, L. W., Hartmann, H. A. 1972. Ultrastructural studies of the nervous system after mercury intoxication. I. Pathological changes in the nerve cell bodies. *Acta Neuropathol. 20*:122-138.
17. Chang, L. W., Hartmann, H. A. 1972. Ultrastructural studies of the nervous system after mercury intoxication. II. Pathological changes in the nerve fibers. *Acta Neuropathol. 20*:316-334.
18. Chang, L. W., Hartmann, H. A. 1972. Blood-brain barrier dysfunction in experimental mercury intoxication. *Acta Neuropathol. 21*:179-184.
19. Cigna-Rossi, L., Clemente, G. F., Santaroni, G. 1976. Mercury and selenium distribution in a defined area and in its population. *Arch. Environ. Health 31*:160-165.

20. Clarkson, T. W. 1971. Epidemiological and experimental aspects of lead and mercury contamination of food. *Food Cosmet. Toxicol.* *9*:229-243.
21. Clarkson, T. W. 1972. Recent advances in the toxicology of mercury with emphasis on the alkylmercurials. *CRC Crit. Rev. Toxicol.* *1*:203.
22. Clarkson, T. W. 1972. The pharmacology of mercury compounds. *Annu. Rev. Pharmacol.* *12*:375-406.
23. Clarkson, T. W., Marsh, D. O. 1976. The toxicity of methylmercury in humans: Dose-response relationships in adult populations, in *Effects and Dose-Response Relationships of Toxic Metals* (Nordberg, G. F., ed.), Elsevier, Amsterdam, pp. 246-261.
24. Danziger, S. J., Pessick, P. A. 1973. Metallic mercury exposure in scientific glass-ware manufacturing plants. *J. Occup. Med.* *15*:15-20.
25. *Documentation of the Threshold Limit Values for Substances in Workroom Air*, 3rd ed., American Conference of Governmental Industrial Hygienists, Ohio, 1971.
26. Elkins, H. B. 1967. Excretory and biologic threshold levels. *Am. Ind. Hyg. Assoc. J.* *28*:305-314.
27. Endo, T., Nakaya, S., Kimura, R., Murata, T. 1984. Gastrointestinal absorption of inorganic mercuric compounds in vivo and in situ. *Toxicol. Appl. Pharmacol.* *74*:223-229.
28. Foà, V., Caimi, L., Amante, L., Antonini, C., Gattinoni, A., Tettamanti, G., Lombardo, A., Giuliani, A. 1976. Pattern of some lysosomal enzymes in the plasma and of proteins in urine of workers exposed to inorganic mercury. *Int. Arch. Occup. Environ. Health* *37*:115-124.
29. Forzi, M., Cassitto, M. G., Gilioli, R., Armeli, G., Foà, V. 1974. Persönlichkeitsfehlentwicklungen in Arbeitern bei der Elektrolytischen Chlor-Alkali Gewinnung, in *II Industrial and Environmental Neurology Congress* (Klinkova-Deutschova, E., Lukas, E., eds.), Univerzita Karlova, Prague, p. 78.
30. Forzi, M., Cassitto, M. G., Gilioli, R., Monzani, G., Armeli, G., Foà, V. 1974. Testpsychologische Leistungsfähigkeit in Quecksilberdampfexponierten Arbeitern, in *II Industrial and Environmental Neurology Congress* (Klinkova-Deutschova, E., Lukas, E., eds.), Univerzita Karlova, Prague, p. 70.
31. Forzi, M., Cossitto, M. G., Bulgheroni, C., Foà, V. 1976. Psychological measures in workers occupationally exposed to mercury vapors. A validation study, in *Adverse Effects of Environmental Chemicals and Psychotropic Drugs*, Vol. 2, Elsevier, Amsterdam, pp. 165-172.
32. Fowler, B. A., Brown, H. W., Lucier, G. W., Beard, M. E. 1974. Mercury uptake by renal lysosomes of rats ingesting methylmercury hydroxide. *Arch. Pathol.* *98*:289-297.
33. Friberg, L. 1961. On the value of measuring mercury and cadmium concentrations in urine. *Pure Appl. Chem.* *3*:289.

34. Friberg, L., Nordberg, G. F. 1972. Inorganic mercury. Relations between exposure and effects, in *Mercury in the Environment. A Toxicological and Epidemiological Appraisal* (Friberg, L., Vostal, J., eds.), CRC Press, Cleveland, pp. 130-140.
35. Friberg, L., Vostal, J. 1972. *Mercury in the Environment. A Toxicological and Epidemiological Appraisal*, CRC Press, Cleveland.
36. Gilioli, R., Forzi, M., Cassitto, M. G., Rota, E., Monzani, G., Foà, V. 1974. Neurophysiological findings in workers chronically exposed to mercury vapours, in *Proceedings of II Industrial and Environmental Neurology Congress* (Klimkova-Deutschova, E., Lukas, E., ed.), Univerzita Karlova, Prague, pp. 397-402.
37. Gilioli, R., Bulgheroni, C., Caimi, L., Foà, V., Filippini, C., Boiardi, A., Bussone, G., Quarti, N., Boeri, R. 1976. Correlations between subjective complaints and objective neurophysiological findings in workers of a chlor-alkali plant, in *Adverse Effects of Environmental Chemicals and Psychotropic Drugs*, Vol. 2, Elsevier, Amsterdam, pp. 157-164.
38. Goldwater, L. J., Ladd, A. C., Jacobs, M. B. 1964. Absorption and excretion of mercury in man. VII. Significance of mercury in blood. *Arch. Environ. Health 9*:735-741.
39. Hamilton, A., Hardy, H. L. 1949. Mercury, in *Industrial Toxicology*, Hoeber, New York, pp. 104-126.
40. Hursh, J. B., Clarkson, T. W., Cherian, M. G., Vostal, J., Vander Mallie, R. 1976. Clearance of mercury (Hg 197, Hg 203) vapor inhaled by human subjects. *Arch. Environ. Health 31*: 302-309.
41. International Committee. 1969. Maximum allowable concentrations of mercury compounds. *Arch. Environ. Health 19*:891-905.
42. Joselow, M. M., Goldwater, L. J. 1967. Absorption and excretion of mercury in man. XII. Relationship between urinary mercury and proteinuria. *Arch. Environ. Health 15*:155.
43. Kark, R. A. P., Poskanzen, D. C., Bullock, J. D., Boylen, G. 1971. Mercury poisoning and its treatment with N-acetyl-DL-penicillamine. *N. Engl. J. Med. 285*:10-15.
44. Kesta, L., Byrne, A. R., Zelenko, V. 1975. Correlation between selenium and mercury in man following exposure to inorganic mercury. *Nature 254*:238-239.
45. Kostyniak, P. J. 1975. An extra-corporeal complexing hemodialysis system for the treatment of methylmercury poisoning. I. In vitro studies of the effects of four complexing agents on the distribution and dialysability of methylmercury in human blood. *J. Pharmacol. Exp. Ther. 192*:260-269.
46. Langolf, G. D., Chaffin, D. B., Henderson, R., Whittle, H. P. 1978. Evaluation of workers exposed to elemental mercury using quantitative tests of tremor and neuromuscular functions. *Am. Ind. Hyg. Assoc. J. 39*:976-984.

47. Levine, S. P., Cavender, G. D., Langolf, G. D., Albers, J. W. 1982. Elemental mercury exposure: peripheral neurotoxicity. *Br. J. Ind. Med. 39*:136-139.
48. Lindsstedt, G., Gottoberg, I., Holmgren, B., Johnsson, T., Karlsson, G. 1979. Individual mercury exposure of chloralkali workers and its relation to blood and urinary mercury levels. *Scand. J. Work Environ. Health. 5*:1-59.
49. Lauwerys, R., Lavenne, F. 1972. *Précis de Toxicologie Industrielle et des Intoxications Professionalles*, Duculot, Gembloux, Belgium.
50. Lovejoy, H. B., Bell, Z. G., Vizena, T. R. 1973. Mercury exposure evaluations and their correlations with urine mercury excretion. IV. Elimination of mercury by sweating. *J. Occup. Med. 15*:590.
51. Magnuson, H. T., Passett, D. W., Gerarde, H. W., Smith, H. T., Stokinger, H. E. 1964. Industrial toxicology in the Soviet Union. Theoretical and applied. *Am. Ind. Hyg. Assoc. J. 23*: 185.
52. Magos, L. 1967. Mercury-blood interaction and mercury uptake by the brain after vapor exposure. *Environ. Res. 1*:323-337.
53. Magos, L. 1968. Uptake of mercury by the brain. *Br. J. Ind. Med. 25*:315.
54. Margel, S., Marcus, L., Mashiah, A., Savin, H., Dalit, M. 1984. Extracorporeal removal of mercury by hemoperfusion through agarose-polymercaptal microsphere beads. *J. Biomed. Mater. Res. 18*:617-629.
55. McFarland, R. B., Reigel, H. 1978. Chronic mercury poisoning from a single brief exposure. *J. Occup. Med. 20*:532-534.
56. Mattiussi, R., Armeli, G., Bareggi, V. 1982. Statistical study of the correlation between mercury exposure (TWA) and urinary mercury concentrations in chloralkali workers. *Am. J. Ind. Med.* 3:335-339.
57. Miettinen, J. K. 1971. Absorption and elimination of dietary mercury (Hg^{2+}) and methylmercury in man, in *Mercury, Mercurialism and Mercaptans* (Miller, M. W., Clarkson, T. W., eds.), Charles C. Thomas, Springfield, Ill., pp. 233-243.
58. Miller, T. B., Clarkson, T. W. 1971. *Mercury, Mercurials and Mercaptans*, Charles C. Thomas, Springfield, Ill.
59. Miller, J. M., Chaffin, D. B., Smith, R. G. 1975. Subclinical psychomotor and neuromuscular changes in workers exposed to inorganic mercury. *Am. Ind. Hyg. Assoc. J. 10*:725-733.
60. Mushak, P. 1983. Mammalian biotransformation processes involving various toxic metalloids and metals, in *Chemical Toxicology and Clinical Chemistry of Metals* (Brown, S. S., Savory, J., eds.), Academic, London, pp. 227-245.

61. Nordberg, G. F. 1976. *Effects and Dose-Response Relationships of Toxic Metals*, Elsevier, Amsterdam.
62. Nordberg, G. F. 1980. Neurotoxic effects of metals and their compounds. A review with special reference to metal uptake in nervous tissue, in *Advances in Neurotoxicology* (Manzo, L., ed.), Pergamon, Oxford, pp. 3-15.
63. Passow, H. 1970. The red blood cells: Penetration, distribution and toxic actions of heavy metals, in *Effects of Metals on Cells, Subcellular Elements, and Macromolecules* (Maniloff, J., Coleman, J. R., Miller, M. W., eds.), Charles C. Thomas, Springfield, Ill., pp. 291-334.
64. Pharmacology Biochemistry Group. 1972. Colloquium on mercury intoxication. *Biochem. J. 130*:59-70.
65. Piotrowski, J. K., Trojanowska, B., Mogilnicka, E. M. 1975. Excretion kinetics and variability of urinary mercury in workers exposed to mercury vapours. *Int. Arch. Occup. Environ. Health 35*:245-256.
66. Roels, H., Lauwerys, R., Buchet, J. P., Bernard, A., Barthels, A., Oversteyns, M., Gaussin, J. 1982. Comparison of renal function and psychomotor performance in workers exposed to elemental mercury. *Int. Arch. Occup. Environ. Health 50*:77-93.
67. Rothstein, A., Hayes, A. 1964. The turnover of mercury in rats exposed repeatedly to inhalation of vapor. *Health Phys. 10*:1099.
68. Scheide, E. P., Taylor, J. K. 1975. A piezoelectric crystal dosimeter for monitoring mercury vapor in industrial atmospheres. *Am. Ind. Hyg. Assoc. J. 12*:897-901.
69. Singerman, A., Catalina, R. L. 1969. Exposure to metallic mercury: Enzymatic studies, in *Proceedings of the XVI International Congress of Occupational Health, Tokyo*, Japan Industrial Safety Assoc., Tokyo, pp. 554-557.
70. Shapiro, I. M., Kornblatt, D. R., Summer, A. J., Uzzel, B., Spitz, L. K., Ship, I. I., Block, P. 1982. Neurophysiological and neuropsychological function in mercury exposed dentists. *Lancet 1*:1147.
71. Sinczuk-Walcza, K., Izyckc, J. 1978. Evaluation of neurological status and EEG tests in workers exposed to metallic mercury vapours. *Med. Pr. 3*:251-257.
72. Skerfving, S., Vostal, J. 1972. Symptoms and signs of intoxication, in *Mercury in the Environment. A Toxicological and Epidemiological Appraisal* (Friberg, L., Vostal, J., eds.), CRC Press, Cleveland, pp. 93-108.
73. Smith, R. G., Vorwals, A. J., Patil, L. S., Mooney, T. F. 1970. Effects of exposure to mercury in the manufacture of chlorine. *Am. Ind. Hyg. Assoc. J. 31*:687.
74. Società Italiana di Medicina del Lavoro—Associazione Italiana degli Igienisti Industriali. 1975. Valori limite ponderati degli

inquinanti chimici e particolati degli ambienti di lavoro per il 1975. *Med. Lav. 66*:361-371.
75. Stoker, H. S., Seager, S. L. 1972. *Environmental Chemistry, Air and Water Pollution*, Scott, Foresman, Glenview, Ill.
76. Suwa, N., Takahata, N. 1969. Clinical studies of chronic inorganic mercury poisoning. *Adv. Neurol. Sci. 13*:89-94.
77. Takahata, N., Hayashi, H., Watanabe, B., Anso, T. 1970. Accumulation of mercury in the brains of two autopsy cases with chronic inorganic mercury poisoning. *Folia Psychiatr. Neurol. Jpn. 24*:59-69.
78. Task Group on Metal Toxicity. 1976. Conceptual considerations: Critical organ, critical concentration in cells and organs, critical effect, subcritical effect, dose-effect and dose-response relationships, in *Effects and Dose-Response Relationships of Toxic Metals* (Nordberg, G. F., ed.), Elsevier, Amsterdam, pp. 10-13.
79. Taylor, A., Marks, V. 1973. Measurements of urinary mercury excretion by atomic absorption in health and disease. *Br. J. Ind. Med. 30*:293-296.
80. Vigliani, E. C., Pennacchietti, M. 1947. Osservazioni neurologiche su dieci casi di mercurialismo cronico professionale. *Rass. Med. Ind. 16*:1-15.
81. Vroom, F. Q., Greer, M. 1972. Mercury vapour intoxication. *Brain 95*:305-318.
82. Ware, R. A., Chang, L. W., Burkholder, P. M. 1974. An ultrastructural study on the blood-brain barrier dysfunction following mercury intoxication. *Acta Neuropathol. 30*:211-224.
83. Watanabe, S. 1969. Mercury in the body 10 years after long-term exposure to mercury. *Proceedings of the XVI International Congress on Occupational Health*, Japan Industrial Safety Assoc., Tokyo, p. 552.
84. Zeglio, P. 1958. Esiti a distanza del mercurialismo cronico. *Lav. Um. 10*:420-422.

16

Human Effects of Methyl Mercury as an Environmental Neurotoxicant

Tadao Takeuchi*
Kumamoto University, Kumamoto City, Japan

I. INTRODUCTION: MINAMATA DISEASE

Minamata disease first occurred more than 25 years ago, between 1953 and 1956. At the time it was a mystery, an unknown illness caused by an unknown factor affecting the nervous system. A study group from the Kumamoto University Medical School (9) found that the disease was induced by eating large amounts of fish contaminated by mercury. The mercury came from a new industrial plant which had been manufacturing acetaldehyde and vinyl compounds on a large scale since 1952. The study group realized that a major factor in the illness was the presence of methyl mercury in fish.

Mercury thus came to be recognized throughout the world as a major environmental neurotoxicant (4,10,13,19,33,34). Further investigations by numerous scientists confirmed that the conversion of mercury into methyl mercury was a two-stage process, the first associated with the acetaldehyde plant's manufacturing process, the second associated with food chains and the presence of inorganic mercury in water (9,22).

---

* *Present affiliation:*
Shokei-Gakuen University, Kumamoto City, Japan

Important lessons were learned from the Minamata findings, not least that there must have been other places in the world contaminated by mercury as a result of industrial pollution.

Methyl mercury's role as an environmental neurotoxicant and its effect on human beings are discussed in this chapter in the light of the Minamata findings.

A. Incidence and Distribution of Victims

In the first outbreak in the Minamata area (1953-1956), 54 victims were recorded. Between then and 1961 a further 34 were recorded, bringing the total to 88. Between 1961 and 1970 a further 33 victims, including 22 infants affected by fetal Minamata disease during the intrauterine life, were added on clinical and pathological grounds, bringing the total to 121 (9,21).

A second outbreak occurred in Niigata (30). Detailed clinical investigations led to recognition of chronic mild cases with incomplete symptoms and signs of the Hunter-Russell syndrome (8). Similar mild cases came to be recognized in widespread areas of Minamata (22).

The number of patients with chronic disease increased until 1973, even though the findings of a second study group from the Kumamoto University Medical School (22) had led to a halt in mercury-based production in 1968. By 1983 some 1800 victims had been officially recorded and another 6000 and more patients were waiting for official diagnosis, following clinical examination. The distribution of victims was unexpectedly wider and more alarming than initially thought (Fig. 1).

B. Mercury Levels

Extremely high mercury concentrations were found in the Minamata Bay sediments in 1959, in the wake of research strongly pointing to an organomercurial cause of the disease (29). Samples analyzed by the dithizone method revealed a maximum concentration of mercury in the sediment of 2010 ppm (wet weight) near the drainage channel running from the chemical plant into the bay The further away samples were taken from this drainage channel's discharge point, the lower the concentration of mercury. There was in fact a very sharp fall: 40 to 59 ppm in the center of the bay, and 22 and 12 ppm at distances of about 1.0 and 1.5 km, respectively, from the discharge point, at the mouth of the bay leading to the Shiranui Sea (Fig. 1). Directly outside Minamata Bay sediment only contained between 0.4 and 3.4 ppm of mercury, and farther out into the Shiranui Sea, between 0.01 and 0.05 ppm. However, fish and shellfish in the Shiranui Sea contained high levels of mercury (9).

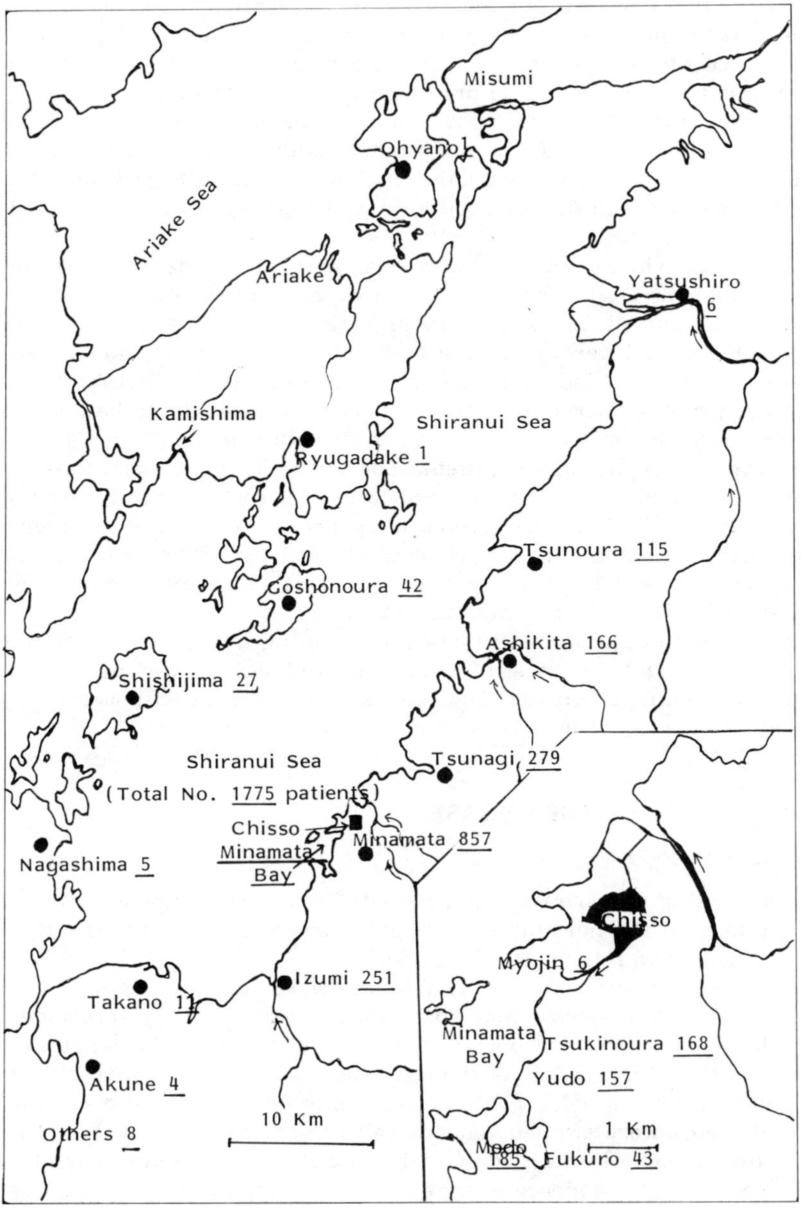

Figure 1 Distribution of victims in the vicinity of the Minamata area in 1981.

In 1972-1973 chemical studies by atomic absorption spectrometry revealed there had been no significant decrease in the mercury levels in the sediments in these areas. In the center of the bay the concentration was between 40 and 50 ppm, and at the mouth of the bay the position was virtually unchanged. What had changed was the concentration in fish, which had gradually but steadily declined following the cessation of acetaldehyde production in 1968, the level of mercury ranging from between 0.5 and 2.0 ppm (wet weight) from 1972 onward (22).

At the time when methyl mercury was shown to be the cause of the disease (1959-1961), the mercury levels in the hair of residents consuming contaminated fish were very high (Fig. 2). Only near Minamata Bay was there less mercury in the residents' hair, because advice from the study group had led to a fishing ban there in 1957. In 1972-1973 the study group (22) reported that Minamata area residents had a maximum concentration of 19.4 ppm mercury in their hair, 11.4 ppm of which was methyl mercury. Patients not originally recognized to be victims were subsequently discovered to have almost identical mercury levels in the hair. Presumably poisoning had occurred over a longer period with lower doses of methyl mercury, but the effects were similar, even though some decline in the mercury levels must have occurred because of the long time span involved (24).

There are no data currently available showing a relationship between mercury levels and the onset of chronic mild illness; nevertheless, autopsies on patients with chronic mild symptoms revealed a mercury level well in excess of normal limits.

II. PATHOLOGY OF THE DISEASE

A. Histopathology

Postmortem studies performed in Minamata* revealed pathological lesions, mainly in the nervous system and predominantly in the brain cortex (Figs. 3 and 4) (5,18,23,25-29).

In acute cases two outstanding features were apparent. One was acute edema in the perivascular space that was occasionally accompanied by small hemorrhages with perivascular demyelination. The other was a lesion in the nerve cells, particularly the small neurons, with changes in the astrocytes and microglial cells or macrophages. Neurons in the cerebral cortex were swollen, and there were severe cellular changes with neuronophagia in both acute and subacute cases. Acute shrinkage of neurons and ischemic changes were also apparent. Neuron loss was already apparent in the brains of victims who had died less than 1

*Over 100 autopsies have been carried out.

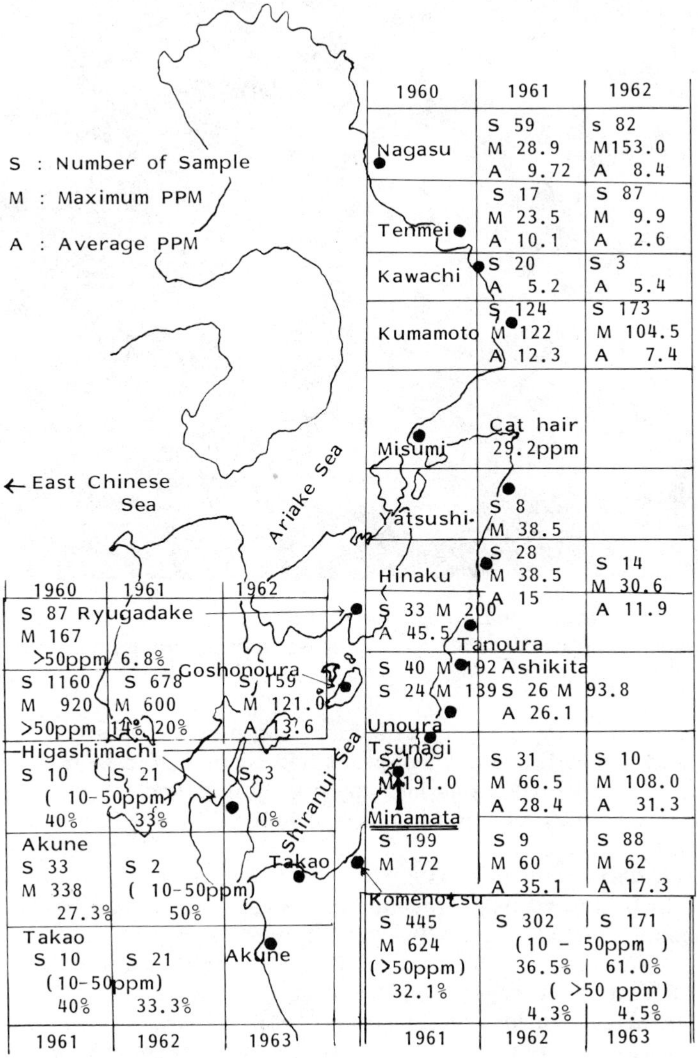

Figure 2 Status of mercury content in the hair from 1960 to 1963.

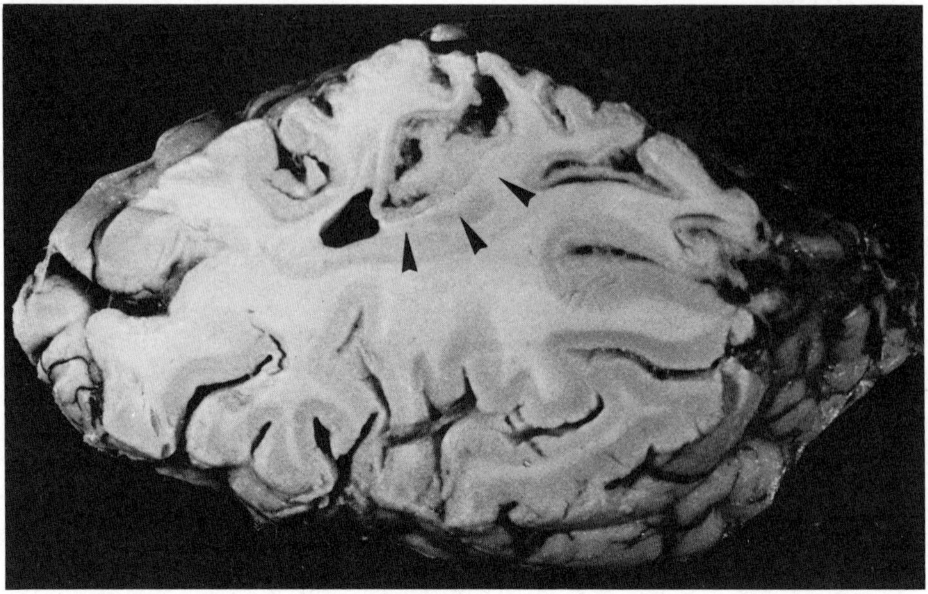

Figure 3 Occipital lobe (56-year-old male, clinical course 48 days, Hg 7.8 ppm). Conspicuous cortical atrophy is seen in the calcarine cortex.

month after the onset of symptoms. The surviving nerve cells were often atrophic and sclerotic. Brown pigment had accumulated in the cell bodies, and the neuronal axons had apparently changed to an abnormal form (e.g., neurofibrillary tangles). Their incidence increased in chronic cases.

Severe neuron damage often caused the affected brain cortex to become spongy, and brain atrophy subsequently occurred. Single nerve cell necrosis occurred focally or diffusely with or without glial cell proliferation. In subjects who had a 2- to 3-month clinical course before death, proliferation of hypertrophic astrocytes was frequently recognized in the affected areas of the brain cortex. In the chronic stage, increase of glial cells with no hypertrophy and phagocytes with lipofuscin and mercury deposits were always present.

B. Grading the Lesions

We decided to establish a six-step grading system in an attempt to classify the distribution and severity of the pathological lesions, which differed in each individual case, presumably because of the differing amount of contaminated fish each patient had consumed (22,25). The

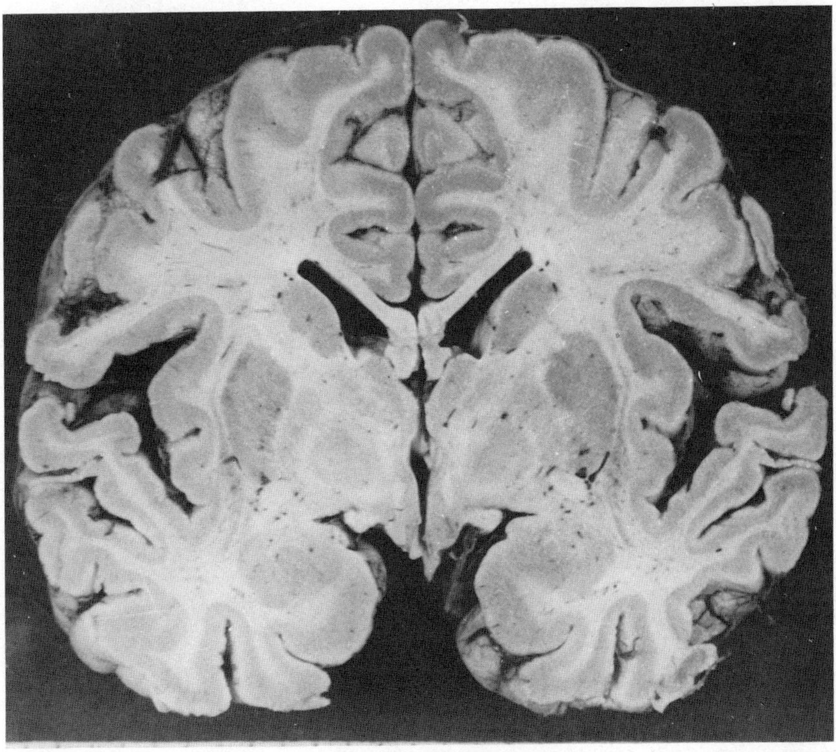

Figure 4 Brain atrophy (26%) in a severe protracted case (8-year-old female, clinical course 2.9 years, Hg 1.3 ppm).

system reflects the differing degree of deterioration in the cerebral cortex and particularly neuron loss in protracted and chronic cases (Fig. 5).

1. *Cerebral Cortex*

The most severe changes in the cerebral cortex (sixth grade) featured a macroscopic spongy state with variously sized cystic cavities which were readily visible to the naked eye (Fig. 5a). Such changes were observed in one male child patient who died after a clinical course of 4 years, passing through the typical Hunter-Russell syndrome as described by Pentschew (8,14) to the decortication syndrome (26). The brain was intensely atrophic and weighed only 600 g (control 1335 ± 15.7 g, at the same age and sex in Japanese).

The second most severe changes (fifth grade) consisted in intense atrophy of the brain (26-37% decrease), with a miscoscopic spongy

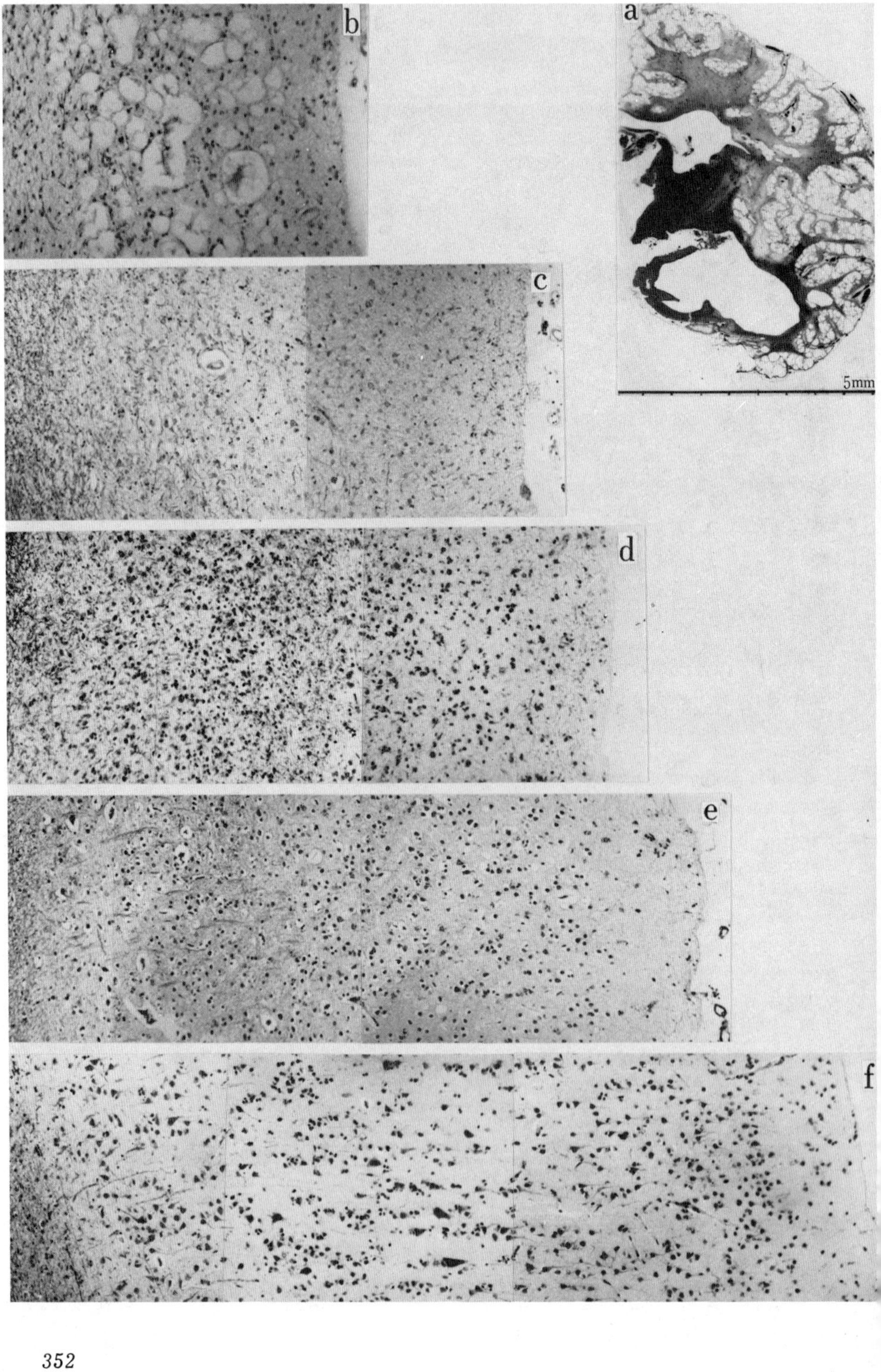
a
b
c
d
e
f
5mm

state in the affected cortices. The spongy state tended to occur from the second to the upper parts of the third or fourth layer, but in severe prolonged cases it was often found to involve the entire cortex (Figs. 5b and 6a).

The third most severe changes (fourth grade) were characterized by a loosened appearance in the cortical tissue as a result of marked neuron loss in the affected layers, where massive laminal necrosis of neurons and proliferation of the glial tissue had occurred. Brain weight was reduced by about 20%, and patients of this type had displayed the typical signs and symptoms of the Hunter-Russell syndrome (Fig. 5c).

Moderate disturbance (third grade) occurred when over 50% of the neurons in the damaged regions of the brain had disappeared (Fig. 5d). In these cases, brain weight reduction was less than 20%.

When the neuron loss rate fell to between 30 and 50% in the affected cortex relatively slight changes (second grade) were observed (Fig. 5e). Neuron loss was not difficult to recognize, but required careful microscope assessment, since the neuronal "thinning out" took place following selective single-cell necrosis with a high proliferation of glial cells. Brain weight reduction was about 10-15%.

Minor changes (first grade) were found mainly in chronic mild cases where the patients had, when first examined, shown minor or doubtful signs and symptoms of the disease. The brain appeared macroscopically to be within normal limits, but neuron loss with a slight increase in glial cells in the affected cortex had occurred. In these cases no more than 30% of the nerve cells were lost. The brain cortex presented a slight "thinning out" of neurons that was difficult to recognize under the microscope without a control (Fig. 5f).

2. *Cerebellum*

Changes in the cerebellum were a predominant feature in all cases examined the distribution and degree of lesions varying from case to case. The granule cells were apparently the most susceptible to methyl mercury, since they had usually begun to disintegrate from the region beneath the Purkinje cell layer. The most severe damage was the entire loss of both the granule cells and Purkinje cells, very rarely with a spongy state in the upper parts of the granular layer. The lesions were usually of the central type, mostly involving the deeper portions of both the neocortex and paleocortex (Fig. 7).

Figure 5 Six degrees of brain lesions in Minamata disease: (a) most severe case and (b-f) from second most severe case (fifth degree) to minor case (first degree) seen in the same calcarine areas (paraffin sections, Klüver-Barrera stain).

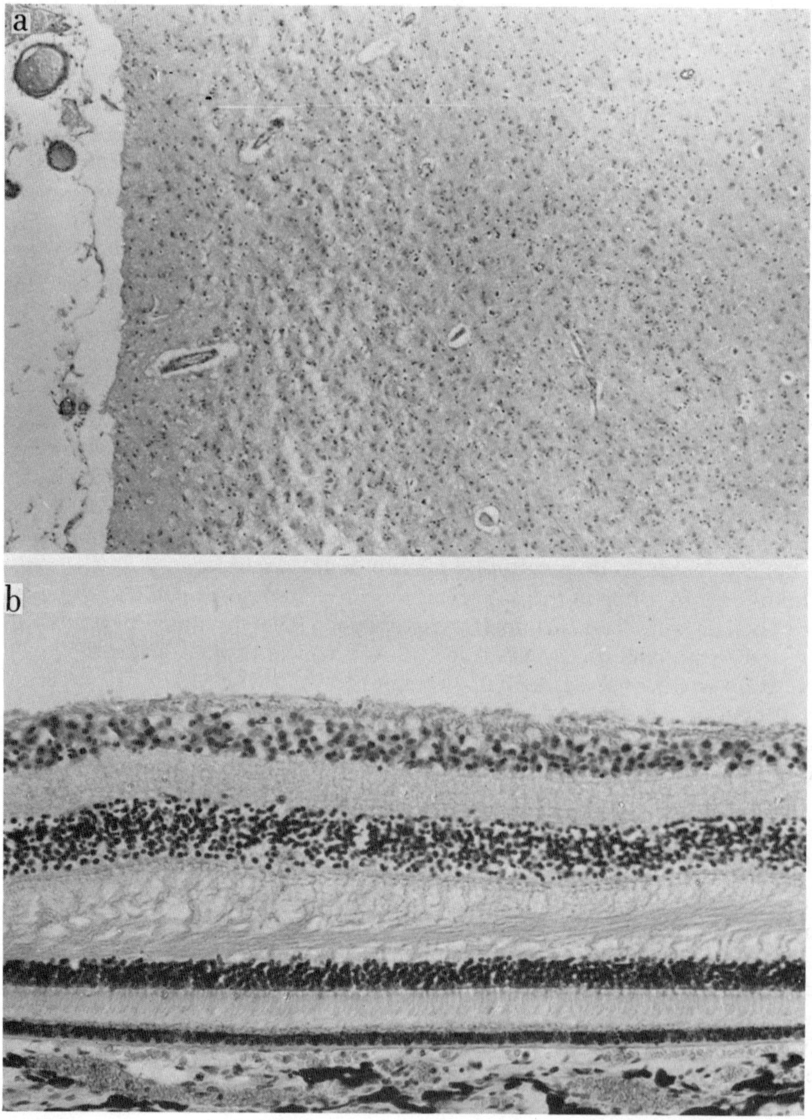

Figure 6 Calcarine cortex (a) and retina (b) in the same case (52-year-old male, clinical course 100 days, Hg 2.6 ppm). There is loss of almost all neurons in the calcarine cortex, with proliferation of astrocytes, in spite of a completely intact retina (paraffin sections, hematoxylin and eosin stain).

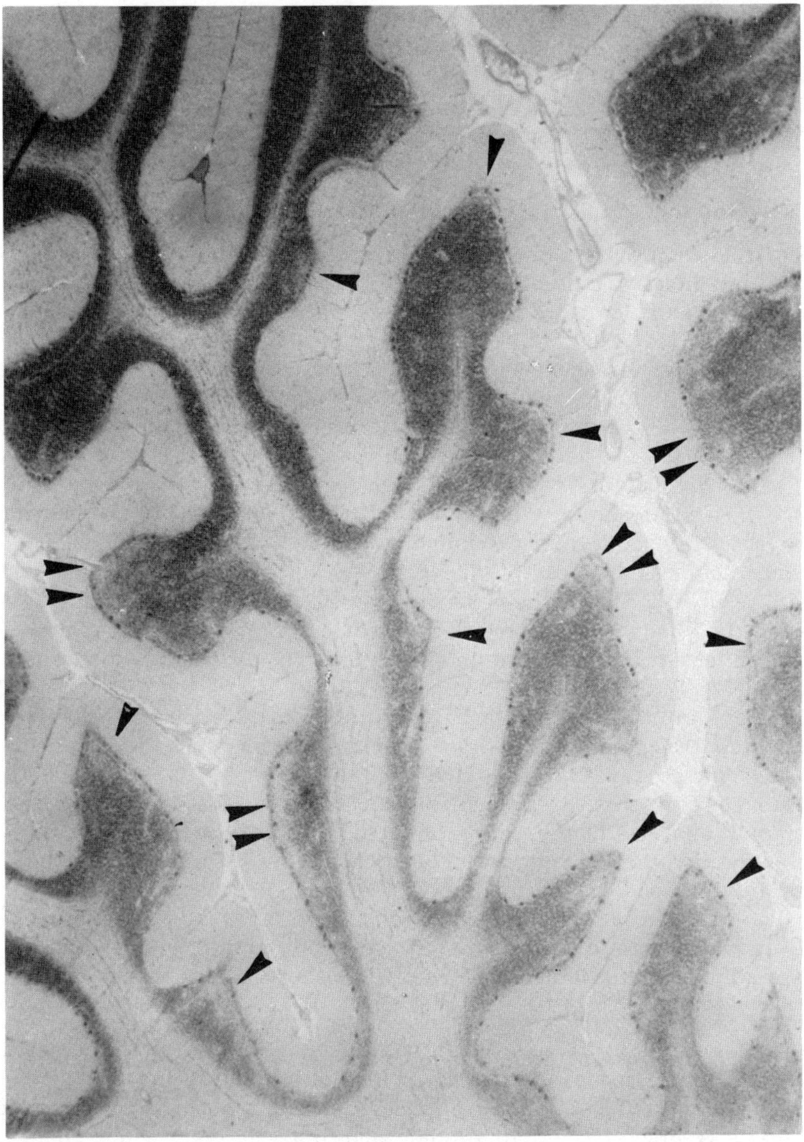

Figure 7 Central-type atrophy of the cerebellum (34-year-old male, clinical course 96 days, Hg 4.6 ppm). Loss of granule cells has occurred from the region beneath the Purkinje cell layers to deeper portions. The Purkinje cells are relatively well spared.

Minor disturbances were characterized by slight loss of granule cells beneath the Purkinje cell layer at the crest portion of the deeper gyri, resulting in "apical scar formation" and often irregularity in the Purkinje cell layer with a proliferation of Bergmann glial cells. Golgi type II cells were relatively preserved.

C. Distribution of Lesions as Related to Clinical Symptoms

The distribution of lesions in the nervous system was highly characteristic in Minamata disease. The cortex both in the cerebrum and cerebellum was much more heavily affected than the brainstem, spinal cord, or peripheral nerves. The calcarine cortex was severely involved in all cases, particularly in the region along the depth of the calcarine fissure or the lateral walls of the sulcus (Fig. 6). The destruction of nerve tissue was prominent in the anterior portions of the calcarine cortex and became less marked toward the occipital pole. The anterior cortical regions receiving projection fibers from the peripheral area of the retina were predominantly affected, and this was probably the cause of the symmetrical concentric constriction of the visual field often observed in Minamata disease. The 18th and 19th visual association areas were less involved than the calcarine cortex, but, when intensely involved, the changes may have been sufficient to give rise to abnormal ocular movements which were detected clinically by electro-ophthalmography (32).

Changes in other cortical areas of the cerebrum such as the pre- and postcentral gyri and superior lateral gyri were similar to those in the calcarine cortex, but as a rule were less severe.

Intense lesions in the precentral cortex caused the development of secondary bilateral degeneration of the pyramidal tracts in the brainstem and spinal cord (Fig. 8). These changes were predominantly found in childhood methyl mercury poisoning. Pathological reflexes and limb paralysis sometimes occurred in severe cases. The pictures showing exaggeration of reflexes, spactic gait, and slow movements presumably indicated a less severe involvement of the precentral cortex.

Figure 8 Lesions in peripheral nervous tissues in a severe protracted case (15-year-old male, clinical course about 10 years, Hg 0.034 ppm): (a) spinal cord (L2) showing secondary degeneration of both pyramidal and Goll's tracts, (b) spinal ganglion showing relatively minor changes with surviving ganglion cells, (c) almost normal motor nerve fibers in the anterior root, and (d) intense disturbance of sensory fibers in the posterior root. This posterior alteration was possibly related to disappearance of neurons in the posterior horns, since ganglion cells in the spinal ganglions survived well (paraffin sections, Klüver-Barrera stain).

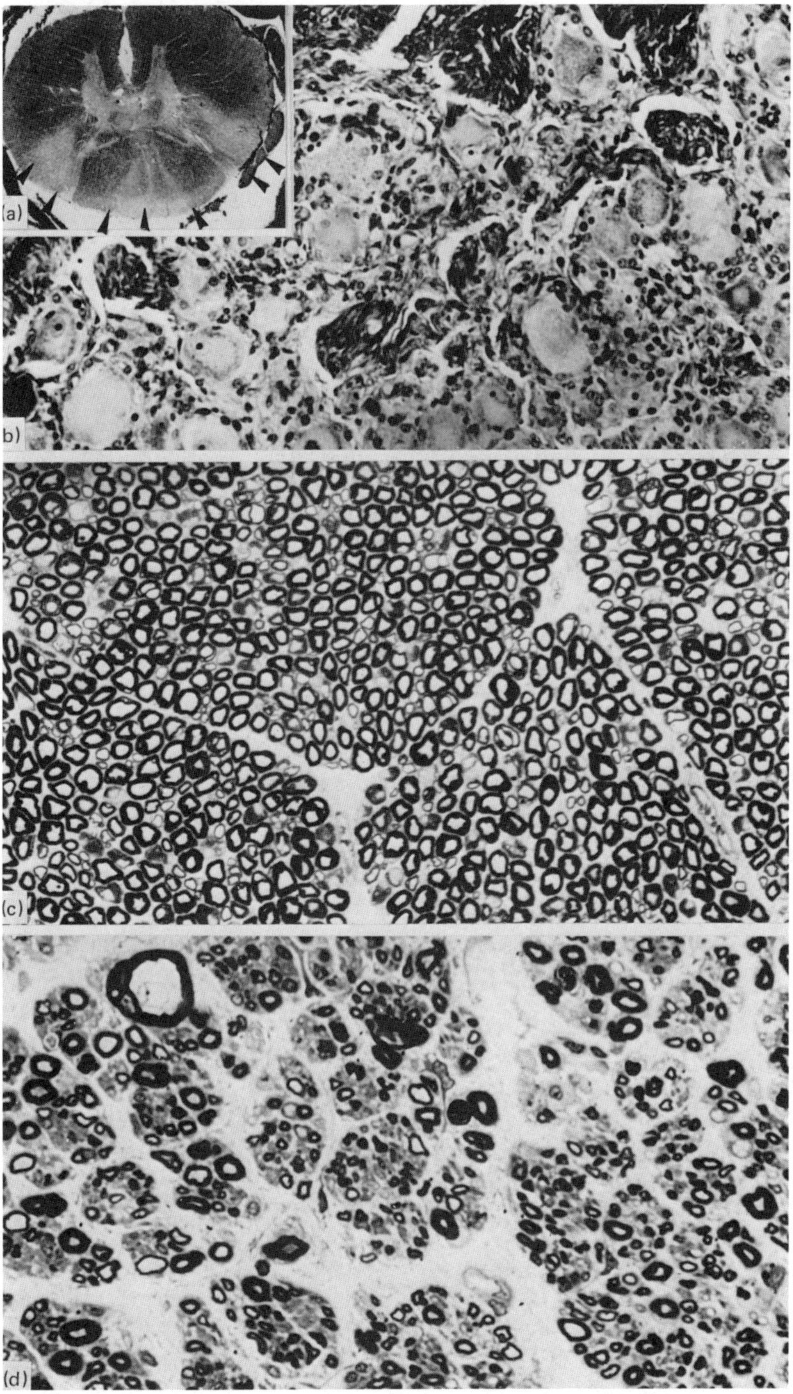
(a)
(b)
(c)
(d)

Impaired superficial sensations, predominantly in the distal parts of the extremities, suggested some involvement of the peripheral nerves, particularly the sensory nerve fibers. In fact, the latter were always more severely involved than the motor nerves, even in chronic cases (Fig. 8). Since such impaired sensations often improved during subsequent stages of the disease, one may assume that the early peripheral nerve disorders must have been partly reversible in nature, although improvement related to regeneration cannot be excluded in the less severe cases. Sometimes impaired sensation remained for long periods, with disturbances in position, vibration, pressure, and two-point discrimination, often accompanied by stereognostic changes.

Lesions in both insular and adjacent temporal cortices were a morphological background to the impairment of auditory acuity, particularly to high tones, as detected by audiograms. Moreover, the mental and psychic disturbances encountered in Minamata disease were consistent with widespread involvement of the brain cortex in areas including the frontal cortex. The characteristic distribution of lesions occurring in severe cases of childhood methyl mercury poisoning caused the decortication syndrome (27) and cerebral palsy. The mental debility in children reported by Harada (7) might be related to less severe cortical changes, but undeniably there was a close relationship between mental disorders and the "thinning out" of nerve cells in the frontal cortex.

Damage to the cerebellar cortex gave rise to cerebellar ataxia, with ataxic gait, dysmetria, and dysarthria. The minor changes occurring in chronic cases presumably caused the slight speech impairment and slow movements.

Muscular rigidity and hyperkinesis occasionally seen in severe victims were possibly due to involvement of the putamen. Tremors, slow movements, and muscle weakness were most common in chronic cases. The pathological correlates of these symptoms remain unclear (31).

III. PERIPHERAL NERVOUS SYSTEM DISORDERS

In our studies (20) it was not possible to determine whether the sensory nerves were involved primarily or secondarily as a result of damage to ganglion cells. Biopsy findings in chronic cases (6,28) did not clarify this problem, except that results revealed damage to the spinal ganglion cells which was less severe than the prominent damage to the peripheral sensory nerve fibers (Fig. 8). Both human studies and animal data suggested that methyl mercury (a) can give rise to neuronal necrosis with subsequent secondary degeneration of nerve fibers in the brain and (b) can also produce primary segmental degeneration. In other words, the neurological effects of methyl mercury differed both quantitatively and qualitatively between the brain and the peripheral nerves. In the brain, poisoning mainly caused neuronal necrosis, whereas in the peripheral nerve it more frequently caused primary degeneration of sensory nerve fibers (6,20,24,26). This conflicts with the experimental results reported by some investigators (1). Only in the rat did peripheral lesions tend to occur more predominantly than in the brain cortex.

IV. EFFECTS OF METHYL MERCURY ON THE FETUS

A total of 22 infants affected by transplacental methyl mercury intoxication had been recorded by 1970. Subsequently a further 18 infantile cases were discovered, giving a total of 40 children who had cerebral palsy (7,9,22). Three of them died, the autopsy revealing both cerebral underdevelopment and toxic regressive changes in the brain (Fig. 9). In two cases, brain weight was reduced by about two-thirds and about one-half, respectively, when compared to that of normal brains of the same age and sex. The brain weight of the third case revealed only a slight reduction. Various forms of hypoplasia of the brain, such as columnar block, abnormal cytoarchitecture, status marmoratus, residual matrix cells, residual nerve cells in the medulla, hypoplasia of the corpus callosum, and hypoplasia of the cerebellum with disorganization of the cellular cortical architecture, were noted. Loss of nerve cells in both the cerebral cortex and the cerebellum was also apparent (7,11).

An infant who suffered intrauterine poisoning during the outbreak of methyl mercury intoxication in Iraq was reported to have shown hypoplastic nerve cells but no loss of neurons in the brain cortex. The infant died at a relatively early age, which was presumably too short a time span to permit loss of neurons (3). Experimentally, loss of neurons in the brain of fetal cases has been demonstrated in the cat and rat (12,22).

V. LEVELS AND HISTOCHEMICAL DISTRIBUTION OF MERCURY IN THE BRAIN

In the first outbreak of Minamata disease, mercury levels were measured in the tissues of 12 severely affected victims who died within 100 days after the onset of symptoms. An average mercury concentration of 10.7 ppm (wet weight) in the brain, 38.1 ppm in the liver, and 46.1 ppm in the kidney was found (29). In protracted cases, the mercury level was lower, but retention of the metal in the brain was clearly documented (Table 1). In chronic mild victims, cerebral concentrations of mercury were considerably lower than in protracted severe cases, but were still higher than in normal controls (Table 2).

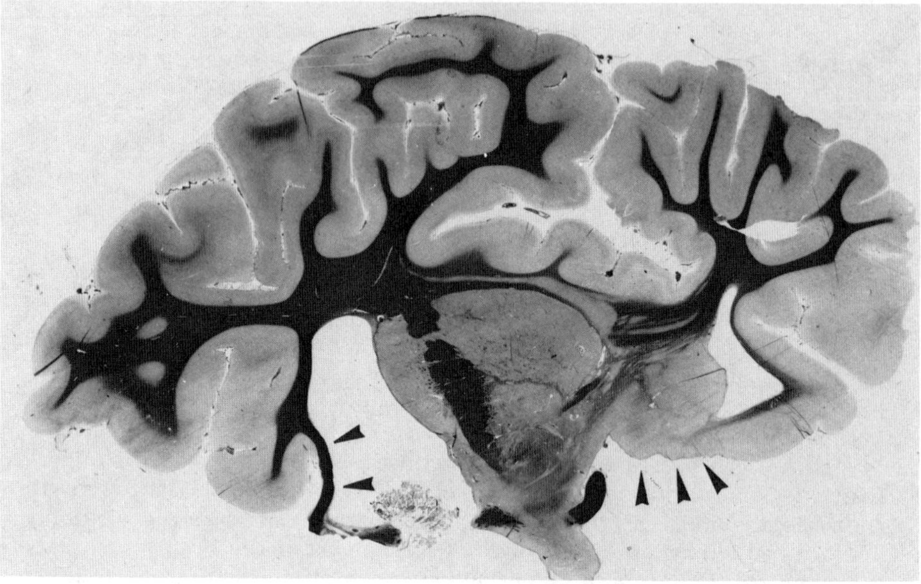

Figure 9 Hypoplasia in the infantile brain (fetal Minamata disease): frontal cut section at the level of the corpus mamillare showing reduction of the corpus callosum, status marmoratus, and hypomyelination (celloidin section, Heidenhein-Woelke stain).

With the use of an autochemography technique (16), mercury deposits were detected in the various organs and tissues, including the kidney, liver, thyroid, testes, pancreas, and skin. In the brain, mercury accumulation occurred in both neurons and glial cells (e.g., astrocytes and microglial cells or macrophages; Fig. 10a and b). It was found to predominate in either the neurons or the glial cells, with differences between individuals for no apparent reasons. The electron microscope revealed accumulation of mercury mainly in lysosomes. Moreover, the concentration of mercury into subcellular organelles was always concomitant with selenium deposits, as demonstrated by electron-microscopic x-ray microanalysis (Fig. 11a and b).

Recent autochemographic investigations (26) have shown the retention of mercury in the brain even in subjects with chronic mild intoxication, which could help in recognizing certain chronic forms of Minamata disease.

Table 1 Total Mercury and Methyl Mercury Levels in the Human Brain in Severe Victims

Time course after onset of symptoms	Number of cases	Total mercury (μg/g)		Methyl mercury (μg Hg/g)
		Dithizone	Atomic absorption	
Days				
19	1	9.6	14.6	2.75
26	1	15.4	21.4	8.42
45±	2	11.3	19.2	5.37
60±	2	15.1	18.4	4.85
70	1	7.8	11.6	5.32
90±	3	6.9	16.4	6.09
100±	2	3.6	8.8	2.46
Years				
1.4±	2	4.0	4.9	0.446
2-2.7	2	0.8	2.1	0.693
7	1	0.68	4.57	0.776
14	1	1.5	4.07	1.015
17-18	2	1.19	4.23	0.700
Control				
1956-1960	15	ND[a]-0.05		
1973	16		ND-0.23	ND-0.02
Average			0.076 ± 0.031	0.009 ± 0.010

[a]Not detectable.

Table 2 Total Mercury and Methyl Mercury Levels in the Human Brain in Chronic Mild Cases[a]

Time course after onset of symptoms	Number of cases	Total mercury (μg/g) (atomic absorption)	Methyl mercury (μg Hg/g)
Victims			
7 years	2	1.64	0.127
15	1	0.48	0.047
17	4	0.487	0.011
21	2	0.260	0.094
27	1	1.380	0.024
Unknown	133	0.20 ± 0.40	0.026 ± 0.030
Residents of Minamata	250	0.10 ± 0.10	0.022 ± 0.030
Control (1981)	16	0.076 ± 0.031	0.009 ± 0.010

[a]Estimated by Dr. Takizawa.

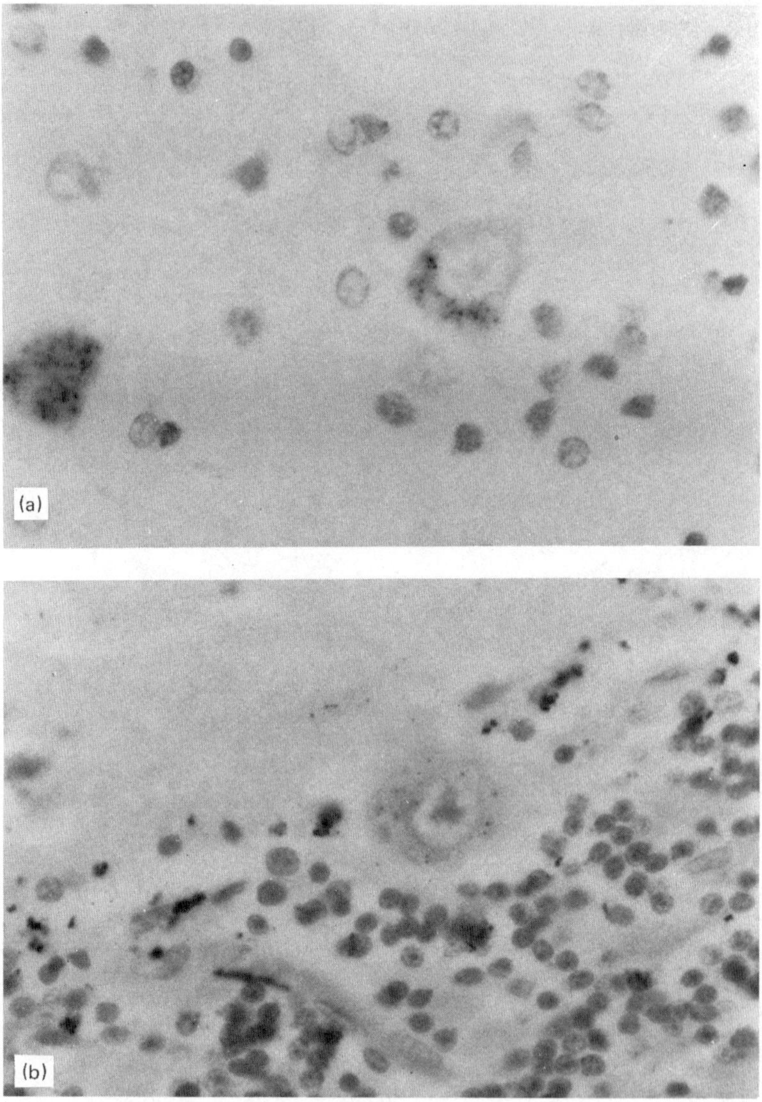

Figure 10 Histochemical staining of mercury deposited by the autochemography technique. (a) Section from the precentral cortex (72-year-old male, clinical course 16 years, Hg 0.73 ppm). Mercury is deposited in the Betz cells and in some of the macrophage cytoplasm. (b) Section from the cerebellum with minor changes (16-year-old male, clinical course unknown, Hg 3.05 ppm) showing deposition of mercury in macrophages and Purkinje cells, particularly notable in lysosomes (paraffin sections, hematoxylin and eosin-mercury double stain).

(a)

(b)

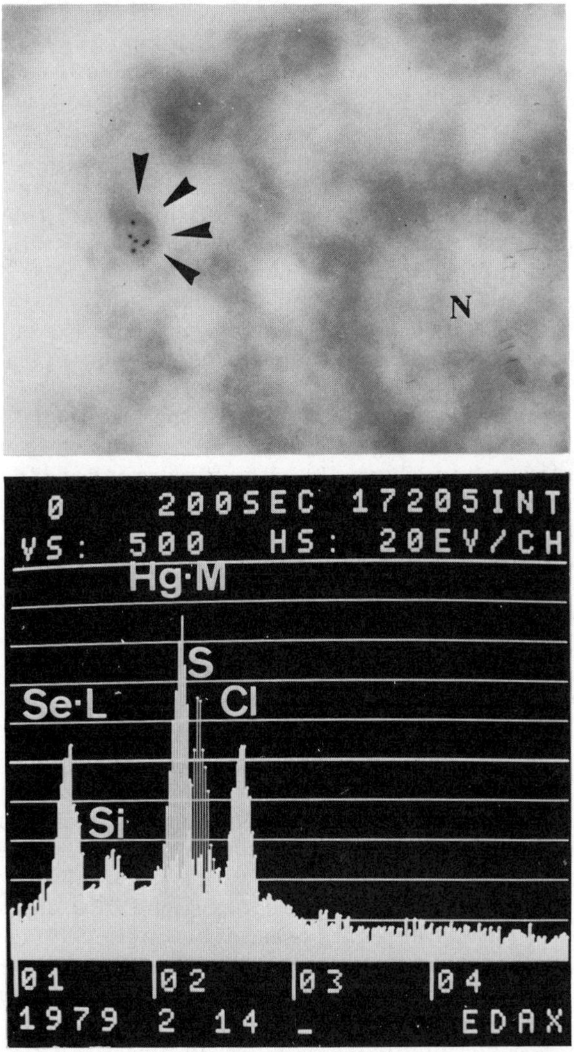

Figure 11 (a) Electron microphotograph of a nerve cell of the occipital lobe in an autopsied brain of Minamata disease. Note the small particles of mercury (arrow) in the lysosome of the cell (10,500 ×). (b) Spectrum of the particles showing distinctive x-ray peaks of mercury, selenium, and sulfur, in addition to concomitant peaks of silicon and chloride. (From T. Shirabe, K. Eto, and T. Takeuchi, *Neurotoxicology* *1*:349-356, 1979).

VI. PATHOGENESIS

Pathogenetic mechanisms of methyl mercury intoxication have been discussed in a recent review by Chang (2). Low concentrations of methyl mercury may lead to (a) irreversible damage to parenchymatous cells, following disintegration of ribosomes and rough endoplasmic reticulum due to reduction of RNA and protein synthesis; (b) nuclear damage with inhibition of DNA synthesis and chromosome disturbances; (c) breakdown of biological membranes; and (d) disruption of enzymatic systems and disintegration of mitochondria in vulnerable cells (15,17, 21).

Disturbance in protein metabolism has been recognized as the only alteration prior to all other biochemical changes caused by methyl mercury intoxication (2). Moreover, mercury was found to impair the blood-brain barrier, probably because of damage of the endothelial and glial membranes leading to leakage of plasma solutes into the nerve parenchyma. As pointed out by Chang (2), "the impairment of the blood-brain barrier, together with the possible inhibition of certain associated enzymes may be responsible for the reduction in uptake of amino acids and other metabolites by the nervous system after methylmercury intoxication." The molecular mechanisms involved in the mercury-induced disorders in protein metabolism are discussed by Omata and Sugano in Chapter 17 of this volume.

It seems of interest that, although the brain contains less mercury than other organs (e.g., the liver and kidney), brain functions are selectively disturbed following exposure to methyl mercury. Moreover, mercury also remains for considerably longer periods of time within the brain as compared with other tissues. One reason for such retention may be that, following single-cell damage, necrotic neuronal cell debris cannot be removed from the brain in the same way as necrotic epithelial cells from the liver or kidney. Presumably, as cell damage and loss of neurons occur, the mercury level in the macrophages and glial cells is progressively increased. It was also considered that single-cell necrosis of neurons with no regenerative compensation may accelerate the damage to the remaining nervous tissue following continuous exposure to mercury over a long period of time (25).

Evidence has been obtained in studies of Minamata disease that the effects of methyl mercury on the nerve cells occurred predominantly in smaller neurons. As recently suggested by Cavanagh (1), the low ribosomal content in small neurons may increase their vulnerability to methyl mercury poisoning. In particular, the ability of these elements to recover from the early inhibition of protein synthesis may be selectively impaired. Anoxic conditions related to disturbances in both the erythrocyte and blood-brain barrier functions (26) may also play a role in the pathogenesis of nervous system lesions in Minamata disease.

REFERENCES

1. Cavanagh, J. B. 1977. Metabolic mechanisms of neurotoxicity caused by mercury, in *Neurotoxicology*, Vol. 1 (Roizin, L., Shiraki, H., Grcevic, N., eds.), Raven Press, New York, pp. 283-288.
2. Chang, L. M. 1982. Pathogenetic mechanisms of the neurotoxicity of methylmercury, in *Mechanisms of Action of Neurotoxic Substances* (Prasad, K. N., Vernadakis, A., eds.), Raven Press, New York, pp. 51-66.
3. Choi, B. H., Lapham, L. W., Amin-Zaki, L., Saleem, T. 1978. Abnormal neuronal migration, deranged cerebral cortical organization, and diffuse white matter astrocytosis of human fetal brain: A major effect of methylmercury poisoning in "utero." *J. Neuropathol. Exp. Neurol. 37*:719-733.
4. Elhassani, S. B. 1983. The many faces of methylmercury poisoning. *J. Toxicol. Clin. Toxicol. 19*:875-906.
5. Eto, K., Takeuchi, T. 1978. A pathological study of prolonged cases of Minamata disease, with particular reference to 83 autopsy cases. *Acta Pathol. Jpn. 28*:565-584.
6. Eto, K., Takeuchi, T. 1977. Pathological changes of human sural nerves in Minamata disease (methylmercury poisoning). Light and electron microscopic studies. *Virchows Arch. B Cell Pathol. 23*:109-128.
7. Harada, Y. 1978. Clinical investigations of Minamata disease. Congenital (fetal) Minamata disease, in *Minamata Disease* (Katsuna, K., ed.), Shuhan, Kumamoto, Tokyo, pp. 73-117.
8. Hunter, D., Russell, D. S. 1954. Focal cerebral and cerebellar atrophy in a human subject due to organic mercury compounds. *J. Neurol. Neurosurg. Psychiatry 17*:235-241.
9. Kumamoto University Study Group. 1968. *Minamata Disease* (Kutsuna, M., ed.), Shuhan, Kumamoto, Japan.
10. Magos, L. 1975. Mercury and mercurials. *Br. Med. Bull. 31*: 241:245.
11. Matsumoto, H., Koya, G., Takeuchi, T. 1965. Fetal Minamata disease. A neuropathological study of two cases of intrauterine intoxication by a methylmercury compound. *J. Neuropathol. Exp. Neurol. 24*:563-674.
12. Morikawa, N. 1961. Pathological studies of organic mercury poisoning. *Kumamoto Med. J. 14*:71-93.
13. National Research Council. 1978. *An Assessment of Mercury in the Environment*, National Academy of Sciences, Washington, D.C.
14. Pentschew, A. 1958. Quecksilbervergiftung. *Handbuch d. spez.Path.Ant.u.Histol.* XIII/2, Vol. B, Springer-Verlag, Berlin, pp. 2007-2024.

15. Sakai, K. 1974. Effects of methylmercury chloride on ascites hepatoma AH 13 cells, especially on cell growth and nucleic acid synthesis. *Kumamoto Med. J. 28*:124-134.
16. Sakai, K., Okabe, M., Eto, K., Takeuchi, T. 1975. Histochemical demonstration of mercury in human tissue cells of Minamata disease by use of autoradiographic procedure. *Acta Histochem. Cytochem. 8*:257-264.
17. Sakai, K., Takeuchi, T. 1973. *Rep. Radioisotope Res. Soc. 7*:55.
18. Shiraki, H. 1979. Neuropathological aspects of organic mercury intoxication including Minamata disease, in *Handbook of Clinical Neurology*, Vol. 36 (Vinken, P. J., Bruyn, G. W., es.), North Holland, Amsterdam, pp. 83-145.
19. Shiraki, H., Takeuchi, T. 1971. Minamata disease, in *Pathology of Nervous System (II)* (Minckler, J., ed.), McGraw-Hill, New York, pp. 1651-1665.
20. Takeuchi, T. 1974. Pathogenesis of Minamata disease from experimental viewpoint. *Clin. Neurol. Tokyo 14*:244-248.
21. Takeuchi, T. 1972. Biological reactions and pathological changes of human beings and animals under the condition of organic mercury contamination, in *Environmental Mercury Contamination* (Hartung, R., Dinman, D. B., eds.), Ann Arbor Scientific, Michigan, pp. 247-289.
22. Takeuchi, T. 1973. Pathological, clinical and epidemiological research about Minamata disease after ten years. *Report of the Kumamoto University 2nd Study Group*.
23. Takeuchi, T. 1977. Neuropathology of Minamata disease in Kumamoto, especially at the chronic stage, in *Neurotoxicology* (Roizin, L., Shiraki, H., Grcevic, N., eds.), Raven Press, New York, pp. 235-246.
24. Takeuchi, T., Eto, K. 1975. Minamata disease, in *Studies on the Health Effects of Alkylmercury in Japan*, Environmental Agency, Tokyo, Japan, pp. 28-62.
25. Takeuchi, T., Eto, K. 1977. Pathology and pathogenesis of Minamata disease, in *Minamata Disease: Methylmercury Poisoning in Minamata and Niigata, Japan* (Tsubaki, T., Irukayama, K., eds.), Kodansha, Tokyo, pp. 103-141.
26. Takeuchi, T., Eto, K. 1984. Neuropathology of Minamata disease in mild, borderline cases recently observed and biochemical detection of mercury. *J. Kumamoto Med. Ass. 58*:41-55.
27. Takeuchi, T., Eto, N., Eto, K. 1979. Neuropathology of childhood cases of methylmercury poisoning (Minamata disease) with prolonged symptoms, with particular reference to the decortication syndrome. *Neurotoxicology 1*:1-20.
28. Takeuchi, T., Eto, K., Oyanagi, S., Miyajima, H. 1978. Ultrastructural changes in human sural nerves in the neuropathy

induced by intrauterine methylmercury poisoning. *Virchows Arch. B Cell Path.* *27*:137-154.
29. Takeuchi, T., Morikawa, N., Matsumoto, H., Shiraishi, Y. 1962. A pathological study of Minamata disease in Japan. *Acta Neuropathol. (Berl.)* *3*:40-57.
30. Tsubaki, T. 1967. Outbreak of intoxication by organic compounds in Niigata prefecture. *Jpn. J. Med.* *6*:132-133.
31. Tsubaki, T., Hiroto, K., Shirakawa, K., Kondo, K., Sato, T. 1978. Clinical, epidemiological and toxicological studies on methylmercury poisoning, in *Toxicology as a Predictive Science* (Plaa, G. L., Duncan, W. A. M., eds.), Academic, New York, pp. 339-357.
32. Tsutsui, J., Fukai, S., Nakamura, Y. 1974. Disturbance of binocular vision in Minamata disease. *Jpn. Rev. Clin. Ophthalmol.* *68*:529-531.
33. Wheatley, B. 1979. *Methylmercury in Canada*, Medical Services Branch, Department of National Health and Welfare, Ottawa, Canada.
34. WHO. 1980. *Health Risk Evaluation for Methylmercury*, EHE 80.22, Geneva.

17

Methyl Mercury: Effects on Protein Synthesis in Nervous Tissue

Saburo Omata and Hiroshi Sugano
University of Niigata, Niigata, Japan

I. INTRODUCTION

Among the neurotoxic agents, methyl mercury has received increasing attention as a major cause of environmental pollution and a potential health hazard for man (6,25) following the outbreaks of methyl mercury poisoning in Japan (see T. Takeuchi in this book) and other countries (31,49).

In recent years, extensive studies have been carried out in humans (25,29) and experimental animals (17,20,32,49) in order to clarify the mechanisms underlying the neurotoxic action of methyl mercury. In this context, a critical role has been attributed to disorders of protein metabolism in the nervous tissue. In this chapter the effects of methyl mercury on protein synthesis will be reviewed.

Protein synthesis in nervous tissue is concerned either with the maintenance of general cellular activities through the production of essential cellular components or with specific expressions of neural processes through supplying protein species which constitute integral parts of functional complexes such as receptors, transport, and conduction systems. Thus alterations in protein synthesis activities in the nervous system will give rise to modifications of both the general and specific nerve functions.

II. EFFECTS OF METHYL MERCURY ON PROTEIN SYNTHESIS ACTIVITIES IN VARIOUS SYSTEMS: AN OVERVIEW

The action of methyl mercury on protein synthesis of nervous tissue was first demonstrated by Yoshino et al. (71) in studies showing significant reduction of [^{14}C]leucine incorporation into protein in slices of rat brain cortex after a single intraperitoneal injection of methyl mercury thioacetamide (75 mg/kg body weight). An important aspect of this pioneering work was the observation that protein synthesis activity was markedly inhibited during the latent period, in which other neurochemical and metabolic parameters, including respiratory and glycolytic activities, were not yet altered by mercury. Yoshino et al. (71) proposed that the disturbance of protein synthesis in the brain may have a direct bearing on methyl mercury poisoning. Thereafter inhibitory actions of methyl mercury on protein synthesis were confirmed in various systems under different experimental conditions.

The amounts of methyl mercury administered by Yoshino et al. (71) elicited neurological symptoms as early as 5 days after starting of the exposure, and the rats died within a week. The administration of methyl mercury chloride with a different dosage schedule (average 27 mg/kg for 3 days, subcutaneously) also led to a considerable reduction in protein synthesis activities in the rat brain during the symptomatic period (20). In keeping with these findings, Cavanagh and Chen (8) reported that the injection of methyl mercury dicyandiamide (5 mg/kg, for 8 days, orally) produced inhibition of [^{14}C]glycine incorporation into the protein of nervous tissue in rats even before the development of neurological symptoms. Using the dosage schedule of 10 mg/kg per day, for 7 days, subcutaneously, Omata et al. (43) also showed early inhibition of protein synthesis activity in the brain of rats before the appearance of symptoms of neurotoxicity. On the other hand, Verity et al. (66), with a dosage schedule of 10 mg/kg per day for 7 days, orally, found that protein synthesis activity in rat brain synaptosomes was significantly reduced by methyl mercury only at the symptomatic period, whereas significant impairment occurred in cerebral and cerebellar slices even during the latent period.

Doses of methyl mercury lower than those mentioned above also modified the pattern of protein synthesis in newborn (28) or 30-day-old (57,60,61) rats. Moreover, a single dose of methyl mercury (10 mg/kg intraperitoneally or subcutaneously) caused persistent depression of protein synthesis activity in cerebellar granule cells from 1 to 10 days, while only transient inhibition was observed in Purkinje cells and cerebral neurons (57,61). In these experiments, however, no neurotoxic signs were detected in the treated animals.

Stimulation of protein synthesis activity after methyl mercury dosing was also reported in the brain (2,28,50) and other nonneural tissues (2,30,43,51), probably reflecting recovery processes including gene activation (2) and proliferation of glial cells (50).

In peripheral nervous tissues, Cavanagh and Chen (8) found marked reduction of [^{14}C]glycine incorporation into the protein of dorsal root ganglia following administration of methyl mercury dicyandiamide (5 mg/kg daily, for 8 days, orally). This effect became apparent at the 7th day of treatment and persisted up to the symptomatic period. These biochemical findings were correlated with morphological changes occurring in intracellular structures involved in protein synthesis (7, 9). The protein synthesis activity of the sciatic nerve was significantly reduced at the 7th day, but showed a sharp increase above normal during the symptomatic period. This latter event was attributed to hyperfunction of Schwann cells (8). Similar results have been obtained by Omata et al. (46) using an improved method (18) in determining the protein synthesis activity. It was also demonstrated that dorsal roots, but not ventral roots, exhibited alterations in the protein synthesis activity resembling those seen in the sciatic nerve (46).

Effects of methyl mercury on protein synthesis were also investigated using mammalian cell lines. Mouse glioma cells showed similar extent of susceptibility to methyl mercury and mercury chloride added to the culture medium ($IC_{50} = 2 \times 10^{-5}$ M), while Yoshida ascite and myeloma cells exhibited stronger inhibition by methyl mercury (39).

III. CRITICAL ASSESSMENT OF EFFECTS OF METHYL MERCURY

A. Effects of Malnutrition

Since reduction in body weight gain associated with reduced food intake is a characteristic feature of experimental methyl mercury intoxication (9), the possibility must be considered that inhibition of protein synthesis merely reflects a nutritional disorder related to low food intake by the methyl mercury-treated animals.

Evidence that reduction in protein synthesis activity in the brain of rats exposed to methyl mercury was not based on malnutrition was obtained in in vitro studies measuring the incorporation of [^{14}C]leucine into proteins in brain slices (66) and cell-free systems (43). On the other hand, the reduction of protein synthesis activity in the liver of symptomatic rats proved to be clearly related to nutritional alterations (43).

B. Changes in the Amino Acid Pool and Uptake of Precursor Amino Acids

To determine the effect of methyl mercury on protein synthesis, most investigators have measured the rate of incorporation of precursor amino acids into protein by using either in vivo or in vitro systems relating the rate of protein labeling with that of protein synthesis. However, precise quantitation is difficult, since protein labeling is related not only to the rate of protein synthesis, but also to biochemical processes involving the precursor amino acids such as the uptake in brain tissue, rate of metabolic conversion, and the actual magnitude of the pool size (19). It seems unlikely that none of these factors other than the rate of protein synthesis is affected by methyl mercury intoxication. Therefore all these parameters should be taken into consideration when effects of methyl mercury on protein synthesis are assessed following administration of trace amounts of labeled amino acids. Omata et al. (44) have determined the protein synthesis rate in the brain of methyl mercury-treated rats using methods developed by Dunlop et al. (18) and Reith et al. (48) by flooding the whole precursor pool following injection of a large amount of [^{14}C]valine of low specific activity. The specific activity of the radiotracer remains constant in the brain during the experimental period, not only in control rats, but also in the methyl mercury-treated rats. In such instances the depression of cerebral protein synthesis was apparently comparable with that seen in in vivo experiments on methyl mercury-treated rats injected with trace amounts of labeled glycine (8) or leucine (20).

Omata et al. (46) further assessed the effect of methyl mercury on peripheral nervous tissues. In rats given a subcutaneous daily injection of 10 mg/kg of methyl mercury for 7 days, protein synthesis activity in the dorsal root ganglia was inhibited to the extent of 60% after 5 days. At this stage, no reduction in body weight or neurotoxic signs were observed. The decrease in protein synthesis persisted during the symptomatic period (day 15), when typical signs of organomercurial poisoning, such as crossing of the hind limbs, also appeared in the animals. On day 5 both the sciatic nerves and dorsal roots showed a slight reduction in valine incorporation. The incorporation of this amino acid returned to normal levels on day 10 and showed a marked increase during the symptomatic period. On the contrary, protein synthesis in the ventral roots exhibited a gradual decrease as the intoxication proceeded, reaching levels similar to those observed in the brain during the symptomatic period. These findings are consistent with the results obtained in morphological studies (11,12).

In recent experiments we found that the level of free amino acids was considerably altered in the brain of the rat after methyl mercury dosing, and the concentration of leucine in cerebrum of the symptomatic rats was significantly reduced (Mochizuki et al., in preparation). So, our previous results showing 44% decrease in [^{14}C]leucine incorporation into brain proteins (43) seem to be exaggerated owing to the greater dilution of [^{14}C]leucine with the preexisting leucine in the brain of the symptomatic rats as compared with that in controls.

C. Physiological Factors Influencing Methyl Mercury Action on Protein Synthesis

It has been observed that male animals are affected by methyl mercury poisoning more severely than female ones after acute treatment of mice (41) or chronic exposure of rats (36). Methyl mercury considerably alters the production of hormones in various animal species, including chickens (62) and rats (3). However, whether methyl mercury differentially affects protein synthesis in tissues of male and female animals is still unknown.

Based on results of lethality studies, old animals proved to be more susceptible to methyl mercury than adult ones (41); moreover neurotoxicity resulting from exposure to low levels of methyl mercury is especially severe in young animals (11,24).

In morphological studies reported by Syversen et al. (59) cerebellar granule cells were more susceptible in adult than in young rats following exposure to a single dose of methyl mercury.

The mechanisms by which the age may affect the vulnerability of the protein synthesic system in methyl mercury-treated animals remain to be investigated. Incorporation of [^{14}C]leucine into brain protein of 48-hr-old rats was not altered by in utero exposure to methyl mercury, 1 mg per pregnant mother (20). Joiner and Hupp (28) estimated the effects of methyl mercury on the rate of [^{14}C]leucine incorporation into brain protein over the critical period of brain development (1-21 days after birth) and found that the normal pattern of protein synthesis during development was considerably modified. These authors also suggested that glial and microneuron proliferation was disturbed by methyl mercury treatment. In this respect, however, the rapid change in amino acid levels that physiologically occurs during brain development (23) must be taken into account in order to appropriately assess the effect of methyl merucry on the free amino acid levels in the brain. The influence of methyl mercury on biochemical events related to protein synthesis activity during the development of the nervous system has been reviewed by Reuhl and Chang (49).

IV. MECHANISM OF ACTION OF METHYL MERCURY

A. Morphological Aspects

It has been suggested that one of the earliest morphological responses to methyl mercury treatment occurs in organelles that are involved in protein synthesis. Disruption of the cytoplasmic protein-synthesizing apparatus with fragmentation of rough endoplasmic reticulum and reduction in the ribosome population and polyribosome structure was observed in the peripheral nervous tissue of rats at an early stage of exposure to methyl mercury (7,26,70). In the central nervous system the appearance of these deleterious effects was somewhat delayed (11,12, 27).

Morphological studies in humans (25) and experimental animals (4,7, 59) also revealed that smaller neurons in the central nervous system are especially susceptible to methyl mercury (see also chapters by Cavanagh and T. Takeuchi in this book). These findings are ascribed to the relatively small ribosome content of these cell types, which therefore seem unable to tolerate the methyl mercury-induced disorganization of intracellular organelles (7). Accordingly, the protein-synthetic activities were most severely impaired in granule neuron cells isolated from the cerebellum of both symptomatic and asymptomatic rats (60,61).

In order to elucidate the sites of action of methyl mercury, knowledge of the patterns of accumulation and localization of this substance in the nervous tissue is necessary (33). In this respect a number of experimental investigations have been carried out at both cellular and subcellular levels (see Ref. 10 for a review). However, as pointed out by Yoshino et al. (72), results of studies dealing with the subcellular distribution of mercury must be interpreted with care. Discrepancy exists between the results of biochemical subcellular distribution and autoradiography studies which can at least in part be accounted for by the blood component contamination of the nuclear fraction obtained by fractionation techniques (see Ref. 10 for details). Indeed, the interaction of $CH_3{}^{203}HgCl$ with purified nuclei was found to be less than 1/100 that with the crude nuclear fraction obtained by conventional fractionation procedures (42). Nevertheless, a typical pattern of cerebral intracellular distribution of mercury has been identified showing the highest content of mercury in the cytosolic fraction, followed by mitochondria (or synaptosomes) and microsomes. The mercury content in ribosomes was found to be rather low (42,45,56). Implications of this subcellular distribution pattern in the events leading to disorders of protein synthesis are not clear at present, although it is known that a fraction of the cytosolic mercury in nervous tissue is associated with glutathione (63,69). Moreover, the low rates of methyl mercury biotransformation in and removal from the brain (42,45,56) clearly imply a rather persistent action of this compound on the toxicity targets in the nervous tissue. In the peripheral nervous system the highest accumulation of methyl mercury occurs in dorsal root ganglia, that is, at the level where inhibition of protein synthesis is the most extensive (46).

B. Biochemical Aspects

Protein synthesis is a complex process involving multiple events such as activation of amino acids, synthesis of aminoacryl transfer RNA, formation of the initiation complex, elongation of peptide chains through peptidyltransferase and translocation factors, and termination of chain elongation. It is unclear at present which step(s) is primarily or selectively blocked by methyl mercury. The putative site should contain active sulfhydryl groups that are known to avidly react with mercurials, causing alterations in a variety of cellular activities including tubulin polymerization (16,33,35). Aryl mercury compounds such as p-hydroxymercuribenzoate and p-chloromercuriphenylsulfonate were found to interact at 0.5-1 mM concentrations with certain protein components in 60 S subunits of liver ribosomes, leading to the inhibition of poly(U)-dependent phenylalanine incorporation (52). Although sulfhydryl groups essential for ribosome activities have been reported to exist exclusively on the larger subunits (1), Tiryaki et al. (64) showed that the in vitro binding of *Escherichia coli* phenylalanyl transfer RNA with rat-liver 40 S subunits was inhibited (I_{50} = 6 μM) by p-chloromercuribenzoate, while neither N-ethylmaleimide nor iodacetamide exerted an appreciable effect, even at 500 μM. According to Walker (68), a mercury derivative of transfer RNA species competitively inhibits amino acylation of the intact transfer RNA, but it is unclear whether such a form of transfer RNA is produced under in vivo conditions. In nonneural tissues (liver) methyl mercury was found to affect protein synthesis reactions by modifying the elongation factors (51). This indicates that the in vitro action of organic mercurials is not a simple one.

It has been suggested (15,65) that changes in cation translocation and decline in adenosine triphosphate levels might explain the decrease in protein synthesis activities in synaptosomal and microsomal preparations from rat brain after methyl mercury exposure. In this respect, it would be important to ascertain whether protein synthesis activities can be restored by addition of an adenosine triphosphate-generating system to the tissue preparation. Energy metabolism in nervous tissue is considerably affected by methyl mercury owing to modifications in the glycolytic pathways (10,22,53), tricarboxylic acid cycle (21,56), and respiratory chains (21,67). However, the primary target of the biochemical action of methyl mercury has not conclusively been determined.

A critical analysis of results obtained in in vitro experiments compared with those of in vivo studies would help in elucidation of the cellular sites involved in methyl mercury toxicity. Richardson and Murphy (50) compared the in vivo and in vitro effects of methyl mercury on brain protein synthesis. When brain mercury concentrations of 23 μg/g were attained in animals dosed with daily subcutaneous injections of 10 mg/kg of methyl mercury for 7 days, a 15% inhibition of the in vivo incorporation of [^{3}H]leucine was observed. Moreover, the in vitro incorporation of [^{3}H]leucine into protein showed about 20% inhibition when methyl mercury was added to brain homogenates at a concentration of 20 μg/g tissue, indicating that the in vivo and in vitro effects of methyl mercury on protein synthesis are quantitatively comparable. However, studies performed on the brain postmitochondrial supernatant produced different results (43,54). In this system, the amount of mercury needed to inhibit [^{14}C]leucine incorporation by about 70% was approximately 12 μg/mg protein. On the other hand, in the brain of symptomatic rats, [^{14}C]leucine incorporation was already reduced to about 70% of control when mercury concentrations in the postmitochondrial supernatant were as low as 0.4 μg/mg protein. These findings suggest that the primary targets of methyl mercury toxicity might be intracellular processes coupled with the formation and regulation of protein synthesis rather than the protein synthesis machinery per se.

Inhibition of protein synthesis resulting from addition of methyl mercury to the rat brain postmitochondrial supernatant is accompanied by the breakdown of the polyribosome structure mediated through activation of ribonuclease and inactivation of ribonuclease inhibitor (54). In contrast, injection of methyl mercury produced no appreciable changes in the cerebral polyribosome profiles in vivo (2,43) under such conditions that protein synthesis activities were considerably affected. These results provide additional support to the concept that the mechanism of action of methyl mercury is different between in vivo and in vitro conditions. Mercury chloride was about 10 times more effective than methyl mercury in suppressing the protein-synthetic activities in the postmitochondrial fraction of the brain homogenate (54,59) and glioma cells (40).

C. Changes in Metabolic Activities Related to Protein Synthesis

One may assume that changes in nucleic acid metabolism caused by methyl mercury would modify translational activities through alterations in the process of gene replication and gene transcription. The sensitivity of nucleic acid metabolism to methyl mercury in neural cells (40, 47) was essentially not different from that seen in nonneural elements, and differences in sensitivity related to the neural cell types (47) and mercurial species (38) have also been described.

Based on investigations on unscheduled DNA synthesis in the cat brain following chronic methyl mercury intoxication, Miller and Miller (34) have proposed that impairment in mitochondrial function due to genetic damage may give rise to a defect in the energy supply required for the maintenance of protein synthesis in neurons.

Reduction of RNA content and changes in its base composition in the neurons of spinal ganglia (13,14) have also been advocated as possible causes for the extensive morphological alterations occurring at an early stage of methyl mercury intoxication (11,12). On the other hand, according to Carmichael and Cavanagh (5), the inhibition of ribosomal RNA synthesis by methyl mercury is not directly responsible for the cytoplasmic ribosome loss in the spinal ganglia neurons as indicated by morphological observations and the results of autoradiographic studies in which the incorporation of [^{3}H]uridine into RNA was measured. The reason for the apparent discrepancy with the result of Chang et al. (13,14) mentioned above has been discussed by Carmichael and Cavanagh (5).

Recently, Nagumo et al. (37) determined RNA-synthetic activity in slices of dorsal root ganglia of methyl mercury-treated rats and concluded that the observed early depression in protein synthesis was not related to the decrease in the transcriptional activity.

Numerous enzyme systems were affected to various extents by methyl mercury (10,34,49). Discussion of these studies is beyond the scope of the present review. Studies have not yet been carried out as to whether alterations in these enzyme activities are related to changes in the synthesis of the individual enzyme molecules.

V. CONCLUDING REMARKS AND SUMMARY

When effects of methyl mercury on the protein synthesis in nervous tissue are evaluated, the basic questions to be answered are the following: (a) Does the observed inhibition of protein synthesis originate primarily from methyl mercury intoxication or represent a secondary phenomenon related to disorders in other cellular activities? (b) Is the apparent inhibition by methyl mercury due to blockade of functioning of the protein synthesis machinery per se or caused by changes in activities closely related to the protein synthesis pathway (i.e., cellular uptake or intracellular metabolism of amino acids) that affects the amino acid pool? (c) How is the inhibition of protein synthesis activity causally related to the neurotoxicity of methyl mercury?

It has been shown that the lowering of protein synthesis activity in methyl mercury-treated animals was not solely due to nutritional disorders derived from the reduced food intake. Moreover, results from the authors' laboratory have demonstrated that the methyl mercury-induced depression of protein synthesis in central and peripheral nervous tissues was due to real inhibition of the protein synthesis machinery, even though certain reported effects of methyl mercury on incorporation of labeled amino acids into the protein must be interpreted with caution. In this respect, determination of amino acid levels has often been carried out for gross brain areas and it is possible that the methyl mercury-associated changes in uptake and metabolism of amino acids might vary, depending on the particular brain areas and different cell types.

Finally, it remains to be clarified at the molecular level which step or site in protein synthesis pathway is most sensitive to methyl mercury intoxication. There is morphological evidence that peripheral neuropathy is a prominent feature in acutely intoxicated rats. In keeping with this finding, the suppression of protein synthesis activity in the dorsal ganglia occurred earlier and was more marked as compared with that in the brain. Whether these features are common to other animal species remains to be established.

It must be pointed out that changes in protein synthesis activity have generally been investigated using the whole brain tissue. Thus the possibility cannot be excluded that a specific mechanism of methyl mercury neurotoxicity involves the defective production of critical protein or enzyme species, leading to disorders in neural function.

REFERENCES

1. Baliga, B. S., Munro, H. N. 1971. Specificity of mammalian transferase II binding to ribosomes. *Nature New Biol. 233*:257-258.
2. Brubaker, P. E., Klein, R., Herman, S. P., Lucier, G. W., Alexander, L. T., Long, M. D. 1973. DNA, RNA, and protein synthesis in brain, liver, and kidneys of asymptomatic methylmercury treated rats. *Exp. Mol. Pathol. 18*:263-280.
3. Burton, G. V., Meike, A. M. 1980. Acute and chronic methylmercury poisoning impairs rat adrenal and testicular function. *J. Toxicol. Environ. Health 6*:597-606.
4. Carmichael, N., Cavanagh, J. B. 1975. Some effects of methylmercury salts on the rabbit nervous system. *Acta Neuropathol. 32*:115-125.
5. Carmichael, N., Cavanagh, J. B. 1976. Autoradiographic localization of ^{3}H-uridine in spinal ganglion neurons of the rat and the effects of methylmercury poisoning. *Acta Neuropathol. 34*:137-148.
6. Cavanagh, J. B. 1969. Toxic substances and the nervous system. *Br. Med. Bull. 25*:268-273.
7. Cavanagh, J. B. 1977. Metabolic mechanisms of neurotoxicity caused by mercury, in *Neurotoxicology* (Roizin, L., Shiraki, H., Grčević, N., eds.), Raven Press, New York, pp. 283-288.
8. Cavanagh, J. B., Chen, F. C. K. 1971. Amino acid incorporation in protein during the "silent phase" before organo-mercury and p-bromophenylacetylurea neuropathy in the rat. *Acta Neuropathol. 19*:216-224.

9. Cavanagh, J. B., Chen, F. C. K. 1971. The effects of methylmercury-dicyandiamide on the peripheral nerves and spinal cord of rats. *Acta Neuropathol.* *19*:208-215.
10. Chang, L. W. 1977. Neurotoxic effects of mercury—A review. *Environ. Res.* *14*:329-373.
11. Chang, L. W., Annau, Z. 1984. Developmental neuropathology and behavioral teratology of methylmercury, in *Neurobehavioral Teratology* (Yanai, J., ed.), Elsevier, Amsterdam, pp. 405-431.
12. Chang, L. W., Hartman, H. A. 1972. Ultrastructural studies of the nervous system after mercury intoxication. II. Pathological changes in the nerve fibers. *Acta Neuropathol.* *20*:316-334.
13. Chang, L. W., Desnoyers, P. A., Hartman, H. A. 1972. Quantitative cytochemical studies of RNA in experimental mercury poisoning. I. Changes in RNA content. *J. Neuropathol. Exp. Neurol.* *31*:489-501.
14. Chang, L. W., Desnoyers, P. A., Hartman, H. A. 1973. Quantitative cytochemical studies of RNA in experimental mercury poisoning. II. Changes in the base composition and ratios. *Acta Neuropathol.* *23*:77-83.
15. Cheung, M., Verity, M. A. 1981. Methylmercury inhibition of synaptosome protein synthesis: Role of mitochondrial dysfunction. *Environ. Res.* *24*:286-298.
16. Clarkson, T. W. 1972. The pharmacology of mercury compounds. *Annu. Rev. Pharmacol.* *12*:375-406.
17. Damstra, T. 1978. Neurochemistry and toxicology: Overview. *Environ. Health Perspect.* *26*:121-124.
18. Dunlop, D. S., Elden, W. V., Lajtha, A. 1975. A method for measuring brain protein synthesis rates in young and adult rats. *J. Neurochem.* *24*:337-344.
19. Dunlop, D. S., Lajtha, A., Toth, J. 1977. Measuring brain protein metabolism in young and adult rats, in *Mechanism, Regulation and Special Function of Protein Synthesis in the Brain*, Vol. 2 (Roberts, S., Lajtha, A., Gispen, W. H., eds.), Elsevier-North Holland, Amsterdam, pp. 79-96.
20. Farris, F. F., Smith, J. C. 1975. In vivo incorporation of ^{14}C-leucine into brain protein of methylmercury treated rats. *Bull. Environ. Contam. Toxicol.* *13*:451-455.
21. Fox, J. H., Patel-Mandlik, K., Cohen, M. M. 1975. Comparative effects of organic and inorganic mercury on brain slice respiration and metabolism. *J. Neurochem.* *24*:757-762.
22. Grundt, I. K., Roux, F., Treich, L., Loriette, C., Raulin, J., Fournier, E. 1982. Effects of methyl mercury and triethyllead on Na^+-K^+ ATPase and pyruvate dehydrogenase activities in glioma C_6 cells. *Acta Pharmacol. Toxicol.* *51*:6-11.
23. Himwich, W. A., Agrawal, H. C. 1969. Amino acids, in *Handbook of Neurochemistry*, Vol. 1 (Lajtha, A., ed.), Plenum, New York, pp. 33-52.

24. Hughes, R., Besler, R., Brett, C. W. 1975. Behavioral impairment produced by exposure to subclinical amounts of methylmercury chloride. *Environ. Res.* *10*:54-58.
25. Hunter, D., Russell, D. S. 1954. Focal cerebral and cerebellar atrophy in human subject due to organic mercury compounds. *J. Neurol. Neurosurg. Psychiatry* *17*:235-254.
26. Jacobs, J. M., Carmichael, N., Cavanagh, J. B. 1975. Ultrastructural changes in the dorsal root and trigeminal ganglia of rats poisoned with methyl mercury. *Neuropathol. Appl. Neurobiol.* *1*:1-19.
27. Jacobs, J. M., Carmichael, N., Cavanagh, J. B. 1977. Ultrastructural changes in the nervous system of rabbits poisoned with methyl mercury. *Toxicol. Appl. Pharmacol.* *39*:249-261.
28. Joiner, F. E., Hupp, E. W. 1979. Developmental and methylmercury effects on brain protein synthesis. *Arch. Environ. Contam. Toxicol.* *8*:465-470.
29. Kurland, L. T., Faro, S. N., Siedler, H. 1960. Minamata disease. *World Neurol.* *1*:370-395.
30. Lapin, C. A., Carter, D. E. 1981. Early indices of methyl mercury toxicity and their use in treatment evaluation. *J. Toxicol. Environ. Health* *8*:767-776.
31. Löfroth, G. 1973. The mercury problem: A review at midway, in *Trace Substances in Environmental Health VI* (Hemphill, D. D., ed.), University of Missouri, Columbia, pp. 63-70.
32. Lund-Larsen, T., Berg, T. 1973. Effect of heavy-metal ions and dexamethasone on the release of messenger-like ribonucleoprotein from liver nuclei of normal and adrenalectomized, cortisol-treated rats. *Hoppe-Seyler's Z. Physiol. Chem.* *354*:1334-1338.
33. MacGregor, J. T., Clarkson, T. W. 1974. Distribution, tissue binding and toxicity of mercurials, in *Advance in Experimental and Medical Biology*, Vol. 48 (Friedman, M., ed.), Plenum, New York, pp. 463-503.
34. Miller, C. T., Miller, D. R. 1979. Biochemical toxicology of methylmercury, in *Effects of Mercury in the Canadian Environment.* National Research Council of Canada, No. 16739, Ottawa, Canada, pp. 106-125.
35. Miura, K., Inokawa, M., Imura, N. 1984. Effects of methylmercury and some metal ions on microtubule networks in mouse glioma cells and in vitro tubulin polymerization. *Toxicol. Appl. Pharmacol.* *73*:218-231.
36. Munro, I. C., Nera, E. A., Charbonneau, S. M., Junkins, B., Zawidzka, Z. 1980. Chronic toxicity of methylmercury in the rat. *J. Environ. Pathol. Toxicol.* *3*:437-447.
37. Nagumo, S., Omata, S., Sugano, H. 1984. Alteration in RNA synthesis in the dorsal root ganglia of methylmercury-treated rats. *Arch. Toxicol.*, in press.

38. Nakada, S., Imura, N. 1980. Stimulation of DNA synthesis and pyrimidine deoxyribonucleotide transport system in mouse glioma and mouse neuroblastoma cells by inorganic mercury. *Toxicol. Appl. Pharmacol.* *53*:24-28.
39. Nakada, S., Nomoto, A., Imura, N. 1980. Effect of methylmercury and inorganic mercury on protein synthesis in mammalian cells. *Ecotoxicol. Environ. Safety* *4*:184-190.
40. Nakada, S., Saito, H., Imura, N. 1981. Effect of methylmercury and inorganic mercury on the nerve growth factor-induced neurite outgrowth in chick embryonic sensory ganglia. *Toxicol. Lett.* *8*:23-28.
41. Nomiyama, K., Matsui, K., Nomiyama, H. 1975. Effects of temperature and other factors on the toxicity of methylmercury in mice. *Toxicol. Appl. Pharmacol.* *56*:392-398.
42. Omata, S., Sakimura, K., Sugano, H. 1975. Intracellular distribution of ^{203}Hg-labelled methylmercury chloride in rat tissues, in *Studies on the Heatlh Effects of Alkylmercury in Japan* (Tsubaki, T., ed.), Environmental Agency, Tokyo, Japan, pp. 137-143.
43. Omata, S., Sakimura, K., Tsubaki, H., Sugano, H. 1978. In vivo effect of methylmercury on protein synthesis in brain and liver of the rat. *Toxicol. Appl. Pharmacol.* *44*:367-378.
44. Omata, S., Horigome, T., Momose, Y., Kambayashi, M., Mochizuki, M., Sugano, H. 1980. Effect of methylmercury chloride on the in vivo rate of protein synthesis in the brain of the rat: Examination with the injection of a large quantity of ^{14}C-valine. *Toxicol. Appl. Pharmacol.* *56*:207-215.
45. Omata, S., Satoh, M., Sakimura, K., Sugano, H. 1980. Time-dependent accumulation of inorganic mercury in subcellular fractions of kidney, liver, and brain of rats exposed to methylmercury. *Arch. Toxicol.* *44*:231-241.
46. Omata, S., Momose, Y., Ueki, H., Sugano, H. 1982. In vivo effect of methylmercury on protein synthesis in peripheral nervous tissues of the rat. *Arch. Toxicol.* *49*:203-214.
47. Prasad, K. N., Nobles, E., Ramanujam, M. 1979. Differential sensitivity of glioma cells and neuroblastoma cells to methylmercury toxicity in cultures. *Environ. Res.* *19*:189-201.
48. Reith, M. E. A., Schotman, P., Gispen, W. H. 1978. Measurement of in vivo rates of protein synthesis in brain, spinal cord, heart, and liver of young versus adult rats, intact versus hypophysectomized rats. *J. Neurochem.* *30*:587-594.
49. Reuhl, K. R., Chang, L. W. 1979. Effects of methylmercury on the development of the nervous system: A review. *Neurotoxicol.* *1*21-55.
50. Richardson, R. J., Murphy, S. D. 1974. Neurotoxicity produced by intracranial administration of methylmercury in rats. *Toxicol. Appl. Pharmacol.* *29*:289-300.

51. Sauvé, G. J., Nicholls, D. M. 1981. Liver protein synthesis during the acute response to methylmercury administration. *Int. J. Biochem.* *13*:981-990.
52. Sjoqvist, A., Hultin, T. 1973. Conformational effects of mercurials on rat liver ribosomes: A comparison between the unmasking of a shielded protein in the 60S subunit by phenyl mercurials and EDTA. *Chem. Biol. Interact.* *6*:131-148.
53. Snell, K., Ashby, S. L., Barton, S. J. 1977. Disturbance of perinatal carbohydrate metabolism in rats exposed to methylmercury in utero. *Toxicology* *8*:277-283.
54. Sugano, H., Omata, S., Tsubaki, H. 1975. Methylmercury inhibition of protein synthesis in brain tissues I. Effects of methylmercury and heavy metals on cell-free protein synthesis in rat brain and liver, in *Studies on the Health Effects of Alkylmercury in Japan* (Tsubaki, T., ed.), Environmental Agency, Tokyo, Japan, pp. 129-136.
55. Suzuki, T. 1979. Dose-effect and dose-response relationships of mercury and its derivatives, in *The Biochemistry of Mercury in the Environment* (Nriagu, J. O., ed.), Elsevier/North-Holland, New York, pp. 399-431.
56. Syversen, T. L. M. 1974. Distribution of mercury in enzymatically characterized subcellular fractions from the developing rat brain after injections of methylmercuric chloride and diethylmercury. *Biochem. Pharmacol.* *23*:2999-3007.
57. Syversen, T. L. M. 1977. Effects of methylmercury on in vivo protein synthesis in isolated cerebral and cerebellar neurons. *Neuropathol. Appl. Neurobiol.* *3*:225-236.
58. Syversen, T. L. M. 1981. Effects of methyl mercury on protein synthesis in vitro. *Acta Pharmacol. Toxicol.* *49*:422-426.
59. Syversen, T. L. M., Totland, G., Flood, P. R. 1981. Early morphological changes in rat cerebellum caused by a single dose of methylmercury. *Arch. Toxicol.* *47*:101-111.
60. Syversen, T. L. M. 1982. Effects of repeated dosing of methyl mercury on in vivo protein synthesis in isolated neurons. *Acta Pharmacol. Toxicol.* *50*:391-397.
61. Syversen, T. L. M. 1982. Changes in protein and RNA synthesis in rat brain neurons after a single dose of methylmercury. *Toxicol. Lett.* *10*:31-34.
62. Thaxton, J. P., Gilbert, J., Hester, P. Y., Brake, J. 1982. Mercury toxicity as compared to adrenocorticotropin-induced physiological stress in the chicken. *Arch. Environ. Contam. Toxicol.* *11*:509-514.
63. Thomas, D. J., Smith, J. C. 1979. Partial characterization of a low-molecular weight methylmercury complex in rat cerebrum. *Toxicol. Appl. Phnrmacol.* 47547-556.

64. Tiryaki, D., Üçer, U., Bermek, E. 1976. The effect of sulfhydryl reagents upon the activity of 40S ribosomal subunits. *Experientia 32*:1270-1271.
65. Verity, M. A., Brown, W. J., Cheung, M. 1975. Organic mercurial encephalopathy: In vivo and in vitro effects of methyl mercury on synaptosomal respiration. *J. Neurochem. 25*:759-766.
66. Verity, M. A., Brown, W. J., Cheung, M., Czer, G. 1977. Methyl mercury inhibition of synaptosome and brain slice protein synthesis: In vivo and in vitro studies. *J. Neurochem. 29*:673-679.
67. Von Verg, R., Lijoi, A., Smith, C. 1979. Oxygen consumption of rat tissue slices exposed to methylmercury in vitro. *Neurosci. Lett. 14*:309-314.
68. Walker, R. T. 1974. The preparation and properties of a stable mercury derivative of transfer RNAs from *Escherichia coli*. *Arch. Biochem. Biophys. 162*:481-486.
69. Winroth, G., Carlstedt, I., Karlson, H., Berlin, M. 1981. Methyl mercury binding substances from the brain of experimentally exposed squirrel monkeys (*Saimiri sciureus*). *Acta Pharmacol. Toxicol. 49*:168-173.
70. Yip, R. K., Chang, L. W. 1981. Vulnerability of dorsal root neurons and fibers toward methylmercury toxicity: A morphological evaluation. *Environ. Res. 26*:152-167.
71. Yoshino, Y., Mozai, T., Nakao, K. 1966. Biochemical changes in the brain in rats poisoned with an alkylmercury compound, with special reference to the inhibition of protein synthesis in brain cortex slices. *J. Neurochem. 13*:1223-1230.
72. Yoshino, Y., Mozai, T., Nakao, K. 1966. Distribution of mercury in the brain and its subcellular units in experimental organic mercury poisonings. *J. Neurochem. 13*:397-406.

18

Neurotoxicity of Selected Metals

Luigi Manzo
University of Pavia Medical School, Pavia, Italy

Kenneth Blum
University of Texas Health Science Center at San Antonio, San Antonio, Texas

E. Sabbioni
Radiochemistry and Nuclear Chemistry Division, EC Joint Research Center, Ispra, Varese, Italy

I. INTRODUCTION

This chapter will discuss biological and clinical aspects of metal-induced diseases of the nervous system other than those caused by lead, mercury, and cisplatin, which are covered in other sections of the book. The neurological effects of arsenic, thallium, tin, manganese, gold, aluminum, and bismuth will be examined in separate sections. Metals which are recognized for their neurotoxic potential and sources of human exposure are listed in Table 1.

II. ARSENIC

Arsenic toxicity can be caused by exposure to a variety of arsenic-based compounds used as pesticides, mordants, paints, wood preservatives, and medicinal agents (49). Arsenic was once a favorite instrument for homicide. Today most cases of intoxication are the result of either children accidentally ingesting arsenic-based insecticides or

Table 1 Metal-Induced Diseases of the Nervous System in Humans

	Sources of exposure			
Element	Occupational	Iatrogenic, medicinal	Environmental, nonoccupational	Accidental, suicidal
Aluminum		*		
Arsenic	*	*	*	*
Bismuth		*		
Boron	*			
Gold		*		
Lead	*		*	*
Manganese	*			
Mercury	*		*	*
Platin		*		
Tellurium	*			
Thallium	*	*	*	*
Tin	*	*		

weed killers, or individuals deliberately taking arsenic in a suicide attempt (42). Major outbreaks of mass accidental poisoning have also occurred following arsenic absorption from contaminated food, water, and alcoholic beverages (15,28).

A. Acute Poisoning

Gastrointestinal disturbances occur immediately after ingestion, and death may follow from circulatory collapse, hepatic failure, or toxic encephalitis. The course of the disease is often complicated by renal failure resulting from direct toxicity of arsenic to the kidney or, more frequently, from prolonged hypotension and renal ischemia leading to acute tubular necrosis. If the acute stage is survived, symptoms of peripheral neuropathy become apparent after a latent interval of 1-3 weeks

In milder cases, numbness and a "pins and needles" sensation usually confined to the lower limbs may be the only symptom. With serious intoxications, sensory involvement may be prominent and, in combination with distal weakness, may even impair ambulation.

Dermatological symptoms including skin pigmentation, hyperkeratosis of the hands and feet, and leukonychia, that is, transverse white bands in the fingernails (Mee's lines), have been described in 50-60% of patients with acute arsenic poisoning (15). Peripheral blood and bone marrow abnormalities may also occur (49).

B. Chronic Intoxication

Occupational exposure to inorganic arsenic has occurred mainly in the smelting industry and in the manufacture and application of arsenic-based pesticides (26,28). Chronic nonoccupational poisoning has also arisen from abuse of arsenic-containing "tonic" preparations (43), as well as from deliberate self-administration of arsenic by autolesionistic individuals (38).

Ingesting as little as 3 mg of inorganic arsenic per day may give rise in a few weeks to symptoms of toxicity. Neurological manifestations are prominent. While encephalopathy has been reported, peripheral neuritis is more common. Numbness, tingling, burning, and less frequently, diplopia, shooting pains, and muscle cramps are noted. Examination may reveal decreases in touch, pain, and temperature sensations; impaired vibratory and position senses; and occasional fasciculations (26). Tendon reflexes are diminished or absent, and grasp weakness and wrist and foot drops are frequent (15). Electrodiagnostic studies indicate a mild or moderate slowing of motor nerve conduction and marked abnormalities of the sensory action potentials (15).

The course of the disease is variable. Patients with mild sensorimotor disturbances recover completely and quickly, but recovery from an acute attack or after cessation of exposure during chronic poisoning may take as long as 3 years, and residual disability is to be expected in those severely affected initially (42).

C. Mechanisms of Neurotoxicity

The mechanisms by which arsenicals exert their harmful effects on the nervous system have been insufficiently studied. There is little information on the metabolism and movements of arsenic in tissue cells in vivo, and no one has succeeded in producing a reasonable animal model of arsenic neuropathy (57). Moreover, most unfortunately, the present knowledge of the arsenic interaction with the nervous tissue is largely based on data obtained in the rat, that is, in an animal species exhibiting patterns of arsenic metabolism quite dissimilar to those seen in humans (61).

The biological effects of arsenicals vary with the chemical form and the oxidation state of the arsenic atom. The trivalent arsenic is considerably more toxic than the pentavalent forms (28).

One manifestation shared by all the arsenicals is the inhibition of cellular respiration. Accumulation into mitochondria with uncoupling of phosphorylation and mitochondrial swelling have been observed in response to both trivalent and pentavalent arsenic (28). It has been suggested that inorganic arsenicals exert their toxic effects following metabolic conversion to arsenous acid, $HAs(OH)_2$, which can react with SH– groups of enzymes to form very stable dithioesters (40). In this respect, the thiol moiety of the pyruvate dehydrogenase complex is thought to be a highly sensitive target because of the strong affinity of arsenous acid for the disulfhydryl lipoic acid component of this system. The reaction generates irreversible arsenic complexes with concurrent inhibition of lipoic acid activity and reduced oxidation of pyruvate (28).

As discussed elsewhere in this book (see Cavanagh in this book), any impairment in energy production or transformation processes is expected to produce adverse consequences on neuronal functioning, particularly in the peripheral nervous system, being the nervous tissue uniquely dependent upon glucose as its substrate for energy metabolism. Thus the biochemical effects of arsenic are consistent with an "energy deprivation" process as an event leading to toxic neuropathy of the distal axonopathy type (67).

III. THALLIUM

A. Acute Poisoning

Thallium toxicosis most commonly occurs following ingestion of rodenticides containing thallium sulfate (10,18,54,59). Severe symptoms may result even with doses as low as 6 mg Tl/kg body weight; 14 mg Tl/kg is often fatal.

When large doses are taken, paresthesias, lethargy, delirium, myocardial abnormalities, convulsion, and coma appear soon after ingestion and are followed within a few days by respiratory failure and death. In the less severe cases, the onset of symptoms may be insidious. Vomiting and abdominal pain are common, but the patient may have only nausea and anorexia. Clinical deterioration is often manifested by rapidly progressive peripheral neuropathy which becomes prominent in the second or third week. Painful paresthesias in the lower limbs and numbness in the fingers and toes with loss of pin-prick and touch sensations are common. Gait may be impaired because of pain in the joints and feet (54). Extrapyramidal manifestations, convulsions, and signs of cranial nerve involvement including ptosis, facial nerve paralysis, disconjugate eye movements, and optic neuritis may develop (10). Sleep disorders, emotional lability, and hallucinatory and paranoid symptoms can also be observed.

Alopecia is characteristic in the late stages of intoxication and, as in arsenic poisoning, may be accompanied by nail dystrophy (Mee's lines). Other nonneurological symptoms include tachycardia and retrosternal pain associated with electrocardiographic abnormalities (63).

Recovery from thallium poisoning requires months and may be incomplete. The most common sequelae are giddiness, lack of concentration, intense headache, memory failure, and emotional disturbances often associated with gradual decay of the patient's intelligence.

B. Chronic Intoxication

Serious complications have resulted in the past from the medicinal use of thallium as a treatment for syphilis, gonorrhea, and dysentery, or as a depilatory for treating ringworm of the scalp (63). Chronic intoxication has also been described in subjects recovering thallium from flue dust of sulfuric acid waters or in plants handling thallium compounds for production of rodenticides, optic glass, dyes, and artificial diamonds (66).

The effects of chronic thallotoxicosis resemble those of acute poisoning. Excitation and insomnia are often the initial symptoms. After exposure for weeks or months leg joint pain, weakness, and sometimes polyneuritis occur. The hair may fall out after a few months. Anorexia, vomiting, weight loss, depression, hysterical laughter, cardiac disturbances, and albuminuria are also noted.

In recent years, increasing attention has also been focused on the potential health hazards of thallium as a trace pollutant (66). Thallium is released into the environment by various anthropogenetic sources, including copper, zinc, and lead smelters, coal mines, and coal-burning power plants. Localized accidents of environmental pollution by thallium have been described (6).

The health effects of the prolonged intake of minute amounts of thallium are unknown, but the results of recent studies in animals have indicated that thallium may act as a cumulative neurotoxic element in the case of continuous low-level absorption (48).

Recently sleep disturbances, headache, and mild polyneuritic symptoms have been described in subjects living around a thallium-emitting cement plant. They also exhibited accumulation of thallium in hair and excessive thallium levels in urine (6).

C. Perinatal Exposure

Thallium easily crosses the placenta (25,62) and the human fetus may suffer from the transplacentally acquired thallotoxicosis (63). Reduced fetal movements and characteristic manifestations of thallium toxicity in the newborn were seen in cases of acute thallium poisoning occurring during pregnancy. No data are available on the neurological consequences of thallium absorption during development. Following injection of minute amounts of ^{201}Tl-labeled thallium sulfate in pregnant rats, the radiotracer was rapidly concentrated in the fetal brain, reaching peak levels similar to those measured in the maternal brain. On the other hand, a faster decline in cerebral thallium concentrations was observed in the fetus as compared with the adult animal (25).

D. Mechanisms of Action

1. *Peripheral Neuropathy*

Thallium can substitute for potassium in the activation of several potassium-dependent enzymes and may interfere with some vital potassium-dependent processes in biological systems (63). Thallium interaction with the membrane (Na,K)-activated adenosinetriphosphatase is thought to play a relevant role in patterning movements of the thallium ion in tissue cells and apparently accounts for the remarkable tendency of thallium to concentrate intracellularly.

There is considerable evidence suggesting that the intracellular accumulation of thallium in the nervous tissue is associated with alterations in biochemical activities related to protein metabolism, transmembrane ion transport, and energy-producing reactions (63). Thallium is actively concentrated into mitochondria, and high concentrations of thallium uncouple the mitochondrial oxidative phosphorylation (52).

In recent years special toxicological significance has been attributed to the tendency of thallium to form insoluble complexes with riboflavin, since this process might result in the depletion of flavine adenine nucleotides and flavoproteins required in steps of pyruvate metabolism and the electron transport chain (9). The disturbance in mechanisms providing energy for the maintenance of the axon's integrity may precipitate nerve fiber damage. This interpretation clearly brings thallium-induced peripheral nervous system disorders into line with arsenic neuropathy.

It is significant that alopecia associated with thallium intoxication has also been attributed to disorders in the energy-transforming and energy-producing processes which are essential for mitotic activity in the hair follicle (9).

2. *Other Neurological Effects*

Tachycardia, hypertension, and certain digestive disorders in thallium poisoning have been interpreted as neurogenic dysfunctions related to toxic vagus nerve damage, but direct action of thallium on the myocardium as well as on vascular and intestinal smooth muscle has also been documented (63).

The mechanisms underlying the central nervous system effects of thallium have been poorly investigated. The rate of lipid peroxidation in the corpus striatum was significantly increased by repeated administration of sublethal doses of thallium acetate in the rat (36). The striatum is particularly abundant with monoamines, and free radical-mediated damage in this area may lead to dopamine depletion (24). Indeed, reduced levels of striatal dopamine and serotonin have been reported in thallium-intoxicated rats (37). The interaction with central neurotransmitter systems may at least in part explain the origin of extrapyramidal manifestations occurring during thallium poisoning. Indirect evidence for a monoaminergic deficit associated with thallium's central nervous system toxicity was provided by various experimental and clinical findings, including the thallium-induced increase in the spontaneous discharge rate of Purkinje neurons in rat cerebellum (51), the high urinary level of catecholamine products in thallium poisoning, and the favorable response to the application of L-dopa in the treatment of choreiform sequellae of human thallotoxicosis (see Ref. 63 for references).

Thallium poisoning was found to inhibit the activity of certain enzymes in the rat striatum, including monoamine oxidase, guanine deaminase, and succinic dehydrogenase, whereas the striatal protein content was increased, probably because of depletion of enzymes responsible for neuronal protein catabolism (35). The possible implications of these biochemical effects of thallium still remain to be clarified.

IV. ORGANOTINS

A. Triethyl Tin

Organotin compounds are increasingly used as biocides, catalysts, preservatives, and polymer stabilizers (46). Their neurotoxicity was dramatically brought home in 1953-1954 by an outbreak of medicinal intoxication in France. A total of 290 people were intoxicated, and 110 of them died after using an oral preparation with the trade name Stalinon used to treat boils and other cutaneous infections. The commercial product consisted of diethyl tin diodide, but contained variable amounts of triethyl tin, a proven neurotoxic agent causing brain edema and myelinopathy (71).

Pathological and clinical findings of the Stalinon accident were reviewed by Foncin and Gruner (27). Nausea, vomiting, vertigo, intense headache, photophobia, altered consciousness, and visual impairment due to papilledema appeared with a sudden onset after a latent period of about 4 days. Convulsive manifestations were recorded in about 10% of the victims. Paraplegia of the flaccid type involving the abdominal and respiratory muscles occurred in most cases in conjunction with the severe encephalic symptoms. Sensory disturbances, hyporeflexia, and loss of sphincter control were constant.

Death occurred after an acute course of usually 4-10 days as the consequence of deep coma or, more frequently, hyperacute intracranial hypertension. About one-third of the patients recovered with sequelae involving the psychic and intellectual sphere or the visual system. Irreversible complete flaccid paraplegia was observed in six cases (27).

1. *Neuropathological Findings*

Postmortem studies showed pathological changes characterized by diffuse edema in the central nervous system white matter without thinning of the cortex. This edema was apparently formed by splitting of myelin lamellae slong the intraperiod lines, with consequent vacuolar development within the myelin sheaths. There were minor axonal changes and vascular alterations. In one single case studied the peripheral nerve was found to be normal (27).

Edema of the white matter and spinal cord with extensive destruction of the myelin sheath was also reported as a prominent feature of triethyl tin neurotoxicity in laboratory animals (71). For this reason, triethyl tin has attracted considerable attention in recent years as a model for brain edema and as a prototype for the effects of a series of neurotoxic agents, such as hexachlorophene, which cause primary noninflammatory demyelination with axon preservation.

2. *Mechanism of Toxicity*

Low concentrations of triethyl tin were found to inhibit mitochondrial adenosinetriphosphatase activity, oxidative phosphorylation, and glucose oxidation in both rat brain slices and in vivo (see Refs. 22 and 29 for references). Selective binding of triethyl tin to purifed rat brain myelin has also been observed, suggesting a primary interaction of this compound with the myelin lamellae in brain. Peripheral nerve myelin proved to be much more resistant (45). The effects of triethyl tin on energy metabolism could produce the osmotic imbalance that gives rise to glial swelling (45).

On the other hand, recent studies have demonstrated the capability of triethyl tin to produce neurophysiological and behavioral changes without induction of cerebral edema. Using a maximal electroshock seizure test Fox and Doctor (29) observed dose-dependent changes in the neuronal excitability in adult rats exposed either acutely or subacutely to triethyl tin. Neonatal rats injected with triethyl tin also exhibited decreases in maximal electroshock seizure severity during development, whereas a long-term increase in seizure severity was observed in the adult animals developmentally exposed to this agent (22).

Based on the observation that triethyl tin can mediate a chloride-hydroxide exchange across biological membranes, it has been proposed that the triethyl tin-induced disorders in neuronal excitability may reflect specific alterations in water, electrolyte, and acid-base balance (29). However, changes involving central neurotransmitter systems may also have a role. Both decreases in the brain levels of norepinephrine and serotonin in triethyl tin-treated animals and the preferential interaction of triethyl tin with the α-adrenergic and GABAergic systems have been recently documented (29). In addition, the uptake of taurine, a neuroactive compound that strongly inhibits neuronal activity, was found to be the most sensitive to triethyl tin (39).

Hemorrhage with patchy neuronal necrosis in the brain and cerebellum (68) and abnormalities in the central nervous system myelination (70) have been reported in rodents as a result of the early postnatal exposure to triethyl tin.

B. Other Tin Compounds

While the neurotoxicity of metallic tin and inorganic tin compounds is negligible (27), several organic tin compounds other than triethyl tin can potentially induce neurological disorders. One of these agents is tetraethyl tin, which is thought to be converted to triethyl tin in vivo (71).

Trimethyl tin is also a potent neurotoxicant, inducing learning deficits, aggressive behavior, tremors, hyperexcitability, and seizures in experimental animals. In human exposure, trimethyl tin caused headache, memory disturbances, loss of vigilance, disorientation, and seizures (see Refs. 13 and 14 for references). Unlike triethyl tin, trimethyl tin is only slightly or not at all myelinotoxic, but mainly causes degeneration and necrosis in various regions of the central nervous system. Mice treated with trimethyl tin, 3 mg/kg intraperitoneally, showed extensive necrotic lesions in the granular layers of the hippocampal fascia dentata that were accompanied by degeneration of some large brainstem neurons (14). In monkeys, the same doses of trimethyl tin proved fatal within 12-18 hr. The intoxicated animals showed neuropathological changes primarily involving the pyramidal cells in both cerebral cortical layers and in Ammon's horn. However, no consistent loss of cells in the fascia dentata was noted (60). It has been postulated (13) that the trimethyl tin-induced lesions in the central nervous system may reflect disruption of the oxidative phosphorylation and adenosine triphosphate synthesis, causing intracellular edema and cell death in the affected neuronal elements. Triphenyl tin acetate is a fungicide which is also used in northern Italy to control algal infestations of rice crops. In two cases of acute occupational intoxication caused by inhaling triphenyl tin acetate the prominent symptoms were nausea, dizziness, persistent headache, photophobia, transient loss of consciousness, and convulsions. Elevated blood tin concentrations were found in one of the affected patients (47).

In animal studies, acute exposure to triaryl tins was shown to cause biochemical changes in the brain, including reduced utilization of glucose and depletion of catecholamines. These effects, however, proved to be less specific as compared to those of triethyl tin, and brain edema did not occur (see Ref. 47 for references).

V. MANGANESE

Severe manganese intoxication (manganism, locura manganica, or "manganese madness") has frequently occurred in the villages of northern Chile and other areas of the world where manganese ore is mined (53). Recently this metal has also attracted considerable attention as a trace pollutant (12). Manganese is added as an antiknock agent in lead-free gasoline and is released into the environment when coal is burned. A possible relationship between high manganese levels in the brain and the etiology of specific inherent disorders of the nervous system has been postulated (4,24).

A. Human Intoxication

The initial symptoms usually become apparent after 1 or 2 years of exposure. They include psychiatric manifestations such as hallucinations, compulsive acts, and emotional instability accompanied by headache, muscular weakness, and sleep disturbances. Symptoms similar in many respects to those of Parkinson's disease then develop. Rigidity, abnormalities of gait, and masklike facies are more obvious than tremor.

At autopsy, brains from manganese-intoxicated patients reveal morphological features resembling those of parkinsonism, with loss of neuromelanin in the substantia nigra, apparently reflecting degeneration in basal ganglia regions (12).

B. Mechanism of Toxicity

There is considerable evidence suggesting that manganese may influence specific processes related to the storage, uptake, and release of neurotransmitters in the brain (24). At high concentrations, manganese ions act as powerful synaptic transmission inhibitors reducing norepinephrine release from the synaptic nerve endings and acetylcholine release at the neuromuscular junctions. These effects are probably brought about by competition between Mn^{2+} and Ca^{2+} ions (3).

Striatal dopaminergic terminals are able to accumulate manganese in storage vesicles, and long-term exposure to manganese may lead to disruption of basal ganglia monoaminergic pathways (12). In the early stages of toxicity, manganese-exposed animals exhibit high levels of dopamine and norepinephrine in the striatum without any evidence of anatomical damage in the central nervous system (12). Whether these neurochemical changes represent correlates of the early psychiatric manifestations of manganese poisoning remains to be clarified (11). In the late stage of experimental manganese intoxication neuromorphological changes become apparent and are accompanied by considerable decrease in the striatal levels of catecholamines (12).

There are similarities in behavioral and neurochemical effects elicited by manganese and the powerful neurotoxin 6-hydroxydopamine in animals (24). It has been postulated that the accumulation of manganese in nervous tissue can promote dopamine autoxidation with generation of dopamine-derived free radicals. Selective neuronal damage and reduced monoaminergic input in the affected regions of the brain may occur (24).

The interaction of manganese with the neuronal membrane adenosinetriphosphatase may be an additional factor contributing to the early disruption of the normal balance of dopamine dynamics in the striatum (23).

The possibility of neurological hazards associated with manganese exposure during development is suggested by recent observations indicating long-term behavioral and neurochemical changes in experimental animals perinatally exposed to manganese compounds (12; see also Nelson in this book).

An aging brain may also be particularly susceptible to the effects of manganese on monoaminergic neurons (65).

VI. GOLD

Neurological complications have occasionally been described following use of organic gold derivatives, such as aurothioglucose and aurothiomalate, in the treatment of rheumatoid arthritis, chronic bronchial asthma, and lupus erythematosus (32,34).

The first symptom is usually paresthesia followed by weakness which may progress rapidly. Although most commonly symmetrical, sudden onset of asymmetrical weakness associated with pain has been observed (42). Axonal degeneration and segmental demyelination are the most prominent findings in most sural nerves examined (42). Psychiatric manifestations may also appear.

Whether the peripheral nerve disorders occur because of a direct effect of gold or through allergic processes is still unclear, but the sudden onset of neuropathy and the association with skin disorders favor an immune mechanism (42).

Administration of aurothioglycose to experimental animals was found to induce damage in the ventromedial nucleus of the hypothalamus associated with hyperphagia and eventually obesity (21). The development of these lesions is inhibited by hypophysectomy and exacerbated by the simultaneous administration of glucose with aurothioglucose. Obesity is not produced by gold compounds unable to reach the ventromedial hypothalamus. However, the depositing of gold probably represents a secondary event in the mechanism causing hypothalamic damage which is currently attributed to the glucose antimetabolite moiety rather than to metallic gold toxicity (32).

VII. ALUMINUM

Aluminum was for a long time considered to be a nonabsorbable nontoxic element whose only health hazard was restricted to the occupational inhalation of aluminum-contaminated dust (72). In recent years, however, this metal has been implicated as an etiological factor in dialysis encephalopathy, a progressive syndrome which is now recognized as a significant cause of death in uremic patients on hemodialysis. Speech disorders, memory deficits, dementia, confusion, myoclonus, focal seizures, electroencephalographic abnormalities have been reported as the prominent clinical features (1,16).

A. Dialysis Encephalopathy

Various considerations based on biological, chemical, and epidemiological data suggest that the development of dialysis encephalopathy in patients with renal insufficiency is related to accumulation of aluminum in the brain (5,16). The source of the high aluminum absorption was originally identified as the aluminum-containing phosphate binders, but later studies also indicated a relationship between encephalopathy and aluminum content in the water used to prepare the dialysis fluid (8,19, 58,72).

Several investigations have been carried out to clarify whether the high aluminum level in the brain of uremic patients is the causative factor in dialysis encephalopathy or merely reflects an alteration in mineral metabolism associated with the neurological disorders.

Recent studies in cats and rabbits demonstrating neurological, behavioral, and electrophysiological changes following intracranial application of soluble aluminum salts (16) point to aluminum being directly involved. The early stages of experimental aluminum encephalopathy are characterized by alterations in learning memory performance. Motor disorders such as dyspraxia, muscle hypertonia, myoclonic jerks, or general convulsion become apparent later. The prominent neuromorphological findings are focal areas of cytoplasmic clearing and widespread, patchy neurofibrillary degeneration in the cerebral cortex, spinal cord, pons, and medulla. In the affected areas the normal network of neurofibrillae is disrupted and dense tangles are formed which acquire a homogeneous appearance (17). As discussed below, a number of physiological implications have been considered related to these changes. On the other hand, the results of recent electrophysiological studies (16) suggest that functional disorders induced by aluminum in susceptible animals are not restricted to neurons with morphological abnormalities.

B. Alzheimer's Dementia

The histological changes associated with experimental aluminum intoxication resemble the morphological features of the brain of subjects with senile or presenile dementia of the Alzheimer type. These patients may also present a moderate accumulation of aluminum in brain regions associated with neurofibrillary degeneration. These findings were used to suggest that aluminum may have a role in the excessive production of the 10-nm filament component of neurofibrillary degeneration (17). Aluminum has also been implicated as a factor in a specific form of amyotrophic lateral sclerosis and parkinsonism dementia in an area with a high content of aluminum in drinking water (56).

The aluminum accumulation upon intranuclear particles in neurons and glia in Alzheimer's affected brain has been regarded as an important noxious factor, since molecular structures involved in genetic processes seem to be a primary site of action of intracellular aluminum toxicity. Thus complexes of aluminum with DNA are formed in vitro (16), and reduced ribosomal protein synthesis was found to occur as a result of the aluminum blockade of neuronal brain RNA initiation sites (64).

On the other hand, as pointed out by Crapper McLachlan et al. (16), certain histopathological findings of human Alzheimer's disease cannot be explained by the known neurotoxic effects of aluminum, and it is clear that the diseases presently related to aluminum (e.g., dialysis encephalopathy and experimental neurointoxications in animals) are not identical to Alzheimer's disease. Subjects with dialysis-associated encephalopathy often exhibit very high aluminum levels in brain (up to 20 times the normal), but this condition is not associated with specific histopathological changes or neurofibrillary tangles. Moreover, in dialysis encephalopathy aluminum is not bonded to intranuclear constituents, but is mainly concentrated in the cytoplasm of neurons (17).

A critical approach to the hypothesis relating senile dementia of the Alzheimer type to the excessive absorption of aluminum from particular sources should take into account the ubiquitous distribution of this element. After oxygen and silicon, aluminum is the most abundant element in the earth's crust (33). Domestic tap water may contain aluminum in high concentrations either naturally or because aluminum has been added as a flocculant in the purification processes. Acid rain also markedly increases the aluminum content of water (72). Presumably, all the individuals from an area where the aluminum content of water is high ingest considerable amounts of this metal in food and water. Moreoever, many aluminum compounds are used as food additives, nonprescription drugs, cosmetics, and deodorants, and all these factors can easily increase the intake of aluminum to several hundred milligrams per day (44). It is therefore unlikely that an aluminum overload condition represents the primary factor in the development of neurofibrillary changes in Alzheimer's disease and senile dementia. It is more probable that in susceptible individuals cellular disorders related to an underlying disease can alter the disposition of aluminum, allowing this metal to accumulate in nuclei and interact with intranuclear targets in the central nervous system.

C. Prevention of Aluminum Toxicity

A rational approach in prophylaxis of aluminum neurotoxicity is based on procedures and controls aimed at reducing the absorption of this metal by dialysis patients. It is now believed that concentrations above 10-20 μg Al/liter in the dialysate may lead to significant blood uptake of aluminum, and it has been proposed that the aluminum levels in the dialysis water should not exceed 10-15 μg Al/liter (5).

Treatment of water by either reverse osmosis or deionization is currently performed in order to lower aluminum levels in the dialysate fluid. Recent data suggest that these measures have been effective in reducing the incidence of encephalopathy cases. This is generally assumed to be due to the removal of aluminum from water, even though one cannot exclude the possibility that, besides aluminum, the water treatment removes other substances, as yet unidentified, which may be of importance in the development of neurological disorders in patients with renal insufficiency (19).

Monitoring of serum aluminum has been proposed as an approach to preventing aluminum neurotoxicity, but the measurement of low levels of this metal may pose serious analytical problems (33). Moreover, serum aluminum levels may not reflect the actual accumulation of aluminum in uremic patients treated with aluminum-free dialysate but also taking aluminum hydroxide to control hyperphosphatemia (31). In this respect, the rate of intestinal absorption of aluminum in subjects receiving aluminum-containing phosphate binders seems to vary with the different types of preparations (57).

The iron-chelating agent desferrioxamine competes for aluminum at binding sites within the body and raises serum aluminum concentrations, allowing the metal's removal by hemodialysis. Desferrioxamine has been receommended as a treatment for reducing the aluminum overload in uremic patients (20).

VIII. BISMUTH

An epidemic of 942 cases of diffuse encephalopathy with 72 deaths in subjects receiving an oral preparation of bismuth subnitrate for treatment of chronic bowel complaints occurred in France in the 1970s (50). In the same period, the Australian Drug Evaluation Committee reported a series of 29 cases of encephalopathy related to the use of an oral preparation of bismuth subgallate by cholostomized patients (7,55). More recently, central nervous system disease attributed to the oral intake of bismuth has been described in Belgium, Switzerland, Spain, and Germany (see Ref. 41 for references).

The affected patients showed neurological disorders only after months or years of therapy without side effects. There was a prodromal phase of psychiatric symptoms with depression, anxiety, irritability, and confusion, followed by clumsiness, difficulty in gait, myoclonic jerks, and marked dysarthria. Electroencephalographic abnormalities have also been documented (30). Animal models of bismuth encephalopathy have not been established and very little is known of the interaction of bismuth with the nervous tissue.

Bismuth is absorbed to some extent from the gastrointestinal tract and considerably high serum and/or brain bismuth levels have been detected in fatal cases of encephalopathy in patients treated with bismuth-containing medications (2). Abnormalities of bismuth metabolism and pharmacokinetics were observed in symptomatic patients (69).

REFERENCES

1. Alfrey, A. C., Le Gendre, G. R., Kaehny, W. D. 1976. The dialysis encephalopathy syndrome: Possible aluminum intoxication. *N. Engl. J. Med. 294*:184-188.
2. Allain, P., Chaleil, D., Emile, J. 1980. L'élevation des concentrations de bismuth dans les tissues des malades intoxiqués. *Therapie 35*:303-304.
3. Arqueros, L., Daniels, A. J. 1981. Manganese as agonist and antagonist of calcium ions: Dual effect upon catecolamine release from adrenal medulla. *Life Sci. 28*:1535-1540.

4. Banta, R. G., Markesbery, W. R. 1977. Elevated manganese levels associated with dementia and extrapyramidal signs. *Neurology 27*:213-216.
5. Berlin, A. 1983. International workshop on the role of biological monitoring in the prevention of aluminum toxicity in man. Memorandum on the summary and conclusions. *J. Toxicol. Clin. Toxicol. 19*:907-910.
6. Brockhaus, A., Dolgner, R., Ewers, U., Soddemann, H., Wiegand, H. 1981. Intake and health effects of thallium among a population living in the vicinity of a cement plant emitting thallium containing dust. *Int. Arch. Occup. Environ. Health. 48*:365-389.
7. Burns, R., Thomas, D. W., Barron, V. J. 1974. Reversible encephalopathy possibly associated with bismuth subgallate ingestion. *Br. Med. J. 1*:220-223.
8. Cannata, J. B., Briggs, J. D., Junor, B. J. R., Fell, G. S., Beastall, G. 1983. Effect of acute aluminum overload on calcium and parathyroid-hormone metabolism. *Lancet 1*:501-503.
9. Cavanagh, J. B. 1979. Metallic toxicity and the nervous system, in *Recent Advances in Neuropathology*, Vol. I (Smith, W. T., Cavanagh, J. B., eds.), Churchill Lvingstone, Edinburgh, pp. 247-275.
10. Cavanagh, J. B., Fuller, N. H., Johnson, H. R. M., Rudge, P. 1974. The effects of thallium salts, with particular reference to the nervous system changes. *Q. J. Med. 43*:293-319.
11. Chandra, S. V. 1983. Psychiatric illness due to manganese poisoning. *Acta Psychiatr. Scand. Suppl. 303*:49-54.
12. Chandra, S. V., Seth, P. K. 1980. Manganese encephalopathy. Effect of manganese exposure on growing versus adult rodents, in *Advances in Neurotoxicology* (Manzo, L., ed.), Pergamon, Oxford, pp. 49-55.
13. Chang, L. W., Tiemeyer, T. M., Wenger, G. R., McMillan, D. E. 1983. Neuropathology of trimethyltin intoxication. III. Changes in the brain stem neurons. *Environ. Res. 30*:399-411.
14. Chang, L. W., Tiemeyer, T. M., Wenger, G. R., McMillan, D. E., Reuhl, K. R. 1982. Neuropathology of trimethyltin intoxication. I. Light microscopy study. *Environ. Res. 29*:435-444.
15. Chhuttani, P. N., Chopra, J. S. 1979. Arsenic poisoning, in *Handbook of Clinical Neurology*, Vol. 36 (Vinken, P. J., Bruyn, G. W., eds.), North Holland, Amsterdam, pp. 199-216.
16. Crapper McLachlan, D. R., Farnell, B., Galin, H., Karlik, S., Eichhorn, G., De Boni, U. 1983. Aluminum in human brain disease, in *Biological Aspects of Metals and Metal-Related Diseases* (Sarkar, B., ed.), Raven Press, New York, pp. 209-218.
17. Crapper, D. R., Quittkat, S., Krishnan, S. S., Dalton, A. J., De Boni, U. 1980. Intranuclear aluminum content in Alzheimer's

disease, dialysis encephalopathy and experimental aluminum encephalopathy. *Acta Neuropathol.* *50*:19-24.

18. Davis, L. E., Standefer, J. C., Kornfeld, M., Abercrombie, D. M., Butler, C. 1981. Acute thallium poisoning: Toxicological and morphological studies of the nervous system. *Ann. Neurol.* *10*:38-44.
19. Davison, A. M., Walker, G. S., Oli, H., Lewins, A. M. 1982. Water supply aluminum concentration, dialysis dementia, and effect of reverse-osmosis water treatment. *Lancet* *2*:785-787.
20. Day, J. P., Ackrill, P., Garstang, F. M., Hodge, K. C., Metcalfe, P. J., O'Hara, M., Romero-Martinez, R. A. 1983. Reduction of the body burden of aluminum in renal patients by desferrioxamine chelation therapy, in *Chemical Toxicology and Clinical Chemistry of Metals* (Brown, S. S., Savory, J., eds.), Academic, London; pp.353-356.
21. Debons, A. F., Das, K. C., Fuhr, B., Siclari, E. 1982. Inhibition by hypophysectomy of the hyperphagia and obesity following gold thioglucose. *Physiol. Behav.* *29*:681-685.
22. Doctor, S. V., Fox, D. A. 1983. Immediate and long-term alterations in maximal electroshock seizure responsiveness in rats neonatally exposed to trimethyltin bromide. *Toxicol. Appl. Pharmacol.* *68*:268-281.
23. Doherty, J. D., Salem, N., Lauther, C. J., Trams, E. G. 1983. Mn^{2+}-stimulated ATPase in rat brain. *Neurochem. Res.* *8*:493-500.
24. Donaldson, J. 1981. The pathophysiology of trace metal: neurotransmitter interaction in the CNS. *Trends Pharmacol. Sci.* *2*:75-78.
25. Edel Rade, J., Marafante, E., Sabbioni, E., Gregotti, C., Di Nucci, A., Manzo, L. 1982. Placental transfer and retention of ^{201}Tl-thallium in the rat. *Toxicol. Lett.* *11*:275-280.
26. Feldman, R. G., Niles, C. A., Kelly-Hayes, M., Sax, D., Dixon, W. J., Thompson, D. J., Landaw, E. 1979. Peripheral neuropathy in arsenic smelter workers. *Neurology* *29*:939-946.
27. Foncin, J. F., Gruner, J. E. 1979. Tin neurotoxicity, in *Handbook of Clinical Neurology* (Vinken, P. J., Bruyn, G. W., eds.), North Holland, Amsterdam, pp. 279-290.
28. Fowler, B. A. 1977. Toxicology of environmental arsenic, in *Toxicology of Trace Elements* (Goyer, R. A., Mehlman, M. A., eds.), Wiley, New York, pp. 79-122.
29. Fox, D. A., Doctor, S. V. 1983. Triethyltin decreases maximal electroshock seizure severity in adult rats. *Toxicol. Appl. Pharmacol.* *68*:260-267.
30. Gastaut, J. L., Jouglard, J. 1980. Confrontation clinique et électroencéphalographique au cours de l'éncephalopathie myoclonique bismuthique. *Therapie* *35*:293-296.

31. Gilli, P., Malacarne, F., Fagioli, F. 1983. Is serum aluminum monitoring useful in evaluating aluminum intoxication? *Lancet 1*: 656.
32. Goetz, C. G., Klawans, H. L. 1979. Neurologic aspects of other metals, in *Handbook of Clinical Neurology* (Vinken, P. J., Bruyn, G. W., eds.), North Holland, Amsterdam, pp. 319-345.
33. Hamilton, E. I. 1982. Aluminium and Alzheimer's disease. A comment. *Sci. Total Environ. 25*:87-91.
34. Hanakago, R. 1980. Severe polyneuritis showed amyotrophy following gold therapy for bronchial asthma, in *Advances in Neurotoxicology* (Manzo, L., ed.), Pergamon, Oxford, pp. 391-395.
35. Hasan, M., Chandra, S. V., Dua, P. R., Raghubir, R., Ali, S. F. 1977. Biochemical and electrophysiologic effects of thallium poisoning on the rat corpus striatum. *Toxicol. Appl. Pharmacol. 41*:353-359.
36. Hasan, M., Ali, S. F. 1981. Effects of thallium, nickel and cobalt administration on the lipid peroxidation in different regions of the rat brain. *Toxicol. Appl. Pharmacol. 57*:8-13.
37. Hasan, M., Ali, S. F., Tariq, M. 1978. Levels of dopamine, norepinephrine and 5-hydroxytryptamine in different regions of the rat brain in thallium toxicosis. *Acta Pharmacol. Toxicol. 43*: 169-173.
38. Hutton, J. T., Christians, B. L., Dippel, R. L. 1982. Arsenic poisoning. *N. Engl. J. Med. 307*:1080.
39. Huxtable, R. A., Barbeau, A. (eds.). 1976. *Taurine*, Raven Press, New York.
40. Knowles, F. C., Benson, A. A. 1983. The biochemistry of arsenic. *Trends Biochem. Sci.*, 178-190.
41. Lagier, G. 1980. Encéphalopathies bismutiques: Situation dans les pays autres que la France. *Therapie 35*:315-317.
42. Le Quesne, P. M. 1982. Metal-induced diseases of the nervous system. *Br. J. Hosp. Med. 28*:534-538.
43. Leslie, A. C. D., Smith, H. 1978. Self poisoning by the abuse of arsenic containing tonics. *Med. Sci. Law 18*:159-164.
44. Lione, A. 1983. The prophylactic reduction of aluminum intake. *Food Chem. Toxicol. 21*:103-109.
45. Lock, E. A., Aldridge, W. N. 1975. The binding of triethyltin to rat brain myelin. *J. Neurochem. 25*:871-876.
46. Luyten, J. G. A. 1972. Applications and biological effects of organotin compounds, in *Organotin Compounds*, Vol. 3 (Sawyer, A. K., ed.), Marcel Dekker, New York, p. 931.
47. Manzo, L., Richelmi, P., Sabbioni, E., Pietra, R., Bono, F., Guardia, L. 1981. Poisoning by triphenyltin acetate. Report of two cases and determination of tin in blood and urine by neurotron activation analysis. *Clin. Toxicol. 18*:1343-1353.

48. Manzo, L., Scelsi, R., Moglia, A., Poggi, P., Alfonsi, E., Pietra, R., Mousty, F., Sabbioni, E. 1983. Long-term toxicity of thallium in the rat, in *Chemical Toxicology and Clinical Chemistry of Metals* (Brown, S. S., Savory, J., eds.), Academic, London, pp. 401-405.
49. Manzo, L., Tonini, M., Sabbioni, E. Toxicology of arsenicals. A review of human effects with special reference to peripheral nervous system. Commission of the European Communities Report (in press).
50. Martin-Bouyer, G., Foulon, G., Guerbois, H., Barin, C. 1981. Epidemiological study of encephalopathies following bismuth administration per os. Characteristics of intoxication patients: Comparison with a control group. *Clin. Toxicol. 18*:1277-1283.
51. Marwaha, J., Freedman, R., Hoffer, B. 1980. Electrophysiological changes at the central noradrenergic synapses during thallium toxicosis. *Toxicol. Appl. Pharmacol. 56*:345-352.
52. Melnick, R. L., Monti, L. G., Motzkin, S. M. 1976. Uncoupling of mitochondrial oxidative phosphorylation by thallium. *Biochem. Biophys. Res. Commun. 69*:68-71.
53. Mena, I. 1979. Manganese poisoning, in *Handbook of Clinical Neurology* (Vinken, P. J., Bruyn, G. W., eds.), North Holland, Amsterdam, pp. 217-237.
54. Moeschlin, S. 1980. Thallium poisoning. *Clin. Toxicol. 17*:133-146.
55. Morrow, A. W. 1973. Request for report. Adverse reactions with bismuth subgallate. *Med. J. Aust. 60*:912-923.
56. Perl, D. P., Gajdusek, D. A., Garruto, R. M., Yanagihara, R. T., Gibbs, C. J. 1982. Intraneuronal aluminum accumulation in amyotrophic lateral sclerosis and Parkinsonism-dementia of Guam. *Science 217*:1053-1055.
57. Politis, M. J., Schaumburg, H. H., Spencer, P. S. 1980. Neurotoxicity of selected chemicals, in *Experimental and Clinical Neurotoxicology* (Spencer, P. S., Schaumburh, H. H., eds.), Williams and Wilkins, Baltimore, pp. 613-630.
58. Randall, M. E. 1983. Aluminium toxicity in an infant not on dialysis. *Lancet 1*:1327-1328.
59. Richelmi, P., Bono, F., Guardia, L., Ferrini, B., Manzo, L. 1980. Salivary levels of thallium in acute human poisoning. *Arch. Toxicol. 43*:321-325.
60. Reuhl, K. R., Mackenzie, B. A., Gilbert, S. G., Rice, D. C. 1983. Neuropathology of trimethyltin intoxication in the monkey, presented at the 2nd International Conference on the Clinical Chemistry and Chemical Toxicology of Metals, Montreal, Canada, July 19-22, 1983.
61. Sabbioni, E., Edel, J., Goetz, L. 1984. Trace metal speciation in environmental biochemical research, in *Proceedings of the Inter-*

national Symposium on Health Effects and Interactions of Essential and Trace Elements, Lund, Sweden, June 13-18, 1983 (in press).

62. Sabbioni, E., Gregotti, C., Edel, J., Marafante, E., Di Nucci, A., Manzo, L. 1982. Organ/tissue disposition of thallium in pregnant rats. *Arch. Toxicol. Suppl.* 5:225-230.
63. Sabbioni, E., Manzo, L. 1980. Metabolism and toxicity of thallium, in *Advances in Neurotoxicology* (Manzo, L., ed.) Pergamon, Oxford, pp. 249-270.
64. Sarkander, H. I., Balss, G., Schlossner, R., Stoltenburg, G., Lux, R. M. 1983. Blockade of neuronal brain RNA initiation sites by aluminum. A primary molecular machanism of aluminium-induced neurofibrillary changes, in *Brain Aging* (Cervos-Navarro, J., Sarkander, H. I., eds.), Raven Press, New York, pp. 259-273.
65. Silbergeld, E. K. 1982. Current status of neurotoxicology basic and applied. *Trends Pharmacol. Sci.* 5:291-296.
66. Smith, I. C. H., Carson, B. L. 1977. *Trace Metals in the Environment, Thallium*, Ann Arbor, Michigan.
67. Spencer, P. S., Sabri, M. I., Politis, M. 1980. Methyl-n-butylketone, carbon disulfide and acrylamide. Putative mechanisms of neurotoxic damage, in *Advances in Neurotoxicology* (Manzo, L., ed.), Pergamon, Oxford, pp. 173-180.
68. Suzuki, K., 1971. Some new observations on triethyltin intoxication in rats. *Exp. Neurol.* *31*:207-214.
69. Thomas, D. W., Sobecki, S., Hartley, T. F., Coyle, P., Alp, M. H. 1983. Variable absorption and excretion of bismuth and its potential for toxicity, in *Chemical Toxicology and Clinical Chemistry of Metals* (Brown, S. S., Savory, J., eds.), Academic, London, pp. 391-394.
70. Toews, A. D., Blaker, W. D., Thomas, D. J., Gaynor J. J., Krigman, M. R., Mushak, P., Morell, P. 1983. Myelin deficits produced by early postnatal exposure to inorganic lead or triethyltin are persistent. *J. Neurochem.* *41*:816-822.
71. Watanabe, I. 1980. Organotins (triethyltin), in *Experimental and Clinical Neurotoxicology* (Spencer, P. S., Schaumburg, H. H., (eds.), Williams and Wilkins, Baltimore, pp. 545-557.
72. Wills, M. R., Savory, J. 1983. Aluminium poisoning: Dialysis encephalopathy, osteomalacia, and anaemia. *Lancet* *2*:29-34.

19

Membrane Effects of Pesticides

John D. Doherty
Office of Pesticide Programs, Environmental Protection Agency, Washington, D.C.

I. INTRODUCTION

It is widely accepted that the plasma membrane and certain intracellular membranes of nerve cells have both general and highly specific functions in regulating cellular tonicity and nerve activity. Agents which are neurotoxic may have highly specific receptors within nerve membranes or they may also become generally incorporated into the cell membranes and affect several membrane functions. Many pesticides are neurotoxic. As chemicals, these agents include a wide variety of molecular types. The insecticides and other invertebrate poisons were by design usually developed to be potent neurotoxic agents. The herbicides and fungicides were designed to attack nonneuronal targets but these agents have often been shown to have neurotoxic effects. The resulting neurotoxicity may refer to either the direct result of killing the pest or to the incidental neurotoxicity in humans or domestic animals due to the various circumstances of exposure. In this chapter the term *expression of neurotoxicity* shall refer to the neurotoxicological consequences in mammals. Alterations in the basic physiological processes such as membrane effects which result in the death of the insect or other pests are often the same processes which result in the expression of neurotoxicity.

Pesticide neurotoxicity in humans includes death resulting from acute intoxication or otherwise transient effects of varying degree, persistent effects due to accumulation of the pesticide in nerve tissue, and in some cases a long-term or permanent injury due to a structural change. The spectrum of neurotoxicity may include convulsions, tremors, changes in electroencephalographic recordings, slurred speech, loss of memory, supersensitivity to sound, and possibly altered mood and mental state and/or restlessness, as well as other symptoms. For examples see the description of symptoms from victims poisoned by Kepone (44) and the persistent effects in humans due to organophosphates (14).

II. MEMBRANE TARGETS OF TOXICITY

Table 1 shows selected representative examples of nerve and muscle membrane components that have been shown to be targets for various pesticides. Table 1 also lists the primary physiological effect and the proposed or proven toxicological consequence of the primary physiological effect.

Figure 1 shows the relationships between the effects of pesticides on cellular membranes and the expression of neurotoxicity. In this scheme pesticides are shown to be able to have direct effects on both the cell plasma and intracellular membranes. Examples of pesticides having direct effects on the plasma membrane are 1,1,1-trichloro-2,2-dichlorophenylethane (DDT), the pyrethroids, cartap, and the nicotine derivatives, as well as others. The carbamates and organophosphates have direct effects on acetylcholinesterase (AChE), an enzyme which is associated with the outer surface of membranes, and the resulting buildup of acetylcholine (ACh) has direct effects on the plasma membrane to produce the primary expression of neurotoxicity. Other pesticides such as those designed to be mitochondrial poisons (i.e., rotenone) attack intracellular membranes. The pathway via mitochondrial inhibition is often important in the expression of neurotoxicity of agents not designed as nerve poisons. This chapter will emphasize that many pesticides may exert their expression of neurotoxicity via both of these pathways.

III. BASIC INTERACTIONS OF PESTICIDES WITH NERVE MEMBRANES

Reviews of the basic structure and function of nerve membranes can be found elsewhere. For the purposes of this chapter, it must be recognized that the nerve plasma membranes consist of a matrix of complex

Table 1 Membrane Components and Selected Examples of Some Known Effects of Pesticides with Their Proposed Neurotoxicological Consequence[a]

Membrane component	Pesticides	Primary effects	Physiological or toxicological consequence (proven or supposed)	References
Na^+ conductance channel	DDT	Prolongs Na^+ influx	Repetitive firing, nerve excitation	35,36
	Pyrethroids	Affect closing of conductance channel	Repetitive firing, nerve excitation, nerve block	33-35
Membrane ATPases (Na^+ and K^+), Mg^{2+}, or (Ca^{2+} and Mg^{2+})	Organochlorines (DDT, Kepone), alkyltins, others	Inhibit ATP hydrolysis and function of the enzyme	Failure to maintain proper ionic distribution and intracellular buffering, depletion of cellular energy supply	See text
Ca^{2+} binding to plasma membrane and intracellular structures (mitochondria and sarcoplasmic reticulum)	Organochlorines, pyrethroids, organphosphates, organotins, lindane	Inhibition of Ca^{2+} binding and release of bound Ca^{2+}	Unstable membranes, rise in intracellular free Ca^{2+} with effects on transmitter release or muscle contraction	See text
ACh receptor	Nicotinic insecticides	Bind ACh receptor	Nerve block and other effects	15
	Cartap and nereistoxin	ACh antagonist (primarily)	Nerve block	15
	Organophosphates and carbamates	Indirect stimulation by ACh due to inhibition of AChE	ACh responses	15

Table 1 (Continued)

Membrane component	Pesticides	Primary effects	Physiological or toxicological consequence (proven or supposed)	References
ACh receptor	Organophosphates	Phosphorylation of the ACh receptor	Change in subsequent ACh response (persistent changes in EEG?)	9,23
	Organophosphates	Decrease in total number of ACh receptors	Tolerance to organophosphates	4,5
Dopamine receptor	Kepone	Changes in total number of dopamine receptors	Not known	41
GABA receptor channel ionophore	Type II pyrethroids	Inhibit binding of specific ligand	Not established	26
L-Glutamate receptor	Pyrethroids	Inhibit binding of specific ligand	Not established	43
Opiate receptor	DDT	Inhibits binding of antagonist (naloxone)	Not known	25
Digitalis receptor	DDT	Increases binding of ligand (ouabain)	Not known	25
Na^+-dependent transmitter uptake	Organochlorines, pyrethroids, alkyltins	Inhibition of uptake	Nerve stimulation (?)	10,13
Cl^- transport system	Lindane, heptachlor epoxide, 2,4-D, Frescon	Inhibition of Cl^- transport	Unstable nerve (?)	18,24,31

[a] ACh, acetylcholine; AChE, acetylcholinesterase; ATP, adenosine triphosphate; ATPase, adenosinetriphosphatase; DDT, 1,1,1-trichloro-2,2-dichlorophenylethane; 2,4-D, 2,4-dichlorophenoxyacetic acid; GABA, γ-aminobutyric acid; and EEG, electroencephalogram.

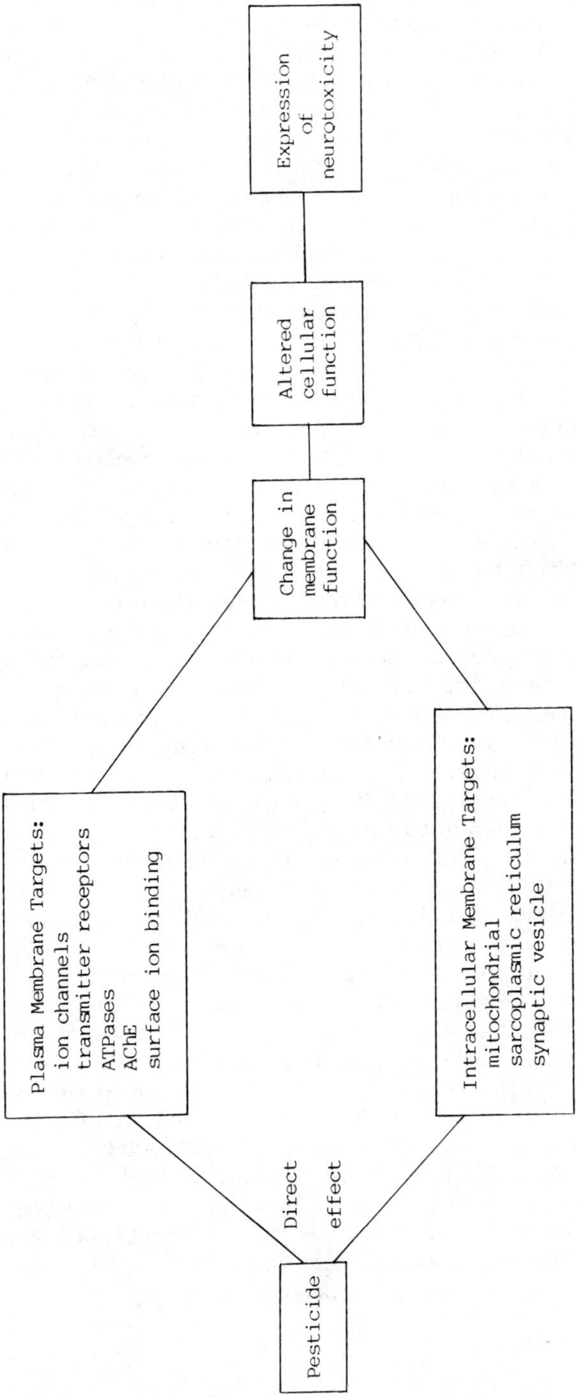

Figure 1 Selected examples of principal membrane targets for pesticides and the expression of neurotoxicity.

lipids and proteins which serve to function as both a structural barrier between the inside and outside of the nerve cell, as well as containing specific entities which serve as transmitter receptors, ionic channel gates, and enzymatic ion pumps. The intracellular membranes of the structures such as the mitochondria, the synaptic vesicles, and the sarcoplasmic reticulum are also composed of complex lipids and/or proteins and have components which attract pesticides. The introduction to this chapter stated that agents can form both general and specific interactions with nerve membranes. An example of a highly specific interaction between an agent and a specific membrane component is that of tetrodotoxin and its ability to selectively block the Na^+ conductance channel on the external membrane surface. This agent is effective at concentrations as low as 10^{-11} M and appears to act on a one-to-one basis with the receptor. Because only a single target is affected, the primary result is that only a single membrane function is directly altered. On the other end of the scale there are agents that become enmeshed into the nerve membrane matrix to cause changes in the physical nature of the membrane. In this case the interaction is considered as a general association and many membrane functions, both in the plasma membrane and in the intracellular structures, may be altered in a variety of ways by the presence of these agents.

There are several examples of insecticides which have been shown to have relatively highly specific associations with nerve membrane components. These include the nicotine derivatives and cartap, which have direct effects on the ACh receptor (see Ref. 15 for a review). These agents mimic the natural substrate or ACh in their affinity for the receptor, with resulting expression of neurotoxicity. Nicotine, however, has been shown to have noncholinergic receptors (2) and thus may not be singularly specific for the ACh receptor. The pyrethroid compounds are thought to have a high degree of specificity for the internal gate of the Na^+ channel. It has been reported that only as little as 1% of the membrane Na^+ channels need to be affected in order to cause nerve excitation due to tetramethrin (28). DDT also has a specificity to alter the Na^+ conductance channel in a characteristic manner. As will be indicated below, other membrane effects may also contribute to the expression of neurotoxicity of DDT and the pyrethroids. Compounds such as the organochlorines and organotins have not been clearly demonstrated to have a single specific molecular receptor within the membranes but have been shown to alter several membrane functions. The expression of neurotoxicity of these chemicals is thought to be the result of their lipid solubility and a generalized incorporation of the insecticide into both plasma and intracellular membranes which may cause several changes in nerve cell physiology. Agents which become incorporated into cellular membranes may still have characteristic expressions of neurotoxicity possibly dependent upon the pattern and degree of incorporation.

A. Experimental Approaches

The neuropharmacology of pesticide interactions at the level of cellular membranes can be viewed as being studied by two major approaches: neurophysiologically and neurochemically. The neurophysiological approach, which utilizes sophisticated electronic equipment to measure the small changes in electric potential that exist about membranes, is, according to Narahashi (35), the most powerful and straightforward manner to study the effects of neuropoisons. Data have been generated by physiologists to show the effects of insecticides on membranes in axons and in neuromuscular junctions. The neurochemical approaches consist of techniques to study the effects of pesticides as inhibitors of enzymes such as AChE and adenosinetriphosphatase (ATPase) and other enzymes of neurological or general cellular importance. Neurochemical approaches also include the use of radiolabeled pesticides and neurotransmitters and isotopes of ions of neurophysiological importance such as $^{22}Na^+$ and $^{45}Ca^{2+}$. More recent developments in the last several years related to the perfection of the use of radioligands which bind specific synaptic receptors for nerve transmitters have been used to assess the effects of insecticides in the synaptic region.

The in vitro neurophysiological and biochemical approaches which test the effects of adding pesticides to isolated preparations of nerves or nerve cell components provide insights as to how the pesticides might affect membranes, and a theoretical basis for the expression of neurotoxicity can be generalized. Neurophysiological and neurochemical approaches can also be used to assess changes in the structure and function of the nervous system in test animals treated with pesticides in vivo after the nerve tissue or subcellular fractions have been removed from the exposed animals. The latter approach is especially important in assessing the chronic exposure to pesticides and for looking for delayed effects following acute exposure. A later section of this chapter will discuss how the latter approach was used to explain tolerance to organophosphates.

B. DDT and Pyrethroid Effects on the Na^+ Channel

It has been recognized for many years that the insecticides DDT and the natural and synthetic pyrethroids have in common an ability to affect the Na^+ conductance channel in both myelinated and nonmyelinated axons (see Ref. 35 for a review). In particular, these two quite different chemical classes have the ability to cause selective effect(s) on the mechanism(s) which regulates the influx of Na^+. The possibility that DDT does have a specific effect within the axon plasma membrane which causes this chemical but not others of similar lipophilic nature (such as polyhalogenated biphenyls and many other organochlorine insecticides) to characteristically alter the Na^+ conductance is a most interesting problem related to the effects of pesticides on nerve membranes. The case that DDT does have some degree of membrane specificity in causing this change in the Na^+ conductance is supported by data with squid giant axons. Several earlier attempts to show changes in the Na^+ conductance in this species failed to demonstrate the expected effects. However, when DDT was perfused internally, the predicted effect was found (27), implying that a specific area within the internal surface of the plasma membrane is affected.

It has been known for many years that nerves bathed in Ca^{2+} free buffer resemble nerves poisoned by DDT and that adding Ca^{2+} to the buffer reverses the effects of DDT on isolated axons (39). DDT has been shown to inhibit several species of membrane-bound ATPase- and adenosine triphosphate- (ATP) supported reactions in axons (11,30) and in some cases to a greater degree than the noninsecticidal analog 1,1-dichloro-2,2-dichlorophenylethylene (DDE), which does not have the same axonic effects as does DDT. A most interesting aspect of these studies is that DDT can prevent ATP-dependent binding of radiolabeled $^{45}Ca^{2+}$ to suspensions of axons (16,17). These findings led to a current hypothesis that DDT causes its characteristic effects in axons by interfering with the association of Ca^{2+} on the cell membrane surface (17). However, such an action would not explain why it is necessary to perfuse DDT internally in the squid to obtain an effect. Moreover, several other agents which affect the Na^+ gate in a manner resembling DDT, such as batrachotoxin and pyrethroids, also are more effective when applied internally. Interestingly, several structural analogs of DDT can block Na^+ conductance, especially when they are less lipophilic than DDT. Other analogs of DDT which are intermediate in polarity have properties of both blocking axonic conduction and causing excitation (46). The specificity which DDT and its analogs have for the Na^+ conductance channel may relate more to a "fit" into the channel rather than to a specific receptor molecule about the channel area (see Ref. 20 and 32 for discussions on the physical nature of DDT action on membranes). As will be indicated in later sections of this chapter, the effects of DDT on membranes in the synaptic and neuromuscular region may also contribute to the expression of neurotoxicity.

The early neurophysiological studies with pyrethroids in isolated nerve bundles led to the conclusion that allethrin at 10^{-6} M (a) resulted in an increase in the negative afterpotential of the action potential, (b) caused repetitive afterdischarges, and (c) at higher concentrations the nerve conduction was blocked. These affects of allethrin were more pronounced when the chemical was applied internally. A number of recent papers have reported the effects of structure-activity relationships of the pyrethroids in efforts to correlate toxicity with effects on the Na^+ channel gating mechanism. One such example (22) utilized nerve cell cultures and reported several novel effects of a series of synthetic pyrethroids. The overall conclusion of these studies was that the pyrethroids apparently have a distinct receptor about the Na^+ channel (internally) which is not shared by veratridine, batrachotoxin, or several other toxins known to have a high affinity for the Na^+ channel internally. This report also described that certain analogs of pyrethroids that had little or no effect on the Na^+ conductance could bind the Na^+ channel pyrethroid receptor and prevent the binding of the more active forms. A similar conclusion was arrived at using axons from invertebrates (29). The synthetic pyrethroids tested thus far have been shown to fall into (roughly) two classes: type I, which produces an expression of neurotoxicity described as the T (tremors) syndrome, and type II, which produces a quite different expression of neurotoxicity described as the CS (choreoathetosis/salivation) syndrome (3,45).

Until recently the only known target for pyrethroids was thought to be on the nerve membrane Na^+ conductance channel but recent data (13) have demonstrated that certain pyrethroids (i.e., resmethrin and permethrin) can prevent the binding of $^{45}Ca^{2+}$ to synaptosome membranes in the absence of Na^+, meaning that there is a synaptic effect of at least some pyrethroids independent of their effects on the Na^+ channel. A correlation between this effect and the toxicity of pyrethroids is not established. Other recent studies which will be discussed below on the effects of pyrethroids on membranes in the synaptic region have shown that these agents can decrease the binding of transmitter receptor ligands. Effects of pyrethroids on membrane targets in the synaptic region and on membrane components other than the Na^+ channel may contribute to explaining the differences in the expressions of neurotoxicity noted for the type I and type II classes of pyrethroids.

C. Inhibition of Membrane ATPases

Most cellular ATPase activity is associated with membranes. The plasma membrane, mitochondria, and the vesicular fractions all have high degrees of specific ATPase activity, and the ATPase species in each membranous fraction have important physiological functions for their respective fractions. It has been established that several of these enzymes with ATPase activity are inhibited by insecticides such as DDT, Kepone, other organochlorine insecticides and organotin pesticides. Some authors report that certain species of ATPases are inhibited to a higher degree than other ATPase species in the same preparation, and that neurotoxicity is primarily the result of inhibition of the most susceptible species of ATPase. Owing to the many critical roles which membrane ATPases play in the dynamics of nerve cells, it is difficult to deny that inhibition of membrane ATPases probably does contribute to the overall expression of neurotoxicity of those pesticides which have been shown to be potent inhibitors of ATPases. Because the differences in the degree of inhibition of the enzyme activity by these insecticides are small (usually about one order of magnitude, but sometimes larger) among the ATPase species, it is also difficult to conclude that insecticidal inhibition of a single membrane ATPase species by any given organochlorine insecticide is solely responsible for the expression of neurotoxicity. Moreover, at the same concentrations of insecticides at which ATPase inhibition occurs there are usually other biochemical systems altered, including changes in other membrane functions. Axonal inhibition of ATPase by DDT has already been discussed. Selected examples of ATPases being inhibited by pesticides with their proposed neurotoxic consequence are discussed below.

1. $(Na^+ + K^+)$ ATPase

The $(Na^+ + K^+)$ ATPase that is responsible for maintaining the proper distribution of Na^+ and K^+ about the plasma membrane in nerve and most other cells is inhibited by several insecticides, but good correlations between this inhibition and the expression of neurotoxicity have not been firmly established. For example, the neurotoxicity would be expected to resemble nerves poisoned by ouabain, a specific inhibitor of this enzyme. In the case of DDT, the spectrum of neurotoxicity in most species is quite different from ouabain, although DDT is a moderately potent inhibitor of this enzyme. In the eel electroplaque, however, correlations between $(Na^+ + K^+)$ ATPase and DDT inhibition of it as well as similarities in DDT and ouabain physiological effects were claimed (37).

2. Mg^{2+} ATPase

The synaptic region and the axons have membranous Mg^{2+} ATPase which is found in the plasma membranes, vesicular fractions, and mitochondria. This enzyme has many physiological functions depending upon the structure with which it is associated. For example, in the mitochondria, Mg^{2+} ATPase appears to be involved with regulation of cellular ATP production and inhibition would result in toxicity to the cell by affecting the energy supply. The hypothesis that mitochondrial ATPase inhibition is the locus of action of organochlorine insecticides has been reviewed elsewhere (8). In the brain synaptic vesicular fraction, the Mg^{2+} ATPase appears to be involved in the packaging of transmitters into the storage vesicles. Inhibition of this vesicular ATPase might then result in disruption of the storage system for transmitters (12).

3. $(Ca^{2+} + Mg^{2+})$ ATPase

The membranous $(Ca^{2+} + Mg^{2+})$ ATPase is believed to be responsible for maintaining low intracellular Ca^{2+} levels. Current theories on transmitter release provide that in response to stimuli from the axons the intracellular Ca^{2+} rises. This event triggers transmitter release in presynaptic terminals and causes other intracellular stimuli coupled responses. The free Ca^{2+} is then removed by ATPase-dependent processes. The role of synaptosomal (Ca^{2+} + Mg^{2+}) ATPase inhibition by organochlorine insecticides (including DDT) has been investigated by several independent laboratories. Important papers regarding this aspect of insecticide interaction with membranes include those of Yamaguchi et al. (47), where it was demonstrated that correlations exist between the effects of heptachlor epoxide on ATPase and Ca^{2+} binding and the release of the transmitter glutamate. The neurochemical basis of the action of Kepone was reviewed recently (10; see also the chapter by Bondy in this volume), and the attractiveness of the hypothesis that this agent exerts neurotoxicity through inhibition of synaptosomal ATPase with resulting effects on intracellular free Ca^{2+} levels was elaborated on. One very interesting observation noted in support of Kepone inhibition of ATPase being involved in its neurotoxicity is that the closely related chemical mirex was reported as not inhibiting membrane ATPase and that it is not neurotoxic.

In muscle, the cytosol Ca^{2+} level increases in response to stimuli and causes muscle contractions. The free Ca^{2+} is sequestered by the membranous sarcoplasmic reticulum mediated by an ATPase to stop the contractions. Disruption of the sarcoplasmic reticulum ATPase in muscle by DDT (19), organophosphates (1,21), and lindane (38) has been suggested as being related to neurotoxicity.

D. Transmitter Receptor Studies

The plasma membrane in the synaptic region differs from the peripheral nerve in that the synaptic region has higher concentrations of specific receptors for the various nerve transmitters on the cell surface of the pre- and postsynaptic membranes. Numerous reports have presented data which indicate that the synaptic region is an important locus for the neurotoxic action for many insecticides, and there are a variety of ways by which the membranes in the synaptic region are affected by these chemicals.

The most well-known example of the effects of a pesticide on a membrane transmitter receptor is the action of ACh due to its accumulation following inhibition of AChE by carbamate and organophosphate insecticides. In this instance, the effects of the insecticides are an indirect effect on the membrane to produce an expression of acute neurotoxicity which subsides after the enzyme is either decarbamylated or dephosphorylated or atropine is administered to protect the membrane receptors by acting as an antagonist to the excess ACh.

There are other aspects of the actions of AChE inhibitors on membranes in the synaptic region which may have important implications in the expression of neurotoxicity. These include the observation of several years ago (9,23) that certain organophosphates can phosphorylate the area immediately about the receptor for ACh in such a way that the response to ACh arriving at the receptor might be modified. The receptor may stay phosphorylated long after the inhibited enzyme is regenerated, and such an altered membrane receptor may be responsible for or related to the lingering effects noted after recovery of the more prominent acute effects due to enzyme inhibition. Such a hypothesis has not been investigated to the extent that phosphorylation of the synaptic membrane has a proven relationship to expression of neurotoxicity.

Another aspect of changes in membrane properties resulting from carbamates and organophosphates regards development of tolerance. It has been known for many years that test animals treated with certain organophosphates or carbamates develop a tolerance to these agents. Often test animals made tolerant to carbamates are also tolerant to organophosphates and vice versa. The tolerant animals are also tolerant to carbachol, an ACh agonist, implying that there are fewer membrane receptors for ACh. That this tolerance was related to changes at the level of the membrane was suggested earlier (40). A more convincing proof of this theory has been presented recently and is based on direct biochemical measurement of the concentration of muscarinic ACh receptors in tolerant and nontolerant rats and mice. These recent studies have demonstrated that rats made tolerant to diisopropylfluorophosphate (DFP), a powerful AChE inhibitor, have in fact membranes that bind fewer molecules of the specific ligand [^{3}H]quinuclidinyl benzylate (see Ref. 4). Similar depressions in the binding of this ligand for the muscarinic ACh receptor were found in rats made tolerant to the organophosphate insecticide disulfoton (4). A later paper from this same laboratory (5) demonstrated that disulfoton also decreased the number of nicotinic binding sites in tolerant rats. This reported change in receptor binding sites in brain tissue is consistent with the theory that increased nerve activity results in changes in membrane receptor density as a mechanism of internal compensation or adaptation. Another group of researchers reported changes in the receptor density for dopamine and γ-aminobutyric acid (GABA) as a response to DFP and that these changes were the result of abnormal ACh activity. The ACh nervous system was shown to interact closely with both the dopaminergic and GABA-ergic systems and the effects on the membrane receptors for dopamine and GABA were considered to be secondary to the action of DFP on the ACh system (42).

There are several other studies which report the use of ^{3}H receptor ligands to demonstrate changes in membrane synaptic receptor density as a response to pesticide poisoning. Treatment with the organochlorine Kepone (41) was shown to decrease in vivo the binding of [^{3}H]-spiroperidol (a dopamine receptor ligand), [^{3}H]muscimol (a GABA receptor ligand), and [^{3}H]quinuclidinyl benzilate, but the binding of [^{3}H]diazepam (a GABA ligand) and [^{3}H]serotonin in the brain cortex was reported as being unaffected. Measurement of the specific protein concentration of the brain areas where receptor binding was altered led to the conclusion that the changes in receptor binding could be accounted for in "terms of a region specific hyperplastic increase in nonreceptor proteins." Another example is the demonstration that DDT inhibits in vitro the specific binding of [^{3}H]naloxone (an opiate receptor ligand) and stimulates the binding of [^{3}H]ouabain, a ligand for the ($Na^+ + K^+$) ATPase; the noninsecticidal analog DDE was reported to be inactive in these receptor assays (25).

The ability of the various synthetic pyrethoids to displace ligands specific for various nerve transmitters has been reported recently and it appears that this class of chemicals has the potential to interfere with the membrane receptors for several transmitters and that the type I and type II pyrethroids differ in their properties of being able to displace these ligands in vitro. Important papers reporting the results of synthetic pyrethroids on interference with the binding of ligands for specific receptors include the study of the displacement of [^{3}H]kainic acid. It was reported (43) that the order of potency for displacement of [^{3}H]kainic acid, a ligand for the transmitter candidate L-glutamate, was decamethrin > cis-permethrin > trans-permethrin, and this order paralleled the order of potency for producing neurotoxicity symptoms in mice. In this study, the brain concentrations of the pyrethroids at times when signs of neurotoxicity were obvious were noted to be in good agreement with the concentrations which could displace kainic acid in vitro.

Another very important paper investigating the effects of pyrethroids on specific transmitter receptor targets in the synaptic region utilized sulfur-35-labeled t-butylbicyclophosphorothionate, a GABA receptor ligand (26). In this study correlations between the toxicity of the type II pyrethroids and affinity for the GABA receptor-ionophore complex (the authors stated that no false positive or negatives appeared in correlating inhibition with toxicity) were reported. The type I pyrethroids phenothrin and permethrin did not effectively inhibit the radioligand binding at test levels as high as 5-10 μM.

Lastly, another approach to study membrane receptors for insecticidal neurotoxicity is to use radiolabeled isomers of the insecticides and determine their binding patterns in nerve. Two recent articles (6,7) illustrate an example of this approach. In this work, the binding of DDT to housefly head membranes was studied. This brain preparation was shown to have a negative temperature correlation for DDT binding, was sensitive to Ca^{2+}, and was also shown to bind permethrin and cypermethrin (although the pyrethroids appeared not to bind the same specific site). Importantly, a strain of flies resistant to DDT bound less radioactivity than susceptible flies and bound little noninsecticidal DDT (7).

IV. SUMMARY

This chapter has attempted to present some of the many aspects of the effects of pesticides on nerve membranes in relation to the expression of neurotoxicity. Several components of the membranes such as the Na^+ channel, Ca^{2+} binding, and transmitter receptors have been indicated as being affected by pesticides. In many cases, especially for the organochlorine and more recently the pyrethroid insecticides, more than one membrane component seems to be affected. All of the affected components may contribute to the expression of neurotoxicity.

REFERENCES

1. Abe, T., Kawal, N., Miwa, A., Fujimoto, Y., Tatsuno, T., Fukami, J. 1983. Effects of paraoxon and fenitrooxon on crustacean muscle membrane. *Comp. Biochem. Physiol. 74C*:249-253.
2. Abood, L. G., Lowy, K., Tometsko, A., Booth, H. 1978. Electrophysiological, behavioral and chemical evidence for a noncholinergic stereospecific site for nicotine in rat brain. *J. Neurosci. Res. 3*:327-333.
3. Casida, J. E., Gammon, D. W., Glickman, A. H., Lawrence, L. J. 1983. Mechanisms of selective action of pyrethroid insecticides. *Annu. Rev. Pharmacol. Toxicol. 23*:413-438.
4. Costa, L. G., Schwab, B. W., Hand, H., Murphy, S. D. 1981. Reduced [^{3}H]quinuclidinyl benzilate binding to muscarinic receptors in disulfoton-tolerant mice. *Toxicol. Appl. Pharmacol. 60*: 441-450.
5. Costa, L. G., Murphy, S. D. 1983. [^{3}H]Nicotine binding in rat brain: Alteration after chronic acetylcholinesterase inhibition. *J. Pharmacol. Exp. Ther. 226*:392-397.
6. Chang. C. P., Plapp, F. W. 1983. DDT and pyrethroids: Receptor binding and mode of action in the housefly. *Pesti. Biochem. Physiol. 20*:76-85.
7. Chang, C. P., Plapp, F. W. 1983. DDT and pyrethroids: Receptor binding in relation to knockdown resistance (kdr) in the housefly. *Pesti. Biochem. Physiol. 20*:86-91.

8. Cutkomp, L. K., Koch, R. B., Desaiah, D. 1982. Inhibition of ATPase by chlorinated hydrocarbons, in *Insecticide Mode of Action* (Coates, J., ed.), Academic, New York, pp. 45-64.
9. Dekin, M. S., Guy, H. R., Edwards, C. 1978. Effects of organophosphate insecticides on the cholinergic receptors of frog skeletal muscle. *J. Pharmacol. Exp. Ther. 205*:319-325.
10. Desaiah, D. 1982. Biochemical mechanisms of chlordecone neurotoxicity. A review. *Neurotoxicology 3*:103-110.
11. Doherty, J. D., Matsumura, F. 1975. DDT effects on certain ATP related systems in the peripheral nervous system of the lobster *Homarus americanus*. *Pesti. Biochem. Physiol. 5*:242-252.
12. Doherty, J. D., Salem, N., Jr., Lauter, C. J., Trams, E. G. 1981. Synaptic vesicle associated ATPase: Stimulation by Mn^{2+} and inhibition by neurotoxic pesticides. *Toxicologist 1*:abstract number 488.
13. Doherty, J. D., Lauter, C. J., Trams, E. G. 1982. Dopamine release and distruption of synaptic Ca^{++} binding by pyrethroids. *Trans. Am. Soc. Neurochem. 13*.
14. Duffy, F. J., Burchfiel, J. L., Bartels, P. H., Gaon, M., Sim, V. M. 1979. Long-term effects of an organophosphate upon the human electroencephalogram. *Toxicol. Appl. Pharmacol. 47*:161-176.
15. Eldefrawi, A. T., Mansour, N. A., Eldefrawi, M. E. 1982. Insecticides affecting acetylcholine receptor interactions. *Pharmacol. Ther. 16*:45-65.
16. Ghiasuddin, S. M., Matsumura, F. 1979. DDT inhibition of Ca^{++} ATPase of the peripheral nerves of the American lobster. *Pesti. Biochem. Physiol. 10*:151-161.
17. Ghiasuddin, S. M., Matsumura, F. 1979. Ca^{++} regulation by Ca^{++}-ATPase in relation to DDT's action on the lobster nerve. *Comp. Biochem. Physiol. 64C*:29-36.
18. Ghiasuddin, S. M., Matsumura, F. 1982. Inhibition of gamma-aminobutyric acid (GABA)-induced chloride uptake by gamma-BHC and heptachlor epoxide. *Comp. Biochem. Physiol. 73C*:141-144.
19. Greenwood, M., Huddart, H. 1975. The effect of an organochlorine insecticide on the fine structure of flight muscle and the implications for the maintenance of contractility. *Comp. Biochem. Physiol. 51A*:475-481.
20. Holan, G., Spurling, T. H. 1974. Mode of action of DDT analogues: Molecular orbital studies. *Experientia 30*:480-481.
21. Huddart, H., Greenwood, M., Williams, M. J. 1974. The effect of some organophosphorus and organochlorine compounds on calcium uptake by sarcoplasmic reticulum isolated from insect and crustacean skeletal muscle. *J. Comp. Physiol. 93*:139-150.

22. Jacques, Y., Romey, G., Cavey, M. T., Kartalovski, B., Lazdunski, M. 1980. Interaction of pyrethroids with the Na^+ channel in mammalian neuronal cells in culture. *Biochem. Biophys. Acta 600*:882-897.
23. Kuba, K., Albuquerque, E. X., Daly, J., Barnard, E. A. 1974. A study of the irreversible cholinesterase inhibitor, diisopropylfluorophosphate, on time course of end-plate currents in frog sartorius muscle. *J. Pharmacol. Exp. Ther. 189*:499-512.
24. Kwiecinski, H. 1981. Myotonia induced by chemical agents. *CRC Crit. Rev. Toxicol. 8*:279-310.
25. LaBella, F. S. 1981. Opiate and digitalis receptor assays distinguish between neuroexcitant and inactive organochlorines and between anesthetics and convulsants. *Brain Res. 219*:166-171.
26. Lawrence, L. J., Casida, J. E. 1983. Stereospecific action of pyrethroid insecticides on the γ-aminobutyric acid-receptor-ionophore complex. *Science 221*:1399-1401.
27. Lund, A. E., Narahashi, T. 1981. Interaction of DDT with sodium channel in squid giant axon membranes. *Neuroscience 6*: 2253-2258.
28. Lund, A. E., Narahashi, T. 1981. Kinetics of sodium channel modification by the insecticide tetramethrin in squid axon membranes. *J. Pharmacol. Exp. Ther. 219*:464-473.
29. Lund, A. E., Narahashi, T. 1982. Dose dependent interaction of the pyrethroid isomers with sodium channels of squid axon membranes. *Neurotoxicology 3*:11-24.
30. Matsumura, F., Narahashi, T. 1971. ATPase inhibition and electrophysiological change caused by DDT and related neuroactive agents in lobster nerve. *Biochem. Pharmacol. 20*:825-837.
31. Moreton, R. B., Gardner, D. R. 1981. Increased intracellular chloride activity produced by the molluscicide, N-(triphenylmethyl)morpholine (Frescon), in *Lymnaea stagnalis* neurons. *Pesti. Biochem. Physiol. 15*:1-9.
32. Mullins, L. J. 1954. Some physical mechanisms of narcosis. *Chem. Rev. 54*:289-323.
33. Narahashi, T. 1962. Nature of the negative after-potential increased by the insecticide allethrin in cockroach giant axons. *J. Cell. Comp. Physiol. 59*:67-76.
34. Narahashi, T. 1962. Effect of the insecticide allethrin on membrane potentials of cockroach giant axons. *J. Cell. Comp. Physiol. 59*:61-65.
35. Narahashi, T. 1976. Effects of insecticides on nervous conduction and synaptic transmission, in *Insecticide Biochemistry and Physiology* (Wilkenson, C. F., ed.), Plenum, New York, pp. 327-352.

36. Narahashi, T., Haas, H. G. 1968. Interaction of DDT with the components of lobster nerve membrane conductance. *J. Gen. Physiol.* *51*:177-198.
37. Niemi, W. D., Webb, G. D. 1980. DDT and sodium transport in the eel electroplaque. *Pesti. Biochem. Physiol.* *14*:170-177.
38. Publicover, S. J., Duncan, C. J., Smith, J. L. 1979. The action of lindane in causing ultrastructural damage in frog skeletal muscle. *Comp. Biochem. Physiol.* *74C*:237-241.
39. Roeder, K. D., Weiant, E. A. 1946. The site of action of DDT in the cockroach. *Science* *103*:304-306.
40. Rider, J. A., Ellinwood, L. E., Coon, J. M. 1952. Production of tolerance in the rat to octamethyl pyrophosphoramide (OMPA). *Proc. Soc. Exp. Biol. Med.* *81*:455-459.
41. Seth, P. K., Agrawal, A. K., Bondy, S. C. 1981. Biochemical changes in the brain consequent to dietary exposure of developing and mature rats to chlordecone (Kepone). *Toxicol. Appl. Pharmacol.* *59*:262-267.
42. Sivam, S. P., Norris, J. C., Lim, D. K., Hoskins, B., Ho, I. K. 1983. Effect of acute and chronic cholinesterase inhibition with diisopropylfuorophosphate on muscarinic, dopamine and GABA receptors in the rat striatum. *J. Neurochem.* *40*:1414-1422.
43. Staatz, C. G., Bloom, A. S., Lech, J. J. 1982. Effects of pyrethroids in [^{3}H]kainic acid binding to mouse forebrain membranes. *Toxicol. Appl. Pharmacol.* *64*:566-569.
44. Taylor, J. R., Selhorst, J. B., Hauff, S. Q., Martinez, A. J. 1978. Chlordecone intoxication in man. *Neurology* *28*:626-630.
45. Verschoyle, R. D., and Aldridge, W. N. 1980. Structure-activity relationships of some pyrethroids in rats. *Arch. Toxicol.* *45*:325-329.
46. Wu, C. H., Bercken, J. van den, Narahashi, T. 1975. The structure-activity relationship of DDT analogs in crayfish giant axons. *Pesti. Biochem. Physiol.* *5*:142-149.
47. Yamaguchi, I., Matsumura, F., Kadous, A. A. 1980. Heptachlor epoxide: Effects on calcium-mediated transmitter release from brain synaptosomes in rat. *Biochem. Pharmacol.* *29*:1815-1823.

20

Biochemical Toxicology of Organophosphorous Compounds

Mohamed B. Abou-Donia
Duke University Medical Center, Durham, North Carolina

I. INTRODUCTION

Organophosphorous compounds were originally developed to be used as agricultural insecticides, and later as potential nerve agents in warfare (Table 1). Most organophosphorous compounds available today are used as agricultural or industrial chemicals. A few of these chemicals have some medicinal use. Many of these compounds, but not all, are organophosphates or are desulfurated to organophosphates. These compounds may be classified into the following chemical groups (42, 59,65):

PHOSPHATE — PHOSPHOROTHIOATE — PHOSPHOROTHIOATE OR PHOSPHOROTHIOLATE — PHOSPHONATE

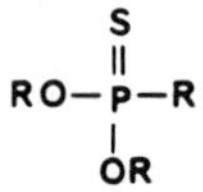
PHOSPHONOTHIOATE

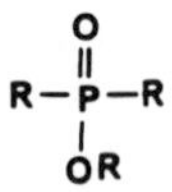
PHOSPHINITE

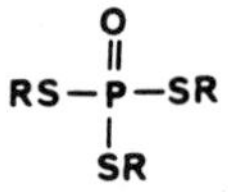
PHOSPHOROTRITHIOATE

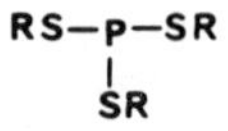
PHOSPHOROTRITHIOITE

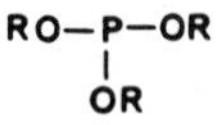
PHOSPHITE

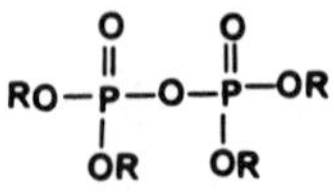
PYROPHOSPHATE

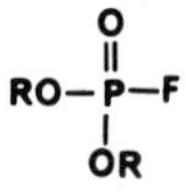
PHOSPHOROFLUORIDATE

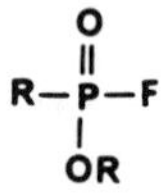
PHOSPHONOFLUORIDATE

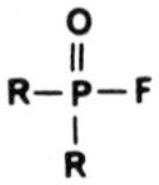
PHOSPHINOFLUORIDATE

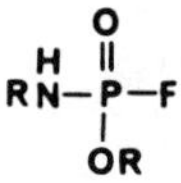
PHOSPHOROAMIDOFLUORIDATE

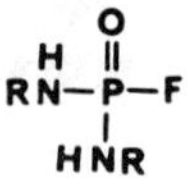
PHOSPHORODIAMIDOFLUORIDATE

II. THE USES OF ORGANOPHOSPHOROUS COMPOUNDS

A. Pesticides

Most organophosphorous compounds belong to one of the pesticide groups. In 1959 it was estimated that over 50,000 organophosphorous compounds had been made, and more have been added since that time. Most of these pesticides are used in agriculture as insecticides, acaricides, fungicides, or cotton defoliants (65).

1. Insecticides and Acaricides

The first organophosphorous insecticide was tetraethylpyrophosphate (TEPP) (78). It was developed as a substitute for nicotine, which was in short supply during World War II (Table 1). Parathion (45) was developed in 1944 to replace TEPP, which is extremely toxic and unstable in moisture. Because of the high toxicity of parathion, other chemicals were developed, such as chlorthion (72) and malathion (64, 88). The selective toxicity of malathion (65), that is, a low toxicity in mammals and high insecticidal effect, is attributed to (a) its oxidation (activation) in insects to the more toxic metabolite malaoxon and (b) its hydrolysis in animals to the less toxic chemical malathion acid.

a. Contact Insecticides: Organophosphorous insecticides are generally contact poisons. Some of these compounds may also have stomach or fumigant actions, for example, dichlorvos (42), trichlorphon (65), and EPN (77).

b. Systemic Insecticides: Systemic organophosphorous insecticides are absorbed by plants. Absorption may take place through the foliage or the roots. These insecticides are distributed throughout the plant,

Table 1 Chemical Designation of Organophosphorous Esters Mentioned in Text

Compound	Designation
Amiprophos	0-Ethyl-0-2-nitro-4-methyl-N-isopropylphosphoramidothionate
Apholate	2,2,4,4,6,6-Hexakis(1-aziridinyl)-2,2,4,4,6,6-hexahydrol-1,3,5,2,4,6-triazatriphosphorine
Bensulide	0,0-Diisopropyl-S-2-phenylsulphonylaminoethylphosphorodithioate
Chlorthion	0-3-Chloro-4-nitrophenyl-0,0-dimethylphosphorothioate
Coroxon	0,0-Diethyl-0-(3-chloro-4-methyl-7-coumarinyl)phosphate
Coumaphos	0,0-Diethyl-0-(3-chloro-4-methyl-7-coumarinyl)phosphorothioate
Cremart	0-Ethyl-3-methyl-6-nitrophenyl-N-sec-butylphosphoramidothionate
Cyanofenphos	0-Ethyl-0-4-cyanophenylphenylphosphonothioate
Cyclophosphamide	2-[Bis(2-chloroethyl)amino]tetrahydro-2H-1,3,2-oxazaphosphorin-2-oxide
DEF (butifos)	S,S,S-Tri-n-butylphosphorothrithioate
DFP	0,0-Diisopropylphosphorofluoridate
Diazinon	0,0-Diethyl-0[2-isopropyl-6-methyl-4-pyrimidinyl]phosphorothioate
Dichlorvos	2,2-Dichloroethenyl-0,0-dimethylphosphate
Echothiophate (phospholine)	0,0-Diethyl-S-choline phosphorothioate
Edifenphos (hinosan)	0-Ethyl-S,S-diphenylphosphorodithioate
EPN	0-Ethyl-0-4-nitrophenylphenylphosphonothioate
Ethephon (ethrel)	2-Chloroethylphosphonic acid
Ethoprophos (mocap)	0-Ethyl-S,S-dipropylphosphorodithioate
Fenchlorphos (ronnel)	0,0-Dimethyl-0-(2,4,5-trichlorophenyl)-phosphorothioate

Table 1 (Continued)

Compound	Designation
Fensulfothion	0,0-Dimethyl-0-4-methylsulphinylphenyl-phosphorothioate
Fyrol FR-2	Tris(1,3-dichloro-2-propyl)phosphate
Glyphosate	N(phosphonomethyl)glycine
Gophacide (phosacetim)	0,0-Di-4-chlorophenyl-N-acetimido-phosphoramidothionate
Haloxon	0,0-Di-(2-chloroethyl)-0-(3-chloro-4-methyl-7-coumarinyl)phosphate
Inezin	S-Benzyl-0-ethylphenylphosphonothioate
Kitazin	S-Benzyl-0,0-diisopropylphosphorothioate
Leptophos	0-4-Bromo-2,5-dichlorophenyl-0-methyl-phenylphosphonothioate
Malathion	0,0-Dimethyl-S-[1,2-(ethoxycarbonyl)-ethyl]phosphorodithioate
Merphos (folex)	S,S,S-Tri-n-butylphosphorotrithioite
Metepa	Tris(2-methyl-1-aziridinyl)phosphine oxide
Mipafox	N,N'-Diisopropylphosphorodiamidic fluoride
Paraoxon	0,0-Diethyl-0-4-nitrophenylphosphate
Parathion	0,0-Diethyl-0-4-nitrophenylphosphorothioate
Phosbutyl	0-Ethyl-S-phenyl-N-butylphosphoramidothiolothionate
Prophos	0-Ethyl-S,S-di-n-propylphosphorodithioate
Round-up	N-(Phosphonomethyl)glycineisopropylamine
Sarin	0-Isopropylmethylphosphonofluoridate
Soman	0-Pinacolylmethylphosphonofluoridate
S-Seven	0-Ethyl-0-2,4-dichlorophenylphenyl-phosphonothioate

Table 1 (Continued)

Compound	Designation
Tabun	0-Ethyl-N,N-dimethylphosphoramido-cyanidate
TEPA	Tris(1-aziridinyl)phosphine oxide
TEPP	0,0,0,0-Tetraethylpyrophosphate
ThioTEPA	Tris(1-aziridinyl)phosphine sulfide
TOCP	Tri-o-cresylphosphate
Trichlorphon	0,0-Dimethyl(1-hydroxy-2,2,2-tri-chloroethyl)phosphonate
Tris-PB	Tris(2,3-dibromopropyl)phosphate
Zinophos (thionazin)	0,0-Diethyl-0-(2-pyrazinyl)phosphoro-thioate
Zytron	0-(2,4-Dichlorophenyl)-0-methyliso-propylphosphoramidothioate

thus protecting it from insects. Examples of systemic insecticides are parathion (42), malathion (65), and diazinon (47).

2. *Veterinary Pesticides*

This group of organophosphorous pesticides is used to control parasites of domestic animals. Parasites include insects, mites, and helminths. These pesticides may be given orally or applied dermally. Examples of these compounds are haloxon (29), coroxon (29), coumaphos (13,42), and fenchlorphos (ronnel) (65).

3. *Nematocides*

Organophosphorous nematocides are introduced into the soil through the water system. They control nematodes through their anticholinesterase action. Examples of these pesticides are prophos (mocap) (15) and zinophos (66).

4. *Insect Chemosterilants*

Organophosphorous insect chemosterilants are alkylating agents. Most of these chemicals are aziridine derivatives of phosphoric acid such as TEPA, thio-TEPA, metepa, and apholate (42). In addition to insect sterilization, these chemicals may also have mutagenic, teratogenic, and carcinogenic effects. Some organophosphorous alkylating agents such as cyclophosphamide have anticancer activity.

Aziridines inhibit the biosynthesis of DNA, an action that is accompanied by an increase of the acid-soluble DNA precursor content (42, 58). This may result from the inhibition of DNA polymerase or the stimulation of DNA degradation. The primary site of alkylation of nucleic acids by alkylating agents is the N-7 of guanine moiety. Alternatively, thio-TEPA may inhibit DNA biosynthesis by inhibiting the conversion of purine ribonucleotides into deoxycompounds for incorporation into DNA (70).

5. *Fungicides*

Several organophosphorous compounds are used as fungicides. Most of these chemicals are phosphorothiolates, phosphonothiolates, or phosphoramidic acid derivatives. Examples of this group of compounds are kitazin (61), inezin (84), edifenphos (65), and phosbutyl (67). The fungicidal action of these organophosphorous compounds does not seem to be related to inhibition of esterases. The fungicide kitazin blocks the biosynthesis of fungal wall chitin by inhibiting the incorporation of glucosamine into the cell wall chitin (42).

6. *Herbicides*

Several organophosphorous compounds are used as herbicides. Examples of this class are bensulide (20), zytron (86), cremart (42), and amiprophos (18). The mechanism of the herbicidal action of these compounds is not well understood. The organophosphorous herbicide ethephon generates ethylene in plants, thus inducing fruit ripening, abscission, and flowering (36). The related compound O-ethylpropylphosphonic acid retards plant growth by reducing the internode elongation. Recently another phosphonic compound, glyphosate, has been used as a nonselective herbicide. It is a postemergence herbicide for control of both annual and perennial weeds. Its herbicidal action may possibly be explained by an interference with the biosynthesis of phenylalanine in plants, perhaps by inhibiting chorismate mutase and/or prephenate dehydratase (52). Two organophosphorous compounds that are useful as cotton defoliants are DEF (butifos) and merphos (folex) (64,65).

7. *Rodenticides*

Only a few organophosphorous compounds are used as rodenticides, probably because the rapid onset of cholinergic effects by organophosphorous compounds prevents the consumption of a lethal dose by a rodent. An example of a compound that is effective is gophacide (38).

8. *Insecticide Synergists*

Some organophosphorous compounds, such as propyl-2-propynylphenylphosphonate (27), are used as synergists of pyrethroids based on their ability to inhibit the microsomal mixed-function oxidases. The cotton defoliant DEF seems to have a potentiating effect for the insecticidal activity of malathion (30).

9. *Insect Repellents*

Some phosphoramidates have repellent activity against insects. An example from this group of compounds is O-n-butyl-O-cyclohexenyl-N,N-diethylphosphoramidate (57).

B. Nerve Gases

During World War II, the highly toxic organophosphorous compounds were developed as potential chemical warfare agents. The extreme toxicity of these compounds results from their irreversible inhibition of acetylcholinesterase (AChE). Examples of these chemicals are sarin (O-isopropylmethylphosphonofluoridate), tabun (O-ethyl-N,N-dimethylphosphoramidocyanidate), and soman (O-pinacolylmethylphosphonofluoridate). The lethal dose for humans of any of these chemicals is less than 0.01 mg/kg. Sarin is the most toxic of the three nerve gases (88).

C. Therapeutic Use

Some organophosphorous compounds, such as DFP (59), are used to treat glaucoma. Acetylcholinesterase inhibitors cause a decrease in intraocular pressure in primary glaucoma. DFP has a long duration of action and is extremely potent. However, the peanut or sesame seed oil solutions needed for local applications are unpleasant to most patients and DFP has been replaced by echothiophate (59).

D. Flame Retardants

In 1972 the United States established flammability standards for children's sleepwear. Since then, chemicals have been used to make fabrics flame retardant. Among these are some organophosphorous chemicals such as tris-PB (59) and fyrol FR-2 (88). The compound tris-PB was found to be carcinogenic in mice, while fyrol FR-2 and some of its metabolites were found to be mutagenic.

E. Other Industrial Uses

Many aliphatic and aromatic organophosphorous esters have industrial uses as plasticizers and industrial fluids. They are used in the application of wall coverings and in plasticized automotive upholstering.

Most of these chemicals have low acute toxicity. Some such as tri-o-cresylphosphate (TOCP) (88), however, produce delayed neurotoxicity (6).

III. MECHANISMS OF ACTION OF ORGANOPHOSPHOROUS COMPOUNDS

A. Inhibition of Acetylcholinesterase

Most organophosphorous esters are direct or indirect inhibitors of acetylcholinesterase. Acetylcholinesterase (EC 3.1.1.7) (AchE), also known as specific or true cholinesterase (ChE), is present in neurons, at the neuromuscular junction, and in some tissues (68). Acetylcholine (ACh), which is released as the result of cholinergic stimulation, is hydrolyzed by AChE. In the peripheral nervous system AChE occurs in high concentrations in the postganglionic parasympathetic, preganglionic autonomic, and somatic motor fibers at the neuromuscular junction; AChE is also present in ganglion cells in the peripheral nervous system. The central nervous system contains high concentrations of AChE. Another enzyme, butyrylcholinesterase (EC 3.1.1.8) (BuChE)—also known as ChE, nonspecific ChE, or pseudo-ChE—is present in plasma and in many tissues, in various types of glia, and in liver and other organs. The physiological function of BuChE is unknown. Also, unlike AChE, inhibition of BuChE at most sites does not result in apparent functional disruption.

Acetylcholine is a neurotransmitter used in the peripheral nervous system and central nervous system. In peripheral nervous system it is used at skeletal neuromuscular junctions; it also functions at the following synapses: parasympathetic preganglionic and postganglionic nerves, sympathetic preganglionic nerves, sweat glands, and a very few blood vessels (sympathetic vasodilator fibers). In the central nervous system ACh serves as a transmitter in many areas. The most important are cerebral cortex and striatum. Choline acetyltransferase catalyzes the synthesis of ACh from choline and acetyl coenzyme A at the cholinergic nerve terminal. There are two types of acetylcholine receptors: muscarinic and nicotinic (59,88). Muscarinic sites are present in smooth muscle, glands, and most cholinergic synapses in the central nervous system, and in the striatum in particular. Nicotinic sites exist in skeletal muscle; all ganglia, including the adrenal medulla; and in some synapses in the central nervous system.

Invasion of the nerve terminal membrane by the nerve action potential initiates neurotransmission. This results in the depolarization of the terminal membrane and entrance of Ca^{2+} through the Ca^{2+} channels (ionophores); this triggers the release of ACh from the nerve terminal (59). Once released, ACh diffuses within the synaptic cleft and interacts specifically with ACh receptor. This transmitter-receptor interaction releases energy, which opens ionophores so that ions can flow into or out of the postsynaptic cells, depending on the concentration gradient. This flow of ions then causes the postsynaptic response. The action of ACh is terminated with hydrolysis by AChE. This enzyme is located on the postsynaptic membrane and on the plasma membrane of the entire cholinergic neuron. Inhibition of AChE may have a greater physiological effect at some cholinergic synapses than at others. Therefore AChE inhibitors may have selective actions. Inhibition of AChE enhances transmission at all cholinergic synapses. Since AChE is present in great excess at cholinergic synapses, however, 60-90% of the enzyme must be inhibited before the onset of cholinergic dysfunction.

The AChE-active center contains two active sites: an anionic site, which interacts with the nitrogen atom and methyl groups of the choline moiety of ACh, and an esteratic site, which binds the carbonyl carbon atom of the acetyl moiety of ACh (59). The interaction between AChE and ACh results in the formation of a covalent bond between the carbonyl carbon and the hydroxyl group of a serine amino acid residue in the esteratic site. The acetylated enzyme is then hydrolyzed with water to release acetate. This hydrolytic step takes place in less than 0.1 msec (51). Organophosphorous compounds only act at the esteratic site, and phosphoric acid ester is formed with the enzyme. The phosphoric ester of some inhibitors is hydrolyzed very slowly, or in other compounds the esterification reaction is virtually irreversible. The duration of enzyme inhibition is therefore very long. In cases of irreversible organophosphorous inhibitors, the duration of action may take months, because it is determined by the time required to synthesize new AChE molecules. The overall reaction of acetylcholinesterase with the substrate acetylcholine and the phosphorous ester inhibitors may be represented according to the following scheme, by considering the phosphorous esters as substrates (59):

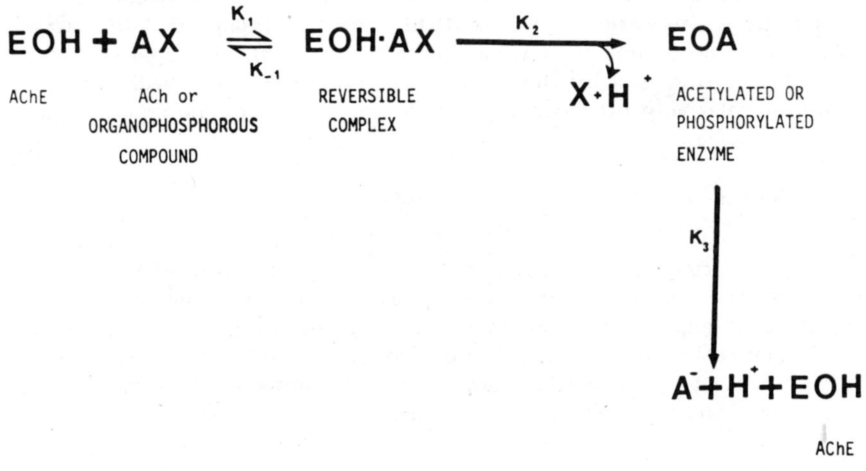

This reaction consists of three steps:

1. Complex formation, which is governed by the affinity constant $K_a = k_{-1}/k_1$
2. Acetylation or phosphorylation, k_2
3. Deacetylation of dephosphorylation, k_3

In all reactions k_3 is the rate-determining step, since it is the slowest in every case. With ACh k_2 and k_3 are very fast, so that the total reaction occurs rapidly (in milliseconds) and new enzyme is generated. The turnover number, which is the number of substrate molecules hydrolyzed per minute by one molecule of enzyme, has been estimated to be 300,000 for ACh. In the case of organic phosphates k_2 is moderately fast, but k_3 is extremely slow ($k_2 \gg k_3$), so the phosphorylated enzyme EOA accumulates, while the amount of EOH · AX is minimal at any time. The turnover number of phosphorylated AChE by dimethylphosphate was estimated to be 0.008 and interpreted to represent irreversible inhibition of AChE.

1. *"Aging" of Phosphorylated Enzyme*

With some organophosphorous compounds dephosphorylation of the inhibited AChE occurs very slowly or not at all. Diisopropyl phosphorofluoridate (DFP) is one of these compounds, and the rate of regeneration of AChE activity of plasma and red cells coincides with the rate of resynthesis of new enzyme. "Aging" involves the loss of one of the alkyl or alkoxy group on the phosphorylated enzyme, which results in the negatively charged monoalkyl- or monoalkoxyphosphorylated enzyme (59).

2. *Reactivation of Phosphorylated AChE*

Oxime derivatives such as 2-PAM (2-pyridine aldoxime methiodide) accelerate regeneration of active AChE from the phosphorylated enzyme (59). It is noteworthy, however, that the aged phosphorylated enzyme is not reversed by the oximes. The aldoxime itself will inhibit AChE at relatively high concentrations by binding to the anionic site of the enzyme.

3. *Tolerance of Acute Sublethal Effects of Organophosphorous Compounds*

When some organophosphorous compounds are administered repeatedly or fed in the diet at sublethal doses for several days, they initially cause acute cholinergic toxicity (28). In time, however, the animals no longer respond with cholinergic signs after each dose, and they appear normal. At sacrifice, however, these apparently normal animals have markedly inhibited AChE activity in blood and nervous tissue and elevated levels of ACh in their brains. It has been suggested that this apparent adaptation involves development of tolerance of the cholinergic receptor site or a decrease in the total number of ACh receptors (see Doherty, this volume).

4. *Signs and Symptoms of Acute Poisoning*

Inhibition of AChE results in accumulation of endogenous ACh in nerve tissue and effector organs with subsequent signs and symptoms that mimic the muscarinic, nicotinic, and central nervous system actions of ACh (59,65).

a. Muscarinic Signs: Stimulation of muscarinic receptors (which are found primarily in the smooth muscle, the heart and exocrine gland) results in the following symptoms:

1. Tightness in the chest and wheezing due to broncoconstriction
2. Increased bronchial secretions, salivation, lacrimation, and sweating
3. Increased gastrointestinal tone with consequent development of nausea, vomiting, abdominal cramps, diarrhea, and involuntary defecation
4. Frequent contraction of smooth muscle of the bladder, resulting in involuntary urination
5. Bradycardia that can progress to heart block
6. Constriction of the pupils

b. Nicotinic Signs: The following signs result from accumulation of ACh at the endings of motor nerves to skeletal muscle and autonomic ganglia:

1. Muscular effects, including easy fatigability and mild weakness followed by involuntary twitching and cramps. Weakness affects the muscles involved in respiration and contributes to dyspnea, hypoxemia, and cyanosis.
2. Nicotinic actions at automonic ganglia may, in severe intoxication, mask some of the muscarinic effects. Thus tachycardia caused by stimulation of sympathetic ganglia may override the usual bradycardia due to muscarinic action on the heart. Elevation of blood pressure and hyperglycemia also reflect nicotinic action at sympathetic ganglia.

c. Central Nervous System Signs: Accumulation of ACh in the central nervous system resulting from mild to moderate poisoning by organophosphorous compounds leads to the following signs: tension, anxiety, restlessness, insomnia, headache, emotional instability, excessive dreaming, and nightmares. Severe poisoning with organophosphorous compounds results in slurred speech, tremor, generalized weakness, ataxia, convulsions, depression of the respiratory and circulatory centers, and coma.

d. Cause of Death: The immediate cause of death in fatal organophosphorous poisoning is asphyxia resulting from respiratory failure with the following contributing factors:

1. The muscarinic bronchoconstriction and increased bronchial secretions
2. Nicotinic action leading to paralysis of respiratory muscles
3. Action on the central nervous system causing depression and paralysis of the respiratory center

5. *Diagnosis of Organophosphorous Poisoning*

The following clinical findings should lead to the diagnosis of organophosphorous poisoning:

1. Evidence of exposure to organophosphorous insecticides within the previous 24 hr.
2. Signs and symptoms of poisoning.
3. Measurement of blood cholinesterase activity indicative of overexposure to organophosphorous insecticides. Erythrocyte AChE is a better reflection of the level of AChE in the central nervous system than is plasma ChE. Blood ChE usually becomes markedly depressed before symptoms appear. There are usually no symptoms or signs until the cholinesterase level reaches about 25% of the normal or preexposure value. A decrease of 60% in red blood cell AChE from the preexposure enzymatic level is an indication for removal from the work place.
4. Detection of organophosphorous insecticide metabolites, for example, p-nitrophenol in the urine after exposure to parathion or EPN.
5. Response to treatment with atropine or 2-PAM.

6. *Treatment of Organophosphorous Poisoning*

A number of treatment measures should be undertaken when organophosphorous poisoning is diagnosed. They include the following:

1. Removal of secretions and maintenance of an unobstructed airway.
2. Initiation of artificial respiration and administration of oxygen.
3. Atropine administration. As soon as cyanosis has been over come, 2-4 mg of atropine should be administered intravenously. (In a state of cyanosis, atropine sulfate may produce ventricular fibrillation in the presence of hypoxia.) This dose is repeated at 5- or 10-min intervals until signs of atropinization

appear (dry flushed skin and tachycardia as high as 140 beats per minute). A mild degree of atropinization should be maintained for at least 48 hr.

4. Pralidoxime administration. Oximes only supplement the atropine therapy and should never replace it. These chemicals act by hydrolyzing the phosphorylated enzyme, resulting in the free AChE. Reactivation of AChE occurs rapidly (within 10-15 min), while plasma ChE requires 12-15 days to return to normal levels. The initial adult dose of 1000 mg in an isotonic solution is given intravenously at a rate not to exceed 500 mg/min. After about 1 hr, a second dose of 1000 mg is given if muscle weakness has not been relieved.
5. Decontamination of skin by washing with alkaline soap and water to accelerate the hydrolysis of the phosphorous ester.
6. Treatment of convulsions by administering sodium thiopental (2.5% solution) given intravenously.

The following drugs should be avoided in cases of organophosphorous poisoning: morphine, theophylline, aminophylline, succinylcholine, and tranquilizers of the reserpine or phenothiazine drug classes.

B. Organophosphorous-Induced Delayed Neurotoxicity

A number of organophosphorous compounds with antiesterase activity also produce a condition known as delayed neurotoxicity (OPIDN) in humans and sensitive species (6,80,81). After a delay period of 6-14 days, clinical signs are observed. There is an initial ataxia followed by paralysis. Neuronal lesions characterized by wallerian-type degeneration of axons and myelin in the central and peripheral nervous systems have been identified in this condition (see Cavanagh, in this volume). This effect was first demonstrated in humans exposed to tri-o-cresylphosphate (TOCP) (6,76). More than 40,000 cases of TOCP-induced delayed neurotoxicity have been documented (49,82,83,85). While being developed as an insecticide, mipafox caused delayed neurotoxicity in three reported persons (21). Leptophos, another experimental insecticide, was implicated as the cause of delayed neurotoxicity seen in some workers handling it (14,89). Certain animal species (cats, dogs, cows, and chickens) are susceptible to OPIDN, while others (rodents and some primates) are less sensitive (14,37,63,80). The adult chicken has become the experimental animal to study this effect (6,80).

Most substituted phenylphosphonothioate insecticides produce delayed neurotoxicity in chickens (4,6). This group of insecticides includes leptophos (6,15,16,86,87), EPN (6,8,39,44,46), cyanophenphos (9), and S-Seven (11). Two aliphatic thiophosphates used as a cotton defoliant, DEF and merphos (6,10,12,85), have been shown to produce delayed neurotoxicity in hens.

The development, severity, and regression of OPIDN are dependent on the chemical used, the dosage, the route of administration, and the frequency and duration of exposure (6). Species sensitivity to delayed neurotoxicity from organophosphorous esters is inversely correlated with the rate of its metabolism and excretion in various species (2,3,4-7, 48,50,89). Some chemicals, such as TOCP, are metabolized to a more active neurotoxic agent (7).

1. *Mechanisms of OPIDN*

Although OPIDN has been known for over half a century, its mechanisms are yet to be defined. The common characteristics of neurotoxic organophosphorous esters are that they are phosphorous compounds and they are direct and indirect inhibitors of esterases.

a. Electrophysiological Studies: Electrophysiological studies following an intra-arterial injection of DFP in cats suggested that an initial functional deficit occurs at the level of the motor nerve ending (6,19). Motor impairment of motor nerve terminals was produced by DFP injection into the femoral artery. Recovery from a subactute DFP dosage was explained by a regeneration of the original motor axons; reinnervation by collateral sprouting was insignificant. Simultaneous impairment of both sensory and motor functions was seen in the cat leg treated with an intra-arterial injection of DFP. Recently the monosynaptic reflex in cats given a single dermal dose of TOCP was studied (60). There was a significant decrease in reflex action in the unconditioned monosynaptic reflex as compared with the control measurements. The dorsal and ventral root compound action potentials were also significantly decreased from control levels.

b. Histopathological Changes: Histopathological lesions in organophosphorous-induced delayed neurotoxicity are identical, irrespective of the organophosphorous compound used. Thus DFP (25,26,43), TOCP (6,17,22,23,31-33,72-74,79), mipafox (34,62), phenolphosphonothioates (4,6,8,9,15,16,39,44,46), and aliphatic esters (53,56,74) all produce similar lesions. Bouldin and Cavanagh (25,26) studied the ultrastructural changes in the peripheral nerves of cats treated with DFP. DFP produced a focal, distal, nonterminal lesion, that is, a "chemical transection" of the axon that precipitates anterograde somatofugal wallerian-type degeneration of the more distal axon (see Cavanagh in this volume).

c. Inhibition of Esterases: It has been suggested that the initial event in organophosphorous-induced delayed neurotoxicity was the phosphorylation of esterases. A role for brain acetylcholinesterase (AChE) (24) in the mechanism of OPIDN was hypothesized. However, Earl and Thompson (40,41) showed that whole hen brain pseudocholinesterase (BuChE) was selectively inhibited by TOCP and AChE was not. They proposed that BuChE was the primary target in OPIDN. This hypothesis was subsequently eliminated. In both in vivo and in vitro studies, Johnson (53-56,75) showed a difference in the inhibitory effect on hen brain esterase activity between known delayed neurotoxic and nonneurotoxic organophosphorous compounds. A small portion (about 6%) of the total phenylvalerate-hydrolyzing activity in hen brain is susceptible to inhibition only by delayed neurotoxic compounds such as mipafox. Compounds not capable of producing delayed neurotoxicity, such as paraoxon, do not inhibit the activity of this enzyme, which Johnson called neurotoxic esterase (NTE). Johnson proposed that NTE is the primary site of the action of the organophosphorous esters causing delayed neurotoxicity. To produce delayed neurotoxicity the enzyme should not only be inhibited, but also has to undergo a process called aging.

Recent studies have shown that dermal administration of DEF effectively inhibited NTE at high dose levels, while NTE was not inhibited with oral doses (61). These results were well correlated with the clinical and histological studies of oral and dermal applications of DEF (10,12). It should be noted, however, that, although there is a good relationship between inhibition and aging of NTE by organophosphorous compounds and their ability to induce OPIDN in chickens, NTE has not been isolated, and its physiological function is not known. Furthermore, NTE is present, becomes inhibited, and ages in treatment with neurotoxic organophosphorous esters in young chicks and in rats without subsequent development of OPIDN (55).

2. *A Working Hypothesis for the Mechanisms of OPIDN*

It is proposed that delayed neurotoxic organophosphorous compounds phosphorylate the active center of a neurotoxicity target protein (e.g., neurotoxic esterase) whose normal physiological function is unknown but could be related to calcium homeostasis, for example, (synaptic) plasma membrane. The ensuing increase in intracellular calcium leads to an increased Ca^{2+}-calmodulin-dependent phosphorylation of tubulin (71). As a result, tubulin is aggregated into insoluble filamentous polymers, as distinct from microtubules (53). Accumulation of such structures leads to the disruption of axoplasmic transport and accumulation of mitochondria at the distal parts of the axons (69). Disrupted mitochondria release their calcium stores into the axoplasm. This overloads and disrupts any axonal membrane control mechanisms for intracellular/extracellular ionic gradients that are still functional. Focal internodal swelling and Ca^{2+}-dependent proteolysis develop, followed by focal degeneration that spreads somatofugally to involve the entire distal axon (25,26). If exposure to organophosphorous compounds continues, this process will extend into the proximal portions of the axon. However, if exposure to these compounds ceases, the axons will begin to be supplied with new, unmodified proteins, allowing some regeneration and restoration of axonal function. Some results from morphological, electrophysiological, biochemical, and neurochemical studies are consistent with this explanation (6).

C. Organophosphorous-Induced Late Acute Effect

Some sulfur-containing organophosphorous compounds such as DEF and merphos produce a condition termed the late acute effect (10,12). It is characterized by a latent period of 1-2 days after the oral administration. Then clinical signs, including general weakness, loss of balance, loss of coordination, diarrhea, loss of appetite, and paralysis, become manifest. Death usually results in severe cases. In hens the color of the comb changes from bright red to dark blue and finally loses color before death. Since these signs were similar to those produced by the DEF metabolite *n*-butyl mercaptan (n-BM), it was hypothesized that the late acute effect resulted from n-BM, which is generated through the hydrolysis of orally administered DEF in the gastrointestinal tract. n-BM produced hematoxicity in treated hens and manifested as reductions in hemoglobin, red blood cell counts, packed cell volume, and glucose-6-phosphate dehydrogenase activity (1). There was an increase in methemoglobin and production of Heinz bodies in the blood of treated hens.

Orally administered DEF drastically reduced spleen weight and size and caused histological changes. Increased medullary bond formation was found in hens treated orally with DEF. Our results suggest that DEF ingestion also disrupts normal egg development, maturation, and the mechanism of laying with a corresponding effect on medullary bone formation.

IV. SUMMARY AND CONCLUSIONS

Organophosphorous compounds are used widely in industry and especially in agriculture. New chemicals of this type are being developed. It is widely accepted that some organophosphorous compounds produce neurological symptoms and probably nerve damage in man. The conditions associated with animal ingestion of or contact with certain organophosphorous compounds are acute poisoning, delayed neurotoxicity, and the late acute effect. The primary source of acute activity in these chemicals is anticholinesterase activity at various sites; is easily recognized, and can be treated. The early signs of OPIDN are subtle, and the duration and type of exposure, as well as the chemical composition of the specific compounds, can determine whether the condition develops and perhaps whether irreversible nerve damage occurs.

ACKNOWLEDGMENTS

The secretarial work of Mrs. Erna S. Daniel is appreciated. This study was supported in part by NIOSH Grants No. OH00823 and OH02003 and NIEHS Grant No. ES02717.

REFERENCES

1. Abdo, K. M., Timmons, P. R., Graham, D. G., Abou-Donia, M. B. 1983. Heinz body production and hematological changes in the hen after administration of a single oral dose of *n*-butyl mercaptan and *n*-butyl disulfide. *Fundam. Appl. Toxicol.* *3*:69-74.
2. Abou-Donia, M. B. 1976. Pharmacokinetics of a neurotoxic oral dose of leptophos in hens. *Arch. Toxicol.* *36*:103-110.
3. Abou-Donia, M. B. 1979. Pharmacokinetics and metabolism of a topically applied dose of *O*-4-bromo-2,5-dichlorophenyl *O*-methyl phenylphosphonothioate in hens. *Toxicol. Appl. Pharmacol.* *51*: 311-328.
4. Abou-Donia, M. B. 1979. Delayed neurotoxicity of phenylphosphonothioate esters. *Science* *205*:713-715.
5. Abou-Donia, M. B. 1980. Metabolism and pharmacokinetics of a single oral dose of *O*-4-bromo-2,5-dichlorophenyl *O*-methyl phenylphosphonothioate (leptophos) in hens. *Toxicol. Appl. Pharmacol.* *55*:131-145.
6. Abou-Donia, M. B. 1982. Organophosphorous ester-induced delayed neurotoxicity. *Annu. Rev. Pharmacol. Toxicol.* *21*:511-548.
7. Abou-Donia, M. B. 1983. Toxicokinetics and metabolism of delayed neurotoxic organophosphorus esters. *Neurotoxicology* *4*: 89-105.
8. Abou-Donia, M. B., Graham, D. G. 1978. Delayed neurotoxicity of *O*-ethyl *O*-4-nitrophenyl phenylphosphonothioate: Subchronic (90 days) oral administration in hens. *Toxicol. Appl. Pharmacol.* *45*:685-700.
9. Abou-Donia, M. B., Graham, D. G. 1979. Delayed neurotoxicity of a single dose of *O*-ethyl *O*-4-cyanophenyl phenylphosphonothioate in the hen. *Neurotoxicology* *2*:449-466.
10. Abou-Donia, M. B., Graham, D. G., Abdo, K. M., Komeil, A. A. 1979. Delayed neurotoxic, late acute and cholinergic effects of *S,S,S*-tributyl phosphorotrithioate (DEF) in hens. *Toxicology* *14*: 229-243.
11. Abou-Donia, M. B., Graham, D. G., Komeil, A. A. 1979. Delayed neurotoxicity of *O*-ethyl *O*-2,4-dichlorophenyl phenylphosphonothioate: Effects of a single oral dose on hens. *Toxicol. Appl. Pharmacol.* *49*:293-303.

12. Abou-Donia, M. B., Graham, D. G., Timmons, P. R., Reichert, B. L. 1979. Delayed neurotoxic and late acute effects of *S,S,S*-tributyl phosphorotrithioate on the hen: Effect of route of administration. *Neurotoxicology 1*:425-447.
13. Abou-Donia, M. B., Makkawy, H. A., Graham, D. G. 1982. Coumaphos: Delayed neurotoxic effect following dermal administration. *J. Toxicol. Environ. Health 10*:87-99.
14. Abou-Donia, M. B., Othman, M. A., Tantawy, G., Khalil, A. Z., Shawer, M. B. 1974. Neurotoxic effect of leptophos. *Experientia 30*:63-64.
15. Abou-Donia, M. B., Preissig, S. H. 1976. Delayed neurotoxicity of leptophos: Toxic effects on the nervous system of hens. *Toxicol. Appl. Pharmacol. 35*:269-282.
16. Abou-Donia, M. B., Preissig, S. H. 1976. Delayed neurotoxicity from continuous low-dose oral administration of leptophos to hens. *Toxicol. Appl. Pharmacol. 38*:595-608.
17. Ahmed, M. M. 1972. A note on the toxic effects of tri-cresylphosphate on spinal ganglion of slow loris (*Nycticebus coucang coucang*). *Anat. Anz. 131*:476-480.
18. Aya, M., Fukazawa, A., Kishino, S., Kume, R. 1972. Tokunol—On the structure and herbicidal activity. *Noyaku-Kenkyu 17*:62.
19. Baker, T., Lowndes, H. E., Johnson, M. K., Sandberg, I. C. 1980. The effects of phenyl methanosulfonyl fluoride on delayed organophosphorus neuropathy. *Arch. Toxicol. 46*:305-311.
20. Below, J. F., Jr. 1969. Stabilized herbicides containing N-[2-(dialkoxyphosphinothioyl-thio)ethyl]arene sulfonamides and a Lewis base. South African patent 6,803,592; CA *71*:100732.
21. Bidstrup, P. L., Bonnell, J. A., Beckett, A. G. 1953. Paralysis following poisoning by a new organic phosphorous insecticide (mipafox). *Br. Med. J. 1*:1068-1072.
22. Bischoff, A. 1967. The ultrastructure of tri-ortho-cresyl phosphate poisoning. I. Studies on myelin and axonal alterations in the sciatic nerve. *Acta Neuropathol. 9*:158-174.
23. Bischoff, A. 1970. Ultrastructure of tri-ortho-cresyl phosphate poisoning: II. Studies on spinal cord alterations. *Acta Neuropathol. 15*:142-155.
24. Bloch, H., Hottinger, A. 1943. Uber die Spezifität der Cholinesterase-hemmung durch Tri-*o*-kresyl Phosphat. *Z. Vitaminforsch. 13*:90.
25. Bouldin, T. W., Cavanagh, J. B. 1979. A teased-fiber study of the spatiotemporal spread of axonal degeneration. *Am. J. Pathol. 94*:241-252.
26. Bouldin, T. W., Cavanagh, J. B. 1979. A fine structural study of the early stages of axonal degeneration. *Am. J. Pathol. 94*:253-269.

27. Bowers, W. S. 1968. Juvenile hormone: Activity of natural and synthetic synergists. *Science 161*:895.
28. Brodeur, J., DuBois, K. P. 1964. Studies on the mechanism of acquired tolerance by rats to *O,O*-diethyl *S*-2(ethylthio) ethyl phosphorodithioate (Di-Syston). *Arch. Int. Pharmacodyn. 149*: 560-570.
29. Brown, N. C., Hollinshead, D. T., Kingsbury, P. A., Maline, J. C. 1962. A new class of compounds showing anthelmintic properties. *Nature 194*:397.
30. Casida, J. E., Baron, R. L., Eto, M., Engel, J. L. 1963. Potentiation and neurotoxicity induced by certain organophosphates. *Biochem. Pharmacol. 12*:73-83.
31. Cavanagh, J. B. 1954. The toxic effects of tri-ortho-cresyl phosphate on the nervous system. *J. Neurol. Neurosurg. Psychiatry 17*:163-172.
32. Cavanagh, J. B. 1964. Peripheral nerve changes in ortho-cresyl phosphate poisoning in the cat. *J. Pathol. Bacteriol. 87*:365-383.
33. Cavanagh, J. D., Patangia, G. M. 1965. Changes in the central nervous system in the cat as the result of tri-*o*-cresyl phsophate poisoning. *Brain 88*:165-180.
34. Davies, D. R. 1963. Neurotoxicity of organophosphorous compounds, in *Handbuch der Exper. Pharmacol. Erganzungswerk*, Vol. 15 (Koelle, G. D., ed.), Springer-Verlag, Berlin, pp. 860-862.
35. DeLorenzo, R. J. 1982. Calmodulin in neurotransmitter release and synaptic function. *Fed. Proc. Fed. Am. Soc. Exp. Biol. 41*: 2265.
36. DeWilde, R. C. 1971. Ethephon, practical application of (2-chloroethyl)phosphonic acid in agricultural production. *Hortic. Sci. 6*:364.
37. Draper, A. H., James, M. F., Johnson, B. C. 1952. Tri-*o*-cresyl phosphate as a vitamin E antagonist for the rat and lamb. *J. Nutr. 47*:583-597.
38. Dubois, K. P., Kinoshita, F., Jackson, P. 1967. Acute toxicity and mechanism of action of a cholinergic rodenticide. *Arch. Int. Pharmacodyn. 169*:108-116.
39. Durham, W. F., Gaines, T. B., Hayes, J. W., Jr. 1956. Paralytic and related effects of certain organic phosphorous compounds. *Arch. Ind. Health 13*:328-330.
40. Earl, C. J., Thompson, R. H. S. 1952. The inhibitory action of tri-ortho-cresyl phosphate on cholinesterase. *Br. J. Pharmacol. 7*:261-269.
41. Earl, C. J., Thompson, R. H. S. 1952. Cholinesterase levels in the nervous system in tri-ortho-cresyl phosphate poisoning. *Br. J. Pharmacol. 7*:685-694.
42. Eto, M. 1974. *Organophosphorous Pesticides: Organic and Biological Chemistry*, CRC Press, Cleveland.

43. Fenton, J. C. B. 1955. The nature of the paralysis in chickens following organophosphorous poisoning. *J. Pathol. Bacteriol. 69*: 181-183.
44. Frawley, J. P., Zwickey, R. W., Fuyat, H. V. 1956. Myelin degeneration in chicks with subacute administration of organic phosphorous insecticides. *Fed. Proc. 15*:424.
45. Gage, J. C. 1958. A cholinesterase inhibition derived from *O,O*-diethyl *O-p*-nitrophenyl thiophosphate *in vivo*. *Biochem. J. 54*: 426.
46. Gaines, T. B. 1969. Acute toxicity of pesticides. *Toxicol. Appl. Pharmacol. 14*:515-534.
47. Gysin, H. 1954. Uber einige neue Insektizide. *Chimia 8*:22.
48. Hassan, A., Abdel-Hamid, F. M., Mohamed, S. I. 1977. Metabolism of ^{14}C-leptophos in the rat. *Arch. Environ. Contam. Toxicol. 6*:447-454.
49. Henschler, D. 1958. Die trikresylphosphatvergiftung: Experimentelle Klarung von problemen der Atiologie und pathogenese. *Klin. Wochenschr. 36*:663-683.
50. Holmstead, R. L., Fukuto, T. R., March, R. B. 1973. The metabolism of *O*-(4-bromo-2,5-dichlorophenyl) *O*-methyl phenylphosphonothioate (leptophos) in white mice and on cotton plants. *Arch. Environ. Contam. Toxicol. 1*:133-147.
51. Igbal, Z. M., Menzer, R. E. 1972. Metabolism of *O*-ethyl *S,S*-dipropylphosphorodithioate in rats and liver microsomal systems. *Biochem. Pharmacol. 21*:1569-1584.
52. Jawarski, E. G. 1972. Mode of action of *N*-phosphonomethylglycine: Inhibition of aromatic amino acid biosynthesis. *J. Agric. Food Chem. 20*:1195.
53. Johnson, M. K., Barnes, J. B. 1970. Age and the sensitivity of chicks to the delayed neurotoxic effects of some organophosphorus compounds. *Biochem. Pharmacol. 19*:3045-3047.
54. Johnson, M. K. 1974. The primary biochemical lesion leading to the delayed neurotoxic effects of some organophosphorus esters. *J. Neurochem. 23*:785-789.
55. Johnson, M. K. 1981. Initiation of organophosphate neurotoxicity. *Toxicol. Appl. Pharmacol. 61*:480-481.
56. Johnson, M. K. 1982. The target for initiation of delayed neurotoxicity by organophosphorus esters: Biochemical studies and toxicological applications. *Rev. Biochem. Toxicol. 4*:141.
57. Kashafutdinow, G. A., Il'ina, N. A. 1971. Repellant activity of cyclohexenyl esters of amidophosphoric acid. *Med. Parazitol. Parazit. Bolezni. 40*:36.
58. Kilgore, W. W., Painter, R. R. 1964. Effect of the chemosterilant apholate on the synthesis of cellular components in developing house fly eggs. *Biochem. J. 92*:353.

59. Koelle, G. B. 1963. Cholinesterase and anticholinesterase agents, in *Handbuch der Experimentellen Pharmakologie*, Vol. 15, Springer-Verlag, Berlin, pp. 187-298.
60. Lapadula, D. M., Abou-Donia, M. B. 1982. Monosynaptic reflex depression in cats with organophosphorous neuropathy: Effects of tri-*o*-cresyl phosphate. *Neurotoxicology 3*:51-62.
61. Lapadula, D. M., Carrington, C. D., Abou-Donia, M. B. 1984. Induction of hepatic microsomal cytochrome P-450 and inhibition of brain, liver, and plasma esterases by an acute dose of *S,S,S*-tri-*n*-butyl phosphorotrithioate (DEF) in the adult hen. *Toxicol. Appl. Pharmacol. 73*:300-310.
62. Majno, G., Karnovsky, M. L. 1961. A biochemical and morphological study of myelination and demyelination. III. Effect of an organophosphorus compound (Mipafox) on the biosynthesis of lipid by nervous tissue of rats and hens. *J. Neurochem. 8*:1-16.
63. Malone, J. C. 1964. Toxicity of haloxon. *Res. Vet. Sci.* 5:17-31.
64. Martin, H., Worthin, C. R. 1977. *Pesticide Manual: Basic Information on the Chemicals Used as Active Components of Pesticides*, 5th ed., British Crop Protection Council, Worchestershire.
65. Matsumura, F. U. 1975. *Toxicology of Insecticides*, Plenum, New York.
66. Menzie, C. M. 1969. *Metabolism of Pesticides*, Bureau of Sport, Fisheries, and Wildlife Special Scientific Report, Wildlife No. 127.
67. Melnikov, N. N. 1971. Chemistry of Pesticides. *Residue Rev. 36*:303.
68. Nachmansohn, D. 1959. *Chemical and Molecular Basis of Nerve Activity*, Academic Press, New York.
69. Padilla, S. S., Lapadula, D. M., Reiter, L. W., Abou-Donia, M. B. 1983. Alteration of slow axonal transport in hens treated with tri-*o*-cresyl phosphate (TOCP). *Fed. Proc. Am. Soc. Exp. Biol. 42*:87.
70. Painter, R. R., Kilgore, W. W. 1967. The effect of apholate and thiotepa on nucleic acid synthesis and nucleotide ratios in house fly eggs. *J. Insect. Physiol. 13*:1105.
71. Patton, S. E., O'Callaghan, J. P., Miller, D. B., Abou-Donia, M. B. 1983. Effect of oral administration of tri-*o*-cresyl phosphate on *in vitro* phosphorylation of membrane and cytosolic proteins from chicken brain. *J. Neurochem. 41*:897-901.
72. Plapp, F. W., Casida, J. E. 1958. Hydrolysis of alkylphosphate bond in certain diaklyl aryl phosphorothioate insecticides by rats, cockroaches and alkali. *J. Econ. Entomol. 51*:800.
73. Prineas, J. 1969. The pathogenesis of dying-back polyneuropathies. An ultrastructural study of experimental TOCP intoxication in the cat. *J. Neuropathol. Exp. Neurol. 28*:571-597.
74. Prineas, J. 1969. Triorthocresyl phosphate myopathy. *Arch. Neurol. 21*:105-156.

75. Richardson, R. J., Davis, C. S., Johnson, M. K. 1979. Subcellular distribution of marker enzymes and neurotoxic esterase in adult hen brain. *J. Neurochem.* *32*:607-615.
76. Roger, H., Recordier, M. 1934. Les polyneurites phosphocréosotiques (phosphate de creosote, ginger paralysis, apiol). *Ann. Med. Paris* *35*:44-63.
77. Shindo, N., Wada, S., Ota, K., Suzuki, F., Ohta, Y. 1967. *O*-Ethyl *O*-(*p*-nitrophenyl)benzenethionophosphate. U.S. Patent. 3,327,026; C.A., 6T,64531.
78. Schrader, G. 1963. *Die Entwicklung Insektizider Phosphorsaure-Ester*, Verlag Chemie, Weinheim.
79. Smith, M. I., Lillie, R. D. 1931. The histopathology of tri-orthocresyl phosphate poisoning. *Arch. Neurol. Psychiatry* *26*:976-992.
80. Smith, M. I., Elvove, E., Frazier, W. H. 1930. The pharmacological action of certain phenol esters, with special reference to the etiology of so-called ginger paralysis. *Pub. Health Rep.* *45*:2509-2524.
81. Smith, M. I., Engel, E. W., Stohlman, F. F. 1932. Further studies on the pharmacology of certain phenol esters with special reference to the relation of chemical constitution and physiologic action. *Nat. Inst. Health Bull.* *160*:1-53.
82. Susser, M., Stein, Z. 1957. An outbreak of tri-ortho-cresyl phosphate (TOCP) poisoning in Durban. *Br. J. Ind. Med.* *14*:111-119.
83. Uesugi, Y., Tomizawa, C., Murai, T. 1972. Degradation of organophosphorus fungicides, in *Environmental Toxicology of Pesticides* (F. Matsumura, D. M. Boush, T. Misato, eds.), Academic, New York, p. 327.
84. Vora, D. D., Dastur, D. K., Braganca, M. B., Parihar, L. M., Iyer, C. G. S. 1962. Toxic polyneuritis in Bombay due to ortho-cresyl phosphate poisoning. *J. Neurol. Neurosurg. Psychiatry* *25*:234-242.
85. Waston, A. J., Leasure, J. K. 1959. Zytron—For crabgrass control. *Down to Earth* (Winter Issue), 1.
86. Waters, E. M., Gerstner, H. B. 1979. *Leptophos, an Overview*, Oak Ridge National Laboratory, Oak Ridge, Tenn.
87. Whitacre, D. M., Badie, M., Schewemmer, B. A., Diaz, L. I. 1976. Metabolism of ^{14}C-leptophos and ^{14}C-bromo-2,5-dichlorophenyl in rats: A multiple dosing study. *Bull. Environ. Contam. Toxicol.* *16*:689-696.
88. Windholz, M. 1983. *The Merck Index: An Encyclopedia of Chemicals and Drugs*, Merck, Rahway, N.J.
89. Xintaras, C., Burg, J. R., Tanaka, S., Lee, S. T., Johnson, B. L., Cottrill, J. 1978. *NIOSH Health Survey of Velsicol Pesticides Workers. Occupational Exposure to Leptophos and Other Chemicals*, NIOSH, Cincinnati.

21
Chlorinated Insecticides

S. C. Bondy
Laboratory of Behavioral and Neurological Toxicology, National Institute of Environmental Health Sciences, Research Triangle Park, North Carolina

I. INTRODUCTION

Polychlorinated pesticides have been widely used in the last 30 years, as they are most effective in destruction of insects. Such compounds are characterized by the presence of several chlorine atoms within aromatic or cage-structured organic molecules. Examples of caged, cuboidal species include chlordecone (Kepone) and mirex, while the cyclodienes (chlordane, aldrin, and dieldrin) can also approximate such an enclosed structure. Aromatic halogenated pesticides include p',p'-dichlorodiphenyltrichloroethane (DDT) and γ-benzenehydrochloride (lindane and hexachlorophene) (Fig. 1). Although these molecules are not obviously closely related, they possess several common characteristics. Neurotoxicity to man is generally manifested in part by tremor, seizures, or other signs of neuronal hyperactivity (30). Another characteristic common to these chemicals is their persistence in the environment. This persistence is due in part to their low water solubility and also to the resistance of the carbon-chlorine bond to degradation by bacteria and multicellular organisms. This stability allows such chemicals to be

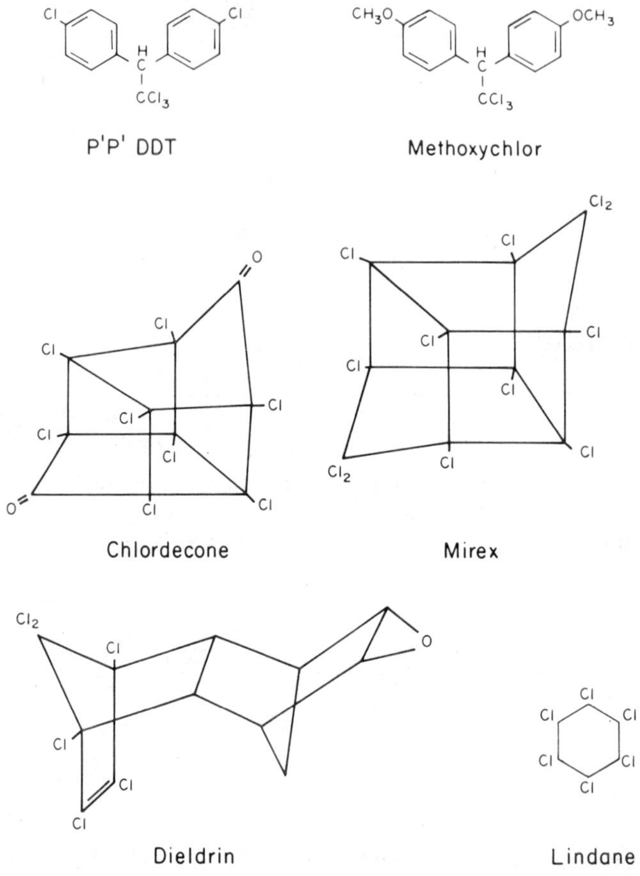

Figure 1 Structure of some organochlorine insecticides.

passed along the food chain, reaching man by way of meat, fish, eggs, or dairy products. Such sequential bioconcentration can be many thousand-fold (20,59).

Organochlorine insecticides tend to persist within the tissues of exposed animals, and their serum half-life tends to be several weeks long (39). This low rate of excretion means that, in addition to the hazards of acute toxicity, the potential for chronic and secondary effects is marked. There is ample time for extensive distribution of these chemicals to varied bodily tissues and organs. It is likely that the potent insecticidal properties of this group of compounds are also partly attributable to such persistence. Exposed insects can carry the toxic agents back to colonies in hives and nests, causing contamination of hitherto unexposed fellows.

II. PRIMARY MECHANISMS OF ACTION

The insecticidal properties of these polychlorinated chemicals may well involve mechanisms similar to those causing neurotoxicity in higher species. In either class of animal, repetitive discharge of neurons has been reported to be caused by those organochlorines which have been studied in detail, namely, DDT and chlordecone (30,64). More detailed analysis has attributed this to an abnormal slowing of sodium channel closure after they have opened during passage of a nerve impulse. By such means, a single action potential can be followed by a repetitive train of impulses (56). This disturbance is similar to the effect of low external calcium and can be mitigated by increasing calcium levels (32). Low external calcium levels also retard the rate of restoration of resting levels of sodium conductance after an action potential (14). From such studies, the concept has arisen that organochlorine insecticides owe their neurotoxicity to a primary interference with the normal equilibrium between intra- and extracellular calcium (12,30).

Another aspect of the neurotoxicity of organochlorines is their steroid-mimicking capacity. Several of these compounds have been found to have antagonist or agonist effects upon gonadal steroids (1, 17,38). Reproductive deficits have been described in man and in experimental animals after exposure to chlordecone (18,51). Since some corticosteroids are capable of enhancing nerve excitability (19), the ability of organochlorines to mimic steroids may in part underlie their capacity to cause neuronal hyperexcitability (31).

A large variety of biochemical, endocrine, and behavioral sequelae have been reported after exposure of experimental animals to organochlorines. Many of these are likely to be indirect effects, but may give clues as to the nature of the underlying primary lesions. The identification of such primary lesions may allow a sequential description of events accounting for many of the secondary but significant phenomena caused by organochlorines.

It should be borne in mind that most organochlorines are not merely neurotoxic, but also damaging to a variety of other organs, including liver, kidney, and cardiac muscle. Some of these compounds have also been shown to be carcinogenic (41). Thus a primary target is likely to be a rather widespread cell component and not merely a very localized tissue-specific feature.

Unfortunately there are few comparative studies involving more than one of these compounds, and thus the generality of the observed effects is not clear. However, a survey of the literature can yield clues as to the important and ubiquitous characteristics of this broad class of insecticides. This review discusses some of the more investigated organochlorines and emphasizes those aspects that may form the basis for the toxic nature of these chemicals.

III. SELECTED EXAMPLES

A. DDT

This is the best-known insecticide and has in the past been very widely used, especially in the period 1950-1970. Its use in the United States has been severely limited since 1972 because of its health hazard to humans and its nonspecific ecological damage. However, DDT has been of enormous benefit to mankind and of dramatic value in the control of insect-borne diseases such as malaria and typhus. DDT has been estimated to have saved more lives than penicillin (64).

The effect of DDT on insects and vertebrates seems similar: at low doses hyperactivity is seen, and at higher doses tremor and convulsions occur. An early event appears to involve a delayed rate of closing of axonal sodium channels following the passage of a nerve impulse (see Doherty in this book for review).

The tremorigenic properties of DDT cannot be solely attributed to the brain, but probably also reflect a direct effect on the spinal cord and peripheral nerves. Spinal preparations of various mammals still show tremor when exposed to DDT (3,46). It is likely that such effects are modulated by the abnormal nervous output of the poisoned brain (30).

Biochemical changes that have been repeatedly observed following exposure of various species to DDT include increased turnover of serotonin (27,58) and inhibition of adenosinetriphosphatases (ATPases) (11).

Another widely reported feature of DDT is its estrogenic and androgenic activity. This seems to be largely due to the o,p'-DDT component of DDT (15,61), which constitutes around 20% of the commercial product (13). Since the largely neurotoxic component is p,p'-DDT (64) (Fig. 1), the estrogenic and neurotoxic components of crude commercial DDT are partially separable. Endocrine effects are thought to be mediated by an interaction of the pesticide with the estrogen receptor system (42). Derangement of reproductive function may reflect a direct effect on target organs (ovary and testis), but may also be due to interference with the hypothalamo-pituitary axis. Such interference with the liberation and recognition of releasing factors and trophic hormones could interfere with the normal feedback loops involved in the maintenance of systemic hormone levels.

Failure of thermoregulatory mechanisms in DDT-treated rats (64) may also imply hypothalamic damage (35). The lethality of DDT is at least partly due to the acute hyperthermia that it can induce (63). The lower sensitivity of neonatal animals to DDT may in part be due to the immaturity of thermoregulatory systems, preventing a major response to DDT.

The cerebellum has been claimed as a primary target for DDT (58), but tremor can occur in spinal and decerebellate animals (3,46), and no clear morphological or electrophysiological evidence of specific damage to the cerebellum of DDT-treated animals exists (16,64).

B. Chlordecone

Chlordecone contains a single carboxyl group which is readily hydrated. This results in its being less lipophilic than DDT or mirex, which contain no oxygen (Fig. 1). For this reason the tissue/blood ratio of chlordecone in exposed animals is higher than that for other organochlorine insecticides (50).

Several of the features that have been described for DDT are also characteristic of the consequences of exposure to chlordecone. This compound also causes an elevation of 5-hydroxyindolacetic acid in several brain regions, indicating an increased rate of serotonin turnover (43). Exposed animals also show poor thermoregulatory capacities, and ATPases have been proposed as primary targets of chlordecone (10).

The estrogenic nature of chlordecone has been extensively studied, and this toxicant can grossly interfere with the normal endocrine cycle of female rats (53). This is paralleled by reproductive disturbances found in exposed human males, such as reduced sperm motility and density (51). Since chlordecone competes weakly for the peripheral estrogen receptor (37), it can act as an antagonist or agonist of physiological estrogens,and its persistence allows it to block the estrus cycle (53). Chlordecone also interacts with the estradiol receptors within the brain (4). This may be related to the estrogen-like effects that chlordecone produces in the hypothalamo-pituitary axis such as the reduction of pituitary met-enkephalin and elevation of circulating levels of prolactin and luteinizing hormones (23,54). The use of dimethylsulfoxide as a solvent in organochlorine insecticide studies is common (36). The use of this vehicle however, may significantly influence results of experiments, since dimethylsulfoxide has marked effects on the endocrine system (55), and also promotes extremely rapid penetration of hydrophobic molecules which could otherwise only be taken up slowly.

In addition to causing selective damage to the nervous system, chlordecone may act as a nonspecific stressor to which the animal must adapt. This has been suggested by comparative endocrine studies on the effect of chlordecone upon stressed and unstressed rats (44). Basal prolactin and corticosterone levels are reduced in chlordecone-treated rats, while the responsivity of these hormones to stress is increased (43). This dual effect implies damage to hypothalamic homeostatic and adaptive mechanisms.

Chlordecone has been reported to cause a variety of changes in neurotransmitter-related parameters including depression of high-affinity γ-aminobutyric acid (GABA) and catecholamines (10). Many of the most dramatic findings of such reports are in vitro findings which do not involve testing of exposed animals, and such data are liable to give false positives. However, data on dopaminergic and GABA parameters have also been derived from in vivo studies (7,15,21,45), even though no clear generalization emerges from transmitter-related data, suggesting that effects on this feature are indirect.

The ability of mitochondria and synaptosomes from chlordecone-treated rats to take up calcium is unusually low (12), as is the level of membrane-bound calcium in such animals (26). Any elevation in intracellular calcium and failure of calcium sequestration could result in repetitive presynaptic neurotransmitter release and deranged mitochondrial electron transport (52).

C. Cyclodienes

The cyclodienes (aldrin, dieldrin, chlordane) are characterized by a much more rapid rate of cutaneous absorption than the other organochlorines, and this can make them especially hazardous (36). This toxicity may be enhanced by the metabolism of cyclodienes to their corresponding epoxides by liver microsomes. Such epoxides are also toxic, lipid soluble, and accumulate in adipose tissues (36). Furthermore, these epoxides are frequently carcinogenic, by virtue of their electrophilic nature (60). Cyclodienes are not significantly estrogenic. They also differ from DDT in that they appear to affect the central nervous system more than peripheral nerves. Effects may be synaptic rather than axonal, since dieldrin (Fig. 1) applied directly to squid axons produces no significant effects on the resting potential, membrane currents, or magnitude of the action potential (57), and spontaneous discharges are not readily induced in isolated insect nerves (30). However, aldrin-transdiol, a metabolite of aldrin, appears to have a direct inhibitory effect in inhibition of the action potential (57). Nonetheless, the major site of action of these compounds appears to be synaptic rather than axonal (29). This view is strengthened by the finding that dieldrin has many properties similar to the convulsant pentylenetetrazol. For example, both are antagonized by the same proportions of anticonvulsants (28), suggesting a common mechanism of action.

Hexachlorocyclohexanes such as lindane (Fig. 1) have many characteristics in common with cyclodienes and exhibit cross-tolerance with them (6). Lindane is the γ-isomer of benzene hexachloride and 50-10,000 times more insecticidal than any of its five isomeric variants (30). This differing toxicity has been attributed to the varying goodness of fit of the isomers into critical areas within neuronal membranes (36).

IV. CONCLUSIONS

A. Relatedness of Observed Effects

Several recurrent features found for organochlorines are apparent:

1. Tremor and hyperexcitability
2. Estrogenic activity
3. Hyperthermia

4. Adverse effects on ATPases
5. Elevated rates of serotonergic activity
6. Altered calcium metabolism

Relations between these repeatedly described effects are unclear, but certain suggestive clues exist.

1. Are neuronal excitability and estrogenicity related?

Estrogens are known to influence monoaminergic activity and act in an antidopaminergic manner in the striatum, perhaps by way of formation of catecholestrogens (65).

While estrogens are not in themselves tremorigenic, any tremor induced in animals treated with these insecticides could be exacerbated by estrogenic mimicry. Corticosterone can be tremorigenic (19,47), and organochlorines may mimic steroids other than sex hormones. Chlordecone and DDT are known to alter levels of steroid hydroxylases in rat liver (62). However, the partial separation of neurotoxic and estrogenic properties in the p,p' and o,p' isomers of DDT, respectively, and the existence of neurotoxic but not estrogenic class of compounds (cyclodienes) suggest that these properties do not always share an underlying common site of action. The release of neurohormones can be induced by lower levels of p,p'-DDT than those required to cause tremors and convulsions in insects, and appears to involve a direct effect upon insect neurosecretory cells (49). Thus even the single p,p'-DDT isomer can have tremorigenic and endocrine activities that occur at separate loci.

2. What is the relation between tremor, calcium levels, and altered ATPase activity?

A connection between damage to the neuronal sodium pump and tremor is obviously a possibility (10). However, the onset of tremor appears to precede significant changes of Na-K ATPase levels after chlordecone exposure in rats (54). It is not clear whether changes in ATPase levels are a cause or a result of tremor. The failure of maintenance of calcium homeostasis implied in data from both DDT and chlordecone could also account for enhanced nerve excitability, and this may be by way of inhibition of calcium ATPase (52).

3. How specific are the biochemical sequelae of exposure to organochlorines?

It is known that injection of serotonin can cause hyperthermia and that hypothalamic serotonergic mechanisms are involved in thermoregulation (35). It is tempting to invoke a relation between the hyperthermia and elevated 5-hydroxyindolacetic acid levels found in DDT-treated animals. The toxicity of DDT can certainly in part be accounted for by elevated body temperature (63). However, several other neurotoxic agents and states that are not known to cause hyperthermia appear to increase serotonin turnover. These include exposure to manganese (24), acrylamide (2), triethyl lead (25), organophosphates (40) and malnutrition (33). Thus this effect may be a rather nonspecific response to stress.

It is not clear whether the many changes reported for neurotransmitter-related parameters represent an adaptive response of an exposed animal or a direct effect of the organochlorines on a specific neuronal circuit. However, in view of their toxicity to a variety of tissues, it is more likely that these compounds act initially as membrane-disrupting agents, perhaps by interference with membrane phospholipids (48).

B. The Future of Organochlorine Insecticides

In view of the beneficial aspects of many of these compounds, it is likely that their use will be continued in some form, on a relatively large scale. Thus it is important to obtain as much data as possible on such compounds so that a cost-benefit analysis can be carried out with a good data base. Also, by such means, modified compounds can be developed, incorporating features designed to optimize efficiency and minimize hazard. In this regard, a promising development in recent years has been the synthesis and testing of a nonchlorinated DDT analog, 2,2-bis(p-ethoxyphenyl)-3,3-dimethyloxetane (EDO; Fig. 2). This compound was designed to incorporate features optimizing lipid-protein interactions combined with the dimethyloxetane site, believed to prevent sodium channels from closing (22). It retains the insecticidal activity of DDT, but is biodegradable and thus less persistent, presumably because of the absence of carbon-chlorine bonds. EDO is less toxic to mammals than DDT and is not degraded by the DDT dehydrochlorinase found in resistant insects (22,64). Such compounds will not be accumulated in the food chain. Another compound that is not as persistent as DDT and almost nontoxic to mammals—presumably because of its rapid metabolism—is methoxychlor (Fig. 1). This compound is rapidly demethylated, conjugated, and excreted by rats, and its half-life in this species is around 20 times less than that of DDT (34).

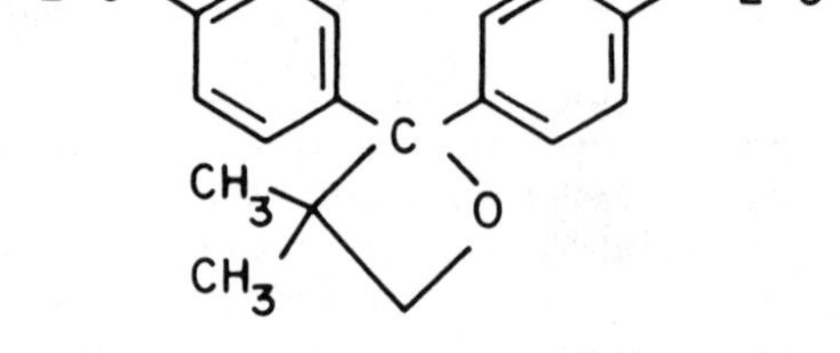

Figure 2 Structure of 2,2-bis-(p-ethoxyphenyl)-3,3-dimethyloxetane (EDO).

Progress has been made in the therapy of individuals accidentally exposed to organochlorine insecticides. This has generally consisted of acceleration of the excretion rate by use of compounds with ion exchange properties such as cholestyramine, which can enhance the rate of fecal excretion of chlordecone considerably (8). Cholestyramine binds to chlordecone and can reduce the half-life of this insecticide in humans from 150 to 83 days.

More recently, the possibility of treatment with specific antisera has been explored. Such antisera have been found to cause a tissue redistribution of chlorinated biphenyls in the mouse (9). This approach promises a selective blockage of the more reactive groups of toxic agents.

REFERENCES

1. Adkins-Regan, E., Hurvitz, E. D. 1982. O,p'-DDT causes growth of an androgen-dependent gland in *Coturnix* quail. *Experientia* *38*:1082-1083.
2. Ali, S. F., Hong, J. S., Wilson, W. E., Uphouse, L. L., Bondy, S. C. 1983. Effect of acrylamide on neurotransmitter metabolism and neuropeptide levels in several brain regions and upon circulating hormones. *Arch. Toxicol.* *52*:35-43.
3. Bromiley, R. G., Bard, P. 1949. Tremor and changes in reflex status produced by DDT in decerebrate, decerebrate-decerebellate and spinal animals. *Johns Hopkins Hosp. Bull.* *84*:414-429.
4. Brown, H., Uphouse, L. L. 1985. Interactions of Chlordecone with the CNS estradiol receptor. In preparation.
5. Bulger, W. H., Muccitelli, R. M., Kupfer, D. 1978. Interactions of chlorinated hydrocarbon pesticides with the 8S estrogen binding protein in rat testes. *Steroids* *32*:165-177.
6. Busvine, J. R. 1954. Houseflies resistant to a group of chlorinated hydrocarbon insecticides. *Nature* *174*:783-785.
7. Chang-Tsui, Y. Y. H., Ho, I. K. 1979. Effects of Kepone^R (chlordecone) on synaptosomal γ-aminobutyric uptake in the mouse. *Neurotoxicology* *1*:357-367.
8. Cohn, W. J., Baylan, J. J., Blanke, R. V., Faries, M. W., Harrell, J. R., Guzelian, P. S. 1978. Treatment of chlordecone (Kepone) toxicity with cholestyramine. *N. Engl. J. Med.* *298*:243-244.
9. Colburn, W. A., White, C. M. 1982. Effect of specific antisera on the redistribution of chlorinated biphenyls in the mouse. *Life Sci.* *24*:169-176.
10. Desiah, D. 1982. Biochemical mechanisms of chlordecone neurotoxicity: A review. *Neurotoxicology* *3*:103-110.
11. Doherty, J. D. 1979. Insecticides affecting ion transport. *Pharmacol. Ther.* *7*:123-151.

12. End, D. W., Carchman, R. A., Dewey, W. L. 1981. Neurochemical correlates of chlordecone neurotoxicity. *J. Toxicol. Environ. Health* 8:707-718.
13. Fry, D. M., Toone, C. K. 1981. DDT-induced feminization of gull embryos. *Science* 213:922-924.
14. Frankenhaeuser, B., Hodgkin, A. L. 1957. The action of calcium on the electrical properties of squid axons. *J. Physiol.* 137:218-244.
15. Fujimori, K., Beret, H., Mehendale, H. M., Ho, I. K. 1982. Comparison of brain discrete area distributions of chlordecone and mirex in the mouse. *Neurotoxicology* 3:125-130.
16. Globus, J. H. 1948. DDT[2,2]bis(p-chlorophenyl)1,1,1-trichloroethane] poisoning. *J. Neuropathol. Exp. Neurol.* 7:418-431.
17. Gelbert, R. J. 1978. Mirex, dieldrin and aldrin: Estrogenic activity and the induction of persistent vaginal estrus and anovulation in rats following neonatal treatment. *Environ. Res.* 16:131-138.
18. Guzelian, P. S. 1982. Comparative toxicology of chlordecone (Kepone) in humans and experimental animals. *Annu. Rev. Pharmacol. Toxicol.* 22:89-113.
19. Hall, E. D. 1982. Glucocorticoid effects on central nervous excitability and synaptic transmission. *Int. Rev. Neurobiol.* 23:165-195.
20. Hansen, D. J., Wilson, A. J., Nimmo, D. R., Schimmel, S. C., Bahner, L. H., Huggett, R. 1976. Kepone: Hazard to aquatic organisms. *Science* 193:528.
21. Ho, I. K., Fujimori, H., Chang-Tsui, H. 1981. Neurochemical evaluation of chlordecone toxicity in the mouse. *J. Toxicol. Environ. Health* 8:701-706.
22. Holan, G. 1971. Rational design of insecticides. *Bull. WHO* 44:355-362.
23. Hong, J. S., Ali, S. F. 1982. Chlordecone (Kepone^R) exposure in the neonate selectively alters brain and pituitary endorphin levels in prepuberal and adult rats. *Neurotoxicology* 3:111-118.
24. Hong, J. S., Hung, C. R., Seth, P. K., Mason, G., Bondy, S. C. 1984. The result of manganese treatment on the levels of neurotransmitters, hormones and neuropeptides. Interaction of stress with such effects. *Environ. Res.* 34:242-249.
25. Hong, J. S., Tilson, H. A., Hudson, P., Ali, S. F., Wilson, W. E., Hunter, V. 1984. Correlation of neurochemical and behavioral effects of triethyl lead chloride in rats. *Toxicol. App. Pharmacol.* 73:336-344.
26. Hoskins, B., Ho, I. K. 1982. Chlordecone-induced alterations in content and subcellular distribution of calcium in mouse brain. *J. Toxicol. Environ. Health* 9:535-544.

27. Hrdina, P. S., Singhal, R. L., Peters, D. A. V., Ling, G. M. 1973. Some neurochemical alterations during DDT poisoning. *Toxicol. Appl. Pharmacol. 25*:276-288.
28. Joy, R. M. 1973. Electrical correlates of preconvulsive and convulsive doses of chlorinated hydrocarbon insecticides in the CNS. *Neuropharmacology 12*:63-70.
29. Joy, R. M. 1976. The alteration by dieldrin of cortical excitability conditioned by sensory stimuli. *Toxicol. Appl. Pharmacol. 38*: 357-368.
30. Joy, R. M. 1982. Chlorinated hydrocarbon insecticides, in *Pesticides and Neurological Diseases* (Ecobichon, D. J., and Joy, R. M., eds), CRC Press, Boca Raton, Fla., pp. 91-150.
31. Mactutus, C. F., Unger, K. L., Tilson, H. A. 1982. Neonatal chlordecone exposure impairs early learning and memory in the rat on a multiple passive avoidance task. *Neurotoxicology 3*:27-44.
32. Matsumura, F., Narahashi, T. 1971. ATPase inhibition and electrophysiological change caused by DDT and related neuroactive agents in lobster nerve. *Biochem. Pharmacol. 20*:825-837.
33. Miller, M., Leaky, J. P., Stern, W. C., Morgane, P. J., Resnick, O. 1977. Tryptophan availability: Relation to elevated brain serotonin in developmentally protein-malnourished rats. *Exp. Neurol. 57*:142-157.
34. Murphy, S. D. 1980. Pesticides, in *Toxicology, The Basic Science of Poisons* (Doull, J., Klaassen, C. D., Amdur, M. O., eds.), Macmillan, New York, pp. 357-408.
35. Myers, R. D. 1980. Hypothalamic control of thermoregulation. Neurochemical mechanisms, in *Handbook of the Hypothalamus*, Vol. 3 (Morgane, P. J., Panksepp, J., eds.), Marcel Dekker, New York, pp. 83-210.
36. O'Brien, R. D. 1967. *Insecticides, Action and Metabolism*, Academic, New York.
37. Palmiter, R. D., Mulvihill, E. R. 1978. Estrogenic activity of the insecticide Kepone on the chicken oviduct. *Science 201*:356-358.
38. Peakall, D. B. 1970. Pesticides and the reproduction of birds. *Sci. Am. 222*:72-78.
39. Politis, M. J., Schaumburg, H. H., Spencer, P. S. 1980. Neurotoxicity of selected chemicals, in *Experimental and Clinical Neurotoxicology* (Spencer, P. S., Schaumburg, H. H., eds.), Williams and Wilkins, Baltimore, pp. 613-630.
40. Prioux-Guyonneau, M., Coudray-Lucas, C., Cog, H. M., Cohen, Y., Wepierre, J. 1982. Modification of rat brain 5-hydroxytryptamine metabolism by sublethal doses of organophosphate agents. *Acta Pharmacol. Toxicol. 51*:278-284.

41. Reuber, M. D. 1978. Carcinogenicity of Kepone. *J. Toxicol. Environ. Health 4*:895-911.
42. Robinson, A. K., Stancel, G. M. 1982. The estrogenic activity of DDT: Correlation of estrogenic effect with nuclear level of estrogen receptor. *Life Sci. 31*:2479-2484.
43. Rosecrans, J. A., Hong, J. S., Squibb, R. E., Johnson, J. H., Wilson, W. E., Tilson, H. A. 1982. Effects of perinatal exposure to chlordecone (Kepone[R]) on neuroendocrine and neurochemical responsiveness to environmental challenges. *Neurotoxicology 3*: 131-142.
44. Rosecrans, J. A., Johnson, J. H., Tilson, H. A., Hong, J. S. 1984. Hypothalamic-pituitary adrenal axis function in adult Fischer-344 rats exposed during development to neurotoxic chemicals perinatally. *Neurobehav. Toxicol. Teratol. 6*:281-288.
45. Seth, P. K., Agrawal, A. K., Bondy, S. C. 1981. Biochemical changes in the brain consequent to dietary exposure of developing and mature rats to chlordecone. *Toxicol. Appl. Pharmacol. 59*:262-267.
46. Shankland, D. L. 1964. Involvement of spinal cord and peripheral nerves in DDT-poisoning syndrome in albino rats. *Toxicol. Appl. Pharmacol. 6*:197-213.
47. Shapiro, S. 1965. Neonatal cortisol administration: Effects on growth, development and the pituitary-adrenal response to stress. *Proc. Soc. Exp. Biol. 120*:771-776.
48. Sharp, C. W., Hunt, D. G., Clements, S. T., Wilson, W. E. 1974. The influence of dichlordiphenyltrichloroethane, polychlorinated biphenyls and anionic amphiphilic compounds on stabilization of sodium and potassium activated adenosine triphosphatases by acidic phospholipids. *Mol. Pharmacol. 10*:119-129.
49. Singh, G. J. P., Orchard, I. 1982. Is insecticide-induced release of insect neurohormones a secondary effect of hyperactivity of the central nervous system? *Pest. Biochem. Physiol. 17*:232-242.
50. Taylor, J. R., Selhorst, J. B., Calabrese, V. P. 1980. Chlordecone, in *Experimental and Clinical Neurotoxicology* (Spencer, P. S., Schaumburg, H. H., eds.), Williams and Wilkins, Baltimore, pp. 407-421.
51. Taylor, J. R. 1982. Neurological manifestations in humans exposed to chlordecone and follow-up results. *Neurotoxicology 3*:9-16.
52. Tilson, H. A., Mactutus, C. F. 1982. Chlordecone neurotoxicity: A brief overview. *Neurotoxicology 3*:1-8.
53. Uphouse, L. L. 1985. A model for the neurotoxicity of Kepone. *Neurotoxicology* (in press).
54. Uphouse, L. L., Maury, W. 1982. Unpublished observations.
55. Uphouse, L. L., Mason, G., Bondy, S. C. 1982. Comments concerning the use of dimethylsulfoxide as a solvent for studies of chlordecone neurotoxicity. *Neurotoxicology 3*:149-154.

56. Van den Berken, J. 1972. The effect of DDT and dieldrin on myelinated nerve fibers. *Eur. J. Pharmacol. 20*:205-214.
57. Van den Berken, J., Narahashi, T. 1974. Effect of aldrin-transdiol—a metabolite of the insecticide dieldrin—on nerve membrane. *Eur. J. Pharmacol. 27*:255-258.
58. Van Woert, M. H., Plaitakis, A., Hwang, E. C. 1982. Neurotoxic effects of DDT [1,1,1,-trichloro-2,2-bis(p-chlorophenyl)-ethane]: Role of serotonin, in *Mechanisms of Actions of Neurotoxic Substances* (Prasad, K., Vernadakis, A., eds.), Raven Press, New York, pp. 143-154.
59. Waters, E. M., Huff, J. E., Gerstner, H. B. 1977. Mirex: An overview. *Environ. Res. 14*:212-222.
60. Weisburger, J. H., Williams, G. M. 1980. Chemical carcinogenesis, in *Cancer Medicine*, 2nd ed. (Holland, J. F., Frei, E., eds.), Lea and Febiger, Philadelphia.
61. Welch, R. M., Levin, W., Conney, A. H. 1969. Effect of chlorinated insecticides on steroid metabolism, in *Chemical Fallout: Current Research on Persistent Pesticides* (Miller, M. W., Gerd, G. G., eds.), Charles C. Thomas, Springfield, Ill., pp. 390-407.
62. Welch, R. M., Levin, W., Conney, A. H. 1967. Insecticide-inhibition and stimulation of steroid hydroxylases in rat liver. *J. Pharmacol. Exp. Ther. 155*:167-173.
63. Woolley, D. E. 1973. Studies on 1,1,1-trichloro-2,2-bis(p-chlorophenyl)-ethane(DDT)-induced hyperthermia. Effects of cold exposure and aminopyrine injections. *J. Pharmacol. Exp. Ther. 184*:261-268.
64. Woolley, D. E. 1982. Neurotoxicity of DDT and possible mechanisms of action in *Mechanisms of Actions of Neurotoxic Substances* (Prasad, K., Vernadakis, A., eds.), Raven Press, New York, pp. 95-141.
65. Yes, S. C. C. 1980. Neuroendrocrine regulation of the estrus cycle, in *Neuroendocrinology* (Krieger, D. T., Hughes, J. C., eds.), Sinauer, Sunderland, Mass., pp. 259-274.

22

Halogenated Hydrocarbons

Anna Maria Seppäläinen
Institute of Occupational Health, Helsinki, Finland

I. INDUSTRIAL USES

Many halogenated hydrocarbons are widely used in industry as solvents, some have been employed for anesthesia and analgesia in surgery and obstetrics, and some have gained use as refrigerants or fumigants. The plastics and rubber industry also uses halogenated hydrocarbons.

Halogenated hydrocarbons are nonflammable at room temperature, relatively stable chemically, highly volatile, and minimally soluble in water, making them very useful solvents. They are used in degreasing, in paints, printing inks, lacquers, varnishes, gas, and tar purification, as a vehicle for adhesives, and for drugs, chemicals, and perfume manufacture. They are also used in dry cleaning (trichloroethylene and tetrachloroethylene) and in solvent soaps. Methyl chloride was widely used as a refrigerant and caused hazards, for example, to unsuspecting occupants of apartments with leaky domestic refrigerators (22). More recently methyl chloride has been used in the production of synthetic rubber (17) and as a foaming agent in the production of foamed plastics (38). Methylene chloride (dichloromethane) was largely used in paint and varnish removers used in homes and hobbies, but was hazardous to subjects with coronary heart disease, because of its metabolism to carbon monoxide (46,49). Methylene chloride is also used as refrigerant in air conditioners, in the production of photographic films, and as a carrier gas for various aerosols. 1,1,1-Trichloroethane (methyl chloroform) has become increasingly used in degreasing, where it has replaced trichloroethylene. It is also used as a propellant in aerosol sprays and in the production of fluorohydrocarbons. It has been tried in anesthesia, but it caused adverse effects on the circulatory system (8,25). Tetrachloromethane (carbon tetrachloride) was tried as an anesthetic and later as an anthelmintic for hookworm, but was soon abandoned because of toxicity (18). Its good solvent properties, volatility, and nonflammability made it popular in dry cleaning and even in dry shampooing at the beginning of the century. It has also been used as a fire extinguisher. The latest use is as a raw material in the production of freons for refrigerators (18).

II. METABOLISM AND EXCRETION

Being highly volatile, these solvents are largely exhaled unaltered. A large portion (up to 70% of the dose) of inhaled methylene chloride is retained in animal tissues (1). A sustained high level of carboxyhemoglobin was noted in subjects exposed to 700 and 1000 ppm of methylene chloride (47). 1,1,1-Trichloroethane is nearly completely eliminated via the lungs unaltered, the metabolized fraction being transformed through trichloracetaldehyde (chloral) to trichloroacetic acid and trichloroethanol (1). Trichloroethylene mainly accumulates in fat and brain tissue. The blood levels rise during exposure and reach the maximum in about 2 hr and then decline gradually. In man and experimental animals, chloral hydrate, trichloroacetic acid, trichloroethane, and urochloralic acid are recognized as metabolites of trichloroethylene (1). Tetrachloroethylene undergoes biotransformation to a much lesser degree than trichloroethylene; it gives rise to several metabolites like di- and trichloroacetic acids, their respective alcohols and oxalic acid (1). Tetrachloromethane is exhaled in 50% of the dose after inhalation exposure, the rest being quickly excreted through urine and feces (1).

III. TRICHLOROETHYLENE

Trichloroethylene (TCE), chemically $CHCl=CCl_2$, is a colorless aliphatic hydrocarbon, liquid at usual temperatures, with a fruity odor. It is widely used in degreasing and dry cleaning, and as an anesthetic in some countries. Lloyd et al. (29) estimated that over 280,000 workers in the United States are exposed to TCE in industry and that approximately a further 5000 medical, dental, and hospital personnel are routinely exposed to it as an anesthetic gas. It has been noted that TCE induces an increased incidence of hepatocellular carcinoma in mice (29). Sudden death has been occasionally reported in young, apparently healthy people exposed to TCE at work, usually at the end of a shift and during mild exercise. These deaths have been attributed to syncope possibly associated with increased sensitivity to catecholamines (43). Minor electrocardiographic changes were noted by Lilis et al. (28), but no disturbance in the cardiac rhythm in the resting electrocardiograms of 57 TCE-exposed workers was found.

Investigations of systemic hemodynamics revealed signs of noradrenergic activation with increased excretion of urinary vanillylmandelic acid (28). As reviewed by Defalque (6), epinephrine has increased arrhythmias after TCE exposure, which has proved fatal in some cases. Huber (19) in his review found very little information on pathological changes in the heart muscle among TCE-exposed workers.

A. Neurotoxicity

The most usual neurotoxic effect of TCE is the psycho-organic syndrome (19). Symptoms like psychic lability, depression, neurotism, alcohol intolerance, and sleep disturbances were most frequent among workers with long-term exposure to TCE (51). These symptoms were found in 301 out of 343 workers clinically studied; moreover, vegetative manifestations like headache, vertigo, sweating, vasomotor, and gastrointestinal disturbances were found in 227. A slight psycho-organic syndrome was diagnosed in 108 workers and mild peripheral nervous system disorders such as paresthesias, paresis of the eye muscles, and taste or smell disturbances were found in 60 workers (51). Similar findings have been reported by many other research groups (5,13,19,26,28).

There appears to be no well-conducted epidemiological study on workers with TCE exposure currently available. Some small groups have been studied by psychological or electrophysiological tests. In a study, two groups of four female TCE workers were compared to control subjects. The results of a choice reaction time test showed an improvement during the working day in control women, while this learning effect was lacking in the TCE-exposed subjects. The group of women with the highest TCE exposure even showed an increase in the reaction time, especially in the afternoon testing (14).

Triebig et al. (53) studied psychological functions in eight workers with long-term occupational exposure to 50 ppm of TCE. The subjects were studied on Friday after a week's work, and 6 weeks later, right after a 15-day holiday. On both occasions the means of the test results were within the normal range, and no significant changes were noted between the two test results. The urinary concentration of trichloroethanol and trichloroacetic acid decreased during the weekend period, but the 15-day holiday interval from the exposed work did not greatly decrease the urinary concentrations from those taken on a Monday morning after a weekend rest.

The results of the psychometric test on a group of 31 workers with variable lengths (1 week to 20 years) of TCE exposure were compared to those of 50 control workers. Statistically significant differences in reaction times and tapping speed revealed some impairment among TCE workers already in the morning before the work shift (23). No further impairment was noted after the shift, nor did experimental exposure to 95 ppm of TCE for 4 hr induce any impairment in the psychological functions of 20 voluntary students (23).

Szulc-Kuberska et al. (50) have reported on perceptive hearing impairments and on vestibular disorders of mixed central-peripheral type among 40 workers with long-term TCE exposure. The results were based on audiograms and electronystagmograms, but no reference population was studied and noise levels in the working areas were not reported.

Experimental studies on short-term human exposure to TCE in concentrations varying from 50 to 300 ppm have mainly yielded negative results. Salvini et al. (35) reported on impairment in the performance of more complex psychological tests after two 4-hr exposures, separated by a 1.5-hr lunch break, to 110 ppm of TCE among six voluntary university students, as well as among six workers who had regularly worked with TCE and other solvents. Stewart et al. (48), however, challenged these results, because of the uncontrolled lunch break, and reported negative psychological findings in a controlled experimental study on subjects exposed to 50 or 110 ppm ot TCE for 8 hr. These data are in keeping with those reported by Ettema and Zielhuis (10), who tested on subjects exposed to 150 or 300 ppm of TCE for 2.5 hr. Alcohol (blood alcohol levels greater than 0.3 g/liter) produced marked effects in the same tests.

With telemetric recordings Konietzko et al. (24) noted electroencephalographic (EEG) changes among 20 volunteers during experimental exposure to 95 ppm of TCE for 4 hr. The change was noted already after 1 hr of exposure. Similar changes, namely, increased length of alpha periods (alpha = EEG activity of 8-13 cps), as well as increase in the amplitude of the alpha activity, was noted among six workers during occupational TCE exposure. Three workers with long-term occupational exposure to a mean of 110 ppm of TCE also exhibited slow waves of delta and theta range during the work shift, although the resting EEG had been normal (24).

The typical finding in acute TCE poisoning has been facial numbness because of lesion in the fifth cranial nerve (2,31). Moreover, several papers have reported trigeminal neuropathy after TCE anesthesia with a closed circuit (11,12). Trichloroethylene has been used as a treatment for tic doloureux, but the specificity of the analgesic effect upon the trigeminal nerve is doubtful. Similar effects have also been reported following administration of chloroform (7).

In one case of fatal TCE poisoning neuropathological findings showed most severe damage in the brainstem, the fifth nerves, tracts, and nuclei, particularly the sensory division. The nucleus and tractus solitarius were also affected, while the changes in the cerebral and cerebellar cortex were less severe and compatible with secondary hypoxemia (2).

A short intensive exposure to TCE while investigating a leak in an overheated degreasing machine caused facial anesthesia in a 26-year-old man and slowing of the distal sensory branches of the ulnar nerve (11). The ulnar nerve conduction velocity returned to normal in 9 weeks, but it was reported to have become slow again at 34 weeks after the intoxication without clinical signs or symptoms referrable to the ulnar nerve. Facial anesthesia receded from the periphery and from the tip of the nose in 4 weeks, but the whole situation improved quite slowly, and 18 months after the accident the patient still had anisocoria and left ptosis, as well as slight dulling to pin prick in a patchy distribution around the face.

Lawrence and Partyka (27) described severe bulbar palsy in a man who had worked for several hours in a closed underground pit with intensive TCE fumes. The following night the man had become dizzy and nauseated. Dysphagia, dysarthria, and dyspnea had soon developed, necessitating hospitalization. Incapacitating bulbar palsy still remained 11 years later, but conduction velocities in the peripheral nerves were normal.

Accidental exposure to several thousand ppm of TCE (as estimated from urinary trichloro compounds) caused a high spinal lesion with poor recovery in a 20-year-old female worker (34). Motor conduction velocities of leg nerves and the EEG were normal, and a biopsy of the cutaneous dorsalis lateralis nerve revealed no pathology.

Triebig et al. (52) found no statistically significant differences in the motor conduction velocities of 7 workers with long-term exposure to TCE in comparison to 12 unexposed persons.

IV. TETRACHLOROETHYLENE

Tetrachloroethylene (CCl_2-CCl_2) is a colorless, nonflammable liquid with a smell similar to that of chloroform. It is mainly used as solvent, in the degreasing of metal surfaces, in dry cleaning, and as a resin solvent. Inhalation is the main route of entry, but absorption through skin and mucosas may add to the occupational toxic effects.

The first symptoms of excessive exposure are mild irritation of the eyes and dysfunction of the central nervous system (33). In humans, one exposure to 106 ppm caused no significant central nervous system effects, whereas a concentration of about 200 ppm produced minimal effects, like light-headedness, thickness of the tongue, and tightness around the mouth. Higher concentrations (600 ppm and 1000 ppm) were tolerated for minutes only because of marked irritation of the eyes and upper respiratory tract (33). Rats, rabbits, and monkeys tolerated repeated exposures to 400 ppm without obvious effects (33).

The clinical picture of chronic tetrachloroethylene poisoning is characterized by neurasthetic syndrome. Chmielewski et al. (4) studied 9 workers occupationally exposed to tetrachloroethylene at concentrations not exceeding the Polish maximum allowable concentration (MAC) value of 200 mg/m^3 (about 30 ppm), and another group of 16 workers whose exposure levels exceeded the MAC value by 2- to 15-fold. Only the latter group underwent neurological and EEG examinations. Abnormal EEGs were found in 4 out of 16; pseudoneurotic syndrome was diagnosed in 6 cases and slight encephalopathy in 4 cases. A pseudoneurotic syndrome was found in four subjects with a short period of exposure (2-15 months). Drowsiness and fatigue were symptoms among workers with short and long exposure, while complaints of headache and dizziness were reported by workers exposed up to 10 years. Complaints of irritation symptoms were made by subjects with 1-10 years of exposure.

Ikeda et al. (20) studied six workers with 10-18 years of exposure to 30-220 ppm of tetrachloroethylene and four with 3 months to 3 years of exposure to 10-40 ppm. All workers with long exposure complained of dizziness, sore throat, shoulder stiffness, or chilly sensations, while only one among the less exposed and none among the controls had such symptoms.

V. 1,1,1-TRICHLOROETHANE

1,1,1-Trichloroethane (CH_3CCl_3, methyl chloroform) is a colorless, nonflammable liquid with a slightly sweet smell, similar to ether; it is partly soluble in water. The temperature stability is poor, which is why commercial products contain stabilizing agents. It is used in increasing amounts as a degreasing agent replacing trichloroethylene. It is also used as a propellant for aerosols.

1,1,1-Trichloroethane has been considered one of the least toxic among chlorinated aliphatic hydrocarbons (44). However, as reviewed by Halevy et al. (15), the literature contains reports on more than 30 fatalities arising from its use. The cause of death has been severe depression of the central nervous system, but damage to the lungs, liver, and kidney has also been detected at autopsy.

Acute intoxication after overexposure (1000-2000 ppm) is characterized by dizziness and unsteadiness (45), but the symptoms usually subside quickly, without any residual damage to the central nervous system or liver.

Long-term effects of occupational exposure to 1,1,1-trichloroethane have not been studied extensively. Maroni et al. (30) studied 22 female workers exposed to 1,1,1-trichloroethane in concentrations varying from 100 to 990 ppm, and 7 control workers. Subjective symptoms like headache, anxiety, restlessness, depression, and insomnia, as well as eye irritation and digestive disorders, were very frequent (present in 30-80%) among both exposed and control workers, with no statistically significant differences between the groups. Results of psychological tests and nerve conduction studies did not differ between the exposed and controls either. The high incidence of subjective symptoms was considered to be based on generally unfavorable environmental and work conditions rather than on specific effects of the chemical.

Some experimental human studies have been conducted on the possible central nervous effects of 1,1,1-trichloroethane. Salvini et al. (36) found no statistically significant effect on a perceptive test, immediate memory, complex reaction times, and manual ability following an 8-hr exposure to 350 ppm of 1,1,1-trichloroethane; neither did an 8-hr exposure to 450 ppm cause statistically significant changes. Savolainen et al. (37) found a slight central nervous system stimulation at 200 ppm of 1,1,1-trichloroethane, while 400 ppm tended to cause depression. Tests used reaction times, critical flicker fusion, body sway, and gaze deviation.

Seppäläinen et al. (42) studied visual evoked potentials in connection with 1,1,1-trichloroethane and xylene exposure. A total of 400 ppm of 1,1,1-trichloroethane decreased significantly the latency of the start of the cortical response, while a later peak (P150) was delayed. A total of 200 ppm of 1,1,1-trichloroethylene or xylene alone was ineffective. The findings indicated some stimulation but, on the other hand, a longer procession time within the central nervous system.

Recent studies have not demonstrated neurotoxic effects of 1,1,1-trichloroethane at the concentrations generally present in work places. However, more comprehensive investigations on long-term effects are necessary.

VI. METHYL CHLORIDE

Methyl chloride (CH_3Cl) is a colorless, almost odorless gas used principally as a chemical intermediate and as a foaming agent in the production of foamed plastics. In the past it was used extensively in refrigerators (22).

Its use as a refrigerant caused several deaths in homes. The autopsy revealed hyperemic and edematous lungs, heart enlargement, and congestion of the liver with fatty degeneration; kidneys were also swollen and congested. Microscope examination of the brain revealed an accumulation of lipoid-filled histocytes in the leptomeninges of the hemispheres, and the cerebral cortex was hyperemic (22). Acute intoxication was characterized by progressive drowsiness, mental confusion, stupor, weakness, nausea, abdominal pain, and vomiting. In more severe cases convulsions and delirium developed later. Most patients also showed tremors and hiccup during the acute stage (22).

Hansen et al. (17) reported on 15 cases with moderately severe intoxications among workers exposed to methyl chloride gas from an aqueous solution or slurry in a synthetic rubber manufacturing plant. The main symptoms were nausea, vomiting, dizziness, weakness, and excessive sleepiness. Poor muscular coordination, hiccups, and low-grade fever were also reported. No permanent effects were manifested.

Scharnweber et al. (38) reported on six foam workers exposed to relatively low levels (200 to 400 ppm) of methyl chloride for at least 2-3 weeks before onset of symptoms. Abnormal behavior, confusion, headache, poor memory, disturbance in balance, tiredness, stuttering, and dizziness were observed. One worker complained of numbness and weakness of his right arm, another diplopia. Symptoms usually disappeared in 6 weeks to 3 months. One worker had increased nervousness and tremor of hands as residual symptoms 3 months after hospitalization.

The reports cited described accidents of acute or subacute poisonings, usually without relevant sequelae. Chronic poisoning after long-term occupational exposure to methyl choride has not come to my knowledge.

VII. METHYLENE CHLORIDE

Methylene chloride (dichloromethane, CH_2Cl_2) is a colorless, nonflammable liquid at room temperature. It is used as a solvent for oils, fats, waxes, glues, and many other chemicals. It is especially suitable for extraction of substances that do not tolerate high temperatures, for example, extraction of fats in the food industry. It is widely used as a paint remover (49). It is also used in degreasing, as a refrigerant in air conditioners, in the manufacture of photographic films, and as a propellant in spray cans.

Hanke et al. (16) studied in detail 14 workers who had been gluing floor mats with a glue containing methylene chloride. The exposure level had been intensive, since air concentrations of 1400-18400 mg/m^3 (400-5300 ppm) were measured in work areas. Expired air contained 610-3600 mg/m^3 of methylene chloride after gluing. All these workers complained of headache, dizziness, sleep and gastric disturbances, and difficulties in concentration and memory after exposure. Four workers had an enlarged liver and exhibited abnormal liver function tests.

Neuropsychiatric and psychological examinations revealed abnormal findings, especially vegetative disorders, and the capacity to concentrate was impaired. Eleven workers had EEG examinations after a weekend without exposure, and all had EEG abnormalities that were more pronounced on the frontal and precentral brain areas.

At concentrations generally found in work places the main hazard from methylene chloride seems to be increased levels of the metabolic product carbon monoxide that can be dangerous to the heart, especially in subjects with coronary heart disease. Accidental death due to myocardial infarction has been attributed to carbon monoxide formed from methylene chloride (49). Impaired performance under difficult or demanding task conditions (32) and changes in the visual evoked potentials (47) noted after 1-hr exposure to methylene chloride have also been attributed to carbon monoxide rather than to the parent compound.

An excess of self-reported neurological symptoms was found among a group of 46 workers exposed to methylene chloride at concentrations below 100 ppm (3), but a follow-up study on 29 subjects and their age-matched controls showed no long-term changes attributable to methylene chloride. The tests used were a sensitive psychological test battery and nerve conduction velocity studies.

VIII. OTHER HALOGENATED HYDROCARBONS

Trichloromethane (chloroform, $CHCl_3$) was once a widely used anesthetic agent, and tetrachloromethane (CCl_4) was also tried for anesthesia. Tetrachloromethane killed some young women, who used it as a dry shampoo in large basins in small bathrooms (18). Later it saw use as a killer of the hookworm. Both tri- and tetrachloromethane are presently used mainly in closed processes in chemical and drug industries and in small amounts in laboratories. Low exposure levels have not caused reported cases of central nervous system intoxications.

IX. GENERAL COMMENTS

Several authors have noted and emphasized central nervous effects of halogenated hydrocarbons; however, neuropathic symptoms have been rare and mainly found in connection with trichloroethylene. The typical pattern of central nervous system disorders is similar to that found among workers exposed to various organic solvents and their mixtures (9,21,39,41). In this respect, studies of solvent-exposed workers have usually dealt with mixed exposure. Solvent mixtures are often used in industry, so that effects of individual compounds after long-term exposure have not been thoroughly investigated. Seppäläinen et al.

(41) suggested that more profound effects found in female workers exposed to solvents might partly be due to their more usual contact with halogenated hydrocarbons in comparison to the male workers studied. In a recent investigation on patients suffering from solvent poisoning, Seppäläinen and Antti-Poika (40) found a tendency for more usual and widespread polyneuropathy among workers who had been exposed to mixtures of solvents and to halogenated hydrocarbons as well, than among subjects exposed solely to halogenated hydrocarbons or solely to solvent mixtures. Electroencephalographic abnormalities also tended to be initially less frequent among workers exposed solely to halogenated hydrocarbons than among those exposed to various solvent mixtures. On the other hand, the EEG changes connected with halogenated hydrocarbons showed less improvement during the follow-up.

Our knowledge about the neurotoxic effects of long-term exposure to halogenated hydrocarbons is still inadequate. Sizable groups of workers with quantified exposure to specific solvents would be a desirable target to future studies.

REFERENCES

1. de Bruin, A. 1976. *Biochemical Toxicology of Environmental Agents*, Elsevier, Amsterdam.
2. Buxton, P. H., Hayward, M. 1967. Polyneuritis cranialis associated with industrial trichloroethylene poisoning. *J. Neurol. Neurosurg. Psychiatry 30*:511-518.
3. Cherry, N., Venables, H., Waldron, H. A., Wells, G. G. 1981. Some observations on workers exposed to methylene chloride. *Br. J. Ind. Med. 38*:351-355.
4. Chmielewski, J., Tomaszewski, R., Glombiowski, P., Kowalewski, W., Kwiatkowski, S. R. Szczekocki, W., Winnicka, A. 1976. Clinical observations of the occupational exposure to tetrachloroethylene. *Bull. Inst. Maritime Trop. Med. Gdynia 27*:197-205.
5. Cotter, L. H. 1950. Trichloroethylene poisoning. *Arch. Ind. Hyg. Occup. Med. 1*:319-322.
6. Defalque, R. J. 1961. Pharmacology and toxicology of trichloroethylene. A critical review of the world literature. *Clin. Pharmacol. Ther. 2*:665-688.
7. Deflaque, R. J. 1961. The "specific" analgesic effect of trichloroethylene upon the trigeminal nerve. *Anesthesiology 22*:379-384.
8. Dornette, W. H. L., Jones, J. P. 1960. Clinical experiences with 1,1,1-trichloroethane. A preliminary report of 50 anesthetic administrations. *Anesth. Analg. Cleveland 39*:249-250.
9. Elofsson, S. A., Gamberale, F., Hindmarsch, T., Iregren, A., Isaksson, A., Johnsson, I., Knave, B., Lydahl, E., Mindus, P.,

Persson, H. E., Philipson, B., Steby, M., Struwe, G., Söderman, E., Wennberg, A., Widén, L. 1980. Exposure to organic solvents. A cross-sectional epidemiologic investigation on occupationally exposed car and industrial spray painters with special reference to the nervous system. *Scand. J. Work Environ. Health 6*:239-273.

10. Ettema, J. H., Zielhuis, R. L. 1975. Effects of alcohol, carbon monoxide and trichloroethylene exposure on mental capacity. *Int. Arch. Occup. Environ. Health 35*:117-132.
11. Feldman, R. G., Mayer, R. M., Taub, A. 1970. Evidence for peripheral neurotoxic effect of trichloroethylene. *Neurology 20*: 599-606.
12. Firth, J. B., Stuckey, R. E. 1945. Decomposition of trilene in closed circuit anaesthesia. *Lancet 1*:814-816.
13. Grandjean, E., Münchinger, R., Turrian, V., Haas, P. A., Knoepfel, H.-K., Rosenmund, H. 1955. Investigations into the effects of exposure to trichloroethylene in mechanical engineering. *Br. J. Ind. Med. 12*:131-142.
14. Gun, R. T., Grygorcewicz, C., Nettelbeck, T. J. 1978. Choice reaction time in workers using trichloroethylene. *Med. J. Aust. 10*:535-536.
15. Halevy, J., Pitlik, S., Rosenfeld, J., Eitan, B.-D. 1980. 1,1,1-Trichloroethane intoxication: A case report with transient liver and renal damage. Review of the literature. *Clin. Toxicol. 16*: 467-472.
16. Hanke, C., Ruppe, K., Otto, J. 1974. Unterschungsergebnisse zur toxischen Wirkung von Dichlormethan bei Fussbodenlegern. *Zentralbl. Gesamt Hyg. Ihre Grenzgeb. 20*:81-84.
17. Hansen, H., Weaver, N. K., Venable, F. S. 1953. Methylchloride intoxication. *Arch. Ind. Hyg. Occup. Med. 8*:328-334.
18. Hardin, B. L. 1954. Carbon tetrachloride poisoning—A review. *Ind. Med. Surg. 23*:93-105.
19. Huber, F. 1969. Zur Klinik und Neuropathologie der Trichloräthylenvergiftung. *Z. Unfallmed. Berufskr. 62*:226-267.
20. Ikeda, M., Koizumi, A., Watanabe, T., Endo, A., Sato, K. 1980. Cytogenetic and cytokinetic investigations on lymphocytes from workers occupationally exposed to tetrachloroethylene, in *Mechanisms of Toxicity and Hazard Evaluation* (Holmstedt, B., Lauwerys, R., Mercier, M., Roberfroid, M., eds.), Elsevier/North-Holland, Amsterdam, pp. 347-350.
21. Juntunen, J., Antti-Poika, M., Tola, S., Partanen, T. 1982. Clinical prognosis of patients with diagnosed chronic solvent intoxication. *Acta Neurol. Scand. 65*:488-503.
22. Kegel, A. M., McNally, W. D., Pope, A. S. 1929. Methyl chloride poisoning from domestic refrigerators. *J. Am. Med. Assoc. 93*:353-358.

23. Konietzko, H., Elster, I., Sayer, H., Weichardt, H. 1975. Zentralnervöse Schäden durch Trichloräthylen. *Staub Reinhalt. Luft 35*:240-241.
24. Konietzko, H., Elster, I., Schomann, P., Weichardt, H. 1976. Feldunterschungen in Lösungsmittelbetrieben. 5. Mitteilung. *Zentralbl. Arbeitsmed. Arbeitsschutz 26*:60-63.
25. Krantz, J. C., Jr., Park, C. S., Ling, J. S. L. 1959. Anesthesia LX: The anesthetic properties of 1,1,1-trichloroethane. *Anesthesiology 20*:635-640.
26. Lachnit, V., Rankl, W. 1950. Die chronische Trichloräthylenvergiftung. *Z. Unfallmed. Berufskr. 43*:334-341.
27. Lawrence, W. H., Partyka, E. K. 1981. Chronic dysphagia and trigeminal anesthesia after trichloroethylene exposure. *Ann. Intern. Med. 95*:710.
28. Lilis, R., Stanescu, D., Muica, N., Roventa, A. 1969. Chronic effects of trichloroethylene exposure. *Med. Lav. 60*:594-601.
29. Lloyd, J. W., Roscoe, M. M., Jr., Breslin, P. 1975. Background information on trichloroethylene. *J. Occup. Med. 17*:603-605.
30. Maroni, M., Bulgheroni, C., Cassitto, M. G., Merluzzi, F., Gilioli, R., Foá, V. 1977. A clinical, neurophysiological and behavioral study of female workers exposed to 1,1,1-trichloroethane. *Scand. J. Work Environ. Health 3*:16-22.
31. Mitchell, A. B. S., Parsons-Smith, B. G. 1969. Trichloroethylene neuropathy. *Br. Med. J. Clin. Res. 1*:422-423.
32. Putz, V. R., Johnson, B. L., Setzer, J. V. 1979. A comparative study of the effects of carbon monoxide and methylene chloride on human performance. *J. Environ. Pathol. Toxicol. 2*:97-112.
33. Rowe, V. K., McCollister, D. D., Spencer, H. C., Adams, E. M., Irish, D. D. 1951. Vapor toxicity of tetrachloroethylene for laboratory animals and human subjects. *AMA Arch. Ind. Hyg. 4*:482-493.
34. Sagawa, K., Nishitani, H., Kawai, H., Kuge, Y., Ikeda, M. 1973. Transverse lesion of spinal cord after accidental exposure to trichloroethylene. *Int. Arch. Arbeitsmed. 31*:257-264.
35. Salvini, M., Binaschi, S., Riva, M. 1971. Evaluation of the psychophysiological functions in humans exposed to trichloroethylene. *Br. J. Ind. Med. 28*:293-295.
36. Salvini, M., Binaschi, S., Riva, M. 1971. Evaluation of the psychophysiological functions in humans exposed to the 'threshold limit value' of 1,1,1-trichloroethane. *Br. J. Ind. Med. 28*:286-292.
37. Savolainen, K., Riihimäki, V., Laine, A., Kekoni, J. 1981. Short-term exposure of human subjects to m-xylene and 1,1,1-trichloroethane. *Int. Arch. Occup. Environ. Health 49*:89-98.

38. Scharnweber, H. C., Spears, G. N., Cowles, S. R. 1974. Chronic methyl chloride intoxication in six industrial workers. *J. Occup. Med. 16*:112-113.
39. Seppäläinen, A. M. 1982. The use of clinical neurophysiological methods in studies of workers exposed to solvents, in *Kyoto Symposia (EEG Suppl. No. 36)* (Buser, P. A., Cobb, W. A., Okuma, T., eds.), Elsevier, Amsterdam, pp. 693-700.
40. Seppäläinen, A. M., Antti-Poika, M. 1983. Time course of electrophysiological findings for patients with solvent poisoning. A descriptive study. *Scand. J. Work Environ. Health 9*:15-24.
41. Seppäläinen, A. M., Lindström, K., Martelin, T. 1980. Neurophysiological and psychological picture of solvent poisoning. *Am. J. Ind. Med. 1*:31-42.
42. Seppäläinen, A. M., Salmi, T., Savolainen, K., Riihimäki, V. 1983. Visual evoked potentials in short-term exposure of human subjects to *m*-xylene and 1,1,1-trichloroethane, in *Application of Behavioral Pharmacology in Toxicology* (Zbinden, G., Cuomo, V., Racagni, G., Weiss, B., eds.), Raven Press, New York, pp. 349-352.
43. Smith, G. F. 1966. Trichloroethylene: a review. *Br. J. Ind. Med. 23*:249-262.
44. Stewart, R. D. 1968. The toxicology of 1,1,1-trichloroethane. *Ann. Occup. Hyg. 11*:71-79.
45. Stewart, R. D. 1971. Methyl chloroform intoxication. Diagnosis and treatment. *J. Am. Med. Assoc. 215*:1789-1792.
46. Stewart, R. D., Fisher, T. N., Hosko, M. J., Peterson, J. E., Baretta, E. D., Dodd, H. C. 1972. Carboxyhemoglobin elevation after exposure to dichloromethane. *Science 176*:295-296.
47. Stewart, R. D., Fisher, T. N., Hosko, M. J., Peterson, J. E., Baretta, E. D., Dodd, H. C. 1972. Experimental human exposure to methylene chloride. *Arch. Environ. Health 25*:342-348.
48. Stewart, R. D., Hake, C. L., Lebrun, A. J., Kalbfleisch, J. H., Newton, P. E., Peterson, J. E., Cohen, H. H., Stuble, R., Busch, K. A. 1974. Effects of trichloroethylene on behavioral performance capabilities, in *Behavioral Toxicology* (Xintaras, C., Johnson, B. J., de Groot, I., eds.), U.S. Department of Health, Education and Welfare, Washington, D. C., pp. 96-129.
49. Stewart, R. D., Hake, C. L. 1976. Paint-remover hazard. *J. Am. Med. Assoc. 235*:398-401.
50. Szulc-Kuberska, J., Tronczynska, J., Latkowski, B. 1976. Oto-neurological investigations of chronic trichloroethylene poisoning. *Minerva Otorinolaringol. 26*:108-112.
51. Trense, E. 1965. Praktische Ergebnisse der Untersuchung von 546 Tri-Arbeitern. *Zentralbl. Arbeitsmed. Arbeitsschutz 15*:114-116.

52. Triebig, G., Reichenbach, T., Flügel, K. A. 1978. Biochemische Untersuchungen und Messungen der Nervenleitgeschwindigkeit bei chronisch Trichloräthylen-belasteten Personen. *Int. Arch. Occup. Environ. Health 42*:31-40.
53. Triebig, G., Schaller, K. H., Erzigkeit, H., Valentin, H. 1977. Biochemische Untersuchungen und psychologische Studien an chronisch Trichloräthylen-belasteten Personen unter Berücksichtigung expositionsfreier Intervalle. *Int. Arch. Occup. Environ. Health 38*:149-162.

23

Alleged Neurotoxic Effects of Chronic Cyanide Exposure

Roger P. Smith
Dartmouth Medical School, Hanover, New Hampshire

I. INTRODUCTION

The contention that single, high doses of cyanide are virulently neurotoxic seems beyond dispute. Its two most prominent effects, activation of the chemoreceptors of the carotid body to produce hyperpnea followed by a rapid asphyxial death due to central respiratory arrest, have been recognized for many years (20). The evidence for a distinct syndrome of "chronic" cyanide poisoning is far more tenuous and rests primarily on epidemiological studies of acquired and/or inherited human diseases such as Leber's optic atrophy and others described below, together with some fragmentary and sometimes controversial animal studies (4,46).

There are hints in the literature that certain effects of cyanide as given by slow infusion over hours to days cannot be duplicated by repeated bolus doses or continuous feeding. Certainly, there is a very different spectrum in the efficacy of cyanide antagonists under conditions of slow infusion. These differences, however, can be explained on pharmacokinetic grounds. It is not clear whether the teratogenic effects of cyanide infusions, which primarily involve neural tube anomalies, represent quantitative or qualitative differences due to the unusual method of administration. Very similar teratogenic effects are produced by single doses of chemicals which release cyanide slowly and continuously over several hours. Unusual effects of cyanide in combination with nitroprusside have been described recently in such diverse isolated tissues as rabbit aorta, human blood platelets, and guinea pig ileum (27,28,31,39). The mechanism for these effects does not seem to involve the well-characterized inhibition of cytochrome c oxidase by cyanide, but their physiological significance remains obscure.

After single doses of cyanide the rates and pathways for its distribution, biotransformation, and excretion are critical in determining whether an animal will live or die. The same factors are no less important, but assume a different perspective in cases of chronic exposure to cyanide. Death in acute cyanide poisoning is undoubtedly due to cyanide itself, but human illness with chronic exposure might well be due to the accumulation of metabolites of cyanide. Cyanide is known or alleged to be metabolized to at least three neurotoxic substances in mammalian species: thiocyanate, cyanate, and formate. Thiocyanate has long been known to accumulate in the body with chronic or repeated exposure to cyanide and to be responsible for at least some toxic effects, particularly on the thyroid gland. No data seem to exist at present to indicate whether or not cyanate and/or formate may accumulate as well. It may prove to be extremely difficult to make a clear distinction between effects due to cyanide per se and neurotoxic effects due to its biotransformation products. Complications of another order of magnitude are introduced when the alleged sources of cyanide exposure are complex mixtures such as tobacco smoke or cassava, which may contain many other neurotoxic chemicals.

II. CYANIDE METABOLISM

Pathways of cyanide biotransformation in species ranging from microorganisms to higher mammals have been reviewed recently (4,46-48). The longest known and best characterized of these is the reaction mediated by rhodanese (thiosulfate cyanide sulfurtransferase), in which the sulfane sulfur of thiosulfate is transferred irreversibly to cyanide to form thiocyanate. Sulfane sulfur is divalent and bonded only to another sulfur (51). Rhodanese is widely distributed in nature, occurring in bacteria, plants, and animals. In mammals the highest activity is found in the liver, where the enzyme is associated with the mitochondrial matrix. Although some evidence suggests that a small metabolic pool of free cyanide exists in man and other animals, it may still be true that the major physiological function of rhodanese is unrelated to cyanide detoxication. One interesting and viable suggestion for the function of rhodanese is synthesis of iron-sulfur centers in such proteins as ferredoxin (50).

Although more than 30 years old, the classic metabolic studies of Boxer and Rickards (5) remain among those most widely cited. Using both ^{14}C-labeled cyanide and thiocyanate salts in beagles and rats, they were able to show that these two substances exist in a state of dynamic equilibrium. The equilibrium is certainly achieved within 24 hr and perhaps in as little as 5-8 hr. Their data suggest that a small pool of rapidly turning over cyanide comes into equilibrium with a larger (100- to 1000-fold) pool of more slowly turning over thiocyanate with a biological half-life of 3.6 days.

Since small amounts of $H^{14}CN$ were detected in the breath of rats after injection of [^{14}C]thiocyanate, it is clear that thiocyanate is metabolized at least to some extent to cyanide. This reaction was at one time ascribed to a "thiocyanate oxidase" in red blood cells. Subsequently it was said to be due to an enzymatic action of hemoglobin. Finally, it was traced to an artifact in the analytical procedure used for measuring cyanide (49). Thiocyanate is not oxidized to cyanide by any constituent of human or rat erythrocytes, although cyanide does accumulate in red cells against high concentration gradients when it is added as such to the media (34). Since, as noted above, the rhodanese reaction is essentially irreversible, the mechanism and anatomical locus for thiocyanate oxidation needs reexploration. Thiocyanate is oxidized by lactoperoxidase as found in milk or saliva, but the product is hypothiocyanite ($OSCN^-$) instead of cyanide (37).

About 1.5% of an administered dose of cyanide in rats is excreted as carbon dioxide, whereas nearly a third of a given dose of thiocyanate is converted to carbon dioxide (5). Although it had a relatively low specific activity, labeled formate was specifically identified as a metabolic product of both, and cyanate was postulated as an intermediate. Radiocarbon appeared in the methyl groups of choline and methionine and in the ureide carbons of allantoin, indicating that both cyanide and thiocyanate enter the single-carbon metabolic pool. Cyanide in particular, however, is not a quantitatively important contributor to that pool (5).

One of the surprising findings in dogs was that despite the alleged importance of thiocyanate as a detoxification product of cyanide metabolism, only 7% of the injected counts were excreted as urinary thiocyanate, even after an entire week (5). Similarly, Doherty et al. (14) could account for only 10-15% of the total dose as urinary thiocyanate in hamsters when sodium cyanide was infused continuously over several days via subcutaneously implanted osmotic minipumps. When examined by whole-body autoradiographic techniques ^{14}C-labeled cyanide and ^{35}S-labeled thiocyanate resulted in almost identical distribution patterns in rats. The highest activity over a 24-hr period was in the stomach, but the submaxillary salivary glands, the thyroid gland, the blood, and the walls of large blood vessels also had significant activity (11). The long-lasting high activity in the stomach which could account for 18% of the total dose is due to active secretion into the stomach and indirectly to an enterohepatic recycling of thiocyanate (35).

Other studies in rats (62) showed that small amounts of cyanide were excreted in the urine in the form of 2-iminothiazolidine-4-carboxylic acid. The final product was believed to have originated through a nonenzymatic reaction of cyanide with cystine first to yield β-thiocyanoalanine, which then underwent tautomerization and isomerization to yield the product found in urine. These same workers insisted that in the rat 80% of a given dose of cyanide is excreted as urinary thiocyanate.

Cyanide also binds avidly to methemoglobin, to such heme enzymes as cytochrome c oxidase and catalase, and to several forms of vitamin B_{12}. Since the hydroxo and cyano forms of vitamin B_{12} are equally effective in the treatment of pernicious anemia, further support is lent to the concept of a small endogenous metabolic pool of cyanide (41). A further link between vitamin B_{12} and cyanide is that the former is involved in at least some one-carbon transfer reactions. Vitamin B_{12} may also be important for the normal synthesis of the myelin sheath, and the deficiency state often involves neurological damage. Possible implications of these links are discussed below.

For many years it has been widely accepted that the rhodanese reaction is the major mechanism for cyanide detoxication in vivo. Although, as noted above, some workers have been able to account for only a small fraction of the total cyanide dose as thiocyanate in urine, thiocyanate is still quantitatively the most important biotransformation product identified (5). In recent years some authorities have begun to question the importance of the rhodanese reaction as a pathway for cyanide metabolism (48,51). Not only is the physiological function for rhodanese being critically reexamined (see above), but questions are again being raised about the normal substrate for that enzyme. Certainly thiosulfate is an effective antagonist in acute cyanide poisoning (20). Even so, thiosulfate penetrates into mitochondria very poorly (51), and it may be effective as a cyanide antagonist primarily because its lack of toxicity permits the administration of huge doses. Thus an enormous molar excess of thiosulfate relative to cyanide may offset the inefficiency with which it penetrates into mitochondria. In addition to a variety of nonendogenous substrates, the rat is able to use the sulfur of methionine and to a lesser degree that of sulfate in cyanide detoxication (2).

Thiosulfate represents only one form of sulfane sulfur which can serve as a substrate for rhodanese (51). Sulfane sulfur may also be formed from cystine by a series of reactions involving transaminases and mercaptopyruvate sulfurtransferase. An alternative pathway has been suggested in the rat (17). A number of sulfhydryl-containing small molecules including 3-mercaptopyruvate, are substrates for mercaptopyruvate sulfurtransferase, which generates various forms of sulfane sulfur. At least some of these can serve as sulfur donors for rhodanese, but a new suggestion (51) is that serum albumin may also function as a carrier of sulfane sulfur groups. As such, these groups are highly available and chemically reactive toward free cyanide. Although additional evidence is needed, the possibility that serum albumin sulfane sulfur may be important in the conversion of cyanide to thiocyanate certainly deserves critical examination (48). The fact that the rhodanese reaction and the sulfane sulfur reaction generate the same product, namely, thiocyanate, may be a coincidence that has misled many investigators.

III. NEUROPATHOLOGICAL LESIONS ASCRIBED TO CYANIDE

A few alleged human cases of "chronic" cyanide poisoning more likely are due to persistent sequelae after one or more acute toxic exposures. Sequelae may follow hypoxic insults irrespective of their cause (20). Demyelinating lesions in brain and peripheral nerves following experimental cyanide administration have been described many times (23), but the fact that they are specific for cyanide has been disputed (3). In rats lesions were said to have been produced by doses that caused no overt signs of toxicity (42), whereas in dogs no pathological changes were found after repeated convulsive doses (52). In monkeys infused cyanide was judged to have produced neuronal damage only secondarily to changes in circulation or respiration (6). Remyelination has been said to occur more rapidly in the peripheral than in the central nervous system (23). The lack of agreement about the pathological effects of cyanide in the brain illustrated above confounds attempts to extrapolate to human disease states as noted below.

IV. NEUROTOXIC CYANIDE METABOLITES

As evaluated after single intraperitoneal doses in mice, the LD_{50} for sodium cyanide is about 0.1 mmol/kg (45). Under the same conditions lethal doses for sodium thiocyanate are between 5 and 8 mmol/kg, and lethal doses for sodium cyanate are around 4-5 mmol/kg (44). When given in single lethal doses to mice, thiocyanate and cyanate produced similar toxic syndromes which were different from that elicited by cyanide and included tremor, hyperreactivity, extensor rigidity, tonic-clonic convulsions, and opisthotonus. Death appeared to be due to respiratory arrest during the tonic phase. Subanesthetic doses of phenobarbital and morphine protected mice against both convulsions and death, but doses of thiosulfate effective against cyanide were not effective against thiocyanate. Thus the conclusion was reached that the acute toxicity of thiocyanate and cyanate was not due to their conversion to cyanide in vivo (44). More likely it is due to the parent ions themselves perhaps acting through a common mechanism.

Because cyanate and thiocyanate are 40- to 80-fold less acutely toxic than cyanide, it is easy to overlook the fact that both are well-known neurotoxins when given chronically to humans and that there is a distinct possibility that some of their effects may be cumulative in nature.

Beginning in the mid-1930s both sodium and potassium thiocyanate were widely used in the United States for the control of hypertension. Doses varied considerably, but often began at 0.1-0.2 g, two to three times daily, with a maintenance regimen of 0.2 -0.4 g/day. It was common practice to adjust the dosage on the basis of periodic determination of serum thiocyanate concentration. The therapeutic range was 1.4-2.1 mmol/liter, and it was generally agreed that serious toxicity and death could occur if the serum thiocyanate rose above 2.6 mmol/liter (19). Normal human subjects have concentrations of 0.007-0.02 mmol/liter, whereas smokers have values about threefold higher (38), but well under toxic levels.

The toxic side effects ascribed to thiocyanate are protean, but those which are probably due to effects on the central nervous system include lassitude, weakness, and drowsiness, which could progress with continued therapy to irritability, nervousness, insomnia, confusion, blurred vision, tinnitus, ataxia, hallucinations, mania, delirium, and convulsions (1). Certainly, some of these manifestations are also present in human diseases ascribed to chronic cyanide poisoning (see below).

Cyanate was tested for a short time as a possible means for preventing crises in patients with sickle cell anemia. Cyanate irreversibly carbamylates the N-terminal valine residues on the globin chains of hemoglobin to result in an increase in the oxygen affinity of hemoglobin. In the case of hemoglobin S the increased affinity helps to prevent deoxygenation to the point where a crisis is precipitated (9). Perhaps nonspecific carbamylation of other proteins, some of which may have occurred in the central nervous system, resulted in the observed neurotoxic side effects. After prolonged administration of daily doses of 35 mg/kg (0.5 mmol/kg) some patients developed hyporeflexia, muscular weakness, and nerve conduction abnormalities (36). On fascicular biopsy, segmental demyelination was found, with neuronal degeneration of the distal axons (21). Again, at least some features of this syndrome are common to some of the alleged conditions caused by chronic exposure to cyanide (below).

Formate as a metabolite of cyanide (5) is of particular interest, since it is also a metabolite of methyl alcohol (20). Following the ingestion of toxic doses of methanol by man, formate accumulates and is responsible for the severe metabolic acidosis and the retinal damage that are characteristic features of the intoxication syndrome. Formate still produces retinal damage in monkeys even when they have been systemically buffered to prevent acidosis (32). Since methanol retinal damage may progress to total and permanent blindness, formate becomes the only known metabolite of cyanide to exhibit one of the prominent features of Leber's disease (see below). Moreover, the ocular lesions are said to be similar, involving primarily the ganglion cells of the retina followed by atrophy (20).

There is a high degree of individual susceptibility to methanol toxicity. As little as 10 ml (about 3.5 mmol/kg) is considered toxic, and the lethal dose in man may lie somewhere between 2 and 8 oz. Lethal blood concentrations in man are on the order of 100-200 mg/dl, and the blood half-life is said to be about 8 hr (20). In two severely poisoned human patients blood formate accumulated initially to levels ranging from 11 to 26 mEq/liter (33), which would be about three orders of magnitude higher than an acute lethal blood level of cyanide. No published studies were found which addressed the issue of the chronic toxicity of either methanol or formate in a relevant species such as susceptible subhuman primates. Formate salts do not have a high degree of acute toxicity, but they may cause convulsions in poisoned animals (20).

V. ANTAGONISM OF CHRONIC CYANIDE POISONING

Cyanide is cited as an unusual example of a chemical possessing a very high acute toxicity and little or no propensity to produce chronic or cumulative effects resulting in death (22). Indeed, few if any controlled studies involving the administration of cyanide as such either in discrete sublethal doses or incorporated into the diet or drinking water purport to show any adverse effects. In contrast, numerous reports involving the ingestion of sorghums, Sudan grasses, and other cyanogenic plants by both ruminant and nonruminant animals describe severe toxicity and death (46). Many animals show an amazing capacity for inactivating cyanide. When administered to golden hamsters continuously over a period of 4 days via subcutaneously implanted osmotic minipumps, the animals were able to detoxify 30-40 single subcutaneous LD_{50} doses (14).

Appropriate single doses of either sodium nitrite or sodium thiosulfate are able to prevent or reverse the course of acute cyanide poisoning in animals. Nitrite acts by converting a tolerable fraction of the circulating hemoglobin to methemoglobin, which binds cyanide in the biologically inactive form of cyanmethemoglobin. Although undeniably effective, nitrite treatment is less efficient than thiosulfate treatment, and especially so during the slow administration of cyanide. Firstly, it has a finite capacity, since there is a limit to the amount of methemoglobin that can be generated before death results from anemic hypoxia. Secondly, the protection is only temporary, since cyanmethemoglobin will eventually dissociate its bound cyanide, which may lead to a recurrence of signs of poisoning and even death. Essentially the same is true of the administration of hydroxycobalamin to inactive cyanide (25).

These points may be illustrated in mice, a species in which the methemoglobin reductase activity in red cells is at least an order of magnitude greater than in human erythrocytes. When a series of aliphatic nitriles, all of which were believed to cause death because of a metabolic release of cyanide, were injected into mice, the mean times to death varied widely, presumably because of varying rates for activation. Thiosulfate given in repeated injections protected mice against death by all nitriles tested. In contrast, repeated injections of nitrite protected only against some of the nitriles, namely, those that produced death within the shortest period of time (55). Thus a balance was struck between the rate of metabolic activation of the nitrile and the rate of release of cyanide from cyanmethemoglobin.

Also in mice nitrite produces a uniquely prolonged methemoglobinemia compared to other agents, such as hydroxylamine. Perhaps nitrite concomitantly inhibits methemoglobin reductase as well as mediating the oxidation of hemoglobin to methemoglobin. In any event, the degree of protection against cyanide mortality was correlated with the duration of the methemoglobinemia when identical peak levels were generated (26). Not only did animals treated with hydroxylamine die at lower doses of cyanide, but the deaths tended to be delayed relative to those in animals given nitrite. The observations are consistent with the hypothesis that cyanmethemoglobin is a substrate for mouse methemoglobin reductase. Hydroxylamine-generated methemoglobin is able to trap and temporarily inactivate a pool of cyanide equivalent to a lethal dose. As methemoglobin reductase acts, that pool of cyanide is released to result in a late death. In this case a balance was struck between the rate of cyanomethemoglobin reduction and endogenous rates of cyanide detoxication (26). The explanation of these phenomena, however, clearly relates to pharmacokinetic processes and not to any qualitative differences in the effects of cyanide.

VI. TERATOGENIC EFFECTS OF CYANIDE

Several aliphatic nitriles have been shown to be teratogenic in the golden hamster, including acrylonitrile, propionitrile (56), succinonitrile (15), acetonitrile (53), and the cyanogenic glycoside amygdalin (54). In each case the teratogenic effects could be prevented or ameliorated by thiosulfate treatment. Although this evidence does not prove that the terata were induced by cyanide, it makes it unlikely that they were due to thiocyanate. No other reports of cyanide teratogenicity were found in the literature. Perhaps the circumstance that the cyanide was slowly and continuously released from these nitriles by metabolic activation was critical for the induction of teratogenic effects. By implanting subcutaneously osmotic minipumps which continuously infused sodium cyanide between days 6 and 9 of gestation, it was possible to induce similar teratogenic effects to those induced by nitriles. These also could be blocked or ameliorated by contemporaneous administration of thiosulfate (14).

The most commonly observed anomalies in hamsters exposed to cyanide or to cyanogenic compounds in utero were neural tube defects such as encephalocoeles and exencephaly, although fetal crown rump length was significantly smaller in the offspring of treated dams than in control dams (14). Whether or not these teratogenic effects bear any relationship to the alleged neurotoxic actions of chronic cyanide can only be regarded for the present as highly speculative.

VII. HUMAN DISEASES ASCRIBED TO CHRONIC CYANIDE EXPOSURE

Evidence associating the following diseases with cyanide has been reviewed recently (4,46). Thus the discussions below are limited primarily to the most relevant or most recent developments.

A. Leber's Optic Atrophy

This rare hereditary disease was characterized in 1871 as a sudden progressive bilateral loss of vision accompanied by scotomas and optic atrophy (29). The disease affects predominantly postadolescent males, but it has occasionally been described in women. The inheritance pattern is generally that of a sex-linked recessive trait. Very few cases have gone to autopsy, but the primary changes may occur in the ganglion cells of the retina. Degeneration of the optic nerve fibers may occur secondarily. In some patients diffuse encephalomyelopathic complications occur, also suggesting that the ophthalmological condition is the minimal manifestation of a more diffuse disorder (16,30,57). Wilson (57) was the first to suggest that the disease probably resulted from an inborn error of metabolism, but that environmental factors such as heavy smoking made important contributions to its manifestation.

In a further expansion of this hypothesis Wilson (58) suggested that Leber's disease may actually represent an inborn error in cyanide metabolism. He noted that both cigarette and pipe smokers have increased plasma concentrations of thiocyanate (0.10-0.11 mmol/liter) and excrete larger amounts of thiocyanate in their urine than do non-smokers (plasma levels of 0.05-0.06 mmol/liter). These observations suggested an increased cyanide exposure by smoke inhalation followed by increased conversion of cyanide to thiocyanate. Increases in thiocyanate in the urine or serum of smokers with Leber's disease were not so large, or totally absent, suggesting a defect in cyanide metabolism to thiocyanate. Also noted were changes in the patterns of vitamin B_{12} metabolism, suggesting that vitamin B_{12} and pathways for converting cyanide to thiocyanate competed for free cyanide and that persons with decreased levels of vitamin B_{12} might be at special risk for the neuropathological effects of chronic cyanide in smoke (60,61).

Most recently Cagianut et al. (7,8) demonstrated a marked reduction in the rhodanese activity of biopsy specimens of liver from two affected adult males in a family exhibiting Leber's disease. This well-known Swiss family had five symptomatic individuals over four generations. Thus the evidence for a role of cyanide in Leber's disease is probably stronger than that for any of the other conditions below. Caution, however, is still indicated in reaching this conclusion. As noted above, the physiological function of rhodanese is still not known. It is conceivable that Leber's disease may be a manifestation of rhodanese deficiency which is unrelated to cyanide metabolism. If environmental factors associated with smoking are critical, cigarette smoke contains many toxic chemical entities in addition to cyanide. Formate is the only cyanide metabolite that has been associated with blindness in man, and no evidence exists at present to suggest that it could accumulate to the very high concentrations necessary for retinal damage as a result of chronic cyanide exposure.

B. Tobacco Amblyopia

As noted above, some authorities regard tobacco amblyopia as the mildest possible expression of Leber's disease. As the name suggests, the condition occurs only in smokers, and its characteristic features are a central scotoma and degeneration of the ganglion cells of the retina and of the optic nerve. The pathological changes are said to resemble demyelinating lesions seen in the central nervous system of animals given repeated small doses of cyanide. As noted above, attempts to reproduce these lesions experimentally have not been uniformly successful (4,46). Similar lesions could also be produced by azide (43), which acts in vivo very differently from cyanide (24).

Again, changes in the patterns of vitamin B_{12} metabolism in smokers have been offered as evidence for cyanide exposure (60,61). A detailed analysis of the cyanide content of smoke from a number of brands of pipe and cigarette tobaccos was more remarkable for the similarities found than for the differences. Tobacco brands associated with amblyopia contained no more cyanide than those that had not, suggesting inherent differences between victims and nonvictims rather than differences in the smoke (13).

Perhaps the most compelling evidence relates to an apparently favorable clinical response of patients with tobacco amblyopia to hydroxocobalamin, but not to cyanocobalamin (10). Some patients who continued to smoke showed full recovery of vision if given the hydroxo form, but only partial recovery with the cyano form. It is still possible, however, that tobacco amblyopia may represent some as yet unappreciated function of vitamin B_{12} that is unrelated to cyanide metabolism or, alternatively, it might represent an unusual manifestation of mild carbon monoxide intoxication.

C. Retrobulbar Neuritis in Pernicious Anemia

Retrobulbar neuritis is a rare but recognized complication of pernicious anemia, as is also tobacco amblyopia. Some authorities make no distinctions between these two conditions (4). There is an overwhelming preponderance of male smokers in both, and both are sometimes associated with vague signs of peripheral neuropathy, myelopathy, or encephalopathy (18,41,59).

D. Tropical Neuropathies and Amblyopias

The full-blown syndrome, also called Nigerian nutritional neuropathy, may include optic atrophy, nerve deafness, and sensory spinal ataxia, but any or all of these may be present in a given patient. The diet of afflicted individuals usually contains large amounts of cassava. Cyanide is liberated from the root tissues of cassava when they are damaged to allow the enzyme linamerase to act on the cyanogenic glycoside linamarin (12). Occasional reports of cases of acute cyanide poisoning have been associated with the ingestion of unprocessed cassava roots. Ordinarily most of the cyanide is washed out, diluted, or otherwise eliminated during elaborate processing procedures. Inevitably, however, questions arise about possible adverse effects of the long-term low-level ingestion of cyanide and subsequent accumulation of thiocyanate, which might be responsible for goiter, cretinism, tropical ataxic neuropathy, and tropical diabetes. The first two conditions may be secondary to an interference in thyroid function by thiocyanate, whereas tropical ataxic neuropathy may require concomitant protein malnutrition and cassava ingestion (12).

Once again the evidence that this or these conditions are due to chronic cyanide ingestion is epidemiological in nature and involves measurements of thiocyanate and vitamin B_{12} in afflicted populations. Cassava may be teratogenic in rats, but the evidence is fragmentary to date (40). Although tropical nutritional neuropathy involves far more patients than the other conditions above, the evidence for a role for chronic cyanide neurotoxicity may well be the least convincing in this case.

IX. SUMMARY

The evidence for novel neurotoxic effects of chronic cyanide exposure in man has been reviewed. Although these represent problems of potentially enormous social significance, the evidence is still fragmentary, incomplete, and at this date unconvincing. It is an area, however, deserving of intensive investigation. Cyanide metabolism and excretion under conditions of chronic administration need careful study, especially with respect to known or alleged neurotoxic metabolites. Some novel effects of cyanide, including perhaps teratogenicity have been reviewed. There appears to be much yet to be learned about this common violent poison.

REFERENCES

1. Barnett, H. J. M., Jackson, M. V., Spaulding, W. B. 1951. Thiocyanate psychosis. *J. Am. Med. Assoc. 147*:1554-1558.
2. Barrett, M. D. P., Alexander, J. C., Hill, D. C. 1978. Utilization of ^{35}S from radioactive methionine or sulfate in the detoxification of cyanide by rats. *Nutr. Metab. 22*:51-57.
3. Bass, N. H. 1968. Pathogenesis of myelin lesions in experimental cyanide encephalopathy. *Neurology 18*:167-177.
4. Baumeister, R. G. H., Schievelbein, H., Zickgraf-Rudel, G. 1975. Toxicological and clinical aspects of cyanide metabolism. *Arzneim. Forsch. 25*:1056-1064.
5. Boxer, G. E., Rickards, J. C. 1952. Studies on the metabolism of the carbon of cyanide and thiocyanate. *Arch. Biochem. 39*:7-25.
6. Brierley, J. B., Prior, P. F., Calverley, J., Brown, A. W. 1977. Cyanide intoxication in *Macaca mulatta*—Physiological and neuropathological aspects. *J. Neurol. Sci. 31*:133-157.
7. Cagianut, B., Rhyner, K., Furrer, W., Schnebli, H. P. 1981. Thiosulfate-sulfur transferase (rhodanese) deficiency in Leber's hereditary optic atrophy. *Lancet 1*:981-982.
8. Cagianut, B., Schnebli, H. P., Rhyner, K., Furrer, W. 1982. Thiosulfat-sulfur-transferase-Mangel bei Lebers hereditärer Optikusatrophie. *Klin. Monatsbl. Augenheilkd. 1*:32-35.
9. Cerami, A., Manning, J. M. 1971. Potassium cyanate as an inhibitor of the sickling of erythrocytes in vitro. *Proc. Nat. Acad. Sci. U.S.A. 68*:1180-1183.

10. Chisholm, I. A., Bronte-Stewart, J., Foulds, W. S. 1967. Hydroxocobalamin versus cyanocobalamin in the treatment of tobacco amblyopia. *Lancet 2*:450-451.
11. Clemedson, C.-J., Sorbo, B., Ullberg, S. 1960. Autoradiographic observations on injected S^{35}-thiocyanate and C^{14}-cyanide in mice. *Acta Physiol. Scand. 48*:382-389.
12. Cock, J. H. 1982. Cassava: A basic energy source in the tropics. *Science 218*:755-762.
13. Darby, P. W., Wilson, J. 1967. Cyanide, smoking, and tobacco amblyopia, observations on the cyanide content of tobacco smoke. *Br. J. Ophthalmol. 51*:336-338.
14. Doherty, P. A., Ferm, V. H., Smith, R. P. 1982. Congenital malformations induced by infusion of sodium cyanide in the golden hamster. *Toxicol. Appl. Pharmacol. 64*:456-464.
15. Doherty, P. A., Smith, R. P., Ferm, V. H. 1983. Comparison of the teratogenic potential of two aliphatic nitriles in hamsters: Succinonitrile and tetramethylsuccinonitrile. *Fundam. Appl. Toxicol. 3*:41-48.
16. Enghoff, E. 1963. Über eine an Leber's Opticusatrophie erinnernde heredodegenerative Krankheit. *Acta Med. Scand. 173*:83-90.
17. Fellman, J. H., Avedovech, N. A. 1982. Cysteine thiosulfonate in cysteine metabolism. *Arch. Biochem. Biophys. 218*:303-308.
18. Freeman, A. G., Heaton, J. M. 1961. The aetiology of retrobulbar neuritis in addisonian pernicious anaemia. *Lancet 1*:908-911.
19. Goodman, L. S., Gilman, A. 1958. *The Pharmacological Basis of Therapeutics*, 2nd ed., Macmillan, New York, pp. 743-747.
20. Gosselin, R. E., Smith, R. P., Hodge, H. C. 1984. *Clinical Toxicology of Commercial Products*, 5th ed., Williams and Wilkins, Baltimore, pp. III-123-130.
21. Graziano, J. H., Thornton, Y. S., Leong, J. K., Cerami, A. 1973. Pharmacology of cyanate II. Effects on the endocrine system. *J. Pharmacol. Exp. Ther. 185*:667-675.
22. Hayes, W. J., Jr. 1967. The 90-dose LD_{50} and a chronicity factor as measures of toxicity. *Toxicol. Appl. Pharmacol. 11*:327-335.
23. Hirano, A., Levine, S., Zimmerman, H. M. 1968. Remyelination in the central nervous system after cyanide intoxication. *J. Neuropathol. Exp. Neurol. 27*:234-245.
24. Kaplita, P. V., Borison, H. L., Smith, R. P. 1982. Cardiovascular and respiratory actions of sodium azide in the anesthetized cat. *Pharmacologist 24*:165.
25. Krapez, J. R., Vesey, C. J., Adams, L., Cole, P. V. 1981. Effects of cyanide antidotes used with sodium nitroprusside infusions; sodium thiosulfate and hydroxocobalamin given prophylactically to dogs. *Br. J. Anaesth. 53*:793-804.

26. Kruszyna, R., Kruszyna, H., Smith, R. P. 1982. Comparison of hydroxylamine, 4-dimethylaminophenol and nitrite protection against cyanide poisoning in mice. *Arch. Toxicol. 49*:191-202.
27. Kruszyna, H., Kruszyna, R., Smith, R. P. 1982. Nitroprusside increases cyclic guanylate monophosphate concentrations during relaxation of rabbit aortic strips and both effects are antagonized by cyanide. *Anesthesiology* 57:303-308.
28. Kruszyna, R., Kruszyna, H., Smith, R. P. 1983. Unpublished observations.
29. Leber, T. 1871. Ueber hereditäre und congenital angelegate Schnervenleiden. *A. Graefes Arch. Klin. Ophthalmol. 2*:249-291.
30. Lees, F., MacDonald, A. M., Turner, J. W. A. 1964. Leber's disease with symptoms resembling disseminated sclerosis. *J. Neurol. Neurosurg. Psychiatry 27*:415-421.
31. Li, J.-G., Burgess, B. K., Corbin, J. L. 1982. Nitrogenase reactivity: Cyanide as substrate and inhibitor. *Biochemistry 21*:4393-4402.
32. Martin-Amat, G., McMartin, K. E., Hayreh, S. S., Hayreh, M. S., Tephly, T. R. 1978. Methanol poisoning: Ocular toxicity produced by formate. *Toxicol. Appl. Pharmacol. 45*:201-208.
33. McMartin, K. E., Ambre, J. J., Tephly, T. R. 1980. Methanol poisoning in human subjects. Role for formic acid accumulation in the metabolic acidosis. *Am. J. Med. 68*:414-418.
34. McMillan, D. E., Svoboda, IV, A. C. 1982. The role of erythrocytes in cyanide detoxification. *J. Pharmacol. Exp. Ther. 221*:37-42.
35. Okoh, P. N., Pitt, G. A. J. 1982. The metabolism of cyanide and the gastrointestinal circulation of the resulting thiocyanate under conditions of chronic cyanide intake in the rat. *Can. J. Physiol. Pharmacol. 60*:381-386.
36. Peterson, C. M., Tsairis, P., Ohnishi, A., Lu, Y. S., Grady, R., Cerami, A., Dyck, P. J. 1974. Sodium cyanate induced polyneuropathy in patients with sickle-cell disease. *Ann. Intern. Med. 81*:152-158.
37. Pruitt, K. M., Tenovuo, J., Andrews, R. W., McKane, T. 1982. Lactoperoxidase-catalyzed oxidation of thiocyanate: Polarographic study of the oxidation products. *Biochemistry 21*:562-567.
38. Schrieber, H. 1925. On thiocyanate in human blood serum. *Biochem. Z. 163*:241-251.
39. Schwerin, F., Rosenstein, R., Smith, R. P. 1982. Reversal by sodium cyanide (CN^-) of sodium nitroprusside (SNP) induced inhibition of platelet aggregation. *Pharmacologist 24*:194.
40. Singh, J. D. 1981. The teratogenic effects of dietary cassava in the pregnant albino rat; a preliminary report. *Teratology 24*:289-291.

41. Smith, A. D. M. 1961. Retrobulbar neuritis in addisonian pernicious anemia. *Lancet 1*:1001-1002.
42. Smith, A. D. M., Duckett, S., Waters, A. H. 1963. Neuropathological changes in chronic cyanide intoxication. *Nature 200*:179-181.
43. Smith, A. D. M., Duckett, S. J. 1965. Cyanide, vitamin B_{12}, experimental demyelination and tobacco amblyopia. *Br. J. Exp. Pathol. 46*:615-622.
44. Smith, R. P. 1973. Cyanate and thiocyanate: Acute toxicity. *Proc. Soc. Exp. Biol. Med. 142*:1041-1044.
45. Smith, R. P., Gosselin, R. E. 1979. Hydrogen sulfide poisoning. *J. Occup. Med. 21*:93-97.
46. Towill, L. E., Drury, J. S., Whitfield, B. L., Lewis, E. B., Galyan, E. L., Hammons, A. S. 1978. *Reviews of the Environmental Effects of Pollutants: V. Cyanide.* ORNL/EIS-81, EPA-600/1-78-027.
47. Vennesland, B., Conn, E. E., Knowles, C. J., Westley, J., Wissing, F. (eds.), 1981. *Cyanide in Biology*, Academic, London.
48. Vennesland, B., Castric, P. A., Conn, E. E., Solomonson, L. P., Volini, M., Westley, J. 1982. Cyanide metabolism. *Fed. Proc. 41*:2639-2648.
49. Vesey, C. J., Wilson, J. 1978. Red cell cyanide. *J. Pharm. Pharmacol. 30*:20-26.
50. Violini, M., Alexander, K. 1981. Multiple forms and multiple functions of the rhodaneses, in *Cyanide in Biology* (Vennesland, B., Conn, E. E., Knowles, C. J., Westley, J., Wissing, F., eds.) Academic, London, pp. 77-91.
51. Westley, J. 1983. Rhodanese and the sulfane pool, in *Enzymatic Basis of Detoxication*, Vol. 2 (Jakoby, W. B., ed.), Academic, New York, pp. 245-262.
52. Wheatley, M. D., Lipton, B., Ward, A. A., Jr. 1947. Repeated cyanide convulsions without central nervous pathology. *J. Neuropathol. Exp. Neurol. 6*:408-411.
53. Willhite, C. C. 1981. Malformations induced by inhalation of acetonitrile vapors in the golden hamster. *Teratology 23*:698.
54. Willhite, C. C. 1982. Congenital malformations induced by laetrile. *Science 215*:1513-1515.
55. Willhite, C. C., Smith, R. P. 1981. The role of cyanide liberation in the acute toxicity of aliphatic nitriles. *Toxicol. Appl. Pharmacol. 59*:589-602.
56. Willhite, C. C., Ferm, V. H., Smith, R. P. 1981. Teratogenic effects of aliphatic nitriles. *Teratology 23*:317-323.
57. Wilson, J. 1963. Leber's hereditary optic atrophy, some clinical and aetiological considerations. *Brain 86*:347-362.
58. Wilson, J. 1965. Leber's hereditary optic atrophy: A possible defect of cyanide metabolism. *Clin. Sci. Mol. Med. 29*:505-515.

59. Wilson, J., Langman, M. H. S. 1966. Relation of sub-acute combined degeneration of the cord to vitamin B_{12} deficiency. *Nature 212*:787-789.
60. Wilson, J., Matthews, D. M. 1966. Metabolic inter-relationships between cyanide, thiocyanate and vitamin B_{12} in smokers and non-smokers. *Clin. Sci. 31*:1-7.
61. Wilson, J., Linnell, J. G., Matthews, D. M. 1971. Plasma cobalamins in neuro-ophthalmological diseases. *Lancet 1*:259-261.
62. Wood, J. L., Cooley, S. L. 1956. Detoxication of cyanide by cystine. *J. Biol. Chem. 218*:449-457.

24

Toxicology of Selected Animal and Marine Neurotoxins

Lawrence Rodichok and Russell Mankes
Albany Medical College, Albany, New York

I. BOTULISM

Botulism is an illness caused by intoxication with the protein neurotoxin produced by the organism *Clostridium botulinum*. Intoxication may occur in three forms: foodborne, wound, and infant botulism. Foodborne botulism is caused by ingestion of the neurotoxin preformed in foods contaminated with the organism. This illness has become relatively rare in most developed countries, although outbreaks continue to be reported, largely caused by home-processed foods or improperly cooked or uncooked marine meats. Wound botulism is extremely rare and occurs by means of absorption of the neurotoxin produced in a wound contaminated by the microorganism. Infant botulism is a recently described illness, having been first proven to occur in 1976 (105). All three forms of botulism represent medical emergencies in which prompt recognition and treatment are crucial to the survival of the patient.

A. Chemistry and Pathophysiology

Botulinum toxins types A through G have been described and are antigenically distinct (92,95). Type A has been most carefully studied. The neurotoxin itself has a molecular weight of approximately 150,000. It is composed of two chains: a larger subunit, the H chain, and a smaller subunit, the L chain, which is approximately one-half the molecular weight of the larger subunit. There are probably two disulfide bonds in each dichain, the reduction of which results in nearly complete loss of toxicity. Activation of toxicity may occur through cleavage ("nicking") of one or more dipeptide bonds at the N and/or C terminus, as well as at one or more additional sites along the chain. Cultures of the microorganism produce proteases which may be responsible for activation of toxicity through cleavage of these dipeptide bonds. However, recent evidence questions a direct cause and effect relationship between the action of proteases and an increase in toxicity.

Each neurotoxin occurs in association with one or more additional proteins, forming a "toxic complex." The total molecular weight of these complexes is about 500,000. These additional associated proteins appear to protect the neurotoxin from inactivation in the gut and enhance toxicity by increasing the opportunity for absorption.

While foodborne botulism occurs by ingestion of the toxin in contaminated foods and wound botulism occurs through absorption of the toxin from a contaminated lesion, infant botulism is caused by release of neurotoxin by microorganisms which have colonized the intestinal tract itself. Infant botulism appears limited to children between the ages of 5 and 25 weeks (6). *Clostridium botulinum* does not readily colonize the adult intestinal tract. Older children and adults regularly ingest *C. botulinum* spores without ill effect. Infant botulism therefore appears to represent a somewhat opportunistic infection of the infant gastrointestinal tract prior to the development of the normal bacterial flora. Experimentally, the gastrointestinal tract of germ-free mice (infant or adult) is more readily colonized than that of normal mice. The resistance of infants under the age of 3-5 weeks may be due to yet additional factors, since it has been shown that even germ-free infant mice less than 3 days old cannot be infected (67).

Botulinum toxins block transmission at cholinergic synapses by preventing the release of acetylcholine. There is no effect on conduction of impulses along the axon into the presynaptic nerve terminal and no effect on the excitability of muscle fibers themselves (20,50). Morphological studies have demonstrated binding of labeled toxin to the presynaptic membrane (52,90,91). The toxin causes a decrease in the frequency of fusion of synaptic vesicles with the presynpatic membrane induced electrically or pharmacologically (51). Electron-microscopic studies have demonstrated that presynaptic terminals poisoned with botulinum toxin do not show morphological evidence of calcium influx after attempted stimulation (51). Thus the effect of the toxin can be antagonized by high concentration of calcium. Intoxication is dependent upon early release of acetylcholine (88). The toxin would therefore appear to bind to the presynaptic membrane and subsequently migrate to an internal receptor within the presynaptic terminal under the influence of synaptic activity. The effect of the toxin can be partially reversed by high extracellular calcium concentration, 4-aminopyridine (which increases the intracellular accumulation of calcium), and black widow spider venom, known to stimulate the release of synaptic vesicles (86,89).

B. Human Intoxication

The vast majority of outbreaks of foodborne botulism have been caused by types A, B, and E. Of 108 outbreaks occurring in the United States between 1970 and 1977, 51% were caused by type A, 21% by type B, and 12% by type E (15). Those due to type A and B are usually related to contamination of home-processed foods. Most type E outbreaks are caused by contaminated marine products and therefore tend to occur in coastal areas. Type E botulism accounted for 23 of 29 outbreaks in Canada between 1961 and 1973 (37), largely occurring in coastal Indians or Eskimos, with whom the custom of eating uncooked marine products is common. Type E is also more common in Japan and the Soviet Union, again related to the ingestion of uncooked fish.

The period from ingestion to the onset of symptoms varies from several hours to 8 days, most commonly 18-36 hr. The shorter incubation periods tend to occur with types B and E and are correlated with more severe illness. Within the first few hours there may be severe abdominal cramping and diarrhea with nausea and vomiting, particularly with types B and E. These early gastrointestinal symptoms are probably due to noxious products in the contaminated food other than the neurotoxin. Early vomiting may in fact lessen mortality. Dry mouth, diplopia, or difficulty focusing, and dysarthria are the earliest neurological symptoms and those that usually prompt a need for medical attention (38). This may be followed by progessive symmetric weakness descending to involve facial, bulbar, respiratory, and limb musculature. Signs on neurological examination nearly always include weakness of extraocular muscles, most commonly the lateral rectus. Approximately half of the patients also have ptosis and dilated, poorly reactive pupils (38,54,98). Weakness of the facial muscles, soft palate, pharynx, and tongue is also common. Limb weakness occurs in 50-75% and is usually symmetrical and predominantly proximal. Muscle stretch reflexes are normal or slightly diminished. There should be normal mentation and sensory function, although some patients do report paresthesias (44). Respiratory involvement occurs in 20-75% of cases. The fatality rate has declined primarily with improvements in mechanical ventilatory care and other types of intensive care. In the United States from 1970 to

1977, 15.7% of cases were fatal. Mortality from types A and E is consistently higher (15-20%) than that for type B (5%) (15).

Differential diagnosis of early symptoms may include a variety of acute abdominal illnesses such as appendicitis. With the onset of neurological involvement, Guillain-Barré syndrome may be considered. Normal pupillary reactions and loss of muscle stretch reflexes with prominent sensory symptoms favor that diagnosis. Acute autoimmune myasthenia gravis could cause a similar picture. Often the diagnosis of botulism is entertained only when a cluster of cases is recognized.

The most effective means of establishing the diagnosis is by demonstrating the toxicity of patient serum injected into mice. The type of neurotoxin is identified by neutralization with specific antitoxins. The diagnosis may also be confirmed by demonstration of the microorganism and/or its neurotoxin in feces, vomitus, or the food source. Electromyography may show augmentation of muscle action potentials with repetitive stimulation at 20-50 per second (21).

Inducing vomiting, gastric lavage, and purgatives may be employed to remove unabsorbed toxin, but may not be safe once gastrointestinal motility is impaired. Antitoxin should be administered as soon as possible. If the type of toxin is known, then either bivalent AB or monovalent E antitoxin is necessary (38). All are of equine origin, and hypersensitivity reactions are relatively common. Nearly 2% have an anaphylactic reaction within 10 min (10). Prompt supportive care, including mechanical ventilatory support, may be crucial. Guanidine, 7-35 mg/kg per day in divided doses, may partially reverse the neuromuscular blockage (21).

All instances of infant botulism reported have been caused by either type A or B toxin (6). The only established source of the organism has been contaminated honey (4). In some cases the presence of *C. botulinum* and/or its toxin in infant fecal specimens was not correlated with any clinical symptoms or with only mild transient gastrointestinal difficulties such as constipation (5). In symptomatic cases the earliest sign is constipation, followed over a period of hours to days by poor suck or gag reflexes manifested by difficulty in feeding. Somatic muscle paralysis causes poor head control and flaccid arm and leg weakness—a "floppy baby." Respiratory distress may appear suddenly. Fulminant cases of infantile botulism are indistinguishable from and a potential cause of sudden infant death syndrome. Treatment of infantile botulism is similar to that for the foodborne variety. Antitoxin has not been administered in most cases. Its potential value is in need of further study.

II. TETANUS

Although an effective toxoid vaccine has been available for nearly four decades, tetanus continues to occur, even in developed countries with widespread immunization programs (58). Mortality remains high in all age groups, but is particularly high in the very young (neonatal) and very old.

A. Chemistry and Pathophysiology

Tetanus toxin is a protein of molecular weight 150,000 produced by the anaerobe *Clostridium tetani*. The molecule contains six sulfhydryl groups and one or two disulfide bridges (71). The disulfide bridges link one heavy chain (molecular weight about 95,000) and one light chain (molecular weight about 55,000) to form the complete toxin (29, 65). The primary clinical manifestations of intoxication are caused by a blockade of inhibitory input to spinal motor neurons (66). The toxin probably gains access to this site via retrograde axonal transport along nerve fibers, and perhaps along perineurium and epineurium as well, originating at the portal of entry (80). The effects of the toxin may remain localized to the motor neuron pool supplying the affected muscles ("local tetanus"); however, the effect becomes generalized if enough toxin is present to enter the bloodstream and affect the nervous system diffusely. The specific effect on motor neurons is probably caused by a blockade of the release of glycine by inhibitory afferents influencing the activity of Renshaw cells (31). It also blocks presynaptic inhibition and the accompanying primary afferent depolarization mediated by γ-aminobutyric acid (GABA) (32). Inhibition of the presynaptic release of neurotransmitters, particularly inhibitory ones, appears to be a general property of tetanus toxin. Electrophysiological and biochemical studies demonstrate inhibition of K^+-stimulated, calcium-dependent release of GABA from spinal afferents (31), from the basket cell projection to Purkinje cells (31), from hippocampal slices in vitro (26), from the striatonigral inhibitory pathway (33), and from developing cell cultures of rat cerebellum (77). Furthermore, the effect is not limited to glycine and GABA, since the release of a variety of inhibitory amino acids from synaptosomes (72) and of dopamine from slices of substantia nigra are similarly blocked (25). Tetanus toxin appears to bind selectively to presynaptic terminals (81), and to the gangliosides G_{T1b} or G_{D1b} in particular (84). Trans-synaptic transfer of intact toxin has also been demonstrated (87).

B. Human Intoxication

The incubation period for tetanus may be as short as a few days in neonates or as long as a month or more and tends to be correlated in adults with the distance the toxin must travel to reach the central nervous system. In 10-30% of cases no portal of entry is discovered (43). Shorter incubation periods are correlated with more severe disease and higher mortality (1). Early manifestations occur in those muscles nearest the portal of entry; however, isolated local tetanus is rare in man. Subsequent involvement usually appears in the head and neck muscles, with trismus ("lockjaw") and spasms of the facial muscles causing involuntary facial mimicry ("risus sardonicus"). These are often the presenting symptoms. Paraspinal muscles are affected early as well, probably because their motor nerves extend only a short distance from the spinal cord. Spasms of hyperextension of the neck and back occur spontaneously or may be induced by the slightest sensory stimulus. These are very painful and may be forceful enough to cause vertebral compression fractures. Eventually, full tetanus may involve the extremity muscles, causing flexion of the arms across the chest, carpopedal spasm, and hyperextension of the legs (46). Autonomic dysfunction is more common than generally appreciated and may in fact be life threatening. Blood pressure is often very labile, with spontaneous periods of hypertension and tachycardia, particularly early in the course of the illness, and later periods of profound hypotension lasting minutes to hours (56). Maximum symptomatology may be sustained for 1-2 weeks before recovery begins. Mortality ranges from 20% to as high as over 90%, depending in part on the age and general health of the patient (46).

Treatment includes careful debridement and cleansing of the wound itself. The role of antibiotics is not clear, but penicillin or tetracycline is usually administered in an effort to eliminate any remaining organisms. Human antitoxin is given to neutralize any circulating toxin. Supportive care, particularly mechanical ventilation, is crucial. Pharmacological control of muscle spasms may be achieved with diazepam (97), baclofen, or even curare. Sensory stimulation should be minimized. A variety of agents have been reported to control autonomic dysfunction, including β-adrenergic blockers and morphine (82,83). During convalescence active immunization with toxoid should be initiated.

III. DIPHTHERIA

Diphtheria has become an uncommon illness in most areas of the world, since the widespread active immunization with toxoid. It continues to appear, however, in unimmunized children, even in more developed countries. It can be an acute, life-threatening illness because of either the local pharyngeal infection or because of the neurotoxicity of diphtheria exotoxin.

A. Chemistry and Pathophysiology

Corynebacterium diphtheriae releases a protein neurotoxin whose mechanism of action has been elucidated in great detail (9,23,74,75). Only strains of *C. diphtheriae* lysogenic for or infected with bacteriophage carrying the toxic gene can produce the toxin. Although synthesis of the toxin protein is directed by the viral genome, expression may be modified by protein factors (repressors) controlled by the bacterial host. In addition, inorganic iron suppresses toxin production.

The toxin is a pure protein with a molecular weight of 62,000-63,000 containing two disulfide bridges. Treatment with a trypsinlike protease followed by exposure to reducing agents such as dithiothreitol leaves two polypeptide fragments. Fragment A, the segment from the N-terminal end, has a molecular weight of 24,000 while the B fragment has a molecular weight of about 38,000.

Whole toxin inhibits protein synthesis in susceptible cells in culture (24,34). It inactivates elongation factor 2 (EF2), an enzyme which mediates the translocation of the polypeptide-tRNA complex from acceptor site to donor site along the ribosome. Fragment A first catalyses the hydrolysis of NAD^+ to nicotinamide (NA) and adenosine diphosphoribose (ADPR). The second step is the adenosine diphosphate ribosylation of EF2, also catalyzed by fragment A (22). The net reaction, which is reversible, is

$$NAD^+ + EF\text{-}2 \rightleftharpoons ADPR\text{-}EF2 + NA + H^+$$

This fragment contains a single binding site for NAD^+ with a K_d of approximately 8 μM (23,62).

Although fragment A is the enzymatically active fraction, it is quite nontoxic to intact cells. Fragment B is essential for initial binding of the toxin to the cell surface, permitting entry (106). Fragment B includes a receptor called the P site because of its affinity for adenosine triphosphate and certain other ribonucleotide phosphates. Although probably not identical to the actual cell surface binding moiety of the toxin, these two receptors are allosterically related. Certain endogenous polyribonucleotides may occupy the P site, perhaps to protect the toxin, and most dissociate from it prior to attachment to a cell (23). Once attached to a susceptible cell, the toxin then probably undergoes a conformational change, causing fragment A to enter the cytosol (11, 70). During this process the disulfide bridges are reduced. Not all cell types are sensitive to diphtheria toxin (68,69). Resistant cell lines lack a surface receptor to mediate toxin entry. Resistance is a stable dominant genetic trait (35).

The predilection of diphtheria toxin for the peripheral nervous system is unexplained. Morphologically affected peripheral nerves show signs of segmental demyelination probably caused by alteration of Schwann cell function (2). The toxin has been shown to inhibit synthesis of myelin proteolipids and basic proteins (78).

B. Human Intoxication

Clinical manifestations are either due to the local effects of the infection, usually nasopharyngeal, or to systemic intoxication (42). Local manifestations may be mild and consist only of a low-grade fever and sore throat. The pharynx is erythematous and a membrane, white to yellow-gray, forms over the tonsils and soft palate. In severe cases the membrane extends into the upper respiratory tract and may obstruct the airway (36).

Systemic intoxication results primarily in myocardial and neurological complications. Myocardial damage may occur acutely or after the first week. In either case, progressive pump failure and conduction disturbances develop leading to systemic hypotension and pulmonary edema.

Neuropathy may occur in 10-75% of cases of oropharyngeal diphtheria (36,45,100). Early manifestations occur in the first 2 weeks and are caused by local spread of toxin to nerves supplying bulbar muscles (64). There is paralysis of soft palate and other oropharyngeal muscles, resulting in dysphagia and potential aspiration. Bulbar muscles may recover, only to be again affected in the later polyneuritic stage. In the fourth or fifth week there may be paralysis of accommodation leading to visual blurring. Multiple neuropathies may follow by the sixth or seventh weeks, again most often affecting the lower cranial nerves to the bulbar muscles. The extraocular muscles may also be weak at this time, but involvement of other cranial nerves such as the facial nerve is rare. Respiratory compromise can occur when the phrenic and accessory respiratory nerves are affected. Sensory disorders are more common than motor disorders, beginning with paresthesias but occasionally leading to a profound deficit of all sensory modalities. Motor involvement is usually mild and largely proximal. Profound limb weakness is uncommon. Loss of bowel and bladder control is occasionally reported. Recovery occurs over weeks to months, but is usually complete. Delayed polyneuropathy may follow cutaneous diphtheria as well.

Other reported neurological complications include rare instances of a diffuse encephalopathy and cases of embolic stroke from cardiac mural thrombi.

IV. SNAKE VENOM NEUROTOXINS

The snake venom toxins are a diverse group of peptide poisons derived from ancestral RNAase enzymes and possessing similarities to some basic phospholipases of pancreatic origin (60). They are composed of 18-108 amino acid residues, with up to 8 disulfide bridges (47). Some are subunit proteins associated with active or inactive "helper" peptide chains (85). These may be covalently bound or hydrophobically attracted to each other. These helper proteins function to depress enzymatic activity and enhance toxicity (104).

The snake venoms may be composed of one or several peptide toxins and a host of other noxious substances, including enzymes, lipids, vasoactive amines, kinins, prostaglandins, and hemolysins (30). The poisonous mixture expelled into the victim depends on the snake; crotalids possess the most complex and elapids the least complex venoms. Except for the Colombrae (or fangless snakes), all species envenomate through specialized teeth through which the venom is injected via a syringelike action (53). Venom production occurs in specialized "salivary" glands with finite storage capacity. Repeated envenomations may lead to depletion of the lethal fractions and lowered toxicity (85).

The lethal peptide neurotoxins to be considered here are classified (on the basis of their activity) into the presynaptic neurotoxins (Elapidae, Hydrophidae), the postsynaptic neurotoxins (Elapidae, Viperidae), and the sodium channel blockers (Crotalidae) (60).

A. Presynaptic Snake Venom Neurotoxins

Presynaptic snake venom neurotoxins block the release of acetylcholine and may be found in association with other peptide toxins (99). Notexin, taipoxin, textilon, and β-bungarotoxin are presynaptic neurotoxins of elapids. Crotoxin has been isolated from the venom of crotalids, and caudotoxin from the Viperidae (39). These families (Crotalidae, Viperidae, and Elapidae) are distributed worldwide and encompass all the fanged terrestrial snakes.

1. *Chemistry and Pathophysiology*

Presynaptic neurotoxins are basic proteins consisting of one or more chains of molecular weight 13,500 usually associated with or bound to "helper" peptides of molecular weight 7000-18,000 (taipoxin, β-bungarotoxin, and crotoxin) (28). The phospholipase activity of these poisons is low and inhibited by cholesterol (55). They are in general potent toxins with LD_{50} values to mice of 0.18-0.02 mg/kg.

Crotoxin from *Crotalus durissus terrificus* is a complex of crotoxin B (basic PLA_2) and the helper protein crotapotin. Crotoxin B is a single 140-amino acid peptide chain with 8 disulfide bridges. Crotapotin consists of 3 peptide chains of 14-40 amino acids each. When crotapotin is added to crotoxin, the lethality increases while phospholipase activity is suppressed.

Notexin and β-bungarotoxin both have 119 amino acids and 7 disulfide bridges. Forty-three of these residues and all seven disulfide bridges of notexin and bungarotoxin are identical. The phospholipase activity in these poisons is low.

The response of postsynaptic receptors at the myoneural junction to acetylcholine is not affected. The release of transmitter from the presynaptic terminal is gradually blocked over a period of 30-60 min. The latency period may be shortened by repetitive synaptic activity (99). Although there may be an early increase in miniature end-plate potentials, eventually the amplitude of miniature end-plate potentials and end-plate potentials declines as paralysis becomes complete. There may be a decrease in the number of synaptic vesicles, perhaps because of an inhibition of choline uptake systems (28,90). Tubocurarine does not affect the action of presynaptic neurotoxins, nor will black widow spider venom reverse the presynaptic block. Alkylation of histidine residues of the basic phospholipase A_2 moiety by p-bromophenacyl abolishes neurotoxicity and myotoxicity of the presynaptic neurotoxins (99).

2. *Human Disease*

Crotalid envenomations are discussed under the sodium channel blockers. Elapid envenomations are discussed under postsynaptic neurotoxins.

Viperid envenomation results in local pain, swelling, and ecchymosis. Systemic signs may include emesis, hypotension, and even cardiovascular collapse (55). Hemorrhage may be significant and contribute to the clinical course of the envenomation. Fatalities are usually associated with direct arterial or venous injection of the poison. Cautery, incision, and suction to limit toxin absorption are considered ineffective and dangerous (47). Antivenins for all common vipers are available and are administered (50-80 ml) for systemic intoxications.

Symptomatic treatment is advisable and may include analgesics (not morphine), clotting factors, correction of fluid and electrolyte imbalance, maintenance of blood pressure, and renal dialysis.

B. Postsynaptic Snake Venom Neurotoxins

The postsynaptic neurotoxins are characteristic of the venoms of the Elapidae (96). These snakes are widely distributed, and venomous genera include coral snakes of North and South America, African and Asian cobras, and the kraits of Asia and Africa (60).

1. *Chemistry and Pathophysiology*

Over 113 venom peptides affecting nicotinic acetylcholine receptors have been isolated and sequenced. They consist of short, 60- to 62-amino acid residue chains, or longer 71- to 78-residue peptides. Four to five disulfide bridges covalently bind the molecule into a characteristic shape. This is similar to erabutoxin (63), with a shallow saucer, a central loop, and a cleft into which the acetylcholine receptor fits. Amino acids at positions 29, 37, and 38 are required for neurotoxicity (27). Position 37 contains a strong guanidino group, and position 31 is an ionizable aspartic acid residue, presumably capable of binding both the cationic and anionic sites of the acetylcholine receptor (103). These neurotoxins cause a flaccid paralysis which is blocked by d-tubocurarine. Cholinergic neurons in the mammalian central nervous system bind elapid neurotoxins without functional impairment in neural transmission (19,61,73,76). The neurotoxin of the cobra (*Naja naja naja*) is pure peptide. The carbohydrate fractions reported by others (28) are probably due to contamination by sugars from the Sephadex or cellulose columns (18). In a later report (16) a 61-amino acid type I neurotoxin which blocked neuromuscular transmission but not muscle contraction was described. Neostigmine was not a useful antagonist, although the toxin caused acetylcholine receptor blockade at the level of the rat diaphram (17).

2. *Human Intoxication*

Elapid envenomation is rare in the United States, and fatalities are extraordinary (30,102). The total number of fatalities in Asia (India), Africa, and Australia may exceed 100,000 per year, or 25% of those bitten. The mortality rate varies with the species responsible. The bite of the black mamba (*Dendroaspis polylepis*) is invariably fatal. It injects 1000 mg of venom per bite, while the lethal dose for humans is only 120 mg. *Micruroides euryxanthes* bites are rarely fatal (47).

Elapid venoms cause little local pain, cardiotoxicity, myotoxicity, or hematotoxicity. Lymphatic involvement is rapid. Systemic symptoms include severe headaches, emesis, abdominal pain, diarrhea, and hypotension. Neurological symptoms appear within 1 hr (47,96). There is rapid ascending paralysis that may ultimately include dysphagia, external ophthalmoplegia, and fatal respiratory failure (85).

First aid is of little help in elapid envenomations, although immobilization of the limb, pressure bandages, and constriction bands to retard venom absorption are recommended by some authors (96). The only treatment available is antivenom. These are available for specific elapid species.

Morphine and alcohol are contraindicated. Diazepam, furosemide, and antibiotics are useful. Digitalis glycosides have been used for cardiac tachyarrhythmias. Tetanus prophylaxis is warranted. Hypotension and renal failure are treated symptomatically. Dehydration may occur, so fluid and electrolyte status must be monitored.

C. SNAKE VENOM NEUROTOXINS AFFECTING SODIUM CHANNELS

Unlike those of the Elapidae, Hydrophidae, and Viperidae, the venom produced by the Crotalidae (pit vipers) is a most complex mixture of poisons. Crotalid venoms include myotoxic, cardiotoxic, hematotoxic, and neurotoxic factors, as well as protease, peptidase, hyaluronidase, and a number of vasoactive substances. Of the neurotoxins, crotoxin is a presynaptic neurotoxin protein, and crotamine is a lower molecular weight (about 5000) peptide which acts on the sodium channel, mimicking tetrodotoxin. The crotalids inhabit the Far East (Asia) and the Americas (47,60).

1. *Chemistry and Pathophysiology*

Crotamines are strongly basic peptides found in the crotalid species *Crotalidae adamanteus*, *Crotalidae horridus horridus*, *Crotalidae viridus viridua*, and *Crotalidae viridus hellisi*. They are composed of 42-43 amino acids with 3 disulfide bridges. The N and C terminals are composed of basic amino acids, while the middle portion is strongly hydrophobic (79).

Crotamine causes an immediate muscular contraction and reduction of muscle membrane resting potential. These effects are inhibited by tetrodotoxin. Crotamine binding to a regulatory site on the membrane sodium channel has been postulated, based on inhibition studies. Procaine and crotamine are noncompetitive inhibitors. High potassium is competitive (60).

2. *Human Disease*

Crotalid venoms are not particularly poisonous (as compared to elapid venoms), but these snakes can efficiently inject large volumes of toxin, often directly into a deep artery or vein. In the United States 7000 envenomations with 15 deaths are reported yearly (30,85). Fatalities are more frequent in South America and the Far East.

Envenomations are painful. Local swelling ensues with tissue destruction and regional lymphatic involvement. This is followed by increasing flaccid weakness with widespread fasciculations. Paralysis may include eye muscles, bulbar muscles, and ultimately ventilatory function (3,48,49). There may also be signs of a coagulopathy manifested by petechial hemorrhages and ecchymoses (49). Active first aid treatment, including incision, suction, constriction bands, and hypothermia, has been advocated by some authors (30); others feel that such measures are ineffective and counterproductive (47). Specific antivenom, available in most areas, should be administered (4-30 or more vials intravenously). Symptomatic treatment may include analgesics (not morphine), respiratory support, correction of coagulation defects and electrolyte imbalances, prophylactic antibiotics, and antitetanus therapy.

Surgical excision of the bite and fasciotomy are controversial and best avoided. This is also true of corticosteroid therapy.

V. MARINE NEUROTOXINS

Toxins of marine animals (excluding sea snakes) represent a wide variety of protein venoms and nonprotein tissue poisons. Intoxication may occur from direct envenomation (coneshell, jellyfish) or via ingestion of a poison stored in host tissue (shellfish, "ciguatera").

Envenomations and intoxications are infrequent, rarely fatal, and usually attributable to human carelessness or inattention. The practice of ingesting raw (or cooked) "fugu" claims many lives each year. Bare-handed collection of coneshells produces painful (and occasionally lethal) envenomations.

Antidotes are made for only a few toxins (Chironex and Conus), and these are not widely available. The most common intoxicants (tetrodotoxin, saxitoxin, ciguatera) are not amenable to antibody production because of their nonprotein nature (12).

A. Saxitoxin: "Paralytic Shellfish Poisoning"

Saxitoxin (Stx) is a nonprotein nitrogenous neurotoxin produced by dinoflagellate protozoa, blue green algae, and scattered species of starfish, crabs, and mollusks. Poisonings were described in the writings of Homer. Periodic "blooms" of the protists gave rise to epidemics of paralytic shellfish poisonings of the seventeenth century and reportedly led to the naming of the Red Sea. These seasonal "red tides" of *Gonyaulax* species have led to restrictions on mussels in California and clams in Alaska from May to October. Many types of shellfish may act as host to the toxin (12,47). Saxitoxin is concentrated in the hepatopancreas and siphon of clams. Unlike other hosts, which are seasonally toxic, *Saxidomus giganteus* retains and stores the ingested Stx and may be toxic year round (7). Detoxification (to 20%) may be accomplished by boiling the suspected shellfish in water containing one tablespoon sodium bicarbonate per quart. This method is used in commercial canning operations, but has the disadvantage of destroying the taste as well as the toxicity of the shellfish (93).

1. *Chemistry and Pathophysiology*

Saxitoxin and related gonyautoxins are low molecular weight (372), highly polar, dibasic nonprotein nitrogenous neurotoxins structurally similar to tetrodotoxin.

Saxitoxin: molecular formula $C_{10}H_{27}N_7O_4$

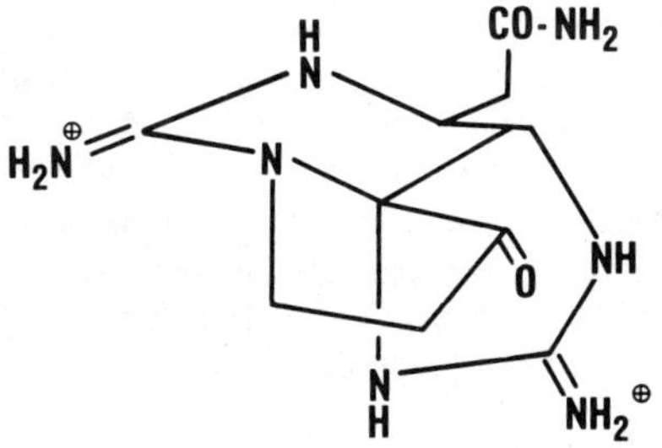

Gonyautoxins (types II and III) are epimers of hydroxy Stx. The guanidine moiety is critical for activity (12,14,40).

Saxitoxin is rapidly absorbed from the gastrointestinal tract and excreted in the urine. A single shellfish mussel may contain up to 50 human lethal doses of Stx (0.3-1.0 mg per person). The mortality rate is 1-10%, death occurring 1-2 hr after ingestion. In mice the Stx oral LD_{50} is 260 μg/kg, and 3 μg/kg if given intravenously (47,93).

Saxitoxin acts as an axonal poison and, like tetrodotoxin, retards the influx of sodium ions during membrane depolarization. It most probably binds to a subunit of the sodium channel (14). Inhibition at the presynaptic terminal and muscle membrane occurs simultaneously. The motor end plate becomes insensitive to acetylcholine. Cardiovascular effects are not noted at levels below 7 μg/kg and are caused by effects on both peripheral conduction, as well as brainstem regulatory centers. Other clinical signs such as alterations of memory suggest effects on the central nervous system (12,14,40).

2. *Human Disease*

Poisonings of epidemic proportion have been reported from Europe and the coasts of America coincident with red tides, such as in 1966 on the Canadian Atlantic coast and in 1972 and 1974 on the U.S. Atlantic coast. Epidemics of high mortality have been reported in England, Belgium, France, Norway, Germany, Japan, Alaska, the United States, Canada, New Zealand, and South Africa. In Western Europe in 1976, 120 cases of Stx poisoning were seen, but none were fatal. Seven deaths occurred in an outbreak in Papua, New Guinea.

Symptoms appear within 30 min of exposure to the toxin. They begin with distal paresthesias, rapidly spreading rostrally, causing diffuse sensory disturbance. Progressive neuromuscular blockade may lead to limb paralysis, involvement of bulbar muscles causing dysarthria and dysphagia, as well as facial weakness and paresis of eye movement causing diplopia. Ultimately respiratory embarrassment may ensue and, without mechanical support, lead to death in 1-24 hr. Involvement of the autonomic nervous system causes diarrhea, sweating, salivation, pupillary disturbances, and cardiac arrhythmias. Mental status usually is unaffected.

No antidote is available and therefore treatment is entirely symptomatic. Saxitoxin and other toxins are adsorbed to charcoal; thus gastrointestinal charcoal lavage is recommended (12). Neostigmine (a cholinesterase inhibitor) and various sympathomimetics have been used in conjunction with assisted ventilation. These may shorten recovery time. The use of digitalis preparations is not advisable.

In fatal cases autopsy findings have included only nonspecific changes attributable to cardiopulmonary collapse.

Many apparent shellfish "poisonings" are actually due to other mechanisms (34):

1. Enteric shellfish poisoning (bacterial) has a longer latency and is characterized by nausea, vomiting, diarrhea, and abdominal pains. Recovery is rapid, and the poisoning is without serious consequences.
2. Allergic shellfish poisoning (erythematous) has a short latency period and is characterized by the usual allergic symptoms and their potential dangers.

Detection of Stx may be accomplished by a standard mouse bioassay or by one of three chemical assays. The mouse bioassay is relatively sensitive, able to detect the total toxicity of as little as 40 μg/100 g of tissue (the maximum safe level for human consumption is considered to be 80 μg/100 g of shellfish). A chemical assay (100 times more sensitive than the mouse bioassay) is available which is based on the oxidation of Stx to purines which are then measured spectrophotometrically (8,13).

A multistep method has been proposed for detection of several biotoxins in fish and shellfish. This procedure is based on three consecutive bioassays: (a) assessment of the overall toxicity of shellfish extracts using the mouse death time assay, (b) in vitro study of the activity of the water-soluble fraction obtained from the extract on the guinea pig phrenodiaphragmatic preparations, and (c) in vivo assessment of the lipid-soluble fraction on the guinea pig phrenodiaphragmatic preparation with simultaneous recording of blood pressure and gastric motility (101).

B. Ciguatera Toxin Fishes

Ciguatera is caused by the ingestion of tropical reef fish containing ciguatoxin. Poisonings occur primarily in the Pacific Islands and on the east and west coasts of the United States. It has been reported in nonendemic areas via transported frozen fish (34). Over 400 species have become ciguatoxic after ingestion of reef algae or microbial flora. These fish range from small herbivores to large carnivores which prey upon them. Ciguatera is produced by an unknown marine algae or by reef microbial flora. It is taken up by the fish in the food and concentrated in the liver, testes, and intestines (47,93). Visceral organs contain 50-100 times the concentration of ciguatoxin as compared to flesh (40). Toxin storage may be very persistent. Snapper (*Liffanus bohar*) may be maintained on nontoxic diets for 30 months and still retain the ciguatoxin.

1. *Chemistry and Pathophysiology*

Ciguatoxin is thought to be a complex high molecular weight neutral lipid. Its presence in foods cannot be determined by chemical analysis. Animal bioassay is the only practical method of detection available. Other methods, including radioimmunoassay, are being developed (57). However, some common laboratory animals such as rats and chickens are resistant to ciguatoxin. The cat and mongoose are commonly employed for ciguatoxin bioassay. The crude extract of suspected contaminated flesh may not be submitted for bioassay. Extraction of ciguatoxin requires polar solvents, dilution followed by washing with nonpolar solvents and finally with diethyl ether. These extracts, when injected intraperitoneally into mice, yield LD_{50} values of about 0.08 mg/kg. No method of food storage or preparation will reduce the ciguatoxin levels in fish. Ciguatoxin causes increased sodium permeability, and therefore impairs generation of the action potential along nerve fibers. This partial depolarization is antagonized by tetrodotoxin and is thought to arise from a competitive inhibition of calcium ions. Some investigators have proposed an effect of ciguatoxin on the neuromuscular junction as well. Ciguatoxin was erroneously considered a cholinesterase inhibitor, but clinical use of pralidoxime led to actual patient deterioration and had to be abandoned. In laboratory animals ciguatoxin causes death by respiratory failure; however, cardiac arrest may also occur. Sublethal intoxication in hens has been reported to cause demyelination of spinal cord and sciatic nerve. Finally, ciguatoxin may have some central effect, as it is reported to be a hallucinogen (40,93).

2. *Human Disease*

Ciguatoxin is fatal to humans in less than 1% of cases. Large outbreaks involving hundreds of victims may occur. Symptoms of ciguatera intoxication include an "inversion of the senses," that is, distorted sensory phenomena such as "hot feels cold." Other symptoms which occur 2-20 hr postingestion may include distal paresthesias, headache, nausea, vertigo, ataxia, metallic taste, and abdominal cramps. In mild cases these symptoms persist for 24 hr and result in weakness and dehydration. Neurological signs, seen in more significant intoxications, may include objective sensory deficits, flaccid motor paralysis, including impaired ventilation, organic mental syndromes with hallucinations, and even seizures (41,47). Autonomic dysfunction may lead to cardiac arrhythmias and hypotension. Respiratory failure and death may occur within 36-48 hr. Recovery from intoxication is slow, with persistent dysesthesias, sensitivity to cold, and pruritus lasting up to 2 weeks. A second intoxication within 6 months is more severe than the first.

There is no antidote and therefore treatment is symptomatic. Ciguatera has been claimed to inhibit cholinesterase, but pralidoxime is not effective and may even cause rapid deterioration and death. Treatment includes removal of ingested fish by gastric lavage or other appropriate measures dependent upon the clinical status. Symptomatic treatment may also include atropine (useful in early poisonings) and calcium gluconate. Cardiovascular and respiratory distress are treated as needed (41). Diazepam, neostigmine, chlorpromazine, hydrocortisone, and pralidoxime have not been useful.

Teas brewed from *Daboisia myoporoides* have successfully treated even severe intoxications. The tepid extracts have been found to contain effective concentrations of tropa and pyridine alkaloids, including nicotine, nornicotine, scopolamine, and atropine (47).

C. Tetrodotoxin

Intoxications and envenomations with tetrodotoxin (Ttx) kill over 150 persons per year. The animals which produce Ttx are widely dissimilar species of marine and wetland creatures inhabiting most temperate and tropical seas, rivers, and inlands. Best known are the tetrodotoxin fish—the puffers, or Japanese "fugu." Ocean sunfish or head fish inhabit warm deep oceans. The California newt and 11 other species of tetrodotoxic *Salamandidae* are found in the United States, Japan, and Europe. Frogs (Atelopia) of Costa Rica and Central America exude Ttx and two additional toxins of similar potency and activity. Tetrodotoxin is also a venom of the blue-ringed octopus of Pacific coral reefs. Toxin content may be uniform or may be concentrated in certain organs (usually the gonads) (12,47).

1. Chemistry and Pathophysiology

Tetrodotoxin, related maculotoxins, and chiriquitoxin are low molecular weight (319.3), highly polar, nonprotein nitrogenous neurotoxins. The caged guanidinium moiety is common to many neurotoxins (e.g., saxitoxin and batrachotoxin). Tetrodotoxin is a highly polar zwitterion.

Tetrodotoxin

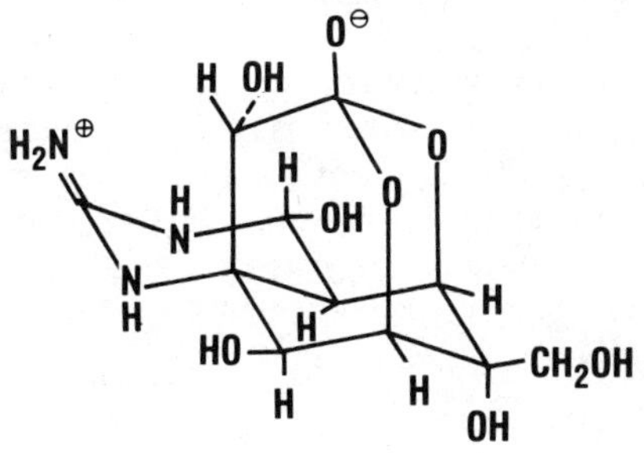

Chiriquitotoxin, found in the skin of the Costa Rican frog *Atelopus Chiriquiensis*, is 6-hydroxymethyl tetrodotoxin. It has an identical toxic and pharmacological profile to that of Ttx (12,40,47,93). An intact molecular periphery is required for binding to the sodium channel of the nerve membrane (14). Purified Ttx exhibits a steep dose-response curve. Its LD_{10} in mice is 8 μg/kg, and its LD_{100} is 12 μg/kg. Tetrodotoxin reversibly blocks the sodium channels by binding to receptors in the outer surface of the neural membrane. It is effective (100%) at concentrations as low as 10^{-8} M (14,59). Tetrodotoxin is an emetic, affecting the chemoreceptor trigger zone of the fourth ventricle. Severe hypotension occurs in humans following doses of 0.5 μg/kg (35 μg per person) due to paralysis of vascular smooth muscle and a consequent drop in total peripheral resistance. At higher doses, blockade of vasomotor nerves occurs, exacerbating the hypotensive effect. The cause of respiratory failure is unclear and may be due to a central effect or to phrenic nerve involvement (12,47).

Two major causes of tetrodotoxin resistance in animals are (a) calcium-based action potential spikes as seen in clams and other mollusks and (b) chemical alteration of the receptor protein of the exterior side of the sodium channel, causing a rejection of Ttx binding. This accounts for the resistance of tetrodotoxic fish (40).

2. *Human Disease*

Except for rare direct envenomation, consumption of raw fugu or puffer fish accounts for the vast majority of tetrodotoxin poisonings. Mortality approaches 60% of those severely intoxicated. The ovaries, testes, liver, and intestines are most toxic, while the skin and flesh have lesser concentrations of Ttx. Greatest risk occurs prior to the spawning season in May to July. In Japan, fugu is prepared by strictly licensed cooks who prepare the musculature cut into small pieces, soaked in water and kneaded for several hours before cooking. A trace of toxin, however, makes the fish more appetizing (12). Early symptoms may appear 10-45 min after ingestion and include nausea, emesis, diarrhea, epigastric pain, and perioral paresthesias. The onset of motor weakness occurs at 1 hr and may include not only limb weakness, but also paralysis of eye movement and ventilatory failure which appear in 6-24 hr. Pupillary reactivity may be lost. Circulatory failure may follow. Consciousness is usually retained until death. In successfully treated patients recovery may be complete within 24 hr. Envenomation with Ttx produces identical symptoms to those of Stx ingestion, but the onset occurs in minutes rather than hours. As no antidote for Ttx exists, treatment is purely symptomatic and supportive. Artificial respiration is often required.

ACKNOWLEDGMENT

We wish to acknowledge the research and editorial assistance of Valerie A. Butler.

REFERENCES

1. Adams, E. B. 1968. The prognosis and prevention of tetanus. *S. Afr. Med. J. 42*:739-743.
2. Allt, G., Cavanagh, J. B. 1969. Ultrastructural changes in the region of the node of Ranvier in the rat caused by diphtheria toxin. *Brain* 459-468.
3. Aragon, F., Gubensek, J. 1981. *Bothrops asper* venom from the Atlantic and Pacific zones of Costa Rica. *Toxicon 19*:797-805.
4. Arnon, S. S., Midura, T. F., Damus, D. 1979. Honey and other environmental risk factors for infant botulism. *J. Pediatr. 94*: 331-336.
5. Arnon, S. S., Midura, R. F., Clay, S. A. 1977. Infant botulism: Epidemiological clinical, and laboratory aspects. *J. Am. Med. Assoc. 237*:1946-1951.
6. Arnon, S. S. 1980. Infant botulism. *Annu. Rev. Med. 31*:541-560.
7. Baden, D. G., Mende, T. J. 1982. Toxicity of two toxins from the Florida red tide marine dinoflagellate, *Ptychodiscus brevis*. *Toxicon 20*:457-461.
8. Bates, H. A., Kostriken, R., Rapoport, H. 1978. A chemical assay for saxitoxin: improvements and modification. *J. Agric. Food Chem. 26*:252.
9. Berry, L. J. 1977. Bacterial toxins. *CRC Crit. Rev. Toxicol. 5*:273-279.
10. Black, R. E., Gunn, R. A. 1980. Hypersensitivity reactions associated with botulinal antitoxin. *Am. J. Med. 69*:567-570.
11. Boquet, P., Pappenheimer, A. M. 1976. Interaction of diphtheria toxin with mammalian cell membranes. *J. Biol. Chem. 251*:5770-5778.
12. Bower, D. J., Hart, R. J., Matthews, P. A. 1981. Non-protein neurotoxins. *Clin. Toxicol. 18*:813-863.
13. Buckley, L. J., Oshima, Y., Shimizu, Y. 1978. Construction of a paralytic shellfish toxin: Analysis and its application. *Anal. Biochem. 85*:157.
14. Catterall, W. A. 1980. Neurotoxins that act on voltage-sensitive sodium channels in excitable membranes. *Annu. Rev. Pharmacol. Toxicol. 20*:15-43.
15. Center for Disease Control. *Botulism in the United States, 1899-1977*, Handbook for Epidemiologists, Clinicians, and Laboratory Workers, issued May 1979.

16. Charles, A. K., Deshpande, S. S. 1981. Peripheral versus central action of a toxin from Indian cobra (*Naja naja naja*) venom. *Toxicon 19*:305-317.
17. Charles, A. K., Gangal, S. V., Deshpande, S. S., Joshi, A. P. 1982. Effects on muscle of a toxin from Indian cobra (*Naja naja naja*) venom. *Toxicon 20*:1019-1035.
18. Charles, A. K., Joshi, A. P. 1981. Source of carbohydrate in a toxic protein from Indian cobra (*Naja naja naja*) venom. *Toxicon 19*:431-436.
19. Chen, Y. H., Tai, J. C., Huang, W. J., Lai, M. Z., Hung, M. C., Lai, M. D., Yang, Y. T. 1982. Role of aromatic residues in the structure-function relationship of β-bungarotoxin. *Biochemistry 21*:2592-2600.
20. Cherington, M., Ginsberg, S. 1971. Type B botulism: Neurophysiologic studies. *Neurology 21*:43-46.
21. Cherington, M., Ryan, D. W. 1970. Treatment of botulism and guanidine: Early neurophysiologic studies. *N. Engl. J. Med. 282*:195-197.
22. Collier, R. J. 1967. Effect of diphtheria toxin on protein synthesis: Inactivation of one of the transfer factors. *J. Mol. Biol. 25*:83-98.
23. Collier, R. J. 1982. From Pap to ApUp. *Cell. Immunol. 66*:17-23.
24. Collier, R. J., Pappenheimer, A. M. 1964. Studies on the mode of action of diphtheria toxin II: Effect of toxin on amino acid incorporation in cell-free systems. *J. Exp. Med. 120*:1019-1039.
25. Collingridge, G. L., Collins, G. G. S., James, T. A. 1980. Effect of tetanus toxin on transmitter release from the substantia nigra and striatum *in vitro*. *J. Neurochem. 34*:540-547.
26. Collingridge, G. L., Thompson, P. A., Davies, J. 1981. *In vitro* effect of tetanus toxin on GABA release from rat hippocampal slices. *J. Neurochem. 37*:1039-1041.
27. Condrea, E., Fletcher, J. E., Rapuano, B. E. 1981. Dissociation of enzymic activity from lethality and pharmacological properties by carbamylation of lysine in *Naja nigricollis* and *Naja naja atra* snake venom phospholipases A_2. *Toxicon 19*:705-720.
28. Coulter, A. R., Broad, A. T., Sutherland, S. K. 1979. Isolation and properties of a high molecular weight neurotoxin from the eastern brown snake (*Pseudonaja textilis*), in *Neurotoxins: Fundamental and Clinical Advances* (Chubb, I. W., ed.), University of Adelaide Press, Adelaide, Australia.
29. Craven, C. J., Dawson, D. J. 1973. The chain composition of tetanus toxin. *Biochem. Biophys. Acta 317*:277-285.
30. Crembor, P., Oehme, F. W. 1981. A literature review of snakes and snake bite therapy. *Vet. Hum. Toxicol. 23*:97-100.
31. Curtis, D. R., DeGroat, W. C. 1968. Tetanus toxin and spinal inhibition. *Brain Res. 10*:208-212.
32. Curtis, D. R., Felix, D., Game, C. J. A. 1973. Tetanus toxin and the synaptic release of GABA. *Brain Res. 51*:358-362.

33. Davies, J., Tongroach, P. 1979. Tetanus toxin and synaptic inhibition in the substantia nigra and striatum of the rat. *J. Physiol. 290*:23-36.
34. Dembert, M. L., Pearn, J. H. 1982. Physicians: Know the ciguatera poisoning symptoms. *Am. J. Public Health 72*:1298.
35. Dendy, P. R., Harris, H. 1973. Sensitivity to diphtheria toxin as a species-specific marker in hybrid cells. *J. Cell Sci. 12*:831-837.
36. Dobie, R. A., Tobey, D. N. 1979. Clinical features of diphtheria in the respiratory tract. *J. Am. Med. Assoc. 242*:2197-2201.
37. Dolman, C. E. 1974. Human botulism in Canada (1919-1973). *Can. Med. Assoc. J. 110*:191-200.
38. Donadio, J. A., Gangarosa, E. J., Faich, G. A. 1971. Diagnosis and treatment of botulism. *J. Infect. Dis. 124*:108-112.
39. Dowdal, M. J., Fohlman, J. P., Watts, A. 1979. Presynaptic action of snake venom neurotoxins on cholinergic systems. *Adv. Cytopharmacol. 3*:63-76.
40. Endean, R. 1979. Neurotoxins occurring in marine animals from Australian waters, in *Neurotoxins: Fundamental and Clinical Advances* (Chubb, I. W., ed.), Adelaide University Press, Adelaid, Australia, pp. 57-73.
41. Engleberg, N. C., Morris, J. G., Lewis, J., McMillan, J. P., Pollard, R. A., Blake, P. A. 1983. Ciguatera fish poisoning: A major common-source outbreak in the U.S. Virgin Islands. *Ann. Intern. Med. 98*:336-337.
42. Fischer, G. W. 1981. Diphtheria, in *Infections in Children*, Vol. 38 (Wedgwood, R. J., Davis, S. D., Ray, C. G., eds.), Harper and Row, Philadelphia, pp. 652-662.
43. Garnier, M. J. 1975. Tetanus in patients three years of age and up: A personal series of 230 consecutive patients. *Am. J. Surg. 129*:459-463.
44. Goode, G. B., Shearn, D. L. 1982. Botulism: A case with associated sensory abnormalities. *Arch. Neurol. 39*:55.
45. Gupta, O. K., Sakensa, P. N., Gupta, N. N. 1975. A clinical study of 856 patients with diphtheria. *Indian J. Pediatr. 40*:93-101.
46. Haberman, E. 1978. Tetanus, in *Handbook of Clinical Neurology*, Vol. 33 (Vinken, P. J., Bruyn, G. W., eds.), North Holland, pp. 491-547.
47. Habermehl, G. G. 1981. *Venomous Animals and Their Toxins*, Springer-Verlag, New York.
48. Hardy, D. L. 1982. Envenomation by the Mexican lance-headed rattlesnake *Crotalus polystictus*—A case report. *Toxicon 20*:1089-1091.
49. Hardy, D. L., Jeter, M., Corrigan, J. J. 1982. Envenomation by the northern blacktail rattlesnake (*Crotalus molossus molossus*): Report of two cases and the *in vitro* effects of the venom on

fibrinolysis and platelet aggregation. *Toxicon 20*:487-493.
50. Harris, A. J., Miledi, R. 1971. The effect of type D botulinum toxin on frog neuromuscular junctions. *J. Physiol. 217*:497-515.
51. Hirokawa, N., Heuser, J. E. 1981. Structural evidence that botulinum toxin blocks neuromuscular transmission by impairing the calcium influx that normally accompanies nerve depolarization. *J. Cell Biol. 88*:160-171.
52. Hirokawa, N., Kitamura, M. 1979. Binding of *Clostridium botulinum* neurotoxin to the presynaptic membrane in the central nervous system. *J. Cell Biol. 81*:43-49.
53. Huang, T. T., Lewis, S. R., Lucas, B. S. 1978. Venomous snakes, in *Dangerous Plants, Snakes, Arthropods and Marine Life: Toxicity and Treatment* (Ellis, M. D., ed.), Drug Intelligence, Hamilton, Ill., pp. 123-142.
54. Hughes, J. M., Blumenthal, J. R., Merson, M. H. 1981. Clinical features of types A and B foodborne botulism. *Ann. Intern. Med. 95*:442-445.
55. Kelly, R. B., Wedel, R. J., Strong, P. N. 1979. Phospholipase-dependent and phospholipase-independent inhibition of transmitter release by β-bungarotoxin. *Adv. Cytopharmacol. 3*:77-85.
56. Kerr, J. H., Travis, K. W., O'Rourke, R. A. 1974. Autonomic complications in a case of severe tetanus. *Am. J. Med.* 57:303-310.
57. Kimura, L. H., Hokama, Y., Abad, M. A., Yama, H. M., Miyahara, Y. T. 1982. Comparison of three different assays for the assessment of ciguatoxin in fish tissues: Radioimmunoassay, mouse bioassay and *in vitro* guinea pig atrium assay. *Toxicon 20*:907-912.
58. LaForce, F. M., Young, L. S., Beenett, J. V. 1969. Tetanus in the United States: Epidemiologic and clinical features. *N. Engl. J. Med. 280*:569-574.
59. Lazdunzki, M., Balerna, M., Chicheportiche, R., et al. 1979. Marine neurotoxins to study the voltage-dependent sodium channel in excitable membranes, in *Neurotoxins: Fundamental and Clinical Advances* (Chubb, I. W., ed.), Adelaide University Press, Adelaide, Australia, pp. 111-119.
60. Lee, C. Y. C. 1979. Recent advances in chemistry and pharmacology of snake toxins. *Adv. Cytopharmacol. 3*:1-16.
61. Lin-Shiau, S. Y., Chen, C. C. 1982. Effects of β-bungarotoxin and phospholipase A_1 from *Naja naja atra* snake venom on ATPase activities of synaptic membranes from rat cerebral cortex. *Toxicon 20*:409-417.
62. Lory, S., Carroll, S. F., Bernard, P. D. 1980. Ligand interactions of diphtheria toxin I: Binding and hydrolysis of NAD. *J. Biol. Chem. 255*:12011-12020.
63. Low, B. W. 1979. Three dimensional structure of erabutoxin B_1 prototype structure of the snake venom post synaptic neurotoxins: Consideration of structure and functions; description of the

reactive site. *Adv. Cytopharmacol. 3*:141-147.

64. Lupton, M. D., Klawans, H. L. 1978. Neurological complications of diphtheria, in *Handbook of Clinical Neurology* (Vinken, P. J., Bruyn, G. W., eds.), North Holland, Amsterdam, pp. 479-498.
65. Matsuda, M., Yoneda, M. 1975. Isolation and purification of two antigenically active, "complementary" polypeptide fragments of tetanus neurotoxins. *Infect. Immun. 12*:1147-1153.
66. Mellanby, J., Green, J. 1981. How does tetanus toxin act? *Neuroscience 6*:281-300.
67. Moberg, L. J., Sugiyama, H. 1979. Microbial ecological basis of infant botulism as studied with germfree mice. *Infect. Immun. 25*: 653-657.
68. Moehring, T. J., Moehring, J. M. 1972. Response of cultured mammalian cells to diphtheria toxin IV: Isolation of KD cells resistant to diphtheria toxin. *Infect. Immun. 6*:487-492.
69. Moehring, J. M., Moehring, T. J. 1976. Comparison of diphtheria intoxication in human and non-human cell lines and their resistant variants. *Infect. Immun. 13*:221-228.
70. Moehring, T. J., Moehring, J. M. 1976. Interaction of diphtheria toxin and its active subunit fragment A, with toxin-sensitivity and tonxin-resistant cells. *Infect. Immun. 13*:1426-1432.
71. Murphy, S. G., Plummer, T. H., Muller, K. D. 1968. Physical and chemical characterization of tetanus toxin. *Fed. Proc. 27*: 268.
72. Osborne, R. H., Bradford, H. F. 1973. Tetanus toxin inhibits amino acid release from nerve endings *in vitro*. *Nature 244*: 157-158.
73. Oswald, R. E., Freeman, J. H. 1981. Analysis of β-bungarotoxin binding in the goldfish central nervous system. *J. Neurochem. 37*:1586-1593.
74. Pappenheimer, A. M. 1977. Diphtheria toxin. *Annu. Rev. Biochem. 46*:69-94.
75. Pappenheimer, A. M., Gill, D. M. 1973. Diphtheria: Recent studies have clarified the molecular mechanisms involved in its pathogenesis. *Science 182*:353-358.
76. Patrick, J. 1979. Snake neurotoxins and nicotinic acetycholine receptors on nerve, in *Neurotoxins: Fundamental and Clinical Advances* (Chubb, I. W., ed.), University of Adelaide Press, Adelaide, Australia, pp. 27-34.
77. Pearce, B. R., Gard, A. L., Dutton, G. R. 1983. Tetanus toxin inhibition of K^+-stimulated (^3H) GABA release from developing cell cultures of the rat cerebellum. *J. Neurochem. 40*:887-880.
78. Pleasure, D. E., Feldmann, B., Prockop, D. J. 1973. Diphtheria toxin inhibits the synthesis of myelin proteolipid and basic proteins by peripheral nerve *in vitro*. *J. Neurochem. 20*:81-90.

79. Powlick, G. D., Geren, C. R. 1981. Fractionation and partial characterization of toxic components of *Agkistrodon delete piscavorus* (eastern cottonmouth moccasin) venom. *Toxicon 19*:867-874.
80. Price, D. L., Griffin, J., Peck, K. 1975. Tetanus toxin: Direct evidence for retrograde intraaxonal transport. *Science 188*:945-947.
81. Price, D. L., Griffin, J. W., Peck, K. 1977. Tetanus toxin: Evidence for binding at presynaptic nerve endings. *Brain Res. 121*:379-384.
82. Prys-Roberts, C., Corbett, J. L., Kerr, J. H. 1969. Treatment of sympathetic overactivity in tetanus. *Lancet 1*:542-546.
83. Rie, M. A., Wilson, R. S. 1978. Morphine therapy controls autonomic hyperactivity in tetanus. *Ann. Intern. Med. 88*:653-654.
84. Rogers, T. B., Snyder, S. H. 1981. High affinity binding of tetanus toxin to mammalian brain membranes. *J. Biol. Chem. 256*:2402-2407.
85. Russell, F. E. 1980. *Snake Venom Poisoning*, Lippincott, Philadelphia.
86. Scaer, R. C., Tooker, J., Cherington, M. 1969. Effect of quanidine on the neuromuscular block of botulism. *Neurology 19*:1107-1110.
87. Schwab, M. E., Suda, K., Thoenen, H. 1979. Selective retrograde transsynaptic transfer of a protein, tetanus toxin, subsequent to its retrograde axonal transport. *J. Cell Biol. 82*:798-810.
88. Simpson, L. L. 1971. Ionic requirements for the neuromuscular blocking action of botulinum toxin: Implications with regard to synaptic transmission. *Neuropharmacology 10*:673-684.
89. Simpson, L. L. 1977. Pharmacological studies on the subcellular site of action of botulinum toxin type A. *J. Pharmacol. Exp. Ther. 206*:661-669.
90. Simpson, L. L., Rapport, M. M. 1971. The binding of botulinum toxin to membrane lipids: Phospholipids and proteolipid. *J. Neurochem. 18*:1761-1767.
91. Simpson, L. L., Rapport, M. M. 1971. The binding of botulinum toxin to membrane lipids: Sphingolipids, steroids and fatty acids. *J. Neurochem. 18*:1751-1759.
92. Smith, L. D. S. 1977. *Botulism: The Organism, Its Toxins, the Disease*, Charles C. Thomas, Springfield, Ill.
93. Southcott, R. V. 1979. Marine envenomation and intoxication in man, in *Neurotoxin, Fundamental and Clinical Advances* (Chubb, I. W., ed.), Adelaide University Press, Adelaide, Australia, pp. 75-84.
94. Strauss, N., Hendee, E. D. 1959. The effect of diphtheria toxin on the metabolism of HeLa cells. *J. Exp. Med. 109*:145-163.

95. Sugiyama, H. 1980. *Clostridium botulinum* neurotoxin. *Microbiol. Rev.* *44*:419-448.
96. Sutherland, S. K., Coulter, A. R., Broad, A. T. 1979. Clinical and experimental aspects of snake bite in Australia, in *Neurotoxins: Fundamental and Clinical Advances* (Chubb, I. W., ed.), University of Adelaide Press, Adelaide, Australia, pp. 9-18.
97. Tempero, K. 1973. The use of diazepam in the treatment of tetanus. *Am. J. Med. Sci.* *226*:4-12.
98. Terranova, W., Palumbo, J. N., Breman, J. G. 1979. Ocular findings in botulism type B. *J. Am. Med. Assoc.* *241*:475-477.
99. Thesleff, S. 1979. Reptile toxins and neurotransmitter release, in *Neurotoxins, Fundamental and Clinical Advances* (Chubb, I. W., ed.), Adelaide University Press, Adelaide, Australia, pp. 19-25.
100. Thomas, K., Chungath, J. P., Philip, E. 1972. Postdiphtheritic paralysis. *Indian Pediatr.* *9*:561-565.
101. Tonini, M., d'Angelo, L., Manzo, L., Crema, A. 1978. Three-step bioassay for detection of marine biotoxins in seafood, in *Chemical Toxicology of Food* (Galli, C. L., Paoletti, R., Vettorazzi, G., eds.), Elsevier, Amsterdam, pp. 375-380.
102. Trestrail, J. H. 1982. The underground zoo: The problem of exotic venomous snakes in private possession in the United States. *Vet. Hum. Toxicol.* 144-145.
103. Tsetlin, V. I., Karlsson, E., Utkin, Y. 1982. Interacting surfaces of neurotoxins and acetylcholine receptor. *Toxicon* *20*:83-93.
104. Tu, A. T. 1977. *Venoms: Chemistry and Molecular Biology*, Wiley, New York.
105. Turner, H. D., Brett, E. M., Gilbert, R. J., Ghosh, A. C., Liebeschuetz, H. J. 1978. Infant botulism in England. *Lancet* *1*:1277-1278.
106. Uchida, T., Pappenheimer, A. M., Greany, R. 1973. Diphtheria toxin and related proteins I: Isolation and properties of mutant proteins serologically related to diphtheria toxin. *J. Biol. Chem.* *248*:3838-3844.

Part IV
ASSESSMENT OF NEUROTOXICITY

25

Neurotoxin-Induced Animal Models of Human Diseases

Edith G. McGeer and Patrick L. McGeer
Kinsmen Laboratory of Neurological Research, University of British Columbia, Vancouver, British Columbia, Canada

I. INTRODUCTION

Glutamate, aspartate, and a number of structurally related amino acids possess the capacity not only to excite neurons in the central nervous system, but also to bring about their destruction if administered at a sufficiently high concentration. For this reason these amino acids have been called excitotoxins, a term coined by Olney and his colleagues in 1974 (see Ref. 74). The neurotoxic effect, unnoticed until recent years, is still a poorly understood phenomenon. Nevertheless, it is being exploited by neuroscientists as a means of providing much new information about the operation of neuronal systems. This review will be concerned with one aspect of such excitotoxic research: the use of such agents to produce animal models of some human degenerative disorders.

The amino acids of present interest in this context are shown in Figure 1. These compounds are found in a variety of plants or animals. Ibotenic acid is obtained from the mushroom *Amanita strobiliformis*. Kainic acid (KA) comes from the Japanese seaweed *Digenea simplex*.

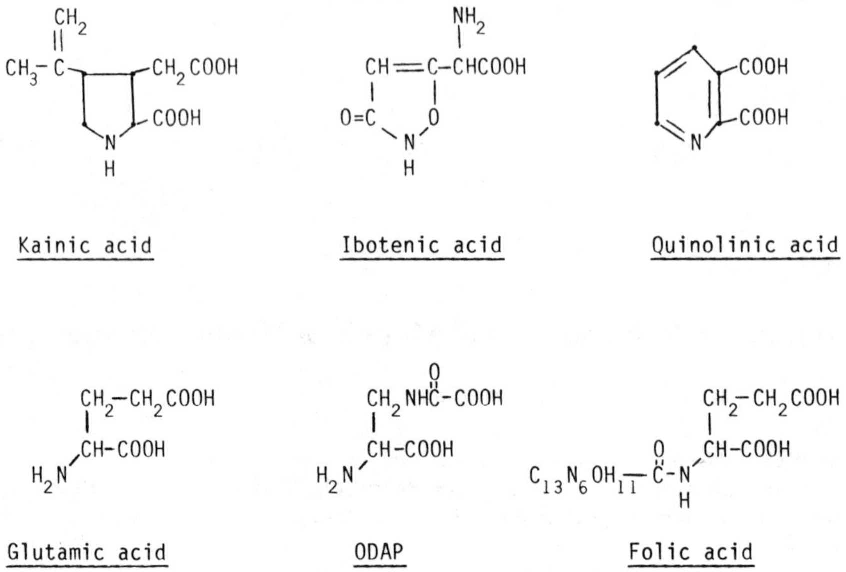

Figure 1 Structures of some excitatory and neurotoxic amino acids

β-N-Oxalyl-L-α,β-diaminopropionic acid (ODAP) is found in chickling pea and may be the toxin responsible for the crippling neurodegenerative illness neurolathyrism (see Refs. 77 and 80). Quinolinic (96) and folic acids both occur widely in animals as well as plants, since quinolinic acid is a metabolite of the essential amino acid tryptophan and folic acid is a required vitamin. Kainic acid is the most active of these derivatives and has been the most widely studied, but differences in the mechanism of actions and effects achieved by the other amino acids suggest that these will also come into widespread use.

Glutamate and aspartate are now widely recognized as important neurotransmitters in the central nervous system. As originally articulated by Olney and his colleagues (see Ref. 74), the excitotoxins activate excitatory receptors on the dendrites and soma of neurons and, when present in high concentrations in the vicinity of these receptors, produce a state of pathological depolarization in which the neuronal plasma membrane permeability is increased for extended periods of time. This causes energy-dependent homeostatic mechanisms to draw heavily on the cell's energy stores in an effort to restore the ionic balance. If these energy sources become irreversibly depleted, or if the ionic exchange reaches a state where cell membrane pumps can no longer function, then cell death can be expected to occur (see Ref. 27). The excitatory receptors exist at much lower concentrations, if at all, on axons and nerve endings. Excitotoxic lesions therefore spare these processes unless the cell bodies and dendrites are also within the injected area. This selective action is critical to the value of these excitotoxins as neurobiological tools.

Later work has made it evident that a number of mechanisms are involved (62,63). Kainic acid seems to act on a receptor which is distinct from, but modulatory on, the glutamate postsynaptic receptor, and its local neurotoxicity depends upon a cooperative reaction with neurotransmitter glutamate and thus on an intact glutamatergic afferent to the injected area. Kainic acid, however, also often causes damage in areas remote from the injection site, and such remote damage seems to depend upon the severe epileptiform activity which KA often induces, depending upon the site, rapidity, and amount injected (63). Ibotenic and quinolinic acids produce local neuronal lesions with the same type of selectivity as found with KA, but have little or no tendency to produce epileptiform activity and remote damage (39,94-96). Folic acid, on the other hand, seems to cause remote but little local damage (65, 76). Unlike KA, ibotenic acid seems to exert its local neurotoxicity by direct action on postsynaptic glutamate receptors, and the effect persists even after lesioning of glutamatergic afferents. The sites of action of quinolinic and folic acids are not yet established. Quinolinic acid has been said to have a more selective excitatory action than glutamic acid (81).

II. MODELS OF HUMAN DISEASE

A. Huntington's Disease and Intrastriatal Injections of Neurotoxins

The extensive neuronal losses seen in the basal ganglia following intrastriatal injections of KA, ibotenate, or quinolinate are very similar to those characteristic of Huntington's disease (HD). Thus it has been proposed that such injections can be used to produce animal models of this condition (23,24,57,75).

Huntington's disease is a hereditary condition, transmitted as an autosomal dominant disease. It is worldwide in distribution, having first been recognized in Scandinavia but later described in more detail by a New York physician after whom the disease is named. It has an incidence of about 5 per 100,000 population in North America, which makes it a relatively rare disorder. It has, nevertheless, attracted considerable clinical and basic research attention because of its severity and unusual time of onset. Typically, it manifests itself in the late 30s or early 40s, which is just beyond the normal child-bearing age. The time of onset accounts for its genetic preservation despite its devastating physical and social characteristics.

The disease is characterized by jerky, uncontrolled movements and severe mental deterioration which progress relentlessly. Death usually occurs within 15 years of the onset of symptoms. There is no known preventive measure, or any effective treatment. Curiously, the mental symptoms can mimic schizophrenia in the early stages, although in the later phases the severe dementia is unmistakable.

The most striking pathological changes in HD occur in the basal ganglia, particularly the caudate, where there is marked atrophy and severe loss of neurons. The putamen is also heavily involved, and damage to the globus pallidus is usually prominent. Negligible histological changes are observed in the substantia nigra, although severe biochemical losses are noted. Loss of cells in the cerebral cortex, particularly the frontal cortex, is also characteristic. The histological changes seen in animals following intrastriatal injections of KA have been repeatedly found to resemble those seen in HD (23), particularly in chronic preparations (51,59,93). If care is taken with the injection, there is little or no remote damage in the acute state (58).

A summary of the biochemical changes reported in HD and in rats given intrastriatal injections of KA is shown in Table 1. As can be seen from Table 1, there are marked decreases in the neostriatum in the levels of neuronal markers for acetylcholine, γ-aminobutyric acid (GABA), enkephalin, substance P, and angiotensin-converting enzyme. These indices are all associated with neurons originating in the basal ganglia. On the other hand, indices of dopamine, noradrenaline, and serotonin, where the cell bodies exist in the substantia nigra, locus ceruleus, and raphe, respectively, are unchanged or elevated. This is consistent with the lack of excitatory receptors on axons and nerve endings, and therefore the relative sparing of these neuronal processes when contacted by KA. Many different types of receptors are lost in the striatum, since these presumably are located on the neurons that degenerate. In both conditions the myelinated axons of the internal capsule are preserved.

In the substantia nigra, there is a loss of neuronal markers for GABA and substance P neurons, while dopamine and acetylcholine markers are normal or increased. The losses result from destruction of GABA and substance P neuronal cell bodies in the striatum that give rise to projections to the substantia nigra. In keeping with such a picture is the fact that stimulation of the KA-lesioned caudates in rats causes a response in only about 58% of nigral neurons, as compared to 92% in intact animals; a pure inhibitory response was decreased from 59% to about 30% (71).

Quantitative data differ somewhat from laboratory to laboratory, in the levels seen in both HD and KA lesions. Qualitatively, however, there would appear to be marked biochemical similarities between HD and the KA model. It must be pointed out, however, that some chemical differences between the human disease and the KA "model" appear in studies of the striatal levels of three peptide neurotransmitters: thyrotropin-releasing hormone, somatostatin, and cholecytokinin. All

Table 1 Biochemical Similarities Between Huntington's Disease and the Kainic Acid "Model"

Biochemical changes	HD	KA model	References
In neostriatum			
Presynaptic GABA indices[a]	Markedly decreased		23[b]
γ-Hydroxybutyrate levels	Increased		1,2
GABA transaminase activity	Normal	Decreased	23[b]
Presynaptic acetylcholine indices[a]	Markedly decreased		23[b]
Presynaptic dopamine indices[a]	Normal or elevated		23[b]
Presynaptic serotonin indices[a]	Normal or elevated		23[b],72,99
Presynaptic noradrenaline indices[a]	Normal or elevated		23[b]
Angiotensin-converting enzyme	Decreased		23[b]
Enkephalin levels	Decreased		23[b],93
Thyrotropin-releasing hormone levels	Increased	Decreased	100,101
Somatostatin levels	Increased	Decreased	15
Cholecytokinin levels	Normal	Decreased	28
Ganglioside levels	Decreased		41
DNA levels	Increased		41
Aspartate transaminase	Decreased		105
Ornithine δ-transaminase	Decreased		105,106
Binding sites ("receptors") for			
Serotonin	Decreased		23[b],32
Dopamine	Decreased		23[b],31
Acetylcholine (muscarinic)	Decreased		8,14,31
Benzodiazepines	Decreased[c]		68,98
Kainic acid	Decreased		5,40,55,103
GABA	Decreased[c]		23[b],109
Noradrenaline (β-adrenergic)	Normal		29,110
Cholecytokinin	Decreased		28
Substantia nigra			
Presynaptic GABA indices[a]	Decreased		23[b]
Substance P levels	Decreased		23[b]
Presynaptic dopamine indices[a]	Normal		23[b]
GABA binding sites	Increased		25,104
Cholecytokinin levels	Decreased		28

[a]Including levels, release, turnover, activity of synthetic enzymes, and/or uptake of transmitter or (in the case of cholinergic systems) the precursor.
[b]And references therein.
[c]At more than 1 month following KA injections; no decrease in acute preparations.

of these have been found decreased in the KA "model" and increased or normal in HD; the KA model is, of course, generally examined in an "acute" phase shortly after the injections of KA, while HD is a chronic condition. The few studies on the chronic model suggest secondary changes may occur (51,59,93).

Pharmacological and behavioral studies have also indicated marked similarities. Following a bilateral injection of KA into the striatum, rats do not display the bizarre choreiform movements seen with HD patients. However, they do show enhanced activity at night, but not during the day (30), abnormal locomotion (44), learning problems (26, 88), and body weight changes (87) reminiscent of those seen in HD. Although some rats given such striatal injections of KA may also show hippocampal damage, others do not, and the behavioral impairments thought comparable to HD are attributable to the loss of striatal neurons (82).

Such rats also show a markedly enhanced response to amphetamine, scopolamine, and pilocarpine, but an attenuated cataleptic response to haloperidol (30,90), sedative effects with apomorphine (89), and some decrease in stereopathy with haloperidol or physostigmine (11). These findings have all been interpreted as possibly akin to the movement and mentation disorders, as well as the pharmacological responses, seen in HD. Thus there is a similarity in pathology, behavioral disorder, and pharmacology, although the time course of neurotoxicity is different and the genetic factor is lacking.

The model is potentially useful for the preclinical testing of potential new therapeutic agents (11,79), as well as being a preparation in which the acute and chronic effects of losses of striatal neurons can be evaluated. For example, the deoxyglucose technique developed by Sokoloff et al. (97) can be applied to study the relative effects of striatal losses on remote areas of brain by quantitative autoradiography. In one study (47) it was shown that 10 days after a unilateral injection of KA into the caudate/putamen, there were marked changes in glucose metabolism, as revealed by the deoxyglucose method, in a number of distant areas. Sharply reduced metabolism was found to occur in the injected striatum, the ipsilateral rostral sulcal cortex, the dentate fascia of the hippocampus, the ventromedial nucleus of the thalamus, and the corticobulbar tract on the affected side. By contrast, there was markedly increased metabolism in the ipsilateral globus pallidus, entopeduncular nucleus, the substantia innominata, the lateral habenular nucleus, and the pars reticulata of the substantia nigra. Frey and Agranoff (35) have found similar increases following injection of ibotenic acid into the rat striatum. These results suggest which pathways may have dominant excitatory or inhibitory mechanisms in the dynamic loops associated with movement. Changes in activity seen in the limbic areas suggest an overlap with the mental symptoms. Animals tested several months after the injection of KA still showed decreased metabolism in the striatum, but do not show some of the acute increases (P. L. McGeer and E. G. McGeer, unpublished results), which may indicate the development of compensatory mechanisms. The KA technique, therefore, when combined with the deoxyglucose technique, indicates the possibility of gaining much significant information about dynamic mechanisms associated with movement.

The KA model has also been used to examine the viability of intracerebral grafts designed to replace destroyed neurons. Cell suspensions from fetal or neonatal rats have been injected into the KA-lesioned striata of adult hosts. Cells survive for at least 2 months and form neuronal masses easily distinguished from the surrounding gliotic tissue. Clear evidence of a significant recovery of choline acetyltransferase and glutamic acid decarboxylase activities has been reported (92), as well as histochemical and autoradiographic evidence for new vascularization, near-normal glucose metabolism, and possible innervation by dopaminergic afferents (48,64).

It has been proposed (57,62,75) that the mechanism of cell death in HD might be an excitotoxic one. This hypothesis does not require the formation in HD of a unique KA-like neurotoxin, and indeed there is evidence against such a possibility (8). Rather, the hypothesis suggests some abnormality in the glutamate bouton—and most probably in the postsynaptic membrane—which would render the neurons more sensitive. Then, just as KA will acutely destroy striatal neurons when directly injected, so might the action of glutamate from the corticostriatal glutamate pathway be excitotoxic in HD. If such a mechanism of cell death is valid, then an obvious protective treatment might be a drug that inhibits glutamate release or blocks its postsynaptic action. Some support for the hypothesis comes from as yet unconfirmed studies with fibroblasts from HD patients and normals. Fibroblasts from HD patients have been said to be more vulnerable in culture to high doses of glutamate than are those from normals (38) and show a difference in affinity in glutamate-binding studies (107). It is on the basis of this hypothesis that clinical trials are presently underway with baclofen (p-chlorophenyl-GABA) which is reported to inhibit glutamate release (34,83) and to have a very minor protective action against KA-induced damage (53,54,61). In a short-term trial of baclofen in HD, no positive results were obtained (3), but, on the postulated mechanism, an agent which inhibited glutamate release could at best slow progression of the disease and could not be expected to affect established symptomatology. Similar reasoning might suggest chronic treatment with naloxone (36, 60), hypotaurine, or taurine (91), which have all been found to inhibit slightly the neurotoxicity of KA. In the future better inhibitors of glutamate release or of its postsynaptic action might be found, and these would deserve evaluation as inhibitors of the progressive deterioration in HD.

B. Epilepsy

Excitotoxic agents in many circumstances induce various types of epileptic phenomena. This is seen both after systemic and intracerebral injections. It is logical, therefore, that they should be considered as a means for producing animal models of epilepsy, particularly status epilepticus. Status epilepticus describes a situation where a continuous seizure, or a series of intermittent seizures, extends over a period of at least an hour without restoration of normal brain function. Following such an episode, nerve cell loss and gliosis occur, primarily in specific sites of the limbic system such as the hippocampal formation. Ben-Ari et al. (6,7) found that KA at a dose of 0.4-1.6 μg injected directly into the amygdaloid nucleus unilaterally will produce status epilepticus. Termination of the episode, and thus prevention of death, can be brought about by intraperitoneal doses of 20 mg/kg of diazepam. As expected, neuronal lesions occur in the amygdala. These, however, extend to the hippocampal field, particularly to the highly vulnerable CA3 field. This secondary brain hippocampal damage is reminiscent of that observed following status epilepticus in man (9). Similar seizure activity and pathological consequences are also seen in baboons given amygdala or temporal pole injections (67). Pretreatment with diazepam before the intracerebral KA injections into the amygdala in rats reduced the hippocampal and other distant damage without reducing the toxic effects of the agent at the site of injection (7).

A role for cholinergic structures in this mechanism of spread has been suggested by Kimura et al. (49). They reported that amygdaloid kindling, a form of epilepsy induced by frequent, pathological electrical stimulation of the amygdala, is enhanced by the acetylcholinesterase inhibitor diisopropyl fluorophosphate and reduced by atropine (an acetylcholine blocker). This implies that the spontaneous seizure activity induced by kindling is mediated, at least in part, by a cholinergic system. Animals with lesions of the nucleus of the diagonal band of Broca, the medial septal area, the interpeduncular nucleus, and the habenula all could be successfully kindled. However, animals with lesions to the substantia innominata (SI) could not be kindled. The SI contains the largest collection of cholinergic cell bodies in the brain (50) and is the main source of afferents to the neocortex (46,52), as well as to the amygdala (70). There is evidence of reciprocal, though noncholinergic, connections from the amygdala to the SI, and these may be involved in amygdaloid kindling.

Injections of either KA or folic acid into the SI also cause convulsions in rats and distant damage to neurons in the pyriform and other cortices, amygdala, and thalamus; some of the neurons destroyed are GABA neurons. Unlike KA, folic acid causes little local damage in the injected SI. The distant damage can be blocked almost completely by prior administration of diazepam. Distant damage in all areas except the thalamus is markedly reduced by prior administration of scopolamine, a cholinergic antagonist, which is consistent with the hypothesis that excitation of the cholinergic system is largely responsible for the effects noted. Injections of folic acid into the SI cause more convulsive activity and far more distant damage than similar injections into either the amygdala or striatum, again emphasizing the probable importance of this widespread cholinergic system in convulsive phenomena (65). The distant damage conforms to that observed in the autopsied brains of human epileptics (13).

These results have led to a reconsideration of older literature in which folic acid was described as having epileptogenic activity in rats (42), enhancing kindling (37) and being elevated in brain during maximum seizure susceptibility following barbiturate withdrawal (22). The possible relationship of folic acid to human epilepsy was first raised when it was realized that prolonged treatment with diphenylhydantoin lowered serum folate levels (10,43). While epileptic activity was not exacerbated in most patients by replacement therapy (43,45, 84), in some patients it clearly had such an effect (4,19,85). Brennan et al. (13) found cerebrospinal fluid folate levels significantly raised (by threefold) following grand mal seizures in untreated patients, and suggested that folates released from nerve endings might interact with receptors to play a role in the spread of seizures. The suggestion has also been put forward on the basis of experiments in rats that, if the blood-brain barrier is breached in epilepsy, then folic acid could concentrate in the epileptic focus itself (66).

High doses (>4 mg/kg) of KA given intravenously to rats will also cause seizures in limbic structures similar to those seen in temporal lobe epilepsy and likewise resistant to conventional anticonvulsant drug therapy (21). The primary action is believed to be in the CA3 region of the hippocampus (56). Electroencephalographic changes have also been recorded following KA injections into the reticular formation (33), a procedure which results in some distant damage to the hippocampus (86). Intrahippocampal injections in both cats (102) and rats (18) result in the long-term in spontaneous recurrent seizures, a finding which further strengthens the clinical relevance of the model, since spontaneity is a characteristic of human epilepsy. This long-term effect seems to depend upon amygdala involvement (17).

These experiments suggest the value of neurotoxins in the study of "models" of epilepsy in man, particularly temporal lobe epilepsy, which is frequently associated with loss of neurons in the hippocampus and other brain areas (69).

C. Interstitial Myocardial Necrosis

Rats injected bilaterally with 2-3 nmol of KA into the thalamus almost invariably show hematuria, elevated blood fibrinogen levels, and acute myocardial necrosis. The type of cardiac damage differs markedly from that observed after myocardial infarction, where there is necrosis of all tissues beyond the thrombosed vessels. Instead, the changes are those of focal myocardial cell injury where the integrity of large vessels is unaffected and where the damage does not follow the perfusion bed of any particular vessel. Cells are often damaged even though they are close to vessels whose integrity is intact. This type of focal myocardial necrosis is seen in humans and animals following strokes, subarachnoid hemorrhage, or traumatic brain damage. It has also been noted following extremely stressful conditions (12). It is not readily detected clinically, but is often detected at autopsy and may be a far more important contributing factor to cardiac failure, and even to the long-term consequences of a classic myocardial infarction, than is commonly recognized. Such cardiac damage has been produced erratically after repetitive electrical stimulation of certain brain regions, but KA has the advantage of being rapid and highly reproducible. Thus it forms a model for investigation of this interesting phenomenon which may have important clinical implications. The hematuria appears to come from tissue destruction in the bladder. Although cases showing gross hematuria all display myocardial damage, the reverse is not true. In contrast to thalamic injections of KA, focal myocardial necrosis is not detected following electrolytic lesions of the thalamus or injections of KA into the cerebellum, cortex, or periphery, even when the doses are as high as 10 mg/kg subcutaneously or 5 mg/kg intraperitoneally.

It cannot be said with certainty whether the myocardial damage is humoral, neurogenic, or both. However, the most likely hypothesis on the basis of current evidence is that circulating catecholamines play an important part. Urinary levels of the catecholamines increase 2- to 10-fold following thalamic injections, and the cardiac effects are partially blocked by pretreatment with reserpine or 6-hydroxydopamine (12); no treatment was found which blocked the effect entirely.

Chelly et al. (20) found that either L-glutamate (10^{-5}-10^{-7} mol/kg) or KA (10^{-8}-10^{-10} mol/kg), injected into the cisterna magna of dogs, produced a dose-dependent increase in blood pressure and slowing of the heart rate. Intravenous injections of larger doses were ineffective. This again suggests the possibility that excitation of central nervous system structures can produce effects in the cardiovascular system.

D. Hypothalamopituitary Disturbances

In infant animals the arcuate nucleus seems particularly sensitive to the neurotoxic effects of systemically administered excitotoxins. Animals treated in this fashion may be extremely useful in studying possible mechanisms underlying various hypothalamopituitary disturbances in man (78). For example, in mice treated in infancy with either single or multiple subcutaneous injections of glutamate, Olney (73) described an obesity syndrome in which treated animals, initially lower in body weight, surpassed the weight of littermate controls at about 45 days of age and thereafter continued to amass considerable carcass fat at a slow but steady pace through adulthood. Despite this, the treated animals were slightly hypophagic compared to controls. All researchers who have attempted to measure food intake in glutamate-treated obese animals have reported them to be either hypophagic or normophagic, but never hyperphagic. Thus the glutamate-obese mouse may be a promising model for studying human obesity.

Cameron and his colleagues (16) have employed the glutamate-obese mouse as a model for studying mechanisms of carbohydrate disturbance in diabetes. Such mice are insulin sensitive, hyperinsulinemic, and mildly hypoglycemic on fasting. In KK mice, an inbred strain with a high genetic susceptibility to diabetes, the disease was unmasked by a glutamate challenge in infancy. Such mice became markedly obese and developed hyperglycemia accompanied by gross hyperinsulinemia, implying a state of insulin resistance. Food restriction restored glucose levels to normal. Cameron et al. (16) considered the hyperglycemia and hyperinsulinemia to be a direct result of the hypothalamic abnormality in this diabetes-prone strain and suggested that further study of this model might shed light on the role of the hypothalamus in obesity and diabetes.

III. SUMMARY

The use of excitotoxic amino acids has already produced interesting animal "models" which may be useful in exploring diseases as diverse as Huntington's, epilepsy, myocardial necrosis, and hypothalamopituitary disturbances. Many other possibilities exist (108). Literature has appeared, for example, suggesting that subthalamic injections produce hemiballismus and that SI injections may produce a model of Alzheimer's disease. As the range of excitotoxic amino acids with different properties is extended, their application to producing modes of human disease states should also expand.

REFERENCES

1. Ando, N., Gold, B. I., Bird, E. D., Roth, R. H. 1979. Regional brain levels of γ-hydroxybutyrate in Huntington's disease. *J. Neurochem. 32*:617-622.
2. Ando, N., Simon, J. R., Roth, R. H. 1979. Inverse relationship between GABA and γ-hydroxybutyrate levels in striatum of rat injected with kainic acid. *J. Neurochem. 32*:623-625.
3. Barbeau, A. 1973. GABA and Huntington's chorea. *Lancet 2*: 1499-1500.

4. Baylis, E. M., Crowley, J. M., Preece, J. M., Sylvester, P. E., Marks, V. 1971. Influence of folic acid on blood-phenytoin levels. *Lancet 1*:62-65.
5. Beaumont, K., Maurin, Y., Reisine, T. D., Fields, J. Z., Spokes, E., Bird, E. D., Yamamura, H. I. 1979. Huntington's disease and its animal model. Alterations in kainic acid binding. *Life Sci. 24*:809-816.
6. Ben-Ari, Y., Lagowska, J., Tremblay, E., Le Gal La Salle, G. 1979. A new model of focal status epilepticus: intra-amygdaloid application of kainic acid elicits repetitive secondarily generalized convulsive seizures. *Brain Res. 163*:176-179.
7. Ben-Ari, Y., Tremblay, E., Ottersen, O. P., Naquet, R. 1979. Evidence suggesting secondary epileptogenic lesions after kainic acid: Pretreatment with diazepam reduces distant but not local brain damage. *Brain Res. 165*:362-365.
8. Beutler, B. A., Noronha, A. B. C., Poon, M. M., Arnason, B. G. W. 1981. The absence of unique kainic acid-like molecules in urine, serum, and CSF from Huntington's disease patients. *J. Neurol. Sci. 5*:355-360.
9. Blackwood, W., Corsellis, J. A. N. (eds.). 1976. *Greenfield's Neuropathology*, Arnold, London.
10. Blakely, R. L. 1969. *The Biochemistry of Folic Acid and Related Pteridines*, North-Holland, Amsterdam.
11. Borison, R. L., Diamond, B. I. 1979. Kainic acid model predicts therapeutic agents in Huntington's disease. *Trans. Am. Neurol.. Assoc. 104*:67-69.
12. Boyko, W. J., Galabru, C. K., McGeer, E. G., McGeer, P. L. 1979. Thalamic injections of kainic acid produce myocardial necrosis. *Life Sci. 25*:87-98.
13. Brennan, M. J. W., Costa, J., Ruff, P., Sutej, P. 1982. Elevation of CSF folate levels following grand mal seizures in untreated patients. *Neurosci. Abstr. 12*:139.11.
14. Briggs, R. S., Redgrave, P., Nahorski, S. R. 1981. Effect of kainic acid lesions on muscarinic agonist receptor subtypes in rat striatum. *Brain Res. 206*:451-456.
15. Burd, G. D., Marshall, P. E., Beal, M. F., Landis, D. M. D., Martin, J. B. 1982. Effects of kainic and ibotenic acid on the neostriatal somatostatin system of the rat. *Neurosci. Abstr. 12*:140.2.
16. Cameron, D. P., Poon, T.K.-Y., Smith, G. C. 1976. Effects of monosodium glutamate administration in the neonatal period on the diabetic syndrome in KK mice. *Diabetologia 12*:621-626.
17. Cavalheiro, E. A., Calderasso Filho, L. S., Riche, D. A., Feldblum, S., Le Gal La Salle, G. 1983. Amygdaloid lesion increases the toxicity of intrahippocampal kainic acid injection and reduces the late occurrence of spontaneous recurrent seizures in rats. *Brain Res. 262*:201-207.

18. Cavalheiro, E. A., Riche, D. A., Le Gal La Salle, G. 1982. Long-term effects of intrahippocampal kainic acid injection in rats: A method for inducing spontaneous recurrent seizures. *Electroencephalogr. Clin. Neurophysiol.* *53*:581-589.
19. Chanarin, I., Laidlaw, J., Loughridge, L. W., Mollin, D. L. 1960. Megaloblastic anaemia due to phenobarbitone. The convulsant action of therapeutic doses of folic acid. *Br. Med. J.* *1*: 1099-1102.
20. Chelly, J., Kouyoumdjian, J. C., Mouille, P., Huchet, A. M., Schmitt, H. 1979. Effects of L-glutamic acid and kainic acid on central α-cardiovascular control. *Eur. J. Pharmacol.* *60*:91-94.
21. Clifford, D. B., Lothman, E. W., Dodson, W. E., Ferrendelli, J. A. 1982. Effect of anticonvulsant drugs on kainic acid-induced epileptiform activity. *Exp. Neurol.* *76*:156-167.
22. Cooke, S., Crossland, J. 1978. Effect of short- and long-term administration of some anticonvulsant drugs on the folate content of rat brain. *Br. J. Pharmacol.* *64*:407P.
23. Coyle, J. T., McGeer, E. G., McGeer, P. L., Schwarcz, R. 1978. Neostriatal injections: A model for Huntington's chorea, in *Kainic Acid As a Tool in Neurobiology* (McGeer, E. G., Olney, J. W., McGeer, P. L., eds.), Raven Press, New York, pp. 139-160.
24. Coyle, J. T., Schwarcz, R. 1976. Lesion of striatal neurons with kainic acid provides a model for Huntington's chorea. *Nature* *263*:244-246.
25. Cross, A. J., Waddington, J. L. 1981. Substantia nigra γ-aminobutyric acid receptors in Huntington's disease. *J. Neurochem.* *37*:321-324.
26. Divac, I., Markowitsch, H. J., Pritzel, M. 1978. Behavioral and anatomical consequences of small intrastriatal injections of kainic acid in the rat. *Brain Res.* *151*:523-532.
27. Duce, I. R., Donaldson, P. L., Usherwood, P. N. R. 1983. Investigations into the mechanism of excitant amino acid cytotoxicity using a well-characterized glutamatergic system. *Brain Res.* *263*:77-87.
28. Emson, D. C., Rehfeld, J. F., Langvin, H., Rossor, M. 1980. Reduction in cholecystokinin-like immunoreactivity in the basal ganglia in Huntington's disease. *Brain Res.* *198*:497-500.
29. Enna, S. J., Bird, E. D., Bennett, J. P., Bylund, D. B., Yamamura, H. I., Iversen, L. L., Snyder, S. 1976. Huntington's chorea: Changes in neurotransmitter receptors in the brain. *N. Engl. J. Med.* *294*:1305-1309.
30. Fibiger, H. C. 1978. Kainic acid lesions of the striatum: A pharmacological and behavioral model of Huntington's disease, in *Kainic Acid As a Tool in Neurobiology* (McGeer, E. G., Olney, J. W., McGeer, P. L., eds.), Raven Press, New York, pp. 161-176.

31. Fields, J. Z., Reisine, T. D., Yamamura, H. I. 1978. Loss of striatal dopaminergic receptors after intrastriatal kainic acid. *Life Sci.* *23*:569-574.
32. Fillion, G., Beaudoin, D., Rousselle, J. C., Deniau, J. M., Fillion, M. P., Dray, F., Jacob, J. 1979. Decrease of [^{3}H]-5-HT high affinity binding and 5-HT adenylate cyclase activation after kainic acid lesion in rat brain striatum. *J. Neurochem.* *33*:567-570.
33. Forchetti, C., Ricciardi, G., Proia, A., Scarnati, E., Pacitti, C. 1981. Neuroexcitatory properties of kainic acid. 1. Electroencephalographic changes following intracerebral microinjections in behavioural rats. Preliminary note. *Boll. Soc. Ital. Biol. Sper.* 57:914-818.
34. Fox, S., Krnjevic, K., Morris, M. E., Puil, E., Werman, R. 1978. Action of baclofen on mammalian synaptic transmission. *Neuroscience* *3*:495-515.
35. Frey, K. A., Agranoff, B. W. 1981. Regional brain glucose metabolism and protein synthesis following ibotenic acid lesions of the striatum. *Neurosci. Abstr.* *11*:774.
36. Fuller, T. A., Olney, J. W. 1979. Effects of morphine or naloxone on kainic acid neurotoxicity. *Life Sci.* *24*:1793-1798.
37. Goff, D., Miller, A. A., Webster, R. A. 1978. Anticonvulsant drugs and folic acid on the development of epileptic kindling in rats. *Br. J. Pharmacol.* *64*:406P.
38. Gray, P. N., May, P. C., Mundy, L., Elkins, J. 1980. L-Glutamate toxicity in Huntington's disease fibroblasts. *Biochem. Biophys. Res. Commun.* *95*:707-714.
39. Guldin, W. O., Markowitsch, H. J. 1981. No detectable remote lesions following massive intrastriatal injections of ibotenic acid. *Brain Res.* *225*:446-451.
40. Henke, H. 1979. Kainic acid binding in human caudate nucleus; effect of Huntington's disease. *Neurosci. Lett.* *14*:247-251.
41. Higatsberger, M. R., Sperk, G., Bernheimer, Shannak, K. S., Hornykiewicz, O. 1981. Striatal ganglioside levels in the rat following kainic acid lesions: Comparison with Huntington's disease. *Exp. Brain Res.* *44*:93-96.
42. Hommes, O. R., Obbens, E. A. M. T. 1972. The epileptogenic action of sodium folate in the rat. *J. Neurol. Sci.* *16*:271-281.
43. Houben, P. F. M., Hommes, O. R., Knaven, P. J. H. 1971. Anti-convulsant drug and folic acid in young mentally retarded epileptic patients. A study of serum folate, fit frequency and I.Q. *Epilepsia* *12*:235-247.
44. Hruska, R. E., Silbergeld, E. K. 1979. Abnormal locomotion in rats after bilateral intrastriatal injection of kainic acid. *Life Sci.* *25*:181-194.
45. Jensen, O. H., Olesen, O. V. 1969. Folic acid and anti-convulsive drugs. *Arch. Neurol. Psychiatry.* *21*:208-214.

46. Johnston, M. V., Coyle, J. T. 1979. Laminar distribution of cholinergic innervation in rat neocortex: Lesions of extrinsic and intrinsic components. *Neurosci. Abstr.* 5:116.
47. Kimura, H., McGeer, E. G., McGeer, P. L. 1980. Metabolic alterations in an animal model of Huntington's disease using the ^{14}C-deoxyglucose method. *J. Neurol. Transmission Suppl.* *16*: 103-109.
48. Kimura, H., McGeer, P. L., Noda, Y., McGeer, E. G. 1980. Brain transplants in an animal "model" of Huntington's disease. *Neurosci. Abstr.* *6*:688.
49. Kimura, H., Kaneko, A., Wada, J. A. 1981. Catecholamine and cholinergic systems and amygdaloid kindling, in *Kindling*, Vol. 2 (Wada, J. A., ed.), Raven Press, New York, pp. 265-287.
50. Kimura, H., McGeer, P. L., McGeer, E. G., Peng, J. H. 1981. The central cholinergic system studied by choline acetyltransferase immunohistochemistry in the cat. *J. Comp. Neurol.* *200*:151-201.
51. Krammer, E. 1980. Anterograde and transynaptic degeneration "en cascade" in basal ganglia induced by intrastriatal injection of kainic acid: An animal analogue of Huntington's disease. *Brain Res.* *196*:209-221.
52. Lehmann, J. C., Nagy, J. I., Atmadja, S., Fibiger, H. C. 1980. The nucleus basalis magnocellularis: The origin of a cholinergic projection to the neocortex of the rat. *Neuroscience* 5:1161-1174.
53. Liebman, J., Barnard, P., Sobiaki, R., Pastor, G., Dawson, K. 1979. Kainic acid-induced neurological syndrome: Partial reversal by baclofen and other putative GABA-mimetics. *Neurosci. Abstr.* 5:1908.
54. Liebman, J. M., Pastor, G., Barnard, P. S., Saelens, J. K. 1980. Antagonism of intrastriatal and intravenous kainic acid by comparison with various anticonvulsants and GABA mimetics. *Life Sci.* *27*:1991-1998.
55. London, E. D., Yamamura, H. I., Bird, E. D., Coyle, J. T. 1981. Decreased receptor-binding sites for kainic acid in brains of patients with Huntington's disease. *Biol. Psychiatry* *16*:155-162.
56. Lothman, E. W., Collins, R. C., Ferrendelli, J. A. 1981. Kainic acid-induced limbic seizures: Electrophysiological studies. *Neurology* *31*:806-812.
57. McGeer, E. G., McGeer, P. L. 1976. Duplication of biochemical changes of Huntington's chorea by intrastriatal injections of glutamic and kainic acids. *Nature* *263*:517-519.
58. McGeer, E. G., McGeer, P. L. 1978. Some factors influencing the neurotoxicity of intrastriatal injections of kainic acid. *Neurochem. Res.* *3*:501-517.

59. McGeer, E. G., McGeer, P. L., Hattori, T., Vincent, S. R. 1979. Kainic acid neurotoxicity and Huntington's disease. *Adv. Neurol. 23*:577-591.
60. McGeer, E. G., McGeer, P. L., Vincent, S. R. 1979. Morphine, naloxone and kainic acid neurotoxicity. *Res. Commun. Chem. Pathol. Pharmacol. 25*:411-414.
61. McGeer, E. G., Jakubovic, A., Singh, E. A. 1980. Ethanol, baclofen and kainic acid neurotoxicity. *Exp. Neurol. 69*:359-364.
62. McGeer, P. L., McGeer, E. G. 1981. Kainate as a selective lesioning agent, in *Glutamate: Transmitter in the Central Nervous System* (Roberts, P. J., Storm-Mathisen, J., Johnston, G. A. R., eds.), Wiley, London, pp. 55-75.
63. McGeer, P. L., McGeer, E. G., Hattori, T. 1978. Kainic acid as a tool in neurobiology, in *Kainic Acid As a Tool in Neurobiology* (McGeer, E. G., Olney, J. W., McGeer, P. L., eds.), Raven Press, New York, pp. 123-138.
64. McGeer, P. L., Kimura, H., McGeer, E. G. 1983. Transplantation of newborn brain tissue into adult kainic acid lesioned neostriatum, in *Neural Transplants* (Sladek, J. R., Jr., Gash, D. M., eds.), Plenum, New York, pp. 361-371.
65. McGeer, P. L., McGeer, E. G., Nagai, T. 1983. GABAergic and cholinergic indices in various regions of rat brain after intracerebral injections of folic acid. *Brain Res. 260*:107-116.
66. Mayersdorf, A., Streiff, R. R., Wilder, B. J., Hammer, R. H. 1971. Folic acid and vitamin B_{12} alterations in primary and secondary epileptic foci induced by metallic cobalt powder. *Neurology 21*:418.
67. Menini, C., Meldrum, B. S., Riche, D., Silva-Comte, C., Stutzman, J. M. 1980. Sustained limbic seizures induced by intraamygdaloid kainic acid in the baboon: Symptomatology and neuropathological consequences. *Ann. Neurol. 8*:501-509.
68. Mohler, H., Okada, T. 1978. The benzodiazepine receptor in normal and pathological human brain. *Br. J. Psychiatry 133*:261-268.
69. Nadler, J. V., Perry, B. W., Gentry, C., Cotman, C. W. 1981. Fate of the hippocampal mossy fiber projection after destruction of its postsynaptic targets with intraventricular kainic acid. *J. Comp. Neurol. 196*:549-569.
70. Nagai, T., Kimura, H., Maeda, T., McGeer, P. L., Peng, F., McGeer, E. G. 1981. Cholinergic projections from the basal forebrain of rat to the amygdala. *J. Neurosci. 2*:513-520.
71. Nakamura, S., Iwatsubo, K., Tsai, C. T., Iwama, K. 1979. Neuronal activity of the substantia nigra (pars compacta) after injection of kainic acid into the caudate nucleus. *Exp. Neurol. 66*:682-691.

72. Neckers, L. M., Neff, N. G., Wyatt, R. J. 1979. Increased serotonin turnover in corpus striatum following an injection of kainic acid: Evidence for neuronal feedback regulation of synthesis. *Naunyn-Schmiedebergs Arch. Pharmacol. 306*:173-177.
73. Olney, J. W. 1969. Brain lesions, obesity and other disturbances in mice treated with monosodium glutamate. *Science 164*: 719-721.
74. Olney, J. W. 1978. Neurotoxicity of excitatory amino acids, in *Kainic Acid As a Tool in Neurobiology* (McGeer, E. G., Olney, J. W., McGeer, P. L., eds.), Raven Press, New York, pp. 95-122.
75. Olney, J. W., de Gubareff, T. 1978. Glutamate neurotoxicity and Huntington's disease. *Nature 271*:557-559.
76. Olney, J. W., Fuller, T. A., de Gubareff, T., Labruyere, J. 1981. Intrastriatal folic acid mimics the distant but not the local brain damage properties of kainic acid. *Neurosci. Lett. 25*:185-191.
77. Olney, J. W., Misra, C. K., Rhee, V. 1976. Brain and retinal damage from lathyrus excitotoxin, β-N-oxalyl-L-α,β-diaminopropionic acid. *Nature 264*:659-661.
78. Olney, J. W., Price, M. T. 1978. Excitotoxic amino acids as neuroendocrine probes, in *Kainic Acid As a Tool in Neurobiology* (McGeer, E. G., Olney, J. W., McGeer, P. L., eds.), Raven Press, New York, pp. 239-264.
79. Owen, R. T. 1980. Intrastriatal kainic acid—A possible model for antidyskinetic/antichoreic agents? *Methods Find. Exp. Clin. Pharmacol. 2*:133-137.
80. Pearson, S., Nunn, P. B. 1981. The neurolathyrogen, β-N-oxalyl-L-α,β-diaminopropionic acid, is a potent agonist at "glutamate preferring" receptors in the frog spinal cord. *Brain Res. 206*:178-182.
81. Perkins, M. N., Stone, T. W. 1983. Quinolinic acid: Regional variations in neuronal sensitivity. *Brain Res. 259*:172-176.
82. Pisa, M., Sanberg, P. R., Fibiger, H. C. 1980. Locomotor activity, exploration and spatial alternations learning in rats with striatal injections of kainic acid. *Physiol. Behav. 24*:11-19.
83. Potashner, S. J. 1979. Baclofen effects on amino acid release and metabolism in slices of guinea pig cerebral cortex. *J. Neurochem. 32*:103-109.
84. Ralston, A. J., Snaith, R. P., Hinley, J. B. 1970. Effects of folic acid on fit frequency and behaviour in epileptics on anticonvulsants. *Lancet 1*:867-868.
85. Reynolds, E. H., Chanarin, I., Matthews, D. M. 1968. Neuropsychiatric aspects of anticonvulsant megaloblastic anemia. *Lancet 1*:394-397.
86. Ricciardi, G., Forchetti, C., Gasbarri, A., Scarnati, E., Pacitti, C. 1981. Neuroexcitatory properties of kainic acid.

II. Neuronal damages following intracerebral microinjections in behavioral rats. *Boll. Soc. Ital. Biol. Sper.* *57*:919-925.

87. Sanberg, P. R., Fibiger, H. C. 1979. Body weight, feeding and drinking behaviors in rats with kainic acid-induced lesions of striatal neurons—With a note on body weight symptomatology in Huntington's disease. *Exp. Neurol.* *66*:444-466.
88. Sanberg, P. R., Lehmann, J. Fibiger, H. C. 1978. Impaired learning and memory after kainic acid lesions of the striatum: A behavioral model of Huntington's disease. *Brain Res.* *149*:546-550.
89. Sanberg, P. R., Lehmann, J., Fibiger, H. C. 1979. Sedative effects of apomorphine in an animal model of Huntington's disease. *Arch. Neurol.* *36*:349-350.
90. Sanberg, P. R., Pisa, M., Fibiger, H. C. 1981. Kainic acid injections in the striatum alter the cataleptic and locomotor effects of drugs influencing dopaminergic and cholinergic systems. *Eur. J. Pharmacol.* *74*:347-357.
91. Sanberg, P. R., Staines, W., McGeer, E. G. 1979. Chronic taurine effects on various neurochemical indices in control and kainic acid-lesioned neostriatum. *Brain Res.* *161*:367-370.
92. Schmidt, R. H., Bjorklund, A., Steveni, U. 1981. Intracerebral grafting of dissociated CNS tissue suspensions: A new approach for neuronal transplantation to deep brain sites. *Brain Res.* *218*:347-356.
93. Schwarcz, R., Fuxe, K., Hökfelt, T., Terenius, L., Goldstein, M. 1979. Effects of chronic striatal kainate lesions on some dopaminergic parameters and enkephalin immunoreactive neurons in the basal ganglia. *J. Neurochem.* *34*:772-778.
94. Schwarcz, R., Hökfelt, T., Fuxe, K., Jonsson, G., Goldstein, M., Terenius, L. 1979. Ibotenic acid-induced neuronal degeneration: A morphological and neurochemical study. *Exp. Brain Res.* *37*:199-216.
95. Schwarcz, R., Kohler, C., Fuxe, K., Hokfelt, T., Goldstein, M. 1979. On the mechanism of selective neuronal degeneration in the rat brain: Studies with ibotenic acid, in *Advances in Neurology*, Vol. 23 (Chase, T. N., Wexler, N. S., Barbeau, A., eds.), Raven Press, New York, pp. 655-668.
96. Schwarcz, R. C., Whetsell, W. O., Jr., Mangano, R. M. 1983. Quinolinic acid: An endogenous metabolite that produces axon-sparing lesions in rat brain. *Science* *219*:316-318.
97. Sokoloff, L., Reivich, M., Kennedy, C., Des Rosiers, M. H., Patlak, C. S., Pettigrew, K. D., Sakurada, O., Shinohara, M. 1977. The [^{14}C]deoxyglucose method for the measurement of local cerebral glucose utilization: Theory, procedure and normal values in the conscious and anesthetized albino rat. *J. Neurochem.* *28*:897-916.

98. Sperk, G., Schlogl, E. 1979. Reduction of number of benzodiazepine binding sites in the caudate nucleus of the rat after kainic acid injections. *Brain Res. 170*:563-567.
99. Sperk, G., Berger, M., Hortnagl, H., Hornykiewicz, O. 1981. Kainic acid-induced changes of serotonin and dopamine metabolism in the striatum and substantia nigra of the rat. *Eur. J. Pharmacol. 74*:279-286.
100. Spindel, E. R., Wurtman, R. J., Bird, E. D. 1980. Increased TRH content of the basal ganglia in Huntington's disease. *N. Engl. J. Med. 303*:1235-6.
101. Spindel, E. R., Pettibone, D. J., Wurtman, R. J. 1981. Thyrotropin-releasing hormone (TRH) content of rat striatum: Modification by drugs and lesions. *Brain Res. 216*:323-331.
102. Tanaka, T., Kaijima, M., Daita, G., Ohgami, S., Yonemasu, Y., Riche, D. 1982. Electroclinical features of kainic acid-induced status epilepticus in freely moving cats. Microinjection into the dorsal hippocampus. *Electroencephalogr. Clin. Neurophysiol. 54*:288-300.
103. Vincent, S. R., McGeer, E. G. 1979. Kainic acid binding to membranes of striatal neurons. *Life Sci. 24*:265-270.
104. Waddington, J. L., Cross, A. J. 1980. The striatonigral GABA pathway: Functional and neurochemical characteristics in rats with unilateral striatal kainic acid lesions. *Eur. J. Pharm. 67*:27-32.
105. Wong, P.-T., McGeer, E. G., McGeer, P. L. 1982. Effects of kainic acid injection and cortical lesion on ornithine and aspartate transaminases in rat striatum. *J. Neurosci. Res. 8*:643-650.
106. Wong, P.-T., McGeer, P. L., Rossor, M., McGeer, E. G. 1982. Ornithine aminotransferase in Huntington's disease. *Brain Res. 231*:466-471.
107. Wong, P. T.-H., Singh, V. K., McGeer, E. G. 1982. Glutamic acid binding in fibroblasts from patients with Huntington's disease. *Neurosci. Abstr. 8*:152.
108. Yamamura, H. I. (ed.). 1984. Chemical neurotoxins as tools to model CNS disease states. *Life. Sci. 35*:1-51.
109. Zaczek, R., Schwarcz, R., Coyle, J. T. 1978. Long-term sequelae of striatal kainate lesion. *Brain Res. 152*:626-632.
110. Zahniser, N. R., Minneman, K. P., Malinoff, P. B. 1979. Persistence of β-adrenergic receptors in rat striatum following kainic acid administration. *Brain Res. 178*:589-595.

26

Morphological Assessment of Neurotoxicity: Disulfiram Neuropathy as an Animal Model of Human Toxic Axonopathies

A. P. Anzil*
Max Planck Institute for Psychiatry, Martinsried, Federal Republic of Germany

I. INTRODUCTION

This chapter contains a critical review of recent publications concerning the pathology of disulfiram neuropathy in humans, a brief account of the neuropathological features of its experimental counterpart in rats, and a general discussion on the nature of this toxic neuropathy. Since the topic was already dealt with in an article (2), in the following, emphasis will be placed on new and not yet documented observations in experimental disulfiram neuropathy viewed as an animal model of human disulfiram neuropathy and related toxic axonopathies.

II. EXPERIMENTAL ASPECTS

It has been shown that neuropathy can be induced in rats by intracutaneous or intraperitoneal implantation of disulfiram. Lately I succeeded in making the rats neuropathic by adding disulfiram to their diet. Disulfiram is tetraethylthiouram disulfide, a slightly fatty, dirty

*Present affiliation:
Medical College of Pennsylvania, Philadelphia, Pennsylvania

white powder which is used in the preparation of several pharmaceutical products sold under a variety of brand names for the treatment of chronic alcoholism. However, disulfiram powder is a commodity which can be purchased in bulk quantities and at a low price from ordinary commercial distributors of chemical products. The substance is not exactly what one would call an appetizer, and it has to be mixed with Altromin chow powder #1321 in order to be made palatable to the animals. The properly blended mixture can be kneaded into a flat cake with the necessary amount of water. The cake can be cut into conveniently sized pellets, and these can be air-dried and served ad lib. In the course of the first experimental series it became clear that feeding the rats cubes containing less than 0.1% (wt/wt) disulfiram was not conducive to any observable clinical sign. These experiments were interrupted after several months of continuous administration. On the other hand, raising the disulfiram contents beyond a 1.0% level changed the organoleptic properties of the diet to such a degree that the rats stopped taking their food altogether. After a series of trials it became apparent that pellets containing 1.0% (wt/wt) disulfiram were both palatable enough to the rats and entirely satisfactory for the experimentalist. After 10 weeks of feeding this diet, an occasional rat came down with clear-cut signs of nervous system disease.

A. Clinical Observations

Disulfiram-fed rats look healthy, gain weight, and later on maintain their weight, all the while developing a very insidious neurological impairment. They can be kept alive for days and possibly longer even after presenting a full-blown flaccid paralysis of their hindlegs. Therefore, the advantages of an experimental model of disulfiram neuropathy based on feeding the substance instead of implanting it are both numerous and obvious. On the other hand, feeding experiments, in general, are unlikely to elicit a clinical picture arising at the same time and progressing at the same pace in all experimental animals. Quite to the contrary, a group of rats following the same feeding schedule is likely to comprise apparently healthy animals and obviously sick animals for quite some time after feeding experiments have begun and signs of incipient intoxication have appeared in one of the experimental animals.

A peculiar gait characterized by swaying of the hind part of the body from side to side is probably the first sign of disulfiram poisoning. This gait may persist for a long time before giving way to foot eversion and later on frank footdrop (Fig. 1). At the beginning, footdrop is sporadic and transient and, as a rule, unilateral. Later on, footdrop is bilateral, symmetrical, and irreversible, even when the animal is chased or helped, as the case may be, by the examiner's hand. However, even a completely paralyzed rat (Fig. 2) with the hindquarters flattened against the ground and the hindlegs dragging

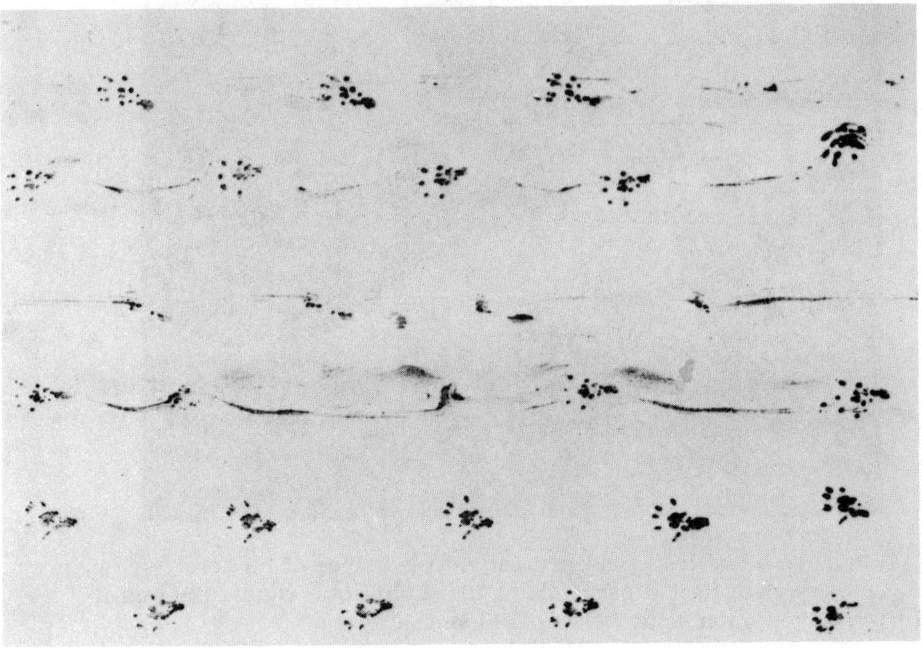

Figure 1 Tracks left by the ink-stained hindfeet of a control rat and two disulfiram-poisoned rats with variable severity of clinical disease. The bottom track is that of a normal rat. The top track reveals only a minor degree of motor impairment. The middle track indicates a major gait disability affecting the right more than the left hindleg.

behind the rump is still able to move about and get food and water as efficiently as its unaffected cage mates do. Other neurological signs, such as, for instance, an obvious motor disability of the forelegs and head, are not observed, at least not at the cutoff points at which the experiments reported here were terminated and the tissues were collected.

B. Morphological Findings

1. *General Remarks*

Tissues were collected from disulfiram-intoxicated animals and ultrathin sections were routinely prepared for electron microscopy. In addition, semithin sections were cut from selected blocks, stained with toluidine blue or paraphenylenediamine, and examined by bright-field or phase-contrast light microscopy. Whenever possible, cross sections were prepared and examined in preference to longitudinal sections.

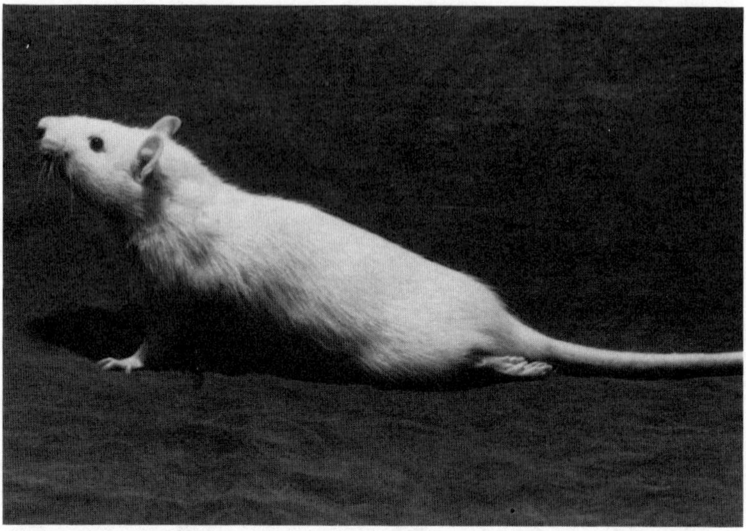

Figure 2 Disulfiram-poisoned rat with evidence of bilateral, symmetrical, complete flaccid paralysis of the hindlimbs. Notice the well-preserved outer appearance of the animal.

From a general viewpoint, the quality of the tissue changes was just about the same in animals poisoned for various lengths of time and affected by a neurological disease of various severity. This is to say that axoplasmic changes, whole-fiber changes, and reparative changes (whenever expected) were present in tissues of all animals studied and only their relative proportion varied (as did the total amount of tissue damage) in a rather loosely positive correlation with the duration of intoxication and severity of clinical disease. Aside from interindividual variations of this type, even greater variations in the amount of tissue pathology present in one and the same animal at any one time during the course of intoxication were observed between different nerve trunks and spinal cord tracts. Furthermore, the amount of pathology generally was greater in distal than in proximal segments of the same affected nerve trunk or central tract, distal and proximal in outstretched nervous structures of this kind being only relative terms having reference to the location of a given point with respect to the cell body.

2. Semithin Sections

Semithin sections were prepared from selected blocks of vibrissa, spinal cord, as well as sciatic, tibial, and plantar nerve samples. Crural nerves and dorsal fasciculi (Fig. 3) of posterior columns of spinal cords were massively affected, and the changes observed were best described as being those of wallerian degeneration. Nerve bundles in the vibrissae were generally unremarkable, except for an occasional myelinated axon with a distinct ground-glass appearance. In some sections, particularly in sections of spinal cord, swollen axons were also seen and were superficially similar to those reported in the neurofilamentous group of toxic axonopathies, except that the cut surface of these swollen axons was apparently more dense and osmiophilic than that of the affected axons in such axonopathies.

3. *Ultrathin Sections*

Ultrathin sections were prepared from all blocks from which semithin sections were cut and from additional blocks, as well taken from the same tissues and locations. The description of the fine structural changes observed in these sections will be brief. Interested readers should consult a previous article (2) for additional information. In order to simplify matters a great deal, I will recall that disulfiram neuropathy is the wallerian degeneration of peripheral nerves, whereas disulfiram myelopathy is wallerian degeneration of the dorsal fasciculi of spinal cord. I will add that wallerian degeneration as occurred in the tissues examined has no special features of its own, and that there is little to be gained by singling out and attributing taxonomic value to an odd feature of a notoriously pleomorphic process, no matter how diffuse and distinctive this feature may be, as, for instance, filamentous or tubulovesicular hyperplasia. In essence, cross sections of tibial nerves and dorsal fasciculi of disulfiram-poisoned rats are liable to reveal a variable assortment of the following profiles (Figs. 4-14). Normal myelinated axons are never entirely missing and, indeed, they can be very numerous. Nerve fibers exhibit essentially two types of axonal changes (Figs. 4-10): one characterized by a granular amorphous degeneration of the axoplasm and another distinguished by a great proliferation of one or more classes of axonal organelles. Profiles divergent from and transitional between these two extremes are quite numerous and make up a large population of nerve fibers with a broad spectrum of primary axonal changes. Nerve fibers of this type, whether they exhibit a normal axon, a profoundly degenerated axon, or an axon with a reactive-dystrophic type of change, have a normal or still essentially normal myelin sheath. When the axon eventually degenerates to the point of disappearance, then myelin sheaths collapse and break down in variously configurated myelin debris (Fig. 11). Up to this point changes are essentially the same in the peripheral as well as in the central nervous system. Past this point the findings are quite different: While in the peripheral nervous system bands of Büngner, axonal sprouting, and remyelination to the point of incipient onion bulb formation are quite prominent; in the central nervous system nothing of this kind is observed. A final word should be said about

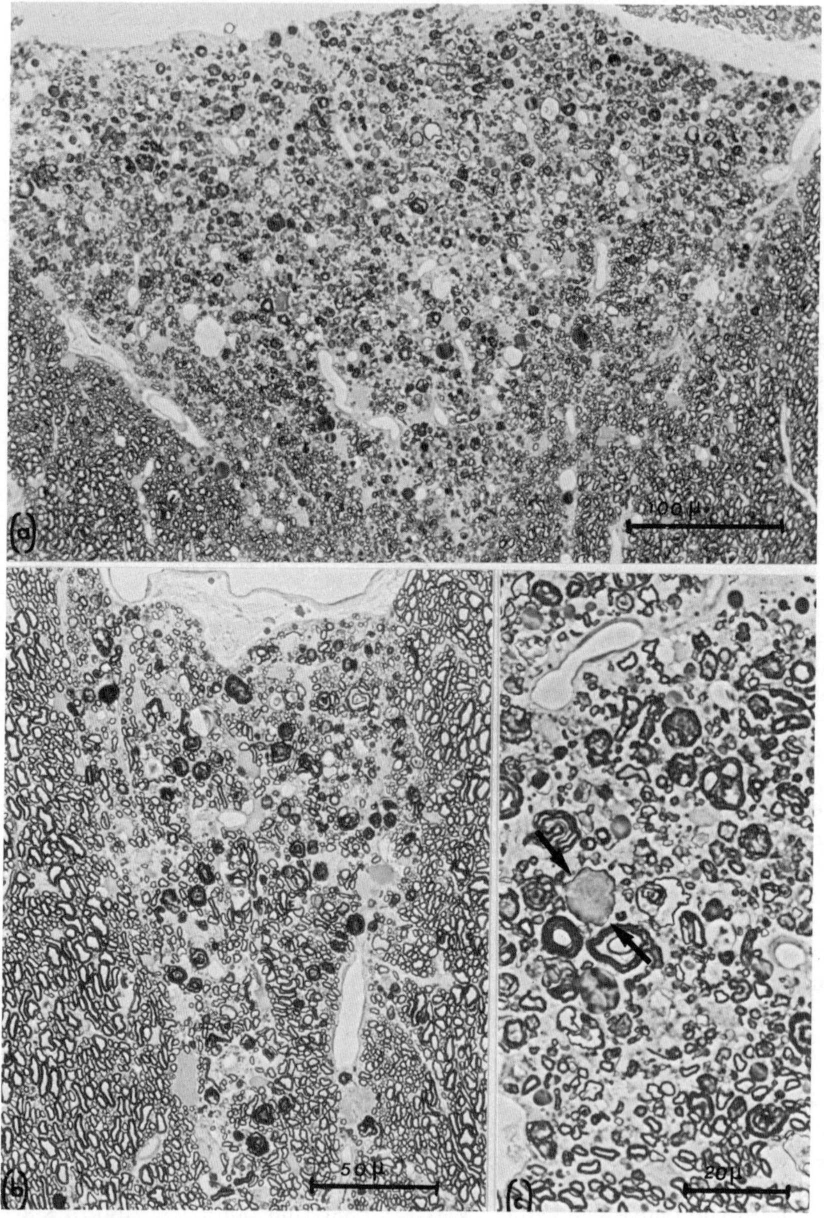
100 μ
50 μ
20μ

the profiles reminiscent of axonal swellings: They contain a motley assortment of axonal organelles and have hardly anything in common with the swollen axon of so-called neurofilamentous neuropathies. On the other hand, regenerating axons with or without a newly formed myelin sheath are fully packed with filaments: They indeed resemble axonal swellings but, at closer inspection and in the context of the surrounding structures, they appear for what they actually are, that is, an expression of a completely different phenomenon (8).

A few additional findings could be mentioned here for the sake of completeness, although they add little and change nothing to the essence of the pathological events I just summarized. With one exception (see below), they seem to be present also in peripheral nerves of normal rats past a certain age (13). Axonal sequestration by attenuated oligoglial processes can also be found in gracile fasciculi of disulfiram-poisoned rats, whereas adaxonal filamentous bodies of peripheral nerves (6) and glycogen bodies are not identified in spinal cord samples of the same rats. A set of remarkable findings already mentioned in a previous article (2) is somewhat at variance with what is commonly known of wallerian degeneration. It consists of some separation of nerve fibers suggestive of interstitial edema and of some irregular occurrence of large, single, or multiple intra-axonal vacuoles, intramyelinic bubbles, and intracytoplasmic vacuoles in Schwann cell profiles (Fig. 5).

III. DISCUSSION

Few papers dealing with the pathology of disulfiram neuropathy of man (1,4,7,9) have appeared in print after the publication of my own review article (2). Only the most recent (1) of these publications strikes a somewhat discordant note and deserves a separate commentary. The authors of the report refer to disulfiram neuropathy as an axonopathy of the neurofilamentous type and purport to supply morphological evidence for that designation. The paper has two electromicrographs

---

Figure 3 Posterior columns of spinal cord: Epon-embedded semithin sections stained with paraphenylenediamine and viewed with phase optics. (a) Cross section of cervical cord: fan-shaped area of degeneration. (b) Cross section of lumbar cord: wedge-shaped area of degeneration. (c) Higher magnification of a small area of spinal cord section. The axon indicated by the arrow has a superficial resemblance to similar profiles of so-called neurofilamentous axonopathies. Axons at the bottom of the photograph are largely spared by the pathological process affecting the axons in the upper part.

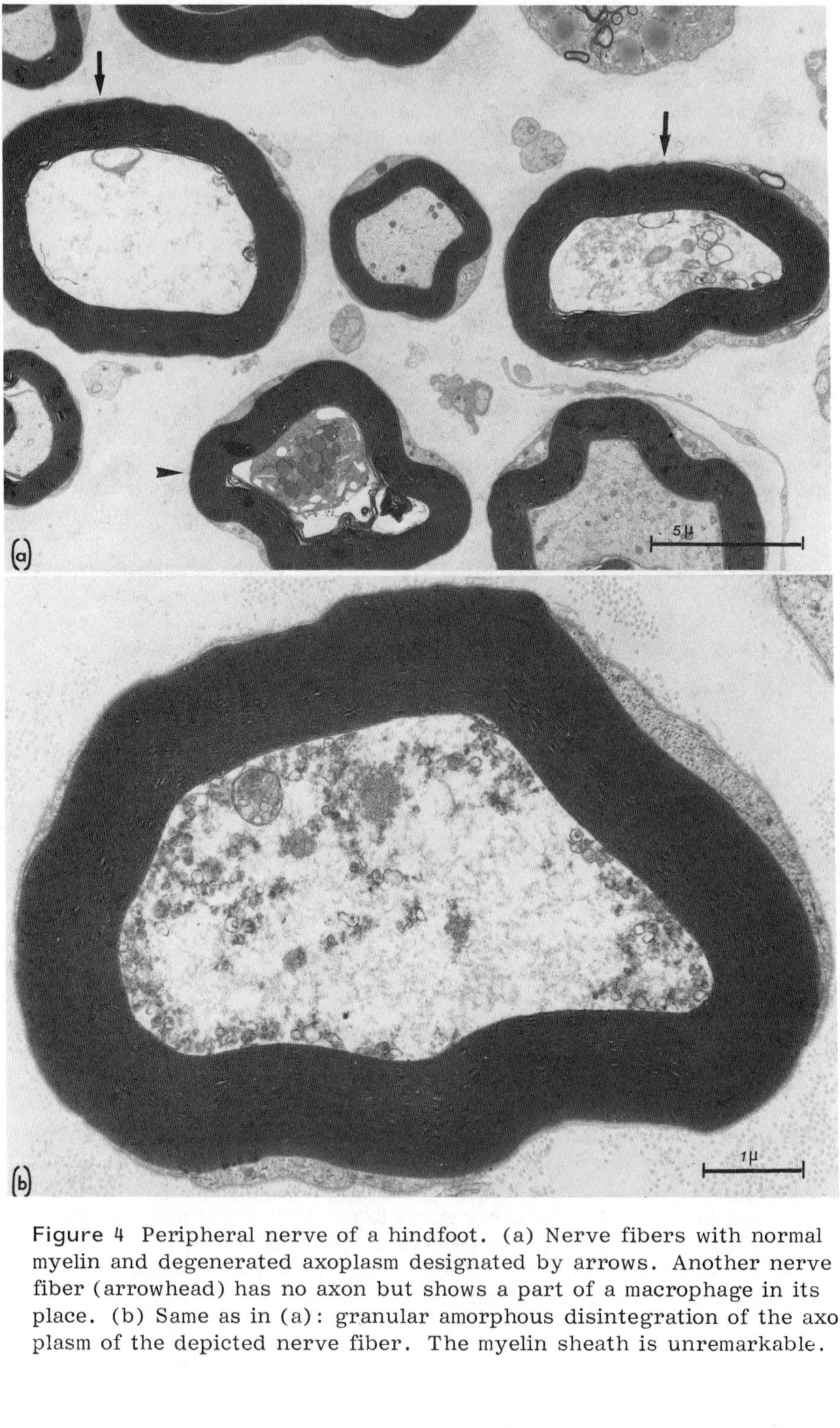

Figure 4 Peripheral nerve of a hindfoot. (a) Nerve fibers with normal myelin and degenerated axoplasm designated by arrows. Another nerve fiber (arrowhead) has no axon but shows a part of a macrophage in its place. (b) Same as in (a): granular amorphous disintegration of the axoplasm of the depicted nerve fiber. The myelin sheath is unremarkable.

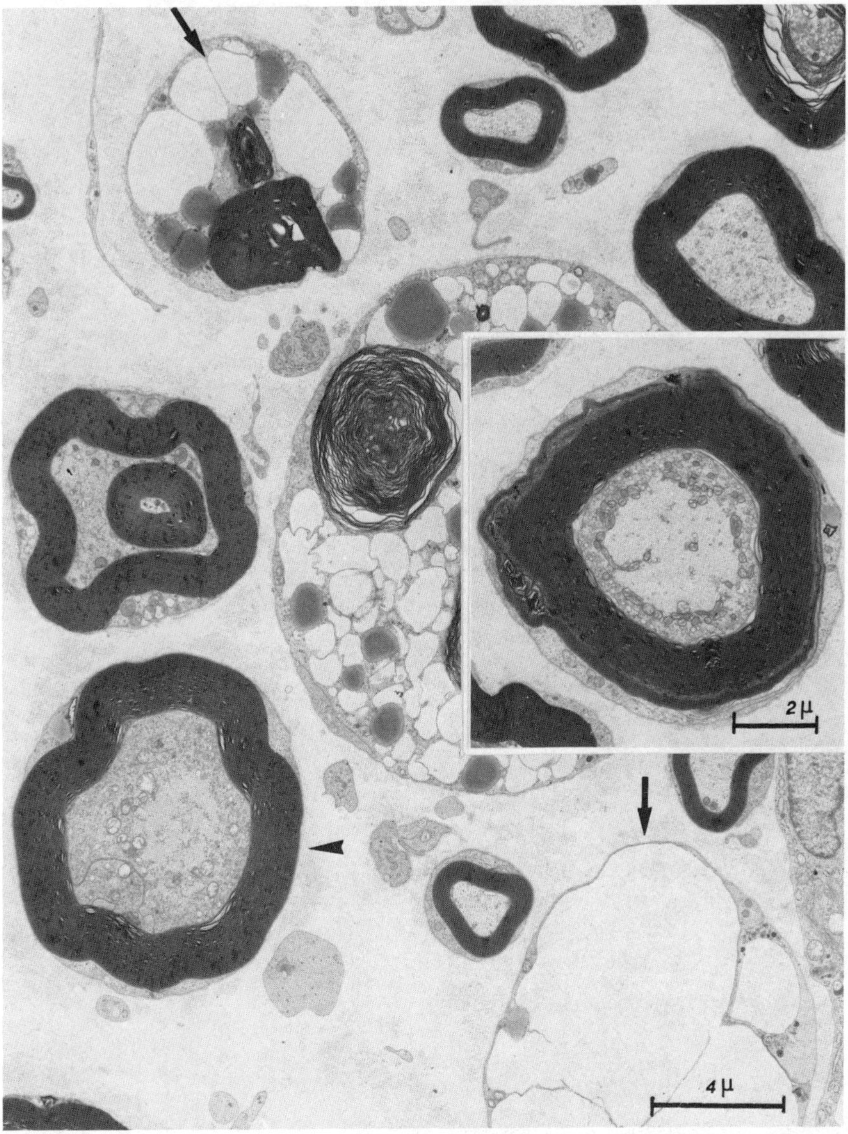

Figure 5 Peripheral nerve of a hindleg. Two vacuolated Schwann cell profiles are designated by arrows. The axoplasm of another nerve fiber (arrowhead) has a core of filaments surrounded by a broad peripheral band crowded with a variety of axonal organelles. Inset: Nerve fiber with peripheral accumulation of axonal organelles around a central core of filaments.

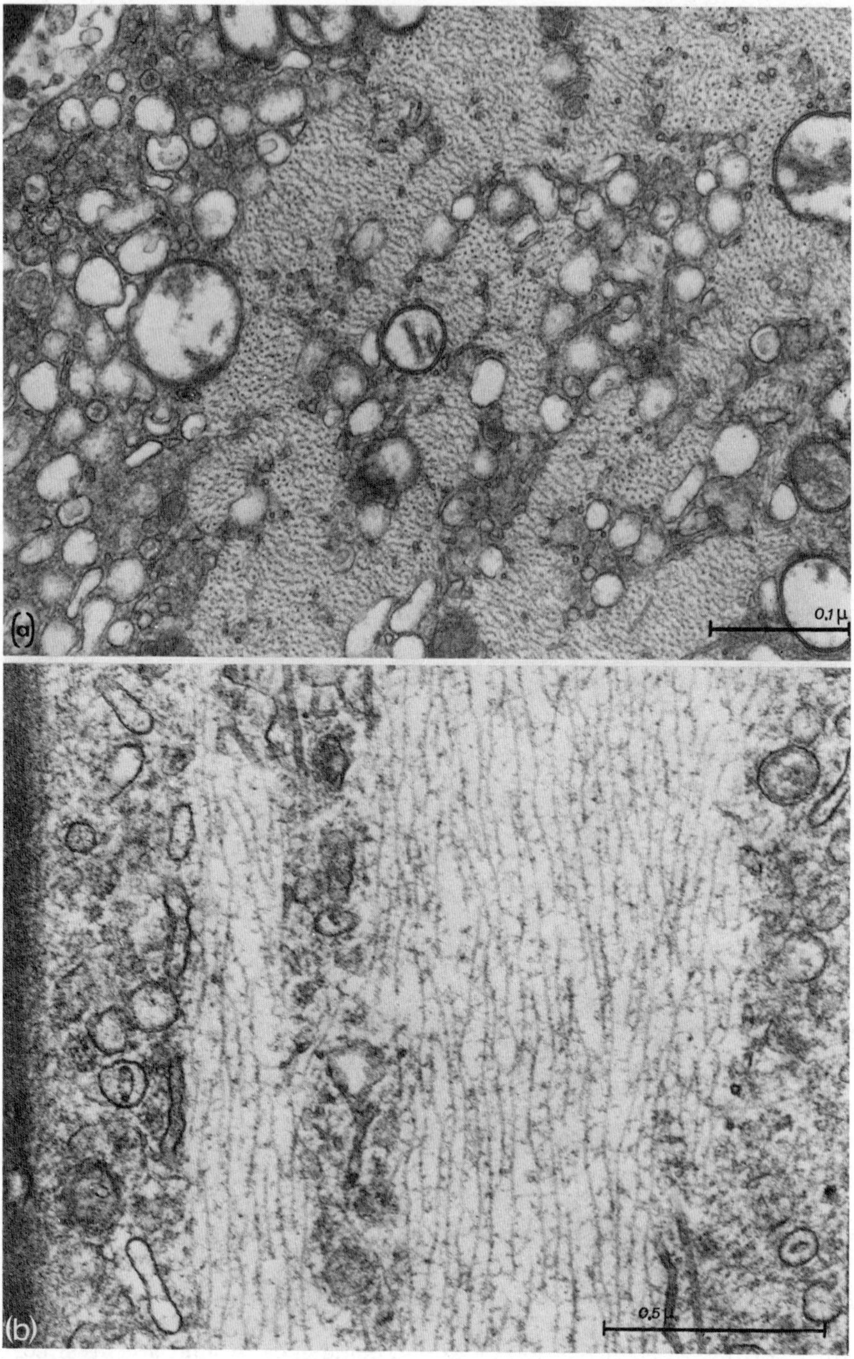
(a)
0,1 μ
(b)
0,5 μ

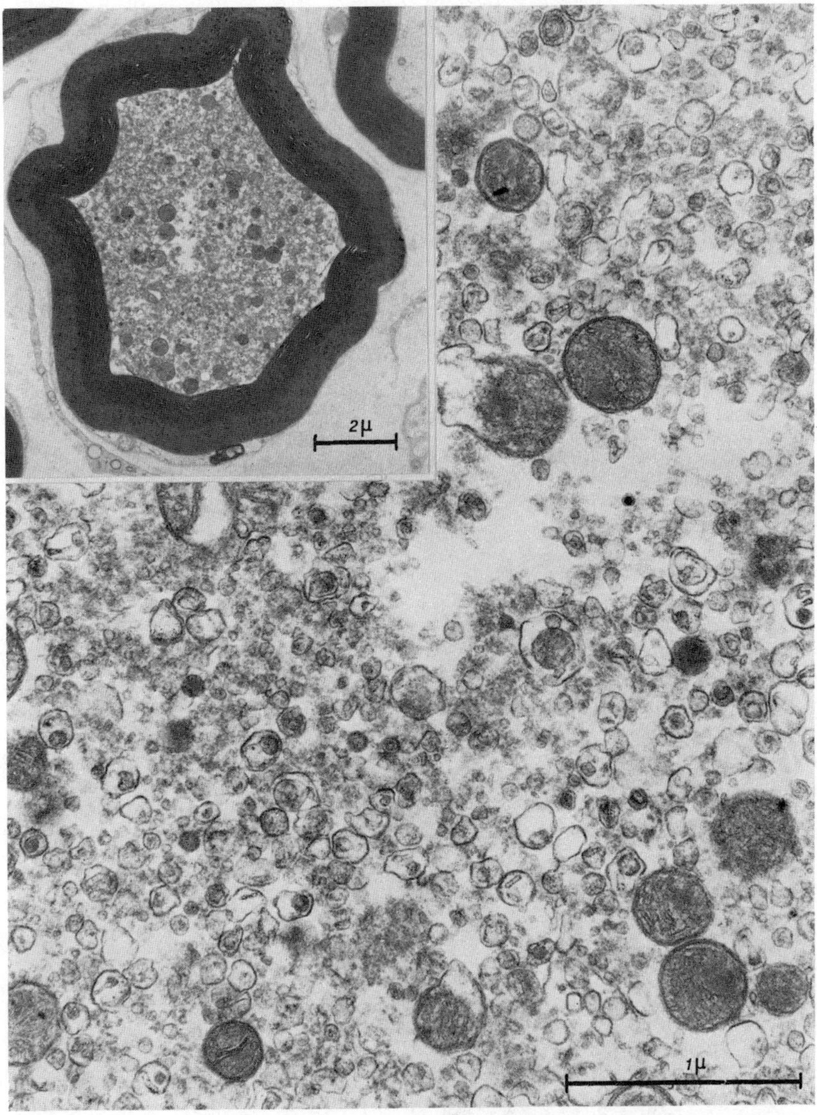

Figure 7 Peripheral nerve of a hindleg. Higher magnification of the degenerating axoplasm of the nerve fiber depicted in the inset. Inset: Transversely cut nerve fiber. The axoplasm is a nondescript collection of degenerating profiles.

Figure 6 Peripheral nerve from a hindlimb. (a) Cross section of a nerve fiber with a filamentous core in the middle of the axon. Miscellaneous vesicular profiles and degenerating mitochondria occupy the periphery and encroach on the center of the axon. (b) Same as in (a): longitudinal section of a nerve fiber showing essentially the same finding.

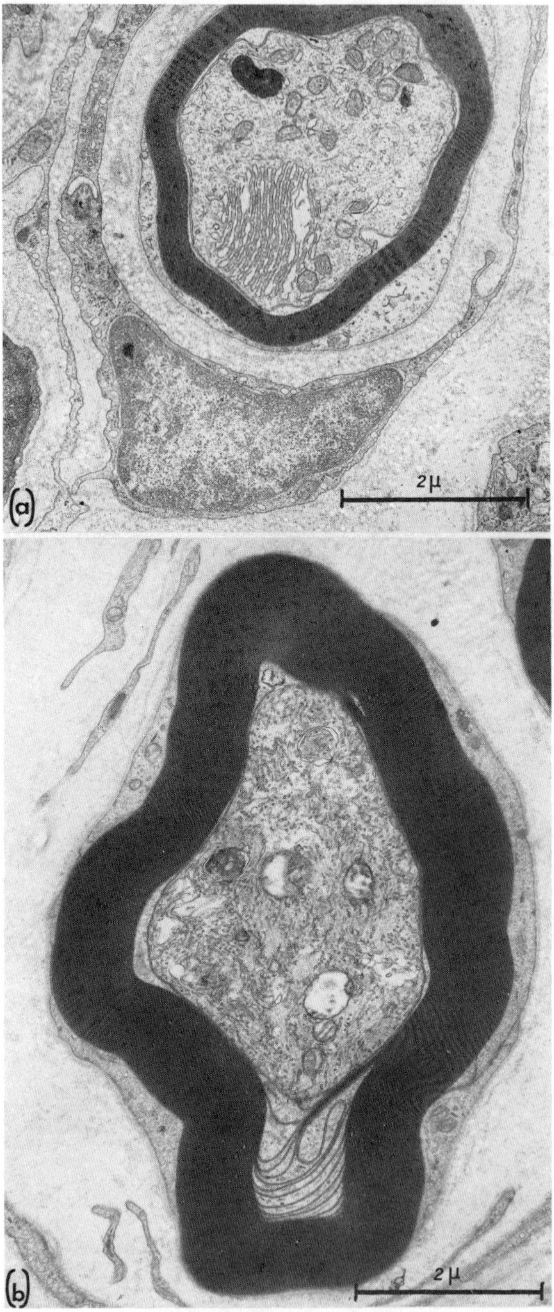

Figure 8 Nerve fascicle of a vibrissa. (a) The depicted nerve fiber has a membranous stack occupying part of the axon; the collection of cytomembranes has little in common with invaginating Schwann cell profiles and the like. (b) Same as in (a). The axoplasm of this axon is rich with elements of the smooth endoplasmic reticulum. Swelling of mitochondria in this particular axon profile is most probably artificial.

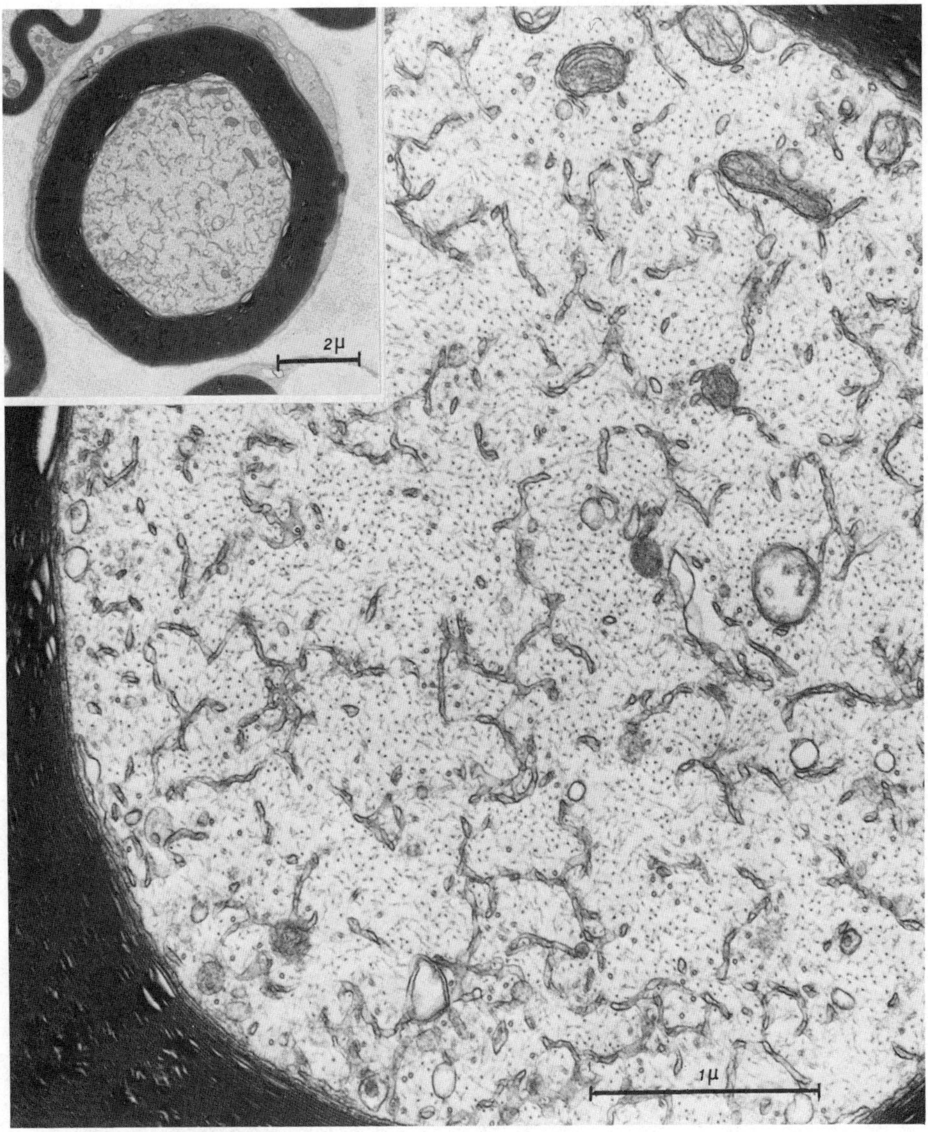

Figure 9 Peripheral nerve of a hindfoot. High magnification of a detail of the nerve fiber illustrated in the inset. The axon is teeming with elements of the smooth endoplasmic reticulum. Inset: At this low magnification a luxuriant proliferation of poorly definable axonal elements is already discernible.

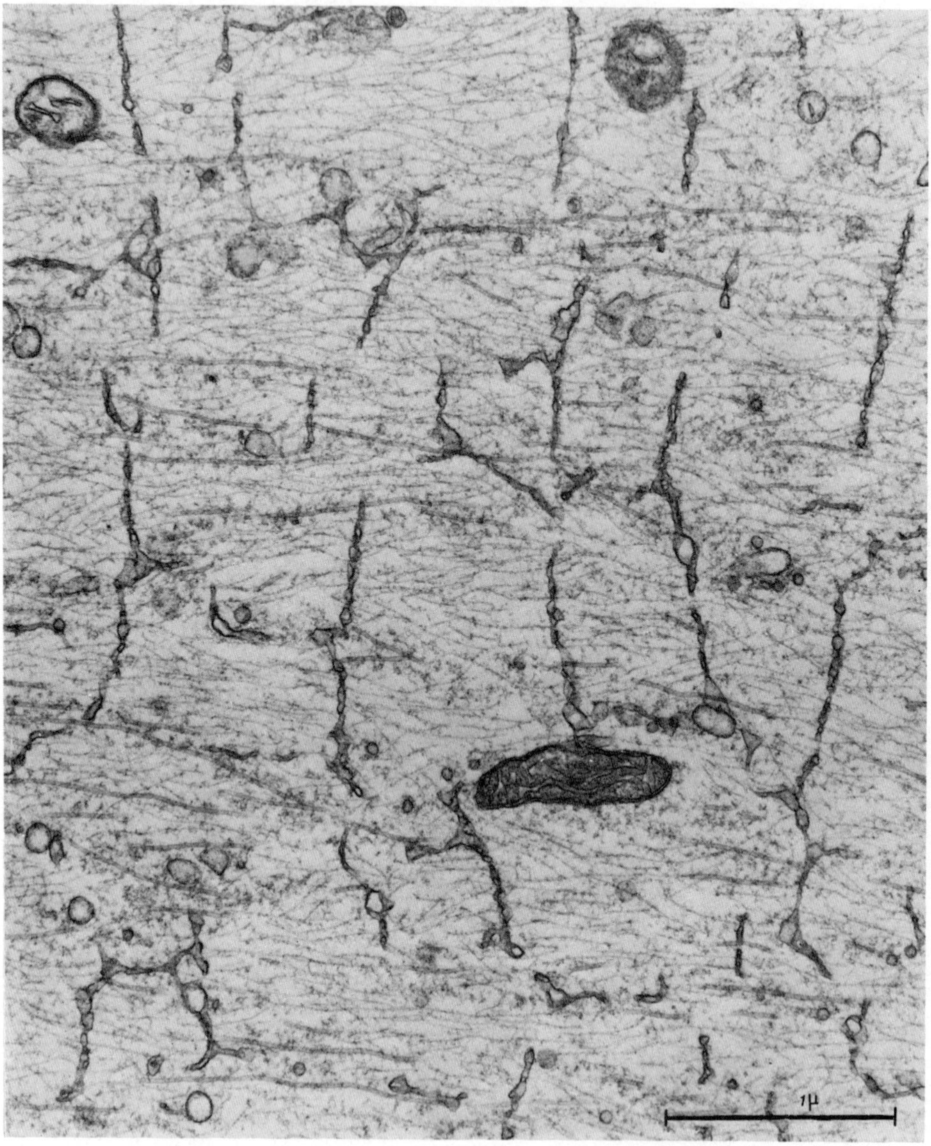

Figure 10 Peripheral nerve of a hindlimb: detail of a nerve fiber cut longitudinally and showing a rich network of smooth endoplasmic reticulum spanning the breadth of the axon.

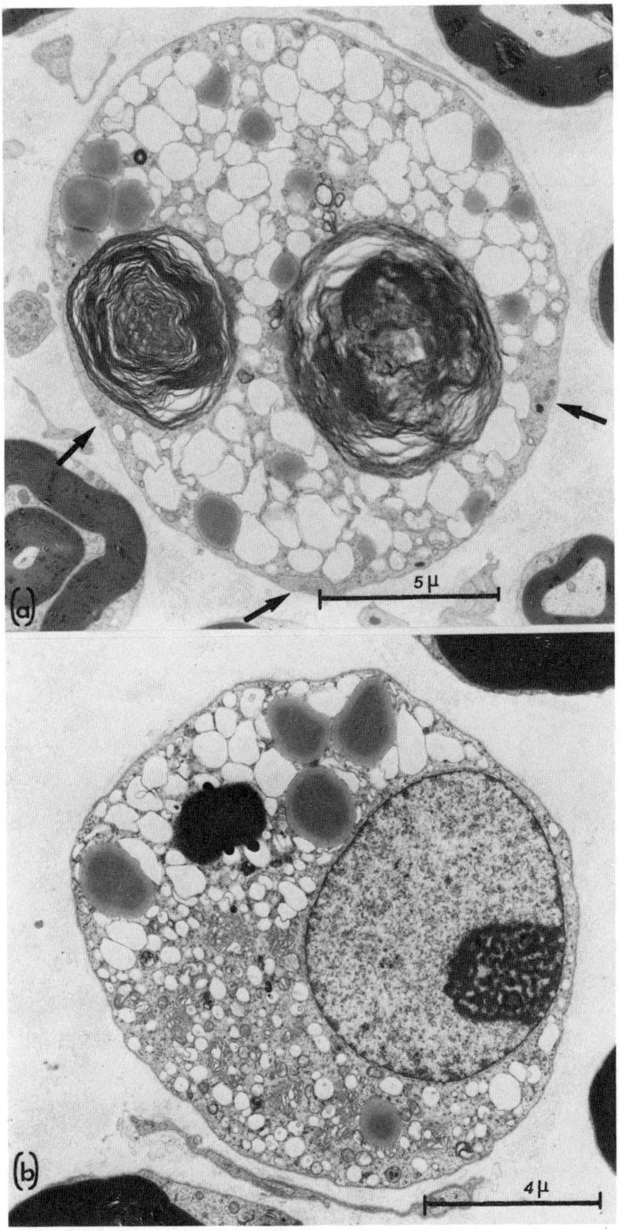

Figure 11 Peripheral nerve of a hindlimb. (a) Schwann cell profile with well recognizable myelin debris and a few fat droplets. The cytoplasm has a sieve-like appearance; the cell processes (arrows) squeezed between cell membrane and basal membrane are probably axonal in nature (axon sprouts). (b) Same as above. Another Schwann cell profile with almost the same appearance except for the lack of identifiable myelin fragments and cell processes under the basal membrane.

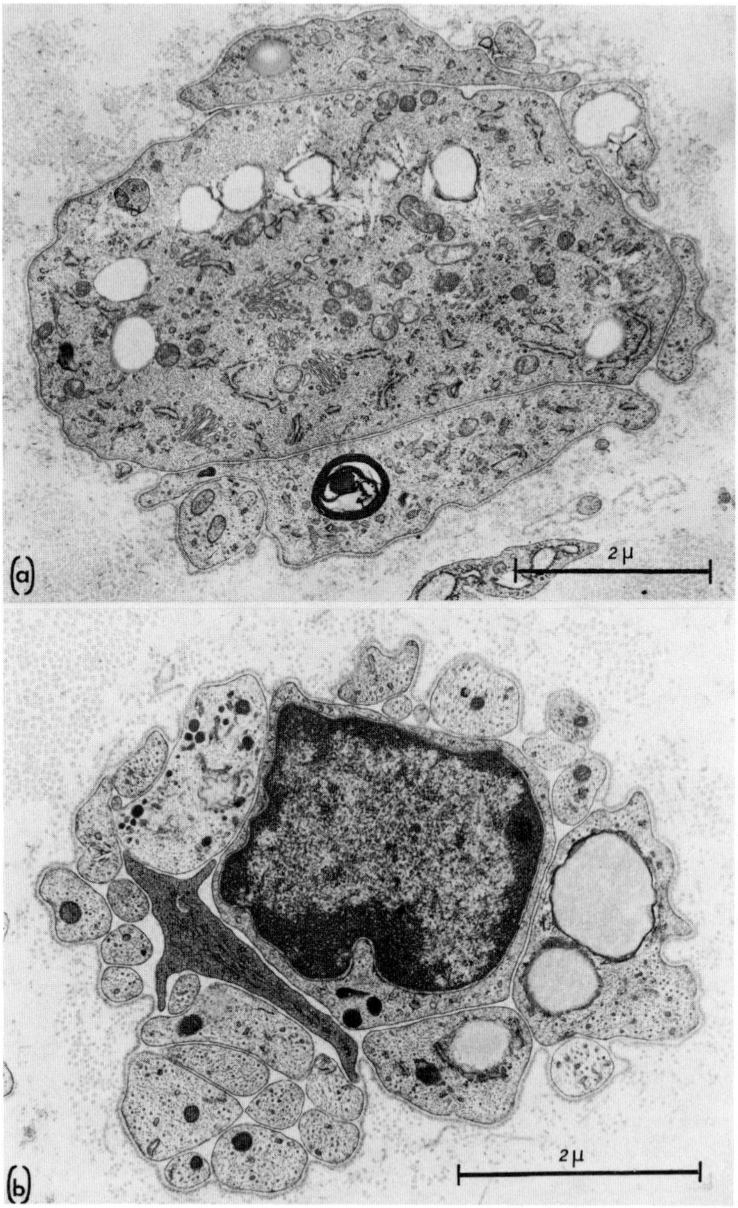

Figure 12 Peripheral nerve of a hindleg. (a) Denervated Schwann cell column with lipid droplets and a myelin rest: two Schwann cell profiles are extremely filament-rich. (b) Same as above. Another collection of Schwan cells bound by a common basal membrane. According to the definition contained in the original description, a denervated Schwann cell complex such as this cannot be called a cross cut profile of a Büngner band since it comprises no axon.

which are supposed to illustrate intraaxonal accumulation of neurofilaments. In fact, the figures only show a somewhat flattened nerve fiber with a normal complement of neurofilaments. The report has also a light micrograph of a plastic-embedded semithin section of a transversely cut nerve fascicle. This picture, at first glance, could be interpreted to lend morphological support to the authors' interpretation. The illustration shows a large, possibly enlarged axon among other myelinated axons of average size. The myelin of the large axon is rather average in thickness and perhaps even slightly attenuated for the size of the axon. All in all, the axon could possibly correspond to one of the all too familiar axonal swellings of so-called filamentous neuropathies. The authors, regretfully, failed to check the ultrastructure of the depicted axon or of any other axon of that same appearance. At any rate, at closer inspection the light micrograph does not seem to uphold the authors' interpretation: Filament-packed axons as seen in plastic-embedded, semithin sections stained with toludine blue are apt to look quite empty, whereas the depicted axon has a ground-glass appearance. To this author, that appearance signifies that the axoplasm contains, besides a number of filaments, a well-stocked assortment of cell organelles bathing in a small amount of amorphous matrix of low electron density. In all my material of experimental disulfiram neuropathy I never came across a single swollen axon which, on electron-microscopic examination, proved to be filled with tightly packed filaments occasionally to the exclusion of all other cell organelles. I have little use for such morphological designations as "neurofilamentous axonopathy," "tubulovesicular axonopathy," and the like. But if neurofilamentous axonopathies characterized by swollen axons packed with filaments and little else do exist—and there is little doubt that they do—then, to be sure, disulfiram neuropathy is not a neurofilamentous type of axonopathy. The question then arises: What is disulfiram neuropathy? How can one define the neuropathic (and myelopathic) lesion induced by disulfiram?

A. Axonal Pathology and the Dying-back Concept

The following points can be considered. The primary lesion induced by disulfiram is one of the axon, that is, an axonopathy. Whole-fiber degeneration, axonal regeneration, and subsequent remyelination up to and including onion bulb formation are entirely secondary phenomena. In trying to define even further this toxic axonopathy, I think that the epithets "peripheral" and "central" fit this type of axonopathy very well. The terms are self-explanatory and make any additional comment appear superfluous. Needless to say, the word *central* today is very much limited; in fact it is restricted to a single tract of the spinal cord.

The next adjective which can be appropriately used in connection with disulfiram neuropathy is *distal*. The clinical picture presented by

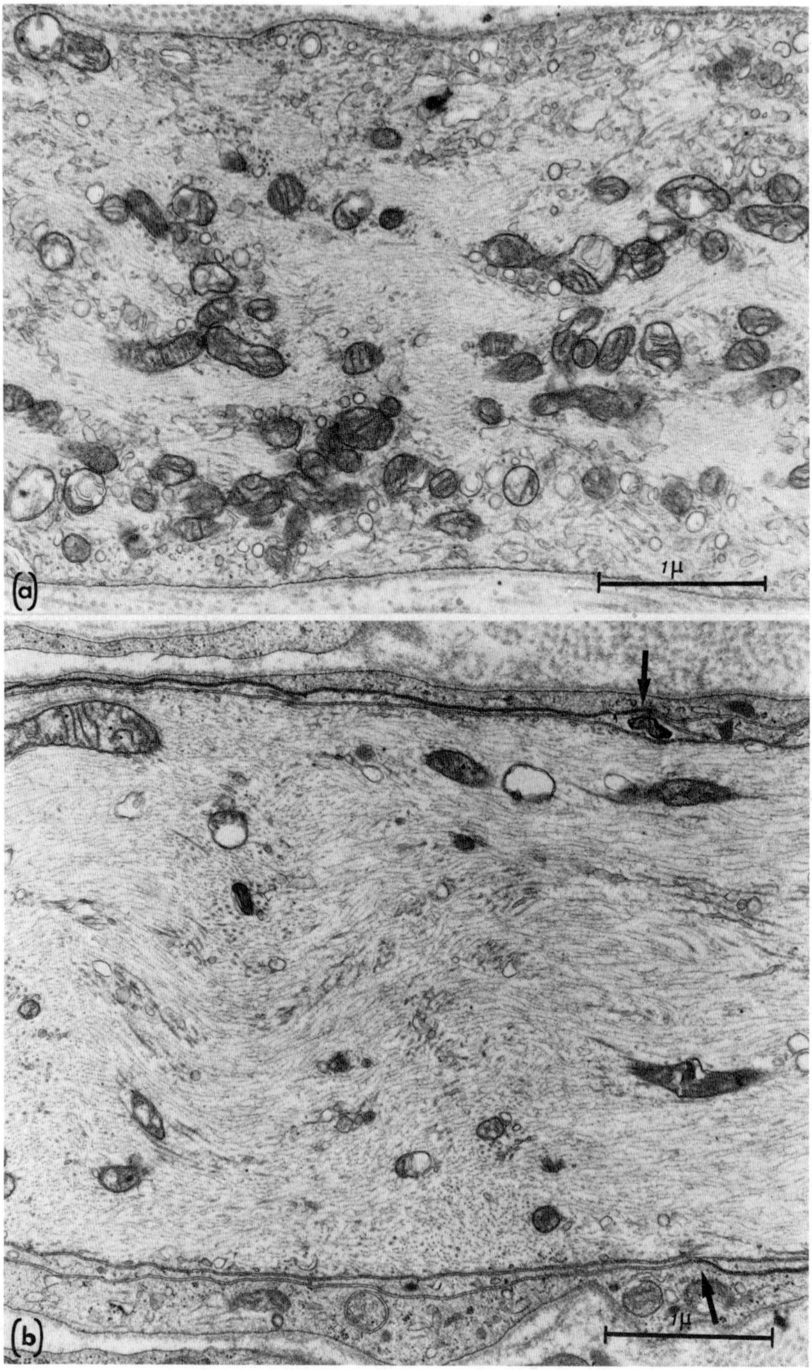
(a)
1μ
(b)
1μ

the rats is an argument for structural changes affecting more the distal than the proximal segments of the hindleg nerve trunks. It further seems that the spatial distribution of the lesions found along crural nerves and dorsal columns speaks for the essential correctness of the term *distal* as applied to disulfiram axonopathy. These are simple, straightforward observations. They do not claim to the completeness or even the thoroughness of high-sounding spatiotemporal studies. Furthermore, in making them, the observer is fully cognizant of the fact that the word *distal* is never absolute and unrestricted, but, on the contrary, always spatially and temporally determined: *distal* in this particular case describes a point along the axon which, at any one particular moment in the lifetime of this axon, is more removed from the cell body than any other point, which is then called proximal. *Distal* is what linguists would call a relative word par excellence, so that what is distal for one axon need not be distal for a neighboring axon, and what is proximal today for a third axon may be distal at a later day for the same axon. Therefore up to this point it seems justified to call disulfiram neuropathy a form of central-peripheral distal axonopathy.

One may next ask: Is disulfiram neuropathy also a dying-back neuropathy? Personally, if someone proposed to drop the question as useless, or to leave it in abeyance as unsolvable or to repel it as disproven, I would favor the first option. Disulfiram neuropathy shows a preferential involvement of the distal segments of a nerve or tract. This preferential susceptibility appears in the forefront at all times, not only in the early days of the clinical disease and the acute stages of the pathological process. The distal character of the neuropathy is not an ephemeral overture: It is a steadfast fixture, regardless of the variable duration of the disease present and of the prevailing quality of the pathology found.

The second point to be kept in mind is that the pathological process of disulfiram neuropathy is not an all-or-none affair across the thickness of a nerve or tract. Quite to the contrary, the neurotoxin affects individual axons in their distal and apparently more exposed region

---

Figure 13 Peripheral nerve of a hindlimb. (a) The longitudinally cut axon is covered with basal membrane without intervening Schwann cell cytoplasm; the axoplasm is rich in organelles, especially mitochondria, and is suggestedly reminiscent of an axon with a reactive type of change; in reality, this is most certainly the profile of a regenerating axon. (b) Same as above. This longitudinally cut profile is also identifiable as a regenerating axon: It is covered with slender Schwann cell processes. At some point Schwann cell membranes (arrows) form major dense lines of myelin sheath.

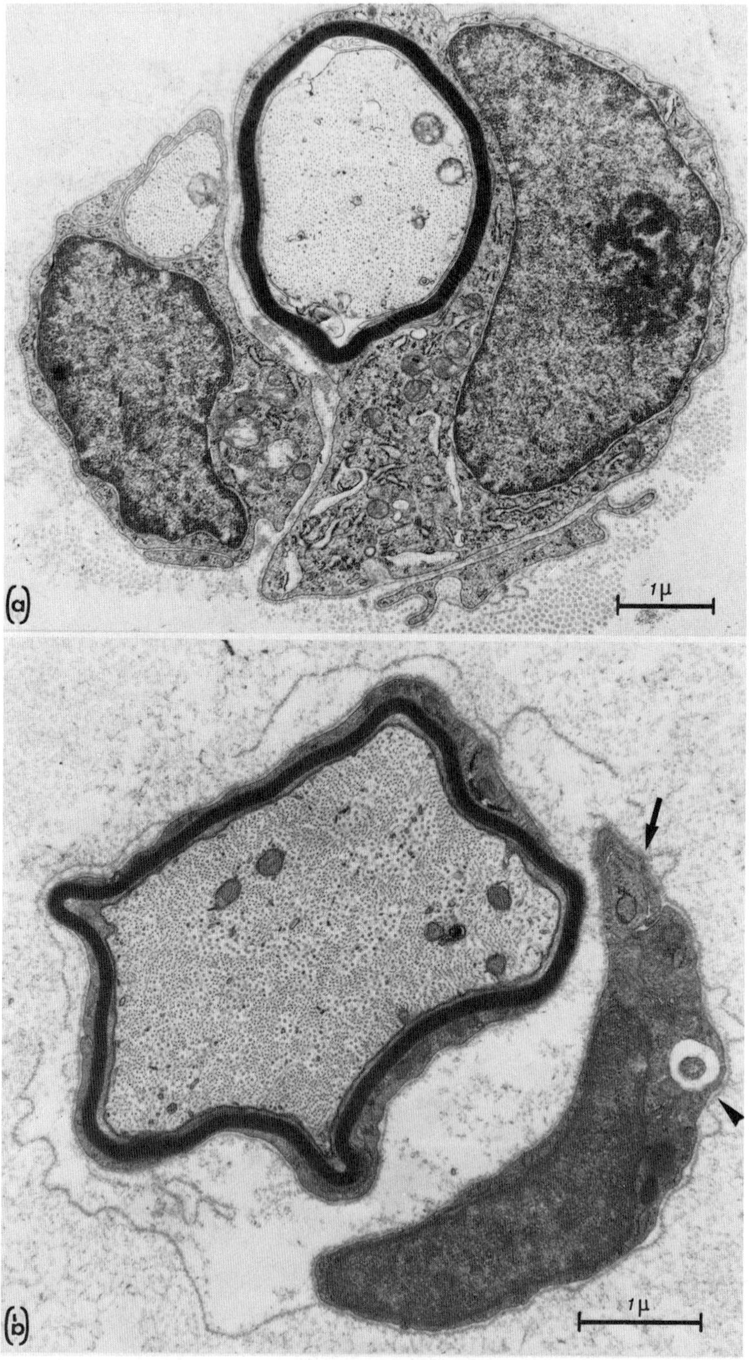
a
1μ
b
1μ

and leaves neighboring axons, likewise in their distal and more vulnerable area, entirely unscathed. To think that all myelinated axons, let alone all axons of a nerve or tract, fall prey simultaneously to the damaging action of disulfiram would be absurd. Accurately and realistically, the case is rather the opposite: Each axon is damaged and degenerates at its own appointed time and pace. In fact, one finds intact axons invariably in all tissue samples studied.

The third point to be considered is that which, for sake of simplicity, one may call the vulgar conception of the dying-back process. On reading some publications, particularly those of an early vintage, one gets the impression that the pathological process affecting the axons in the dying-back type of toxic neuropathies operates like a biological fuse of some sort: It ignites somewhere close to the end and burns all the way up near to the explosive charge, that is, the apparently unaffected cell body. This, of course, would be a highly unusual suspension—indeed, a deviation—of all that is known about wallerian degeneration and the like. Axons hit by wallerian degeneration die forth, if anything. They never die back except for those few millimeters lying immediately above and proximal to the point of physical or chemical transection. To think that wallerian degeneration, no matter how distinctive, may head upward, that is, backward to the cell body, instead of downward, that is, forward to the cell periphery, would run against current understanding of this phenomenon—indeed against many tenets of present-day neurobiology. The only conceivable way for a degeneration of this kind to extend upward along the axon in the direction of the perikaryon would be in a *saltatory, discontinuous, skipping* fashion (*multifocal* would be a very poor substitute for the adjectives I have just used). It would be as if a new round of degeneration would set in on an already affected axon, and then a third, a fourth, and so on, would follow suit at progressively less distal, that is, more proximal points along the axon (Fig. 15).

The sum of the evidence at hand seems to favor a preferential distal vulnerability over a saltatory centripetal spreading of the pathological process, although an involvement strictly limited to the more distal areas is definitely out of the question and skip lesions in the more proximal areas, no matter how few and rare, are certainly the rule. By the way, the fact that an axonopathic process spreading endlong a nerve fiber must move from more distal to more proximal regions stands

---

Figure 14 Peripheral nerve of a hindleg. (a) Schwann cell profiles with regenerating axons; one of these is thinly remyelinated. (b) Same as above. The unmyelinated axon is arrowed; a cilium (arrowhead) stands out in the nucleated Schwann cell profile; a common basal membrane loosely encircles both Schwann cells.

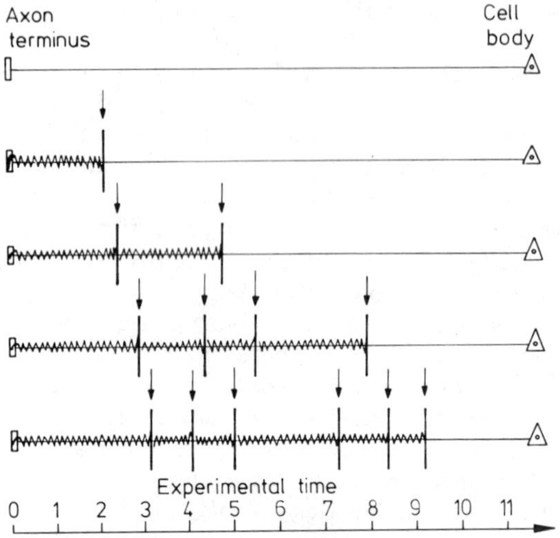

Figure 15 This drawing can be looked upon as a sort of time-table of outbound axonal pathology in disulfiram-poisoned rats. Time units in the bottom line are purely arbitrary. Absence or presence of axonal pathology and especially interaxonal differences in the amount of it are likewise arbitrary in their relative magnitude. The first axon from the top has no structural changes; the second shows one round of wallerian degeneration going out from the point shown by the arrow and off at the time shown in the bottom line; the third axon has two rounds of degeneration indicated by as many arrows; the fourth has four (arrows), and the fifth has six (arrows). Notice that every bout of degeneration taken separately is a dying-forth event, whereas all rounds considered collectively make up a dying-back phenomenon. It is possible that in vivo a situation like the one represented here in a highly schematic form is more apt to be found in the central than in the peripheral limb of disulfiram axonopathy.

only to reason. After all, any *new* pathology has no other way to go but back in the direction of the cell body. Under these circumstances one may question the biological importance of the dying-back concept without denying its clinical significance and heuristic value. But for the reader who may still be ill at ease with this conclusion, I would like to quote from a publication (11) of two well-known workers of long standing in the field of toxic neuropathies: [The term *dying-back*] "does not reflect accurately the spatio-temporal evolution of pathological events in short lengths of isolated *nerve fibers*" [italics in the original].

IV. CONCLUSIONS

On rounding off these comments, I would like to emphasize that the structural changes observed in the peripheral and central nervous system of disulfiram-poisoned rats are entirely compatible with those of wallerian degeneration. There is not a single feature in the material studied which could be taken out and looked on as a distinctive marker of yet a new morphological variant of wallerian degeneration. To the exclusion of a few minor details, everything found in disulfiram-intoxicated animals can also be found in wallerian degeneration, and nothing found in wallerian degeneration is far apart from and alien to what can be observed in rats given disulfiram. A point which is all too often a bit understated and thus tends to go under in the vast literature on toxic neuropathies is certainly the following: The common denominator of the pathological changes seen in *all* toxic axonopathies is wallerian degeneration, regardless of whether the axons are swollen with filaments (5,12), overstocked with smooth endoplasmic reticulum (3), or laid waste with amorphous material (10).

ACKNOWLEDGMENTS

The author is indebted to Dr. L. Palmucci for taking some of the electromicrographs and reading the manuscript, to P. Becker for looking after the feeding experiments, and to S. Luh for work with most of the semithin sections. The photography of V. Heinzinger and the typing of U. Qreini are also gratefully acknowledged.

REFERENCES

1. Ansbacher, L. E., Bosch, E. P., Cancilla, P. A. 1982. Disulfiram neuropathy: A neurofilamentous distal axonopathy. *Neurology 32*: 424-428.
2. Anzil, A. P. 1980. Selected aspects of experimental disulfiram neuromyopathy, in *Advances in Neurotoxicology* (Manzo, L., ed.), Pergamon, Oxford, pp. 359-366.
3. Bischoff, A. 1967. The ultrastructure of tri-orthocresyl phosphate poisoning. I. Studies on myelin and axonal alterations in the sciatic nerve. *Acta Neuropathol. 9*:158-174.
4. Bouldin, T. W., Hall, C. D., Krigman, M. R. 1980. Pathology of disulfiram neuropathy. *Neuropathol. Appl. Neurobiol. 6*:155-160.
5. Cavanagh, J. B. 1982. The pathokinetics of acrylamide intoxication: A reassessment of the problem. *Neuropathol. Appl. Neurobiol. 8*:315-336.

6. Jacobs, J. M., Cavanagh, J. B. 1972. Aggregations of filaments in Schwann cells of spinal roots of the normal rat. *J. Neurocytol.* *1*:161-167.
7. Mokri, B., Ohnishi, A., Dyck, P. J. 1981. Disulfiram neuropathy. *Neurology 31*:730-735.
8. Morris, J. H., Hudson, A. R., Weddell, G. 1972. A study of degeneration and regeneration in the divided rat sciatic nerve based on electron microscopy. II. The development of the regenerating unit. *Z. Zellforsch. 124*:103-130.
9. Olney, R. K., Miller, R. G. 1980. Peripheral neuropathy associated with disulfiram administration. *Muscle Nerve 3*:172-175.
10. Schröder, J. M. 1970. Zur Pathogenese der Isoniazid-Neuropathie. I. Eine feinstrukturelle Differenzierung gegenüber der Wallerschen Degeneration. *Acta Neuropathol. 16*:301-323.
11. Spencer, P. A., Schaumburg, H. H. 1977. Ultrastructural studies of the dying-back process. III. The evolution of experimental peripheral giant axonal degeneration. *J. Neuropathol. Exp. Neurol. 36*:276-299.
12. Suzuki, K., Pfaff, L. D. 1972. Acrylamide neuropathy in rats. An electron microscopic study of degeneration and regeneration. *Acta Neuropathol. 24*:197-213.
13. Thomas, P. K., King, R. H. M., Sharma, A. K. 1980. Changes with age in the peripheral nerves of the rat. An ultrastructural study. *Acta Neuropathol. 52*:1-6.

27

Neural Culture: A Tool to Study Cellular Neurotoxicity

Antonia Vernadakis and David L. Davies
University of Colorado School of Medicine, Denver, Colorado

Fulvia Gremo
University of Cagliari School of Medicine, Cagliari, Italy

I. INTRODUCTION

For several years pharmacology and toxicology have been concerned primarily with studies using whole animals, isolated organs, or tissue preparations. In recent years it has become apparent that it is desirable to work at a lower integrative level of the nervous system than can be provided by whole animals. This is particularly important for experiments requiring days and weeks to study development or to monitor pharmacological or toxicological effects of agents on neurogenesis or synaptogenesis. Moreover, considerable difficulties have frequently been encountered because of the poor accessibility of the central nervous system in vivo. The method of neural tissue and cell culture has gained considerable importance during the last few years and has provided insight into cellular physiology, as well as into the reactions of cells to certain external stimuli such as drugs and neurotoxins.

In this chapter we will briefly review studies in which the use of neural culture has provided clues to the cellular neurotoxic actions of central nervous system drugs, such as phenytoin, ethanol, and opiates, and of hormones when administered exogenously. Toxicological effects observed in culture include cellular activities such as cell growth and differentiation, cell adhesion, neurite outgrowth, synapse formation, and receptor development.

II. NEURAL CULTURE SYSTEMS

We will review briefly the various types of culture systems which have been used by us and others to study pharmacological effects of drugs. For detailed reviews the reader is referred to Murray (65), Dimpfel (29), Vernadakis and Culver (95), and Schrier (78).

Brain tissue explants maintained in *organ* and *organotypic* culture systems have been used extensively to study biochemical, morphological, electrophysiological changes induced by drug treatment. Harrison (43) and his associates pioneered the "hanging-drop" culture method for explants. Harrison placed fragments of embryonic tadpole tissue into a culture made with clotted lymph from adult frog. The cultures were viable for several weeks, long enough to observe the outgrowth of neuronal fibers from the bodies of individual cells. Murray and Stout in 1942 (66) and 1947 (67) described the adaptation of the Maximow double coverslip assembly (56) for cultivating neural tissue. The characteristic advantage of this method over the organ culture is the longer time that the neural tissue can be maintained viable in culture.

Dissociated neural cell cultures obtained by disruption of immature central nervous system tissue using lytic enzymes or mechanical procedures have been used to address a variety of problems concerning the biochemical, physiological, and morphological organization of the central nervous system. Since dissociation procedures disrupt the original tissue organization and abolish cell-cell contacts, it is possible to examine cellular performance and characteristics that reflect inherent cell properties rather than those expressed by an organized tissue. Cells from different central nervous system regions and from a variety of species have been studied in culture since Cavanaugh (13) first reported an attempt to dissociate and culture neural cells from chick embryo spinal cord. Basically, two types of culture systems have been used to maintain dissociated neural cells for study: the aggregate culture system and the monolayer culture system (10,80). Neuron-enriched and glia-enriched cultures can be obtained by using the appropriate animal age (fetal versus postnatal) and plating substratum (e.g., polylysine-coated culture dishes).

The search for methods to obtain neurons and glial cells separately and in large quantities in order to do biochemical analyses has led investigators to use neoplastic tissue as a potential source of neuronal and glial elements. Most of the studies carried out thus far have involved principally two neoplastic neural sources: the mouse neuroblastoma C1300 (3) and the C6 cell strain from a rat astrocytoma (4).

III. CELL GROWTH AND DIFFERENTIATION

Fundamental concepts and experimental evidence for cell recognition, aggregation, and differentiation will be briefly introduced in order to address possible interference by exogenous factors. For an extensive treatment of the subject, the reader is referred to a review by Moscona (62).

Embryonic morphogenesis depends on the aggregation and organization of individual cells and cell groups into characteristic multicellular patterns which give rise to distinct tissues and organs. During the early development of the nervous system, many of its precursor cells migrate from their sites of origin toward their final locations, associate with similar or other cell types into characteristic patterns, and eventually become functionally linked. Evidence (not reviewed here) has established that the complex organization of the nervous system depends on cell-cell recognition or cellular affinities. This capacity arises and evolves in the cells, along with their progressive differentiation. Failure of the mechanisms of cell recognition prevents cells from becoming organized correctly and may lead to defective morphogenesis in the affected parts of the nervous system (82). Cell recognition affinities involve not only those between identical cells (homotypic cell recognition), but also between different cells that cooperate developmentally (allotypic cell recognition; see Ref. 62), such as the contacts between glia and neurons, between neurons and muscle cells, and among different kinds of neurons.

In the following sections of this chapter evidence will be presented that xenobiotics may interfere with cell aggregation by changing the cell surface and thus altering cell recognition and cell-cell interactions such as those between neurons and glial cells.

A. Changes in Cell Growth Induced by Antiepileptic Drugs

Phenytoin is generally regarded as a very useful and safe anticonvulsant drug; however, on occasion it has been associated with acute and chronic toxic effects, the most serious of which are attributable to its effects on the central nervous system. Neurological dysfunctions, including a specific cerebellar symptomatology, have been reported in both epileptic children and adults who were treated with large doses of phenytoin (24,25,36,51,81; see also Bittencourt et al. in this book). However, in some animal studies (19) toxic changes in the nervous system, especially in the cerebellum, have not been demonstrated. Dam (18,19) and Dam and Nielsen (20) believe that phenytoin is not itself neurotoxic, but that associated seizures, deranged metabolic states, and modes of administration contribute to the cerebellar alterations. In animal experiments, problems include absorption, organ

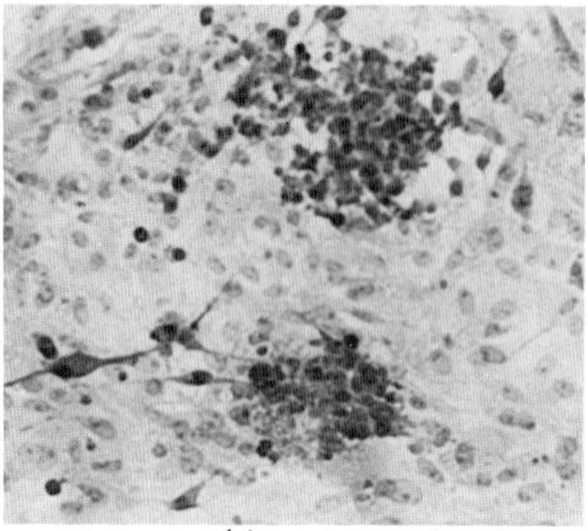

(a)

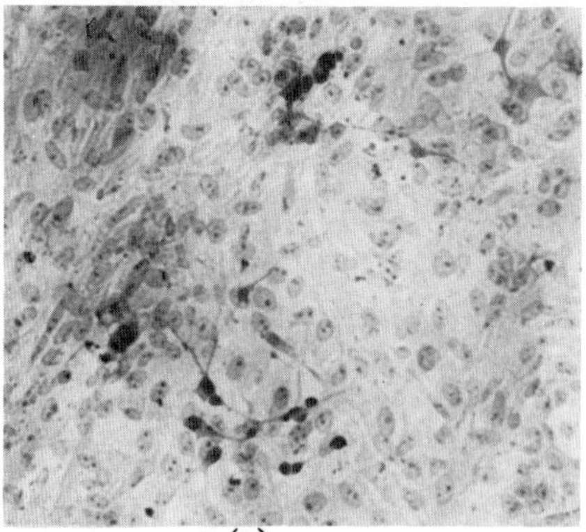

(b)

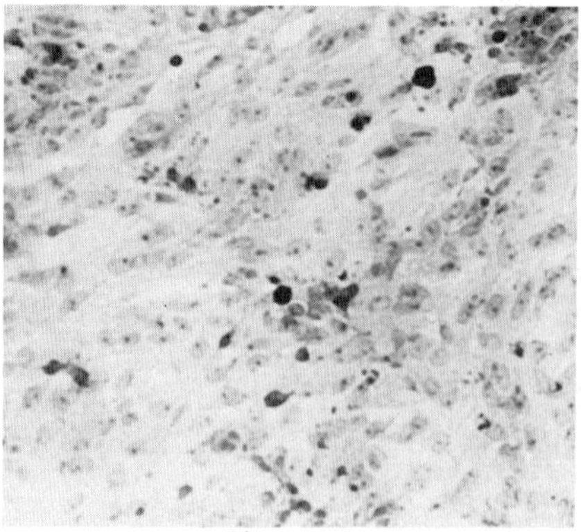

(c)

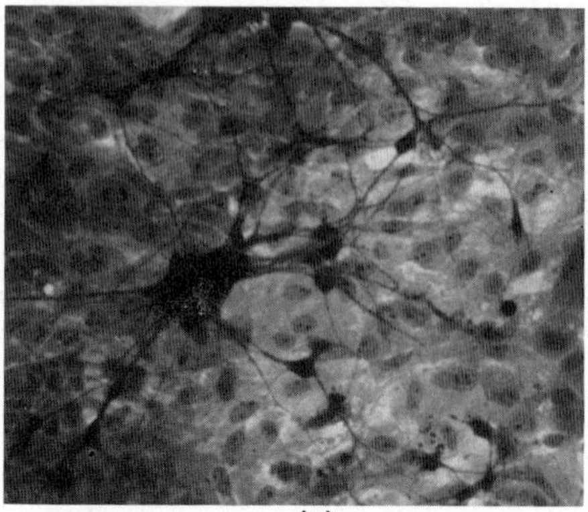

(d)

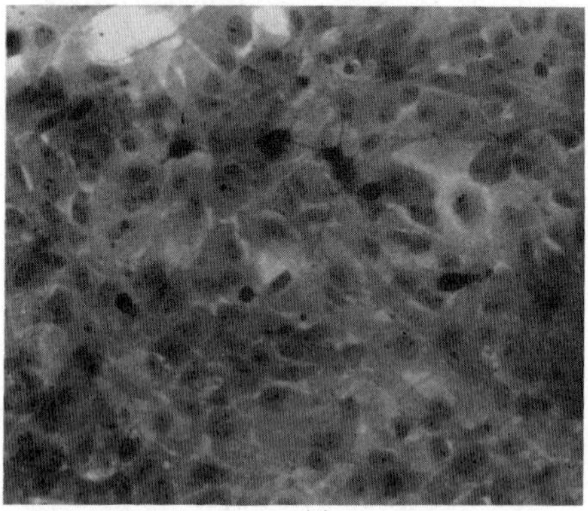

(e)

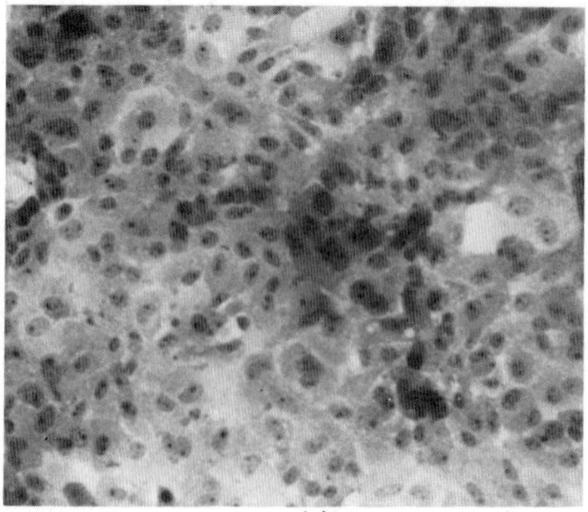

(f)

metabolism, the blood-brain barrier, hypoxia, and seizures; these problems can be circumvented by the use of tissue culture.

Studies using neural cultures have provided evidence that phenytoin directly affects cell growth. We have used dissociated brain cell cultures obtained from cerebral hemispheres and cerebella of chick embryos to examine the consequences of phenytoin, its metabolite 5-(p-hydroxyphenyl)-5-phenylhydantoin (HPPH), and phenobarbital on developing neuronal and glial cells in culture (17). Drugs are administered to cultures daily for a period of three consecutive days. Phenytoin added to dissociated neural cell cultures for this period of time produced a dose-related decrease in the number of neural cell aggregates (Figs. 1 and 2). Moreover, cultures treated with phenytoin, particularly with the higher concentration (5×10^{-4} M), contained fewer neurons, and also the neurons had fewer prominent neuronal processes. The reduction in number of neurons probably reflects the reduction of cell aggregates which consist predominantly of neuroblasts. In contrast to these striking effects on neurons, glial cells appeared to be relatively unaffected by phenytoin. Cultures treated with HPPH exhibited toxicity qualitatively and quantitatively similar to those treated with phenytoin (Fig. 3).

The toxicity produced by phenytoin was not observed with phenobarbital, even in concentrations as high as 1×10^{-3} M; phenobarbital did not alter aggregation or neurite outgrowth after the 3-day treatment period. These findings are also reflected in results comparing the effects of phenytoin and phenobarbital on protein synthesis as evidenced by the incorporation of [^{14}C]l-leucine into protein of neural cells (17). While leucine incorporation was not altered by phenobarbital, 3-day treatment of cultures with 5×10^{-4} M, but not with 1×10^{-4} M, phenytoin produced a significant decrease in the incorporation of [^{14}C]l-leucine.

Cultures dissociated from the cerebellum appeared to be somewhat more sensitive to the neurotoxic effects of phenytoin than cultures dissociated from the cerebral hemispheres. Comparisons of cultures treated with the low dose of phenytoin (1×10^{-4} M) relative to control cultures revealed that the effects of the 3-day drug treatment were more pronounced in cerebellum than in cerebral hemisphere cultures.

Figure 1 Dose-related effect of diphenylhydantoin (DPH) on the morphology of dissociated chick enbryo brain cell cultures. (a-c) Cultures from the cerebella of 10-day-old embryos on day 7 in culture: (a) control, (b) 1×10^{-4} M DPH, and (c) 5×10^{-4} M DPH treatment on days 4-6 in culture (toluidine blue stain, 200×). (d-f) Cultures from the cerebral hemispheres of 8-day-old embryos on day 9 in culture: (d) control, (e) 1×10^{-4} M DPH, and (f) 5×10^{-4} M DPH treatment on days 6-8 in culture (Bodian stain, 200×). (From Ref. 17.)

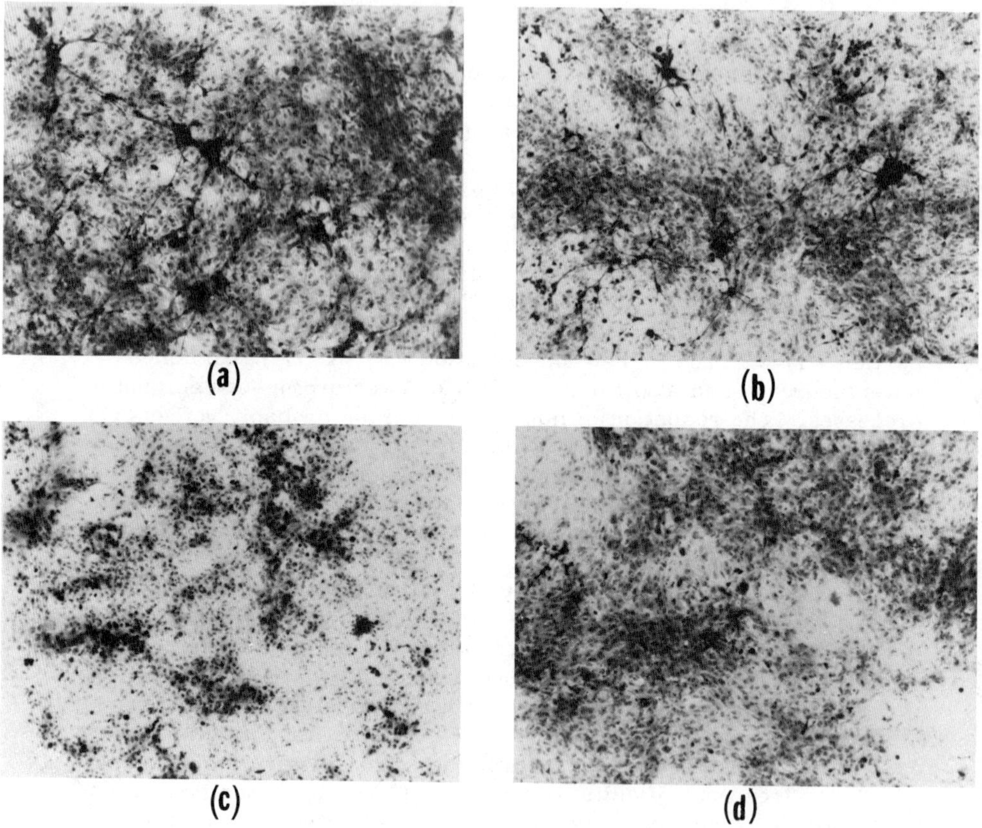

Figure 2 Brain cells from the cerebral hemispheres of 8-day-old chick embryos on day 9 in culture: (a) control, (b) phenobarbital (1×10^{-3} M), (c) DPH (5×10^{-4} M), and (d) HPPH (5×10^{-4} M) treatment on days 6-8 in culture (Bodian stain, 100×). (From Ref. 17.)

Evaluation of these findings is difficult because dissociated cells from different brain regions exhibit different developmental patterns. For example, dissociated cerebellar cells form fewer cell aggregates in culture compared to cells dissociated from cerebral hemispheres. However, treatment of the cultures at different stages of cultivation (1 or 2 weeks) resulted in similar findings. It is also of interest that large neurons resembling Purkinje cells in cultures from cerebellum were absent after treatment with phenytoin.

The susceptibility of the cerebellum to phenytoin neurotoxicity has recently been substantiated by Blank et al. (9) using cerebellar

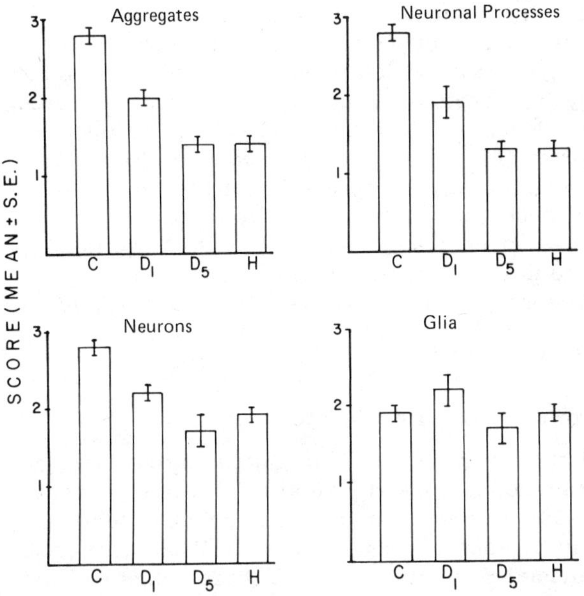

Figure 3 Evaluation of morphological parameters in control (C), 1 × 10^{-4} M DPH (D_1), 5 × 10^{-4} M DPH (D_5) and 5 × 10^{-4} M HPPH (H) cultures. Dissociated cerebral hemisphere cells from 8-day-old chick embryos were in culture for 9 days. Cultures were treated daily on days 6-8 of cultivation. Bars represent the means ± SE of the scores for 8-13 cultures. (From Ref. 17.)

explants from newborn Swiss Webster mice maintained in organotypic culture (Maximow assembly). Cerebellar cultures exposed at explantation to 30 μg/ml or more of phenytoin for a minimum of 5 days showed loss of large cortical neurons. The Purkinje cells became granular and pyknotic; at the ultrastructural level somatic and dendritic cytoplasm was dark and contained dense bodies and many alveolar vesicles, expecially adjacent to the Golgi apparatus. Degenerating Purkinje cell dendrites within the neuropil were numerous, and surviving presynaptic elements, the parallel fibers, were opposed to them. Impaired axonal development was also present. At higher concentrations, Purkinje cells were completely destroyed, and severe changes were seen in other cortical neurons, macroglia, and lastly in intracerebellar nucleus neurons.

The molecular mechanisms involved in the direct cellular toxicity of phenytoin are not understood. We have proposed that the decrease in total cellularity and in the number of aggregates of neural cells observed in phenytoin-treated cultures may be at least partially due to an increase in cell death. An alternative explanation is that phenytoin may alter cell-cell interaction or migration of cells from the aggregates. One possible mechanism by which phenytoin may influence cell aggregation could be by its effects on calcium ions. It has been recognized that reduction of calcium provides an effective tool for dissociating brain and other tissues (61). Conversely, studies on HeLa cells have shown that calcium ions promote aggregation of the cells (26). Moreover, it has been proposed that phenytoin may interact competitively with calcium at the neuronal membrane (88). Thus additional studies to investigate the possibility that an alteration in calcium dynamics plays a role in the morphological effects observed with phenytoin treatment of neural cell cultures seem to be warranted.

B. Changes in Cell Growth Induced by Opiates

It has been established that in humans and in animals narcotic drugs such as morphine and methadone markedly retard brain development (see the review in Ref. 98). Recent studies from our laboratory using both the chick embryo in ovo and dissociated brain cell cultures from chick embryos have provided some evidence that these drugs administered during early embryogenesis and neurogenesis interfere with cell maturation and differentiation (96).

Ornithine decarboxylase (ODC) has been implicated in the growth process in a wide range of living cells and has been used as an index of cell proliferation, hypertrophy, and differentation (16,44,83,84). It has been shown to decline with brain maturation in several animal species, including the chick (1,27,72,76). We also found a decline in ODC activity with neuronal maturation in cell cultures from dissociated cerebral hemispheres of chick embryos (72). Using ODC as a cell maturation index and 2',3'-cyclic nucleotide phosphohydrolase (CNP) as an oligodendrocytic marker (74), we have examined possible differential sensitivity of neurons and glial cells to methadone. Our findings suggest that a neuronal-glial interaction may be involved in the response of neural cells to methadone (96). Cultures were prepared from cerebral hemispheres of 8-day-old chick embryos for mixed neuronal-glial cultures and neuron-enriched cultures, using polylysine-coated Petri dishes; cerebral hemispheres from 15-day-old chick embryos were used for glia-enriched cultures.

Activity of ODC was significantly higher in mixed neuronal-nonneuronal cell cultures treated with methadone as compared to controls. This effect was not observed in either neuron- or glia-enriched cell cultures treated with methadone. In addition, the activity of CNP was higher in the methadone-treated mixed neuronal-nonneuronal cell cultures, but was not affected in the glia-enriched cultures as compared to controls. Thus both ODC and CNP activities were affected by methadone only in the mixed-cell cultures.

As discussed earlier, high ODC activity is associated with growing cells still in an undifferentiated state of maturation. Thus the high ODC activity in the mixed neuronal-nonneuronal cell cultures treated with methadone has been interpreted to reflect cells in the state of maturation and differentiation. That this effect was observed only in the mixed cell cultures and not in the neuron-enriched or glia-enriched cultures implicates a neuronal-glial interaction in the response to methadone. The high ODC activity may reflect either proliferation of glial cells or immature differentiating neuronal cells. The higher activity in CNP in the mixed-cell cultures suggests that either the number or the activity of glial cells may be affected by methadone.

The implication that neuronal changes produced by methadone influence glial cell activity as shown by CNP changes in the mixed-cell cultures should be considered. Nonneuronal cell responsiveness to neurons has been shown in the peripheral nervous system (11,12). However, information concerning the influence of neurons on glial proliferation is sparse, although there is extensive evidence on functional responses of glial cells to neuronal depolarization (48,49). Hanson and Partlow (42) have reported that neuronal sonicate or homogenate of neurons purified from sympathetic ganglia of 12-day-old chick embryos stimulated the incorporation of [^{3}H]methylthymidine in nonneuronal cells. The authors suggest that neurons contain one or more substances which are mitogenic for nonneuronal cells. Opiates have been shown to affect the release of neurotransmitter substances in various nervous tissues (59,85,86,92). Whether in the present study the increase in glial cells is a result of neurotransmitter substances released in the microenvironment remains to be investigated. It has been shown that addition of norepinephrine to C6 glial cells produces a twofold increase in the specific activity of CNP (60). We have found that addition of norepinephrine to cerebellar explants in culture increases the activity of butyrylcholinesterase, an enzyme predominantly localized in glial cells (97). Thus the possibility that glial cells are influenced by a change in the microenvironment produced in culture by methadone treatment remains to be further investigated.

C. Changes in Cell Growth Induced by Ethanol

Numerous types of physiological and behavioral alcohol studies have been conducted in mature organisms; however, physiological studies of ethanol and developing organisms have mostly concentrated on the teratogenic effects of the compound. The effects of ethanol on the human fetus have been described as the fetal alcohol syndrome and have received considerable attention because of their implications in the later life of the child (15,32). Ethanol interacts with various organ systems to produce malformations, but central nervous system involvement is the most characteristic. However, the mechanisms by which ethanol may affect fetal brain growth are complex and may include secondary effects resulting from the influence of alcohol on the maternal organism. Neural cell culture is a useful model with which to delineate direct cellular effects of ethanol that may contribute to the aberrant brain growth and development typical of the fetal alcohol syndrome.

The impact of ethanol on the proliferation, differentiation, and metabolic capabilities of neural cells has been studied in a diversity of cell culture models. Delineation of specific toxin-induced cellular changes must encompass subtle biochemical and metabolic alterations as well as cytological changes. Aberrant metabolism may interfere with the ability of cells to (a) synthesize cellular components, for example, enzymes, transmitters, and trophic factors; (b) alter membrane composition; or (c) respond to neighboring cells, for example, glial cells, and the extracellular milieu.

The morphological and electrophysiological effects of chronic physiological doses of ethanol on neural cell cultures have been studied by Scott and Edwards (79). In cell cultures derived from the dorsal root ganglia of adult mice, nonneuronal cells were observed to be more susceptible to ethanol than neurons. After 8 days, exposure to ethanol concentrations of 0.17 and 0.34 g % did not alter neuronal survival, and only a 10% reduction in the neuronal population resulted from exposure to 0.51 g %; whereas 0.17 g % ethanol concentrations produced marked decreases relative to controls in cell counts on nonneuronal cells. Phase-microscopic examination of the cultures indicated a reduction in nonneuronal cell density in cultures exposed to 0.51 and 0.68 g % ethanol. The nonneuronal cell population in these cultures consisted of fibroblasts, capsule, Schwann, and endothelial cells; the extent to which these different cell types were vulnerable was not determined.

Cellular toxicity resulting from ethanol has been characterized in homogeneous populations of clonal cell lines (33,55) or through comparisons drawn between primary and transformed cell lines (53). Waziri et al. (102) were the first to examine the susceptibility of differentiating cells in culture to ethanol toxicity. Morphological and enzymatic differentiation was induced in C6 glioma cells by applying dexamethasone (10 μg/ml) to cultures. Ethanol concentrations of 1% diminished morphological differentiation, and 1.5% eliminated it. Additionally, exposure of the cells to ethanol also reduced the dexamethasone induction of 2',3'-cyclic nucleotide 3'-phosphohydrolase, an enzyme associated with myelination (52) and an oligodendroglial marker (74). In the absence of a maturation-stimulating agent, the glioblast (C6) and astroblast (NN) cell lines were morphologically indistinguishable when exposed to ethanol (100 mM) from untreated control cultures (55). Ethanol produces a variety of changes in the biosynthetic capacity of cells in the absence of changes in cytological appearance of cell density. Whole-cell synthesis of polyribosomal RNA and protein is reduced in Cox astrocytoma cells exposed to 100 mM ethanol for 10 days (33).

Transformed cell lines differ in their responsiveness to ethanol exposure. The membrane-bound enzymes Mg^{2+} adenosinetriphosphatase (ATPase) and Na^+-K^+ ATPase are differentially affected in various cell lines (53,90). In one neuroblastic and two glioblastic cell lines chronic exposure to ethanol produces a transitory increase in high-affinity choline uptake (55). However, the length of time required to increase this system varied among the clonal cell lines. The changes in ATPase activity and high-affinity uptake of choline may be related to ethanol-induced changes in the properties of plasma membranes. The presence of ethanol tends to fluidize membranes by disordering the acyl chains of the phospholipids embedded in the interior of the membrane (see the review in Ref. 75). Chronic exposure to this drug is thought to result in adaptive changes in the composition of membranes which may in turn affect receptors and membrane-bound enzymes (14). Increased enzymatically releasable sialic acid from the surface of cultured astroblastic clone NN cells also suggests a possible ethanol-induced membrane lesion (68).

The role of glial cells in the neurotoxicity of xenobiotics is not understood. In view of the fact that glial cells make up a large part of the brain tissue and the evidence that they regulate the neuronal microenvironment, their responsiveness and sensitivity to ethanol warrant investigation. We have used primary glia-enriched cultures to study the effects of ethanol on glial growth and differentiation and the possible involvement of glial cells in ethanol neurotoxicity (21,22). Glutamine synthetase (GS) activity was measured, since it has been used as an astrocyte marker (70) implicated in compartmentation of glutamate and glutamine (6) and plays a role in detoxification of ammonia (see the review in Ref. 69). Glia-enriched cultures prepared from cerebral hemispheres of 15-day-old embryonic chicks were exposed to ethanol at four dose levels (0.1, 0.5, 1, and 2% wt/vol) between culture days 6 and 10. Cell number as assessed by DNA content was lower in cultures treated with 1 or 2% ethanol than for controls; the response was dose dependent, indicating that ethanol exposure levels above 1% impair cell proliferation. Glutamine synthetase activity decreased markedly with treatment levels of 1 and 2% ethanol. The decrease in GS activity per milligram protein in ethanol-exposed cultures suggests that GS may be selectively susceptible to ethanol. Alteration in glutamate and glutamine metabolism may be a potential consequence of ethanol-induced changes in GS activity. Alternatively, ethanol may selectively inhibit the proliferation of a subpopulation of astrocytic cells high in GS activity.

D. Changes in Cell Growth Induced by Steroid Hormones

The role of hormones as intrinsic regulators of brain growth has been investigated extensively. Studies in vivo and in culture have established that the effects of steroid hormones depend on the stage of maturation of neural tissue at the time of hormone treatment and that the specific hormones act on specific neural structures (45,99,100). Using cerebellar explants from 15-day-old chick embryos in organ neural tissue culture (for details see Ref. 94), we found that DNA content was significantly higher in explants cultured for 24 hr in the presence of cortisol (2.76×10^{-5} M) or estradiol (2.65×10^{-5} M). The increase in DNA was interpreted to reflect an increase in cell proliferation, and specifically in the number of glial cells, which are known to proflierate actively in the 15-day-old chick embryonic brain. This view was supported by a study using organotypic neural tissue culture (for details see Ref. 94). Cerebellar explants from 15-day-old chick embryos were cultured in the presence of vairous steroid hormones (10^{-8} M) for 5 days. The rate of migration of glial cells from the explants was calculated during the growth period between the first and fifth days in culture. We found that cortisol, corticosterone, estradiol, and progesterone, but not testosterone, enhanced the neuroglial migration rate. In the migration zone, astrocytes appear earlier in the hormone-treated explants as compared to controls. We interpreted these findings to reflect an increase in glial proliferation and differentiation produced by steroid hormones.

Both inhibitory and stimulatory effects on cell proliferation have been reported, using various cell lines as experimental models. Grasso (40) has reported proliferation of C6 glioma cells treated with cortisol, 3×10^{-6} M, during the log growth phase in culture; then cell proliferation ceased as growth in control cultures continued into a stationary phase. A 2-day period of growth inhibition followed, and growth subsequently resumed. It appears from these findings that the proliferative capacity of cortisol-treated cells is not irreparably damaged. A transient cell growth inhibitory effect by a synthetic steroid, dexamethasone, was also observed by Mealy et al. (58) in cell cultures of human glioblastomas. Cortisol significantly increased the growth of primary monolayer cultures of lung cells from rabbit fetuses at 20 days of gestation; however, when cultures were prepared from lung cells from fetuses at 28 days, the effect was reversed and cortisol reduced growth by a factor of 2 (87). Thus, as discussed earlier, with respect to neural tissue, the responsiveness of lung cells to cortisol depends on the age of the fetus at the time of exposure.

A primary effect of corticoid hormones appears to be related to enzyme synthesis and induction. Glycerolphosphate dehydrogenase synthesis is reversibly induced by hydrocortisone (5,57). Numerous reports have described the induction of glutamine synthetase either in C6 glial cells (46), primary astrocyte cultures (50), rat glioma cells (73), or retina Müller cells (63). Other enzymes such as tyrosine hydroxylase (103) or tyrosine aminotransferase (89) or synthesis of enzymes related to lipid metabolism (101) is increased by glucocorticoids. In some cases, enzyme induction by glucocorticoids is only indirect. In adrenergic neuronal cultures, hormones enhance response to nerve growth factor (71). Much less information is available on the effects of thyroid hormones on enzyme induction. Triiodothyronine has been shown to influence the development of 2',3'-cyclic nucleotide 3'-phosphohydrolase in dissociated embryonic mouse brain cells (8). Triiodothyronine increases the activity of choline acetyltransferase and acetylcholinesterase, as well as the developmental rise of glutamic acid decarboxylase in aggregating fetal rat brain cells (47). The activity of Na^+-K^+ ATPase was negligible when neuroblastoma cells were cultured in fetal calf serum deficient in triiodothyronine or thyronine, whereas the activity of C6 glioma cells was not affected (30). The decrease in Na^+-K^+ ATPase in neuronal cells parallels the low activity of this enzyme observed in the brain of hypothyroid rats during development (93).

The studies discussed demonstrate that hormones modulate cell growth and differentiation in culture. Their actions can be cell growth promoting, but also cell inhibitory. Thus in the intact developing organism changes in the hormonal cellular environment by exogenously administered hormones, drugs acting on hormonal regulation, and changes in the maternal homeostatic regulation may contribute to normal or abnormal central nervous system development, depending on the direction of the hormonal influence.

IV. CELL SURFACE AND RECEPTORS

A. Changes in Opiate Receptors

It has been established that infants born to mothers addicted to opiates exhibit physiological dependency on narcotics and develop symptoms of withdrawal at birth. Numerous pharmacological studies have demonstrated that the effects of opiates in vitro and in vivo reflect their interaction with specific neuronal receptors. We have undertaken to examine the hypothesis that interaction of the opiate-binding sites with endogenous opioid peptides during early embryogenesis could be critical for the normal development of the nervous system and the embryo as a whole. We found stereospecific binding activity using [^{3}H]etorphine as a ligand in the chick embryo brain as early as 4 days of incubation (37). Not only was binding activity present in the brain at 4 days of incubation, but the same amount of stereospecific binding activity per milligram of protein was present in body tissue at 4-7 days of incubation. By 10 days of incubation the amount of binding activity had increased substantially in brain tissue and was not detected in body tissue. The number of opiate-binding sites per milligram of protein continued to increase in brain tissue until day 20 of incubation. The possibility that these early opiate receptors are functional is supported by reports that endogenous opiates are present very early in chick embryos; enkephalin-like immunoreactivity has been observed at

5 days of incubation in the chick embryo telencephalic vesicles (23), and leu- and metenkephalin detected by radioimmunoassay in chick embryo brain and gut at 5 days of incubation (31).

In view of the early presence of opiate binding, we were interested in investigating the effects of the narcotic drug L-α-acetylmethadol (LAAM) on opiate receptor binding in the developing chick embryo. Chick embryos were exposed to LAAM at specific periods of embryonic development, and chick viability and hatchability were considered to be indices of the general toxicity of LAAM. In addition, the maximum number of [^{3}H]etorphine-binding sites (B_{max}) and affinity of [^{3}H]etorphine binding (K_D) in the chick brain following treatment with LAAM, methadone, and morphine sulfate were measured as indices of neurotoxicity. The chick embryo was most sensitive to LAAM between days 3 and 6 of incubation, and more sensitive to LAAM than methadone or morphine. The LAAM consistently decreased B_{max} values and increased K_D values for [^{3}H]etorphine in all ages investigated (38).

We have further examined neuronal sensitivity to opiates, using neuron-enriched cultures (Fig. 4) from dissociated chick embryonic brain (39). We found opiate-specific binding sites in dissociated cell cultures from 6.5-day-old chick embryo whole brain 9 days in culture. As in vivo, in embryonic brain tissue preparations Mn^{2+} increased both B_{max} and K_D, and both Na^+ and guanosine triphosphate (GTP) decreased B_{max} (38). Cultures were exposed to L-acetylmethadone, 10^{-6} M, from day 6 to day 7 in culture, and opiate binding activity was assayed either immediately after treatment or 24 hr later. We observed that N-LAAM decreased B_{max} and increased K_D. In view of the findings reported earlier in this chapter that ODC is higher in dissociated brain cell cultures of mixed neurons and glial cells (96), we interpret these findings to mean that N-LAAM influences neuronal maturation and thus produces a decrease in binding sites. Furthermore, the higher K_D supports the view that LAAM may have a general effect on neuronal growth and its microenvironment.

B. Changes in Benzodiazepine Receptors

Several reports have suggested a relationship between convulsant-anticonvulsant mechanisms and alterations in brain benzodiazepine receptors (see references in Ref. 35). Gallager et al. (35) studied the effects of chronic exposure (7 days) of primary dissociated cerebral cortical cells in culture to the anticonvulsant drug phenytoin, using benzodiazepine binding. Phenytoin treatment was associated with increases in nonneuronal-type benzodiazepine-binding sites and with decreases in neuronal-type benzodiazepine-binding sites. These drug effects may involve the differential survival and/or proliferation of neuronal and nonneuronal cell populations containing benzodiazepine-binding sites. Such changes observed in culture are consistent with

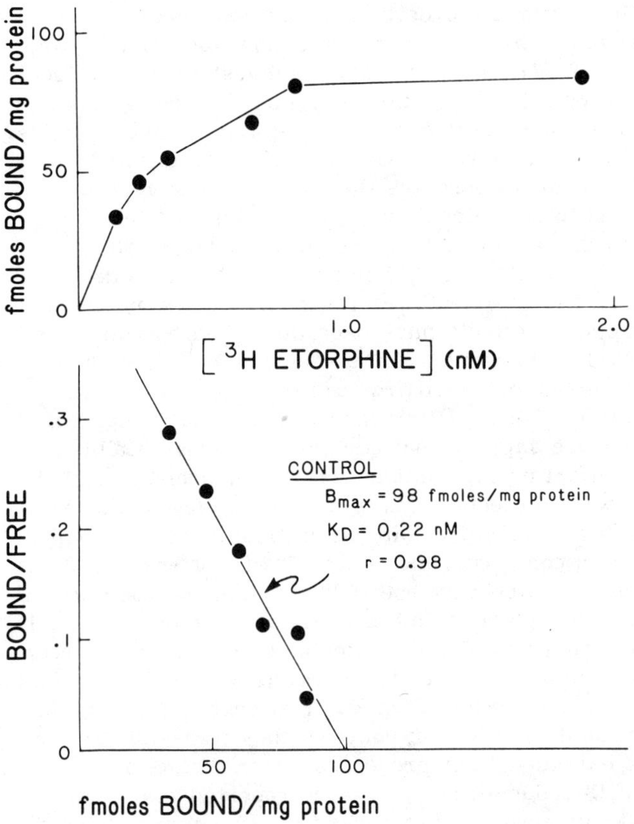

Figure 4 Stereospecific [^{3}H]etorphine binding in neuron-enriched cell cultures dissociated from 6.5-day-old chick embryo whole brain and plated on polylysine-coated dishes: 9 days in culture. (Top) Saturation curve. (Bottom) Scatchard analysis. (From Ref. 38.)

phenytoin-induced changes observed in the fetal and early postnatal animal (34).

C. Changes in Cell Surface Induced by Hormones

It is known that the cell surface is directly involved in a variety of cell functions, including immunological phenomena, cancer proliferation, reactions to hormones and toxins, cell-cell interaction, and adhesion. In this section we will briefly discuss some of the hormonal effects which can be linked with cell surface changes.

Morphological differentiation of both glial and neuronal clone cell lines has been reported to be induced by glucocorticoids and thyroid and sex hormones (7,28,77). Recent studies suggest that the effects of glucocorticoids on cell differentiation may be mediated through their effects on cell surface components, for example, glycoproteins. Arenander and de Vellis (2) have shown that administering hydrocortisone to C6 glial cell cultures decreases the synthesis and release of particulate compartments of proteins and glycoproteins. This effect is selective and inducible at physiological concentrations of the hormone ranging from 10^{-9} to 10^{-6} M. Other authors (54) have demonstrated an increase in the fibronectin extracellular matrix by rat hepatocytes in vitro. Opposite effects have been described by Murota et al. (64), who detected an inhibition of the secretion of intercellular protein in cloned fibroblasts derived from rat carrageenin granuloma.

Recent studies by our laboratory using retina monolayer cell cultures from chick embryos suggest that corticosterone may inhibit neuronal growth by affecting interactions between neurons and glial cells (41). Retinas were dissected from 8-day-old embryos, dissociated, and cells plated on salt-precipitated collagen. At day 5 cultures were treated with corticosterone (from 10^{-9} to 10^{-7} M) for 24 hr. Controls received either Dulbecco's modified essential medium plus 10% fetal calf serum or Dulbecco's modified essential medium only. Inhibition of neuronal sprouting was already detectable after 24 hr of treatment (Fig. 5). When higher doses of the hormone were used (10^{-7} M), process development was severely affected. Flat cells were fewer in number and nonconfluent, whereas in controls they formed a "carpet" over which neurons extended their processes. Our results are in accord with those of Unsicker et al. (91), who reported that corticosterone impaired the outgrowth of processes in cultured sympathetic neurons and adrenal chromaffin cells.

Neuronal sprouting could be directly inhibited by corticosterone, but it could also be a consequence of a hormone-induced interaction between the substratum and flat cells. The lack of flat-cell confluency in hormone-treated cultures may be a result of lack of flat-cell migration from the aggregates due to alterations in adhesiveness induced by the hormone. The number and orientation of flat cells could be due to a change in microtubule assembly, as has already been shown in glial cell lines (7). Also, cell surface modification could account for this phenomenon. Wu and Sato (104) have shown that hydrocortisone alters the cell surface in several lines of rat, mouse, and human cells, as indicated by changes in cell morphology, rate of detachment from the dish, and pattern of surface proteins. The peculiar behavior of flat cells, including their lack of confluency, may be partially responsible for the delay in differentiation of immature neurons. At present, it can only be speculated that if glial cells are influenced by guidance from neuronal growth cones, the absence of interaction between flat cells and neurons may interfere with development of neuronal processes.

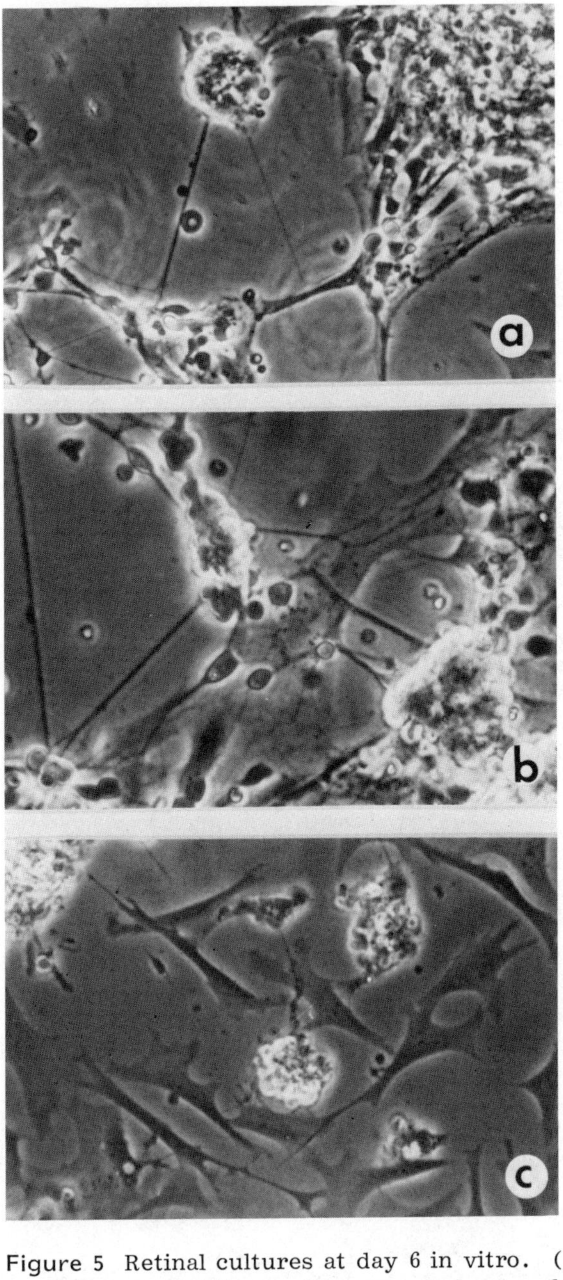

Figure 5 Retinal cultures at day 6 in vitro. (a) Representative picture of control culture. The neuronal network is well developed and neuronal aggregates can be observed (520×). (b) Serum-free cultures after 24 hr of serum deprivation. Neuronal sprouting is more evident than in the controls (750×). (c) Hormone-treated cultures 24 hr after corticosterone treatment (10^{-7} M). The development of neuronal processes is minimal; flat cells are more numerous and not confluent as in controls and serum-free cultures (520×). (From Ref. 41.)

As discussed earlier in this chapter, hormones play a vital role in normal brain maturation. In culture studies have contributed to our understanding of the cellular actions of hormones and have shown that these intrinsic substances have the potential to produce neurotoxic effects.

V. CONCLUSIONS

The studies briefly reviewed in this chapter demonstrate that some central nervous system drugs produce neurotoxic effects by acting directly on neural cellular activity. The effects observed in culture can be related to those reported in in vivo studies. For example, cerebellar neurotoxicity of phenytoin is observed in cultured cerebellar cells, as well as in vivo. The differential sensitivity of brain tissue to hormones is also illustrated in cultured cells. It must be noted, however, that direct extrapolation of neurotoxic effects of drugs observed in culture must take into consideration possible alterations in drug disposition in vitro. Thus dose-response curves in vitro may not be directly correlated to in vivo findings.

The advantage of the neural culture system as an experimental tool to study central nervous system drug effects is that it offers the opportunity to observe the drug sensitivity of individual cell types, cell-cell interactions altered by drugs, and molecular cellular components that are involved in drug action.

ACKNOWLEDGMENTS

This review was partially supported by USPHS Research Grand DA 02131, Training Grant T32 DA 07043, NATO Grant No. 148-80, and the Developmental Psychobiology Research Endowment Fund.

REFERENCES

1. Anderson, T. R., Schanberg, S. M. 1972. Ornithine decarboxylase activity in developing rat brain. *J. Neurochem. 19*:1471-1481.
2. Arenander, A. T., de Vellis, J. 1980. Glial-released proteins in clonal cultures and their modulation by hydrocortisone. *Brain Res. 200*:401-419.
3. Augusti-Tocco, G., Sato, G. 1969. Establishment of functional clonal lines of neurons from mouse neuroblastoma. *Proc. Nat. Acad. Sci. U.S.A. 64*:311-315.
4. Benda, P., Lightbody, J., Sato, G., Levine, L., Sweet, W. 1968. Differentiated rat glial strain in tissue culture. *Science 161*:370-371.

5. Bennett, K., McGinnis, J. F., de Vellis, J. 1977. Reversible inhibition of the hydrocortisone induction of glycerol phosphate dehydrogenase by cytochalasin B in rat glial C6 cells. *J. Cell. Physiol.* *92*:247-260.
6. Berl, S., Takagaki, G., Clarke, D. D., Waelsch, J. 1962. Metabolic compartments *in vivo*: Ammonia and glutamic acid metabolism in brain and liver. *J. Biol. Chem.* *237*:2562-2569.
7. Berliner, J. A., Bennett, K., de Vellis, J. 1978. Effect of hydrocortisone on cell morphology in C6 cells. *J. Cell. Physiol.* *94*:321-334.
8. Bhat, N. R., Shanker, G., Pieringen, K. A. 1981. Investigations on myelination "*in vitro*": Regulation of 2',3'-cyclic nucleotide 3'-phosphohydrolase by thyroid hormone in cultures of dissociated brain cells from embryonic mice. *J. Neurochem.* *37*:695-701.
9. Blank, N. K., Nishimura, K. N., Seil, F. J. 1982. Phenytoin neurotoxicity in developing mouse cerebellum in tissue culture. *J. Neurol. Sci.* *55*:91-97.
10. Booher, J., Sensenbrenner, M. 1972. Growth and cultivation of dissociated neurons and glial cells from embryonic chick, rat and human brain in flask cultures. *Neurobiology* *2*:97-105.
11. Bunge, R. P., Bunge, M. B. 1978. Evidence that contact with connective tissue matrix is required for normal interaction between Schwann cells and nerve fibers. *J. Cell Biol.* *78*:943-950.
12. Bunge, R. P., Bunge, M. B., Cochran, M. 1978. Some factors influencing the proliferation and differentiation of myelin-forming cells. *Neurology* *28*:59-67.
13. Cavanaugh, M. W. 1955. Neuron development from trypsin-dissociated cells of differentiated spinal cord of the chick embryo. *Exp. Cell Res.* *9*:42-48.
14. Chin, J. H., Parsons, L. M., Goldstein, D. B. 1978. Increased cholesterol content of erythrocyte and brain membranes in ethanol-tolerant mice. *Biochim. Biophys. Acta* *513*:358-363.
15. Clarren, S. K., Smith, D. W. 1978. The fetal alcohol syndrome. *N. Engl. J. Med.* *298*:1063-1067.
16. Costa, M. 1979. Levels of ornithine decarboxylase activation used as a simple marker of metal induced growth arrest in tissue culture. *Life Sci.* *24*:705-714.
17. Culver, B., Vernadakis, A. 1979. Effects of anticonvulsant drugs on chick embryonic neurons and glia in culture. *Dev. Neurosci.* *2*:74-85.
18. Dam, M. 1970. Number of Purkinje cells after diphenylhydantoin intoxication in pigs. *Arch. Neurol.* *22*:64-67.
19. Dam, M. 1972. The density and ultrastructure of the Purkinje cells following diphenylhydantoin treatment in animals and man. *Acta Neurol. Scand. Suppl.* *49*:3-65.

20. Dam, M., Nielsen, M. 1970. Purkinje cell density after diphenylhydantoin intoxication in rats. *Arch. Neurol. 23*:555-557.
21. Davies, D. L., Vernadakis, A. 1983. Effects of ethanol on the astrocytic marker glutamine synthetase in the cerebral hemispheres of chick embryos, in *Proceedings of the 4th International Congress, International Society for the Development of Neuroscience*, Salt Lake City, Utah, July 3-7, 1983, Pergamon, Elmsford, N.Y., p. 37.
22. Davies, D. L., Vernadakis, A. 1983. Ethanol effects on glial cell cultures: Cell proliferation and glutamine synthetase activity. *Soc. Neurosci. Abstr. 9*:450.
23. Davis, B. M., Brecha, N., Karten, H. J. 1980. Enkephalin-like immunoreactivity in developing avian basal ganglia and nucleus spiriformis lateralis, in *Proceedings of the 10th Annual Meeting, Society for Neuroscience*, Abstract 250:2.
24. Del Cerro, M. P., Snider, R. S. 1967. Studies on Dilantin intoxication. Part 1: Ultra-structural analogies with the lipidoses. *Neurology 17*:452-466.
25. Del Cerro, M. P., Snider, R. S. 1970. Cerebellar alterations resulting from Dilantin intoxication: An ultrastructural study, in *The Cerebellum in Health and Disease* (Fields, W. S., Willis, W. D., eds.), Warren H. Green, St. Louis, pp. 380-408.
26. Deman, J. J., Bruyneel, E. A., Mareel, M. M. 1974. A study of the mechanism of intercellular adhesion. Effects of neuraminidase, calcium and trypsin on the aggregation of suspended HeLa cells. *J. Cell Biol. 60*:641-652.
27. DeMello, F. G., Bachrach, U., Nirenberg, M. 1976. Ornithine and glutamic acid decarboxylase activities in the developing chick retina. *J. Neurochem. 27*:847-851.
28. de Vellis, J., Inglish, D., Cole, K., Mocson, J. 1971. Effects of hormones on the differentiation of cloned lines of neurons and glial cells, in *Influence of Hormones on the Nervous System* (Ford, D. H., ed.), S. Karger, Basel, pp. 25-39.
29. Dimpfel, W. 1980. Rat nerve cell cultures in pharmacology and toxicology. *Arch. Toxicol. 44*:55-62.
30. Draves, D. J., Timiras, P. S. 1980. Differential effects of altered thyroid hormone states on nervous tumor cells, in *Multidisciplinary Approach to Brain Development* (Di Benedetta, C., Balazs, R., Gombos, G., Porcellati, G., eds.), Elsevier/North Holland, New York, pp. 313-315.
31. Epstein, M., Lindberg, I., Dahl, J. L. 1980. Development of enkephalin in chick brain, gut, adrenal and Remak's ganglion, in *Proceedings of the 10th Annual Meeting, Society for Neuroscience*, Abstract 211:14.
32. Erb, L., Andresen, B. D. 1978. The fetal alcohol syndrome (FAS). *Clin. Pediatr. 17*:644-649.

33. Flemming, E. W., Woodson, M. E., Tewari, S. 1981. Ethanol and cycloheximide alter protein and RNA synthesis of Cox astrocytoma cells in culture. *J. Neurosci. Res.* *6*:511-524.
34. Gallager, D. W., Mallorga, P. 1980. Diphenylhydantoin: Pre and postnatal administration alters diazepam binding in developing rat cerebral cortex. *Science* *208*:64-66.
35. Gallager, D. W., Mallorga, P., Swaiman, K., Neales, E. A., Nelson, P. G. 1981. Effects of phenytoin on ^{3}H-diazepam binding in dissociated primary cortical cell culture. *Brain Res.* *218*:319-330.
36. Ghatak, N. R., Santoso, R. A., McKinney, W. M. 1976. Cerebellar degeneration following long-term phenytoin therapy. *Neurology* *26*:818-820.
37. Gibson, D. A., Vernadakis, A. 1982. ^{3}H-Etorphine binding activity in early chick embryos: Brain and body tissue. *Dev. Brain Res.* *4*:23-29.
38. Gibson, D. A., Vernadakis, A. 1983. Critical period for LAAM in the chick embryo: Toxicity and altered opiate receptor binding. *Dev. Brain Res.* *8*:61-69.
39. Gibson, D. A., Vernadakis, A. 1983. Effects of N-LAAM on ^{3}H-etorphine binding in neuronal-enriched cell cultures. *Neurochem. Res.* *8*:1197-1202.
40. Grasso, R. J. 1976. Transient inhibition of cell proliferation in rat glioma monolayer cultures by cortisol. *Cancer Res.* *36*:2408-2414.
41. Gremo, F., Porru, S., Vernadakis, V. 1984. Effects of corticosterone on chick embryonic retinal cells in culture. *Dev. Brain Res.* *15*:45-52.
42. Hanson, G. R., Partlow, L. M. 1978. Stimulation of non-neuronal cell proliferation *in vitro* by mitogenic factors present in highly purified sympathetic neurons. *Brain Res.* *159*:195-210.
43. Harrison, R. G. 1907. Observations on the living developing nerve fiber. *Proc. Soc. Exp. Biol. Med.* *4*:140-143.
44. Heby, O. 1981. Role of polyamines in the control of cell proliferation and differentiation. *Differentiation* *19*:1-20.
45. Heim, L. M., Timiras, P. S. 1963. Gonad-brain relationship: Precocious brain maturation after estradiol in rats. *Endocrinology* *72*:598-606.
46. Holbrook, N. J., Grasso, R. J., Hackney, J. F. 1981. Glucocorticoid receptor properties and glucocorticoid regulation of glutamine synthetase activity in sensitive C6 and resistant C6H glial cells. *J. Neurosci. Res.* *6*:75-88.
47. Honegger, P., Lenoir, D. 1980. Triiodothyronine enhancement of neuronal differentiation in aggregating fetal rat brain cells cultured in a chemically defined medium. *Brain Res.* *199*:425-434.

48. Hösli, L., Andres, P. F., Hösli, E. 1979. Effects of 4-aminopyridine and tetraethylammonium on the depolarization of GABA of cultured satellite glial cells. *Neurosci. Lett.* *11*:193-196.
49. Hösli, L., Andres, P. F., Hösli, E. 1979. Neuron-glia interactions: Indirect effect of GABA on cultured glial cells. *Exp. Brain Res.* *33*:425-434.
50. Juurlink, B. H. J., Schousboe, A., Jorgensen, O. S., Hertz, L. 1981. Induction by hydrocortisone of glutamine synthetase in mouse primary astrocyte cultures. *J. Neurochem.* *36*:136-142.
51. Kogenge, R., Kutt, H., McDowell, F. 1965. Neurological sequelae following Dilantin overdose in a patient and in experimental animals. *Neurology* *15*:823-829.
52. Kurihara, T., Tsukada, Y. 1968. 2',3'-Cyclic nucleotide 3'-phosphohydrolase in the developing chick brain and spinal cord. *J. Neurochem.* *15*:827-832.
53. Mandel, P., Ledig, M., M'Paria, J.-R. 1980. Ethanol and neuronal metabolism. *Pharmacol. Biochem. Behav. Suppl.* *1*:175-182.
54. Marceau, N., Goyette, R., Valet, J. P., Deschenes, J. 1980. The effect of dexamethasone on formation of a fibronectin extracellular matrix by rat hepatocytes *in vitro*. *Exp. Cell Res.* *125*: 497-502.
55. Massarelli, R., Syapin, P. J., Noble, E. P. 1976. Increased uptake of choline by neural cell cultures chronically exposed to ethanol. *Life Sci.* *18*:397-404.
56. Maximow, A. 1925. Tissue-cultures of young mammalian embryos. *Contrib. Embryol.* *16*:47-113.
57. McGinnis, J. F., de Vellis, J. 1978. Glucocorticoid regulation in rat brain cell cultures. *J. Biol. Chem.* *253*:8483-8492.
58. Mealy, J., Jr., Chen, T. T., Schanz, G. P. 1971. Effects of dexamethasone and methylprednisolone on cell cultures of human glioblastomas. *J. Neurosurg.* *34*:324-334.
59. Moleman, P., Brainvels, J. 1979. Morphine-induced striatal dopamine efflux depends on the activity of nigrostriatal dopamine neurons. *Nature* *281*:686-687.
60. Morris, F. A. 1977. Norepinephrine induces glial-specific enzyme activity in cultured glioma cells. *Proc. Nat. Acad. Sci. U.S.A.* *74*:4501-4504.
61. Moscona, A. A. 1965. Recombination of dissociated cells and the development of cell aggregates, in *Cells and Tissues in Culture*, Vol. 1 (Wilmer, E., ed.), Academic, New York, pp. 489-529.
62. Moscona, A. A. 1977. Cell recognition in embryonic morphogenesis and the problem of neuronal specificities, in *Neuronal Recognition* (Barondes, S. H., ed.), Plenum, New York, pp. 205-226.
63. Moscona, A. A., Mayerson, P., Moscona, M. 1980. Induction of glutamine synthetase in the neural retina of the chick embryo:

Localization of the enzyme in Müller fibers and effects of BrdU on cell separation, in *Tissue Culture in Neurobiology* (Giacobini, E., Vernadakis, A., Shahar, L., eds.), Raven Press, New York, pp. 111-129.

64. Murota, S., Kashihara, Y., Tsurufuji, S. 1976. Effects of cortisol and tetrahydrocortisol on the cloned fibroblast derived from rat carrageenin granuloma. *Biochem. Pharmacol.* *25*:1107-1113.

65. Murray, M. R. 1965. Nervous tissues *in vitro*, in *Cells and Tissues in Culture*, Vol. 2 (Willmer, E. N., ed.), Plenum, New York, pp. 373-455.

66. Murray, M. R., Stout, A. P. 1942. Characteristics of human Schwann cells *in vitro*. *Anal. Rec.* *84*:275-293.

67. Murray, M. R., Stout, A. P. 1947. Distinctive characteristics of the sympatheticoblastoma cultivated *in vitro*: A method for prompt diagnosis. *Am. J. Pathol.* *23*:429-441.

68. Noble, E. P., Syapin, P. J., Vigran, R., Rosenberg, A. 1976. Neuraminidase-releasable surface sialic acid of cultured astroblasts exposed to ethanol. *J. Neurochem.* *27*:217-221.

69. Norenberg, M. D. 1981. The astrocyte in liver disease, in *Advances in Cellular Neurobiology*, Vol. 2 (Fedoroff, S., Hertz, L., eds.), Academic, New York, pp. 303-352.

70. Norenberg, M. D., Martinez-Hernandez, A. 1979. Fine structural localization of glutamine synthetase in astrocytes of rat brain. *Brain Res.* *161*:303-310.

71. Otten, U., Thoenen, H. 1977. Effect of glucocorticoids on nerve growth factor-mediated enzyme induction in organ culture of rat sympathetic ganglia: Enhanced response and reduced time requirement to initiate enzyme induction. *J. Neurochem.* *29*:69-75.

72. Parker, K., Vernadakis, A. 1980. Stimulation of ornithine decarboxylase activity in neural cell culture: Potential role of insulin. *J. Neurochem.* *35*:155-163.

73. Pisnak, M. R., Phillips, A. T. 1980. Glucocorticoid stimulation of glutamine synthetase production in cultured rat glioma cells. *J. Neurochem.* *34*:866-872.

74. Poduslo, S. E., Norton, W. T. 1972. Isolation and some chemical properties of oligodendroglia from calf brain. *J. Neurochem.* *19*:727-736.

75. Rubin, E., Rottenberg, H. 1982. Ethanol-induced injury and adaptation in biological membranes. *Fed. Proc.* *41*:2465-2471.

76. Russell, D., Snider, S. H. 1968. Amine synthesis in rapidly growing tissues: Ornithine decarboxylase activity in regenerating rat liver, chick embryo and various tumors. *Proc. Nat. Acad. Sci. U.S.A.* *60*:1420-1427.

77. Sandquist, D., Williams, T. H., Sahu, S. K., Kataoka, S. 1978. Morphological differentiation of the murine neuroblastoma clone

in monolayer culture induced by dexamethasone. *Exp. Cell. Res.* *113*:375-381.

78. Schrier, B. K. 1982. Nervous system cultures as toxicological test systems, in *Nervous System Toxicology* (Mitchell, C. L., ed.), Raven Press, New York, pp. 337-348.
79. Scott, B. S., Edwards, B. A. V. 1981. Effect of chronic ethanol exposure on the electric membrane properties of DRG neurons in cell culture. *J. Neurobiol.* *12*:379-390.
80. Seeds, N. W. 1971. Biochemical differentiation in reaggregating brain cell culture. *Proc. Nat. Acad. Sci. U.S.A.* *68*:1858-1861.
81. Selhorst, J. B., Kaufman, B., Horowitz, S. J. 1972. Diphenylhydantoin-induced cerebellar degeneration. *Arch. Neurol.* *27*:453-456.
82. Sidman, R. L. 1974. Contact interaction among developing mammalian brain cells, in *The Cell Surface in Development* (Moscona, A. A., ed.), Wiley, New York, pp. 221-253.
83. Slotkin, A. T. 1979. Ornithine decarboxylase as a tool in developmental neurobiology. *Life Sci.* *24*:1623-1630.
84. Slotkin, A., Lau, C., Bartolome, M. 1976. Effects of neonatal or maternal methadone administration on ornithine decarboxylase activity in brain and heart of developing rats. *J. Pharmacol. Exp. Ther.* *199*:141-148.
85. Slotkin, T. A., Lau, C., Bartolome, M., Seidler, F. J. 1976. Alterations by methadone of catecholamine uptake and release in isolated rat adrenomedullary storage vesicles. *Life Sci.* *19*:483-492.
86. Slotkin, T. A., Whitmore, W. L., Salvaggio, M., Seidler, F. J. 1979. Perinatal methadone addiction affects brain synaptic development of biogenic amine systems in the rat. *Life Sci.* *24*:1223-1230.
87. Smith, B. T., Torday, J. S., Giroud, C. J. P. 1974. Evidence for different gestation-dependent effects of cortisol on cultured fetal lung cells. *J. Clin. Invest.* *53*:1518-1526.
88. Sohn, R. S., Ferendelli, J. A. 1973. Inhibition of Ca^{++} transport into rat brain synaptosomes by diphenylhydantoin (DPH). *J. Pharmacol. Exp. Ther.* *185*:272-275.
89. Sorimachi, K., Niwa, A., Yasumura, Y. 1981. Hormonal regulation of tyrosine aminotransferase and phenylalanine hydroxylase in rat hepatoma cells continuously cultured in a serum-free medium: Effect of serum, dexamethasone and insulin. *Cell Struct. Funct.* *6*:61-68.
90. Syapin, P. J., Stefanovic, V., Mandel, P., Noble, E. P. 1976. The chronic and acute effects of ethanol on adenosine triphosphatase activity in cultured astroblast and neuroblastoma cells. *J. Neurosci. Res.* *2*:147-155.
91. Unsicker, K., Krisch, B., Otten, U., Thoenen, H. 1978. Nerve

growth factor-induced fiber outgrowth from isolated rat adrenal chromaffin cells: Impairment by glucocorticoids. *Proc. Nat. Acad. Sci. U.S.A.* *75*:3498-3502.

92. Urwyler, S., Tabakoff, B. 1981. Stimulation of dopamine synthesis and release by morphine and D-Ala^{2}-D-Leu^{5}-enkephalin in the mouse striatum *in vivo*. *Life Sci.* *28*:2271-2286.

93. Valcana, T., Timiras, P. S. 1969. Effect of hypothyroidism on ionic metabolism and Na-K activated ATP phosphohydrolase activity in the developing rat brain. *J. Neurochem.* *16*:935-943.

94. Vernadakis, A. 1971. Hormonal factors in the proliferation of glial cells in culture, in *Influence of Hormones on the Nervous System* (Ford, D. H., ed.), S. Karger, Basel, pp. 42-55.

95. Vernadakis, A., Culver, B. 1980. Neural tissue culture: A biochemical tool, in *Biochemistry of Brain* (Kumar, S., ed.), Pergamon, Oxford, pp. 407-452.

96. Vernadakis, A., Estin, C., Gibson, D. A., Amott, S. 1982. Effects of methadone on ornithine decarboxylase and cyclic nucleotide phosphohydrolase in neuronal and glial cell cultures. *J. Neurosci. Res.* *7*:111-117.

97. Vernadakis, A., Gibson, D. A. 1974. Role of neurotransmitter substances in neural growth, in *Perinatal Pharmacology: Problems and Priorities* (Dancis, J., Hwang, J. C., eds.), Raven Press, New York, pp. 65-77.

98. Vernadakis, A., Parker, K. 1980. Drugs and the developing central nervous system. *Pharmacol. Ther.* *11*:593-647.

99. Vernadakis, A., Woodbury, D. M. 1963. Effect of cortisol on the electroshock seizure thresholds in developing rats. *J. Pharmacol. Exp. Ther.* *139*:110-113.

100. Vernadakis, A., Woodbury, D. M. 1971. Effects of cortisol on maturation of the central nervous system, in *Influence of Hormones on the Nervous System* (Ford, D. H., ed.), S. Karger, Basel, pp. 85-97.

101. Volpe, J. J., Marasa, J. C. 1976. Regulation of palmitic acid synthesis in cultured glial cells: Effects of glucocorticoid on fatty acid synthetase, acetyl-GA carboxylase, fatty acid and sterol synthesis. *J. Neurochem.* *27*:841-845.

102. Waziri, R., Kamath, S. H., Sahu, S. 1981. Alcohol inhibits morphological and biochemical differentiation of C6 glial cells in culture. *Differentiation* *18*:55-59.

103. Williams, L. R., Sandquist, D., Black, A. C., Jr., Williams, T. H. 1981. Glucocorticoids increase tyrosine hydroxylase activity in cultured murine neuroblastoma. *J. Neurochem.* *36*:2057-2062.

104. Wu, R., Sato, G. H. 1978. Replacement of serum in cell culture by hormones: A study of hormonal regulation of cell growth and specific gene expression. *J. Toxicol. Environ. Health* *4*:427-448.

28

Central Nervous System Toxicity Evaluation in Vitro: Neurophysiological Approach

Michael J. Rowan
Trinity College, University of Dublin, Dublin, Ireland

I. INTRODUCTION

It is only recently that interest in in vitro techniques has expanded in neurotoxicology. One reason for this is the growing belief that if toxicology is to develop theories of the sites and mechanisms of action of neurotoxic compounds, it will have to use in vitro techniques. The other major reason is the belief that in vitro techniques will be of major use in neurotoxicity screening. Whereas the former approach considers in vitro techniques as analytical tools for apparently scientific reasons, the latter approach considers them as predictive tools and is based mainly on ethical and economical considerations. Both are controversial, especially the second one.

It is often assumed that the simpler the test system one uses, the easier it should be to determine the mechanism of action of neurotoxic compounds. This should certainly be true if one is thinking purely in terms of interactions at the molecular level. However, even at this level there are difficulties for in vitro analyses. Once one has isolated a part of a system which presumably contains the site of action, it is important to know and show that it is behaving in a manner similar to that in situ. The closer one gets to the molecular site of action, the further one is away from the change in behavior one is trying to explain and the more difficult it is to prove that the interaction one is studying is functionally relevant. This is especially the case when one considers the complexity of many of the behavioral or psychological changes that are often the result of neurotoxic insult. There are some who believe that the reductionist approach and assumptions which underly the use of in vitro techniques may in fact not be easily applied to human behavior. Many of these people, however, would even question the value of any in situ analyses based in the biological sciences. There may indeed be a need for some form of "psychotoxicology," but the molecular and cellular mechanisms of neurotoxicity also need to be solved. Approaches using isolated neural tissue may suffer certain theoretical difficulties, but they certainly possess many practical advantages over those carried out in situ (Table 1). These practical advantages greatly facilitate the development of sound theories regarding sites and mechanisms of toxicity in such a complex and integrated system as the nervous system.

The apparent success of in vitro toxicity tests for mutagenicity and carcinogenicity screening has led many to seek similar models for neurotoxicology. This approach, however, has not in general been very successful to date, for both practical and theoretical reasons. The only fast, accurate, and inexpensive in vitro neurotoxicity tests that might be suitable for screening purposes use simple systems which on their own cannot provide the multiplicity of potential sites for neurotoxic interactions which are present in the human nervous system. Isolated neural tissue by its very nature is cut off from its normal environment which may protect it (e.g., blood-brain barrier, metabolic inactivation) or make it more vulnerable to insult (e.g., metabolic activation, selective uptake). Other factors such as the lack of its usual inputs and outputs may make the isolated tissue respond at times in a manner which is abnormal. The problem of deciding on whether or not any observed change has any toxicological significance is immense when one is not using a behaving organism. The chance of obtaining false positives or false negatives must therefore be quite high, as is the chance of obtaining conflicting results from different test systems. In several cases it may be difficult to distinguish nonspecific cytotoxic effects from specific neurotoxic ones. Behavioral tests in combination with in vivo neurobiological measurements seem to provide the best means of screening for neurotoxic compounds at present.

A number of books and reviews concerning the application of in vitro techniques in neurotoxicology have been published recently (17,27,30,36, 42,44,49,65,79,83,95,97,110,127). There are many approaches, including chemical, morphological, and physiological. This review considers the latter approach as applied to neural tissue derived from the central nervous system, since this has so far been largely neglected in the

Table 1 Advantages and Limitations of the in Vitro Approach for Neurotoxicology

Advantages
1. Simplified systems and procedures: aid execution and analysis of experiments.
2. Controlled medium: eases standardization and modification of experimental conditions (e.g., pH, pO_2, temperature, ionic environment).
3. Direct access of chemicals to central nervous system: Dose-response relationships may be rapidly and accurately measured in absence of blood-brain barrier and hepatic metabolism; eases structure-activity studies.
4. Interdisciplinary work facilitated: morphology, chemistry, and physiology can be studied on the same preparation.

Limitations
1. Complexity ignored: reconstitution difficult; "the whole may be greater than the sum of its parts."
2. Distribution and metabolism usually ignored when screening.
3. Mainly limited to short-term tests unless exposure is carried out in vivo.
4. Toxic end points often difficult to define: difficult to extrapolate to in vivo.

literature. This is not to imply that the other approaches are any less valuable. In fact, for a proper evaluation of neurotoxicity in vitro, a combination of as many techniques as possible is usually necessary.

The neurophysiological approach generally involves the use of electrophysiological techniques. These make it possible to directly elicit and monitor physiological activity with only minor perturbations of the central nervous system tissue. There are numerous such techniques and central nervous system preparations available, but most of them are difficult and very time-consuming in preparation and execution and probably not suited to general neurotoxicology. The in vitro technique makes it possible to visually position electrodes, thus avoiding the need for stereotaxic techniques. It also allows stable recording from very small cells without requiring anesthesia. This review emphasizes three isolated central nervous system preparations where the application of electrophysiological techniques is comparatively easy. The first, an invertebrate preparation, the ventral nerve cord of the cockroach, involves the use of a sucrose gap-type recording technique. The second, a mammalian preparation, the rat hippocampal slice, is very suitable for extracellular field potential recording. The third, cultured central nervous system of mammalian origin, makes intracellular recording from identified neurons relatively easy.

II. INVERTEBRATE CENTRAL NERVOUS SYSTEM: COCKROACH ABDOMINAL NERVE CORD

A. Physiology

Since the first observations of Adrian (1) in 1930 on the isolated caterpillar central nervous system, the invertebrate central nervous system has been extensively studied in vitro using electrophysiological techniques (61,69). A major advantage of these preparations is their relative simplicity, where the same particular neuron can be located repeatedly in each individual organism of a species. Many invertebrates have a rapid reproduction rate and are inexpensive to maintain throughout the year. Because they are cold-blooded animals, in vitro central nervous system experiments can be carried out at room temperature. At temperatures of about 20°C the respiration of the neurons is quite low, and since their cell bodies are usually located near the outer surface of the central nervous system, there is normally no problem with oxygenation. Although these preparations are admittedly further removed from man than preparations from vertebrates, it is encouraging to find that in the vast majority of cases the mechanisms and types of both synaptic and axonal transmission are quite similar across phyla. Indeed, invertebrate preparations often offer the opportunity of carrying out neurophysiological experiments on the central nervous system which, in animals higher on the evolutionary scale, would be extremely difficult, if not impossible.

The interest of toxicologists in isolated invertebrate central nervous system preparations has mainly focused on insects or closely related arthropods (5,19,81,116). This is because a large proportion of pesticides in use at present are compounds which are thought to act principally by disrupting insect central nervous system function.

One particular preparation that has been widely used in this area is the abdominal part of the ventral nerve cord of the cockroach *Periplaneta americana* (103). There is an extensive literature dealing with morphological, physiological, biochemical, pharmacological, and toxicological aspects of neural transmission in the giant interneurons present in this preparation (Fig. 1) (12,25,48,90,105,107-109,117). The cell bodies and dendritic trees of these neurons are situated in the terminal (or sixth) abdominal ganglion. Their axons extend from this ganglion along the ventral nerve cord at least to the level of the thorax. The main input to these giant interneurons is from a pair of sensory organs called cerci at the tip of the abdomen. Mechanical stimulation (sound, air puff, touch, pressure) of these causes impulses to be transmitted down the cercal nerves and results in excitation or inhibition at

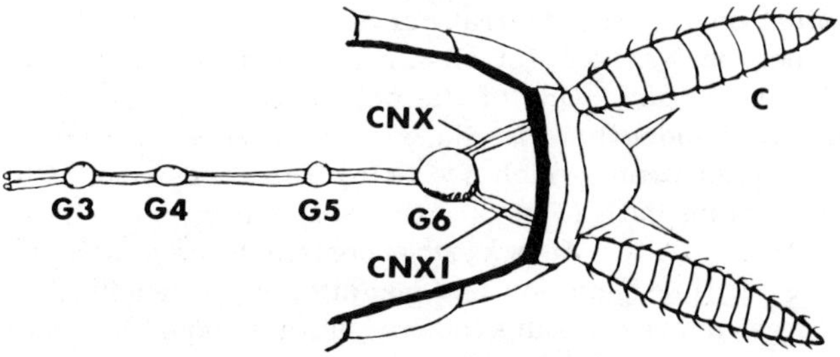

Figure 1 Lower abdomen of the American cockroach viewed from the dorsal side. The alimentary canal has been removed in order to expose the ventral nerve cord (G3-G6, abdominal ganglia 3-6; CNX and CNXI, cercal nerves X and XI; C, cercus). (From Ref. 105.)

the giant interneuron dendrites. Excitation due to the firing of a single cercal nerve is manifest electrophysiologically as a short-duration depolarization of the postsynaptic membrane and is termed a unitary excitatory postsynaptic potential (EPSP). Electrically evoked EPSPs can be elicited by stimulation of cercal nerve XI. The excitatory transmitter at this synapse is believed to be acetylcholine. The postsynaptic receptors seem to be predominantly of a nicotinic type. Short-duration hyperpolarizations termed unitary inhibitory postsynaptic potentials (IPSPs) may also be detected in the giant interneuron dendrites. Electrically evoked IPSPs may be obtained by stimulating cercal nerve X. Here γ-aminobutyric acid (GABA) is thought to be the inhibitory transmitter. The inhibitory pathway from cercal nerve to giant interneuron is probably a bisynaptic route.

B. Methods

A relatively simple in vitro method for studying the effects of chemicals on synaptic and axonal transmission (14,93,94,105) is available for this preparation. Transmembrane postsynaptic electrical activity at the giant interneurons in the sixth abdominal ganglion can be monitored using extracellular electrodes. Rather than attempt to impale the dendrites of the giant interneurons with microelectrodes, one measures the electrical potential difference between the outside of the sixth abdominal ganglion and the cut ends of the giant axons. A mannitol solution which is isotonic with the physiological solution is perfused between the recording electrodes in order to prevent the transmembrane potential being short-circuited; hence the method is called a "mannitol gap" technique.

After the abdominal part of the ventral nerve cord has been dissected free from a cockroach, the outer sheath is removed in order to disrupt the blood-brain barrier. One of the pair of connectives between the sixth and fifth abdominal ganglia is severed so as to reduce the number of giant interneurons which are being recorded from. On transferral to the recording bath (Fig. 2), the compartments are carefully sealed with petroleum jelly. It is very important to have stable drainage through the terminal ganglion and mannitol compartments. Stimulation of the cercal nerves is achieved by platinum hook electrodes or by silver wire suction electrodes. The physiological solution contains 208.6 mM NaCl, 3.1 mM KCl, 5.4 mM $CaCl_2$, and 2.0 mM $NaHCO_3$ in distilled water at pH 7.0.

C. Example

In order to demonstrate the use of this technique for the assessment of neurotoxicity the effects of the convulsant phenol hydroquinone will be described (18,56,104) (Fig. 3). Concentrations of ⩾1 mM were found

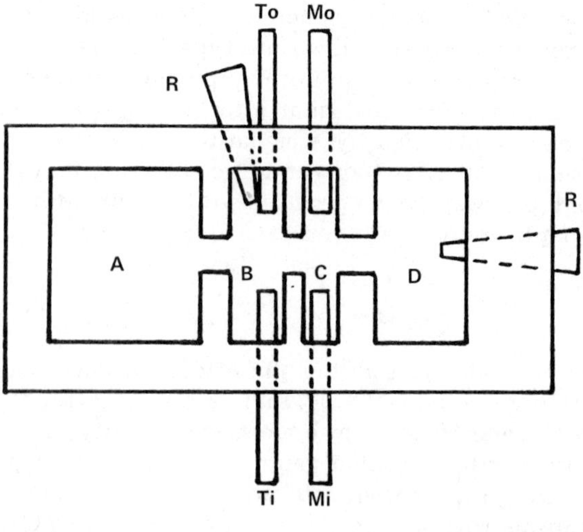

Figure 2 Chamber for mannitol gap recording of giant interneuron synaptic activity in the sixth abdominal ganglion of the cockroach ventral nerve cord. Bathing media were led to the chamber via a gravity feed system and drained away using a suction pump. (R, recording electrodes; Ti and To, physiological or test solution inflow and outflow; Mi and Mo, mannitol solution inflow and outflow; A-D, compartments for cercus and cercal nerves, sixth ganglion, mannitol, and cut end of nerve cord, respectively). (From Ref. 105.)

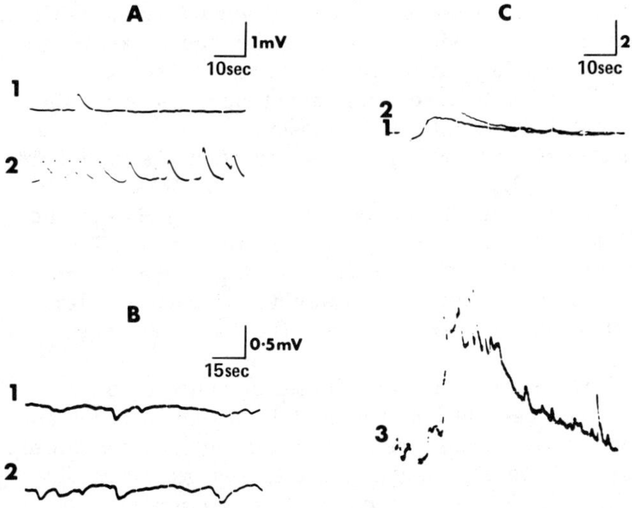

Figure 3 Effects of hydroquinone on transient giant interneuron postsynaptic potentials in the sixth abdominal ganglion of the cockroach: (A) unitary EPSPs; (B) unitary IPSPs, and (C) electrically evoked EPSPs (1, controls; 2, in the presence of 0.5 mM hydroquinone; and 3, in the presence of 1.0 mM hydroquinone). (From Ref. 18.)

to produce a gradual depolarization of the postsynaptic membrane which was dose dependent. With these depolarizing concentrations there was much seizure-like activity, especially when cercal nerve XI was electrically stimulated. With depolarizations greater than about 10 mV, blockade of synaptic transmission occurred. The effects of a nondepolarizing concentration (0.5 mM) were then studied. This was shown to produce an increase in the amplitude of the electrically evoked EPSP and both unitary EPSPs and IPSPs. The frequency of both unitary EPSPs and IPSPs was also increased. It was concluded that hydroquinone probably acts primarily on the presynpatic nerve terminals to increase the release of transmitter.

D. Discussion

In the study of the acute neurophysiological activity of large numbers of chemicals either for screening or structure-activity analyses of neurotoxic compounds, it is necessary to use relatively rapid and inexpensive techniques. The cockroach nerve cord technique appears to partly fulfill these criteria. It not only allows one to detect neurotoxicity, but it also enables one to make inferences regarding probable sites and mechanisms of toxicity. It has proved very suitable not just for insecticides, but also for a wide range of other neurotoxic compounds. Tests can be carried out in the presence or absence of the blood-brain barrier in order to assess its importance. More detailed studies on both axonal and synaptic transmission can be carried out using an oil gap method for recording from single identified giant interneurons (12,13,15,90,91,93).

Takeno et al. (121) have used the spontaneous firing rate of the excised crayfish abdominal nerve cord as a rapid test system for structure-activity studies of neurotoxic compounds. It was found to be very sensitive to many insecticides; however, it seems to offer little advantage, other than larger size, over the cockroach nerve cord technique (87).

Can one extrapolate the results from studies of neurotoxic compounds on the cockroach central nervous system to that of other animals, including man? Clearly this preparation can only be considered as a simplified model, where cholinergic and GABA-ergic processes predominate. It will be selectively more sensitive to certain types of neurotoxic interactions than others. For example, in this test chemicals with a high affinity for, or activity at, nicotinic cholinergic receptors will appear more toxic than chemicals which preferentially interact with muscarinic ones. On the other hand, there are many fundamental processes in inhibitory and excitatory synaptic transmission and axonal transmission which appear to be very similar across all animals. Chemicals which affect these processes should be easily detected using an invertebrate preparation such as this. For example, there appear to be very few species differences in the cellular sites and mechanisms of the acute action of neurotoxic insecticides. It should also be remembered that the terminal abdominal ganglion contains and is sensitive to many biogenic amines, amino acids, and peptides (32,51,92) which provide sites for neurotoxic action that have yet to be properly exploited with the present technique.

III. VERTEBRATE CENTRAL NERVOUS SYSTEM: RAT HIPPOCAMPAL SLICE

A. Physiology

Isolated central nervous system preparations from cold-blooded vertebrates have been extensively investigated using electrophysiological techniques (60,66) since the original work of Kato on the frog spinal cord (58). Although it is possible to obtain high-quality, stable recordings from the isolated central nervous systems of fish, amphibia, and reptiles, the more recently developed (71,126) electrophysiological approaches to mammalian central nervous systems in vitro are now much more popular. This is probably because of the close similarity in both structural and functional organization of mammalian and human central nervous systems. The similar temperature dependence of both physiological and pharmacological phenomena is also a major factor.

At present, the most widely used source of slices is the hippocampus, a large comma-shaped structure which lies along the walls of the lateral ventricle. It is part of the limbic system and is thought to be involved in a wide variety of functions, including cognition, memory, and emotions. The particular attraction of the hippocampus for both in vivo (101) and in vitro (118) electrophysiological work lies in its laminar structure, which is nearly all in a single plane (Fig. 4). Layers of fiber tracts, cell bodies, and dentritic regions can be distinguished both visually and electrically in the isolated transverse hippocampal slice. The neurons studied are cortical cells, but, unlike those in the neocortex, here the pyramidal cells and granule cells are anatomically separate, the former type in the hippocampus proper and the latter type in the dentate gyrus. The close-layered packing of the cells and their inputs in the CA1 region and dentate gyrus, combined with the relative paucity of other neuron cell bodies, make it possible to record stable and large field potentials which can be meaningfully interpreted. Both excitatory and inhibitory synaptic responses can be evoked by electrically stimulating afferent and efferent pathways. The excitatory transmitter at these synapses is thought to be glutamate and/or aspartate, while the inhibitory one is probably GABA. The receptors for the excitatory transmitter at the granule cells are probably of the quisqualate/kainate type (24), whereas those at the CA1 pyramidal cells may be of a different type (64). Both the granule cells and pyramidal cells are sensitive to agonists and antagonists for a wide variety of neurotransmitter and neuromodulator receptors.

B. Methods

There are numerous different methods for isolating hippocampal slices and recording from them (28,61,123). No matter what method is used, the most important factors for success appear to be extreme care during the dissection and adequate oxygenation. What follows is a brief description of the methods used in this laboratory. An adult rat is decapitated and the brain is removed rapidly from the skull. A slab of the brain containing the hippocampus is cut out and the neocortex is removed. Transverse sections of about 350 μm are cut using a vibroslice (55). Slices are then placed between two nylon nets and submerged in a superfusion bath maintained at 34°C (Fig. 5). Although most workers superfuse only one surface of the slice, complete submersion allows smooth and rapid changes in the solution bathing the preparation. This is very useful if different concentrations of different chemicals are to be tested. The normal incubation medium is a modification of the Krebs-Henseleit solution, which mimics the compo-

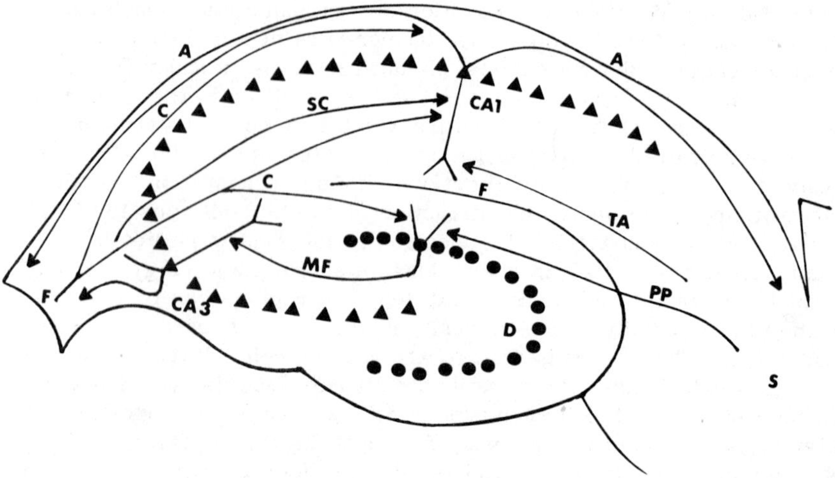

Figure 4 Schematic diagram of some of the pathways in the transverse hippocampal slice. Pyramidal cells (▲) are found in the hippocampus proper (e.g., CA1 and CA3 regions), whereas granule cells (●) are present in the dentate gyrus (D). The granule cells receive a major input from the perforant path (PP) which comes from the subiculum (S). The granule cells send out mossy fibers (MF) which synapse with the pyramidal cells in CA3. These cells send out axons to the fimbria (F), and other ones (Schaffer collaterals, SC) to the pyramidal cells in CA1. The CA1 cells also receive inputs from commissural (C) and temporoammonic (TA) tracts. The axons of the CA1 cells run along the outer surface of the hippocampus, the alveus (A).

sition of the cerebrospinal fluid: NaCl, 120 mM; KCl, 2.5 mM; NaH_2PO_4, 1.25 mM; $MgSO_4$, 1.0 mM; $CaCl_2$, 2.0 mM; $NaHCO_3$, 26.0 mM; and glucose, 10 mM. Glass microelectrodes (about 1 M Ω resistance) filled with 20 mM NaCl in agar are used both to stimulate and record.

With an extracellular recording microelectrode placed in the dendritic region of either granule or pyramidal cells, a synaptic potential can be recorded in response to stimulation of afferent pathways in the same layer. This negative-going potential (Fig. 6, lower traces) is considered to be a population EPSP, but it usually also contains an IPSP component. With an electrode in the cell body region, this EPSP is reversed in polarity because of current spread. If the EPSP is large enough, the cells generate action potentials which are recorded extracellularly at the cell bodies as a sharp downward deflection superimposed on the EPSP (Fig. 6, upper traces). This is termed a population spike.

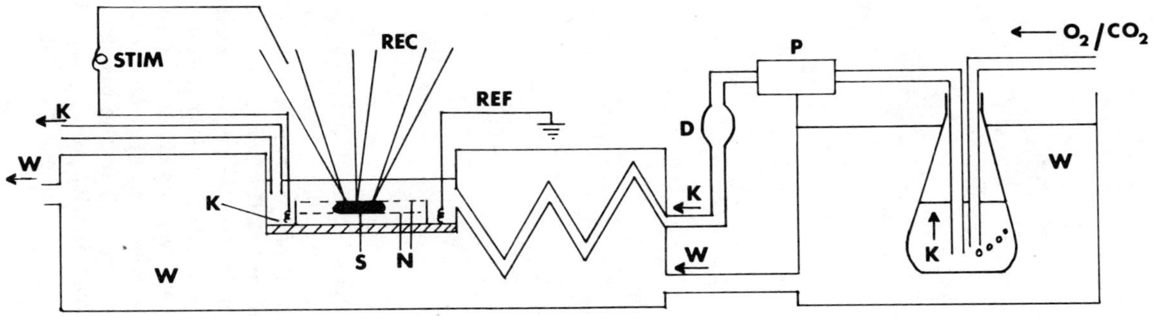

Figure 5 Schematic diagram of recording system for rat hippocampal slice. Warmed (34°C), oxygenated (95% O_2/5% CO_2) artificial cerebrospinal fluid (K) is pumped (P) from a reservoir, through a dropper D, then through the outer chamber into the inner recording chamber, and then removed using a suction pump. The slice (S) is held in position using a pair of fine nylon nets (N) which are held down with pins stuck in a Sylgard base. The chamber is illuminated from below so that the slice can be viewed using a stereomicroscope with a long working distance. Temperature is measured both in the recording chamber and the water reservoir (STIM, stimulating electrode; REC, recording electrodes; REF, reference electrode; and W, circulating heated water).

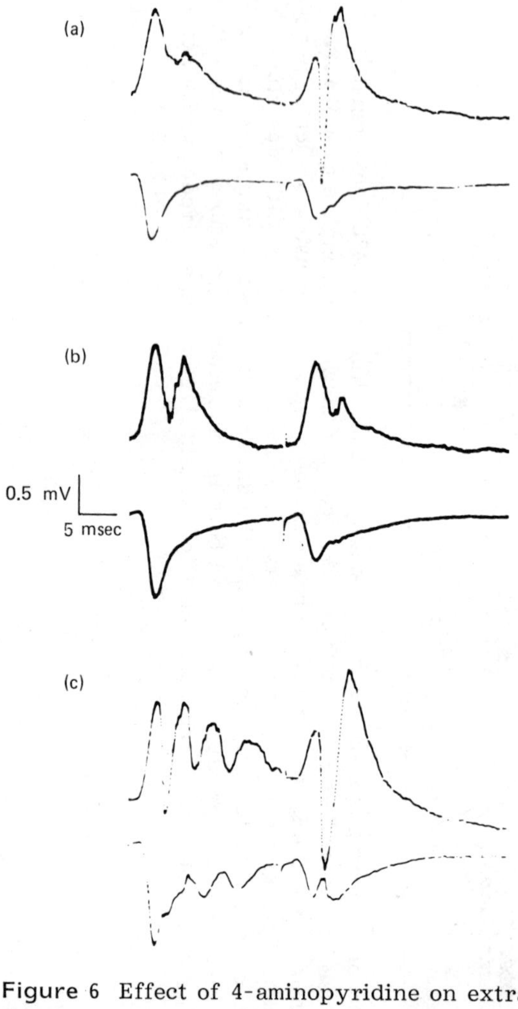

Figure 6 Effect of 4-aminopyridine on extracellular field potentials in the dentate gyrus of the rat hippocampal slice. The upper traces were recorded with an electrode in the cell body layer, where EPSPs generate population spikes (sharp downward deflection). The lower traces were simultaneously recorded in the dendritic region, where EPSPs (gradual downward deflects) are generated by stimulating afferent nerves in the perforant path. Stimuli were delivered as paired pulses (20msec interstimulus interval) every 90 sec: (a) control, (b) in the presence of 10 μM 4-aminopyridine, and (c) in the presence of 50 μM 4-aminopyridine. (From M. J. Rowan and R. Anwyl, unpublished observations.)

C. Example

The use of this technique will be illustrated with a study of the effects of the convulsant 4-aminopyridine (8,70,124). This compound is known to act presynaptically to enhance transmitter release at many different synapses. Only recently have its actions on the mammalian central nervous system been examined in vitro (10,41,45,46,54,63,125). In this example the perforant path-granule cell pathway was studied. Responses to stimulation of the medial perforant path were recorded in both the dendritic layer and the cell body layer of the dentate gyrus. Paired-pulse stimulation was used to illustrate the plasticity of the preparation. Whereas there was inhibition of the second EPSP, there was facilitation of the second population spike (Fig. 6a). At the lowest effective concentration, 10 μM in this preparation, 4-aminopyridine produced an increase of the first EPSP and population spike (Fig. 6b). The increased population spike appeared to be a direct result of the increase in the EPSP size and may be due to increased transmitter release. While paired pulse inhibition of the EPSP was slightly increased, there was a dramatic reversal of paired-pulse facilitation of the population spike. This may be due to increased collateral inhibition. At the higher concentration of 50 μM it is interesting to note the appearance of secondary and tertiary population spikes indicative of seizurelike activity (Fig. 6c). At this concentration of 4-aminopyridine there is a marked increase in the excitability of the preparation. This is illustrated by the observation that EPSPs that previously were well below threshold for inducing a population spike now produce multiple ones (Fig. 7). The delayed positivity which was observed at this concentration in the

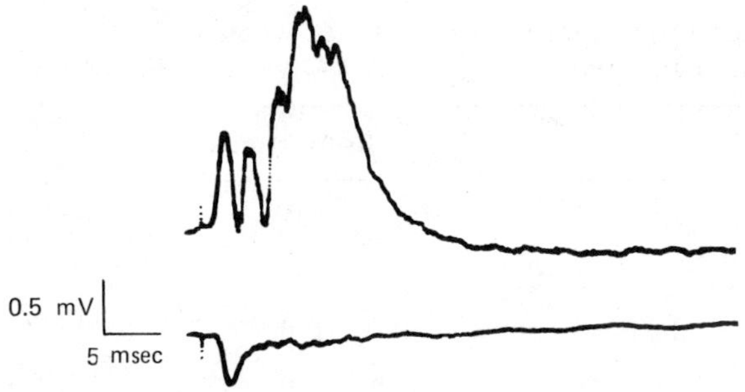

Figure 7 Effect of 50 μM 4-aminopyridine on extracellular field potentials in the dentate gyrus of the rat hippocampal slice. Conditions here are the same as for Figure 6c, but in response to very low level afferent input. Note the large positivity following the population spikes in the cell body layer. (From M. J. Rowan and R. Anwyl, unpublished observations.)

cell body layer may be due to an enhanced afterhyperpolarization similar to that observed by Alger and Nicoll (2).

D. Discussion

The choice of the hippocampal slice seems highly appropriate for toxicological work. A large number of environmental/industrial neurotoxic compounds produce effects which mimic hypoxia (110). The hippocampal slice should provide a good model system to test such compounds because of its extreme sensitivity to oxygen tension changes when maintained at or near body temperature (7,62,72,78). As previously noted, invertebrate neurons are not very sensitive to hypoxia. Thus a property of mammalian slices which has made them difficult to work with is in fact an added attraction to the toxicologist. Another factor is the well-established sensitivity of the hippocampus to a wide range of neurotoxic compounds in vivo (23,80; see Table 2). It is not known why the hippocampus is so sensitive to neurotoxic agents, anoxia, and ischemia (29,43), or why it has such a low seizure threshold. A major factor is probably the functional organization of the region, since the individual cells are apparently not much different from ones of the same type in other regions of the brain. Another contributary factor in some cases is the wide surface area exposed to the lateral ventricles. This should facilitate absorption of compounds from the cerebrospinal fluid. It is interesting to note that certain metals such as lead and zinc appear to accumulate preferentially in the hippocampus (20,26, 35,100,106).

Table 2 Some Environmental/Industrial Compounds Known to Affect the Hippocampus in Vivo

Compound	References
Carbon disulfide	31
Carbon monoxide	68
Dieldrin	120
Lead	20
Methyl methacrylate	52
Tetraethylead and triethyl lead	86,115
Trimethyl tin	9

It would clearly be desirable to study the effects of compounds on slices from the area of the central nervous system where they are most active. This, however, requires that detailed preliminary work be carried out in vivo. Even though it now appears that viable slices can be cut from nearly all regions of the central nervous system, most of them would be unsuitable for field potential studies.

Kuroda (67) has cogently advocated the use of the isolated guinea pig olfactory cortex slice as a system for neurotoxicity testing and evaluation. Using this technique, he demonstrated how neurotoxic agents acting primarily on synaptic transmission and axonal transmission, such as DDT and toluene, respectively, can be readily distinguished using field potential recordings. The hippocampal slice should provide similar information. It should be especially useful in the detection and analysis of the actions of convulsant compounds (34,88,96) and anesthetic-like agents (102).

Up to now neurotoxic compounds have mainly been used on the hippocampal slice for pharmacological reasons. For example, divalent cations such as Cd^{2+}, Co^{2+}, and Mn^{2+} have been used to block Ca^{2+}-mediated processes, including synaptic transmission, Ca^{2+} spikes and Ca^{2+}-dependent K^+ conductances (6,47,75,76,112,122). These observations may be used as a guide for future toxicological work.

In a similar vein, most of the published neurotoxicity studies on hippocampal slices using animals treated subacutely have investigated the effects of drugs (16,21,39). Subacute exposure in vivo to morphine, for example, produced tolerance to the excitatory effect of opiates on pyramidal cells in hippocampal slices. When the opiate was removed from the perfusion medium or naloxone was added, an inhibitory effect was observed. These effects may provide a model for some aspects of the withdrawal syndrome seen in vivo. In a rather more toxicological study, Farnell et al. (33) showed that about 2 weeks after a single intraventricular injection of aluminum the orthodromically evoked population spike in the CA1 region was uncoupled from the EPSP, and long-term potentiation was inhibited before morphological changes occurred in this region.

Other potentially valuable studies include the examination of the effect of animal age on the slice response to neurotoxic compounds, since this is known to be an important variable in vivo. Whereas it is extremely difficult to record from the central nervous systems of fragile young animals in situ, it is probably easier to obtain reliable responses in hippocampal slices from young animals than in those from adult ones (111).

There is obviously much scope for more detailed work with intracellular recordings from the pyramidal and granule cell bodies using fine microelectrodes. Even the single-electrode voltage clamp has been successfully applied to this preparation (e.g., see Ref. 45).

IV. CULTURED CENTRAL NERVOUS SYSTEM

Any part of the central nervous system that is kept alive for more than 1 or 2 days can be termed cultured central nervous system. Although it is probably easier to culture central nervous system tissue from invertebrates and cold-blooded animals, the general tendency is to use a mammalian source. Indeed, one of the big attractions of these techniques is the possibility of testing compounds on human tissue and thereby bypass the problem of considering possible species differences from man. Apart from possessing many of the advantages of other in vitro techniques, culture is the only available method of examining the neurotoxic effects of subacute, subchronic, and chronic exposure to chemicals totally in isolation from the rest of the body. Since the developing nervous system has been shown to be highly sensitive to neurotoxic insult in situ, it is highly appropriate to examine neurotoxicity in developing cultured central nervous system. In fact, this approach may provide a powerful tool for the study of neuroteratology (85; see also the chapter by Nelson in this book).

Although most of the mysteries regarding the composition of the culture medium are being resolved (4), there are still many technical difficulties in the maintenance of stable predictable cultured neuron populations. They are also expensive and time-consuming to keep. Their use at present is therefore limited to detailed studies of well-known neurotoxic agents (see Vernadakis, Davies and Gremo in this book).

Electrophysiological techniques have been used to advantage in three main types of cultured central nervous system neurons: (a) explants, (b) dissociated or dispersed primary cultures, and (c) neuronal cell lines (113). Emphasis here is placed on intracellular recordings from primary cultures, since they possess many of the advantages of both explants and cell lines.

A. Explants

Small tissue fragments (about 1 mm^3) have been explanted from many regions of prenatal or perinatal central nervous system and kept functional for months. Extracellular recordings can be made with ease, whereas intracellular work is difficult because thin layers of connective tissue usually form over the surface of the explant. The activity recorded is usually characteristic of the central nervous system region from which the explant has been removed, and excitatory and inhibitory synaptic activity can be clearly detected (22,40).

An example of the use of explants in neurotoxicology is the study of Raybourn et al. (99). They recorded spontaneous firing in roller-tube cultured cerebellar Purkinje cells explanted from 3 to 4-day-old rats. The disruptive effect of carbon monoxide on firing activity was partly prevented or reversed by exposure of the explant to white light, whereas the somewhat similar effect of hypoxia was not affected. It was concluded that carbon monoxide was having a direct effect independent of hypoxia, possibly by binding to iron-heme groups in the cytochrome systems of the cells. Such a study, where light is an experimental variable, would be impossible in vivo. One unexpected phenomenon was the finding that some cells appeared to be resistant to hypoxia. This could be due to a number of factors, but may mean that the results cannot be easily extrapolated to the intact animal.

B. Dispersed Primary Cultures

Neurons from embryonic or neonatal central nervous system can be completely dissociated by mechanical or enzymatic means without losing their ability to survive in tissue culture. They have been used to address a variety of problems concerning biochemical and morphological changes following exposure to neurotoxic chemicals (see Vernadakis, Davies, and Gremo in this book). Moreover, it is comparatively easy to record with intracellular electrodes from visually identifiable single neurons in these cultures because the neuronal surface membranes are in direct contact with the culture medium. The neurons can differentiate in culture, developing synaptic activity, and are sensitive to a variety of neurotransmitters (84,98).

The technique has been used extensively by MacDonald and Barker (74) to study convulsant compounds, and their work will be used by way of illustration. They dissect out spinal cords from 12- to 14-day-old mouse embryos, mechanically dissociate the cells, and culture them for several weeks on collagen-coated dishes in modified Eagle's medium containing fetal calf serum and heat-inactivated horse serum. The growth of nonneural cells is suppressed by the temporary addition of uridine and 5'-fluoro-2'-deoxyuridine. Large-diameter (20-50 μm) multipolar neurons can be seen using an inverted phase-contrast microscope and are impaled with CH_3COOK- or KCl-filled high-resistance glass microelectrodes. It is now common practice to use beveled pipets and a stepping microdrive in order to make recording and cell penetration easier. Stable recordings can be made for several hours and chemicals applied by iontophoresis, pressure ejection (miniperfusion), or via the perfusion medium. In the case of iontophoresis compounds are applied from the tips of single or multibarreled micropipets which are visually guided to a position about 2 μm from the neuronal cell body. Pressure ejection is carried out using pipets of about 2-10 μm tip diameter. Although less quantitative than bath application, both these techniques allow the localized effect of compounds to be studied. These authors have shown (73) that the structurally diverse convulsants pentylenetetrazol, penicillin, picrotoxin, and bicuculline can produce random, abrupt, and prolonged depolarizations with superimposed spikes. These are similar to paroxysmal depolarizing shifts seen in vivo during epileptiform discharges. The compounds were found to selectively antagonize the inhibitory effect of GABA without altering the inhibitory effect of β-alanine or glycine or the excitatory response to glutamate. Bicuculline and penicillin were also shown to have non-synaptic actions, namely, a depolarizing effect on the cell bodies, mediated by a decrease in membrane potassium conductance, and a prolongation of calcium-dependent action potentials (50).

One particular problem of this and related culture techniques is that all the neurons are not in the same physiological or developmental state at any given time. A very large proportion of neurons derived from fetal central nervous system tissue die, especially in the early stages of culture. This means that the remaining neurons are not a reliable representation of that particular region's neuron types. Cultured neuron heterogeneity is also due to the fact that some cells are differentiating while others may be simultaneously dedifferentiating. The use of adult neurons in culture seems to overcome many of these limitations. Unfortunately, these techniques are at the moment mainly confined to neurons derived from the peripheral nervous system. Scott (114), for example, has applied ethanol for 12 days to adult mouse dorsal root ganglion cells in culture. On withdrawal of the ethanol from the medium, membrane electrical properties were dramatically changed. It should be possible to carry out similar experiments on cultured adult retina (77). These cells should be more representative of central nervous system tissue and should be of great potential use in neurotoxicology.

C. Neuronal Cell Lines

These are groups of cultured neurons which are capable of indefinite cell division as a result of some genetic transformation, be it spontaneously or after exposure to radiation, chemicals, or viruses, or which originate from tumors (e.g., neuroblastoma). High-quality intracellular recordings from visually identified single neurons can be made (82, 119). Even though these cells are structurally very different from normal neurons, many cell lines have been shown to behave remarkably well in terms of excitability and synaptic responsiveness. Some cell lines behave in a manner intermediate between glial cells and neurons, and are therefore usually considered to be hybrids (89). Although cell lines are the simplest and most reproducible neuronal culture systems available and are comparatively easy to obtain and maintain, they have not been used widely for electrophysiological studies of neurotoxicity. An example of the potential of these preparations is the study of Jacques et al. (53) concerning the effects of pyrethroids on the C 1300 mouse neuroblastoma clone NIE-115. They showed that compounds which produced a depolarizing afterpotential following the action potential could be antagonized by inactive pyrethroids and the Na^+ channel-blocking agent tetrodotoxin. This was taken to imply that active pyrethroids interact with the channel in two steps, first binding to some site on the channel and only then affecting the channel gate. Since the active pyrethroids were potentiated by toxins which keep sodium channels open, it was concluded they have a preference for open channels.

V. NONCENTRAL NERVOUS SYSTEM PREPARATIONS

There are many other in vitro neurophysiological techniques involving the peripheral nervous system which have been studied in their own right or as models of how the central nervous system may respond to neurotoxic compounds. It is obviously important to study the effects of a compound on the site which is thought to be most sensitive to it. Practical considerations often prevent one from doing this, especially if the presumed site is in the central nervous system. There are many sensitive preparations which are often more accessible and easier to work with than those derived from the appropriate central nervous system region. These include sensory organs (the retina is technically part of the central nervous system), neuromuscular junctions (both vertebrate and invertebrate at skeletal, smooth, and cardiac muscle),

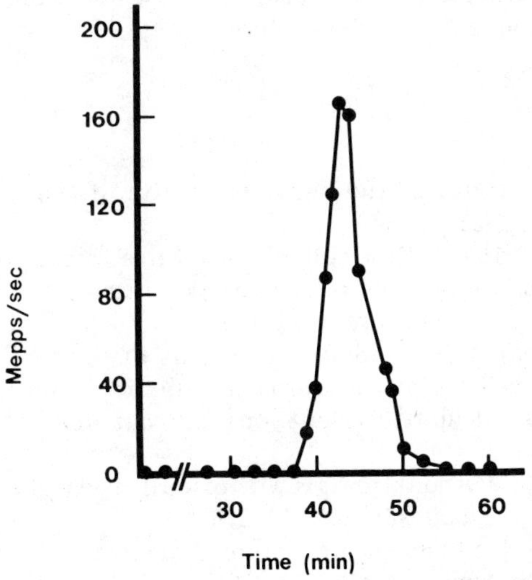

Figure 8 Effect of 0.5 mM lead on miniature end-plate potential (mepp) frequency at the rat diaphragm. Records were taken from a single junction in order to avoid missing the very rapid increase and decrease in frequency. (From Ref. 3.)

autonomic and sensory ganglia, or myelinated and unmyelinated nerve fibers (e.g., see Ref. 11, 37, 38, 57, and 59).

In order to demonstrate the use of one such preparation, the effect of lead on spontaneous transmitter release at the skeletal neuromuscular junction will be illustrated (3). Miniature end-plate potentials were recorded intracellularly from rat diaphragm using glass microelectrodes of 10-20 M Ω filled with CH_3COOK. In the absence of Ca^{2+} in the control medium there was little apparent spontaneous quantal release of acetylcholine. On the addition of lead there was a delayed transient rapid increase in miniature end-plate potential frequency followed by a rapid block (Fig. 8). The effect was prevented by raising the K^+ level from 5 to 10 mM. It was concluded that since the total number of quanta released (77,000) in the presence of lead was much less than the total number of vesicles available for release (about 320,000), lead was not only activating the release mechanisms, but also almost simultaneously inactivating it. It is possible that the inhibitory effect of raised K^+ levels may be due to a direct prevention of the entry of lead into the nerve terminal. Whether or not similar mechanisms operate in the central nervous system is highly debatable.

ACKNOWLEDGMENTS

The author wishes to thank R. Anwyl, P. L. Chambers, N. Collender, L. Cullen, and T. Molloy for invaluable help and advice.

REFERENCES

1. Adrian, E. D. 1930. The activity of the nervous system of the caterpillar. *J. Physiol.* *70*:34-35.
2. Alger, B. E., Nicoll, R. A. 1980. The epileptiform burst after hyperpolarization: A calcium dependent potassium potential in hippocampal pyramidal cells. *Science* *210*:1122-1124.
3. Anwyl, R., Kelly, T., Sweeney, F. 1982. Alterations of spontaneous quantal transmitter release at the mammalian neuromuscular junction induced by divalent and trivalent ions. *Brain Res.* *246*:127-132.
4. Barnes, D., Sato, G. 1980. Methods for growth of cultured cells in serum-free medium. *Anal. Biochem.* *102*:255-270.
5. Beeman, R. W. 1982. Recent advances in mode of action of insecticides. *Annu. Rev. Entomol.* *27*:253-282.
6. Bernardo, L. S., Prince, D. A. 1982. Ionic mechanisms of cholinergic excitation in mammalian hippocampal pyramidal cells. *Brain Res.* *249*:333-344.
7. Bingmann, D., Kolde, G. 1982. PO_2 profiles in hippocampal slices of the guinea pig. *Exp. Brain Res.* *48*:89-96.

8. Bowman, W. C., Savage, A. O. 1981. Pharmacological actions of aminopyridines and related compounds. *Rev. Pure Appl. Pharmacol. Sci.* *2*:317-371.
9. Brown, A. W., Aldridge, W. N., Street, B. W., Verschoyle, R. D. 1979. The behavioural and neuropathological sequelae of intoxication by trimethyltin compounds in the rat. *Am. J. Pathol.* *97*:59-82.
10. Buckle, P. J., Haas, H. L. 1982. Enhancement of synaptic transmission by 4-aminopyridine in hippocampal slices of the rat. *J. Physiol.* *326*:109-122.
11. Bureš, J., Petraň, M., Zachar, J. 1967. *Electrophysiological Methods in Biological Research*, 3rd ed., Academic, New York.
12. Callec, J. J. 1974. Synaptic transmission in the central nervous system of insects, in *Insect Neurobiology* (Treherne, J. E., ed.), North-Holland, Amsterdam, pp. 119-184.
13. Callec, J.-J., David, J. A., Sattelle, D. B. 1982. Iontophoretic application of acetylcholine on to the dendrites of an identified giant interneurone (GI 1) in the cockroach *Periplaneta americana*. *J. Insect Physiol.* *28*:1003-1008.
14. Callec, J.-J., Sattelle, D. B. 1973. A simple technique for monitoring the synaptic actions of pharmacological agents. *J. Exp. Biol.* *59*:725-738.
15. Callec, J. J., Sattelle, D. B., Hue, B., Pelhate, M. 1980. Central synaptic actions of pharmacological agents in insects: Oil-gap and mannitol-gap studies, in *Insect Neurobiology and Pesticide Action* (Sherwood, M., ed.), Society of Chemical Industry, London, pp. 93-100.
16. Carlen, P. L., Corrigall, W. A. 1980. Ethanol tolerance measured electrophysiologically in hippocampal slices and not in neuromuscular junctions from chronically ethanol-fed rats. *Neurosci. Lett.* *17*:95-100.
17. Ceccarelli, B., Clementi, F. (eds.) 1979. *Neurotoxins: Tools in Neurobiology*, Raven Press, New York.
18. Chambers, P. L., Rowan, M. J. 1980. An analysis of the toxicity of hydroquinone on central synaptic transmission. *Toxicol. Appl. Pharmacol.* *54*:238-243.
19. Coats, J. R. (ed.). 1982. *Insecticide Mode of Action*, Academic, New York.
20. Collins, M. F., Hrdina, P. D., Whittle, E., Singhal, R. L. 1982. Lead in blood and brain regions of rats chronically exposed to low doses of the metal. *Toxicol. Appl. Pharmacol.* *65*:314-322.
21. Corrigall, W. A., Linseman, M. A., Lucato, R., Elliott, M. 1981. Differential tolerance to the effects of morphine on evoked activity in the hippocampal slice. *Life Sci.* *28*:1613-1620.
22. Crain, S. M. 1976. *Neurophysiologic Studies in Tissue Culture*, Raven Press, New York.

23. Cremer, J. E. 1981. Specific toxic effects on the nervous system, in *Organ Directed Toxicity. Chemical Indices and Mechanisms* (Brown, S. S., Davies, D. S., eds.), Pergamon, Oxford, pp. 213-217.
24. Crunelli, V., Forda, S., Collingridge, G. L., Kelly, J. S. 1982. Intracellular recorded synaptic antagonism in the rat dentate gyrus. *Nature 300*:450-452.
25. Daley, D. L., Vardi, N., Appignan, B., Camhi, J. M. 1981. Morphology of the giant interneurons and cercal nerve projections of the American cockroach. *J. Comp. Neurol. 196*:41-52.
26. Danscher, G., Fjerdingstad, E. J., Fjerdingstad, E., Fredens, K. 1976. Heavy metal content in subdivisions of the rat hippocampus (zinc, lead, and copper). *Brain Res. 112*:442-446.
27. Dewar, A. J. 1981. Neurotoxicity testing, in *Testing for Toxicity* (Gorrod, J. W., ed.), Taylor and Francis, London, pp. 199-217.
28. Dingledine, R., Dodd, J., Kelly, J. S. 1980. The in vitro brain slice as a useful neurophysiological preparation for intracellular recording. *J. Neurosci. Methods 2*:323-363.
29. Ekström von Lubitz, D. K. J., Diemer, N. H. 1982. Complete cerebral ischaemia in the rat: An ultrastructural and stereological analysis of the distal stratum radiatum in the hippocampal CA-1 region. *Neuropathol. Appl. Neurobiol. 8*:197-215.
30. Eldefrawi, A. T., Mansour, N. A., Eldefrai, M. E. 1982. Insecticides affecting acetylcholine receptor interactions. *Pharmacol. Ther. 16*:45-65.
31. Euler, C. von. 1962. On the significance of the high zinc content in the hippocampal formation, in *Physiologie de l'hippocampe* (Passouant, P., ed.), Editions du Centre National de la Recherche Scientifique, Paris, pp. 135-146.
32. Evans, P. D. 1980. Biogenic amines in the insect nervous system. *Adv. Insect Physiol. 15*:317-473.
33. Farnell, B. J., De Boni, U., Crapper, D. R. 1982. Aluminum neurotoxicity in the absence of neurofibrillary degeneration in CA1 hippocampal pyramidal neurons *in vitro*. *Exp. Neurol. 78*:241-258.
34. Fisher, R. S., Alger, B. E. 1984. Electrophysiological mechanisms of kainic acid-induced epileptiform activity in the rat hippocampal slice. *J. Neurosci. 4*:1312-1323.
35. Fjerdingstad, E. J., Danscher, G., Fjerdingstad, E. 1974. Hippocampus: Selective concentration of lead in the normal rat brain. *Brain Res. 80*:350-354.
36. Fournier, E. P., Roux, F. 1980. Limits of the use in neurotoxicology of cellular models developed in neurobiology, in *Advances in Neurotoxicology* (Manzo, L., ed.), Pergamon, Oxford, pp. 307-318.

37. Fox, D. A., Lowndes, H. E., Bierkamper, G. G. 1982. Electrophysiological techniques in neurotoxicology, in *Nervous System Toxicology* (Mitchell, D. L., ed.), Raven Press, New York, pp. 299-335.
38. Fox, D. A., Sillman, A. J. 1979. Heavy metals affect rod, but not cone, photoreceptors. *Science 206*:78-80.
39. French, E. D., Zieglgänsberger, W. 1982. The excitatory response of in vitro hippocampal pyramidal cells to normorphine and methionine-enkephalin may be mediated by different receptor populations. *Exp. Brain Res. 48*:238-244.
40. Gähwiler, B. H. 1981. Organotypic monolayer cultures of nervous tissue. *J. Neurosci. Methods 4*:329-342.
41. Galvan, M., Grafe, P., ten Bruggencate, G. 1982. Convulsant actions of 4-aminopyridine on the guinea-pig olfactory cortex. *Brain Res. 241*:75-86.
42. Goldberg, A. M. 1980. Mechanisms of neurotoxicity as studied in tissue culture systems. *Toxicology 17*:201-208.
43. Greenfield, J. G., Blackwood, W., McMenemey, W. H., Meyer, A., Norman, R. M. 1961. *Neuropathology*, Edward Arnold, London.
44. Gryder, R., Frankos, V. (ed.). 1980. *Fifth FDA Science Symposium on the Effects of Foods and Drugs on the Development and Function of the Nervous System*, U.S. Department of Health and Human Services, Washington, D.C.
45. Gustafsson, B., Galvan, M., Grafe, P., Wigstrom, H. 1982. A transient outward current in a mammalian central neurone blocked by 4-aminopyridine. *Nature 299*:252-254.
46. Haas, H. L., Wieser, H. G., Yasargil, M. G. 1983. 4-Aminopyridine and fiber potentials in rat and human hippocampal slices. *Experientia 39*:114-115.
47. Haliwell, J. V., Dolly, J. O. 1982. Preferential action of β-bungarotoxin at nerve terminal regions in the hippocampus. *Neurosci. Lett. 30*:321-327.
48. Harrow, I. D., David, J. A., Sattelle, D. B. 1982. Acetylcholine receptors of identified insect neurons, in *Neuropharmacology of Insects. Ciba Foundation Symposium 88* (Everard, D., O'Connor, M., Whelan, J., eds.), Pitman, London, pp. 12-27.
49. Hayes, A. W. (ed.). 1982. *Principles and Methods in Toxicology*, Raven Press, New York.
50. Heyer, E. J., Nowak, L. M., MacDonald, R. L. 1982. Membrane depolarization and prolongation of calcium-dependent action potentials of mouse neurons in cell culture by two convulsants: bicuculine and penicillin. *Brain Res. 232*:41-56.
51. Hue, B., Pelhate, M., Chanelet, J. 1979. Pre- and postsynaptic effects of taurine and GABA in the cockroach central nervous system. *Can. J. Neurol. Sci. 6*:243-250.

52. Innes, D. L., Tansy, M. F. 1981. Central nervous system effects of methyl methacrylate vapor. *Neurotoxicology 2*:515-522.
53. Jacques, Y., Romey, G., Cavey, M. T., Kartalovski, B., Lazdunski, M. 1980. Interaction of pyrethroids with Na^+ channel in mammalian neuronal cells in culture. *Biochim. Biophys. Acta 600*:882-897.
54. Jefferys, J. G. R. 1981. Aminopyridines modify granule cell responses in guinea-pig hippocampal slices. *J. Physiol. 312*:17P.
55. Jefferys, J. G. R. 1981. The Vibroslice a new vibrating-blade tissue slicer. *J. Physiol. 324*:2P.
56. Kaila, K. 1982. Cellular neurophysiological effects of phenol derivatives. *Comp. Biochem. Physiol. 73C*:231-241.
57. Kaneko, A. 1979. Physiology of the retina. *Annu. Rev. Neurosci. 2*:169-191.
58. Kato, G. 1934. *The Microphysiology of Nerve*, Maruzen, Tokyo.
59. Kerkut, G. A. (ed.). 1967-1973. *Experiments in Physiology and Biochemistry*, Vols. 1-6, Academic, London.
60. Kerkut, G. A. 1982. The development of isolated central nervous system preparations. *Comp. Biochem. Physiol. 72C*:161-169.
61. Kerkut, G. A., Heal, H. V. (eds.). 1981. *Electrophysiology of Isolated Mammalian CNS Preparations*, Academic, London.
62. King, G. L., Parmentier, J. L. 1983. Oxygen toxicity of hippocampal tissue *in vitro*. *Brain Res. 260*:139-142.
63. Klee, M. R., Lux, H. D., Speckman, E. J. (eds.). 1982. *Physiology and Pharmacology of Epileptogenic Phenomena*, Raven Pres, New York.
64. Koerner, J. F., Cotman, C. W. 1982. Response of Schaffer collateral-CA1 pyramidal cell synapses of the hippocampus to analogues of acidic amino acids. *Brain Res. 251*:105-115.
65. Kolber, A. R., Wong, T. K., Grant, L. D., De Woskin, R. S., Hughes, T. J. (eds.). 1982. *In Vitro Toxicity Testing of Environmental Agents. Current and Future Possibilities*, Elsevier, Amsterdam.
66. Kudo, Y. 1978. The pharmacology of the amphibian spinal cord. *Prog. Neurobiol. 11*:1-76.
67. Kuroda, Y. 1980. Brain slices, assay systems for the neurotoxicity of environmental pollutants and drugs on mammalian central nervous system, in *Mechanisms of Toxicity and Hazard Evaluation* (Holmstedt, B., Lauwerys, R., Mercier, M., Roberfroid, M., eds.), Elsevier/North Holland, Amsterdam, pp. 59-62.
68. Lapresle, J., Fardcan, M. 1964. The central nervous system and carbon monoxide poisoning II. Anatomical study of brain lesions following intoxication with carbon monoxide (22 cases). *Prog. Brain Res. 24*:31-74.
69. Leake, L. D., Walker, R. J. 1980. *Invertebrate Neuropharmacology*, Blackie, Glasgow.

70. Lechat, P., Bowman, W. C., Thesleff, S. (eds.). 1982. *Aminopyridines and Related Drugs. Advances in the Biosciences*, Vol. 35, Pergamon, Oxford.
71. Li, C.-L., McIlwain, H. 1957. Maintenance of resting membrane potentials in slices of mammalian cerebral cortex and other tissues *in vitro*. *J. Physiol. 139*:178-190.
72. Lipton, P., Whittingham, T. S. 1982. Reduced ATP concentration as a basis for synaptic transmission failure during hypoxia in the *in vitro* guinea pig hippocampus. *J. Physiol. 325*:51-65.
73. MacDonald, R. L., Barker, J. L. 1978. Specific antagonism of GABA-mediated postsynaptic inhibition in cultured spinal mammalian neurons: A common mode of convulsant action. *Neurology 28*:325-330.
74. MacDonald, R. L., Barker, J. L. 1981. Neuropharmacology of spinal cord neurons in primary dissociated cell culture, in *Excitable Cells in Tissue Culture* (Nelson, P. G., Lieberman, M., eds.), Plenum, New York, pp. 81-109.
75. MacVicar, B. A., Weir, G., Riexinger, K., Dudek, F. F. 1981. Paradoxical effects of lithium on field potentials of dentate granule cells in slices of rat hippocampus. *Neuropharmacology 20*:489-496.
76. Madison, D. V., Nicoll, R. A. 1982. Noradrenaline blocks accommodation of pyramidal cell discharge in the hippocampus. *Nature 299*:636-638.
77. Messing, A., Kim, S. U. 1979. Long-term culture of adult mammalian central nervous system neurones. *Exp. Neurol. 65*:293-300.
78. Misgeld, U., Frotscher, M. 1982. Dependence of the viability of neurons in hippocampal slices on oxygen supply. *Brain Res. Bull. 8*:95-100.
79. Mitchell, D. L. (ed.). 1982. *Nervous System Toxicology*, Raven Press, New York.
80. Nadler, J. V. 1981. Structure-activity relationships for neurotoxicity. Hippocampus. *Neurosci. Res. Program Bull. 19*:360-362, 364-367, 415-427.
81. Narahashi, T. (ed.). 1970. *Neurotoxicology of Insecticides and Pheromones*, Plenum, New York
82. Nelson, P. G. 1977. Neuronal cell lines, in *Cell Tissue and Organ Cultures in Neurobiology* (Federoff, S., Hertz, L., eds.), Academic, London, pp. 347-365.
83. Nelson, P. G. 1978. Neuronal cell cultures as toxicological test systems. *Environ. Health Perspect. 26*:125-133.
84. Nelson, P. G., Neale, E. A., MacDonald, R. L. 1981. Electrophysiological and structural studies on neurons in dissociated cell cultures of the central nervous system, in *Excitable Cells in Tissue Culture* (Nelson, P. G., Liberman, M., eds.), Plenum, New York, pp. 39-80.

85. Neubert, D. 1982. The use of culture techniques in studies of prenatal toxicity. *Pharmacol. Ther. 18*:397-434.
86. Niklowitz, W. J. 1980. Neurotoxicology of lead, in *Advances in Neurotoxicology* (Manzo, L., ed.), Pergamon, Oxford, pp. 27-34.
87. Nishimura, K., Ueno, A., Nakagawa, S., Fujita, T., Nakajima, M. 1982. Quantitative structure-activity studies of substituted benzylcrysanthemates 3. Physicochemical substituent effects and the spontaneous neuroexcitatory activity on the crayfish abdominal nerve cords. *Pesti. Biochem. Physiol. 17*:271-279.
88. Oliver, A. P., Hoffer, B. J., Wyatt, R. J. 1977. The hippocampal slice: A system for studying the pharmacology of seizures and for screening anticonvulsant drugs. *Epilepsia 18*:543-548.
89. Patrick, J., Heinemann, S., Schubert, D. 1978. Biology of cultured nerve and muscle. *Annu. Rev. Neurosci. 1*:417-443.
90. Pelhate, M., Sattelle, D. B. 1982. Pharmacological properties of insect axons: A review. *J. Insect. Physiol. 28*:889-903.
91. Pichon, Y. 1974. Axonal conduction in insects, in *Insect Neurobiology* (Trehern, J. E., ed.), Academic, London, pp. 73-117.
92. Pichon, Y. 1974. The pharmacology of the insect nervous system, in *The Physiology of Insecta*, Vol. 4, 2nd ed. (Rockstein, M., ed.), Academic, London, pp. 101-174.
93. Pichon, Y., Callec, J.-J. 1970. Further studies on synaptic transmission in insects. I. External recording of synaptic potentials in a single giant axon of the cockroach, *Periplaneta americana. J. Exp. Biol. 52*:257-265.
94. Pichon, Y., Treherne, J. E. 1970. Extraneuronal potentials and potassium depolarization in cockroach giant axons. *J. Exp. Biol. 53*:485-493.
95. Prasad, K. N., Vernadakis, A. (eds.). 1982. *Mechanisms of Actions of Neurotoxic Substances*, Raven Press, New York.
96. Prince, D. A. 1978. Neurophysiology of epilepsy. *Annu. Rev. Neurosci. 1*:395-415.
97. Proceedings of a satellite symposium to the 1st World-Congress of IBRO, on Environmental Neurotoxicology, 1982. *Neurobehav. Toxicol. Teratol. 4*:597-746.
98. Ransom, B. R., Neale, E., Henkart, M., Bullock, P. N., Nelson, P. G. 1977. Mouse spinal cord in cell culture. I. Morphology and intrinsic neuronal electrophysiological properties. *J. Neurophysiol. 40*:1132-1150.
99. Raybourn, M. S., Cork, C., Schimmerling, W., Tobias, C. A. 1978. An *in vitro* electrophysiological assessment of the direct cellular toxicity of carbon monoxide. *Toxicol. Appl. Pharmacol. 46*:769-779.
100. Record, I. R., Freosti, I. E., Tulsi, R. S., Fraser, F. J., Bulkely, R. A., Manuel, S. J. 1982. Postnatal accumulation of zinc by the rat hippocampus. *Biol. Trace Elem. Res. 4*:279-288.

101. Renshaw, B. A., Forbes, A., Morrison, B. R. 1940. Activity of isocortex and hippocampus: electrical studies with microelectrodes. *J. Neurophysiol.* *3*:74-105.
102. Richards, C. D., White, A. E. 1975. The actions of volatile anaesthetics on synaptic transmission in the dentate gyrus. *J. Physiol.* *252*:241-257.
103. Roeder, K. D., Roeder, S. 1939. Electrical activity in the isolated ventral nerve cord of the cockroach. 1. The action of pilocarpine, nicotine, eserine and acetylcholine. *J. Cell. Comp. Physiol.* *14*:1-9.
104. Rowan, M. J. 1981. Aspects of the pharmacological actions and effects of electron donors and electron acceptors with special reference to hydroquinone, p-benzoquinone and pyridine, Ph.D. thesis submitted to the University of Dublin, Ireland.
105. Rowan, M. J., Chambers, P. L. 1982. The assessment of neurotoxicity using the cockroach nerve cord. *Neurobehav. Toxicol. Teratol.* *4*:605-612.
106. Sato, S. M., Frazier, J. M., Goldberg, A. M. 1984. The distribution and binding of zinc in the hippocampus. *J. Neurosci.* *4*:1662-1670.
107. Sattelle, D. B. 1978. The insect nervous system as a site of action of neurotoxicants, in *Pesticide and Venom Neurotoxicity* (Shankland, D. L., Hollingsworth, R. M., Smyth, K., eds.), Plenum, New York, pp. 7-26.
108. Sattelle, D. B. 1980. Acetylcholine receptors of insects. *Adv. Insect. Physiol.* *15*:215-315.
109. Sattelle, D. B. 1981. Acetylcholine receptors in the central nervous system of an insect (*Periplaneta americana* L.), in *Advances in Physiological Sciences*, Vol. 22 (Rozsa, K. S., ed.), Pergamon, Oxford, pp. 31-55.
110. Savolainen, H. 1982. Neurotoxicity of industrial chemicals and contaminants: Aspects of biochemical mechanisms and effects. *Arch. Toxicol. Suppl.* 5:71-83.
111. Schwartzkroin, P. A., Altschuler, R. J. 1977. Development of kitten hippocampal neurons. *Brain Res.* *134*:429-444.
112. Schwartzkroin, P. A., Szawsky, M. 1977. Probable calcium spikes in hippocampal neurons. *Brain Res.* *135*:157-161.
113. Schrier, B. K. 1982. Nervous system cultures as toxicologic test systems, in *Nervous System Toxicology* (Mitchell, C. L., ed.), Raven Press, New York, pp. 337-348.
114. Scott, B. S. 1982. Adult neurons in cell cultures: Electrophysiological characterization and use in neurobiological research. *Prog. Neurobiol.* *19*:187-211.
115. Seawright, A. A., Brown, A. N., Aldridge, W. N., Verschoyle, R. D., Street, B. W. 1980. Neuropathological changes caused

by trialkyl lead compounds in the rat, in *Mechanisms of Toxicity and Hazard Evaluation* (Holmstedt, B., Lauwerys, R., Mercier, M., Roberfroid, M., eds.), Elsevier/North Holland, Amsterdam, pp. 71-74.

116. Shankland, D. L., Hollingsworth, R. M., Smyth, T. (eds.). 1978. *Pesticide and Venom Neurotoxicity*, Plenum, New York.

117. Shankland, D. L., Rose, J. A., Donniger, C. 1971. The cholinergic nature of the cercal nerve-giant fibre synapse in the sixth abdominal ganglion of the American cockroach, *Periplaneta americana* (L.). *J. Neurobiol.* *2*:247-262.

118. Skrede, K. K., Westgaard, R. H. 1971. The transverse hippocampal slice: A well defined cortical structure maintained *in vitro*. *Brain Res.* *35*:589-593.

119. Spector, I. 1981. Electrophysiology of clonal cell lines, in *Excitable Cells in Tissue Culture* (Nelson, P. G., Lieberman, M., eds.), Plenum, New York, pp. 247-277.

120. Swanson, K. L., Woolley, D. E. 1980. Dieldrin induced changes in hippocampal evoked potentials in the rat. *Proc. West. Pharmacol. Soc.* *23*:81-84.

121. Takeno, K., Nishimura, K., Parmentier, J., Narahashi, T. 1977. Insecticide screening with isolated nerve preparations for structure-activity relationships. *Pesti. Biochem. Physiol.* *7*:486-499.

122. Taylor, C. P., Dudek, F. E. 1982. Synchronous neural after discharges in rat hippocampal slices without active chemical synapses. *Science* *218*:810-812.

123. Teyler, T. J. 1980. Brain slice preparation: Hippocampus. *Brain Res. Bull.* *5*:391-403.

124. Thesleff, S. 1980. Aminopyridines and synaptic transmission. *Neuroscience* *5*:1413-1419.

125. Van Harreveld, A. 1984. Effects of 4-aminopyridine on the field potentials of hippocampal slices. *Neurosci. Lett.* *50*:283-288.

126. Yamamoto, C., McIlwain, H. 1966. Electrical activities in thin sections from the mammalian brain maintained in chemically-defined media *in vitro*. *J. Neurochem.* *13*:1333-1343.

127. Zbinden, T., Gross, F. (eds.). 1979. Pharmacological Methods in Toxicology. *International Encyclopaedia of Pharmacology and Therapeutics, Section 102*, Pergamon, Oxford.

29

Electrophysiological Methods for the in Vivo Assessment of Neurotoxicity

Yasuhiro Takeuchi and Yasuo Koike*
Nagoya University School of Medicine, Nagoya, Japan

I. INTRODUCTION

Electrophysiological methods have often been used to detect functional changes of the nervous system affected by neurotoxicants such as mercury, manganese, carbon disulfide, carbon monoxide, organic solvents, organophosphorous compounds, and chlorinated hydrocarbons. Neurophysiological studies would be a highly valuable approach for objective diagnosis of the neurotoxic disorders in which organic or functional impairment of the nervous system is accompanied by corresponding changes in electrophysiological phenomena. Electrophysiological procedures also have been used to obtain more sensitive and objective findings at an early stage of poisoning, because early neurological symptoms caused by neurotoxic substances are often ambiguous and not so specific.

The electrophysiological methods have a number of advantages in neurotoxicological research. They can continuously follow dynamic functional changes of the nervous system without seriously damaging the subject examined. The biopotentials of the nervous system can be

Present affiliation:
Nagoya University Hospital, Nagoya, Japan

recorded even in the microvolt range using recently developed instruments. In addition, rapidly developing computer techniques can be used to measure and characterize low-amplitude biopotentials. The methods can qualify and standardize the data and statistically compare them in terms of amplitude, duration, frequency, and any other parameter.

There are, however, some disadvantages and limitations. These methods usually require expensive equipment and trained staff for their use, and they only can partially measure the comprehensive functions of the highly integrated nervous system. The central nervous system, in particular, has such enormous functional reserves and compensatory mechanisms that electrophysiological data would be of little value in toxicological studies unless combined with the assessment of other parameters, such as symptoms and biochemical, morphological, and behavioral changes.

This chapter is not concerned with a general analysis of electrophysiological techniques which have been successful in neurotoxicology, but is restricted to procedures commonly employed in the in vivo assessment of neurotoxic effects.

II. CENTRAL NERVOUS SYSTEM FUNCTION

A. Electroencephalography

Electroencephalography (EEG) is taken to mean here the spontaneous electrical activity of brain tissue recorded with either scalp electrodes or depth electrodes in specific brain structures. Changes in EEG amplitude, frequency, variability, and pattern are thought to reflect the instantaneous integrated synaptic activity of the central nervous system, which is assumed to be related to underlying biochemical events (12).

The amplitude of the human EEG with scalp electrodes is in the range of 20-100 μV, and the frequency is usually recorded in a range of 0-50 Hz. The electrical activity recorded in the scalp has traditionally been analyzed within a specific band—δ (2-4 Hz), θ (4-8 Hz), α (8-13 Hz), β_1 (13-20 Hz), and β_2 (20-30 Hz) (20)—but various instruments for quantifying EEG activity within selected frequency bands are commercially available. Activation or evocative techniques such as hyperventilation, photic stimulation, or sleep can increase the amount of information obtained from EEG. In animal experiments, the effects of toxicants on various brain structures can be assessed by means of depth electrodes.

Many investigations have demonstrated the increased incidence of abnormal EEG patterns in workers exposed to a variety of toxic agents, including carbon disulfide (14), n-hexane (36), organic solvents (24), organophosphorous compounds (8), styrene (15,30,34), and other chemicals. The clinical experience indicated that EEG is an easily applicable and effective method when detecting the effects of toxic substances on the central nervous system. In addition, many researchers have tried to elucidate the mechanisms of certain neurotoxic effects by means of EEG analysis. The acute and long-term effects of carbon monoxide (38,44), 1,1,1-trichloro-2,2-dichlorophenylethane (DDT) (45), petroleum (22), methyl methacrylate (19), organic solvents (3,23,28), organochlorine and organophosphorous pesticides (2) have been especially investigated.

Woolley and Barron (45) implanted bipolar electrodes in various cortical areas and in the olfactory bulb, dorsal hippocampus, cerebellar vermis, neocerebellum, medial geniculate body, and midbrain reticular formation of rats. The bipolar electrodes were made of two No. 32 stainless steel wires cemented together to provide a 1-mm longitudinal tip separation and insulated except for about 0.5 mm at each tip. After administration of 100 mg/kg DDT to rats by gavage, spontaneous electrical activity of the cerebellum showed a drastic increase in amplitude, while that recorded from the occipital cortex, reticular formation, and medial geniculate body revealed few changes (Fig. 1). Analysis of average sound-evoked potentials in the DDT-treated animals also demonstrated that the most drastic and consistent changes occurred in the cerebellum.

Peréz et al. (28) implanted permanent bipolar stainless steel electrodes in various regions of the central nervous system, including the anterior lobe of the cerebellum, basolateral amygdala, and mesencephalic reticular formation of 15 cats. The animals were exposed to benzene, toluene, or thinner in different experimental sessions. The concentrations of the solvents ranged from 50 to 200 mg/liter. A basolateral amygdaloid discharge appeared initially, and the amygdala showed the most prominent effects in all three cases. Based on these findings, the authors proposed that the solvent under examination might cause convulsive seizures originating within the limbic system. This kind of study can be regarded as a relevant example of a new approach in the assessment of effects and sites of action of the environmental neurotoxicants. Such studies are also intended to reveal the relationship between changes in encephalographic patterns and behavioral or biochemical effects induced by toxic chemicals in the central nervous system.

1. *The Sleep-Wakefulness Cycle*

Electroencephalography can be used to monitor the sleep-wakefulness cycle in experimental animals, since EEGs from different brain areas have characteristic voltage and frequency patterns in the different stages of the cycle. The results of recent studies indicate that analysis of changes in the sleep-wakefulness cycle can provide a useful index of the central nervous system effects of toxicants.

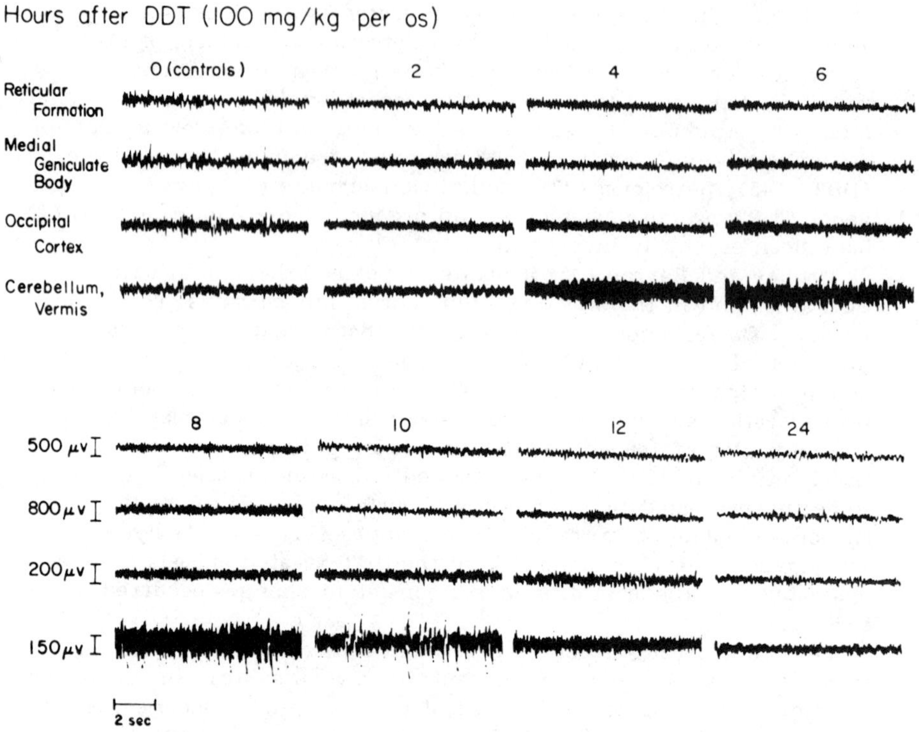

Figure 1 Sequence of changes in the spontaneous electrical activity of four brain areas of an awake, unanesthetized rat after administration of 100 mg/kg DDT. (Reproduced from Ref. 45.)

Fordor et al. (9) recorded the bioelectrical activity from the brain cortex and neck muscles of rats chronically implanted with electrodes and were able to obtain an amplitude-related differentiation of the three distinct stages of the sleep-wakefulness cycle (NREM, REM, and awake). Using this model, the effects of toluene (1000 ppm), dichloromethane (3000 ppm), and carbon monoxide (250 ppm) were compared. The REM showed an increase during an 8-hr exposure to toluene, but decreased in the animals exposed to dichloromethane or carbon monoxide.

Takeuchi and Hisanaga (41) chronically implanted electrodes in the cortex, hippocampus, and cervical muscles of rats. By EEG, electromyography (EMG), and pulse rate, five stages of the sleep-wakefulness cycle were classified: wakeful, spindle, slow wave, preparadoxical, and paradoxical. The animals were exposed to 4000, 2000, or 1000 ppm of toluene for 4 hr. Exposure to 1000 ppm toluene increased the spindle and decreased the paradoxical stage. Exposure to 2000 or 4000 ppm toluene decreased all four sleeping stages. In an additional study (42), the long-term exposure to toluene (2000 ppm, 4 hr/day for 24 weeks) increased the wakeful stage during the 6 hr of the main sleeping time. The sleep-wakefulness cycle was often interrupted in the exposed rats (Fig. 2).

Sǔšić et al. (39) used rats implanted with chronic electrodes for sleep recording to test the effects of diethyldithiocarbamate (DDC) on sleep. The EEG and EMG recordings were carried out continuously for 12 hr during 12 consecutive days. With a daily injection of DDC (500 mg/kg, intraperitoneally), there was an increase in wakefulness, no changes in slow-wave sleep, and suppression or drastic reduction in paradoxical sleep. There was no paradoxical sleep rebound during the DDC withdrawal days.

B. Evoked Potentials

Evoked potentials (EPs) are recorded from cortical and subcortical areas of the brain in response to external stimuli. A series of EPs

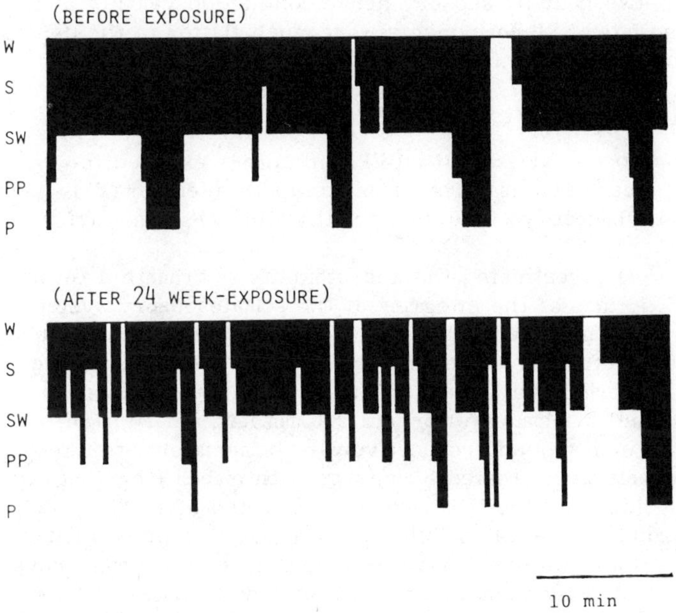

Figure 2 Interruption of the sleep cycle in a rat exposed to 2000 ppm toluene for 25 weeks (W, wakeful; S, spindle; SW, slow wave; PP, preparadoxical; and P, paradoxical). (Reproduced from Ref. 42.)

are usually averaged to extract the evoked potentials from the background EEG (4), and the amplitude and latency of each individual wave component are analyzed. The EPs include a somatosensory evoked potential (SEP) elicited by electrical or other stimulation of the peripheral nerve, a visual evoked potential (VEP) elicited by photic stimulation, and an auditory evoked potential (AEP) elicited by sonic stimulation. The early components of the EPs are generally considered to represent the arrival of inputs in the primary sensory cortex via specific sensory pathways, and the later components to be produced by inputs in the sensory cortex from nonspecific brain regions.

1. *Studies in Humans*

Recently EPs have been used to assess the neurotoxicity potential of certain chemicals in humans (27,36). Mutti et al. (27) investigated somatosensory evoked potentials in 15 female workers from a shoe factory. These subjects were exposed to n-hexane and other organic solvents for 2-8 years (mean 4.5 years). The SEPs were obtained by stimulation of the median nerve at the wrist by silver cup electrodes. Evoked potentials were recorded by two monopolar needle electrodes inserted in the scalp. The latencies of the early SEP components (P15 and N20) were significantly increased, and a negative linear relationship was found between distal sensory nerve conduction velocity and P15 latency. The later SEP components were much flatter in the exposed than in the reference subjects.

2. *Studied in Animals*

Much research has been carried out on EPs in animals exposed to neurotoxic substances such as carbon monoxide (5), dieldrin (21), lead (10,11), methylmercury (6,25,46), triethyl tin (7), trimethyl tin (17), and others.

Howell et al. (17) investigated the neurotoxicity of trimethyl tin in rats as reflected by loss of the integrity of the somatosensory system. The animals were implanted with skull electrodes to record somatosensory evoked response. Stainless steel screws were threaded into the skull at 2.5 mm posterior and 2.5 mm left of bregma (primary somatosensory cortex), and 2.0 mm anterior and 2.0 mm left and right of bregma (reference and ground, respectively). Immediately after testing on day 0, animals were treated by gavage with either 0 or 7 mg/kg trimethyl tin chloride, and the treatment was repeated after 1, 4, and 16 days. The caudal nerve was stimulated with a square pulse (1 Hz, 0.1 msec). Sixty-four responses were used to derive an average wave form. The SEP elicited by caudal nerve stimulation consisted of three negative peaks and two positive peaks. Significant dose by day interactions for N1 and P1 latencies and N1P1 amplitude were demonstrated, and a significant effect of the dose was observed for P2 latency.

Analysis within a single day revealed that N1 and P1 latencies were significantly increased, and N1P1 amplitude was decreased on day 4. From these results, it was proposed that trimethyl tin can influence both specific and nonspecific somatosensory systems.

In a recent study of triethyl tin (TET) (7) rats were implanted with skull screws overlaying the visual cortex (5.5 mm posterior to bregma, 3.0 mm lateral to the midline), and bipolar depth electrodes were implanted in the dentate gyrus of the hippocampal formation (2.5 mm posterior to bregma, 2.5 mm lateral to the midline, and 3.0 mm below the cortical surface, incisors 5 mm above interaural line) under anesthesia. After a week's recovery, TET was intraperitoneally injected at daily doses of 0.188, 0.375, 0.75, or 1.5 mg/kg for 6 days. One hour prior to each visual evoked potential test session, the animals' pupils were dilated with 1% atropine drops. During the test session, four consecutive averaged evoked potentials were obtained, each average consisting of the responses to 64 flashes presented at 0.5 Hz. As illustrated in Fig. 3, alterations in visual evoked potentials were characterized by increased latencies. Triethyl tin continued to increase N1 latency until day 10. Only the N3 latency was affected at a dosage below 1.5 mg/kg per day. This peak appeared most sensitive to TET, since it was affected at the lowest dosage. The visual evoked potential data provide physiological correlates of TET-induced optic nerve demyelination. They also suggest that the optic nerve is not the most sensitive visual structure, since N3 latency, which presumably reflects cortical activity, was affected by TET at lower dosages than those causing changes in direct measures of optic nerve function (e.g., P1 and N1 latency).

The neurophysiological effects of the radiosensitizing agent misonidazole were studied by Rebert (29). Ten rats were implanted with stainless steel screws in midline frontal bone 8-10 mm anterior to bregma for reference, 3 mm posterior to bregma and lateral to the midline for recording auditory brainstem responses, and in the bone over the frontal lobe for ground. During testing the animals were placed in a restraint device which was housed in a sound-attenuating chamber. The auditory evoked potentials were recorded with a pass band of 400 Hz to 10 kHz, averting 1000 responses to clicks presented at 20 per second. Three intensities, corresponding approximately to 5, 30, and 60 dB, were produced by a stimulator. At daily doses of 250 mg/kg, misonidazole produced no obvious signs of behavioral toxicity before 2 weeks, when slight ataxia and lethargy appeared. All the treated animals exhibited a stable auditory evoked potential during the first 2 weeks, followed by a period when the latency of later components increased significantly. The first component of the evoked potential was only slightly affected. These results indicate that misonidazole has only slight effects on the peripheral transducer, or eighth nerve, but a conspicuous effect on central tract conduction velocity.

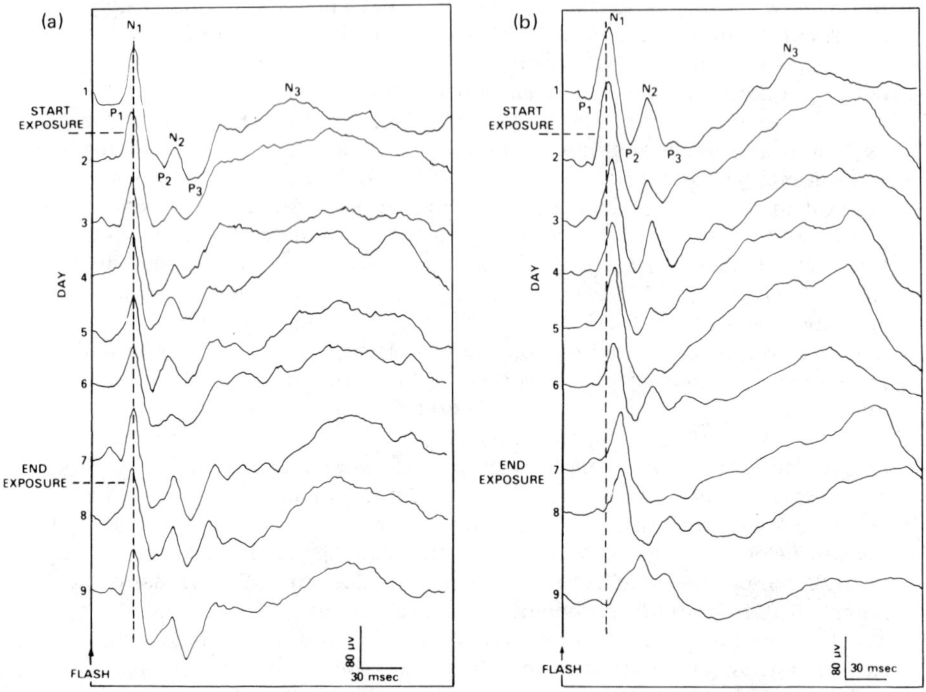

Figure 3 Visual evoked potentials recorded from (a) one animal in the control group and (b) one animal treated with trimethyl tin, 1.5 mg/kg per day intraperitoneally, during the first 9 days of the study. (Reproduced from Ref. 7.)

III. PERIPHERAL NERVE FUNCTION

A. Peripheral Nerve Conduction Velocity

Measuring peripheral nerve conduction velocity is a relatively simple and rapid procedure. The methods in animal experiments are basically the same as in human examination, and the animal data can be easily extrapolated to clinical situations. Moreover, the procedure is not too invasive, so that the same animal can be used for repeated determinations throughout long-term experiments. However, the peripheral nerve in an intact animal does not run linearly, so that accurate measurement of the conduction distance is not very easy. The temperature of the nerve must be kept within a narrow range during measurement, because the lower the temperature of the nerve, the slower the nerve conduction velocity. Moreover, nerve conduction velocities vary with the fiber diameter. The larger fibers conduct action potentials faster than the smaller ones. Therefore, when large-diameter nerve fibers are damaged by toxicants, maximal nerve conduction velocity is decreased, but when large fibers are unaffected, it never decreases, even though smaller fibers are impaired.

1. *Maximal Motor Nerve Conduction Velocity*

Maximal motor nerve conduction velocity is measured by stimulating a nerve trunk with electrodes at two separate points, recording the evoked muscle action potentials and determining the two latencies between the stimuli and the responses. Then the maximal motor nerve conduction velocity is calculated by determining the distance between the two stimulation points and dividing the distance by the difference in the two latencies.

2. *Distal Motor Latency*

Distal motor latency is measured by stimulating the distal portion of the motor nerve and recording the evoked muscle potential. This parameter includes both the distal motor nerve conduction time and neuromuscular junction delay. Thus delayed distal motor latency implies impairment of the distal portion of the motor nerve and/or the neuromuscular junction.

3. *Sensory Nerve Conduction Velocity*

Sensory nerve conduction velocity can be measured orthodromically or antidromically with skin or needle electrodes. The nerve is stimulated with an electrode at one point of a nerve and the evoked nerve action potentials are recorded with an electrode placed at another point along the nerve. A series of evoked nerve potentials is usually averaged out with a computer to detect clearly low-voltage nerve action potentials. Then the sensory nerve conduction velocity is calculated by dividing the distance between the stimulating and recording electrodes by the latency.

4. *Slower Motor Nerve Fiber Conduction Velocity*

Slower motor nerve fiber conduction velocity is measured by using paired stimuli based on the antidromic blocking technique. It can detect lesions to smaller-diameter nerve fibers, in contrast to maximal motor nerve conduction velocity, which can detect only large-diameter fiber impairment. The principle of slower motor nerve fiber conduction velocity is as follows: If the nerve is simultaneously stimulated at two different points, the antidromic volley from the distal stimulus points stops the orthodromic volley from the proximal point on its way to the muscle. Thus only one muscular response is obtained. If the proximal stimulus is sufficiently delayed, two successive muscular responses can be recorded. The interval between the stimuli is then gradually shortened until the second response begins to decrease. Some fibers are now being inhibited by the antidromic volley, although most of them are still conducting the impulse. The slower motor nerve fiber conduction velocity can be calculated by dividing the distance by the latency which is obtained by subtracting 1 msec (the refractory period of the nerve fibers) from the shortest interval between paired stimuli with full response (33).

5. *Muscle Action Potential and Nerve Action Potential*

Muscle and nerve action potentials are sometimes simultaneously applied when peripheral nerve conduction velocity is measured. The changes in their amplitude and duration are at times more sensitive than maximal motor or sensory nerve conduction velocity in subjects poisoned by toxicants (34,40).

6. *Clinical Data*

Clinical investigations have proved that the peripheral nerve conduction velocity decreases in workers exposed to neurotoxic substances such as acrylamide (13,40), allyl chloride (16), carbon disulfide (32), n-hexane (18), lead (31), and methyl n-butyl ketone (MnBK) (1).

Seppäläinen and Hernberg (31) investigated 39 lead workers with no clinical signs of neurological impairment. The maximal motor nerve conduction velocities of the median, ulnar, and lateral popliteal nerves were measured. Slower motor nerve fiber conduction velocities were measured using a partial antidromic block. The mean maximal motor nerve conduction velocities of all three nerves studied were slightly slower in the exposed group. The mean distal latency of the median nerve was significantly prolonged in the exposed group. The marked difference between the exposed groups and controls was the reduction in the slower motor nerve fiber conduction velocity of the ulnar nerve among the lead workers. The authors suggest that subclinical nerve damage can be detected in the lead workers with these methods.

7. *Animal Studies*

Seppäläinen and Linnoila (35) exposed rats to 750 ppm carbon disulfide for 6 hr daily, 12 rats being exposed for 22 weeks, and 15 rats from 2 to 5 weeks. The development of neuropathy was monitored by measuring maximal motor nerve conduction velocity in the sciatic nerve before exposure and twice monthly during the exposure and recovery periods. After 8 weeks of exposure, a slight but statistically slowing of the maximal motor nerve conduction velocity was demonstrated concomitantly with the occurrence of difficulties in voluntary control of the hindlimbs. After 22 weeks of exposure, four rats were allowed to recover for a further 12 weeks. Maximal motor nerve conduction velocity was slightly improved, but remained at a much lower level than in control.

Takeuchi et al. (43) investigated the neurotoxicity of n-pentane, n-hexane, and n-heptane in rats exposed to 3000 ppm for 12 hr daily for 16 weeks. The conduction velocity of the peripheral nerve was measured in the rat's tail, which has four large nerve trunks, as shown in Fig. 4. The tail nerves run along so linearly that the distance between stimulating and recording electrodes can be measured in the intact animal relatively accurately. The electrode used was a stainless steel needle, 0.34 mm in diameter. After insertion of the electrodes, the tail was immersed in a paraffin bath at a temperature kept at between 37 and 38°C. The tail nerve was stimulated by supramaximal 0.3-msec square pulses. The maximal motor nerve conduction velocity was measured by stimulation at two separate points of the ventral caudal nerve, and the evoked muscle potentials were picked up at its distal portion. Mixed nerve conduction velocity (equivalent to sensory nerve conduction velocity here) was also measured by stimulation at the distal portion of the tail, and the evoked nerve action potentials were picked up at two separate proximal points of the nerve. A total of 100 evoked potentials were averaged. n-Hexane caused changes in both maximal motor and mixed nerve conduction velocities and prolonged the distal motor latency, while n-pentane and n-heptane were uneffective (Fig. 5).

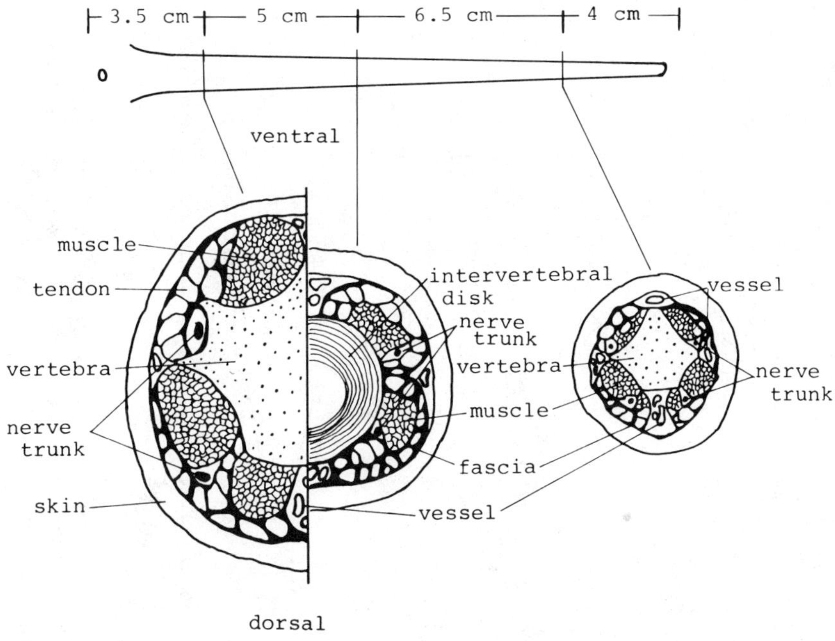

Figure 4 A sketch of rat's tail anatomy. (Drawn by Dr. J. Kitoh.)

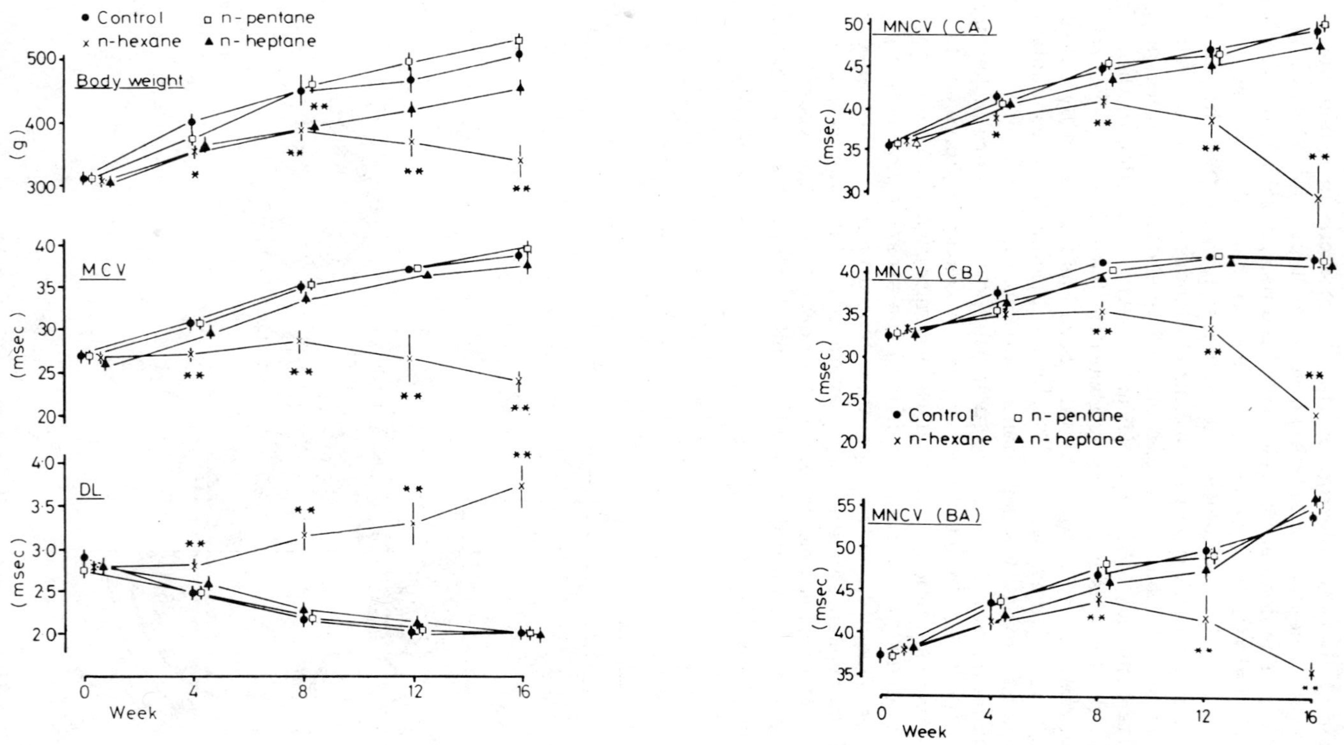

Figure 5 Changes in body weight, maximal motor nerve conduction velocity (MCV), distal motor latency (DL), and mixed nerve conduction velocity (MNCV) (mean ±SE; significance level: *, $p < 0.05$; **, $p < 0.01$). (Reproduced from Ref. 43.)

In studies reported by Seppäläinen and Savolainen (37), the maximal motor nerve conduction velocity of the tail nerve was significantly decreased in rats exposed for 12 weeks to 100 or 300 ppm vinyl toluene (6 hr daily for 5 days a week). The amplitude of the evoked motor action potential was also decreased. The neurotoxic effects of vinyl toluene seemed more pronounced than those of styrene, and the authors suggested that the former substance should not be used as a substitute for styrene.

B. Electromyography

1. *Clinical Studies*

Electromyography (EMG) has been extensively used in human investigations to diagnose toxic myopathies and neuropathies. Electromyograms are usually recorded with a needle electrode inserted into muscle. The electrical activity evoked by insertion and movement of the needles, the electrical activity of the resting muscle, and that of motor units during voluntary muscle contraction are measured. Electromyographic abnormalities are indicated by typical findings such as unusually prolonged discharge or absence of any discharge of electrical activity response to the needle insertion in the muscle, occurrence of spontaneous electrical activity in the muscle at rest, and presence of abnormalities in the shape, amplitude, duration, number or rate of the motor unit action potentials observed during muscle contraction (20).

Impairment of motor units was demonstrated by EMG examination in workers exposed to neurotoxic substances. Iida et al. (18) investigated 44 workers poisoned by n-hexane. The EMG was recorded by a concentric needle electrode in four muscles in the extremities. Electromyographic findings showed fibrillation voltage and positive sharp waves (a sign of denervation), reduced interference voltage and low amplitude (the neurogenic atrophy pattern), and high-amplitude neuromuscular unit (NMU) voltage (a sign of renervation) in a high percentage of n-hexane-exposed workers, including individuals with latent polyneuropathy.

2. *Animal Experiments*

Mendell et al. (26) exposed cats to 600 ppm MnBK, 24 hr daily, for 7 days a week, and EMG examination was performed weekly. Teflon-coated recording electrodes with a 0.1-mm bare tip were used. The reference electrode was located in the subcutaneous tissue. The interference pattern was obtained in the limb muscle by inducing a flexor reflex. Recording of the electrical activity of the muscle at rest and during insertion of the electrode was carried out under general anesthesia. The first EMG findings consisted of pathological insertional activity with positive waves which appeared in all cats after 4-6 weeks of exposure. Subsequently trains of positive sharp waves were observed. After 9-10 weeks, fibrillation potentials appeared in muscles at rest, with insertional activity. The amplitude of motor unit action potentials did not significantly change from control. The EMG abnormalities were apparent in all muscles tested, with more marked changes occurring distally. These findings are consistent with muscle denervation due to distal axonopathy.

IV. CONCLUSIONS AND SUMMARY

This chapter gave a brief overview of the in vivo electrophysiological methods commonly used in neurotoxicology. The description was restricted to electroencephalography (EEG), evoked potentials (EPs), electroneurography (maximal motor and sensory nerve conduction velocities, distal motor latency, slower motor nerve fiber conduction velocity, and muscle and nerve action potentials), and electromyography (EMG). The experimental approaches vary with purpose, and many techniques have been used (12).

Electrophysiological methods are rapidly developing in the field of neurotoxicology, and normative data for experimental animals are also being accumulated. It is hopefully expected that the individual techniques will be improved in the near future and contribute much more to the objective diagnosis of toxic disorders of the nervous system, early detection of neurological changes, and a better understanding of the mechanisms of neurotoxicity.

REFERENCES

1. Allen, N., Mendell, J. R., Billmaier, D. J., Fontaine, R. E., O'Neill, J. 1975. Toxic polyneuropathy due to methyl n-butyl ketone. An industrial outbreak. *Arch Neurol.* *32*:209-218.
2. Burchfiel, J. L., Duffy, F. H., Sim, V. M. 1976. Persistent effects of sarin and dieldrin upon primate electroencephalogram. *Toxicol. Appl. Pharmacol.* *35*:365-379.
3. Contreras, C. M., González-Estrada, T., Zarabozo, D., Fernández-Guardiola, A. 1979. Petit mal and grand mal seizures produced by toluene or benzene intoxication in the cat. *Electroencephalogr. Clin. Neurophysiol.* *46*:290-301.
4. Dawson, G. D. 1954. A summation technique for the detection of small evoked potentials. *Electroencephalogr. Clin. Neurophysiol.* *6*:65-84.
5. Dyer, R. S., Annau, Z. 1977. Carbon monoxide and flash evoked potentials from rat cortex and superior colliculus. *Pharmacol. Biochem. Behav.* *6*:461-465.

6. Dyer, R. S., Eccles, C. U., Annau, Z. 1977. Evoked potential alterations following prenatal methylmercury exposure. *Pharmacol. Biochem. Behav. 8*:137-141.
7. Dyer, R. S., Howell, W. E. 1982. Acute triethyl tin exposure; effects on the visual evoked potential and hippocampal afterdischarge. *Neurobehav. Toxicol. Teratol. 4*:259-266.
8. Duffy, F. H., Burchfiel, J. L., Bartels, P. H., Gaon, M., Sim, V. M. 1979. Long-term effects of an organophosphate upon the human electroencephalogram. *Toxicol. Appl. Pharmacol. 47*: 161-176.
9. Fordor, G. C., Schlipköter, H. W., Zimmerman, M. 1973. The objective study of sleeping behaviour in animals as a test of behavioral toxicology, in *Adverse Effects of Environmental Chemicals and Psychotropic Drugs*, Vol. 1 (Horvath, M., ed.), Elsevier, Amsterdam, pp. 115-123.
10. Fox, D. A., Lewkowski, J. P., Cooper, G. P. 1977. Acute and chronic effects of neonatal lead exposure on development of the visual evoked responses in rats. *Toxicol. Appl. Pharmacol. 40*: 449-461.
11. Fox, D. A., Lewkowski, J. P., Cooper, G. P. 1979. Persistent visual cortex excitability alterations produced by neonatal lead exposure. *Neurobehav. Toxicol. Teratol. 1*:101-106.
12. Fox, D. A., Lowndes, H. E., Bierkamper, G. G. 1982. Electrophysiological techniques in neurotoxicology, in *Nervous System Toxicology* (Michell, C. L., ed.), Raven Press, New York, pp. 299-335.
13. Fullerton, P. M. 1969. Electrophysiological and histological observations on peripheral nerves in acrylamide poisoning in man. *J. Neurol. Neurosurg. Psychiatry 32*:186-192.
14. Harada, M., Kabashima, K., Ooshima, S., Sugimura, K., Tatesu, S. 1980. EEG and CT-scan on chronic carbon disulfide poisoning. *Rinsho Noha 22*:420-426.
15. Härkönen, H. 1977. Relationship of symptoms to occupational styrene exposure and to the findings of electrophysiological and psychological examinations. *Int. Arch. Occup. Environ. Health 40*:231-239.
16. He, F., Shen, D., Guo, Y., Lu, B. 1980. Toxic polyneuropathy due to chronic allyl chloride intoxication; a clinical and experimental study. *Chin. Med. J. 93*:177-182.
17. Howell, W. E., Walsh, T. J., Dyer, R. S. 1982. Somatosensory dysfunction following acute trimethyl tin exposure. *Neurobehav. Toxicol. Teratol. 4*:197-201.
18. Iida, M., Yamamura, Y., Sobue, I. 1969. Electromyographic findings and conduction velocity of n-hexane polyneuropathy. *Electromyography 9*:247-261.

19. Innes, D. L., Tansy, M. F. 1981. Central nervous system effects of methyl methacrylate vapor. *Neurotoxicology 2*:515-522.
20. Johnson, B. L. 1980. Electrophysiological methods in neurotoxicity test, in *Experimental and Clinical Neurotoxicology* (Spencer, P. S., Schaumburg, H. H., eds.), Williams and Wilkins, Baltimore, pp. 726-742.
21. Joy, R. M. 1976. The alteration by dieldrin of cortical excitability conditioned by sensory stimuli. *Toxicol. Appl. Pharmacol. 38*:357-368.
22. Komura, S., Ueda, M., Fujimura, K. 1973. Electroencephalographic studies on liquified petroleum gas poisoning in rabbits. *Tohoku J. Exp. Med. 111*:33-40.
23. Konietzko, H., Elster, I., Bencsath, A., Drysch, K., Weichardt, H. 1975. EEG-Veränderungen unter definierter Trichlorathylene-Exposition. *Int. Arch. Occup. Environ. Health 35*:257-264.
24. Mabuchi, C., Takagi, S., Takeuchi, Y., Koike, Y., Yamauchi, K., Shibata, T. 1974. Neurological symptoms in chronic intoxication by organic solvents. *Igaku No Ayumi 88*:97-106.
25. Mattsson, J. L., Miller, E., Alligood, J. P., Koering, J. E., Levin, S. G. 1981. Early effects of methylmercury on the visual evoked response of the dog. *Neurotoxicology 2*:499-514.
26. Mendell, J. R., Saida, K., Ganansia, M. F., Jakson, D. B., Weiss, H., Gardier, R. W., Chrisman, C., Allen, N., Couri, D., O'Neill, J., Marks, B., Hetland, L. 1974. Toxic polyneuropathy produced by methyl n-butyl ketone. *Science 185*:787-789.
27. Mutti, A., Ferri, F., Lommi, G., Lotta, S., Lucertini, S., Franchini, I. 1982. n-Hexane-induced changes in nerve conduction velocities and somatosensory evoked potentials. *Int. Arch. Occup. Environ. Health 51*:45-54.
28. Peréz, C. M. C., González-Estrada, M. T., Paz, C., Fernândez-Guardiola, A. 1978. Electroencephalographic and behavioral aspects of chronic exposure with industrial solvents to cats, in *Voluntary Inhalation of Industrial Solvents* (Sharp, C., Carroll, L., eds.), National Institute on Drug Abuse, Rockville, Md., pp. 226-245.
29. Rebert, C. S. 1980. The brain stem auditory evoked response as a tool in neuro-behavioral toxicity and medicine. *Prog. Brain Res. 54*:458-462.
30. Rosén, I., Haeger-Aronsen, B., Rehnström, S., Welinger, H. 1978. Neurophysiological observation after chronic styrene exposure. *Scand. J. Work Environ. Health Suppl. 2*:184-194.
31. Seppäläinen, A. M., Hernberg, S. 1972. Sensitive technique for detecting subclinical lead neuropathy. *Br. J. Ind. Med. 29*:443-449.
32. Seppäläinen, A. M., Tolonen, M. 1974. Neurotoxicity of long-term exposure to carbon disulfide in the viscose-rayon industry. *Scand. J. Work Environ. Health 11*:145-153.

33. Seppäläinen, A. M. 1975. Applications of neurophysiological methods in occupational medicine. *Scand. J. Work Environ. Health 1:*:1-14.
34. Seppäläinen, A. M., Härkönen, H. 1976. Neurophysiological findings among workers occupationally exposed to styrene. *Scand. J. Work Environ. Health 3*:140-146.
35. Seppäläinen, A. J., Linnoila, I. 1976. Electrophysiological findings in rats with experimental carbon disulfide neuropathy. *Neuropathol. Appl. Neurobiol. 2*:209-216.
36. Seppäläinen, A. M., Raitta, C., Huuskonen, M. 1979. n-Hexane-induced changes in visual evoked potentials and electroencephalograms of industrial workers. *Electroencephalogr. Clin. Electrophysiol. 47*:492-498.
37. Seppäläinen, A. M., Savolainen, H. 1982. Impaired nerve function in rats after prolonged exposure to vinyltoluene. *Arch. Toxicol. Suppl.* 5:100-102.
38. Stewart, R. D., Peterson, J. E., Fisher, T. N., Hosko, M. J., Baretta, E. D., Dodd, H. C., Herrman, A. A. 1973. Experimental human exposure to high concentrations of carbon monoxide. *Arch. Environ. Health 26*:1-7.
39. Sŭšić, V., Kovačević, R., Maširević, G. 1980. Sleep-waking cycle behaviour after diethyldithiocarbamate in the rat. *Arch. Int. Physiol. Biochem. 88*:37-45.
40. Takahashi, M., Ohara, T., Hashimoto, K. 1971. Electrophysiological study of nerve injuries in workers handling acrylamide. *Int. Arch. Arbeitsmed. 28*:1-11.
41. Takeuchi, Y., Hisanaga, N. 1977. The neurotoxicity of toluene; EEG changes in rats exposed to various concentrations. *Br. J. Ind. Med. 34*:314-324.
42. Takeuchi, Y., Hisanaga, N., Ono, Y. 1979. Changes of sleep cycles and EEG in rats chronically exposed to toluene vapor. *Arh. Hig. Rada Toksikol. Suppl. 30*:467-475.
43. Takeuchi, Y., Ono, Y., Hisanaga, N., Kitoh, J., Sugiura, Y. 1980. A comparative study on the neurotoxicity of n-pentane, n-hexane, and n-heptane in the rat. *Br. J. Ind. Med.* 37:241-247.
44. Yukitake, A. 1973. An experimental study of carbon monoxide poisoning; an appallic-like syndrome. *Folia Psychiatr. Neurol. Jpn. 27*:341-349.
45. Woolley, D. E., Barron, B. A. 1968. Effects of DDT on brain electrical activity in awake, unrestrained rats. *Toxicol. Appl. Pharmacol. 12*:440-454.
46. Zenick, H. 1976. Evoked potential alterations in methylmercury chloride toxicity. *Pharmacol. Biochem. Behav.* 5:253-255.

30

Behavioral Toxicology: Animal Experimental Models

Zoltan Annau
The Johns Hopkins University, Baltimore, Maryland

I. INTRODUCTION

The first conference on Behavioral Toxicology held at the University of Rochester, and subsequently published as a book by that title (6), marks the beginning of this new field as a recognized subdiscipline of the fields of toxicology and behavioral science. The recognition that some of the chemicals released into the environment by man cause damage to the nervous system and alter behavior led to the necessity of studying these chemicals with rigorous scientific methods. Since the appearance of the above book the new discipline has grown rapidly and beyond the initial boundaries of establishing behavioral indices of adverse effects. Investigators have recognized that the behavioral end point is just one index of toxicity and that in order to understand the mechanisms of toxicity, the behavioral measures have to be correlated with morphology, neurochemistry, analytical measures, and many other techniques that have become the standard armamentarium of neurotoxicology. While these new techniques have expanded our understanding of the complexities of neurotoxicity, the ultimate question of how neurotoxic chemicals affect behavior remains the central theme of of this field, and perhaps the central question relevant to humanity.

II. THE PROBLEM AREA

It is clear that all chemicals will have an adverse effect when used in sufficiently large doses (12), but what has become equally clear in recent times is that adverse effects can occur after exposure to infinitely small doses as well. The field of behavioral toxicology has to deal, therefore, not only with the study of the effects of those chemicals that alter nervous system function, but also with developing control procedures that allow us to differentiate between neurotoxic and nonneurotoxic chemicals that simply make the animal ill. It has become clear during the past 20 years that this is not an easy process and that the behavioral toxicologist will never be able to depend on any one test to determine the toxicity of a chemical. Rather than hope for the behavioral "Ames test," we have to develop a test battery that takes into account the major behavioral variables that we now know are affected by toxic chemicals. These variables can be characterized as consisting of tests of motor function, tests of sensory function, and tests of cognitive functions. This chapter will deal with these three areas as well as their subclasses in describing the effects of three neurotoxic chemicals: inorganic lead, carbon monoxide, and methyl mercury.

A. Motor Function

The development of locomotion in an animal is one of the earliest forms of "behavior" that can be measured readily. The measures of locomotor activity, therefore, are not only useful as neonatal measures of normal versus abnormal development, but also as measures of the general state of health of the adult organism. In the neonatal rodent there seems to be a well-characterized pattern of locomotor development in which the rat pups show increasing levels of motor activity that reaches a peak around 15 days of age and then subsequently declines toward adult levels (10). This pattern has been observed by many investigators using a variety of devices for measuring activity (18,20); however, as Reiter and MacPhail have pointed out, this pattern may depend somewhat on the apparatus employed to make the measurements (40). More specifically, some observers have reported that activity peaks can be observed at later ages than those reported by Campbell if activity is recorded in different settings. Nevertheless, there seems to be a potential correlation between the development of locomotor activity in the neonatal rat and the development of specific neurochemical systems. It has been postulated that the development of catecholaminergic systems associated with arousal occurs earlier than the development of systems associated with inhibition such as those involving serotonin, γ-aminobutyric acid (GABA), or acetylcholine (10). Thus the enhancement of locomotor activity may be initially under the influence of excitatory systems, and as the inhibitory systems mature, the rate of motor activity slowly decreases. When the development of these systems is altered by the administration of catecholaminergic poisons such as 6-hydroxydopamine, the activity levels of the animals remain high up to 28 days postpartum (20).

In a series of elegant reviews, Reiter and co-workers have described the organismic and experimental variables affecting measures of motor activity (39,40). These reviews show clearly that motor activity is not a unitary behavioral phenomenon but, rather, that there are many types of motor activity and that differences reported by investigators using this behavior as a dependent variable in drug studies are usually due to the simplistic assumption that all motor activities are the same. This problem, of course, becomes even more acute in the study of neurotoxic agents, since in many instances these agents cause irreversible alterations in nervous system function and therefore the changes in motor activity reported following the treatment have to be described in the context of the above considerations in order to lead to usable and testable hypotheses regarding the nature of the alterations. As we shall see in the review of many of these studies, this is rarely the case, and the outcome of much of the research therefore remains in doubt.

1. *Lead Toxicity*

Interest in using locomotor activity in behavioral toxicology was aroused considerably by a series of articles dealing with lead toxicity. In these articles it was shown that mice exposed to lead acetate through the mother's milk became hyperactive during development (44-46; see also the chapter by Silbergeld in this book). This hyperactivity continued in the mice to adulthood, and these researchers proposed that the increased activity level seen in the lead-exposed mice was a good animal model of human lead toxicity, and perhaps of the human hyperkinetic syndrome. The reasons for this interesting hypothesis were that the animals were exposed to lead only during the neonatal period, for example, through the mother's milk, just as lead-poisoned children are characteristically exposed to lead in the neonatal period. The mice, when challenged by a series of psychotropic drugs such as amphetamine or phenobarbital responded paradoxically to these drugs just as the hyperactive children, and, finally, some of the neurochemical measures taken on these animals appeared similar to the neurochemical alterations seen in lead-exposed children (47). While these experiments aroused a great deal of interest, careful examination of the data revealed some serious flaws in their design and execution. Because of the high concentration of lead acetate used in the mother's drinking water (0.5%), the females reduced their water intake and their milk production. As a consequence, the growth rate of the mouse offspring was severely retarded. This early malnutrition became an insurmountable obstacle in interpreting the data. A second problem was that the paradoxical responses to drugs could be interpreted as a shift in the dose-response curve, rather than an altered response to drugs. The interest that these studies aroused, however, had the effect of a remarkable increase in lead toxicity-oriented research. The subsequent studies on this topic soon demonstrated that malnutrition alone without lead exposure could increase locomotor activity in rodents (11,29), and that the paradoxical responses to psychotropic drugs could also be generated by malnutrition alone.

While there were problems with the interpretation of these early experiments on lead toxicity and motor activity, the subsequent studies on this topic did not clear up the controversy. In fact, in the summary of these studies published by the United States Environmental Protection Agency (1), it is clear that the effects of lead exposure on motor activity range from no effect, to increased activity, to decreased activity. This state of affairs unfortunately arose from the fact that different research groups did not replicate each others' work but, rather, chose new protocols for administering the lead (prenatal versus postnatal), the doses were almost always different, and in many instances no lead tissue levels were determined. Moreover very rarely was there any neuropathology described in conjunction with the behavioral results.

Even if all experimenters used the same apparatus for recording activity, there is no reason why the direction of the activity change should be the same under different lead doses or different durations of lead exposure. This is because lead may interfere with the development of different groups of neurons or myelination in different parts of the nervous system during the development of the neonate. Alterations in motor activity therefore may be due to the selective destruction of particular components of the brain and lead to either decreases or increases in activity. Since lead is not excreted particularly rapidly from the brain during development, it can have a detrimental effect for a prolonged period. As we shall see later, when more rapidly acting agents are used, a somewhat similar pattern of toxin-induced alteration in motor activity is seen as a function of time of administration during gestation.

2. *Carbon Monoxide*

In order to study the effects of severe hypoxia on motor activity, rats were exposed to lethal carbon monoxide (CO) concentrations until respiratory failure at either 5 days of age or at adulthood (15). The animals that survived exposure were tested in groups in a residential maze. Exposure to CO produced hyperactivity in the young animals that persisted up to 5 months of age. In a subsequent study from this group, 5-day-old, 16 to 20-week old, or adult (18 months old) rats were exposed to lethal (0.42%) CO until respiratory arrest and then removed from the chambers (43). In this experiment the animals were tested singly in the maze. Hyperactivity was found in the 16 to 20-week-old female animals that had been exposed to CO, but not in the males. Motor activity was found to be correlated with duration of exposure. The female animals were found to be significantly more resistant to the lethal effects of CO than the male animals. While these experiments indicate that increased motor activity is seen after severe hypoxia, the data tend to be confounded by the sexually determined sensitivity factor which leads to uncontrolled exposure durations.

Prenatal exposure to CO, on the other hand, has been reported to lead to a decrement in motor activity during the neonatal period (23). In these experiments pregnant rats were exposed to 150 ppm CO throughout gestation and removed from CO after the birth of the litter. Motor activity of individual pups was measured in an open field device for 1 hr on days 1, 4, 14, and 21 postpartum. Since the activity of the 1- and 4-day-old animals was negligible, these animals were injected with 100 mg/kg of L-dopa to increase their base-line activity. The activity of the CO-exposed animals was lower than the activity of controls at all ages tested. The animals that had been exposed to CO showed a significant decrease in brain protein levels, as well as reduced ability to convert L-dopa to dopamine. These experiments showing reductions in motor activity following CO exposure again suggest the usefulness of activity measures as an early index of toxicity.

3. *Methyl Mercury*

With the prenatal exposure model again, pregnant rats were intubated with 5 or 8 mg/kg of methyl mercury chloride on day 8 or 15 of gestation (19). The 8-mg dose administered on day 8 of gestation caused some females to resorb their litters, but those females that gave birth delivered normal-sized litters and normal-weight pups. Measures of activity in an electronic activity monitor were taken on days 4, 8, 15, and 21 for 60 min on individual pups. Activity was significantly elevated on day 4 in pups that had received 5 mg/kg on day 8 of gestation, on postnatal day 8 in pups treated with 8 mg/kg, and on postnatal days 8 and 15 in pups treated with 5 mg/kg on day 15 of gestation. The results of these experiments indicated that mercury produced an enhancement of activity only, and that the peak of the enhancement was influenced by the time of mercury administration during gestation. The peak motor activity of the rat is usually observed around day 15 postnatally, and thus a displacement of the peak could be used as an index of alterations in normal nervous system development. This effect of mercury is of course not entirely surprising in view of the previous discussion, since peak mercury levels in the brains of the fetuses are reached at different times according to the intubation period.

Because the concentration of the toxic material injected either in the mother or the offspring remains high for considerable periods of time, a more direct approach to study the effect of cytotoxic agents on brain-behavior relationships can be obtained with the short-lived chemical 5-azacytidine (42). This toxic agent destroys proliferating cells in the fetal but not the maternal brain and is rapidly cleared from the circulation. The results of a series of studies by these authors show that activity measures taken when the animals had reached adulthood following exposure on gestational days 12, 14, 15, 16, 18, and 19 were not consistent, in that activity sometimes increased and sometimes decreased.

A survey of the literature dealing with perinatal neurotoxicology (see the chapter by Nelson in this book) reveals that while activity measures are very useful in revealing subtle damage to the nervous system, the abnormal response may not be associated with readily measurable neuropathology or overt biological indices of toxicity, and the direction of the activity change cannot at this time be used as an index of neural damage. In many instances the behavioral changes seen postnatally are transient, and subsequent measures of activity in adulthood fail to show a difference from normal patterns of activity. In this sense, these measures are at best an index of ongoing perturbations of the animal's nervous system and have to be used with caution. On the other hand, since during the immediate postnatal period it is difficult to elicit complex behaviors from rodents, the measure of locomotor activity and its development remains a useful technique in assessing toxic damage. Of course, activity measures are useful not only in the neonate, but in the adult animal as well. In this case alterations in motor activity patterns can be used as early indicators of peripheral nervous system dysfunction. As an example, rats placed in activity wheels were exposed to low doses of acrylamide. Other animals had electrodes implanted in the brain and evoked potentials were recorded from their visual cortex (8). The first indicator of toxicity was recorded by the reduction in wheel running, before alterations in electrophysiological measures were noted. Similar results were obtained by other investigators using a variety of chemicals and motor activity as one of the behavioral measures (38,50). This measure may therefore prove to be a very useful index of damage to the peripheral motor system, although it does not lend itself readily to mechanistic interpretations.

Other tests that involve the motor system have been published in the literature in the evaluation of toxic chemicals. One of these models, developed by Tilson and co-workers, consists of measuring the forelimb and hindlimb grip strength of the rats. In this procedure the animals are allowed to grip a metal ring with their forepaws and are pulled away from this ring by the experimenter. The force exerted by the animal before its grip breaks is recorded by a transducer (49). A similar procedure is carried out with the hindlimb. The advantage of this procedure is that it can be quantified with the force transducers, although there is some potential for experimental bias in that the animal has to be held by the experimenter. This test has been used primarily with agents that affect the peripheral nervous system.

Rather than measuring locomotion or grip strength, some laboratories have tried to quantitate the pattern of locomotion following peripheral injury. In this procedure the feet of the rat are painted with ink and the animal is made to walk on paper, usually in a narrow corridor (27). The length of the gait and the distance between the fore- and hindlimbs is measured as an index of toxicity, or, more specifically, peripheral neuropathy. While some laboratories have found this procedure useful, it has not gained wide acceptance because of the awkwardness of the procedure and the variability of the data.

As we can see from the above considerations, tests of motor activity and motor function can reveal alterations in central and peripheral pathways. Because of the relative ease of these measurements and the availability of electronic activity monitors from a variety of manufacturers, they have become part of the routine neurotoxicological screen of most laboratories. Further refinements in these measures consist of measuring vertical movements (rearing), as well as patterns of movement in the open field. Whether the addition of these factors will provide data that will enhance our understanding of the mechanisms of toxicity as well as perhaps increase the sensitivity of the measures remains to be seen.

B. Sensory Function

Alterations in sensory function following exposure to neurotoxic chemicals has been an important area of investigation for behavioral toxicologists. As in the investigation of motor functions, a variety of techniques have been developed that rely not only on purely behavioral approaches, but also on a combination of behavioral and electrophysiological techniques. Not too surprisingly, the visual and auditory systems have been the primary targets of these investigations, not only because of their importance to the normal functioning of most organisms studies in the laboratory, but also because of their importance in controlling behavioral events.

Following the outbreak of methyl mercury poisoning in Minamata, Japan (see the chapter by T. Takeuchi in this volume), and the discovery that one of the target organs of this compound was the visual cortex, many investigators attempted to develop laboratory models of mercury toxicity. In order to approximate the human effects, the most elegant studies in this area were performed on primates. Typically the animals were dosed with a series of high "priming" doses, 1 mg/kg at 5-day intervals, and then blood mercury levels were maintained by weekly doses (21,22,32). The animals were trained before the mercury exposures on a complex visual discrimination task in which they were required to monitor two video displays. These displays were controlled by a computer and a flickering visual stimulus was displayed on one. Both the luminance level and the flicker frequency were controlled independently. The animal was required to make a response to the flickering stimulus by pressing a button. The performance of the animals was tested throughout the mercury exposure as well as following cessation of exposure. The results, while varying from animal to animal in terms of obsolute sensitivity to mercury, indicated that as the luminance decreased, performance declined early in the exposure period when no changes could be detected at high luminance levels. Performance also was more disruptible at high flicker frequencies as compared to low frequencies, regardless of the level of luminance.

The usefulness of such tests may be in the transposability of animal research techniques to human populations. It is often difficult to screen large populations for sensory function when chemical contamination is suspected. The relative ease with which modern microcomputers can be incorporated into test equipment may foretell the development of portable testing devices whose origins lie in the animal behavioral laboratories.

Another approach to the evaluation of sensory functions originates in the neurophysiology laboratory. In this approach the animals are prepared surgically with recording electrodes implanted either in the skull or into brain tissue. The electrodes can be aimed to record from any level of the sensory pathway that the investigator chooses, for example, from the optic tract, in the case of the visual system, to the visual cortex. The animals are presented with the appropriate sensory stimulus and the evoked potentials (EP) recorded from the electrodes are averaged as a rule by a computer. Alterations in averaged evoked potentials have been found following prenatal exposure to a variety of agents. Methyl mercury administered to pregnant rats at doses that had no measurable toxic effects on the mother has been reported to alter the visual EPs of the progeny (16,52). The alterations in the EPs subsequently have been used to (a) measure the reversibility of the toxicity and (b) determine the site and nature of the mercury-induced lesion.

Using the EP technique is, of course, not restricted to prenatal exposures, but can also be used to measure toxicity in adult animals. (See the chapter by Takeuchi and Koike in this book.) Examples of this can again be found in the methyl mercury literature, where mature dogs have been used (30).

Although in the early stages of mercury exposure there was a slight alteration in the EP, it remained very stable for almost 2 months of treatment, at which point it underwent rapid deterioration in conjunction with overt behavioral signs of toxicity. This result is, of course, not atypical of the adult exposure literature, where either large doses or prolonged treatment is necessary to cause overt toxicity, and the detection is very often closely followed by severe intoxication.

In a similar vein, the visual evoked potential has been used in animals intoxicated with trimethyltin. In this instance it was not known prior to the experiments that this compound was toxic to the visual system and the EP technique was used as part of the nervous system screen (17). The alterations in the EP led to a thorough examination of the visual system of the exposed animals and to the discovery that this compound causes severe lesions in the retina. This experiment is therefore a demonstration of the utility of this technique in screening for damage of sensory systems caused by chemicals of unknown neurotoxicity.

Finally, it must be mentioned that these techniques have been used very successfully in lead-exposed children, not only to detect lead-induced alteration in the EP, but also to suggest that the accepted levels of safe exposure to toxic compounds may have to be revised in light of the experimental data (35,36).

The auditory system has been less well explored in neurobehavioral toxicology, simply because fewer ototoxic chemicals have been identified. In order to detect such toxicity, however, very elegant behavioral techniques have been developed by Moody and Stebbins (34). In the procedure used with monkeys, the animals are trained to respond to the presence of a tone delivered through earphones by breaking contact with a switch. The intensity of the tones presented to the animals are varied between clearly audible and below threshold. In a "tracking" procedure, the animal's auditory threshold is determined. Following the description of the normal response of an animal, it is exposed to either high-intensity noise or ototoxic chemicals and its auditory function reexamined. Similar procedures have been adopted for a variety of other animals in this laboratory, and the results show this to be a very useful albeit labor-intensive method for detecting auditory system damage.

Another approach that is gaining popularity is the use of simple reflexive behaviors to determine sensory function. This technique has been described recently by several authors using rats as experimental subjects (26,51). The inhibition of the auditory startle reflex is the basis of this measure. Animals are placed in a chamber suspended on springs and are exposed to brief intense auditory stimuli. The amplitude of the startle reflex is recorded. Stimuli preceding this startle-eliciting tone may inhibit the amplitude of the reflex and thus can be used to measure sensory function in the animal. Experiments using ototoxic antibiotics seem to validate this procedure for measuring auditory damage. The advantage of this technique would be that no training of the animal is required, and there is the possibility that the same procedures could be used with humans as with rats in order to obtain interspecies comparisons on the effects of neurotoxic chemicals.

This brief overview of the literature dealing with sensory function shows that several techniques have emerged during the past few years that allow us to determine damage caused by neurotoxic chemicals in a variety of ways and in many diverse species, including man. These techniques are providing data that will allow us to refine our understanding of the mode of action of toxic chemicals, as well as to provide better guidelines for human exposure levels.

C. Cognitive Function

The largest part of the behavioral toxicology literature deals with the effects of toxic chemicals on learning and performance. Because of the great wealth of literature in this aspect of toxicology, we shall simply make an attempt to provide a survey of the major trends that have emerged during the last 10 years. There is no doubt that when one examines this part of the behavioral toxicology literature, it becomes evident that this field originated from the field of behavioral pharmacology, which itself developed from experimental psychology. The technology developed by operant conditioning to measure, quantify, and control behavior was tailor-made for behavioral toxicology when it finally emerged in the 1960s.

In order to simplify the presentation of material that might be somewhat repetitive because of the similarity of many of the experiments, we shall divide this section into three areas: (a) the effects of carbon monoxide on schedule-controlled behavior in adult animals, (b) the effects of lead on learning and schedule-controlled behavior in postnatally exposed animals, and (c) the effects of mercury in animals exposed in utero. In this simplified scheme we will of course neglect large parts of the behavioral toxicology literature dealing with other chemicals, but hope that the survey is representative of the experimental approach taken in the field.

1. *Exposure to Carbon Monoxide*

In most of the experiments dealing with CO, the animals (rats in most instances), were either preexposed to the gas or both preexposed and exposed throughout the experimental session. The levels of carboxyhemoglobin rise relatively slowly under these conditions, reaching saturation approximately at 90 min (2). Thus in many of these experiments the animals were not tested at saturation levels, but under constantly changing blood carboxyhemoglobin levels.

In one of the first such studies, four animals were trained on a differential reinforcement of low rates schedule and exposed to 100, 250, 500, 750, and 1000 ppm of CO for 90 min (4). Carbon monoxide caused extended pausing or complete cessation of responding at 750 ppm or higher concentrations. Until this pausing occurred, however, there was no significant effect on the performance of the animals.

A somewhat similar effect was seen in four rats trained on a fixed consecutive-number performance (48). In this schedule, the rats had to emit a fixed number of responses on one lever in order to obtain a reinforcement by responding once on a second lever. The major effect was increased pausing. In yet another study from the same laboratory (33), a progressive-ratio schedule was used in which the number of responses the animals had to make increased by either five or seven responses after each reinforcement. At the highest level of CO exposure (700 ppm), the animals stopped responding or slowed down significantly.

In a detailed analysis of reinforcement contingencies and their interaction with CO, animals were trained on either a fixed-ratio (30 min) or fixed-interval (3 min) schedule or a multiple schedule consisting of both (5). The fixed-ratio component was more easily disrupted by CO than the fixed-interval component, and in some cases under the multiple schedule the disrupting effect of CO on the fixed-ratio schedule was attenuated. As before, significant disruption in responding seemed to occur at 700 ppm of CO or above. In contrast, a previous report found no differences between the fixed-ratio and fixed-interval schedules' sensitivity to CO in pigeons (31). The pigeon proved to be extremely insensitive to CO, and responding only ceased at 1700 ppm. In this experiment the birds were injected with either d-amphetamine or pentobarbital at different doses and exposed to the range of CO concentrations. The effect of the interactions was to enhance the rate-lowering effects of CO at all doses of both drugs on both schedules.

As we can see from these studies, the effect of CO on schedule-controlled behaviors is to decrease the response rates mostly by increasing pausing between responses or following reinforcements. The CO exposures used in these studies were not toxic to the animals, in the sense that no neuropathology developed as a consequence of the exposure, since the exposures were brief in duration and the CO concentrations were not too high. A more convincing argument for the lack of neuropathology is the fact that the animals recovered to their normal base-line response rates after each exposure. As we shall see later, when animals are exposed to methyl mercury, this is not always the case, and the behavioral alterations can be more persistent.

An overview of the CO literature suggests that the rat or, for that matter, the pigeon is not a sensitive species for modeling behavioral effects in man (28), and that in order to study the effects of CO more thoroughly, primate models would have to be used.

2. Postnatal Lead Exposure

As we pointed out at the beginning of this chapter, the protocol for exposing animals to lead, first described by Pentschew and Garro (37), consisted of feeding the lactating mother a diet containing 4% lead carbonate. The pups were therefore exposed to the lead via the mother's milk. While this model is the basis of almost all animal studies, in some the exposure to lead ceases at weaning, while in others it may continue through life.

In one of the first behavioral studies using this model the dams were administered lead acetate in the drinking water (10 mg Pb/100 ml) or by gavage (25-35 mg/kg per day) on either days 1-10 postpartum or days 11-20 (9). The offspring were trained at 8-10 weeks of age in a T maze. Offspring exposed to lead on days 1-10 of life learned the maze significantly more slowly than controls; however, offspring exposed to lead on days 11-20 were not different from controls in the T maze. The results suggested that the rat was very sensitive at an early age of development, but not after the 10th day of life.

Contrary to these positive results on learning tasks, another group of investigators found no effect of lead treatment (0.02% and 0.1% lead acetate) on the acquisition and subsequent reversal of a brightness discrimination task. A positive effect was found in that lead-exposed animals appeared to be less aggressive than controls in a shock-elicited aggression test (24). In a departure from the usual lead exposure regimen, two studies evaluated the effects of postweaning lead exposure in rats on schedule-controlled behaviors (13,14). In the first study the animals were trained on a fixed-interval 30-sec schedule after being exposed to 50, 300, or 1000 ppm lead acetate in their drinking water. The lower doses of lead seemed to increase response variability and decreased time to response latency. In the second study the animals were trained on a minimum response duration schedule, in which the animal had to hold the lever a certain time in order to obtain reinforcement. The effect of the lead treatment was to push the animals into shorter response durations. Both of these studies are of interest because they indicate that the postweaning rat is still quite sensitive behaviorally to lead treatment, contrary to earlier reports, and that this sensitivity can be revealed by operant techniques. A subsequent report using monkeys treated from day 1 of life with 500 μg/kg per day lead acetate and tested at 2-3 years of age showed similar effects on a fixed-interval schedule, that is increased response rates (41). Because interspecies comparisons are so rarely made, the findings of this study are particularly important in revealing that under similar schedules and similar toxic exposure histories, rats and monkeys respond in the same manner to the toxic insult.

This survey of the postnatal lead exposure literature reveals that both learning tasks and schedule-controlled behaviors can be used effectively to detect subtle effects of exposure to neurotoxic chemicals.

As we shall see in the next section, with prenatal exposures the results are very similar.

3. *Prenatal Methyl Mercury Exposure*

When animals are exposed to neurotoxic chemicals during gestation, the outcome of these exposures can be (a) resorption of the litter, (b) gross teratogenesis, and (c) no apparent ill effects. It is only in the latter instance, when the offspring are born without apparent ill effects, that behavioral testing procedures are necessary. These tests can reveal whether the offspring are developing normally, as we saw earlier in this chapter, or whether there are delayed and permanent consequences of this exposure. The general design of these studies is similar. Pregnant rodents are administered methyl mercury during the first, second, or third trimester, either in a single dose or several doses. After parturition, the offspring are tested in developmental tests and later in a learning task. In one of the first of these studies, mice were given 1, 2, 3, 5, or 10 mg/kg of methyl mercury on day 8 of gestation (25). At the 3- and 5-mg/kg doses, the offspring were significantly retarded in their acquisition of a two-way avoidance task when tested at 56 days of age. The 10-mg/kg dose was sufficiently toxic to reduce the litter sizes significantly, and these animals were not tested behaviorally. In this study a cross-fostering experiment was also performed and the results showed that the mercury effects could be attributed to the prenatal environment exclusively.

A more careful evaluation of the significance of gestational period on the behavioral toxicity of mercury was carried out in a similar study with rats (19). Pregnant rats were treated either on day 8 or 15 of gestation with 5 or 8 mg/kg of methyl mercury. When the adult offspring were tested in a two-day avoidance task, both dose and time of treatment were significant variables. The animals that had been exposed to 8 mg/kg on day 8 of gestation showed a significant increase in the number of trials to reach criterion on the second acquisition (following extinction after meeting criterion in the initial acquisition) of the task. Rats treated on day 15 of gestation, however, were severely affected on initial acquisition of the task, and this persisted during the second acquisition also. The results of these studies show that toxic exposures late in gestation are potentially much more hazardous in terms of their behavioral consequences.

Prenatal mercury exposure at extremely low doses (0.01 mg/kg on days 6-9 of gestation) has also been shown to cause measurable effects in rats trained on a differential reinforcement of high rates schedule (7). This interesting schedule has not been used widely in behavioral toxicology and may prove to be useful in that, by making the animals increase their response rates to their limits, the schedule might reveal latent behavioral alterations that the more commonly used schedules would not reveal.

III. CONCLUSIONS

This brief survey of the experimental approaches used in behavioral toxicology shows that the techniques that have been adopted by investigators have been useful in detecting subtle alterations in nervous system function caused by neurotoxic agents. While the diverse approaches to measuring toxicity have not come to a standardized behavioral screen, there is general agreement that certain types of measures have to be used in order to provide the range of data that will result in either the detection of toxicity or a demonstration of lack of toxicity. In the real world, where industry is releasing large numbers of new chemicals every year, both types of data are needed. As the field develops further, it will become more heavily involved in devising tests that will reveal the mechanisms of neurotoxicity, as well as simply demonstrate its presence. This will necessitate the closer identification of the field with the rapidly expanding field of neurobiology, which deals with brain-behavior relationships as its principal subject (3). This type of collaboration would be fruitful to both sides (neurobiology and neurotoxicology), since the study of toxic chemicals often reveals as much about the normal function of the nervous system as it reveals about its function following injury.

REFERENCES

1. Air Quality Criteria for Lead. 1977. U.S. Environmental Protection Agency, Office of Research and Development, Washington, D.C. EPA-600/8-77-017.
2. Annau, Z. 1975. The comparative effects of hypoxic and carbon monoxide hypoxia on behavior, in *Behavioral Toxicology* (Weiss, B., Laties, V. G., eds.), Plenum, New York, pp. 105-126.
3. Annau, Z. 1982. Screening strategies, in *Application of Behavioral Pharmacology in Toxicology* (Zbinden, G., Cuomo, V., Racagni, G., Weiss, B., eds.), Raven Press, New York, pp. 87-96.
4. Ator, N. A., Merigan, W. H., McIntire, R. W. 1976. The effects of brief exposures to carbon monoxide on temporally differentiated responding. *Environ. Res. 12*:81-91.
5. Ator, N. A. 1982. Modulation of the behavioral effects of carbon monoxide by reinforcement contingencies. *Neurobehav. Toxicol. Teratol. 4*:51-61.
6. *Behavioral Toxicology*. 1975. (Weiss, B., Laties, V. G., eds.), Plenum, New York.
7. Bornhausen, M., Musch, H. P., Greim, H. 1980. Operant behavior changes in rats after prenatal methyl mercury exposure. *Toxicol. Appl. Pharmacol. 56*:305-310.

8. Boyes, W. K., Laurie, R. D., Cooper, G. F. 1980. Acrylamide toxicity: Effects on cortical evoked potentials and locomotor activity in rats. *Soc. Neurosci. Abstr.* 246.7.
9. Brown, D. R. 1975. Neonatal lead exposure in the rat: Decreased learning as a function of age and blood lead concentrations. *Toxicol. Appl. Pharmacol.* *32*:628-637.
10. Campbell, B. A., Lytle, L. D., Fibiger, H. C. 1969. Ontogeny of adrenergic arousal and cholinergic inhibitory mechanisms in the rat. *Science* *160*:637-638.
11. Castellano, C., Oliverio, A. 1976. Early malnutrition and postnatal changes in brain and behavior in the mouse. *Brain Res.* *101*:317-325.
12. *Casarett and Doull's Toxicology*. 1980. (Doull, J., Klaassen, C. D., Amdour, M. O., eds.), Macmillan, New York.
13. Cory-Slechta, D. A., Bissen, S. T., Young, A. M., Thompson, T. 1981. Chronic postweaning lead exposure and response duration performance. *Toxicol. Appl. Pharmacol.* *60*:78-84.
14. Cory-Slechta, D. A., Thompson, T. 1979. Behavioral toxicity of chronic postweaning lead exposure in the rat. *Toxicol. Appl. Pharmacol.* *47*:151-159.
15. Culver, B., Norton, S. 1976. Juvenile hyperactivity in rats after acute exposure to carbon monoxide. *Exp. Neurol.* *50*:89-98.
16. Dyer, R. S., Eccles, C. U., Annau, Z. 1978. Evoked potential alterations following prenatal methyl mercury exposure. *Pharmacol. Biochem. Behav.* *8*:137-141.
17. Dyer, R. S., Howell, W. E., Wonderlin, W. F. 1982. Visual system dysfunction following acute trimethyltin exposure in rats. *Neurobehav. Toxicol. Teratol.* *4*:191-196.
18. Eccles, C. U., Annau, Z. 1982. Prenatal methyl mercury exposure: I. Alterations in neonatal activity. *Neurobehav. Toxicol. Teratol.* *4*:371-376.
19. Eccles, C. U., Annau, Z. 1982. Prenatal methyl mercury exposure: II. Alterations in learning and psychotropic drug sensitivity in adult offspring. *Neurobehav. Toxicol. Teratol.* *4*:377-382.
20. Erinoff, L., MacPhail, R. C., Heller, A., Seiden, L. S. 1979. Age dependent effects of 6-hydroxydropamine on locomotor activity in the rat. *Brain Res.* *64*:195-205.
21. Evans, H. L. 1978. Behavioral assessment of visual toxicity. *Environ. Health Perspect.* *26*:53-57.
22. Evans, H. L. 1982. Assessment of vision in behavioral toxicology, in *Nervous System Toxicology* (Mitchell, C. L., ed.), Raven Press, New York, pp. 81-107.
23. Fechter, L. D., Annau, Z. 1977. Toxicity of mild prenatal carbon monoxide exposure. *Science* *197*:680-682.

24. Hastings, L., Cooper, G. P., Bornschein, R. L., Michaelson, I. A. 1977. Behavioral effects of low level neonatal lead exposure. *Pharmacol. Biochem. Behav.* 7:37-42.
25. Hughes, J. A., Annau, Z. 1976. Postnatal behavioral effects in mice after prenatal exposure to methyl mercury. *Pharmacol. Biochem. Behav.* *4*:386-391.
26. Ison, J. R., Hammond, G. R. 1971. Modification of the startle reflex of the rat by changes in the auditory and visual environments. *J. Comp. Physiol. Psychol.* *75*:435-452.
27. Jolicoeur, F. B., Rondeau, D. B. Barbeua, A., Wayner, N.J. 1979. Comparison of neurobehavioral effects induced by various experimental models of ataxia in the rat. *Neurobehav. Toxicol. Teratol. Suppl.* *1*:175-178.
28. Laties, V. G., Merigan, W. H. 1979. Behavioral effects of carbon monoxide on animals and man. *Annu. Rev. Pharmacol. Toxicol.* *19*:357-392.
29. Loch, R. K., Rafales, L. S., Michaelson, I. A., Bornschein, R. L. 1978. The role of undernutrition in animal models of hyperactivity. *Life. Sci.* *22*:1963-1970.
30. Mattsson, J. L., Miller, E., Alligood, J. P., Koering, J. E., Levin, S. G. 1981. Early effects of methylmercury on the visual evoked response of the dog. *Neurotoxicology* *2*:499-514.
31. McMillan, D. E., Miller, A. T., Jr. 1974. Interactions between carbon monoxide and d-amphetamine or pentobarbital on schedule-controlled behavior. *Environ. Res.* *8*:53-63.
32. Merigan, W. H. 1979. Effects of toxicants on the visual system. *Neurobehav. Toxicol. Suppl.* *1*:15-22.
33. Merigan, W. H., McIntire, R. W. 1976. Effects of carbon monoxide on responding under a progressive ratio schedule in rats. *Physiol. Behav.* *16*:407-412.
34. Moody, D. B., Stebbins, W. C. 1982. Detection of the effects of toxic substances on the auditory system by behavioral methods, in *Nervous System Toxicology* (Mitchell, C. L., ed.), Raven Press, New York, pp. 109-131.
35. Otto, D., Benignus, V., Muller, K., Barton, C. 1981. Effects of age and body lead burden on CNS function in young children. I. Slow cortical potentials. *Electroencephalogr. Clin. Neurophysiol.* *52*:229-239.
36. Otto, D., Benignus, B., Muller, K., Barton, C., Seiple, K., Prah, J., Schroeder, S. 1982. Effects of low to moderate lead exposure on slow cortical potentials in young children: Two year follow up study. *Neurobehav. Toxicol. Teratol.* *4*:733-737.
37. Pentschew, A., Garro, F. 1966. Lead encephalopathy-myelopathy of the suckling rat and its implications on the porphirinopathic nervous diseases. *Acta Neuropathol.* *6*:266-278.

38. Pryor, G. T., Uyano, E. T., Tilson, H. A., Mitchell, C. L. 1983. Assessment of chemicals using a battery of neurobehavioral tests: A comparative study. *Neurobehav. Toxicol. Teratol. 5*: 91-117.
39. Reiter, L. W., MacPhail, R. C. 1979. Motor activity: A survey of methods with potential use in toxicity testing. *Neurobehav. Toxicol. Teratol. 1*:52-66.
40. Reiter, L. W., MacPhail, R. C. 1982. Factors influencing motor activity measurements in neurotoxicology, in *Nervous System Toxicology* (Mitchell, C. L., ed.), Raven Press, New York, pp. 45-65.
41. Rice, D. C., Gilbert, S. G., Willes, R. F. 1979. Neonatal low-level lead exposure in monkeys: Locomotor activity, schedule-controlled behavior, and the effects of amphetamine. *Toxicol. Appl. Pharmacol. 51*:503-513.
42. Rodier, P. M., Reynolds, S. S., Roberts, W. N. 1979. Behavioral consequences of interference with CNS development in the early fetal period. *Teratology 19*:327-336.
43. Schellenberger, N. K., Norton, S. 1980. Factors influencing the persistent effects of acute carbon monoxide exposure on rat motor activity. *Neurotoxicology 1*:541-550.
44. Silbergeld, E. K., Goldberg, A. M. 1973. A lead induced behavioral disorder. *Life. Sci. 13*:1275-1283.
45. Silbergeld, E. K., Goldberg, A. M. 1974. Hyperactivity: A lead induced behavior disorder. *Environ. Health Perspect. 7*: 227-232.
46. Silbergeld, E. K., Goldberg, A. M. 1974. Lead-induced behavioral dysfunction: An animal model of hyperactivity. *Exp. Neurol. 42*:146-157.
47. Silbergeld, E. K., Chisolm, J. J. 1976. Lead poisoning: Altered urinary catecholamine metabolites as indicators of intoxication in mice and children. *Science 192*:153-155.
48. Smith, M. D., Merigan, W. H., McIntire, R. W. 1976. Effects of carbon monoxide on fixed-consecutive-number performance in rats. *Pharmacol. Biochem. Behav.* 5:257-262.
49. Tilson, H. A., Cabe, P. A. 1979. The effects of acrylamide given acutely or in repeated doses on fore- and hindlimb function in rats. *Toxicol. Appl. Pharmacol. 47*:355-362.
50. Tilson, H. A., Mitchell, C. L., Cabe, P. A. 1979. Screening for neurobehavioral toxicity: The need for validation of testing procedures. *Neurobehav. Toxicol. Suppl. 1*:137-140.
51. Young, J., Fechter, L. D. 1983. Reflex inhibition procedures for animal audiometry: A technique for assessing ototoxicity. *J. Acoust. Soc. Am. 73*:1686-1693.
52. Zenick, H. 1976. Evoked potential alterations in methylmercury chloride toxicity. *Pharmacol. Biochem. Behav.* 5:253-255.

Index

A

Acaricides, 424
Acetaldehyde, 206-207
Acetaldehyde dehydrogenase
polymorphism, 206-207
Acetonitrile, 480
Acetophenazine, 70
Acetylcholine
as a neurotransmitter in cockroach abdominal nerve cord, 589
inhibition of release
botulinum toxins, 490
lead, 310
manganese, 395
presynaptic snake venom toxins, 497
kainic acid and, 518,519
Acetylcholine receptors
effect of pesticides on, 407,408, 410
organophosphorous compounds, 408,416,430,432
and snake venom toxins, 498, 499
Acetylcholinesterase, 430-432
aging, 432
effect of lead on, 310
and manganese ion, 173,174
[Acetylcholinesterase]
and organophosphorous compounds, 15,210,430-432
reactivation of, 430,435
and thyroid hormones, 571
L(α)-Acetylmethadol, 572
N-Acetylpenicillamine, 337
N-Acetyltransferase polymorphism and neurotoxicity, 209-210
Acrylamide
behavioral toxicity, 636
disturbance to smooth membrane function by, 34,35
effect on nerve conduction velocity, 622
neuropathy, 14,34-38
biochemical markers of, 278, 279
and serotonin turnover, 452
Acrylonitrile, 480
ACTH (*see* Adrenocorticotropic hormone)
Addiction
neuropeptides and, 135-152
from prenatal exposure to opiates, 571
Adenosine deaminase and manganese toxicity, 173
Adenosinetriphosphatases
effect of manganese on, 173,395

[Adenosinetriphosphatases]
inhibition by lead, 311
in neural culture systems, 569, 571
and pesticides, 407,413-415, 448,451
and organotins, 392,407
and thallium ion, 390
Adenosinetriphosphate (ATP), 312,313,375
Adenylate cyclase and lead toxicity, 309,311,312
β-Adrenergic blocking drugs
secretory glandular effects, 74, 75
sleep disturbances induced by, 104-105
use in lithium intoxication, 225
use in tetanus, 393
Adrenergic neuronal cultures, 570,574
Adrenocorticotropic hormone, 85, 138,139
neuroleptic drugs and, 87
neuropeptides related to, 139-141
and experimental addictions, 145-146
Akathisia, drug-induced
amoxapine, 224
imipramine withdrawal, 224
neuroleptics, 222
ALA-dehydroase, 308
Alcohol (*see* Ethanol)
Alcohol dehydrogenase polymorphism, 206-207
Alcoholics
electroencephalogram in, 207
endorphin levels in CSF, 147
Alcoholism, genetic determinants in, 205-207
Aldosterone, and neuroleptic drugs, 87
Aldrin, 445,450
Aldrin-transdiol, 450
Aliphatic nitriles, 480
Alkyllead
effects on glia, 309,312
gasoline additives and, 306
hippocampal effects, 598
mutagenicity, 309
neurotransmitter changes induced by, 309,452
toxicity in humans, 300
Allethrin, 412
Allotypic cell recognition, 561
Allyl chloride, 622
Allylglycine, 244
Alopecia, thallium-induced, 398, 390
Aluminum, 396-399
in dialysis fluid, 398
hippocampal effects, 599
sources of exposure, 386
Alzheimer's disease, 11,525
aluminum toxicity and, 397-398
Amanita strobiliformis, 515
Amantadine, 107
Amblyopias, cyanogenic agents and, 482-484
γ-Aminobutyric acid (GABA)
in cockroach abdominal nerve cord, 589
effects of chlordecone on, 449
hippocampal neurotransmission and, 599
kainic acid and, 518,519
and lead toxicity, 309-311,315, 316
lithium-induced changes in metabolism of, 71
in neural culture systems, 602
pimozide and, 88
receptors
effect of kainic acid on, 519
muscimol binding sites, chlordecone interaction with, 417
effect of organophosphorous compounds on, 416
type II pyrethroids and, 417

[γ-Aminobutryric acid (GABA)]
synaptosomal uptake of, inhibition by organotins, 393
synthesis, inhibition by isoniazid, 12
tetanus toxin and, 493
δ-Aminolevulinic acid in urine, and lead toxicity, 302,304
Aminopterin, 257
4-Aminopyridine, 596,597
Amiprophos, 425,428
Amitriptyline (*see also* Antidepressants)
cardiovascular effect, 70
esophageal effects, 72
sedation induced by, 224
Amobarbital
cerebellar effects, 238
sleep distrubances induced by, 102
Amphetamine
autonomic nervous system effects, 75
and brain stimulation reward, 114-115,116
and carbon monoxide toxicity, 461
effect in lead-exposed animals, 315,633
and kainic acid toxicity, 520
Amprolium, 14
Amygdala, 615
and kindling, 522
and trimethyl tin toxicity, 393
Amygdalin, teratogenicity, 480
Amyotrophic lateral sclerosis, 305,397
Angiotensin-converting enzyme, and kainic acid, 518,519
Animal neurotoxins, 496-506
Antabuse (*see* Disulfiram)
Antianxiety agents, 226-227
Antiarrhythmic drugs, sleep disturbances induced by, 106
Anticancer drugs, 251-262
Anticholinergic drugs
ophthalomological effects, 73,74
overdosage, 51,57,58,227
cerebral function monitoring, 58
Antidepressants (*see* also individual agents)
monoamine oxidase inhibitors, 224
autonomic nervous system toxicity, 71
hypertensive crises induced by, 224-225
tricyclics, 70,71,72,73
akathisia induced by, 224
and cognitive impairment, 224
effects on the genitourinary tract, 74-76
nocturnal myoclonus and, 106-107
overdosage, 56,57,60,71,223
sedative effects, 224
Antiepileptics (*see also* individual agents)
acute central nervous system effects, 234
cerebellar toxicity, 233-244
effects in neural culture systems, 561-566
and porphyrias, 213
Antiparkinsonian drugs
sleep disturbances induced by, 107-1-8
urological side effects, 74,75
use in neuroleptic-induced akathisia, 222
withdrawal symptoms, 228
Antitubulin drugs, 253-255
Apholate, 425,427
Apomorphin, 88,116,315
action in kainic acid-treated animals, 520
ARA-C (*see* Cytosine arabinoside)
Arsenic, 385-388
chronic intoxication, 19
neuropathy, 14,19,386,387

[Arsenic]
sources of exposure, 386
Aspartate, as a neurotransmitter, 516,599
Aspartate transaminase, kainic acid and, 519
Astrocytes, 19,350
Ataxia
in acrylamide intoxication, 34
in ciguatera, 504
drug-induced
anticholinergic agents, 227
benzodiazepines, 238
carbamazepine, 237
5-fluorouracil, 259
methotrexate, 257
metronidazole, 272
phenytoin, 235
primidone, 237
procarbazine, 262
thiocyanate, 478
in methylmercury intoxication, 358
in nigerian neuropathy, 483
in poisoning by organophosphorous compounds, 15,434, 435
Atelopus chiriquiensis, 506
Atropine, 51,434,505
Auditory evoked potentials (*see* Evoked potentials)
Auditory system disorders
toxin-induced (*see* Ototoxicity)
assessment in experimental toxicology, 288,637,639-640
Aurothioglucose, 395
Aurothiomalate, 395
Autonomic nervous system disorders
acrylamide-induced, 35
in ciguatera, 504
drug-induced, 69-77,102
in neuroleptic malignant syndrome, 221
and methylene chloride exposure, 467
[Autonomic nervous system disorders]
in paralytic shellfish poisoning, 502
in poisoning by organophosphorous compounds, 433
in puffer fish poisoning, 506
trichloroethylene-induced, 461
in tetanus, 494
Axon
degeneration, 2-4
biochemical changes, 278-280
energy metabolism disorders in, 19-23
morphological factors in, 2-4
retrograde, 37
dying-back process (*see* Neuropathy)
growth in neural cultures, 565
terminals, entry of chemicals, 34
toxin-induced damage
cyanate, 478
disulfiram, 535-537,539
lead, 4-7
methylmercury, 538-539
metronidazole, 272
misonidazole, 275,276,281
Axon hillock, 27
Axonal transport, toxin-induced disorders of, 262
anticancer drugs, 252,254
hexacarbons, 29,31
organophosphorous compounds, 437
tetanus toxin, 493
vincristine, 34
5-Azacytidine, 636
Azide, 482
Aziridines, 428

B

Baclofen, 493
in Huntington's disease, 521

Barbiturates
abuse, 111
and brain reward systems, 114,117
cerebellar effects, 238
overdosage, 51,53,56,57
cerebral function monitoring in, 58
and porphyrias, 213
sleep disturbances induced by, 102
withdrawal reactions to, 522
Batrachotoxin, 412,413,505
Baygon, 175
Behavioral techniques in toxicity assessment, 631-644
abuse potential, 111,119-120
cognitive function, 640-644
developmental studies, 163-191
influence of malnutrition, 634
motor function, 632-637
sensory function, 637-640
Behavioral toxicity
acrylamide, 636
Baygon, 175
benzene, 185
t-butanol, 176-187
cadmium, 171-172,185
carbon disulfide, 177,186
carbon monoxide, 182,634-635, 640-641
chlordecone, 175
DDE, 175,185
diazinon, 174,185
2-ethoxyethanol, 178,186
formaldehyde, 186
halothane, 182
lead, 639,642
developing animals, 169-171, 184-185
hyperactivity, 633-634
minimal brain dysfunction in children, 301-302,304
neuropsychological changes in workers, 305
[Behavioral toxicity]
maneb, 175
mercury, inorganic, 330,332, 333
2-methoxyethanol, 179,186
methyl-n-butyl ketone, 197,186
methylene chloride
animal studies, 180,186
human studies, 467
methylmercury, 168-169,184-185,635,638,643
mirex, 176
misonidazole, 287,288
neuroleptic drugs, 220
organotins, 172-173,185
physical agents, 187-189,190
polychlorinated biphenyls, 181
trichloroethylene, 461
Belladonna poisoning, 51
Bensulide, 425,428
Benzamides, 84
Benzhexol, 76
Benzene, 176,186
electroencephalographic effects, 615
Benzene hexachloride, 450
Benzodiazepines, 234 (*see also* Antianxiety agents)
abuse, 111
action on brain reward systems, 114,117
acute cerebellar effects, 238
overdosage, 51,56,57,226
receptors, 572-573
kainic acid and, 519
sleep disturbances induced by, 103-105,108
tolerance to, 103
use in neuroleptic-induced akathisia, 222
withdrawal reactions to, 226, 227
Benztropine, 74,75,76
Bergmann cells, 241,243,356
Bethanidine, 72,73,76
Bicuculline, 244,601

Binding assays in neurotoxicology (*see* Experimental neurotoxicology)
Bismuth, 386,399
Biopterins, 257
Biperiden, 75
Black mamba, 499
Blood-brain barrier, neurotoxic agents and
  anticancer drugs, 252
  elemental mercury, 326-327
  methotrexate, 256
  methylmercury, 364
  nitroimidazoles, 284
Boron, 386
Botulinum toxinc, 52
  antagonism by calcium ion, 491
  antitoxin, 492
  chemistry and physiopathological effects, 490-491
  mechanism of action, 490
  metabolic activation, 490
Botulism, 37,489-492
  diagnosis, 492
  foodborne, 491
  human intoxication, 491
  infant disease, 490,492
Brain
  atrophy, in methylmercury intoxication, 350,351
  cultures, 559-576,600-603
  development, 569-571
    and ethanol toxicity, 567-569
    and opiates, 566-567
  edema, 348,391,392
  failure, in acute poisoning, 45-66
    classification, 59-66
    clinical signs, 47-52
    diagnosis, 46,49,53,65
    scoring system, 59,61-63
    specific semiotics, 47
  lysosomal enzymes, 279
  reward circuitry, 112-115
  transplants, 521
  vasculopathy, 3
Brainstem reflexes, 50,64,65
  cold-caloric test, 49
  corneal, 49-52,64
  pupillary, 49-51,64-66
  and severity of toxic brain failures, 49,63,65
  vestibulo-ocular, 49-52,64-66
Bretylium, 74
Bromocriptine, 88,107
Buccolingual-masticatory syndrome, 222
β-Bungarotoxin, 497
t-Butanol, 176,186
Butaperazine, 70
Butifos (*see* DEF)
n-Butylmercaptan, 438
Butyrylcholinesterase, 430,567
Butyrophenones (*see also* Neuroleptics), 70,87
  sexual dysfunction induced by, 76

C

C6 glial cell culture, 560,570,574
  effect of ethanol in, 568
  effect of opiates in, 567
(Ca^{2+}, Mg^{2+})-dependent adenosinetriphosphatase (*see* Adenosinetriphosphatases)
Cadmium, 171-172,185
  effect in hippocampal slice preparations, 599
Caffein test in malignant hyperthermia, 208
Calcium
  and botulinum toxins, 490,491
  and cell aggregation, 566
  pesticides affecting movements of, 407,447,451
  interaction with bivalent metals, 599
  and lead toxicity, 306,311
Calcium gluconate, in ciguatera, 505

Calomel, 14,24
Cannabinoids, 119
Carbachol, 77
Carbamates, 406,415
 behavioral teratogenic effects, 185
Carbamazepine, 234,237
Carbon disulfide
 behavioral toxicity, 177,186
 effect on motor nerve conduction velocity, 622
 hippocampal effects, 598
 neuropathy, 31,32
  neurofilament disorders in, 14
Carbon monoxide
 cerebral protein changes in animals exposed to, 635
 behavioral toxicity, 634-635, 640-641
  developmental changes, 182
  disorders in brain neurochemistry and, 182,190
 effect on cerebellar Purkinje cells, 600
 effect on evoked potentials, 618
 formation from methylene chloride, 467
 hippocampal effects, 598
 sleep pattern alteration induced by, 616
Carbon tetrachloride, 460,467
Cardiovascular drugs, sleep disorders induced by, 104-106
Cartap, 406,407,410
Cassava, 483
Catatonic-like states, neuroleptic-induced, 220-221
Catecholamine system, 113
 brain development and, 632
 toxins inducing changes in:
  chlordecone, 449
  2-ethoxyethanol, 178
  lead, 302-305,309,310
  manganese, 173-174,394-395
[Catecholamine system]
  2-methoxyethanol, 179
  trichloroethylene, 461
Catecholestrogens, 451
Caudotoxin, 497
Cerebellum, 234-235
 and DDT toxicity, 448,615
 effect of antiepileptic drugs on, 233-244
 explants, 570
  effect of carbon monoxide on firing activity, 600-601
  phenytoin toxicity in, 564-565
 and 5-fluorouracil toxicity, 259
 and mercury toxicity, 328, 353
  changes in protein synthesis, 371
  neuropathology (human), 353,355,356,358
  studies in developing animals, 169
 and misonidazole toxicity, 279
Cerebral function monitor (CFM), 52-59,65,66
Cerebral palsy, in methylmercury poisoning, 358
Chiriquitoxin, 505,506
Chloral hydrate, 103,460
Chlordane, 175,185,445
 metabolic activation, 450
Chlordecone, 445-446,449-450
 action on ATPases, 413,415
 behavioral teratogenic effects, 175
 and dopamine receptors, 408
 effect on intracellular calcium homeostasis, 450
 endocrine effects, 449
 reproductive system disorders induced by, 447
 toxicity targets of, 447
Chlordiazepoxide, withdrawal reactions to, 226

Chloride transport system,
effect of pesticides on, 408
Chlorinated biphenyls, 181,
453
Chlorinated insecticides, 445-
453 (*see also* individual
agents)
Chloroform (*see* Trichloromethane)
Chloromethylparaoxon, 210
Chlorpromazine, 73,74,220-223
effect on gonadotropins, 88
epileptogenic effects, 221
Chlorthion, 424,425
Chlecystokinin, kainic acid
toxicity and, 518,519
Choline acetyltransferase, 430
and lead toxicity, 310
in neural cultures, 571
Cholinergic system
in cockroach abdominal nerve
cord, 589
development, 632
and kindling, 522,523
toxins inducing changes in:
botulinum toxins, 490
2-ethoxyethanol, 178
lead, 309,310,311
2-methoxyethanol, 179
organophosphorous com-
pounds, 430-432
Cholinesterase, 430
erythrocyte, measurement of,
336,434
Chromatolysis, neuronal
in chronic thiamine deficiency,
19
toxin-induced, 5,19,21,35,37
Ciguatera, 503-505
Ciguatoxin, 503-504
Cis-platin, 260-261,386
ototoxicity, 260-261
Clioquinol, 14
Chlomipramine, 107
Clinidine, sexual disorders
induced by, 76
Clozapine, 88,90
Cobalt, 599
Cocaine, 136
and reward systems in the brain,
155-116
neuropeptides and abuse of, 151
Cockroach nerve cord preparation,
588-592
Cognitive impairment, lithium-
induced, 225
Colchicine, 34,256
Coma, drug overdosage induced,
59-66
Conduction velocity (*see* Nerve)
Coneshell, 500
Convulsions, toxic
4-aminopyridine, 597
anticholinergic drugs, 227
barbiturate withdrawal, 111
benzodiazepine withdrawal, 111,
226
chlorinated insecticides, 445,
448,450
folic acid, 522
hydroquinone, 590
isoniazid, 23,210
kainic acid, 522
lead, 301,305
methotrexate, 257
methyl chloride, 466
misonidazole, 275
neuroleptics, 220,221
organotins, 391,393,394
thallium, 388
thiocyanate, 478
tricyclic antidepressants, 223
Coral snakes, 498
Coroxon, 425,427
Corticosteroids, 174,175,185,451,
449
action in cerebellar cultures,
570
action in retinal monolayer cell
cultures, 574,575
effect on membrane excitability,
447
sleep disturbances induced by,
106

Corticotropic-releasing factor (CRF), 85
Coumaphos, 425,427
Crayfish abdominal nerve cord preparation, 592
Cremart, 425,428
Crotalid venoms, 499-500
Crotamine, 499-500
Crotapotin, 497
Crotoxin, 497,499
Cuprizone, encephalopathy induced by, 279
Cyanate, 474,477-478
Cyanide, 473-484
  biotransformation, 474-477
  in cassava, 483
  diseases ascribed to chronic exposure to, 481-484
  metabolite-associated toxicity, 477-479
  neuropathology, 477
  poisoning, antagonism of, 479, 480
  teratogenic effects, 480-481
Cyanogenic agents, 480-481,483
Cyanophenphos, 425
  delayed neurotoxicity, 435
  2',3'-cyclic nucleotide phosphohydrolase, 571
  as an oligodendroglial marker, 566,568
Cyclodienes, 450
Cyclophosphamide, 165,425,427
Cyclopropane, 77,208
Cytosine arabinoside, 258-259

D

Daboisia myoporoides, and ciguatera, 505
Dapsone neuropathy, 14
  genetic factors in, 209
DDE, 174,185
DDT, 446,448
  action on opiate receptors, 408
[DDT]
  action on ouabain-binding sites, 408
  effect in olfactory cortex slice preparation, 599
  electrophysiological effects, 615, 616
  endocrine changes induced by, 448
  inhibition of ATPases by, 407, 413-414
  membrane targets of toxicity, 406
  and sodium channel, 410,411-413
  tremorigenic activity, 448
Debrisoquin, 72,73
Decamethrin, 417
Decortication syndrome in methylmercury intoxication, 358
DEF, 425,428,429
  delayed neurotoxicity, 435
  late acute effects, 438
Delayed neurotoxicity, organophosphate-induced (OPIDN), 435-438
Demyelination toxin-induced
  cyanate, 478
  cyanide poisoning, 477
  diphtheria, 7,495
  lead, 5,301
  methotrexate, 257,258
  triethyl tin, 392
Dendroaspis polylepis, 499
Deoxycytidine, 259
Deoxyglucose technique, 521
  kainic acid toxicity assessed by, 520-521
Desferrioxamine, in dialysis encephalopathy, 399
Desipramine, 70,107
Desmethylmisonidazole, 272
  effect on cerebral glucose utilization, 283
  neuropathy, 285,289
Development and neurotoxicity, 3,561,632

[Development and neurotoxicity]
carbon monoxide, 634-635
ethanol, 567-569
lead, 169-171,301-305,633-634,639,642
manganese, 395
methylmercury, 359,373
neurobehavioral toxicants, 163-191
opiates, 571-572
thallium, 389-390
triethyl tin, 392
DEP, 416,425,429
Dialysis encephalopathy, 396-397,399
Diazepam, 226
cerebellar effects, 238
and kainic acid-induced status epilepticus, 522
use in tetanus, 494
Diazinon, 425,427
behavioral teratogenic effects, 174,185
Dichloromethane (*see* Methylene chloride)
Dichloropehnoxyacetic acid, 408
Dichlorvos, 424,425
Dieldrin, 445,446
changes in evoked potentials induced by, 618
hippocampal effects, 598
metabolic activation, 450
Diethyldithiocarbamate, 617
Dihydrobenzperidol, 87
Diisopropylfluorophosphate (*see* Phenytoin)
Diphtheria, 494
neuropathy, 4,7,496
toxin, 9,494-495
Disopyramide, 76
Disulfiram, 535-557
Disulfoton, 416
L-DOPA, 107-108
and carbon monoxide toxicity, 635
and neuroleptic drugs, 88,91
[L-DOPA]
use in thallium poisoning, 311
Dopamine, 178,179,181
brain levels in manganese intoxication, 173,174,395
indices, and kainic acid toxicity, 518,519
release, inhibition by tetanus toxin, 493
striatal levels in thallium intoxication, 391
Dopamine receptors, 204-206
effect of chlordecone on, 408
DFP-induced changes in, 416
supersensitivity during neuroleptic therapy, 86
Dopaminergic system
dysfunction, trimethyl tin-induced, 393
and endorphins, 142,147
and intracranial self-stimulation, 145
and lead toxicity, 309,310
and neurohypophyseal hormone-related peptides, 148
role in drug reward processes, 113-114,122,145,147,148
nicotine, 117
opiates, 116-117,147
psychomotor stimulants, 115-116
and steroid hormones, 84,451
Dorsal root ganglia
cell cultures, effect of ethanol in, 568,602
edema, misonidazole-induced, 281
as a target of mercury toxicity, 27-28
disorders in protein synthesis, 26,371,372,373
metal accumulation, 374
Doxepin, 224
Drugs of abuse, 111,136
action on reward systems in the brain, 111-124

[Drugs of abuse]
dependence, reinforcement models of, 118-122
motivational toxicity, 112
self-administration techniques, 136,144
Duchenne muscular dystrophy, 209
Dying-back process (*see* Neuropathy)
Dynorphins, 141
Dysarthria, antiepileptic drug-induced, 324

E

Echothiophate, 425,429
Electroencephalogram
in alcoholics, 207
in dialysis encephalopathy, 395
Electroencephalography
bismuth, 399
carbon disulfide, 614
carbon monoxide, 615
DDT, 615
n-hexane, 614
lead, 305
mercury, 330,331
methylene chloride, 467
methylmethacrylate, 615
neuroleptic drugs, 221
neurotoxicity assessment, 614-617
organochlorine pesticides, 615
organophosphorous compounds, 614-615
petroleum, 615
solvents, 468,614,615
styrene, 615
trichloroethylene, 462
Electromyography
botulism, 492
n-hexane, 625
malignant hyperthermia, 207
methyl-n-butyl ketone, 625,626
[Electromyography]
neurotoxicity, 625-626
Electrophysiological measurements (*see also* Electroencephalography, Electromyography, Evoked potentials, Nerve)
in nerve cell cultures, 600-603
in hippocampal slice, 592-599
in invertebrate preparations, 588-592
studies in vitro, 585-604
studies in vivo, 613-626
Encephalopathy
aluminum, 396
bismuth, 399
cuprizone, 279
lead, 301-302,305
methotrexate, 256,257
misonidazole, 275,276
triethyl tin, 392
Endoplasmic reticulum, as a target of neurotoxicants
rough, effect of methylmercury on, 25,28,364,374,375
smooth, 34-35
acrylamide, 14,34-35
isoniazid, 24
organophosphorous compounds, 19
Endorphins, 138,139,140
behavioral effects, 141-143
content in CSF of alcoholics, 147
experimental addiction and, 141, 146-148
self-administration in animals, 146
Energy metabolism disorders, and neurotoxicity, 4,14,19-23
arsenic, 19,388
anticancer drugs, 262
lead, 312
methylmercury, 375
nitrofurantoin, 21
nitroimidazoles, 21,23,283-284
thallium, 21,390

Enkephalins, 146
  kainic acid toxicity and, 518, 519
Enzymes as neural markers
  2,3 cyclic nucleotide phosphohydrolase, 566,568,571
  β-galactosidase, 278,279
  β-glucuronidase, 278-280
  glutamine synthetase, 569,570
  ornithine decarboxylase, 566
Epilepsy
  animal models of, 522-523
  neuroleptic drugs and, 221
EPN as a contact insecticide, 424, 425
  delayed neurotoxicity, 435
  diagnosis of poisoning by, 434
Erabutoxin, 498
Erythredema polyneuritis (pink disease), 24
Estazolam, 105
Estradiol, action in neural culture, 570
Estrogens
  porphyrias and, 213
  receptors, interaction of chlordecone with, 449
Ethanol
  action in neural culture systems, 567-569,602
  dependence, 111,136
    genetic determinants, 206-207
    naltrexone and, 147
  fetal alcohol syndrome, 567
  intolerance after exposure to trichloroethylene, 461
  membrane effects, 569
  self-administration in moneky, 147
  stimulation of brain reward system by, 114-115,117
  vasopressin-related peptides and, 144,150
Ethchlorvynol, 103,226
Ethephon (Ethrel), 425,428
Ethionamide, 14,23
Ethoprophos (Mocap), 425,427
Ethorphine, 571,572
Ethosuximide, 238
2-Ethoxyethanol, 178,186
  interaction with ethanol, 178-179
bis(p-Ethoxyphenyl)-3,3-dimethyloxetane (EDO), 452
Ethrel (*see* Ethephon)
Etoposide, 256
Evoked potentials
  auditory
    misonidazole, 619
    nitroimidazoles, 287,619
  neurotoxicity assessment and, 617-620
  somatosensory
    n-hexane, 618
    trimethyl tin, 618-619,620
  visual
    acrylamide, 636
    carbon monoxide, 618
    lead, 305,618
    methylene chloride, 467
    methylmercury, 618,638
    1,1,1 trichloroethane, 465
    triethyl tin, 619
    trimethyl tin, 639
Excitatory postsynaptic potential (EPSP)
  4-aminopyridine, effect on, 596-597
  in cockroach abdominal nerve cord preparation, 589
  in hippocampal slice, 594
  hydroquinone, effect on, 590-591
Excitotoxins, 515-525
Experimental neurotoxicology
  behavioral toxicity assays, 631-644
  drug-self-administration procedures, 136,144
  morphological approach, 1-38, 535-557

[Experimental neurotoxicology]
neural culture systems, 559-576,600-603
neurophysiological methods in vitro, 585-604
neurophysiological methods in vivo, 613-626
receptor binding assays, 415-418,571-572
toxin-induced animal models of human disease, 515-525
Extrapyramidal disorders
in manganese intoxication, 394
neuroleptic induced, 220,222
in thallium poisoning, 388

F

Fenchlorphos (Ronnel), 425,427
Fensulfothion, 426
Fetal alcohol syndrome, 567-569
FK33-824, 147
Flagyl (*see* Metronidazole)
Flame retardants, 429
Flavoproteins, thallium toxicity and, 21,390
Floxuridine, 260
Flunitrazepam, 105
Fluoride number, pseudocholinesterase variants and, 211-212
Fluoroalkane propellants, 72
5-Fluorouracil, 259-260
Flupentixol, 88
Flufenazine decanoate, prolactin secretion and, 86
Flurazepam, sleep disorders induced by, 103,104,105,108
Flurbiprofen, 286
Folex (*see* Merphos)
Folic acid, 516-517
epileptogenic activity, 523
methotrexate toxicity and, 256
Follicle-stimulating hormone (FSH), 85
Formaldehyde, 179,186
Formate, 478-479
as a metabolite of cyanide, 474-475
visual toxicity, 482
Frescon, 408
Friederich's disease, 11
Fugu (puffer fish poisoning), 505,506
Fungicides, organophosphorous compounds as, 428
Fyrol FR-2, 426,429

G

GABA (*see* γ-aminobutyric acid)
GABA-transaminase, 519
β-Galactosidase, 278,279
Ganglionic blocking agents, 73
Gangliosides
and kainic acid, 519
tetanus toxin bindings, 493
Glasgow coma scale, 48,59
Glaucoma, 74,429
Glioblastoma, nitroimidazole radiosensitizers and, 274
β-Glucuronidase, 278-280
Glutamate, 516
Huntington's disease and, 521
lithium toxicity and, 71
as a neurotransmitter in hippocampus, 593
receptors, effect of pesticides on, 408,417
release, inhibition by pesticides, 415
Glutamate-obese mouse, 525
Glutamine sinthetase, 569,570
Glutethimide, 103,226
des-Glycinamide vasopressin, 144, 148
Glycine
lead toxicity and, 309
and tetanus toxin, 493
Glyphosate, 426,328

Gold compounds, 386,395-396
Gonadotropins, neuroleptic drugs and, 88
Gonyaulax species, 501
Gonyautoxins, 501-502
Gophacide (Phosacetim), 426,428
Grip-strength test, 636-637
Growth hormone, neuroleptic drugs and, 87-88
Guanethidine, 73,74,76
Guanidine, 492

H

Hallucinatory states in poisoning
  anticholinergic drugs, 108,227
  manganese, 394
  thallium, 389
  thiocyanate, 478
  tricyclic antidepressants, 223
Halogenated hydrocarbons, 459-468
  behavioral teratogenic effects, 181-182
Haloperidol, 84,205
  action in kainic acid-treated animals, 520
  neuropeptides and, 147,148
Halothane, 182,190,208
Haloxon, 426,427
Harderian gland, 176
Hatter shakes syndrome, 26
Hemoperfusion in mercury intoxication, 337
Heptachlor epoxide, 408,415
n-Heptane, 623,624
Herbicides, organophosphorous compounds as, 428
Heroin (*see also* Opiates)
  motivational toxicity, 119
  neuropeptides, 144,146,147
  overdosage, 53,54
  self-administration in animals, 144-147
Hexacarbon intoxication, 29-31
Hexachlorophene, 392,445
2,5-Hexandione, 2,31
n-Hexane
  electroencephalographic effects, 614
  neuropathy, 29,625
    electromyographic findings, 625
    nerve conduction velocity, 618,622,623-624
    somatosensory evoked potentials, 618
2,5-Hexanediol, 30,31
Hinosan (*see* Edifenphos)
Hippocampal slice, 592-599
Hippocampus, 593-594
  and kindling, 522
  and opiate withdrawal, 599
  toxic substances and
    aluminum, 599
    carbon disulfide, 598
    carbon monoxide, 598
    DDT, 615
    dieldrin, 598
    kainic acid, 522
    lead, 598
    methylmethacrylate, 598
    morphine, 599
    toluene, 616
    trimethyl tin, 393,598
Homotypic cell recognition, 561
Homovanillic acid, in urine, lead toxicity and, 302-305,310
Horner's syndrome, phenothiazines and, 74
Horseradish peroxidase, 6
Hunter-Russell syndrome, 346, 351,353
Huntington's disease, excitotoxins and, 517-521
HVA (*see* Homovanillic acid)
Hydralazine
  cardiovascular side effects, 72
  neuropathy, 14,209
    disturbances to pyridoxal phosphate metabolism in, 23

Hydroquinone, 590,591
hydroxycobalamin in cyanide intoxication, 480,483
6-Hydroxydopamine, 395,633
  drug self-administration in animals, effect of, 16
  and kainic acid induced myocardial damage, 524
Hydroxylamine, 480
p-Hydroxymercury benzoate, 375
5(p-Hydroxyphenyl)-5-pehnylhydantoin, 563-565
5-Hydroxytriptamin (*see* Serotonin system)
Hyperactivity, toxin-induced
  carbon monoxide, 634-635
  DDT, 448
  lead, 633
Hypotaurine, Huntington's disease and, 521
Hypothalamic-pituitary axis, chlorinated insecticides and, 448-449
Hypoxic cell sensitizers (*see* Nitroimidazole drugs)

I

Ibotenic acid, 515,516
  action on glutamate receptors, 517
  and cerebral glucose metabolism, 520
Ileus in amitriptyline overdosage, 73
β,β-Iminodipropionitrile, 14,31
Imipramine (*see also* Antidepressants)
  autonomic side effects, 70,72, 74,75
  porphyrias and, 213
  sedative effects, 224
  sleep disturbances induced by, 107
Inezin, 426,528
Infant botulism (*see* Botulism)
Inhibitory postsynaptic potential (IPSP), 589,594
Insect repellents, organophosphorous compounds as, 429
Insecticides (*see also* Pesticides, and individual agents)
  and acetylcholine receptors, 407,410
  and ATPase, 413-415
  synergists of, 428
Interstitial myocardial necrosis, 523-524
Intracranial electrical self-stimulation, 145,146
Invertebrate neural preparation, neurotoxicity assessment in, 588-592
Ionizing radiation, 187,190
Iron, lead toxicity and, 306
Isoniazid
  interaction with phenytoin, 212
  metabolism, genetic variation in, 209-210
  neuropathy, 14,23-24,25
    disturbances to pyridoxal phosphate metabolism in, 23
    smooth membrane function, disorders in, 34
Isonicotinic acid hydrazide (*see* Isoniazid)
Isoproterenol, 72

J

Jellyfish, 500

K

Kainic acid, 515-525
  animal models of epilepsy and, 522-523
  binding sites, action of pyrethroids on, 417

[Kainic acid]
and hippocampal neurotransmission, 593
Huntington's disease and, 517-521
hypothalamic-pituitary axis and, 524-525
interaction with glutamatergic pathways, 517
interstitial myocardial necrosis induced by, 523-524
receptor-neurotransmitter changes induced by, 519
Kepone (*see* Chlordecone)
Kitazin, 426,428
Klüver-Bucy-like syndrome, 221

L

Lathyrus odoratus, 31
Lead, 299-316
behavioral toxicity in animals, 462
developmental studies, 169-171,184-185,633-634,642
effect of parental exposure, 165
and hippocampus, 598
human toxicology
adult toxicity, 305-306
behavioral effects, 301,302, 304,305
blood levels, 303,304,307
chelation therpay, 302-303
childhood disease, 301-305
dentine levels, 304
encephalopathy, 301-302,305
electrophysiological effects, 301,305,622,639
hematological disorders, 301, 306
hyperkinetic syndrome, 633
minimal brain dysfunction, 185,314,315
neuropathy, 4-5,305
[Lead]
sources and extent of exposures, 306,307,386
interaction with calcium ions, 306,311,312,313
mechanisms of toxicity, 308-314
alteration of neurotransmitter indices, 309-314
membrane effects, 311,312
mitochondrial changes, 312
postsynaptic effects, 311
neuropathy, experimental, 5-7
neurophysiological changes, 603-604,618
organic (*see* Alkyllead)
protein-bound, 6
reproductive toxicity, 306
vascular bed lesions induced by, 3,4
Leber's optic atrophy, 481-482
Leptophos, 426,435
Lergotrile, 91
Leu-enkephalin, 141,572
Leukonychia in arsenic intoxication, 387
Linamarin, 483
Linamerase, 483
Lindane, 445,446,450
calcium homeostasis and, 407
chloride transport system, inhibition by, 408
Lipofuscin, 350
Lipoic acid, arsenic toxicity and, 388
β-Lipotropin, 138
Lisuride, 91
Lithium, 225
cardiac effects, 71,72
endocrine toxicity, 84
overdosage, 57,58
Lorazepam, withdrawal reaction to, 226
Lorcainide, 106
Lormetazepam, 105
LSD, 205
Luteinizing hormone (LH), 85,88

[Luteinizing hormone (LH)]
chlordecone-induced changes in levels of, 449
Lysosomal enzymes, 278-280

M

Maculotoxins, 505
Malaoxon, 424
Malathion, 424,426,427,429
Malignant hyperthermia, 207-209
Malnutrition, locomotor activity and, 634
Mandrake herb, 256
Maneb, 175
Manganese, 394-395
accumulation in dopaminergic terminals, 395
action in hippocampal slice, 599
effect on opiate binding sites, 572
human disease, 394
neurobehavioral teratogenicity, 173-174,185
neurotransmitter changes induced by, 394-395,452
sources of exposure, 386
Mannitol gap technique, 589-592
Marine biotoxins, 500-506
Mecamylamine, 73
Mee's lines, 387,389
Melanocyte-stimulating hormone (MSH), 138
phenothiazines and, 90
Melatonin, neuroleptics and, 90
Memory
disorders in toxic states
dialysis encephalopathy, 395
lithium, 225
mercury, 330
methotrexate, 257
methylene chloride, 466
thallium, 389
tricyclic antidepressants, 224
trimethyl tin, 393
[Memory]
vasopressin-related peptides and, 143
Mental retardation ana lead toxicity, 302
Meprobamate, withdrawal reaction to, 226
Mercury, inorganic, 14,24,223-237,376
antidotes, 337
blood levels, 332,335-336
calomel, 14,24
distribution in nervous tissue, 328,333
dose-response relationships, 332-333
elemental, 325
environmental surveillance, 333-334
exposure, 323,386
biological monitoring of, 334-336
human intoxication, 24,329-332
hatter shakes syndrome, 26
psychodiagnostic tests in, 330,332-333
levels in urine, 334-335
lung retention, 326
metabolism, 325-329
metabolism, 325-329
nephropathy, 329
neuropathy, 14,24,331
ototoxicity, 332
pink disease, 24
psellism, 331
threshold limit values (TLV), 330,334,335
uses, 323-325
Mercury, organic (*see* Methylmercury)
Merphos, 426,428
delayed neurotoxicity, 435
late acute effects, 438
Metals (*see* individual elements)
Metaproterenol, 72
Met-enkephalin, 141

Metepa, 426,427
Methadone, 151
  effect in neural culture systems, 566-567
Methamphetamine psychosis, 205
Methanol, 478-479
Methaqualone, 103,226
Methotrexate, 256-258
Methotrimeprazine, 70
Methoxychlor, 446,452
2-Methoxyethanol, 165
  behavioral teratogenic effects, 179,186
Methoxyflurane, 208
Methyl alcohol (*see* Methanol)
Methyl-n-butyl ketone (MnBK), 29
  behavioral teratogenic effects, 179-180,186
  effect on nerve conduction velocity, 622
  electromyographic effects, 625,626
  neuropathy, 29
Methy chloride, 459,460,465-466
Methyl chloroform, 460,464-465
  behavioral teratogenic effects, 177,186
α-Methyldopa, 71,73
  behavioral teratogenic effects, 180,186
  changes in sleep-wakefulness cycle induced by, 616
  metabolic activation, 460,467
Methylmercury, 26-28,345-364
  behavioral teratogenic effects, 168-169,184-185,635,638,643
  cerebellar toxicity, 353,355, 356,368,371
  childhood poisoning, 358
  effect on the visual system, 618,638-639
  mechanism of action, 364
    effect on protein synthesis in nervous tissue, 26,369-378
[Methylmercury]
  metabolism and disposition
    intracellular distribution in cerebellum, 374
    levels in brain tissue, 359
    levels in hair, 348,349
  Minamata disease, 345-348
  peripheral nervous system toxicity, 14,26,358-359
    biochemical markers, 278,279
    experimental intoxication, 26-28
    neuropathology, human, 348-359
  transplacental intoxication, 359
Methylmercury dicyandiamide, 370
Methylmethacrylate, 598
Methylparaoxon, 210
Methylphenidate, 88
Methyprylon, 103
Metopyron, 87
Metronidazole, 271,272,273
  neuropathy, 14,21,274,279
Mg^{2+}-ATPase (*see* Adenosinetriphosphatases)
Mianserin, 71
Microbial neurotoxins, 489-496
Microtubules
  assembly in neural cultures, 574
  and organophosphate-induced delayed neuropathy, 437
  proteins associated with, 255
  stability, 255
    disturbances in toxic neuropathies, 14,33-34
  as a target of anticancer drugs, 252,253,262
Micruroides euryxanthenes venom, 499
Midazolam, 103
Miniature end-plate potentials
  and lead toxicity, 603,604
  and presynaptic snake venom neurotoxins, 498

Minamata disease (*see* Methylmercury)
Minoxidil, 72
Mipafox, 426
  delayed neurotoxicity, 435,436
Mirex, 176,445,446
Misonidazole, 272,274
  bone marrow damage, 286
  central nervous system toxicity, human 275-277,281
  effect on auditory evoked potentials, 619
  glutathione depletion by, 282
  interaction with phenobarbitone, 285,286
  interaction with phenytoin, 285,286
  maximum tolerable dose, 277
  neurobehavioral effects in mice, 287-288
  neuropathy, 14,21,274,275
    biochemical changes, 278-280
    clinical effects, 273-278
    experimental, 23,281,282
    flurbiprofen in, 286
    mechanism, 282-284
    nerve conduction disorders in, 287
    neuropathology, animal, 23, 281,282
    neuropathology, human, 275-277
    prevention of, 284-286
    tests for, 287-288
  ototoxicity, 276,277,281
    assessment in mice, 288
  pharmacokinetics, 285,286
Mocap (*see* Ethoprophos)
Monoamine oxidase:
  inhibition by thallium ion, 391
  inhibitors, 261 (*see also* Antidepressants)
  and lead toxicity, 310
Monolayer culture systems, 560
Morphine
  action in hippocampal slice, 599
[Morphine]
  addiction, 136
  behavioral teratogenic effects, 165
  and reward systems in the brain, 114-115
  self-administration in animals and neuropeptides, 150
  use in tetanus, 494
Motivational toxins, 123
Motoneuritis multiplex, 5
Motor activity, 632
  disorders, toxin-induced (*see* Hyperactivity, toxin-induced)
  effect of malnutrition on, 634
  measurements, 632-637
    experimental variables, 633
Morot neuron disease, 305
γ_2-MSH, 139
  and heroin self-administration in animals, 146
  opiate withdrawal-like effects, 141
Multiple sleep latency test, 104, 105
Muscimol binding assay, 417
Muscle
  denervation atrophy, toxin-induced, 5,21
  electrophysiology, 622,625-626
  spindles, 235
    and organophosphate-delayed neuropathy, 15
Mushroom poisoning, 51
Myelin, as a target of toxicity, 5
  diphtheria toxin, 7
  lead, 5
  organophosphorous compounds, 435
  perhexilene, 11
  triethyl tin, 392
Myelination, 568
  developmental lead toxicity, 171
Myelopathy, cytosine arabinoside-induced, 258
Myotonia congenita, 209

N

Na^+,K^+-dependent ATPase (*see* Adenosine triphosphatase)
Naja Naja Naja, 499
Naloxone
binding sites, 311,417
and intracranial electrical self-stimulation, 154
hippocampal effects, 599
in Huntington's disease, 521
and smoking, 147
Naltrexone, 142,146
effect on ethanol intake in monkeys, 147
Narcolepsy, 106
Narcotics (*see* Opiates, and individual drugs)
Nematocides, organophosphorous compounds as, 427
Neostigmine, 502
Nereistoxin, 407
Nerve
action potential, 622
conduction velocity, toxin-induced disorders, 620-625
acrylamide, 622
allyl chloride, 622
arsenic, 387
carbon disulfide, 31,622
cyanate, 478
diphtheria toxins, 7
n-hexane, 29,618,622
lead, 5,301,306,322
methyl-n-butyl ketone, 622
nitroimidazoles, 287
perhexilene, 11
trichloroethylene, 462
vinyl toluene, 625
excitation, pesticide-induced, 407,410,457
fiber
access of toxic chemicals, 2
damage, lysosomal markers (*see* Enzymes as neural markers)
[Nerve]
regeneration, and acrylamide, 37
terminals, tubulin in, 255
membrane function, toxin-induced disorders in
chlorinated insecticides, 447
ciguatoxin, 504
ethanol, 569
excitotoxins, 516
lead, 311-312
pesticides, 405-418
saxitoxin, 502
Nerve agents, 423,429
Nerve growth factor, 571
Nerve train analysis, 287
Neural culture systems, 559-576
cerebellum, 600
dorsal root ganglia, 568,602
electrophysiology, 600-603
growth and differentiation, 561
marker enzymes, 566,567,569
retina, 602
Neural tube defect, cyanogenic compounds-induced, 481
Neuroblastoma cell model, 560
pyrethroids, 602
thyroid hormones, 571
Neuroendocrine system
excitotoxins and, 524-525
neuroleptic drugs and, 83-92
pesticides and, 185
Neurofibrillary degeneration, aluminum toxicity and, 397
Neurofilaments, 14,31
disturbances in toxic neuropathies, 14,28-33
acrylamide, 35
carbon disulfide, 31
hexacarbons, 29-31
β,β-iminodiproionitrile, 33
protein cross-linking, 31
Neurolathyrism, 516
Neuroleptic malignant syndrome, 209,221
Neuroleptics, 83-92

[Neuroleptics]
adverse central nervous system effects, 220-223
effect on ovarian steroids, 89
electroencephalographic effects, 221
neuroendocrine disorders induced by, 83-92
ocular effects, 73,74
and prolactin, 223
sexual dysfunction induced by, 76,85
toxicity, genetic determinants, 204
Neuromuscular junction blocking agents, 52
Neuropathy, 1-38
acrylamide, 34-38
arsenic, 19,386
biochemical markers, 278-280
carbon disulfide, 31
cis-platin, 261
classifications of, 4-38
dapsone, 209
desmethylmisonidazole, 285, 289
diphtheria, 7,9-10,495,496
dying-back process, 11-19, 551-556
electrophysiological assessment of, 620-626
energy supply disorders in, 19-23
ethionamide, 14,23
etoposide, 256
gold, organic, 395
2,5-hexandione, 31
n-hexane, 31,625
hydralazine, 14,23,209
β,β'-iminodipropionitrile, 31
isoniazid, 23-24
lead, 4-7,305
mercury, inorganic, 331
metronidazole, 274
methylmercury, 26,358-359
misonidazole, 275
[Neuropathy]
morphological factors in the development of, 2-4
neurofilament disorders in, 28-33
nitrofurans, 21,273
organophosphorous compounds, 14-19,435,436
perhexilene, 7,11
in pernicious anemia, 483
procarbazine, 262
protein synthesis disorders in, 24-28
pyridoxal phosphate metabolism, alterations in, 23-24
smooth membrane function disorders in, 34-38
trichloroethylene, 461,462
tropical diseases, 483-484
Vinca alkaloids, 14,254,255
Neuropeptides and addiction, 135-161
experimental studies, 144-151
human studies, 151-152
Neuroteratogenicity, assessment of, 167,600
environmental and industrial agents, 163-191
Neurotoxic esterase, 437
Neurotransmitters (*see also* the individual systems):
development, 183,632
and lead toxicity, 309-314
and pesticides, 415-418,451
in teratology studies, 167
Nicotine, 117,152
Nicotine-derived insecticides, 407
Nitrazepam, 103,209,238
p-Nitroacetophenone, 273
Nitrofen, 176
Nitrofurans, 273
Nitrofurantoin, 14,21,279
Nitroimidazole drugs, 14,271-290 (*see also* individual drugs)
p-Nitrophenol, 210,434
Nitroprusside, 474

Nitrous oxide, malignant hyperthermia and, 208
Nocturnal myoclonus, tricyclic antidepressants and, 106-107
Noise, behavioral teratogenic effects, 188,190
Nomifensine, 75
Norepinephrine
  kainic acid and, 518,519
  and lead toxicity, 309,310
  and manganese, 395
Notexin, 497
NTE (*see* Neurotoxic esterase)
Nucleic acids and neurotoxicity
  aluminum, 397
  kainic acid, 519
  methylmercury, 375-377
  nitroimidazoles, 282,283
  organophosphorous compounds, 428
  studies on neural culture systems, 570
Nucleus accumbens, 142
  and psychomotor stimulatns, 115,116
Nucleus arcuatus, 139
Nystagmus
  benzodiazepines, 238
  carbamazepine, 237
  5-fluorouracil, 259
  phenytoin, 235,236
  primidone, 237

O

ODAP (*see* β-N-Oxalyl-L-α,β-diaminopropionic acid)
Olfactory cortex slice preparation, 599
Opiates, 111,119-122,566-567
  action on reward systems in the brain, 116-117
  receptors, 141
    effect of pesticides on, 408
[Opiates]
    in neural cultures systems, 571-572
  withdrawal, assessment in hippocampal slice, 599
ORG 2766, 139,140,146
Organochlorine insecticides
  behavioral teratogenic effects, 185
  bioconcentration, 446
  effect on intracellular calcium, 407
  estrogenicity, 447
  inhibition of ATPase, 413,414, 415
  nonneural effects, 447,451
Organophosphates (*see* Organophosphorous compounds)
Organophosphorous compounds, 423-438
  acute poisoning, 433-435
  behavioral teratogenic effects, 185
  classification, 423-430
  delayed neurotoxicity, 435-438
    experimental animal models, 435
    mechanism, 436-437
    neuropathology, 14-19, 436
  effect on cholinergic systems, 415
  effect on intracellular calcium ion, 407
  electroencephalographic effects, 408
  late acute effects, 438
  mechanism of action, 430-432
  metabolism, 210-211
  phosphorylation of acetylcholine receptors, 408
  and serotonin turnover, 452
  smooth endoplasmic reticulum, disorders induced by, 34
  tolerance to, 408,416,432
  uses, 424-430

Organotins (*see also* the individual agents), 391-394
behavioral teratogenic effects, 172-173,185
and calcium ion, 407
exposure to, 386
inhibition of ATPase by, 413
Ornithine decarboxylase, 566
Ornithine-d-transaminase, kainic acid and, 519
Orphenadrine, overdosage, 57
Ototoxicity, 639
cis-platin, 260-261
lead, 302
mercury, inorganic, 332
misonidazole, 276,277,281
trichloroethylene, 462
β-N-Oxalyl-L-α,β-diaminopropionic acid (ODAP) and lathyrism, 516
Ouabain, 408
Ovarian steroids, neuroleptic drugs and, 89
Oxazepam, 105
Oxytocin, 140,143,148

P

Paint thinner, 180,186
Palmar skin reflex, 75
2-PAM, 432,434
Paralytic shellfish poisoning, 501-502
Paraoxon, 210,426
Paraoxonase polymorphism, 210-211
Parathion, 424,426,427,434
metabolism, 210
Penfluridol, 88
n-Penicillamine, in mercury intoxication, 337
Penicillin, 600
n-Pentane, 623,624
Pentobarbitone, 238
Pentylenetetrazol, 601
Peptidergic pathways, 138-139
Perchloroethylene, 180-181,186
Perhexilene neuropathy, 7,11,14, 15
Periaqueductal gray, 120
Periplaneta americana, 588,589
Permethrin, 413,417
Perphenazine, 70,86
Pesticides (*see also* individual agents)
behavioral teratogenic effects, 174-176,188
carbamates, 185,415
chlorinated agents, 445-453
membrane effects, 405-418, 452
organophosphorous compounds, 423-438
organotins, 393-394
pyrethroids, 417
and receptor-neurotransmitter systems, 415-418
veterinary use, 427
Pharmacogenetics and neurotoxicity, 203-213
Phencyclidine, 117
Phenelzine, 71,209,210
Phenobarbitone, 234,238
and carbon monoxide toxicity, 641
effect in lead-exposed animals, 633
and misonidazole, 285,286
Phenothiazines (*see* Neuroleptics)
Phenothrin, 317
Phenoxybenzemine, 76
Phentolamine, 75
Phenytoin, 212,234,236
and benzodiazepine receptors, 572-573
cerebellar toxicity, 235-236,239-244,561
competition with calcium ion, 566
metabolism, genetic variation in, 212-213

[Phenytoin]
and misonidazole pharmacokinetics, 285,286
neural culture systems, 563-566
smooth pursuit abnormalities induced by, 236
Phosacetim (*see* Gophacide)
Phosbutyl, 426,428
Phospholine (*see* Echothiophate)
Physostigmine, kainic acid and, 520
Physostigmine salicylate, in anticholinergic drug overdosage, 57-59,227
Picrotoxin, 601
Pilocarpine, 77,520
Pimozide, effects on the neuroendocrine system, 84,88,89
Platin (*see* Cis-platin)
Podophyllotoxin, 256
Porphyria, 213,306
Practolol, 76
Pressinamide, 148,149
Primidone, 234
acute cerebellar effects, 237
Procainamide, 209
Procarbazine, 71,261-262
Prochlorperazine, 74
Prodynorphin, 141
Proenkephalin, 141
Progesterone
effect in cerebellar culture, 570
and neuroleptic drugs, 89
Prolactin, 449
and neuroleptic drugs, 84-87, 223
Prolyl-leucyl-glycinamide, 144
Pro-opiomelanocortin, 138,139, 141
Propantheline, 74
Prophos, 426,427
Propionitrile, 480
Propranolol, 76,77
in lithium-induced tremor, 225
[Propranolol]
sleep disturbances induced by, 105
Protein synthesis disorders and neurotoxicity
ethanol, 568
mercury, inorganic, 14,24,26
methylmercury, 26-28,364,369-378
nitroimidazole drugs, 283
phenytoin, 563
Protoporphyrin in red cells, and lead toxicity, 304
Protriptyline, 72,107
Pseudocholinesterase variants, 211-212
Pseudoparkinsonism, neuroleptics and, 222
Psellism, in mercury intoxication, 331
Psychomotor stimulants, 115-116, 119,122
Psychoorganic syndrome, and trichloroethylene poisoning, 461
Psychosis, 143,220
drug-induced, 205-205
isoniazid, 210
lead, 302
phenelzine, 210
Puffer fish poisoning, 505-506
Purkinje cells, as toxicity targets
acrylamide, 34,35
carbon monoxide, 600
antiepileptic drugs, 239-244, 564-565
β,β'-iminodipropionitrile, 33
mercury, 328,353,371
thallium, 391
Pyrethroids, 413
action in neuroblastoma cell culture, 602
effect on glutamate receptors, 408,417
effect on calcium ion homeostasis, 407

[Pyrethroids]
and sodium channels, 406,407, 411-413
synergists of, 429
Pyridoxal phosphate, 14,209
procarbazine toxicity and, 262
and toxic neuropathies, 23-24
Pyridoxine, 210,285
Pyrithiamine, 14

Q

Quazepam, 105
Quinalbarbitone, 238
Quinidine, 106
Quinuclidinyl benzylate binding assays, 416,417,516,517

R

Rauwolfia alkaloids, 72,73
Rebound insomnia, 102-104
Red tides, 501
Reserpine, 72,524
Resmethrin, 413
Retina, 603
culture systems, 570,602
effect of steroid hormones, in, 574,575
damage
in Leber's disease, 481
in methanol poisoning, 479
in tobacco amblyopia, 482
from trimethyl tin toxicity, 639
Reward circuitry in brain, 112-114
action of abusable drugs on, 112-124
endorphins and, 146
Rhodanase, 474,476
Riboflavin, 21,285,390
RO-03-8799, 290
RO-05-9963 (*see* Desmethyl-misonidazole)
RO-07-0582 (*see* Misonidazole)
Rodenticides, 388,428
Ronnel (*see* Fenchlorphos)
Rotenone, 406
Round-up, 426

S

Sarin, 426,429
Saxidomus giganteus, 501
Saxitoxin, 501-502
bioassay of, 503
Schwann cell, 6,7,15,372,494
Scopolamine, 520
Selenium, 360,363
Serotonin system, 632
toxins inducing changes in
acrylamide, 452
chlorinated insecticides, 448, 449,451
2-ethoxyethanol, 178,179
kainic acid, 518,519
manganese, 173-174,452
organophosphorous compounds, 452
organotins, 393
triethyllead, 452
Sexual dysfunction, drug-induced, 78,88
Sleep, 101-102
apnea syndrome, 108
disorders, drug-induced, 101-109
manganese, 394
methyl chloride, 466
methylene chloride, 466
thallium, 389
trichloroethylene, 461
electroencephalography, 614
and toxicity assessment, 615-617
Smoking, 117
naloxone and, 147

[Smoking]
tobacco amblyopia, 482-483
and vasopressin-related neuropeptides, 152
SMON (*see* Subacute myelo-optico neuropathy)
Snake venom toxins, 496-500
Sodium channel, toxins affecting
organochlorine pesticides, 407, 411-413,447,448
pyrethroids, 410,411-413,602-603
sacitoxin, 502
snake venom toxins, 499-500
tetrodotoxin, 410,506
Sodium nitrite, 479-480
Sodium thiosulfate, 479-480
Solvent-refined hydrocarbons, 183
Solvents
behavioral teratogenic effects, 176-180,186
effect on evoked potentials, 618
halogenated hydrocarbons, 459-468
mixtures, 467-468
Soman, 426,429
Somatosensory evoked potentials (*see* Evoked potentials)
Somatostatin, kainic acid and, 518,519
Spinal cord damage, 14,328,350
in experimental disulfiram intoxication, 538,539,541
Spindle apparatus, 253
Spiroperidol binding sites, 116, 311,417
SR-2508, 290
S-Seven, 426
Startle reflex, 639
Status epilepticus, kainic acid-induced, 522
Steroid hormones, neuronal growth and, 569-571
Styrene, 625
Subacute myelo-optico neuropathy (SMON), 14
Substance P, 518,519
Substantia innominata, 522
Succinic dehydrogenase, 173,391
Succinonitrile, 480
Succinylcholine, 208,211-212
Sulfamethazine, 209
Sulpiride, 311
neuroendocrine effects, 84,87, 89,90

T

Tabun, 427,429
Taipoxin, 497
Tardive dyskinesia, neuroleptic-induced, 223-224
genetic determinants in, 204-205
Taurine, 393
in Huntington's disease, 521
Tellurium, 386
Temazepam, 238
Tepa, 427
TEPP, 424,427
Testosterone, neuroleptic drugs and, 89
Tetanus, 492-494
human intoxication, 493-494
toxin, 493
axonal transport, 493
binding to presynaptic terminals, 493
chemistry and pathophysiology, 493
effect on neurotransmitter release, 493
and inhibitory aminoacids, 493
Tetrachloroethylene, 459,460,463-464
Tetraethyllead (*see* Alkyllead)
Tetraethylpyrophosphate (*see* TEPP)

Tetraethylthiuram disulfide (*see* Disulfiram)
Tetraethyl tin (*see* Organotins)
Tetrahydrobiopterin, 308
Tetramethrin, 410
Tetrodotoxin, 505-506
  action on sodium channels, 410
  action in nerve cell culture, 603
Textilon, 497
Thalidomide, 166
Thallium, 388-391
  acute poisoning, 388-389
  blood-nerve barrier, 2
  cardiovascular effects, 390
  levels in brain, 390
  long-term effects, 389
  mechanism of action, 390-391
  neuropathy, 14,21,388
  neurotransmitter imbalance induced by, 391
  placental transfer, 389
  sources of exposure, 396
  as a trace pollutant, 389
Theophylline, 72
Thermoregulation disorders
  chlordecone-induced, 449
  and serotonergic systems, 451
Thiamine deficiency, toxic neuropathy and, 282,283,285
  alcoholism, 19,206
Thiocyanate, 474
  effects in animals, 477-478
  metabolism, 475-476
  and smoking, 481-482
Thionazine (*see* Zinophos)
Thiopental, 77
Thioridazine, 71,221
  neuroendocrine effects, 88,89
Thiotepa, 427,428
Thioxantenes, 70,84
Thymidylate synthetase, 259
Thymidylic acid, 257
Thyroid hormones, action in neural culture, 571
Thyrosine hydroxylase, 257
Thyrotropin-releasing hormone, 85
  kainic acid and, 518,519
Tin, organic (*see* Organotins, and individual agents)
Tobacco amblyopia, 482-483
Tocinamide, 149
TOPC (*see* Tri-ortho-cresylphosphate)
Toluene, 615,616,617
  effect in olfactory cortex slice preparation, 599
Tranylcypromine, 71
Trazodone, 71
Triazolam, 103,105
1,1,1 Trichloro-2,2-dichlorophenylethane (*see* DDT)
1,1,1-Trichloroethane (*see* Methyl chloroform)
Trichloroethylene, 459,460-463
  autonomic nervous system effects, 461
  behavioral effects, 461-462
  electroencephalographic changes in experimental exposure, 462
  metabolism, 460
  noradrenergic activation by, 461
  poisoning, neuropathology, 463
  psychoorganic syndrome in workers, 461
Trichloromethane (chloroform), 177,186,467
Trichlorophenoxyacetic acid, 175, 185
Trichlorphon, 424
Triethyllead (*see* Alkyllead)
Triethyl tin, 391-393 (*see also* Organotins)
  behavioral effects, 392,393
  disease outbreak, 391
  effect on visual evoked potentials, 619
  exposure, 386,391
  mechanism of action, 392-393
Trifluoperazine, 70,90

Trifluperidol, 73
Trigeminal neuropathy, trichloroethylene-induced, 462
Trihexyphenidyl, 74
Trimethyl tin, 393 (*see also* Organotins)
  effect on somatosensory evoked responses, 618-619,620
  effect on the visual system, 639
  hippocampal effects, 598
  neurotransmitter imbalance induced by, 393
  poisoning, 393
Tri-ortho-cresylphosphate (TOPC), 430
  delayed neuropathy, 16-19,435, 436
Tripelennamine, 74
Tri-PB, 427,429
Tropical neuropathies, 483-484
Tryptophan hydroxylase, 172
Tubulin disorders in toxic neuropathies, 14
  organophosphorous compounds, 437
  vincristine, 34,255
Tyrosine hydroxylase, 310
  and cadmium toxicity, 172
  effect of manganese on, 173
  in neural culture, 570

U

Ultrasound, 189,190

V

Valproate, sodium, 238
Vanillyl mandelic acid in urine, and lead exposure, 302
Vascular bed permeability, and neurotoxicity, 3,4,6,284
Vasopressin, 90,143-144
  addiction and peptides related to, 140,148-151
Veratridine, 413
Vestibulotoxins
  mercury, 332
  misonidazole, 281
  trichloroethylene, 462
Vestibulo-ocular reflex, 235 (*see also* Brainstem reflexes)
Viloxazine, 71
Vinblastine, 14,34,253
Vinca alkaloids, 253-256
Vincristine, 14,34,253-255
Vindesine, 253
Vinyl toluene, 625
Visual system, toxins inducing disorders in
  cytosine arabinoside, 258
  diphtheria toxins, 496
  formate, 482
  lead, 302
  methanol, 479
  methylmercury, 356,638
  organotins, 391,639
  thallium, 388
Visual evoked potentials (*see* Evoked potentials)
Vitamin B_{12}, cyanide interaction with, 476
Voltage clamp technique, 599

W

Wernicke's encephalopathy, 281

Z

Zinc, 598
Zinophos (Thionazine), 427
Zytron, 427,428